# PRINCIPLES OF
# ENGINEERING
# GRAPHICS

# PRINCIPLES OF

# ENGINEERING GRAPHICS

## Frederick E. Giesecke

M.E., B.S. in ARCH., C.E., Ph.D.,
Late Professor Emeritus of Drawing, Texas A & M University

## Alva Mitchell

B.C.E., Late Professor Emeritus of Engineering Drawing,
Texas A & M University

## Henry Cecil Spencer

A.B., B.S. in ARCH., M.S., Late Professor Emeritus of
Technical Drawing; Formerly Director of Department,
Illinois Institute of Technology

## Ivan Leroy Hill

B.S., M.S., Professor Emeritus of Engineering Graphics,
Formerly Chairman of Department,
Illinois Institute of Technology

## Robert Olin Loving

B.S., M.S., Professor Emeritus of Engineering Graphics,
Formerly Chairman of Department,
Illinois Institute of Technology

## John Thomas Dygdon

B.S., M.B.A., Associate Professor of Engineering Graphics,
Director of Division of Academic Services and Office of
Educational Services, Illinois Institute of Technology

MACMILLAN PUBLISHING COMPANY
NEW YORK

Collier Macmillan Publishers
LONDON

Editor: John Griffin
Production Supervisor: Elisabeth Belfer
Production Manager: Sandra E. Moore
Text Designer: Patrice Fodero
Cover Designer: Brian Sheridan
Cover illustration: Brian Sheridan

This book was set in 10/12 Caledonia by Waldman Graphics Inc. Printed by Von Hoffman Press, Inc., and bound by Von Hoffman Press, Inc. The cover was printed by The Lehigh Press, Inc.

Macmillan Publishing Company
866 Third Avenue, New York, New York 10022

Collier Macmillan Canada, Inc.

**LIBRARY OF CONGRESS CATALOGING IN PUBLICATION DATA**

Principles of engineering graphics / Frederick E. Giesecke . . . [et
   al.].
      p.   cm.
   ISBN 0-02-342810-4
   1. Engineering graphics.   I. Giesecke, Frederick Ernest, 1869-
   II. Engineering graphics.
T353.P847   1990
604.2--dc20                  89-39693
                                    CIP

Printing: 1 2 3 4 5 6 7 8     Year: 0 1 2 3 4 5 6 7 8 9

# Preface

*Principles of Engineering Graphics* is our response to the latest developments in engineering and technical education. Our goals in writing this new text were to

1. Produce a concise and affordable textbook that can be used either for a one or two semester course in technical drawing and design, descriptive geometry, graphs and diagrams, and computer graphics.

2. Include a thorough introduction to computer graphics.

3. Retain the high standard of accuracy and excellence established in eight editions of *Technical Drawing* and four editions of *Engineering Graphics*.

4. Provide the student with a text that will cover the foundations of the subject and serve as a valuable reference book long after graduation.

For those instructors teaching an introductory course on the theory of engineering graphics including manual and computer graphics techniques, the contents of this text will be sufficient. For those wishing to spend additional time developing the manual or computer drafting skills of their students, the book has been priced so a supplemental manual workbook or computer software manual may be required without imposing too great a financial burden.

*Principles of Engineering Graphics* meets the needs of today's curriculum. Much of this text is adapted or condensed from *Engineering Graphics*, 4th Edition, by the same authors and published by Macmillan Publishing Company. The purpose of this book is *to teach the language of the engineer*. This goal has prompted the authors to illustrate and explain the basic principles from the standpoint of the student— that is, to present each principle so clearly that the student is certain to understand it, and to make the text interesting enough to encourage all students to read and study on their own initiative. By this means the authors hope to free the instructor from the repetitive labor of teaching each student individually the subject matter that the textbook can teach. Thus more class time can be given to the special requirements of individual programs—such as explaining the features of your school's brand of computer graphics software—or in giving more attention to those students having real difficulties.

### Features of This Text

Technical sketching is emphasized throughout the text as well as in an early chapter devoted specifically to sketching. This unique chapter integrated the basic concepts of views with freehand rendering so that multiview drawing can be introduced through the medium of sketches.

The increased use of computer technology for drafting, design work, and manufacturing processes is reflected in many chapters. Two chapters are specifically devoted to this new technology. Chapter 3 presents a generic introduction to computer-aided design and drafting and a survey of hardware and software of current CAD systems. Chapter 8 includes a general discussion of the use and operation of a CAD system. These chapters emphasize the relationship between fundamental drafting techniques and computer graphics. In addition, a comprehensive glossary of CAD/CAM terms and concepts is included in the Appendix.

Thus the growing importance of the engineer's design function is emphasized, especially in the chapter on design and working drawings. The chapter is designed to give the student an understanding of the fundamentals of the design process.

The book consistently reflects the latest trends and practices in education, industry, and especially the various current sections of the ANSI Y14 American National Standards Drafting Manual and other relevant ANSI standards.

The chapters on manufacturing processes, dimensioning, tolerancing, and threads and fasteners have been extensively reviewed to ensure their conformity with the latest ANSI standards. Every effort has been made to ensure that this book is completely abreast of the many technological developments of recent years.

The high quality of drafting in the illustrations and problems that appear in *Engineering Graphics*, 4th Edition, has been maintained in *Principles of Engineering Graphics*. A large number of drawings include the approved system of metric dimensions, now that the metric system is more widely used internationally. The current editions of ANSI standards also indicate a preference for the use of metric units. Many problems, especially in the chapter on design and working drawings, provide an opportunity for the student to convert dimensions to either the decimal-inch system or the metric system.

### Supplements

In addition to the numerous problems in this text, a complete workbook is available for use with this text. It is expected that the instructor who uses this text with the workbook will supplement the problems sheets with assignments from the text, to be drawn on blank paper. Many of the text problems are designed for Size A4 or Size A sheets, the same size as the easily filed problem sheets.

### Acknowledgments

The chapter on graphs and diagrams was prepared by Eugene J. Mysiak, Engineering Manager, Phoenix Company of Chicago, Wood Dale, IL.

The chapters on computer graphics were prepared by Gary R. Bertoline, Assistant Professor of Engineering Graphics, The Ohio State University, Columbus, Ohio. The chapter on manufacturing processes was revised and condensed by Stephen A. Smith, Tool Development Engineer, Packaging Corporation of America/

EKCO Products, A Tenneco Company, Wheeling, IL. Mr. Smith also reviewed the chapter on threads, fasteners, and springs.

The authors also wish to acknowledge the assistance and contributions of James W. Zagorski, Kelly High School, Chicago Public Schools, for the preparation of selected text material on computer graphics in the book. The authors also express their thanks to James E. Novak, Associate Director/Executive Officer of Educational Services, Illinois Institute of Technology for his helpful suggestions and cooperation.

The authors also wish to express their thanks to the numerous persons and firms who have generously contributed their services and materials to the production of this book. Special thanks are due to Mrs. Frances A. Koller for her assistance in typing the manuscript for this book, to our editor John Griffin, and to our Production Supervisor Elisabeth Belfer.

Students, teachers, and engineers, designers, and drafters are invited to write concerning any questions that may arise. All comments and suggestions will be welcome.

Ivan Leroy Hill
Clearwater, FL

Robert Olin Loving
Chicago, IL

John Thomas Dygdon
Illinois Institute
of Technology

# Contents

# The Graphic Language and Design

The old saying "necessity is the mother of invention" continues to hold, and a new machine, structure, system, or device is the result of that need. If the new device, machine, system, or gadget is really needed or desired, people will buy it, providing it does not cost too much. Then, naturally, these questions may arise: Is there a wide potential market? Can this device or system be made available at a price that people are willing to pay? If these questions can be answered satisfactorily, then the inventor, designer, or officials of a company may elect to go ahead with the development of production and marketing plans for the new project or system.

A new machine, structure, or system, or an improvement thereof, must exist in the mind of the engineer or designer before it can become a reality. This original concept or idea is usually placed on paper, or as an image on a cathode ray tube (CRT), and communicated to others by the way of the *graphic language* in the form of freehand *idea sketches*, Figs. 1.1 and 6.1. These idea or design sketches are then followed by other sketches, such as *computation sketches*, for developing the idea more fully.

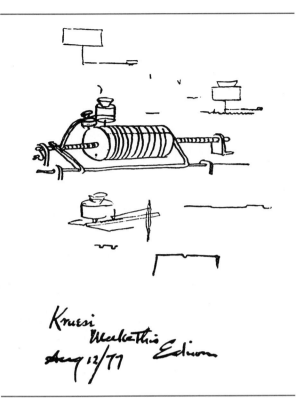

## 1.1 The Young Engineer*

The engineer or designer must be able to create idea sketches, calculate stresses, analyze motions, size the parts, specify materials and production methods, make design layouts, and supervise the preparation of drawings and specifications that will control the numerous details of production, assembly, and maintenance of the product. In order to perform or supervise these many tasks, the engineer makes liberal use of freehand sketches. He or she must be able to record and communicate ideas quickly to associate and support personnel. Facility in freehand sketching (Chapter 6) or the ability to work with computer-controlled drawing techniques, §16.19, requires a thorough knowledge of the graphic language. The engineer or designer who

*Henceforth in this text, all conventional titles such as student, drafter, designer, engineer, engineering technician, engineering technologist, and so on are intended to refer to all persons, male and female.

**Fig. 1.1**   Edison's Phonograph. *Original sketch of Thomas A. Edison's first conception of the phonograph; reproduced by special permission of Mrs. Edison.*

**Fig. 1.2**   Computer-Aided Design and Drafting Section of an Engineering Department. *Courtesy of Jervis B. Webb Co.*

*Fig. 1.3*   Engineering Drafting Department. *Courtesy of AT&T Bell Laboratories.*

uses a computer for drawing and design work must be proficient in drafting, designing, and conceptualizing.

Typical engineering and design departments are shown in Figs. 1.2 and 1.3. Many of the staff have considerable training and experience; others are recent graduates who are gaining experience. There is much to be learned on the job, and it is necessary for the inexperienced person to start at a low level and advance to more responsibility as experience is gained.

## 1.2 The Graphic Language

Although people around the world speak different languages, a universal graphic language has existed since the earliest of times. The earliest forms of writing were through picture forms, such as the Egyptian hieroglyphics, Fig. 1.4. Later these forms were simplified and became the abstract symbols used in our writing today.

A drawing is a *graphic representation* of a real thing, an idea, or a proposed design for later manufacture or construction. Drawings may take many

*Fig. 1.4*   Egyptian Hieroglyphics.

forms, but the graphic method of representation is a basic natural form of communication of ideas that is universal and timeless in character.

## 1.3 Two Types of Drawings

Graphic representation has been developed along two distinct lines, according to the purpose: (1) artistic and (2) technical.

From the beginning of time, artists have used drawings to express aesthetic, philosophic, or other abstract ideas. People learned by listening to their elders and by looking at sculptures, pictures, or drawings in public places. Everybody could understand pictures, and they were a principal source of information. The artist was not just an artist in the

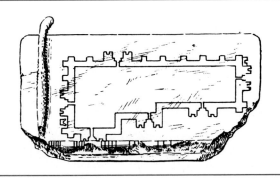

*Fig. 1.5*    Plan of a Fortress.   This stone tablet is part of a statue now in the Louvre, in Paris, and is classified in the earliest period of Chaldean art, about 4000 B.C. *From Transactions ASCE, May 1891*

aesthetic sense, but also a teacher or philosopher, a means of expression and communication.

The other line along which drawing has developed has been the technical. From the beginning of recorded history, people have used drawings to represent the design of objects to be built or constructed. Of these earliest drawings no trace remains, but we definitely know that drawings were used, for people could not have designed and built as they did without using fairly accurate drawings.

## 1.4 Earliest Technical Drawings

Perhaps the earliest known technical drawing in existence is the plan view for a design of a fortress drawn by the Chaldean engineer Gudea and engraved upon a stone tablet, Fig. 1.5. It is remarkable how similar this plan is to those made by modern architects, although "drawn" thousands of years before paper was invented.

In museums we can see actual specimens of early drawing instruments. Compasses were made of bronze and were about the same size as those in current use. As shown in Fig. 1.6, the old compass resembled the dividers of today. Pens were cut from reeds.

The theory of projections of objects upon imaginary plans of projection (to obtain *views*, Chapter 7) apparently was not developed until the early part of the fifteenth century—by the Italian architects Alberti, Brunelleschi, and others. It is well known that Leonardo da Vinci used drawings to record and transmit to others his ideas and designs for mechanical constructions, and many of these drawings are still in existence, Fig. 1.7. It is not clear whether Leonardo ever made mechanical drawings showing orthographic views as we now know them, but it is probable that he did. Leonardo's treatise on painting, published in 1651, is regarded as the first book ever printed on the theory of projection drawing; however, its subject was perspective and not orthographic projection.

The scriber-type compass gave way to the compass with a graphite lead shortly after graphite pencils were developed. At Mount Vernon we can see the drawing instruments used by the great civil engineer George Washington, bearing the date 1749. This set, Fig. 1.8, is very similar to the conventional drawing instruments used today, consisting of a divider and a compass with pencil and pen attachments plus a ruling pen with parallel blades similar to the modern pens.

## 1.5 Early Descriptive Geometry

The beginnings of descriptive geometry are associated with the problems encountered in designs for building construction and military fortifications of France in the eighteenth century. Gaspard Monge (1746–1818) is considered the "inventor" of descrip-

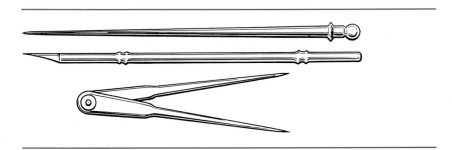

*Fig. 1.6*   Roman Stylus, Pen, and Compass. *From Historical Note on Drawing Instruments, published by V & E Manufacturing Co.*

*Fig. 1.7* An Arsenal, by Leonardo da Vinci. *The Bettmann Archive*

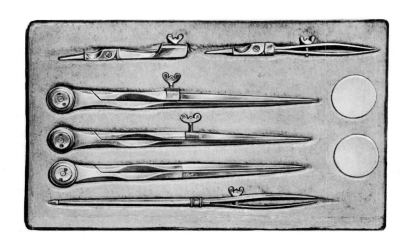

*Fig. 1.8* George Washington's Drawing Instruments. *From Historical Note on Drawing Instruments, published by V & E Manufacturing Co.*

tive geometry, although his efforts were preceded by publications on stereotomy, architecture, and perspective in which many of the principles were used. It was while he was a professor at the Polytechnic School in France near the close of the eighteenth century that Monge developed the principles of projection that are now the basis of our technical drawing. These principles of descriptive geometry were soon recognized to be of such military importance that Monge was compelled to keep his principles secret until 1795, following which they became an important part of technical education in France and Germany and later in the United States. His book, *La Géométrie Descriptive*, is still regarded as the first text to expound the basic principles of projection drawing.

Monge's principles were brought to the United States from France in 1816 by Claude Crozet, an alumnus of the Polytechnic School and a professor at the United States Military Academy at West Point. He published the first text on the subject of descriptive geometry in the English language in 1821. In the years immediately following, these principles became a regular part of early engineering curricula at Rensselaer Polytechnic Institute, Harvard University, Yale University, and others. During the same period, the idea of manufacturing interchangeable parts in the early arms industries was being developed, and the principles of projection drawing were applied to these problems.

## 1.6 Modern Technical Drawing

Perhaps the first text on technical drawing in this country was *Geometrical Drawing*, published in 1849 by William Minifie, a high school teacher in Baltimore. In 1850 the Alteneder family organized the first drawing instrument manufacturing company in the United States (Theo. Alteneder & Sons, Philadelphia). In 1876 the blueprint process was introduced at the Philadelphia Centennial Exposition. Up to this time the graphic language was more or less an art, characterized by fine-line drawings made to resemble copper-plate engraving, by the use of shade lines, and by the use of water color "washes." These techniques became unnecessary after the introduction of blueprinting, and drawings gradually were made less ornate to obtain the best results from this method of reproduction. This was the beginning of modern technical drawing. The graphic language now became a relatively exact method of representation, and the building of a working model as a regular preliminary to construction became unnecessary.

Up to about 1900, drawings everywhere were generally made in what is called first-angle projection, §7.38, in which the top view was placed under the front view, the left-side view was placed at the right of the front view, and so on. At this time in the United States, after a considerable period of argument pro and con, practice gradually settled on the present *third-angle projection* in which the views are situated in what we regard as their more logical or natural positions. Today, third-angle projection is standard in the United States, but first-angle projection is still used throughout much of the world.

During the early part of the twentieth century, many books were published in which the graphic language was analyzed and explained in connection with its rapidly changing engineering design and industrial applications. Many of these writers were not satisfied with the term "mechanical drawing" because they recognized that technical drawing was really a graphic language. Anthony's *An Introduction to the Graphic Language*, French's *Engineering Drawing*, and Giesecke et al., *Technical Drawing* were all written with this point of view.

## 1.7 Drafting Standards

In all of the previously mentioned books there has been a definite tendency to standardize the characters of the graphic language, to eliminate its provincialisms and dialects, and to give industry, engineering, and science a uniform, effective graphic language. Of prime importance in this movement in the United States has been the work of the American National Standards Institute (ANSI) with the American Society for Engineering Education, the Society of Automotive Engineers, and the American Society of Mechanical Engineers. As sponsors they have prepared the *American National Standard Drafting Manual—Y14*, which is comprised of several separate sections that were published as approved standards as they were completed over a period of years. See Appendix 1.

These sections outline the most important idioms and usages in a form that is acceptable to the majority and are considered the most authoritative guide to uniform drafting practices in this country today. The Y14 Standard gives the characters of the graphic language, and it remains for the textbooks to explain the grammar and the penmanship.

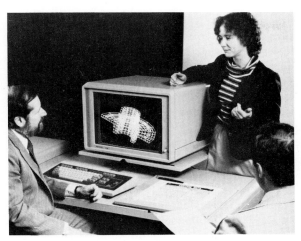

*Fig. 1.9* CAD Workstation. *Courtesy of Control Data Corporation*

## 1.8 Definitions

After this brief survey of the historical development of the graphic language, and before we begin a serious study of theory and applications, a few terms need to be defined.

*Descriptive geometry.* The grammar of the graphic language; it is the three-dimensional geometry forming the background for the practical applications of the language and through which many of its problems may be solved graphically.

*Instrumental* or *mechanical drawing.* Properly applies only to a drawing made with drawing instruments. The use of "mechanical drawing" to denote all industrial drawings is unfortunate not only because such drawings are not always mechanically drawn but also because that usage tends to belittle the broad scope of the graphic language by naming it superficially for its principal mode of execution.

*Computer graphics.* The application of conventional computer techniques with the aid of one of many graphic data processing systems available to the analysis, modification, and the finalizing of a graphical solution. The use of computers to produce technical drawings is called computer-aided design or computer-aided drafting (CAD) and also computer-aided design and drafting (CADD). A typical CAD workstation is shown in Fig. 1.9.

*Engineering drawing* and **engineering drafting.** Broad terms widely used to denote the graphic language. However, since the language is used not only by engineers but also by a much larger group of people in diverse fields who are concerned with technical work or with industrial production, these terms are still not broad enough.

*Technical drawing.* A broad term that adequately suggests the scope of the graphic language. It is rightly applied to any drawing used to express technical ideas. This term has been used by various writers since Monge's time at least and is still widely used, mostly in Europe.

*Engineering graphics* or *engineering design graphics.* Generally applied to drawings for technical use and has come to mean that part of technical drawing that is concerned with the graphical representation of designs and specifications for physical objects and data relationships as used in engineering and science.

*Technical sketching.* The freehand expression of the graphic language, whereas **mechanical drawing** is the instrumental expression of it. Technical sketching is a most valuable tool for the engineer and others engaged in technical work because through it most technical ideas can be expressed quickly and effectively without the use of special equipment.

*Blueprint reading.* The term applied to the "reading" of the language from drawings made by others. Actually, the blueprint process is only one of many forms by which drawings are reproduced today, but the term "blueprint reading" has been accepted through usage to mean the interpretation of all ideas expressed on technical drawings, whether the drawings are blueprints or not.

## 1.9 What Engineering, Science, and Technology Students Should Know

The development of technical knowledge from the dawn of history has been accompanied, and to a large extent made possible, by a corresponding graphic language. Today the intimate connection between engineering and science and the universal graphic language is more vital than ever before, and the engineer, scientist, or technician who is ignorant of or deficient in the principal mode of expression in his or her technical field is professionally illiterate. Thus training in the application of technical drawing is required in virtually every engineering school in the world.

The old days of fine-line drawings and of shading and "washes" are gone forever; artistic talent is no longer a prerequisite to learning the fundamentals of the graphic language. Instead, today's student of graphics needs precisely the aptitudes and abilities that will be needed in the science and engineering courses that are studied concurrently and later.

The well-trained engineer, scientist, or technician must be able to make and read correct graphical representations of engineering structures, designs, and data relationships. This means that the individual must understand the fundamental principles, or the *grammar*, of the language and be able to execute the work with reasonable skill, which is *penmanship*.

Graphics students often try to excuse themselves for inferior results (usually caused by lack of application) by arguing that after graduation they do not expect to do any drafting at all; they expect to have others make any needed drawings under their direction. Such a student presumptuously expects, immediately after graduation, to be the accomplished engineer concerned with bigger things and forgets that a first assignment may involve working with drawings and possibly revising drawings, either on the board or with computerized aids, under the direction of an experienced engineer. Entering the engineering profession via graphics provides an excellent opportunity to learn about the product, the company operations, and the supervision of others.

Even a young engineer who has not been successful in developing a skillful penmanship in the graphic language will have use for its grammar, since the ability to *read* a drawing will be of utmost importance. See Chapter 16.

Furthermore, the engineering student is apt to overlook the fact that, in practically all the subsequent courses taken in college, technical drawings will be encountered in most textbooks. The student is often called upon by instructors to supplement calculations with mechanical drawings or sketches. Thus, a mastery of a course in technical drawing will aid materially not only in professional practice after graduation but more immediately in other technical courses, and it will have a definite bearing on scholastic progress.

Besides the direct values to be obtained from a serious study of the graphic language, there are a number of very important training values that, though they may be considered by-products, are as essential as the language itself. Many students learn the meaning of neatness, speed, and accuracy for the first time in a drawing course. These are basic habits that every successful engineer, scientist, and technician must have or acquire.

All authorities agree that the ability to *think in*

*three dimensions* is one of the most important requisites of the successful scientist and engineer. This training to visualize objects in space, to use the constructive imagination, is one of the principal values to be obtained from a study of the graphic language. The ability to *visualize* is possessed to an outstanding degree by persons of extraordinary creative ability. It is difficult to think of Edison, De Forest, or Einstein as being deficient in constructive imagination.

With the increase in technological development and the consequent crowding of drawing courses by the other engineering and science courses in our colleges, it is doubly necessary for students to make the most of the limited time devoted to the language of the profession, to the end that they will not be professionally illiterate, but will possess an ability to express ideas quickly and accurately through the correct use of the graphic language.

## 1.10 Projections

Behind every drawing of an object is a space relationship involving four imaginary things.

1. The *observer's eye,* or the *station point.*
2. The *object.*
3. The *plane* or *planes of projection.*

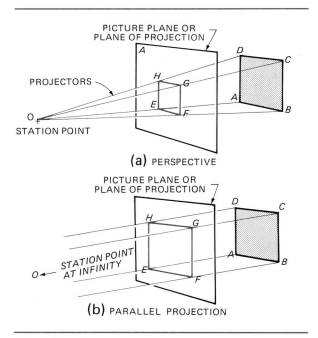

**(a)** PERSPECTIVE

**(b)** PARALLEL PROJECTION

*Fig. 1.10* Projections.

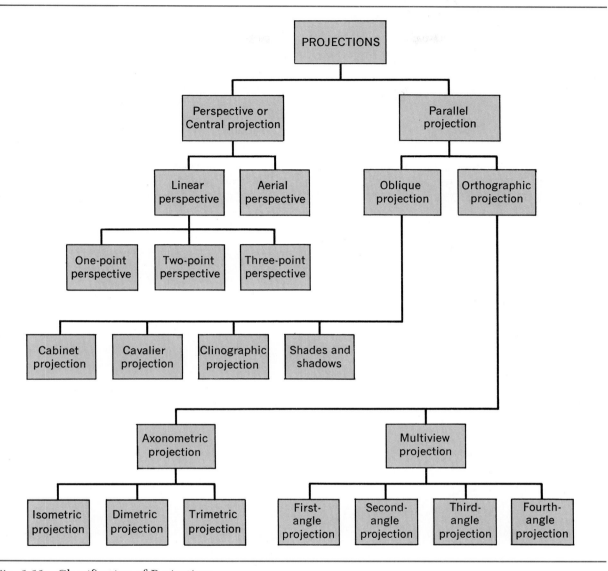

*Fig. 1.11* Classification of Projections.

4. The *projectors*, also called visual rays and lines of sight.

For example, in Fig. 1.10 (a) the drawing **EFGH** is the projection, on the plane of projection **A**, of the square **ABCD** as viewed by an observer whose eye is at the point **O**. The projection or drawing upon the plane is produced by the points where the projectors pierce the plane of projection (piercing points). In this case, where the observer is relatively close to the object and the projectors form a "cone" of projectors, the resulting projection is known as a perspective.

If the observer's eye is imagined as infinitely distant from the object and the plane of projection, the projectors will be parallel, as shown in Fig. 1.10 (b); hence, this type of projection is known as a *parallel* projection. If the projectors, in addition to being parallel to each other, are perpendicular (normal) to the plane of projection, the result is an *ortho-graphic,* or right-angle, projection. If they are parallel to each other but oblique to the plane of projection, the result is an *oblique* projection.

These two main types of projection—perspective and central or parallel projection—are further broken down into many subtypes, as shown in Fig. 1.11,

**Table 1.1** *Classification by Projectors*

| Classes of Projection | Distance from Observer to Plane of Projection | Direction of Projectors |
|---|---|---|
| Perspective | Finite | Radiating from station point |
| Parallel | Infinite | Parallel to each other |
| Oblique | Infinite | Parallel to each other and oblique to plane of projection |
| Orthographic | Infinite | Perpendicular to plane of projection |
| Axonometric | Infinite | Perpendicular to plane of projection |
| Multiview | Infinite | Perpendicular to planes of projection |

and will be treated at length in the various chapters that follow.

A classification of the main types of projection according to their projectors is shown in Table 1.1.

# Instrumental Drawing

For many years the items of equipment essential to students in technical schools, and to engineers and designers in professional practice, remained unchanged. One needed a drawing board, T-square, triangles, an architects' or engineers' scale, and a professional quality set of drawing instruments. Recently, however, there has been a shift toward greater use of the drafting machine, the parallel-ruling straightedge, the technical fountain pen, and other modern equipment—not to mention the significant increase in the use of the computer as a drafting tool.

The basic items of equipment are shown in Fig. 2.1. To secure the most satisfactory results, the drawing equipment should be of high grade. When drawing instruments (item 3) are to be purchased, the advice of an experienced drafter or designer, or a reliable dealer,* should be sought because it is difficult for beginners to distinguish high-grade instruments from those that are inferior.

*Keuffel & Esser Co., Morristown, NJ; Eugene Dietzgen Co., Des Plaines, IL; Charles Bruning Co., Mt. Prospect, IL; Teledyne Post, Des Plaines, IL; Vemco Corp., Pasadena, CA; Staedtler Mars, Elk Grove Village, IL; Koh-I-Noor Rapidograph, Inc., Bloomsbury, NJ; and Tacro, Brooklyn, NY, are some of the larger distributors of this equipment; their products are available through local dealers.

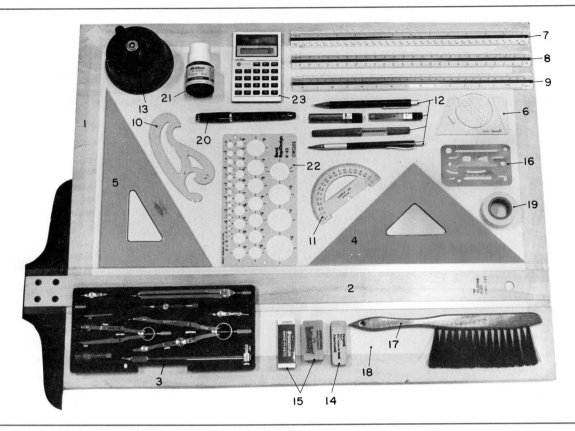

*Fig. 2.1*  Principal Items of Equipment.

## 2.1 Typical Equipment

A complete list of equipment for students of technical drawing follows. The numbers refer to the equipment illustrated in Fig. 2.1.

1. Drawing board (approx. 20″ × 24″), drafting table, or desk, §2.4.

2. T-square (24″, transparent edge), drafting machine, or parallel ruling edge, §§2.5, 2.56, and 2.57.

3. Set of instruments, §§2.33 and 2.34.

4. 45° triangle (8″ sides), §2.16.

5. 30° × 60° triangle (10″ long side), §2.16.

6. Ames Lettering Guide or lettering triangle, §4.16.

7. Architects' triangular scale, §2.28.

8. Engineers' triangular scale, §2.27.

9. Metric triangular scale, §2.25.

10. Irregular curve, §2.53.

11. Protractor, §2.18.

12. Mechanical pencils and/or thin-lead mechanical pencils and HB, F, 2H, and 4H to 6H leads, or drawing pencils. See §§2.8 and 2.9.

13. Lead pointer and sandpaper pad, §2.10.

14. Pencil eraser, §2.12.

15. Plastic drafting eraser or Artgum cleaning eraser, §2.12.

16. Erasing shield, §2.12.

17. Dusting brush, §2.12.

18. Drawing paper, tracing paper, tracing cloth, or films as required. Backing sheet (drawing paper—white, cream, or light green) to be used under drawings and tracings.

19. Drafting tape, §2.7.

20. Technical fountain pens, §2.48.

21. Drawing ink, §2.50.

22. Templates, §2.55.

*Fig.* 2.2 Orderliness Promotes Efficiency and Accuracy.

23. Calculator.

24. Cleansing tissue or dust cloth (not shown).

## 2.2 Objectives in Drafting

On the following pages, the correct methods to be used in instrumental drawing are explained. The student who practices and learns correct manipulation of the drawing instruments will eventually be able to draw correctly by habit, thus giving his or her full attention to the problems at hand.

The following are the important objectives the student should strive to attain:

1. *Accuracy.* No drawing is of maximum usefulness if it is not accurate. The student cannot achieve success in a college career or later in professional employment if the habit of accuracy is not acquired.

2. *Speed.* "Time is money" in industry, and there is no demand for the slow drafter, technician, or engineer. However, speed is not attained by hurrying; it is an unsought by-product of *intelligent and continuous work*. It comes with study and practice.

3. *Legibility.* The drafter, technician, or engineer should remember that the drawing is a means of communication to others, and that it must be clear and legible in order to serve its purpose well. Care should be given to details, especially to lettering, Chapter 4.

4. *Neatness.* If a drawing is to be accurate and legible, it must also be clean; therefore, the student should constantly strive to acquire the habit of neatness. Untidy drawings are the result of sloppy and careless methods, §2.13, and will be unacceptable to an instructor or employer.

## 2.3 Drafting at Home or School

In the school drafting room, as in the industrial drafting room, the student is expected to give thoughtful and continuous attention to the problems at hand.

Technical drawing requires headwork and must be done in quiet surroundings without distractions. The efficient drafting student sees to it that the correct equipment is available and refrains from borrowing—a nuisance to everyone. While the student is drawing, the textbook—the chief source of information—should be available and in a convenient position, Fig. 2.2.

When questions arise, first use the index of the text and endeavor to find the answer for yourself.

Try to develop self-reliance and initiative, but when you really need help, ask your instructor. Students who go about their work intelligently, with a minimum waste of time, first study the assignment carefully to be sure that they understand the principles involved; second, make sure that the correct equipment is in proper condition (such as sharp pencils); and third, make an effort to dig out answers for themselves (the only true education).

One of the principal means of promoting efficiency in drafting is orderliness. Efficiency, in turn, will produce accuracy in drawing. All needed equipment and materials should be placed in an orderly manner so that everything is in a convenient place and can readily be found when needed, Fig. 2.2. The drawing area should be kept clear of equipment not in direct use. Form the habit of placing each item in a regular place outside the drawing area when it is not being used.

When drawing at home, if possible work in a room by yourself. Place a book under the upper portion of the drawing board to give the board a convenient inclination, or you can pull out the study table drawer and use it to support the drawing board at a slant.

It is best to work in natural north light coming from the left and slightly from the front. Never work on a drawing in direct sunlight or in dim light, as either may be injurious to the eyes. If artificial light is needed, the light source should be such that shadows are not cast where lines are being drawn and such that there will be little reflected glare from the paper as possible. Special drafting fluorescent lamps are available with adjustable arms so that the light source may be moved to any desired position. Drafting will not hurt eyes that are in normal condition, but the exacting work will often disclose deficiencies not previously suspected.

***Left-handers*** Place the head of the T-square on the right, and have the light come from the right and slightly from the front.

## 2.4 Drawing Boards

If the left edge of the drafting table top has a true straight edge and if the surface is hard and smooth (such as masonite), a drawing board is unnecessary, provided that drafting tape is used to fasten the drawings. It is recommended that a backing sheet of heavy drawing paper be placed between the drawing and the table top.

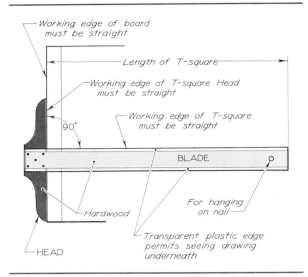

*Fig. 2.3* The T-square.

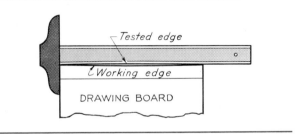

*Fig. 2.4* Testing the Working Edge of the Drawing Board.

However, in most cases a drawing board will be needed. These vary from 9″ × 12″ (for sketching and field work) up to 48″ × 72″ or larger. The recommended size for students is 20″ × 24″, Fig. 2.1, which will accommodate the largest sheet likely to be used.

Drafters use drafting tape, which in turn permits surfaces such as hardwood, masonite, or other materials to be used for drawing boards.

The left-hand edge of the board is called the *working edge* because the T-square head slides against it, Fig. 2.3. This edge must be straight, and you should test the edge with a T-square blade that has been tested and found straight, Fig. 2.4. If the edge of the board is not true, it should be replaced.

## 2.5 T-square

The T-square, Fig. 2.3, is composed of a long strip, called the *blade*, fastened rigidly at right angles to a shorter piece called the *head*. The upper edge of the blade and the inner edge of the head are *working edges* and must be straight. The working edge of the head must not be convex or the T-square will rock when the head is placed against the board. The blade should have transparent plastic edges and should be free of nicks along the working edge. Transparent edges are recommended, since they permit the draftsman to see the drawing in the vicinity of the lines being drawn.

*Do not use the T-square for any rough purpose. Never cut paper along its working edge, as the plastic is easily cut and even a slight nick will ruin the T-square.*

## 2.6 Testing and Correcting the T-square

To test the working edge of the head, see if the T-square rocks when the head is placed against a straight edge, such as a drawing board working edge that has already been tested and found true. If the working edge of the head is not straight, the T-square should be replaced.

To test the working edge of the blade, Fig. 2.5, draw a sharp line very carefully with a hard pencil along the entire length of the working edge; then turn the T-square over and draw the line again along the same edge. If the edge is straight, the two lines will coincide; otherwise, the space between the lines will be twice the error of the blade.

It is difficult to correct a crooked T-square blade, and if the error is considerable, it may be necessary to discard the T-square and obtain another.

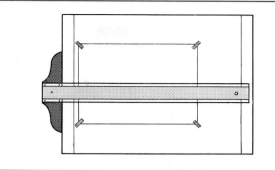

*Fig. 2.6*   Placing Paper on Drawing Board.

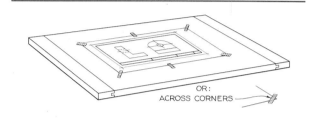

*Fig. 2.7*   Positions of Drafting Tape.

## 2.7 Fastening Paper to the Board

The drawing paper should be placed close enough to the working edge of the board to reduce to a minimum any error resulting from a slight "give," or bending of the blade of the T-square, and close enough to the upper edge of the board to permit space at the bottom of the sheet for using the T-square and supporting the arm while drawing, Fig. 2.6.

Drafting tape is preferred for fastening the drawing to the board, Fig. 2.7, because it does not damage the board and it will not damage the paper if it is removed by *pulling it off slowly toward the edge of the paper.*

To fasten the paper in place, press the T-square head firmly against the working edge of the drawing board with the left hand, while the paper is adjusted with the right hand until the top edge coincides with the upper edge of the T-square. Then move the T-square to the position shown and fasten the upper left corner, then the lower right corner, and finally the remaining corners. Large sheets may require additional fastening, whereas small sheets may require fastening only at the two upper corners.

Tracing paper should not be fastened directly to the board because small imperfections in the surface

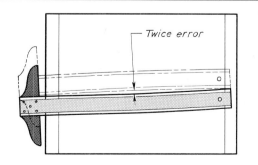

*Fig. 2.5*   Testing the T-square.

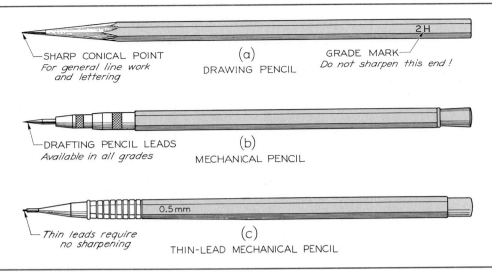

SHARP CONICAL POINT
*For general line work and lettering*

(a)
DRAWING PENCIL

GRADE MARK
*Do not sharpen this end !*

DRAFTING PENCIL LEADS
*Available in all grades*

(b)
MECHANICAL PENCIL

*Thin leads require no sharpening*

(c)
THIN-LEAD MECHANICAL PENCIL

*Fig. 2.8*   Drawing Pencils.

## 2.8 Drawing Pencils

High-quality drawing pencils, Fig. 2.8 (a), should be used in technical drawing—never ordinary writing pencils.

Many makes of mechanical pencils are available, Fig. 2.8 (b), together with refill drafting leads of conventional size in all grades. Choose the holder that feels comfortable in the hand and that grips the lead firmly without slipping. Mechanical pencils have the advantages of maintaining a constant length of lead while permitting the use of a lead practically to the end, of being easily refilled with new leads, of affording a ready source for compass leads, of having no wood to be sharpened, and of easy sharpening of the lead by various mechanical pencil pointers now available

Thin-lead mechanical pencils, Fig. 2.8 (c), are available with 0.3, 0.5, 0.7, or 0.9 mm diameter drafting leads in several grades. These thin leads produce uniform width lines without sharpening, providing both a time savings and a cost benefit.

Mechanical pencils are recommended as they are less expensive in the long run.

## 2.9 Choices of Grade of Pencil

Drawing pencil leads are made of graphite with the addition of a polymer binder or of kaolin (clay) in varying amounts to make 18 grades from 9H, the hardest, down to 7B, the softest. The uses of these different grades are shown in Fig. 2.9. Note that small diameter leads are used for the harder grades, whereas large diameter leads are used to give more strength to the softer grades. Hence, the degree of hardness in the wood pencil can be roughly judged by a comparison of the diameters.

Specifically formulated leads of carbon black particles in a polymer binder are also available in several grades for use on the polyester films now found quite extensively in industry. See §2.62.

To select the grade of lead, first take into consideration the type of line work required. For light construction lines, guide lines for lettering, and for accurate geometrical constructions or work where accuracy is of prime importance, use a hard lead, such as 4H to 6H.

For mechanical drawings on drawing paper or tracing paper, the lines should be **black,** particularly for drawings to be reproduced. The lead chosen must be soft enough to produce jet black lines, but hard enough not to smudge too easily or permit the point to crumble under normal pressure. The same comparatively soft lead is preferred for lettering and arrowheads.

of the board will interfere with the line work. Always fasten a larger backing sheet of heavy drawing paper on the board first, then fasten the tracing paper over this sheet.

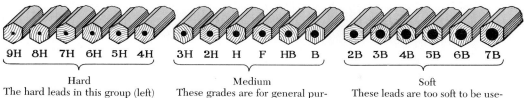

| 9H 8H 7H 6H 5H 4H | 3H 2H H F HB B | 2B 3B 4B 5B 6B 7B |
|---|---|---|
| Hard | Medium | Soft |
| The hard leads in this group (left) are used where extreme accuracy is required, as on graphical computations and charts and diagrams. The softer leads in this group (right) are used by some for line work on engineering drawings, but their use is restricted because the lines are apt to be too light. | These grades are for general purpose work in technical drawing. The softer grades (right) are used for technical sketching, for lettering, arrowheads, and other freehand work on mechanical drawings. The harder leads (left) are used for line work on machine drawings and architectural drawings. The H and 2H leads are widely used on pencil tracings for reproduction. | These leads are too soft to be useful in mechanical drafting. Their use for such work results in smudged, rough lines which are hard to erase, and the lead must be sharpened continually. These grades are used for art work of various kinds, and for full-size details in architectural drawing. |

*Fig. 2.9*  Lead Grade Chart.

This lead will vary from F to 2H, roughly, depending on the paper and weather conditions. If the paper is hard, it will be necessary generally to use harder leads. For softer surfaces, softer leads can be used. The weather factor to consider is the humidity. On humid days the paper absorbs moisture from the atmosphere and becomes soft. This can be recognized because the paper expands and becomes wrinkled. It is necessary to select softer leads to offset the softening of the paper. If you have been using a 2H lead, for example, change to an F until the weather clears up.

## 2.10 Sharpening the Pencil

*Keep your lead sharp!* This is certainly the instruction needed most frequently by the beginning student. A dull lead produces fuzzy, sloppy, indefinite lines. Only a sharp lead is capable of producing clean-cut black lines that sparkle with clarity.

If a good mechanical pencil, Fig. 2.8 (b), is used, much time may be saved in sharpening, since the lead can be fed from the pencil as needed. Two excellent lead pointers for mechanical pencils are shown in Fig. 2.10. Each has the advantage of one-hand manipulation, and of collecting the loose graphite particles inside, where they cannot soil the hands, the drawing, or other equipment.

If thin-lead mechanical pencils are used Fig. 2.8 (c), no sharpening is required since the lead di-

(a) TRU-POINT

(b) BEROL TURQUOISE

*Fig. 2.10*  Pencil Lead Pointers. *Courtesy of Keuffel & Esser Co./Kratos*

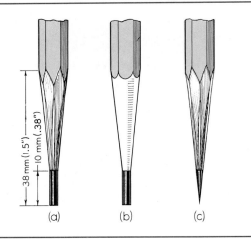

**Fig. 2.11**    Pencil Points.

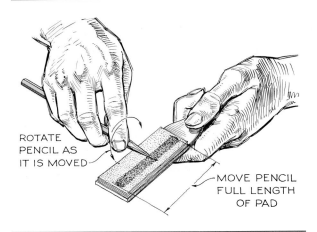

**Fig. 2.13**    Shaping the Lead.

ameter determines the line width. Hence, several thin-lead mechanical pencils are required for the various line widths used in technical drawing. Each thin-lead mechanical pencil will accommodate only one diameter of lead.

If a wood drawing pencil is used, Fig. 2.8 (a), sharpen the unlettered end in order to preserve the identifying grade mark. First, the wood is removed with a knife or a special drafting pencil sharpener starting about 38 mm (1.5″) from the end and about 10 mm (.38″) of uncut lead is exposed, Fig. 2.11 (a) or (b). Next the lead is shaped to a sharp conical point, and the point is wiped clean with cloth or paper tissue to remove loose particles of graphite.

Mechanical devices for removal of the wood are shown in Fig. 2.12 (a) and (b). The procedures for shaping the lead are illustrated in Fig. 2.13.

*Never sharpen your pencil over the drawing or any of your equipment.*

*Keep the pencil pointer close by, as frequent pointing of the pencil will be necessary.*

When the sandpaper pad is not in use, it should be kept in a container, such as an envelope, to prevent the particles of graphite from falling upon the drawing board or drawing equipment, Fig. 2.12 (c).

Many drafters burnish the point on a piece of hard paper to obtain a smoother, sharper point. However, for drawing visible lines the point should not be needle-sharp, but very slightly rounded. First sharpen the lead to a needle point, then stand the pencil vertically, and with a few rotary motions on the paper, wear the point down slightly to the desired shape.

## 2.11 Alphabet of Lines

Each line on a technical drawing has a definite meaning and is drawn in a certain way. The line conventions endorsed by the American National Standards Institute, ANSI Y14.2M–1979 (R1987), are presented in Fig. 2.14, together with illustrations of various applications.

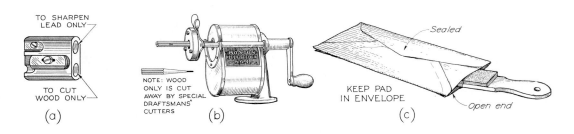

**Fig. 2.12**    Pencil Sharpeners.

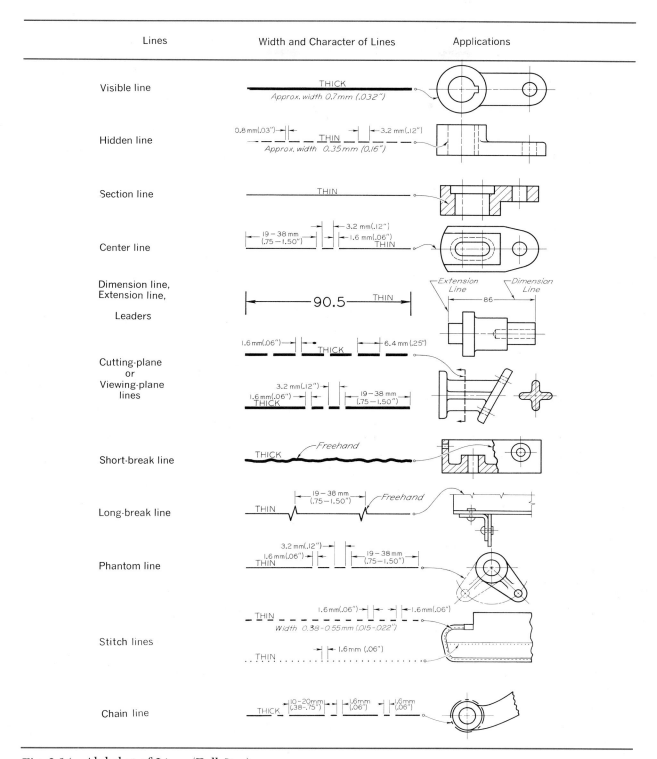

*Fig. 2.14*  Alphabet of Lines (Full Size).

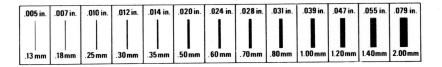

| .005 in. | .007 in. | .010 in. | .012 in. | .014 in. | .020 in. | .024 in. | .028 in. | .031 in. | .039 in. | .047 in. | .055 in. | .079 in. |
|---|---|---|---|---|---|---|---|---|---|---|---|---|
| .13 mm | .18 mm | .25 mm | .30 mm | .35 mm | .50 mm | .60 mm | .70 mm | .80 mm | 1.00 mm | 1.20 mm | 1.40 mm | 2.00 mm |

*Fig. 2.15*    Line Gage. *Courtesy of Koh-I-Noor Rapidograph, Inc.*

Two widths of lines are recommended for use on drawings. All lines should be clean-cut, dark, uniform throughout the drawing, and properly spaced for legible reproduction by all commonly used methods. Minimum spacing of 1.5 mm (.06″) between parallel lines is usually satisfactory for all reduction and/or reproduction processes. The size and style of the drawing and the smallest size to which it is to be reduced govern the actual width of each line. The contrast between the two widths of lines should be distinct. Pencil leads should be hard enough to prevent smudging, but soft enough to produce the dense black lines so necessary for quality reproduction.

When photoreduction and blowback are not necessary, as is the case for most drafting laboratory assignments, three weights of lines may improve the appearance and legibility of the drawing. The "thin lines" may be made in two widths—regular thin lines for hidden lines and stitch lines and a somewhat thinner version for the other secondary lines such as center lines, extension lines, dimension lines, leaders, section lines, phantom lines, and long-break lines.

For the "thick lines"—visible, cutting plane, and short break—use a relatively soft lead such as F or H. All thin lines should be made with a sharp medium-grade lead such as H or 2H. All lines (except construction lines) must be *sharp* and *dark*. Make construction lines with a sharp 4H or 6H lead so thin that they barely can be seen at arm's length and need not be erased.

The high-quality photoreduction and reproduction processes used in the production of this book permitted the use of three weights of lines in many illustrations and drawings for increased legibility.

In Fig. 2.14, the ideal lengths of all dashes are indicated. It would be well to measure the first few hidden dashes and center-line dashes you make and then thereafter to estimate the lengths carefully by eye.

The line gage, Fig. 2.15, is convenient when referring to lines of various widths.

## 2.12 Erasing

Erasers are available in many degrees of hardness and abrasiveness. For general drafting the Pink Pearl or the Mars-Plastic is suggested, Fig. 2.16. These erasers are suitable for erasing pencil or ink line work. Best results are obtained if a hard surface, such as a triangle is placed under the area being erased. If the surface has become badly grooved by the lines, the surface can be improved by burnishing the back side with a hard smooth object or with the back of the fingernail.

The erasing shield, Fig. 2.17, is used to protect the lines near those being erased.

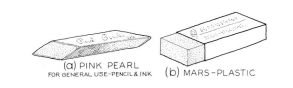

(a) PINK PEARL
FOR GENERAL USE-PENCIL & INK     (b) MARS-PLASTIC

*Fig. 2.16*    Erasers.

*Fig. 2.17*    Using the Erasing Shield.

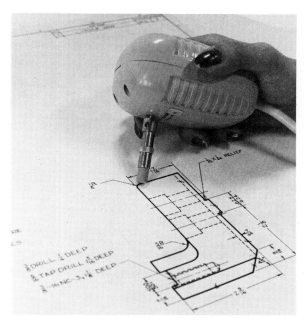

*Fig. 2.18*   Electric Erasing Machine.

*Fig. 2.19*   Dusting Brush.

The electric erasing machine, Fig. 2.18, saves time and is essential if much drafting is being done.

A dusting brush, Fig. 2.19, is useful for removing eraser crumbs without smearing the drawing.

## 2.13 Keeping Drawings Clean

Cleanliness in drafting is very important and should become a habit. Cleanliness does not just happen; it results only from a conscious effort to observe correct procedures.

First, the drafter's hands should be clean at all times. Oily or perspiring hands should be frequently washed with soap and water. Talcum powder on the hands tends to absorb excessive perspiration.

Second, all drafting equipment, such as drawing board, T-square, triangles, and scale, should be wiped frequently with a clean cloth. Water should be used sparingly and dried off immediately. A soft eraser may also be used for cleaning drawing equipment.

Third, the largest contributing factor to dirty drawings is *not dirt, but graphite* from the pencil; hence, the drafter should observe the following precautions:

1.  Never sharpen a lead over the drawing or any equipment.

2.  Always wipe the lead point with a clean cloth or cleansing tissue, after sharpening or pointing, to remove small particles of loose graphite.

3.  Never place the sandpaper pad or file in contact with any other drawing equipment unless it is completely enclosed in an envelope or similar cover, Fig. 2.12 (c).

4.  Never work with the sleeves or hands resting upon a penciled area. Keep such parts of the drawing covered with clean paper (not a cloth). In lettering a drawing, always place a piece of paper under the hand.

5.  Avoid unnecessary sliding of the T-square or triangles across the drawing. Pick up the triangles by their tips and tilt the T-square blade upward slightly before moving. A very light sprinkling of powdered Artgum or drafting powder on the drawing helps to keep the drawing clean by picking up the loose graphite particles as you work. It should be brushed off and replaced occasionally.

6.  Never rub across the drawing with the palm of the hand to remove eraser particles; use a dust brush, Fig. 2.19, or flick—don't rub—the particles off with a clean cloth.

If the foregoing rules are observed, a completed drawing will not need to be cleaned. The practice of making a pencil drawing, scrubbing it with a soft eraser, and then retracing the lines is poor technique and a waste of time, and this habit should not be acquired.

At the end of the period or of the day's work, the drawing should be covered with paper or cloth to protect it from dust.

If the drawing must be removed from the board before it is complete, it can be carried flat in a draw-

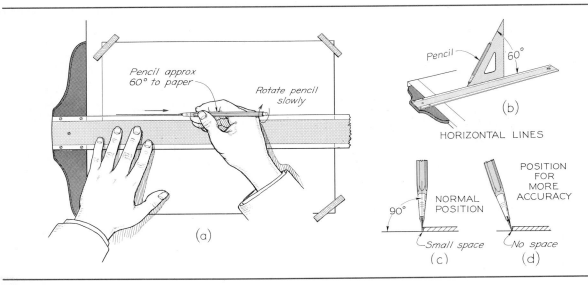

*Fig. 2.20*    Drawing a Horizontal Line.

ing portfolio or gently rolled and carried in a card-board or plastic tube, preferably one with ends that can be closed.

## 2.14 Horizontal Lines

To draw a horizontal line, Fig. 2.20 (a), press the head of the T-square firmly against the working edge of the board with the left hand; then slide the left hand to the position shown, so as to press the blade tightly against the paper. Lean the pencil in the direction of the line at an angle of approximately 60° with the paper, (b), and draw the line from left to right. Keep the pencil in a vertical plane, (b) and (c); otherwise, the line may not be straight. While drawing the line, let the little finger of the hand holding the pencil glide lightly on the blade of the T-square, and rotate the pencil slowly, except for the thin-lead pencils, between the thumb and fore-finger so as to distribute the wear uniformly on the lead and maintain a symmetrical point.

When great accuracy is required, the pencil may be "toed in" as shown at (d) to produce a perfectly straight line.

Thin-lead pencils should be held nearly vertical to the paper and not rotated. Also, pushing the thin-lead pencil from left to right, rather than pulling it, tends to minimize lead breakage.

*Left-handers*    In general, reverse the procedure just outlined. Place the T-square head against the right edge of the board, and with the pencil in the left hand, draw the line from right to left.

## 2.15 Vertical Lines

Use either the 45° triangle or the 30° × 60° triangle to draw vertical lines. Place the triangle on the T-square with the *vertical edge on the left* as shown in Fig. 2.21 (a). With the left hand, press the head of the T-square against the board, then slide the hand to the position shown where it holds both the T-square and the triangle firmly in position. Then draw the line upward, rotating the pencil slowly be-tween the thumb and forefinger.

Lean the pencil in the direction of the line at an angle of approximately 60° with the paper and in a vertical plane, (b). Meanwhile, the upper part of the body should be twisted to the right as shown at (c).

See §2.14 regarding the use of thin-lead pencils.

*Left-handers*    In general, reverse the above pro-cedure. Place the T-square head on the right and the vertical edge of the triangle on the right; then, with the right hand, hold the T-square and triangle firmly together, and with the left hand draw the line upward.

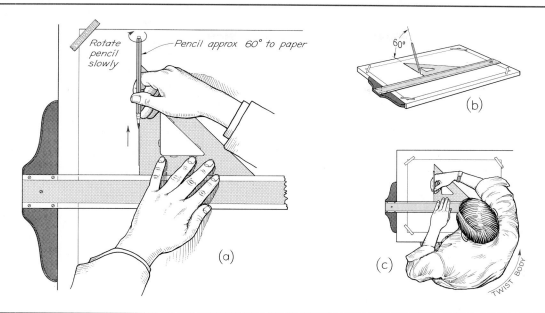

*Fig. 2.21*   Drawing a Vertical Line.

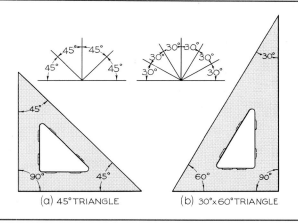

*Fig. 2.22*   Triangles.

The only time it is advisable for right-handers to turn the triangle so that the vertical edge is on the right is when drawing a vertical line near the right end of the T-square. In this case, the line would be drawn downward.

## 2.16 The Triangles

Most inclined lines in mechanical drawing are drawn at standard angles with the *45° triangle* and the *30° × 60° triangle*, Fig. 2.22. The triangles are made of transparent plastic so that lines of the draw-

ing can be seen through them. A good combination of triangles is the 30° × 60° triangle with a long side of 10″ and a 45° triangle with each side 8″ long.

## 2.17 Inclined Lines

The positions of the triangles for drawing lines at all of the possible angles are shown in Fig. 2.23. In the figure it is understood that the triangles in each case are resting upon the blade of the T-square. Thus, it is possible to divide 360° into twenty-four 15° sectors with the triangles used singly or in combination.

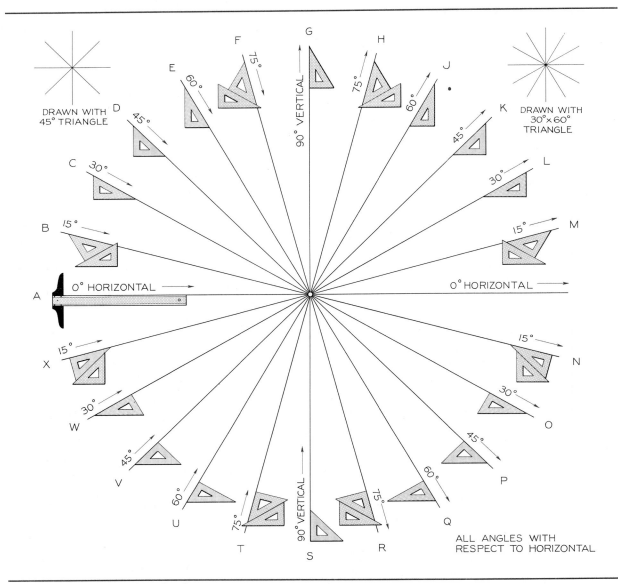

*Fig. 2.23*   The Triangle Wheel.

Note carefully the directions for drawing the lines, as indicated by the arrows, and that all lines in the left half are drawn *toward the center*, while those in the right half are drawn *away from the center*.

## 2.18 **Protractors**

For measuring or setting off angles other than those obtainable with the triangles, the *protractor* is used. The best protractors are made of nickel silver and are capable of most accurate work, Fig. 2.24 (a). For ordinary work the plastic protractor is satisfactory

and much cheaper, (b). To set off angles with greater accuracy, use one of the methods presented in §5.20.

## 2.19 **Drafting Angles**

A variety of devices combining the protractor with triangles to produce great versatility of use are available, one type of which is shown in Fig. 2.25.

## 2.20 **To Draw a Line Through Two Points**

To draw a line through two points, Fig. 2.26, place the pencil vertically at one of the points, and move

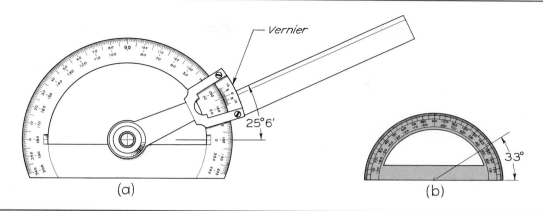

*Fig. 2.24* Protractors.

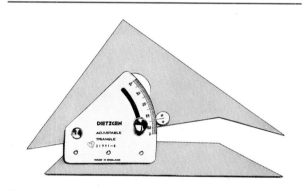

*Fig. 2.25* Adjustable Triangle.

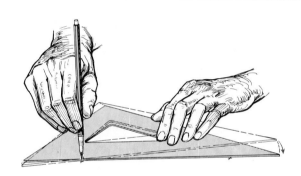

*Fig. 2.26* To Draw a Pencil Line Through Two Points.

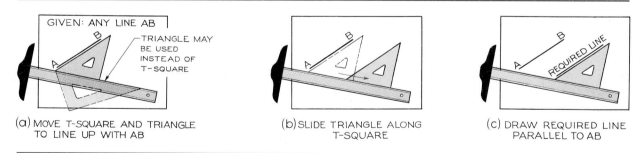

*Fig. 2.27* To Draw a Line Parallel to a Given Line.

the straightedge about the pencil point as a pivot until it lines up with the other point; then draw the line along the edge.

## 2.21 Parallel Lines

To draw a line parallel to a given line, Fig. 2.27, move the triangle and T-square as a unit until the hypotenuse of the triangle lines up with the given line, (a); then, holding the T-square firmly in position, slide the triangle away from the line, (b), and draw the required line along the hypotenuse, (c).

Obviously any straightedge, such as one of the triangles, may be substituted for the T-square in this operation, as shown at (a).

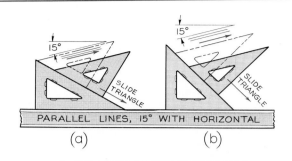

**Fig. 2.28**  Parallel Lines.

To draw parallel lines at 15° with horizontal, arrange the triangles as shown in Fig. 2.28.

### 2.22 Perpendicular Lines

To draw a line perpendicular to a given line, move the T-square and triangle as a unit until one edge of the triangle lines up with the given line, Fig. 2.29 (a); then slide the triangle across the line, (b), and draw the required line, (c).

To draw perpendicular lines when one of the lines makes 15° with horizontal, arrange the triangles as shown in Fig. 2.30.

### 2.23 Lines at 30°, 60°, or 45° with Given Line

To draw a line making 30° with a given line, arrange the triangle as shown in Fig. 2.31. Angles of 60° and 45° may be drawn in a similar manner.

### 2.24 Scales

A drawing of an object may be the same size as the object (full size), or it may be larger or smaller than the object.

The ratio of reduction or enlargement depends upon the relative sizes of the object and of the sheet of paper upon which the drawing is to be made. For example, a machine part may be half size; a building may be drawn $\frac{1}{48}$ size; a map may be drawn $\frac{1}{1200}$ size; or a printed circuit board, Fig. 2.32, may be drawn four times size.

Scales, Fig. 2.33, are classified as the *metric scale* (a), the *engineers' scale* (b), the *decimal scale* (c), the *mechanical engineers' scale* (d), and the *architects' scale* (e).

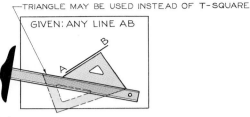

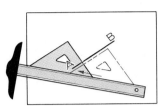

(a) MOVE T-SQUARE AND TRIANGLE TO LINE UP WITH AB

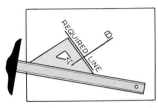

(b) SLIDE TRIANGLE ALONG T-SQUARE

(c) DRAW REQUIRED LINE PERPENDICULAR TO AB

**Fig. 2.29**  To Draw a Line Perpendicular to a Given Line.

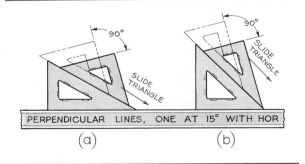

**Fig. 2.30**  Perpendicular Lines.

A full-divided scale is one in which the basic units are subdivided throughout the length of the scale, Fig. 2.33, except for the lower scale at (e). An open divided scale is one in which only the end unit is subdivided, as in the lower scale at (e).

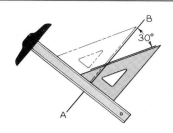

**Fig. 2.31**   Line at 30° with Given Line.

**Fig. 2.32**   Printed Circuit Board. *United Nations/Guthrie*

Scales are usually made of plastic or boxwood. The better wood scales have white plastic edges. Scales are either triangular, Fig. 2.34 (a) and (b), or flat, (c) to (f). The triangular scales have the advantage of combining many scales on one stick, but the user will waste much time looking for the required scale if a *scale guard*, (g), is not used. The flat scale is almost universally used by professional drafters because of its convenience, but several flat scales are necessary to replace one triangular scale, and the total cost is greater.

## 2.25 Metric Scales Fig. 2.33 (a)

The metric system is an international language of measurement that, despite modifications over the past 200 years, has been the foundation of science and industry and is clearly defined. The modern form of the metric system is "The International System of Units," commonly referred to as SI (from the French name, Le Système International d'Unités). It is important to remember that SI differs in several respects from former metric systems. For example, cc was previously an accepted abbreviation for cubic centimeter; in SI the symbol used is $cm^3$. In the past, degree centigrade was accepted as an alternative name for degree Celsius; in SI only degree Celsius is used. Probably the most important characteristic of SI is that it is a unique system; each quantity has only one unit. The SI system was established in 1960 by international agreement and is now considered the standard international language of measurement.

The metric scale is used when the meter is the standard for linear measurement. The meter was established by the French in 1791 with a length of one ten-millionth of the distance from the Earth's equator to the pole. The meter is equal to 39.37 inches or approximately 1.1 yards.

The metric system for linear measurement is a decimal system similar to our system of counting money. For example,

$$1 \text{ mm} = 1 \text{ millimeter } (\tfrac{1}{1000} \text{ of a meter})$$
$$1 \text{ cm} = 1 \text{ centimeter } (\tfrac{1}{100} \text{ of a meter})$$
$$= 10 \text{ mm}$$
$$1 \text{ dm} = 1 \text{ decimeter } (\tfrac{1}{10} \text{ of a meter})$$
$$= 10 \text{ cm} = 100 \text{ mm}$$
$$1 \text{ m} = 1 \text{ meter} = 100 \text{ cm} = 1000 \text{ mm}$$
$$1 \text{ km} = 1 \text{ kilometer} = 1000 \text{ m}$$
$$= 100\,000 \text{ cm} = 1\,000\,000 \text{ mm}$$

The primary unit of measurement for engineering drawings and design in the mechanical industries is the millimeter (mm). Secondary units of measure are the meter (m) and the kilometer (km). The centimeter (cm) and the decimeter (dm) are rarely used.

In recent years, the auto and other industries have used a dual dimensioning system of millimeters and inches; see Fig. 16.22. The large agricultural machinery manufacturers have elected to use all metric dimensions with the inch equivalents given in a table on the drawing, Fig. 16.22.

Many of the dimensions in the illustrations and the problems in this text are given in metric units. Dimensions that are given in the customary units (inches and feet, either decimal or fractional) may be converted easily to metric values. In accordance with standard practice, the ratio 1 in. = 25.4 mm is

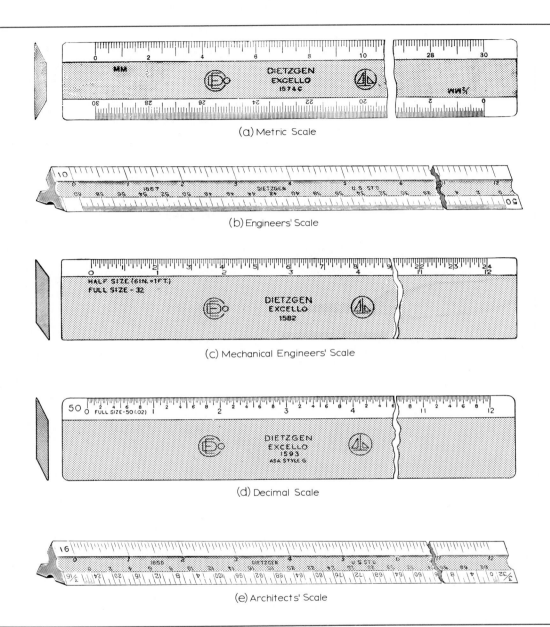

*Fig.* 2.33   Types of Scales.

used. Decimal equivalents tables can be found inside the front cover, and conversion tables are given in Appendix 31.

Metric scales are available in flat and triangular styles with a variety of scale graduations. The triangular scale illustrated in Fig. 2.35 has one full-size scale and five reduced-size scales, all full divided. By means of these scales a drawing can be made full size, enlarged size, or reduced size. To specify the scale on a drawing, see §2.31.

### Full Size Fig. 2.35 (a)

The 1 : 1 scale is full size, and each division is actually 1 mm in width with the numbering of the calibrations at 10 mm intervals. The same scale is convenient also for the ratios of 1 : 10, 1 : 100, 1 : 1000, and so on.

### Half Size Fig. 2.35 (a)

The 1 : 2 scale is one-half size, and each division equals 2 mm with the calibration numbering at 20-

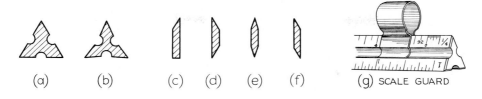

**Fig. 2.34** Sections of Scales and Scale Guard.

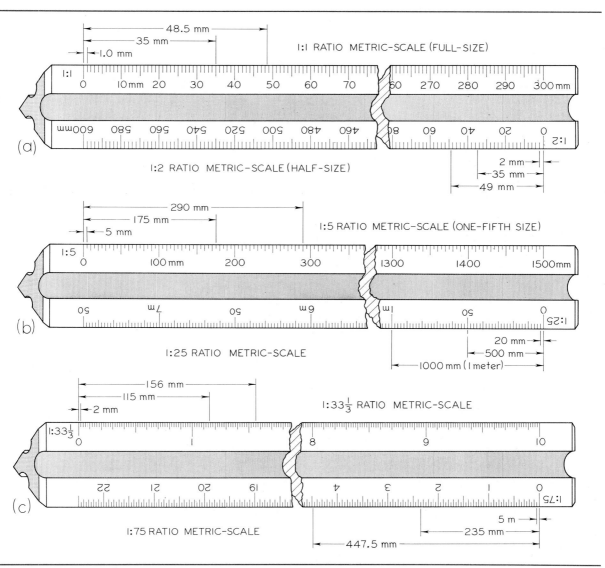

**Fig. 2.35** Metric Scales.

unit intervals. In addition, this scale is convenient for ratios of 1 : 20, 1 : 200, 1 : 2000, and so on.

The remaining four scales on this triangular metric scale include the typical scale ratios of 1 : 5, 1 : 25, 1 : $33\frac{1}{3}$, and 1 : 75, as illustrated at (b) and (c). These ratios also may be enlarged or reduced as desired by multiplying or dividing by a factor of 10. Metric scales are also available with other scale ratios for specific drawing purposes.

The metric scale is also used in map drawing and in drawing force diagrams or other graphical constructions that involve such scales as 1 mm = 1 kg, and 1 mm = 500 kg.

## 2.26 Inch-Foot Scales

Several scales that are based upon the inch-foot system of measurement continue in domestic use today along with the metric system of measurement that is accepted worldwide for science, technology, and international trade.

## 2.27 Engineers' Scale Fig. 2.33 (b)

The *engineers' scale* is graduated in the decimal system. It is also frequently called the *civil engineers' scale* because it was originally used mainly in civil engineering. The name *chain scale* also persists because it was derived from the surveyors' chain composed of 100 links, used for land measurements. The name "engineers' scale" is perhaps best, because the scale is used generally by engineers of all kinds.

The engineers' scale is graduated in units of one inch divided into 10, 20, 30, 40, 50, and 60 parts. Thus, the engineers' scale is convenient in machine drawing to set off dimensions expressed in decimals. For example, to set off 1.650″ full size, Fig. 2.36 (a), use the 10-scale and simply set off one main division plus $6\frac{1}{2}$ subdivisions. To set off the same dimension half size, use the 20-scale, (b), since the 20-scale is exactly half the size of the 10-scale. Similarly, to set off a dimension quarter size, use the 40-scale.

The engineers' scale is used also in drawing maps to scales of 1″ = 50″, 1″ = 500′, 1″ = 5 miles, and so on and in drawing stress diagrams or other graphical constructions to such scales as 1″ = 20 lb and 1″ = 4000 lb.

## 2.28 Architects' Scale Fig. 2.33 (e)

The *architects' scale* is intended primarily for drawings of buildings, piping systems, and other large

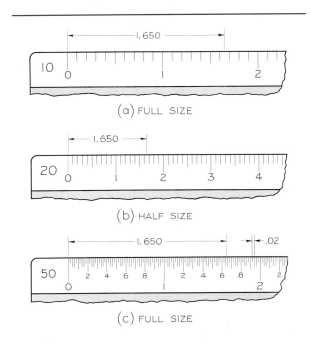

*Fig. 2.36*    Decimal Dimensions.

structures that must be drawn to a reduced scale to fit on a sheet of paper. The full-size scale is also useful in drawing relatively small objects, and for that reason the architects' scale has rather general usage.

The architects' scale has 1 full-size scale and 10 overlapping reduced-size scales. By means of these scales a drawing may be made to various sizes from full size to $\frac{1}{128}$ size. *Note particularly, in all of the reduced scales the major divisions represent feet, and their subdivisions represent inches and fractions thereof.* Thus, the scale marked $\frac{3}{4}$ means $\frac{3}{4}$ inch = 1 foot, not $\frac{3}{4}$ inch = 1 inch; that is, one-sixteenth size, not three-fourths size. And the scale marked $\frac{1}{2}$ means $\frac{1}{2}$ inch = 1 foot, not $\frac{1}{2}$ inch = 1 inch, that is, one-twenty-fourth size, not half-size.

All of the scales, from full size to $\frac{1}{128}$ size, are shown in Fig. 2.37. Some are upside down, just as they may occur in use. These scales are described as follows:

### *Full Size* Fig. 2.37 (a)

Each division in the full-size scale is $\frac{1}{16}″$. Each inch is divided first into halves, then quarters, eighths, and finally sixteenths, the division lines diminishing in length with each division. To set off $\frac{1}{32}″$, estimate

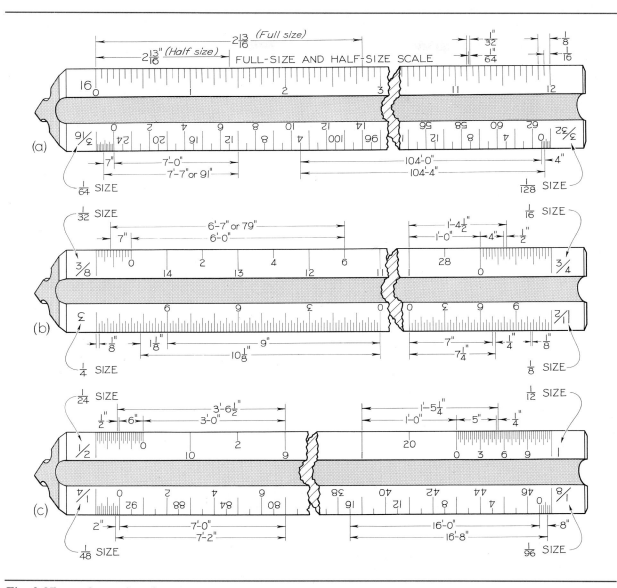

*Fig. 2.37* Architects' Scales.

visually one half of $\frac{1}{16}''$; to set off $\frac{1}{64}''$, estimate one-fourth of $\frac{1}{16}''$.

## Half Size Fig. 2.37 (a)

Use the full-size scale, and divide every dimension mentally by two (do not use the $\frac{1}{2}''$ scale, which is intended for drawing to a scale of $\frac{1}{2}'' = 1'$, or one-twenty-fourth size). To set off $1''$, measure $\frac{1}{2}''$; to set off $2''$, measure $1''$; to set off $3\frac{1}{4}''$, measure $1\frac{1}{2}''$ (half of $3''$), then $\frac{1}{8}''$ (half of $\frac{1}{4}''$); $6\frac{1}{2}$ to set off $2\frac{13}{16}''$ (see figure), measure $1''$, then $\frac{13}{32}''$ ($\frac{6\frac{1}{2}}{16}''$ or half of $\frac{13}{16}''$).

## Quarter Size Fig. 2.37 (b)

Use the $3''$ scale in which $3'' = 1'$. The subdivided portion to the left of zero represents 1 foot compressed to actually $3''$ in length and is divided into inches, then half inches, quarter inches, and finally eighth inches. Thus the entire portion representing 1 foot would actually measure 3 inches; therefore, $3'' = 1'$. To set off anything less than $12''$, start at zero and measure to the left.

To set off $10\frac{1}{8}''$, read off $9''$ from zero to the left, and add $1\frac{1}{8}''$ and set off the total $10\frac{1}{8}''$, as shown. To set off more than $12''$, for example, $1'-9\frac{3}{8}''$ (see your scale), find the 1′ mark to the right of zero and the $9\frac{3}{8}''$ mark to the left of zero; the required distance is the distance between these marks and represents $1'–9\frac{3}{8}''$.

### *Eighth Size* Fig. 2.37 (b)

Use the $1\frac{1}{2}''$ scale in which $1\frac{1}{2}'' = 1'$. The subdivided portion to the right of zero represents 1′ and is divided into inches, then half inches, and finally quarter inches. The entire portion, representing 1′, actually is $1\frac{1}{2}''$; therefore, $1\frac{1}{2}'' = 1'$. To set off anything less than 12″, start at zero and measure to the right.

### *Double Size*

Use the full-size scale, and multiply every dimension mentally by 2. To set off 1″, measure 2″; to set off $3\frac{1}{4}''$, measure $6\frac{1}{2}''$; and so on. The double-size scale is occasionally used to represent small objects. In such cases, a small actual-size outline view should be shown near the bottom of the sheet to help the shop worker visualize the actual size of the object.

### *Other Sizes* Fig. 2.37.

The other scales besides those just described are used chiefly by architects. Machine drawings are customarily made only double size, full size, half size, one-fourth size, or one-eighth size.

### 2.29 Decimal Scale Fig. 2.33 (d)

The increasing use of decimal dimensions has brought about the development of a scale specifically for that use. On the full-size scale, each inch is divided into fiftieths of an inch, or .02″, as shown in Fig. 2.36 (c), and on the half- and quarter-size scales, the inches are compressed to half size or quarter size, and then are divided into 10 parts, so that each subdivision stands for .1″.

The complete decimal system of dimensioning, in which this scale is used, is described in §13.10.

### 2.30 Mechanical Engineers' Scale
Fig. 2.33 (c)

The objects represented in machine drawing vary in size from small parts, an inch or smaller in size, to machines of large dimensions. By drawing these objects full size, half size, quarter size, or eighth

size, the drawings will readily come within the limits of the standard-size sheets. For this reason the mechanical engineers' scales are divided into units representing inches to full size, half size, quarter size, or eighth size. To make a drawing of an object to a scale of one-half size, for example, use the mechanical drafter's scale marked half size, which is graduated so that every $\frac{1}{2}''$ represents 1″. Thus, the half-size scale is simply a full-size scale compressed to one-half size.

These scales are also very useful in dividing dimensions. For example, to draw a $3\frac{11}{16}''$ diameter circle full size, we need half of $3\frac{11}{16}''$ to use as radius. Instead of using arithmetic to find half of $3\frac{11}{16}''$, it is easier to set off $3\frac{11}{16}''$ on the half-size scale.

Triangular combination scales are available that include the full- and half-size mechanical engineers' scales, several architects' scales, and an engineers' scale.

### 2.31 To Specify the Scale on a Drawing

For machine drawings the scale indicates the ratio of the size of the drawing of the part or machine to its actual size irrespective of the unit of measurement used. The recommended practice is to letter **FULL SIZE** or 1:1; **HALF SIZE** or 1:2; and similarly for other reductions. Expansion or enlargement scale are given as $2:1$ or $2\times$; $3:1$ or $3\times$; $5:1$ or $5\times$; $10:1$ or $10\times$; and so on.

The various scale calibrations available on the metric scales and the engineers' scale provide almost unlimited scale ratios. The preferred metric scale ratios appear to be 1 : 1, 1 : 2, 1 : 5, 1 : 10, 1 : 20, 1 : 50, 1 : 100, and 1 : 200. For examples of how scales may be shown on machine drawings, see Figs. 16.24 and 16.25.

Map scales are indicated in terms of fractions, such as Scale $\frac{1}{62500}$, or graphically, such as 400   0   400   800 Ft.

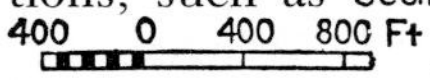

### 2.32 Accurate Measurements

Accurate drafting depends considerably upon the correct use of the scale in setting off distances. Do not take measurements directly off the scale with the dividers or compass, as damage will result to the scale. Place the scale on the drawing with the edge parallel to the line on which the measurement is to be made and, with a sharp pencil having a conical point, make a short dash at right angles to the scale

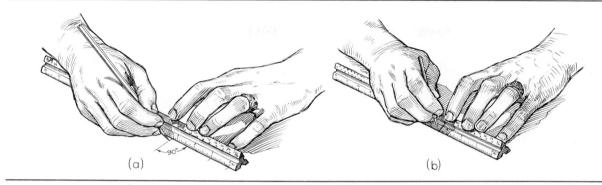

*Fig. 2.38*  Accurate Measurements.

and opposite the correct graduation mark, as shown in Fig. 2.38 (a). If extreme accuracy is required, a tiny prick mark may be made at the required point with the needle point or stylus, as shown at (b), or with one leg of the dividers.

*Avoid cumulative errors* in the use of the scale. If a number of distances are to be set off end-to-end, all should be set off at one setting of the scale by adding each successive measurement to the preceding one, if possible. Avoid setting off the distances individually by moving the scale to a new position each time, since slight errors in the measurements may accumulate and give rise to a large error.

## 2.33 Drawing Instruments

Drawing instruments are generally sold in sets, in cases, but they may be purchased separately. The principal parts of high-grade instruments are usually made of nickel silver, which has a silvery luster, is corrosion-resistant, and can be readily machined into desired shapes. Tool steel is used for the blades of ruling pens, for spring parts, for divider points, and for the various screws.

In technical drawing, accuracy, neatness, and speed are essential, §2.2. These objectives are not likely to be obtained with cheap or inferior drawing instruments. For the student or the professional drafter, it is advisable, and in the end more economical, to purchase the best instruments that can be afforded. Good instruments will satisfy the most rigid requirements, and the satisfaction, saving in time, and improved quality of work that good in-

struments can produce will more than justify the higher price.

Unfortunately, the qualities of high-grade instruments are not likely to be recognized by the beginner, who is not familiar with the performance characteristics required and who is apt to be attracted by elaborate sets containing a large number of shiny low-quality instruments. Therefore, the student should obtain the advice of the drafting instructor, of an experienced drafter, or of a reliable dealer.

## 2.34 Giant Bow Set

Formerly it was general practice to make pencil drawings on detail paper and then to make an inked tracing from it on tracing cloth. As reproduction methods and transparent tracing papers were improved, it was found that a great deal of time could be saved by making drawings directly in pencil with dense black lines on the tracing paper and making prints or photocopies therefrom, thus doing away with the preliminary pencil drawing on detail paper. Today, though inked tracings are still made when a fine appearance is necessary and where the greater cost is justified, the overwhelming proportion of drawings are made directly in pencil on tracing paper, vellum, polyester films, or pencil tracing cloth.

These developments have brought about the present giant bow sets that are offered now by all the major manufacturers, Fig. 2.39. The sets contain various combinations of instruments, but all feature a large bow compass in place of the traditional large compass. The large bow instrument is much sturdier and is capable of taking the heavy pressure neces-

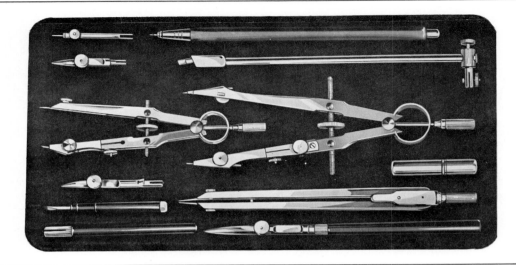

*Fig. 2.39*   Giant Bow Set. *Courtesy of Frank Oppenheimer*

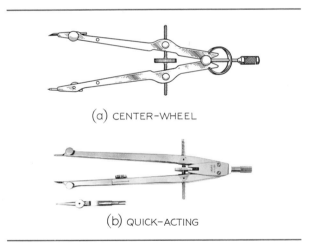

(a) CENTER-WHEEL

(b) QUICK-ACTING

*Fig. 2.40*   Giant Bow Compass. *(b) Courtesy of Frank Oppenheimer*

sary to produce dense black lines without losing the setting.

Most of the large bows are of the center-wheel type, Fig. 2.40 (a). Several manufacturers now offer different varieties of quick-acting bows. The large bow compass shown at (b) can be adjusted to the approximate setting by simply opening or closing the legs in the same manner as for the other bow-style compass.

## 2.35 The Compass

The giant bow compass, Figs. 2.39 and 2.41, has a socket joint in one leg that permits the insertion of either pencil or pen attachments. A *lengthening bar* or a *beam attachment* is often provided to increase the radius. For production drafting, in which it is necessary to make dense black lines to secure clear legible reproductions, the giant bow, or an appropriate template, Figs. 2.68 and 2.69, is preferred.

## 2.36 Using the Compass

These instructions apply generally both to the old style and the giant bow compasses. The compass, with pencil and inking attachments, is used for drawing circles of approximately 25 mm (1″) radius or larger, Fig. 2.41. Most compass needle points have a plain end for use when the compass is converted into dividers and a shoulder end for use as a compass. Adjust the needle point with the shoulder end out and so that the small point extends *slightly* farther than the pencil lead or pen nibs, Fig. 2.43 (d).

To draw a penciled circle, Fig. 2.41: (1) set off the required radius on one of the center lines, (2) place the needle point at the exact intersection of the center lines, (3) adjust the compass to the required radius (25 mm or more), and (4) lean the compass forward and draw the circle clockwise while rotating the handle between the thumb and fore-

Use shoulder–
end of
needle point

**Fig. 2.41** Using the Giant Bow Compass.

finger. To obtain sufficient weight of line, it may be necessary to repeat the movement several times.

Any error in radius will result in a doubled error in diameter; hence, it is best to draw a trial circle first on scrap paper or on the backing sheet and then check the diameter with the scale.

On drawings having circular arcs and tangent straight lines, draw the arcs first, whether in pencil or in ink, as it is much easier to connect a straight line to an arc than the reverse.

For very large circles, a beam compass, §2.38, is preferred, or use the lengthening bar to increase the compass radius. Use both hands, as shown in Fig. 2.42, but be careful not to jar the instrument and thus change the adjustment.

When using the compass to draw construction lines, use a 4H to 6H lead so that the lines will be very dim. For required lines, the arcs and circles must be black, and softer leads must be used. However, since heavy pressure cannot be exerted on the compass as it can on a pencil, it is usually necessary to use a compass lead that is about one grade softer than the pencil used for the corresponding line work. For example, if a 2H lead is used for visible lines drawn with the pencil, then an F lead might be found suitable for the compass work. The hard leads supplied with the compass are usually unsatisfactory for most line work except construction lines. In summary, use leads in the compass that will produce arcs and circles that *match* the straight pencil lines.

It is necessary to exert pressure on the compass to produce heavy "reproducible" circles, and this tends to enlarge the compass center hole in the paper, especially if there are a number of concentric circles. In such cases, use a horn center, or center tack, in the hole, and place the needle point of the compass in the center of the tack.

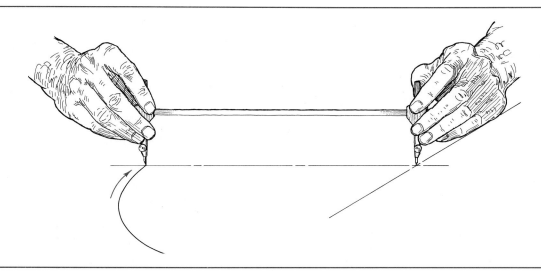

**Fig. 2.42** Drawing a Circle of Large Radius with the Beam Compass.

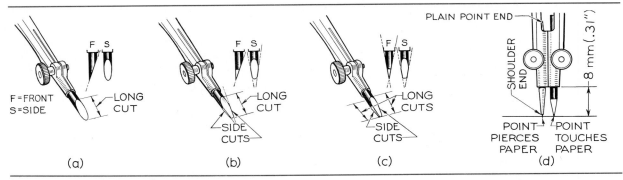

*Fig. 2.43* Compass Lead Points.

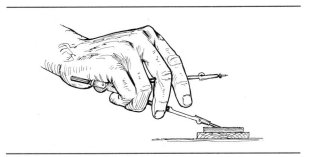

*Fig. 2.44* Sharpening Compass Lead.

## 2.37 Sharpening the Compass Lead

Various forms of compass lead points are illustrated in Fig. 2.43. At (a), a single elliptical face has been formed by rubbing on the sandpaper pad, as shown in Fig. 2.44. At (b), the point is narrowed by small side cuts. At (c), two long cuts and two small side cuts have been made so as to produce a point similar to that on a screwdriver. At (d), the cone point is prepared by chucking the lead in a mechanical pencil and shaping it in a pencil pointer. Avoid using leads that are too short to be exposed as shown.

In using the compass, *never use the plain end of the needle point.* Instead, use the shoulder end, as shown in Fig. 2.43 (d), adjusted so that the tiny needle point extends about halfway into the paper when the compass lead just touches the paper.

## 2.38 Beam Compass

The beam compass, or trammel, Fig. 2.45, is used for drawing arcs or circles larger than can be drawn with the regular compass and for transferring distances too great for the regular dividers. Besides steel points, pencil and pen attachments are provided. The beams may be made of nickel silver, steel, aluminum, or wood and are procurable in var-

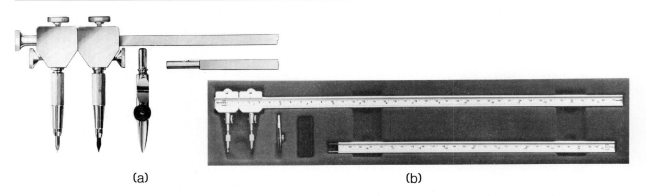

*Fig. 2.45* Beam Compass Sets. *(a) Courtesy of Frank Oppenheimer; (b) Courtesy of Tacro, Div. of A&T Importers, Inc.*

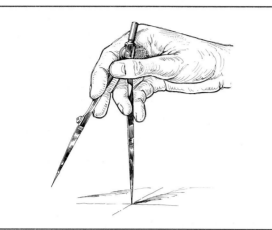

*Fig. 2.46*  Adjusting the Dividers.

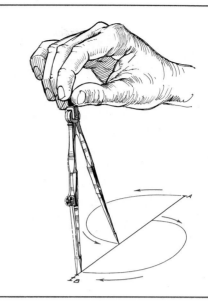

*Fig. 2.47*  Using the Dividers.

ious lengths. A square nickel silver beam compass set is shown in Fig. 2.45 (a), and at (b) a set with the beam graduated in millimeters and inches.

## 2.39 Dividers

The dividers are similar to the compass in construction and are made in square, flat, and round forms.

The friction adjustment for the pivot joint should be loose enough to permit easy manipulation with one hand, as shown in Fig. 2.46. If the pivot joint is too tight, the legs of the compass tend to spring back instead of stopping at the desired point when the pressure of the fingers is released. To adjust tension, use a small screwdriver.

Many dividers are made with a spring and thumbscrew in one leg so that minute adjustments in the setting can be made by turning the small thumbscrew.

## 2.40 Using the Dividers

The dividers, as the name implies, are used for *dividing* distances into a number of equal parts. They are used also for *transferring distances* or for *setting off* a series of equal distances. The dividers are used for spaces of approximately 25 mm (1″) or more. For less than 25 mm spaces, use the bow dividers, Fig. 2.50 (a). *Never use the large dividers for small spaces*

*when the bow dividers can be used; the latter are more accurate.*

To divide a given distance into a number of equal parts, Fig. 2.47, the method is one of trial and error. Adjust the dividers with the fingers of the hand that holds them, to the approximate unit of division, estimated by eye. Rotate the dividers counterclockwise through 180°, and so on, until the desired number of units has been stepped off. If the last prick of the dividers falls short of the end of the line to be divided, increase the distance between the divider points proportionately. For example, to divide the line **AB**, Fig. 2.47, into three equal parts, the dividers are set by eye to approximately one-third the length **AB**. When it is found that the trial radius is too small, the distance between the divider points is increased by one-third the remaining distance. If the last prick of the dividers is beyond the end of the line, a similar decreasing adjustment is made.

The student should avoid *cumulative errors*, which may result when the dividers are used to set off a series of distances end to end. To set off a large number of equal divisions, say, 15, first set off 3 equal large divisions and then divide each into 5 equal parts. Wherever possible in such cases, use the scale instead of the dividers, as described in §2.32, or set off the total and then divide into the parts by means of the parallel-line method, §§5.14 and 5.15.

*Fig. 2.48* Proportional Dividers.

## 2.41 Proportional Dividers

For enlarging or reducing a drawing, proportional dividers, Fig. 2.48, are convenient. They may be used also for dividing distances into a number of equal parts, or for obtaining a percentage reduction of a distance. For this purpose, points of division are marked on the instrument so as to secure the required subdivisions readily. Some instruments are calibrated to obtain special ratios, such as $1 : \sqrt{2}$, the diameter of a circle to the side of an equal square, and feet to meters.

## 2.42 The Bow Instruments

The bow instruments are classified as the *bow dividers, bow pen,* and *bow pencil.* A combination pen and pencil bow, usually with center-wheel adjustment, Fig. 2.49, and separate instruments with either side-wheel or center-wheel adjustment, Fig. 2.50, are available. The choice is a matter of personal preference.

## 2.43 Using the Bow Instruments

The bow pencil and bow pen are used for drawing circles of approximately 25 mm (1″) radius or smaller. The bow dividers are used for the same purpose as the large dividers, but for smaller (approximately 25 mm or less) spaces and more accurate work.

Whether the center-wheel or side-wheel instrument is used, the adjustment should be made with the thumb and finger of the hand that holds the instrument, Fig. 2.51 (a). The instrument is manipulated by twirling the head between the thumb and fingers, (b).

The lead is sharpened in the same manner as for the large compass, §2.37, except that for small radii, the inclined cut may be turned *inside* if preferred, Fig. 2.52 (a). For general use, the lead should be turned to the outside, as shown at (b). In either case, always keep the compass lead sharpened. *Avoid stubby compass leads,* which cannot be properly sharpened. At least 6 mm ($\frac{1}{4}$″) of lead should extend from the compass at all times.

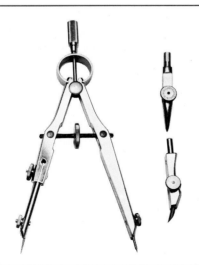

*Fig. 2.49* Combination Pen and Pencil Bow.
*Courtesy of Frank Oppenheimer*

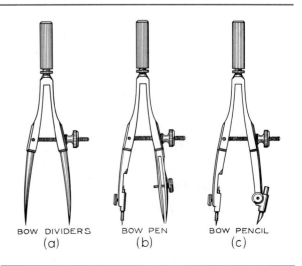

BOW DIVIDERS (a)    BOW PEN (b)    BOW PENCIL (c)

*Fig. 2.50* Bow Instruments with Side Wheel.

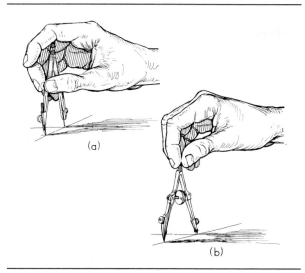

**Fig. 2.51** Using the Bow Instruments.

In adjusting the needle point of the bow pencil or bow pen, be sure to have the needle extending slightly longer than the pen or the lead, Fig. 2.43 (d), the same as for the large compass.

In drawing small circles, greater care is necessary in sharpening and adjusting the lead and the needle point, and especially in accurately setting the

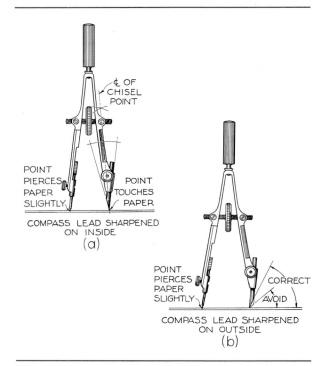

**Fig. 2.52** Compass-Lead Points.

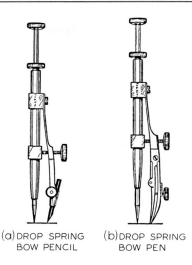

**Fig. 2.53** Drop Spring Bow Instruments.

desired radius. If a 6.35 mm ($\frac{1}{4}$ ") diameter circle is to be drawn, and if the radius is "off" only 0.8 mm ($\frac{1}{32}$ ") the total error on diameter is approximately 25 percent, which is far too much error.

Appropriate templates may be used also for drawing small circles. See Figs. 2.68 and 2.69.

## 2.44 Drop Spring Bow Pencil and Pen

These compasses, Fig. 2.53, are designed for drawing multiple identical small circles, such as drill holes or rivet heads. A central pin is made to move easily up and down through a tube to which the pen or pencil unit is attached. To use the instrument, hold the knurled head of the tube between the thumb and second finger, placing the first finger on top of the knurled head of the pin. Place the point of the pin at the desired center, lower the pen or pencil until it touches the paper, and twirl the instrument clockwise with the thumb and second finger. Then lift the tube independently of the pin, and finally lift the entire instrument.

## 2.45 To Lay Out a Sheet

After the sheet has been attached to the board, as explained in §2.7, proceed as follows, Fig. 2.54 (see also Layout A–2, inside back cover).

I.   Using the T-square, draw a horizontal *trim line* near the lower edge of the paper, and then using the triangle, draw a vertical trim line near the left

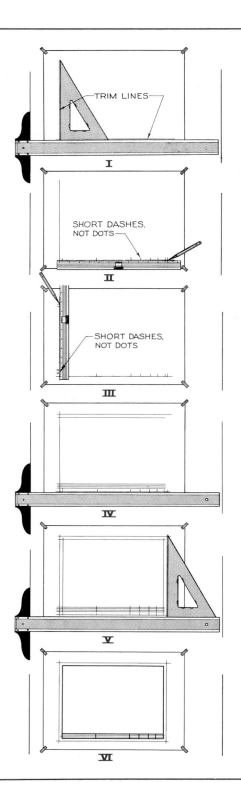

*Fig. 2.54* To Lay Out a Sheet.
Layout A–2; see inside back cover.

edge of the paper. Both should be *light construction lines*.

II.    Place the scale along the lower trim line with the full-size scale up. Draw short light dashes *perpendicular* to the scale at the required distances. See Fig. 2.38 (a).

III.    Place the scale along the left trim line with the full-size scale to the left, and mark the required distances with short light dashes perpendicular to the scale.

IV.    Draw horizontal construction lines with the aid of the T-square through the marks at the left of the sheet.

V.    Draw vertical construction lines, *from the bottom upward,* along the edge of the triangle through the marks at the bottom of the sheet.

VI.    Retrace the border and the title strip to make them heavier. Notice that the layout is made independently of the edges of the paper.*

## 2.46 Technique of Pencil Drawing

By far the greater part of commercial drafting is executed in pencil. Most prints or photocopies are made from pencil tracings, and all ink tracings must be preceded by pencil drawings. It should therefore be evident that skill in drafting chiefly implies skill in pencil drawing.

*Technique* is a style or quality of drawing imparted by the individual drafter to the work. It is characterized by crisp black linework and lettering. Technique in lettering is discussed in §4.11.

### Dark Accented Lines

The pencil lines of a finished pencil drawing or tracing should be very dark, Fig. 2.55. Dark crisp lines are necessary to give punch or snap to the drawing. Ends of lines should be accented by a little extra pressure on the pencil, (a). Curves should be as dark as other lines, (b). Hidden-line dashes and center-line dashes should be carefully estimated as to length and spacing and should be of uniform width throughout their length, (c) and (d).

Dimension lines, extension lines, section lines, and center lines also should be dark. The difference

*In industrial drafting rooms the sheets are available, cut to standard sizes, with border and title strips already printed. Drafting supply houses can supply such papers, printed to order, to schools for little extra cost.

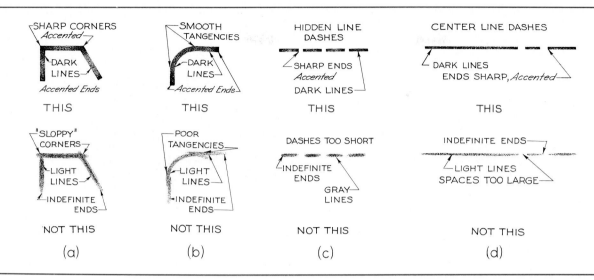

*Fig. 2.55* Technique of Lines (Enlarged).

between these lines and visible lines is mostly in width—there is very little difference, if any, in blackness.

A simple way to determine whether your lines on tracing paper or cloth are dense black is to hold the tracing up to the light. Lines that are not opaque black will not print clearly by most reproduction processes.

Construction lines should be made with a sharp, hard lead and *should be so light that they need not be erased* when the drawing is completed.

### Contrast in Lines

Contrast in pencil lines should be similar to that of ink lines; that is, the difference beween the various lines should be mostly in the *widths* of the lines, with little if any difference in the degree of darkness, Fig. 2.56. The visible lines should contrast strongly

with the thin lines of the drawing. If necessary, draw over a visible line several times to get the desired thickness and darkness. A short retracing stroke backward (to the left), producing a jabbing action, results in a darker line.

### 2.47 Pencil Tracing

While some pencil tracings are made of a drawing placed underneath the tracing paper (usually when a great deal of erasing and changing is necessary on the original drawing), most drawings today are made directly in pencil on tracing paper, pencil tracing cloth, films, or vellum. These are not tracings but pencil drawings, and the methods and technique are the same as previously described for pencil drawing.

In making a drawing directly on a tracing medium, a smooth sheet of heavy white drawing

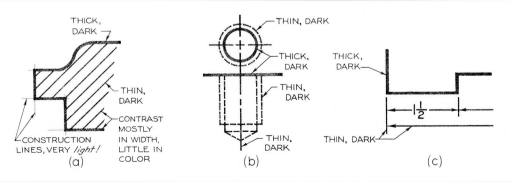

*Fig. 2.56* Contrast of Lines (Enlarged).

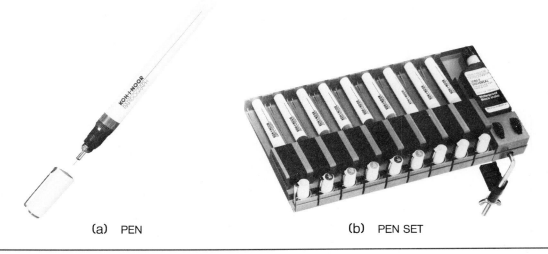

(a) PEN                    (b) PEN SET

***Fig. 2.57***   Technical Fountain Pen and Pen Set. *Courtesy of Koh-I-Nor Rapidograph, Inc.*

paper, a backing sheet, should be placed underneath. The whiteness of the backing sheet improves the visibility of the lines, and the hardness of the surface makes it possible to exert pressure on the pencil and produce dense black lines without excessive grooving of the paper.

Thus all lines must be dark and cleanly drawn when drawings are intended to be reproduced.

### 2.48 Technical Fountain Pens
The technical fountain pen, Figs. 2.57 and 2.58, with the tube and needle point, is available in several line widths. Many prefer this type of pen, for the line widths are fixed and it is suitable for freehand or mechanical lettering and line work. The pen requires an occasional filling and a minimum of skill to use. For uniform line work, the pen should be used perpendicular to the paper. For best results, follow the manufacturer's recommendations for operation and cleaning.

### 2.49 Ruling Pens
The ruling pen, Fig. 2.59, should be of the highest quality, with blades of high-grade tempered steel sharpened properly at the factory. The nibs should be sharp, but not sharp enough to cut the paper.

The *detail pen*, capable of holding a considerable quantity of ink, is extremely useful for drawing long

***Fig. 2.58***   Using the Technical Fountain Pen.

heavy lines, (b). This type of pen is preferred for small amounts of ink work as the pen is easily adjusted for the desired line width and is readily cleaned after use.

For methods using the ruling pen, see §2.51.

### 2.50 Drawing Ink
Drawing ink is composed chiefly of carbon in colloidal suspension and gum. The fine particles of carbon give the deep, black luster to the ink, and the

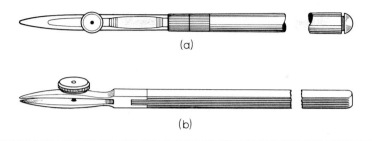

*Fig. 2.59*  Ruling Pens.

gum makes it waterproof and quick to dry. The ink bottle, or flask, should not be left uncovered, as evaporation will cause the ink to thicken.

Special drawing ink is available for use on acetate and Mylar films. Such inks should not be used in technical fountain pens unless the pens are specifically made for acetate-based inks.

For removing dried waterproof drawing ink from pens or instruments, pen-cleaning fluids are available at dealers.

## 2.51 Use of the Ruling Pen

The ruling pen, Fig. 2.59, is used to ink lines drawn with instruments, never to ink freehand lines or freehand lettering. The proper method of filling the pen is shown in Fig. 2.60.

The pen should lean at an angle of about 60° with the paper in the direction in which the line is being drawn and in a vertical plane containing the line, Fig. 2.61.

The ruling pen is used in inking irregular curves,

*Fig. 2.60*  Filling the Ruling Pen.

as well as straight lines, as shown in Figs. 2.65 and 2.66. The pen should be held more nearly vertical when used with an irregular curve than when used with the T-square or a triangle. The ruling pen should lean only slightly in the direction in which the line is being drawn.

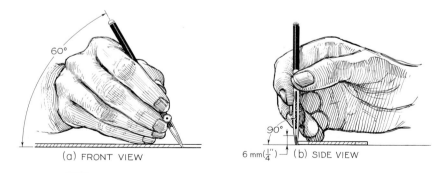

*Fig. 2.61*  Position of Hand in Using the Ruling Pen.

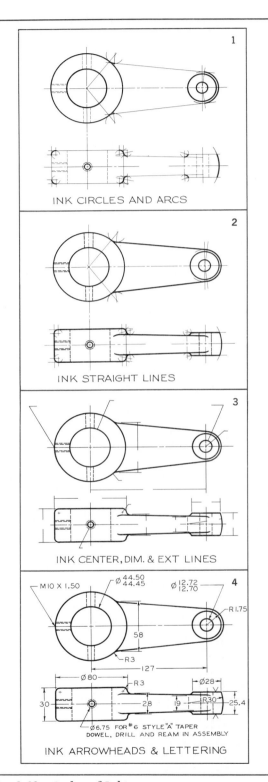

**Fig. 2.62**  Order of Inking.

## 2.52 Order of Inking Fig. 2.62
A definite order should be followed in inking a drawing or tracing, as follows.

1. (a) Mark all tangent points in pencil directly on tracing.
   (b) Indent all compass centers (with pricker or divider point).
   (c) Ink visible circles and arcs.
   (d) Ink hidden circles and arcs.
   (e) Ink irregular curves, if any.

2. (1st: horizontal; 2nd: vertical; 3rd: inclined)
   (a) Ink visible straight lines.
   (b) Ink hidden straight lines.

3. (1st: horizontal; 2nd: vertical; 3rd: inclined) Ink center lines, extension lines, dimension lines, leader lines, and section lines (if any).

4. (a) Ink arrowheads and dimension figures.
   (b) Ink notes, titles, etc. (pencil guide lines directly on tracing.)

Some drafters prefer to ink center lines before indenting the compass centers because ink can go through the holes and cause blots on the back of the sheet.

When an ink blot is made, the excess ink should be taken up with a blotter, or smeared with the finger if a blotter is not available, and not allowed to soak into the paper. When the spot is thoroughly dry, the remaining ink can be erased easily.

For cleaning untidy drawings or for removing the original pencil lines from an inked drawing, the Pink Pearl or the Mars-Plastic eraser is suitable if used lightly. Pencil lines or dirt can be removed from tracing cloth by rubbing lightly with a cloth moistened with carbon tetrachloride or benzine. Use either with care in a well-ventilated area.

When erasure on cloth damages the surface, it may be restored by rubbing the spot with soapstone and then applying pounce or chalk dust. If the damage is not too great, an application of the powder will be sufficient.

When a gap in a thick ink line is made by erasing, the gap should be filled in with a series of fine lines that are allowed to run together. A single heavy line is difficult to match and is more likely to run and cause a blot.

In commercial drafting rooms, the electric erasing machine, Fig. 2.18, is usually available to save the time of the drafter.

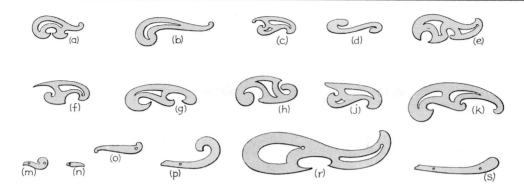

*Fig. 2.63* Irregular or French Curves.

## 2.53 Irregular Curves

When it is required to draw mechanical curves other than circles or circular arcs, an irregular or French curve is generally employed. Many different forms and sizes of curves are manufactured, as suggested by the more common forms illustrated in Fig. 2.63.

The curves are composed largely of successive segments of the geometric curves, such as the ellipse, parabola, hyperbola, and involute. The best curves are made of highly transparent plastic. Among the many special types of curves that are available are hyperbolas, parabolas, ellipses, logarithmic spirals, ship curves, and railroad curves.

Adjustable curves, Fig. 2.64, are also available. The curve shown at (a) consists of a core of lead, enclosed by a coil spring attached to a flexible strip. The one at (b) consists of a spline, to which "ducks" (weights) are attached. The spline can be bent to form any desired curve, limited only by the elasticity of the material. An ordinary piece of solder wire

can be used very successfully by bending the wire to the desired curve.

## 2.54 Using the Irregular Curve

The irregular curve is a device for the *mechanical drawing of curved lines and should not be applied directly to the points* or used for purposes of producing an initial curve. The proper use of the irregular curve requires skill, especially when the lines are to be drawn in ink. After points have been plotted through which the curve is to pass, a light pencil line should be sketched freehand smoothly through the points.

To draw a mechanical line over the freehand line with the aid of the irregular curve, it is only necessary to match the various segments of the irregular curve with successive portions of the freehand curve and to draw the line with pencil or ruling pen along the edge of the curve, Fig. 2.65. It is very important

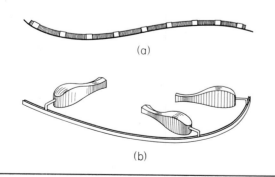

*Fig. 2.64* Adjustable Curves.

*Fig. 2.65* Using the Irregular Curve.

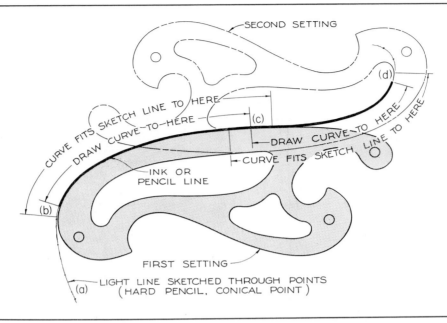

*Fig. 2.66*   Settings of Irregular Curve.

that the irregular curve match the curve to be drawn for some distance at each end beyond the segment to be drawn for any one setting of the curve, as shown in Fig. 2.66. When this rule is observed, the successive sections of the curve will be tangent to each other, without any abrupt change in the curvature of the line. In placing the irregular curve, the short-radius end of the curve should be turned toward the short-radius part of the curve to be drawn; that is, the portion of the irregular curve used should have the same curvilinear tendency as the portion of the curve to be drawn. This will prevent abrupt changes in direction.

The drafter should change position with the drawing when necessary, to avoid working on a near side of the curve.

When plotting points to establish the path of a curve, it is desirable to plot more points, and closer together, where sharp turns in the curve occur.

Free curves may also be drawn with the compass, as shown in Fig. 5.42.

For symmetrical curves, such as an ellipse, Fig. 2.67, use the same segment of the irregular curve in two or more opposite places. For example, at (a) the irregular curve is matched to the curve and the line drawn from 1 to 2. Light pencil dashes are then drawn directly on the irregular curve at these points (the curve will take pencil marks well if it is lightly "frosted" by rubbing with a hard pencil eraser). At (b) the irregular curve is turned over and matched so that the line may be drawn from 2 to 1. In similar manner, the same segment is used again at (c) and

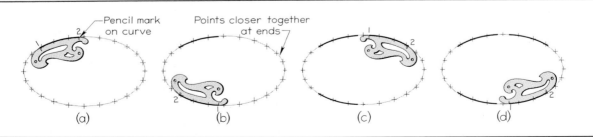

*Fig. 2.67*   Symmetrical Figures.

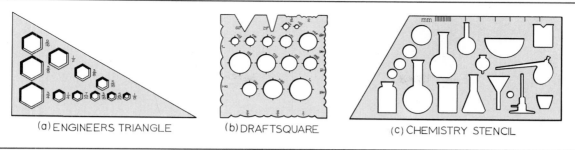

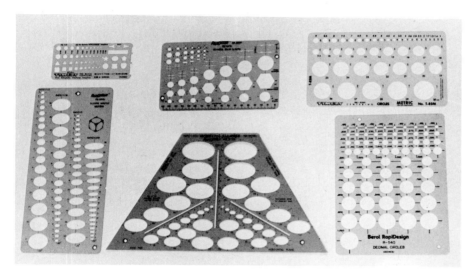

Fig. 2.68   Drafting Devices.

Fig. 2.69   Templates.

(d). The ellipse is completed by filling in the gaps at the ends by using the irregular curve or, if desired, the compass.

### 2.55 Templates

Templates are available for a great variety of specialized needs. A template may be found for drawing almost any ordinary drafting symbols or repetitive features. The engineers' triangle, Fig. 2.68 (a), is useful for drawing hexagons or for bolt heads and nuts; the draftsquare, (b), is convenient for drawing the curves on bolt heads and nuts, for drawing circles, thread forms, and so forth; and the chemistry stencil, (c), is useful for drawing chemical apparatus in schematic form.

Ellipse templates, §5.56, are perhaps more

widely used than any other type. Circle templates are useful for drawing small circles quickly and for drawing fillets and rounds, and are used extensively in tool and die drawings. Some of the more commonly used templates are shown in Fig. 2.69.

### 2.56 Drafting Machines

The drafting machine, Fig. 2.70, is an ingenious device that replaces the T-square, triangles, scales, and protractor. The links, or bands, are arranged so that the controlling head is always in any desired fixed position regardless of where it is placed on the board; thus, the horizontal straightedge will remain horizontal if so set. The controlling head is graduated in degrees (including a vernier on certain machines), which allows the straightedges, or scales, to

*Fig. 2.70*   Drafting Machine. *Courtesy of Keuffel & Esser Co./Kratos*

be set and locked at any angle. There are automatic stops at the more frequently used angles, such as 15°, 30°, 45°, 60°, 75°, and 90°.

Drafting machines and drafting tables, Fig. 2.71, have been greatly improved in recent years. The chief advantage of the drafting machine is that it

*Fig. 2.71*   Adjustable Drafting Table with Track Drafting Machine. *Courtesy of Keuffel & Esser Co./Kratos*

speeds up drafting. Since its parts are made of metal, their accurate relationships are not subject to change, whereas T-squares, triangles, and working edges of drawing boards must be checked and corrected frequently. Drafting machines for left-handers are available from the manufacturers.

### 2.57 Parallel-Ruling Straightedge

For large drawings, the long T-square becomes unwieldy, and considerable inaccuracy may result from the "give" or swing of the blade. In such case the parallel-ruling straightedge, Fig. 2.72, is recommended. The ends of the straightedge are controlled by a system of cords and pulleys which permit the straightedge to be moved up or down on the board while maintaining a horizontal position.

### 2.58 The Computer as a Drafting Tool

The development of CAD workstations for minicomputers and the availability of drafting software for microcomputers have made the computer a valuable drafting instrument. The use of the computer in drafting began in 1963 when Ivan Sutherland, of M.I.T., developed "Sketchpad," a program for engineers and drafters. The use of the computer continued to grow and by the late 1970s was widespread. All drafters and engineers today need to become familiar with computer systems in drafting and design.

Most CAD systems, regardless of their size, consist of an input device such as a joystick or digitizer pad, a cathode-ray tube (CRT—much like a television screen) on which to "build" the drawing, and a plotter that prints the final results. These devices are discussed in Chapters 3 and 8. It is appropriate to point out that this new approach is developing rapidly and that, in the future, much routine drafting can be done on CAD systems.

It should be noted that a student will need to learn the basic concepts of orthographic and pictorial projections as well as dimensioning before a CAD system can be effectively used.

### 2.59 Drawing Papers

Drawing paper, or *detail paper*, is used whenever a drawing is to be made in pencil but not for reproduction. For working drawings and for general use, the preferred paper is light cream or buff in color, and it is available in rolls of widths 24″ and 36″ and

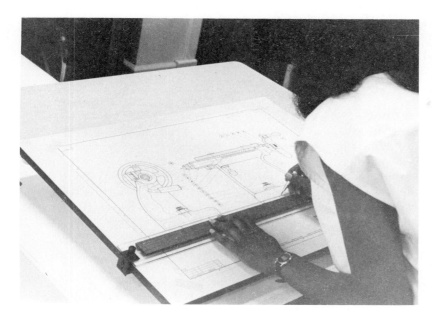

*Fig. 2.72* Parallel-Ruling Straightedge.

in cut sheets of standard sizes such as 8.5″ × 11″, 11″ × 17″, 17″ × 22″, and so on. Most industrial drafting rooms use standard sheets, §2.63, with printed borders and title strips, and since the cost for printing adds so little to the price per sheet, many schools have also adopted printed sheets.

The best drawing papers have up to 100 percent pure rag stock, have strong fibers that afford superior erasing qualities, folding strength; and toughness; and will not discolor or grow brittle with age. The paper should have a fine grain or tooth that will pick up the graphite and produce clean, dense black lines. However, if the paper is too rough, it will wear down the pencil excessively and will produce ragged, grainy lines. The paper should have a hard surface so that it will not groove too easily when pressure is applied to the pencil.

For ink work, as for catalog and book illustrations, white papers are used. The better papers, such as Bristol Board and Strathmore, come in several thicknesses such as 2-ply, 3-ply, and 4-ply.

## 2.60 Tracing Papers

Tracing paper is a thin transparent paper upon which drawings are made for the purpose of reproducing by blueprinting or by other similar processes. Tracings are usually made in pencil, but may be made in ink. Most tracing papers will take

pencil or ink, but some are especially suited to one or to the other.

Tracing papers are of two kinds: (1) those treated with oils, waxes, or similar substances to render them more transparent, called *vellums;* (2) those not so treated, but which may be quite transparent, owing to the high quality of the raw materials and the methods of manufacture. Some treated papers deteriorate rapidly with age, becoming brittle in many cases within a few months, but some excellent vellums are available. Untreated papers made entirely of good rag stock will last indefinitely and will remain tough.

## 2.61 Tracing Cloth

Tracing cloth is a thin transparent muslin fabric (cotton, not "linen" as commonly supposed) sized with a starch compound or plastic to provide a good working surface for pencil or ink. It is much more expensive than tracing paper. Tracing cloth is available in rolls of standard widths, such as 30″, 36″, and 42″, and also in sheets of standard sizes, with or without printed borders and title forms.

For pencil tracings, special pencil tracing cloths are available. Many concerns make their drawings in pencil directly on this cloth, dispensing entirely with the preliminary pencil drawing on detail paper, thus saving a great deal of time. These cloths generally have a surface that will produce dense black lines when hard pencils are used. Hence, these drawings do not easily smudge and will stand up well with handling.

## 2.62 Polyester Films and Coated Sheets

The polyester film is a superior drafting material available in rolls and sheets of standard size. It is made by bonding a mat surface to one or both sides of a clear Mylar polyester sheet. The transparency and printing qualities are very good, the mat drawing surface is excellent for pencil or ink, erasures leave no ghost marks, and the film has high dimensional stability. Its resistance to cracking, bending, or tearing makes it virtually indestructible, if given reasonable care. The film has rapidly replaced cloth and is competing with vellum in some applications. Some companies have found it more economical to make their drawings directly in ink on the film.

Large coated sheets of aluminum, which provides a good dimensional stability, are often used in the aircraft and auto industry for full-scale layouts that are scribed into the coating with a steel point rather than a pencil. The layouts are reproduced from the sheets photographically.

## 2.63 Standard Sheets

Two systems of sheet sizes together with length, width, and letter designations are listed by ANSI as follows.

| Nearest International Size[a] (millimeter) | Standard USA Size[a] (inch) |
|---|---|
| A4 210 × 297 | A  8.5 × 11.0 |
| A3 297 × 420 | B 11.0 × 17.0 |
| A2 420 × 594 | C 17.0 × 22.0 |
| A1 594 × 841 | D 22.0 × 34.0 |
| A0 841 × 1189 | E 34.0 × 44.0 |

[a]ANSI Y14.1–1980 (R1987).

The use of the basic sheet size, 8.5″ × 11.0″ or 210 mm × 297 mm and multiples thereof, permits filing of small tracings and of folded prints in standard files with or without correspondence. These sizes can be cut without waste from the standard rolls of paper, cloth, or film.

For layout designations, title blocks, revision blocks, and list of materials blocks, see inside the back cover of this book. See also §16.5.

# Instrumental Drawing Problems

All of the following constructions, Figs. 2.73 to 2.83, are to be drawn in pencil on Layout A2 (see inside the back cover of this book). The steps in drawing this layout are shown in Fig. 2.54. Draw all construction lines *lightly*, using a hard lead (4H to 6H), and all required lines dense black with a softer lead (F to H). If construction lines are drawn properly—that is, *lightly*—they need not be erased.

If the layout is to be made on the A4 size sheet, width dimensions for title-strip forms will need to be adjusted to fit the available space.

The drawings in Figs. 2.78 to 2.83 are to be drawn in pencil, preferably on tracing paper or vellum; then prints should be made to show the effectiveness of the student's technique. If ink tracings are required, the originals may be drawn on film or on detail paper and then traced on vellum or tracing cloth. For any assigned problem, the instructor may require that all dimensions and notes be lettered in order to afford further lettering practice.

The problems at the end of Chapter 5, "Geometric Constructions," provide excellent additional practice to develop skill in the use of drawing instruments.

Since many of the problems in this chapter are of a general nature, they can also be solved on most computer graphics systems. If a system is available, the instructor may choose to assign specific problems to be completed by this method.

Problems in convenient form for solution may be found in *Principles of Engineering Graphics Problems* by Spencer, Hill, Loving, and Dygdon, a workbook designed to accompany this text that is also published by Macmillian Publishing Company.

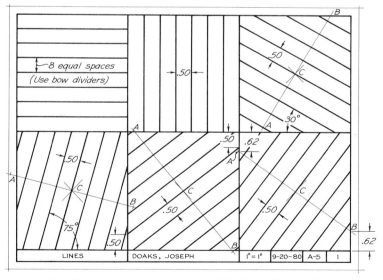

*Fig. 2.73*   Using Layout A–2 or A4–2 (adjusted), divide working space into six equal rectangles, and draw visible lines, as shown. Draw construction lines **AB** through centers **C** at right angles to required lines; then along each construction line, set off 0.50″ spaces and draw required visible lines. Omit dimensions and instructional notes.

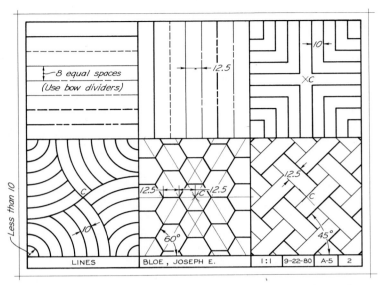

*Fig. 2.74* Using Layout A–2 or A4–2 (adjusted), divide working space into six equal rectangles, and draw lines as shown. In first two spaces, draw conventional lines to match those in Fig. 2.14. In remaining spaces, locate centers **C** by diagonals, and then work constructions out from them. Omit the metric dimensions and instructional notes.

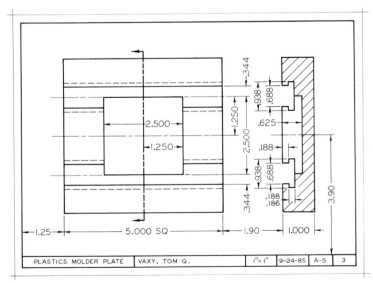

*Fig. 2.75* Using Layout A–2 or A4–2 (adjusted), draw views in pencil, as shown. Omit all dimensions.

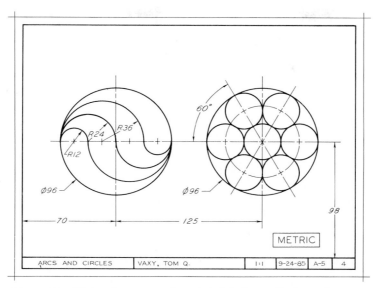

**Fig. 2.76** Using Layout A–2 or A4–2 (adjusted), draw figures in pencil, as shown. Use bow pencil for all arcs and circles within its radius range. Omit all dimensions.

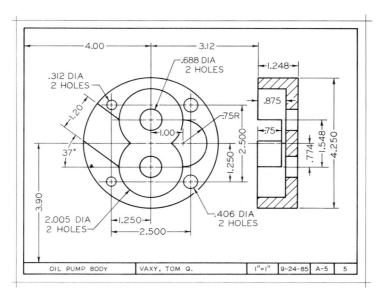

**Fig. 2.77** Using Layout A–2 or A4–2 (adjusted), draw views in pencil, as shown. Use bow pencil for all arcs and circles within its radius range. Omit all dimensions.

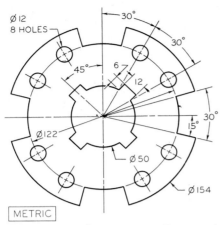

**Fig. 2.78** Friction Plate. Using Layout A–2 or A4–2 (adjusted), draw in pencil. Omit dimensions and notes.

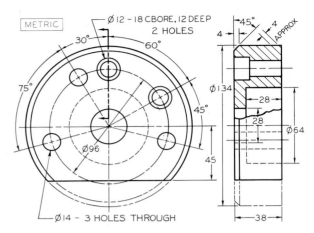

**Fig. 2.79** Seal Cover. Using Layout A–2 or A4–2 (adjusted), draw views in pencil. Omit dimensions and notes. See §9.8.

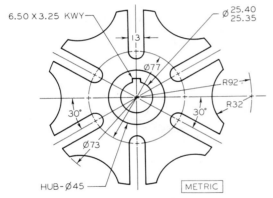

**Fig. 2.80** Geneva Cam. Using Layout A–2 or A4–2 (adjusted), draw in pencil. Omit dimensions and notes.

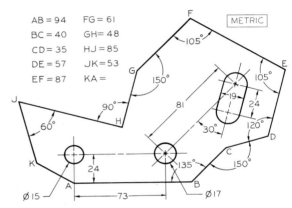

AB = 94    FG = 61
BC = 40    GH = 48
CD = 35    HJ = 85
DE = 57    JK = 53
EF = 87    KA =

**Fig. 2.81** Shear Plate. Using Layout A–2 or A4–2 (adjusted), draw accurately in pencil. Give length of **KA**. Omit other dimensions and notes.

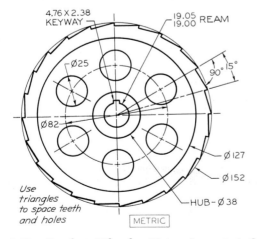

**Fig. 2.82** Ratchet Wheel. Using Layout A–2 or A4–2 (adjusted), draw in pencil. Omit dimensions and notes.

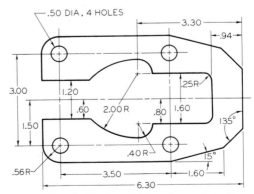

**Fig. 2.83** Latch Plate. Using Layout A–2 or A4–2 (adjusted), draw in pencil. Omit dimensions and notes.

# Introduction to Computer Graphics

by Gary R. Bertoline*

Engineering and technical drawing have a rich history that spans thousands of years. During this time, of course, the tools used to communicate graphically have evolved slowly. However, the changes and improvements have been relatively minor compared to the introduction of the computer to assist in the preparation of engineering and technical drawings.

The introduction of computer-aided design, or drafting, CAD, can be considered revolutionary rather than evolutionary. CAD should be seen as a tool or an aid in the design process. Although it is a very powerful tool in the hands of a trained designer, CAD does not replace skilled designers. CAD enhances their creativity by simplifying and automating the graphic communication of designs.

An entire CAD system consists of *hardware* and *software*. The various pieces of physical equipment comprising a computer system are known as hardware. The programs, instructions, and other documentation that permit the computer system to operate are classified as software. Computer programs include application programs, operating system programs, and languages.

*Assistant Professor of Engineering Graphics, The Ohio State University, Columbus, Ohio.

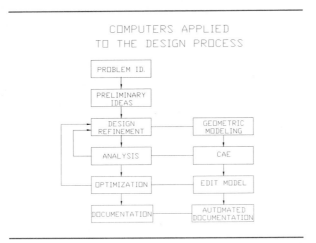

COMPUTERS APPLIED
TO THE DESIGN PROCESS

*Fig. 3.1*    Integration of CAD into the Design Process.

## 3.1 Computer-Aided Drafting

The use of traditional tools for engineering and technical graphics has been explained in the preceding chapter. The term CAD refers to any graphic computer software program that assists the user in design and documentation. It supplements the use of traditional tools and can be integrated into the design process as shown in Fig. 3.1. Some CAD software programs can be used very early in the design process to assist in the geometric definition of the part. The design is analyzed and then modified with the CAD system.

Initially, CAD was developed to assist the designer in the documentation, or stage 5, Fig. 3.1. As computers became more powerful and less expensive, CAD software developed that assisted the designer in the early stages. Three-dimensional computer models of design ideas can be created and modified with CAD. Once the 3D model is created, it can be used for analysis, and orthographic views can be selected and edited. Dimensions are added for detail drawings, and assemblies are created along with parts lists. Working drawings can be created faster and easier.

## 3.2 Advantages of Using CAD

CAD provides many advantages over traditional tools. However, it becomes advantageous only if the user has a thorough understanding of the software and is aware of its strengths and weaknesses. The use of CAD requires some preplanning before starting on a drawing or design. Preplanning is an effort to maximize the CAD software for a particular task. Doing so will result in many advantages such as increased speed of creating drawings. CAD will produce extremely accurate drawings that can be used with numerical control (NC) tools for manufacturing. Drawings are revised by using the CAD editing commands. Ease of revisions is one of the greatest advantages of using CAD. Using the right type of output device will result in drawings that are neat and legible to the designer. This is especially true for lettering, dimensions, title blocks, and parts lists. CAD can be a great advantage on drawings with repetitive features or symmetrical objects, as shown in Fig. 3.2.

## 3.3 Applications of CAD

CAD has a wide variety of applications. Wherever graphic communication is necessary, CAD can be used, but most CAD software is specialized for a limited number of applications. The focus here will be on CAD software used in engineering design.

### Electronic Applications

CAD is useful for the design and documentation of electrical and electronic systems. Designing electrical and electronic systems requires the use of many symbols depicting electronic components. Symbols can be created and used many times with CAD. Once a symbol is created, it can be placed on any drawing in any orientation or at any scale, making CAD a powerful tool for the creation of schematic drawings.

An even more powerful feature is the ability to design circuits and, through artificial intelligence, to create the circuit logic automatically along with drill patterns for component leads. One of the earliest applications of CAD was in electronics, and it remains the most popular application, Fig. 3.3.

### Mechanical Applications

The designing of mechanical devices and systems is the second most popular application of CAD. It is materially assisted by special software commands such as rotate, copy, move, and scale. Assemblies, Fig. 3.4, can be sectioned or exploded to reveal internal features and possible interferences. Three-dimensional models can be viewed from any direction, and orthographic views can be produced for

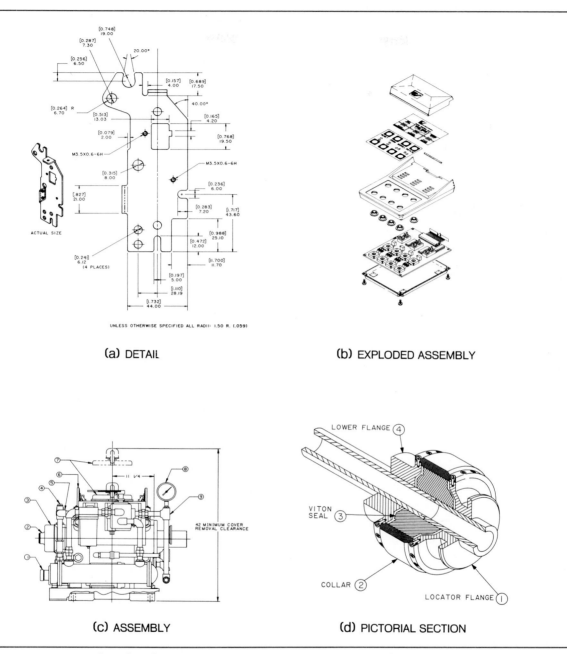

**Fig. 3.2** Drawings Produced with CAD. *Courtesy of Computervision Corporation*

working drawings. The model also is used to create the NC program to run machine tools.

## Civil Applications

Some CAD software is specialized to assist the civil engineer in the design of transportation systems such as bridges and piping systems and preparation of contour maps using surveying data input into the computer. Predrawn details of curbs, decks, bridges, ramps, wall, and other construction elements can be used as symbols and placed on any drawing. Structural steel detailing and table calculations are automatically done with specialized CAD software for civil applications, Fig. 3.5.

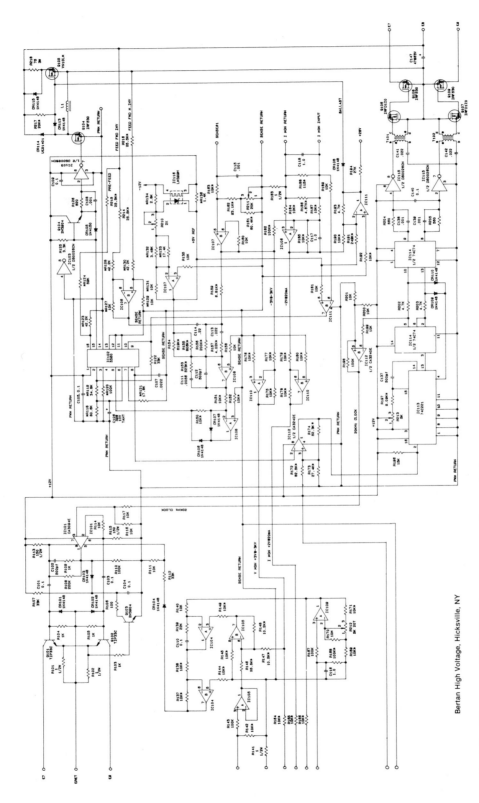

**Fig. 3.3** Electronic CAD Drawing. *Courtesy of VersaCAD*

Bertan High Voltage, Hicksville, NY

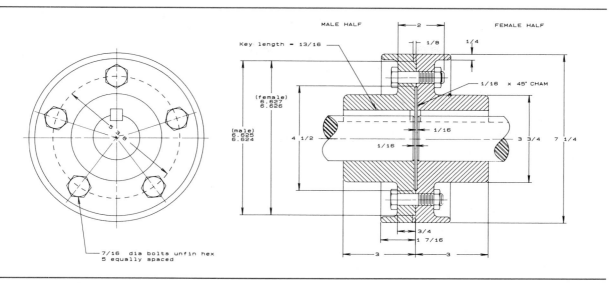

**Fig. 3.4** Mechanical CAD Drawing. *Courtesy of VersaCAD*

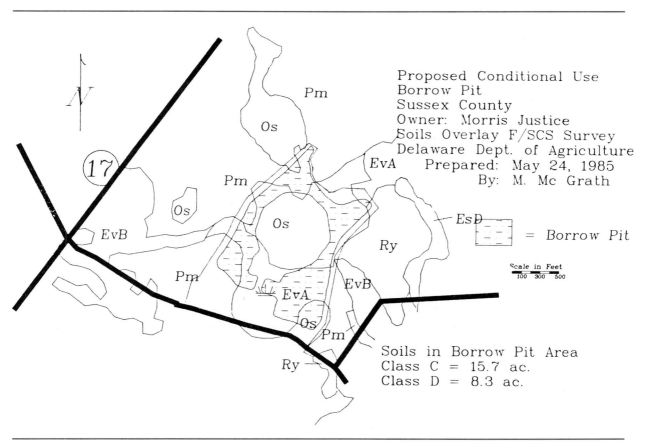

**Fig. 3.5** Civil CAD Drawing. *Courtesy of Autodesk*

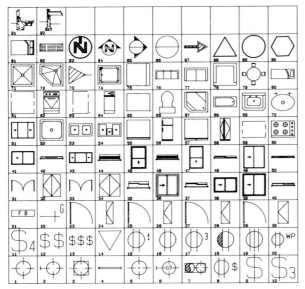

**Fig. 3.6** Architectural Symbols Library. *Courtesy of VersaCAD*

### Architectural Applications

Architectural drawings rely heavily on repetitive features and symbols. Special CAD software has been developed specifically for architectural applications. A floor plan drawing can be reflected to create ceiling plans, mechanical, electrical, and other types of drawings. Symbol libraries, Fig. 3.6, are available for structural details: fixtures, plumbing, furniture, stairs, etc. The schedules for doors, windows, fixtures, and other details can be extracted from the drawings and are updated with each change.

### Technical Illustration

After a design has been created and the engineering drawings produced, there is a need for further documentation. Technical illustrations are necessary for technical manuals, advertising, and sales literature. Many times the drawings produced on a CAD system for the initial design of the product can be used for technical illustrations, Fig. 3.7. Three-dimensional models can be viewed from any direction and color shaded to produce high resolution drawings, suitable for technical illustrations. One of the greatest advantages of using CAD is that once the 3D model is produced it becomes the source for documentation, technical illustration, NC (numerical control) machining, and many other related tasks.

### Charts and Graphs

When used with multi-pen plotters, CAD can create beautiful multicolor charts and graphs. Some software is capable of extracting information from a drawing and automatically creating the type of chart or graph you specify. CAD systems can produce different type fonts and hatching patterns and can control color plotter pens to produce charts and graphs, as shown in Fig. 3.8.

There are many other applications of CAD in engineering design, such as those shown in Figs. 3.9 and 3.10. Because CAD is a tool used to communicate graphically, it can be used for many types of drawings. The user must determine if the type of

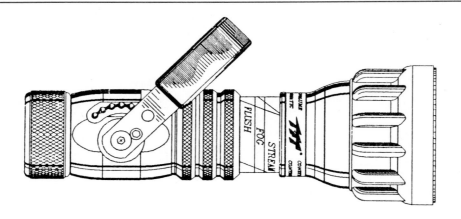

**Fig. 3.7** Technical Illustration Produced with CAD. *Courtesy of Autodesk*

| EVENT | SEP | OCT | NOV | DEC | JAN | FEB | MAR | APR | MAY |
|---|---|---|---|---|---|---|---|---|---|
| SELECT PROJECT | | | | | | | | | |
| WRITE OBJECTIVE | | | | | | | | | |
| LITERATURE SEARCH | | | | | | | | | |
| GATHER INFORMATION | | | | | | | | | |
| PROGRESS REPORT | | | | | | | | | |
| WRITE PROPOSAL | | | | | | | | | |
| PROPOSAL DUE | | | | | | | | | |
| ORDER PARTS | | | | | | | | | |
| FINALIZE DESIGN | | | | | | | | | |
| BUILD HARDWARE | | | | | | | | | |
| TEST SYSTEM | | | | | | | | | |
| PREPARE SPEECH | | | | | | | | | |
| REPORT DUE | | | | | | | | | |
| WRITE REPORT | | | | | | | | | |

*Fig. 3.8*   Chart Drawing—A Project Schedule. *Courtesy of Hewlett Packard*

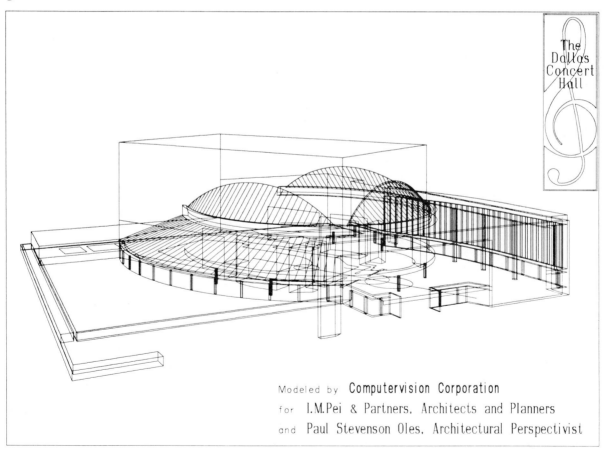

*Fig. 3.9*   Architectural Application of CAD. *Courtesy of Computervision Corporation*

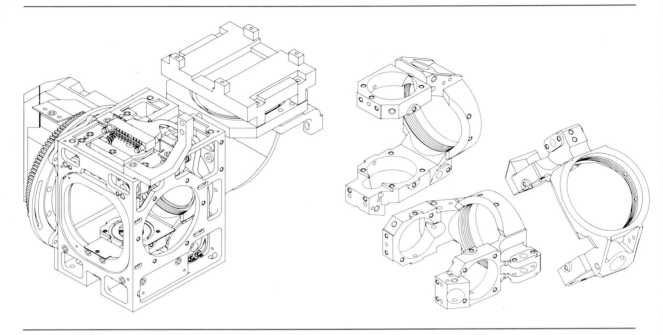

Fig. 3.10 Mechanical Application of CAD. *Courtesy of Computervision Corporation*

drawing to be created should be made with CAD or some other tool.

## 3.4 CAD and Factory Automation

CAD was originally intended to be an aid in creating production drawings. As hardware and software improved, it became apparent that a 3D computer model of a design could assist in the manufacture of the part. Machine tool automation was also developing rapidly. Soon it was realized that the CAD data base could be used for CAM (computer-aided manufacturing). *CAM* is the automation and control of manufacturing processes, including machine tools, robotics, and automated vehicles.

One of the most important principles a user of CAD should realize is the tremendous potential of a CAD-produced drawing compared to one created with hand tools. If the design is created as a 3D model, the graphics data base can have many other applications. The 3D model becomes the central focus and source for manufacturing, illustrations, documentation, and animation, as shown in Fig. 3.11. Three-dimensional CAD models are used for the automated manufacture of mechanical components and assemblies and electronic circuit boards.

## 3.5 Hardware Components of a CAD System

There are many types of CAD *software* programs. These programs can run on many different types of computer *hardware*. It is important to have a general understanding of the instruments (computer hardware) that are commonly used with CAD software. There are four hardware devices commonly used with CAD software: (1) CPU (central processing unit), (2) display device, (3) input device, and (4) output device.

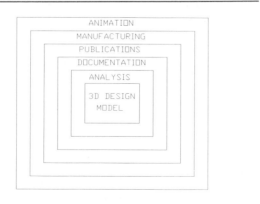

Fig. 3.11 CAD Input Model.

*Fig. 3.12* Mainframe CAD Workstation. *Courtesy of Hewlett Packard*

## 3.6 The Central Processing Unit

The CPU is the brains of the CAD system. It controls all other computer devices and follows the instructions of the CAD software. Generally, two types of CPU are used with CAD software: mainframe and microcomputer.

A mainframe CPU is a powerful computer housed in an environmentally controlled area. It is relatively expensive to purchase and operate. However, one mainframe computer can be used as the CPU for many CAD workstations. A *workstation* Fig. 3.12, is a group of hardware devices used to run CAD software programs. A microcomputer CAD workstation, Fig. 3.13, uses the microcomputer as the CPU, each workstation having its own CPU. Many different CAD software programs have been

*Fig. 3.13* Microcomputer CAD Workstation. *Courtesy of International Business Machines Corporation*

Fig. 3.14    Keyboard. *Courtesy of Hewlett Packard*

developed for both types of CPU. Typically, CAD software that runs on a mainframe computer has more powerful engineering analysis features than does microcomputer software.

## 3.7 Display Devices

Display devices are used to display alphanumeric and graphic images produced by the computer. A display device actually is an output device, but it is important enough to CAD systems to be considered alone. Display devices are categorized by the num-

Fig. 3.15    Optical Mouse. *Courtesy Summagraphics Corp.*

ber of colors and the resolution of the screen, called pixels. *Pixels* are the individual dots that are lined up horizontally and vertically on the screen. The higher the resolution—that is, the smaller and more numerous the dots in a given area—the better the displayed image and the smaller the jaggies or stair-stepping. *Jaggies* or *stair-stepping* is a phenomenon associated with the resolution of the screen and is characterized by angled lines or arcs appearing as a series of short lines like stair-steps. Some CAD systems use dual display devices, one screen to display text and the other to display graphics.

## 3.8 Input Devices

An *input device* is used to enter data and/or control some features of the CAD software. Most CAD software programs use more than one input device. Usually there are two input devices: the keyboard, used to enter alphanumeric data and make menu selections, and the cursor control device, such as a mouse or puck and tablet.

### Keyboard

The keyboard, Fig. 3.14, is the most commonly used device for inputting text and numerics, controlling the software through menu selections, and controlling the cursor. The specific functions depend on the particular CAD software being used. Some CAD software is much more keyboard dependent then others. Although the cursor can be controlled with the keyboard, most CAD programs will use a cursor control device.

### Mouse

One popular cursor control device is called a mouse. A *mouse* is an input device that controls the screen cursor, locates lines and other geometry, and is used as an interactive device for making menu selections. Optical and mechanical mice are the most common, Fig. 3.15.

When the mouse is moved, the screen cursor moves correspondingly. The screen cursor is usually a large or small plus (+) on the screen. The mouse has two or more buttons located on its surface. These buttons are assigned certain functions. For example, one button is the hot or "pick" button used to make menu selections or to locate the end points of lines and centers of circles and arcs. Another but-

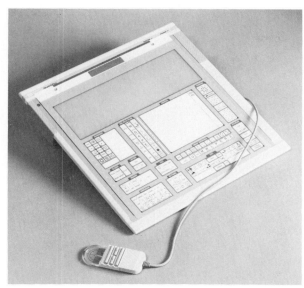

*Fig. 3.16*   Tablet and Menu. *Courtesy of Hewlett Packard*

ton might be assigned a keyboard function, such as RETURN or ENTER.

### Tablet

A *tablet*, Fig. 3.16, is an electronic device used to control screen cursor movement and to select CAD software commands. Typically, a thin plastic overlay on the surface of the tablet lists most of the menu commands outside of an area that represents the computer screen used for cursor control. A stylus or puck attached to the tablet is used to control movement of the screen cursor and to select menu commands from the surface of the tablet.

Menu selections from a tablet are made by moving the stylus or puck over the menu command on the tablet and pushing the "pick" button. A *stylus* looks like a pen and is used with a tablet to control the cursor and to make menu selections. Menu selections are made by pressing on the tip of the stylus or by pressing a small button located on its surface near the bottom. A *puck* looks like a mouse except it usually has a set of crosshairs that is placed over menu items for selection. A puck has many buttons located on its surface that can be assigned different functions.

Some specialized tablets are used primarily to enter drawings created with hand tools into the computer. Such tablets are a means of converting

traditional drawings to CAD drawings. This process is referred to as *digitizing* a drawing.

### Other Input Devices

Other less common input devices are the joystick, programmed function board, light pen, and dials. Most of these devices are for specialized functions, such as a programmed function board used to make some menu selections.

## 3.9 Output Devices

An *output device* is used to produce a hard copy of the graphics produced with the CAD system. For engineering graphics, output is usually in the form of plots on standard size engineering paper. Slides or photographs can also be produced if desired.

### Plotters

*Plotters* are of three types: (1) pen, (2) ink jet, and (3) electrostatic. There are two types of pen plotters, drum and flatbed. A *pen plotter*, Fig. 3.17, uses different sizes, types, and colors of pens to produce hard copy of CAD drawings on standard engineering paper.

*Fig. 3.17*   Pen Plotter. *Courtesy of Hewlett Packard*

*Fig. 3.18*   Ink Jet Plotter. *Courtesy of Hewlett Packard*

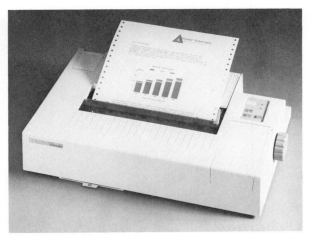

*Fig. 3.19*   Electrostatic Plotter. *Courtesy of Hewlett Packard*

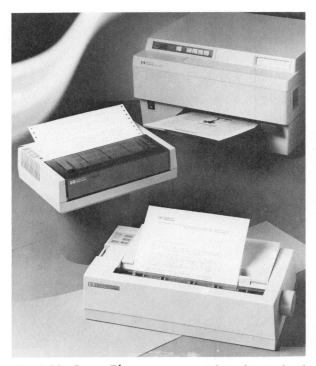

*Fig. 3.20*   Laser Plotter. *Courtesy of Hewlett Packard*

*Ink jet* plotters, Fig. 3.18, spray a fine stream of ink on the paper to produce hard copy. Ink jet plotters are relatively slow, but can produce multicolored shaded plots of assemblies, solid models, and other illustrations.

*Electrostatic* plotters, Fig. 3.19, discharge a printing medium onto electrostatically charged paper to produce hard copy of CAD drawings. Typically, electrostatic plotters are fast and a single color, but multicolor plotters are available.

### Other Output Devices

Laser printers, Fig. 3.20, produce single color plots of drawings. Dot matrix printers are used for drawing output but do not produce high quality output. Normally dot matrix printers are used for check prints. Photoplotters are used to produce photographic negatives for printed circuit boards, color slides or color negatives for presentation graphics.

# *Introduction to CAD Problems*

The following problems are given to examine your retention and understanding of the subject matter presented in this chapter. When necessary, refer to the appropriate section of the chapter to check your answers.

Computer graphics problems in convenient form for solution may be found in *Principles of Engineering Graphics Problems* by Spencer, Hill, Loving, and Dygdon, a workbook designed to accompany this text that is also published by Macmillan Publishing Company.

---

**Prob. 3.1**   Define CAD, CAM, input device, output device, and display device.

**Prob. 3.2**   List five applications of CAD.

**Prob. 3.3**   Describe how CAD is used in factory automation.

**Prob. 3.4**   List the four basic components of a CAD system.

**Prob. 3.5**   List two different methods of controlling cursor movement.

**Prob. 3.6**   List three display options.

# CHAPTER

# 4

# Lettering

The designs of modern alphabets had their origin in Egyptian hieroglyphics, Fig. 1.4, which were developed into a cursive hieroglyphic or hieratic writing. This was adopted by the Phoenicians and was developed by them into an alphabet of 22 letters. This Phoenician alphabet was later adopted by the Greeks, but it evolved into two distinct types in different sections of Greece: an Eastern Greek type, used also in Asia Minor, and a Western Greek type, used in the Greek colonies in and near Italy. In this manner the Western Greek alphabet became the Latin alphabet about 700 B.C. The Latin alphabet came into general use throughout the Old World.

Originally the Roman capital alphabet consisted of 22 characters, and these have remained practically unchanged to this day. The numerous modern styles of letters were derived from the design of the original Roman capitals.

ABCDEFGH
abcdefgh

**GOTHIC** *All letters having the elementary strokes of even width are classified as Gothic*

ABCDEFGH
abcdefghij

*Roman All letters having elementary strokes "accented" or consisting of heavy and light lines, are classified as Roman.*

*ABCDEFGHI*
*abcdefghijklm*

*Italic- All slanting letters are classified as Italics~ These may be further designated as Roman- Italics. Gothic Italics or Text Italic.*

𝕬𝕭𝕮𝕯𝕰𝕱𝕲
𝖆𝖇𝖈𝖉𝖊𝖋𝖌𝖍𝖎𝖏𝖐𝖑

*Text—This term includes all styles of Old English. German text, Bradley text or others of various trade names ~ Text styles are too illegible for commercial purposes.*

*Fig. 4.1*   Classification of Letter Styles.

## 4.1 Lettering* Styles

Before the invention of printing by Gutenberg in the fifteenth century, all letters were made by hand, and their designs were modified and decorated according to the taste of the individual writer. In England these letters became known as Old English. The early German printers adopted the Old English letters, and they are still in limited use. The early Italian printers used Roman letters, which were later introduced into England, where they gradually replaced the Old English letters. The Roman capitals have come down to us virtually in their original form. A general classification of letter styles is shown in Fig. 4.1. If the letters are drawn in outline and filled in, they are referred to as *filled-in* letters. The plainest and most legible style is the GOTHIC, from which our single-stroke engineering letters are derived. The term ROMAN refers to any letter having wide downward strokes and thin connecting strokes, as would result from the use of a wide pen, while the ends of the strokes are terminated with spurs called *serifs*. Roman letters include Old Roman and Modern Roman and may be vertical or inclined. Inclined letters are also referred to as *italic*, regardless of the letter style; those shown in Fig. 4.1 are inclined Modern Roman. *Text* letters are often loosely referred to as Old English, although these letters as well as the other similar let-

ters, such as German Text, are actually Gothic. The Commercial Gothic shown at the top of Fig. 4.1 is a relatively modern development, which originates from the earlier Gothic forms. German Text is the only commercially used form of medieval Gothic in use today.

## 4.2 Extended and Condensed Letters

To meet design or space requirements, letters may be narrower and spaced closer together, in which case they are called compressed or condensed letters. If the letters are wider than normal, they are referred to as extended letters, Fig. 4.2.

## 4.3 Lightface and Boldface Letters

Letters also vary as to the thickness of the stems or strokes. Letters having very thin stems are called LIGHTFACE, while those having heavy stems are called **BOLDFACE,** Fig. 4.3.

## 4.4 Single-Stroke Gothic Letters

During the latter part of the nineteenth century the development of industry and of technical drawing in the United States made evident a need for a simple legible letter that could be executed with single strokes of an ordinary pen. To meet this need, C. W. Reinhardt, formerly chief draftsman for the *Engineering News*, developed alphabets of capital and lowercase inclined and "upright" letters,* based

---

*Lettering*, not "printing," is the correct term for making letters by hand. *Printing* means the production of printed material on a printing press.

*Published in the *Engineering News* about 1893 and in book form in 1895.

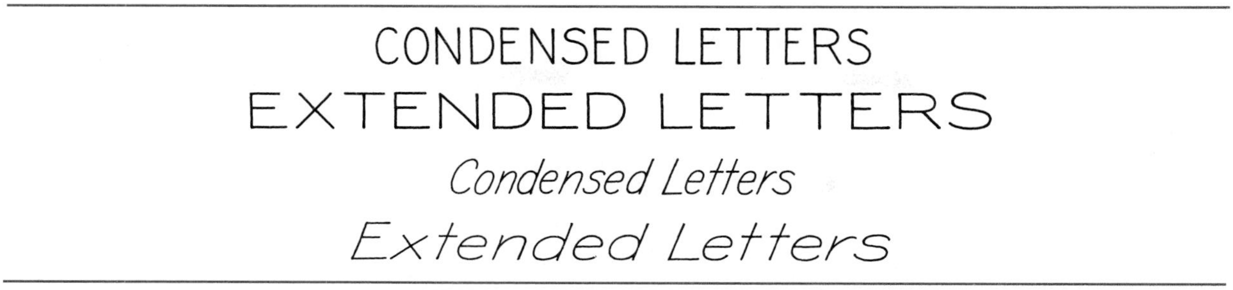

CONDENSED LETTERS
EXTENDED LETTERS
*Condensed Letters*
*Extended Letters*

*Fig. 4.2*   Condensed and Extended Letters.

LIGHTFACE

**BOLDFACE**

*Fig. 4.3*   Lightface and Boldface Letters.

RELATIVELY

Relatively — Letters not uniform in style.

RELATIVELY
RELATIVELY — Letters not uniform in height.

RELATIVELY
*RELATIVELY* — Letters not uniformly vertical or inclined.

RELATIVELY
RELATIVELY — Letters not uniform in thickness of stroke.

RELATIVELY — Areas between letters not uniform.

NOW IS THE TIME FOR EVERY GOOD MAN TO COME TO THE AID OF HIS COUNTRY — Areas between words not uniform.

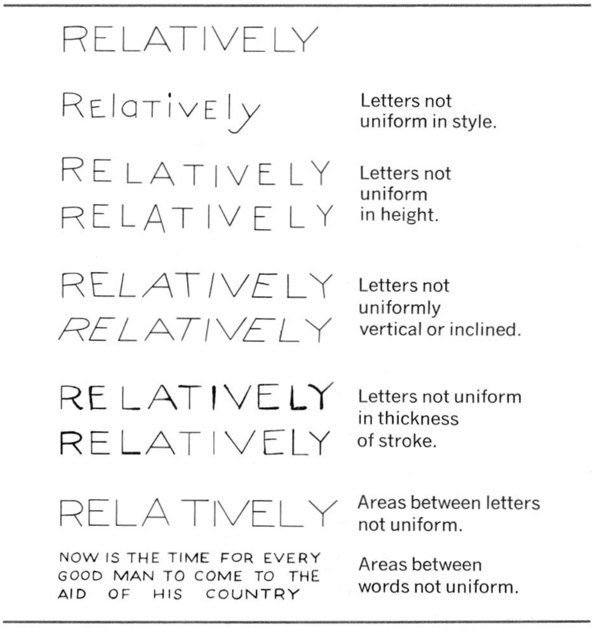

*Fig. 4.4*   Uniformity in Lettering.

upon the old Gothic letters. For each letter he worked out a systematic series of strokes. The single-stroke Gothic letters used on the technical drawings today are based upon Reinhardt's work.

### 4.5 Standardization of Lettering

The first step toward standardization of technical lettering was made by Reinhardt when he developed single-stroke letters with a systematic series of strokes, §4.4. However, since that time there has been an unnecessary and confusing diversity of lettering styles and forms, and the American National Standards Institute in 1935 suggested letter forms that are now generally considered as standard. The lettering forms given in the present standard [ANSI Y14.2M–1979 (R1987)] are practically the same as those given in 1935, except that lowercase forms have since been added.

The letters in this chapter and throughout this text conform to the American National Standard. Vertical letters are perhaps slightly more legible than inclined letters, but are more difficult to execute. Both vertical and inclined letters are standard, and the engineer or drafter may be called upon to use either.

Lettering on drawings must be legible and suitable for easy and rapid execution. The single-stroke Gothic letters shown in Figs. 4.19 and 4.20 meet these requirements. Either vertical or inclined lettering may be used but only one style should be used throughout a drawing. Drawings for microfilm reproduction require well-spaced lettering to prevent "fill-ins." Background areas between letters in words should appear approximately equal, and words should be clearly separated by a space equal to the height of the lettering. Only when special emphasis is necessary should the lettering be underlined.

It is not desirable to vary the size of the lettering according to the size of the drawing except when a drawing is to be reduced in reproduction.

### 4.6 Uniformity

In any style of lettering, uniformity is essential. Uniformity in height, proportion, inclination, strength of lines, spacing of letters, and spacing of words insures a pleasing appearance, Fig. 4.4.

Uniformity in height and inclination is promoted by the use of light guide lines, §4.13. Uniformity in

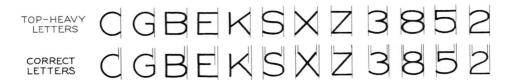

*Fig. 4.5*  Stability of Letters.

strength of lines can be obtained only by the skillful use of properly selected pencils and pens, §§4.9 and 4.10.

## 4.7 Optical Illusions

Good lettering involves artistic design, in which the white and black areas are carefully balanced to produce a pleasing effect. Letters are designed to *look* well and some allowances must be made for errors in perception. Note that in Fig. 4.19 the width of the standard H is less than its height to eliminate a square appearance, the numeral 8 is narrower at the top to give it stability, and the width of the letter W is greater than its height for the acute angles in the W give it a compressed appearance. Such acute angles should be avoided in good letter design.

## 4.8 Stability

If the upper portions of certain letters and numerals are equal in width to the lower portions, the characters appear top-heavy. To correct this, the upper portions are reduced in size where possible, thereby producing the effect of stability and a more pleasing appearance, Fig. 4.5.

If the central horizontal strokes of the letters B, E, F, and H are placed at midheight, they will appear to be below center. To overcome this optical illu-

sion, these strokes should be drawn slightly above the center.

## 4.9 Lettering Pencils

Pencil letters can be best made with a medium-soft lead with a conical point, Fig. 2.11 (c), or with a suitable thin-lead pencil, Fig. 2.8 (c).

Today the majority of drawings are finished in pencil and reproduced. To reproduce well by any process, the pencil lettering must be **dense black,** as should all other final lines on the drawing. The right lead to use depends largely upon the amount of tooth or grain in the paper, the rougher papers requiring the harder pencils. The lead should be soft enough to produce jet-black lettering, yet hard enough to prevent excessive wearing down of the point, crumbling of the point, and smearing of the graphite.

## 4.10 Lettering Pens

The choice of a pen for lettering is determined by the size and style of the letters, the thickness of stroke desired, and the personal preference of the drafter. Fig. 4.6 shows a variety of the best pen points in a range from the *tit quill*, the finest, to the *ball-pointed*, the coarsest. The widths of the lines made by the several pens are shown full size. Letters more than $\frac{1}{2}''$ (12.7 mm) in height generally require a special pen, Fig. 4.7.

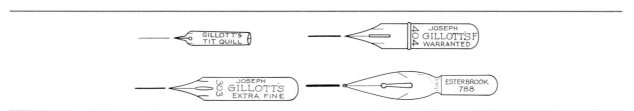

*Fig. 4.6*  Pen Points (Full Size).

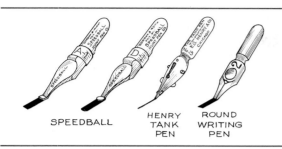

SPEEDBALL     HENRY TANK PEN     ROUND WRITING PEN

*Fig. 4.7* Special Pens for Freehand Lettering.

*Fig. 4.8* Technical Fountain Pen. *Courtesy of Keuffel & Esser Co./Kratos*

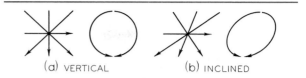

(a) VERTICAL      (b) INCLINED

*Fig. 4.9* Basic Lettering Strokes.

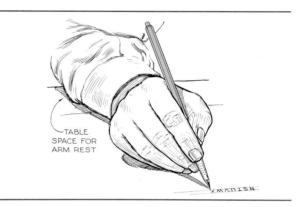

TABLE SPACE FOR ARM REST

*Fig. 4.10* Position of Hand in Lettering.

The technical fountain pen, Fig. 4.8, is a newer instrument that is now used by drafters for lettering and line work. The point is a small tube in which an automatic plunger rod keeps the ink flowing, and the pen has a reservoir cartridge for the storage of ink. The pen point produces a uniform thickness of line and makes the task of inking much simpler. These pens are available in sets of different point sizes and may be used with lettering instruments and templates.

*Any lettering pen must be kept clean.* All of these pens should be frequently cleaned with cleaning fluid to keep them in service.

## 4.11 Technique of Lettering

*Any normal person can learn to letter if a persistent and intelligent effort is made.* While it is true that "practice makes perfect," it must be understood that practice alone is not enough; it must be accompanied by *continuous effort to improve.*

Lettering is freehand drawing and not writing. Therefore, the six fundamental strokes and their direction for freehand drawing are basic to lettering, Fig. 4.9. The horizontal strokes are drawn to the right, and all vertical, inclined, and curved strokes are drawn downward.

Good lettering is always accomplished by conscious effort and is never done well otherwise, though good muscular coordination is of great assistance. Ability to letter has little relationship to writing ability; excellent letterers are often poor writers.

There are three necessary steps in learning to letter.

1. Knowledge of the proportions and forms of the letters and the order of the strokes. No one can make a good letter who does not have a clear mental image of the correct form of the letter.

2. Knowledge of composition—the spacing of letters and words. Rules governing composition should be thoroughly mastered, §4.23.

3. Persistent practice, with *continuous effort to improve.*

First, sharpen the pencil to a needle point; then dull the point *very slightly* by marking on paper while holding the pencil vertically and rotating the pencil to round off the point.

Pencil lettering should be executed with a fairly soft pencil, such as an F or H for ordinary paper, and the strokes should be *dark* and *sharp*, not gray and blurred. In order to wear the lead down uniformly and thereby to keep the lettering sharp, turn the pencil frequently to a new position.

The correct position of the hand in lettering is shown in Fig. 4.10. In general, draw vertical strokes

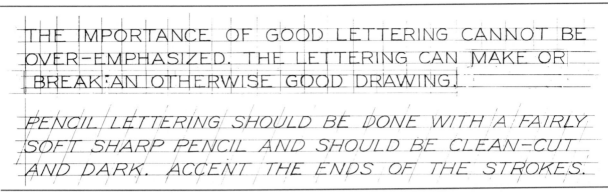

**Fig. 4.11** Pencil Lettering (Full Size).

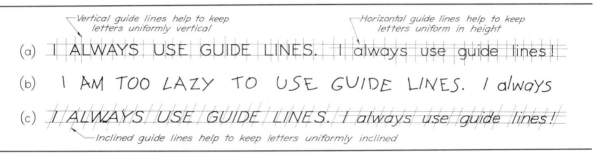

**Fig. 4.12** Guide Lines.

downward or toward you with a finger movement, and draw horizontal strokes from left to right with a wrist movement without turning the paper.

Since practically all pencil lettering will be reproduced, the letters should be **dense black.** Avoid hard pencils that, even with considerable pressure, produce gray lines. Use a fairly soft pencil and keep it sharp by frequent dressing of the point on the sandpaper pad or file. An example (full size) of pencil lettering exhibiting correct technique is shown in Fig. 4.11.

### 4.12 Left-handers
All evidence indicates that the left-handed drafter is just as skillful as the right-hander, and this includes skill in lettering. The most important step in learning to letter is learning the correct shapes and proportions of letters, and these can be learned as well by the left-hander as by anyone else. The left-hander does have a problem of developing a system of strokes that seems personally most suitable. The strokes shown in Figs. 4.19 and 4.20 are for right-handers. The left-hander should experiment with each letter to find out which strokes are best. The habits of left-handers vary so much that it is futile

to suggest a standard system of strokes for all left-handers.

### 4.13 Guide Lines
Extremely light horizontal guide lines are necessary to regulate the height of letters. In addition, light vertical or inclined guide lines are needed to keep the letters uniformly vertical or inclined. Guide lines are absolutely essential for good lettering and should be regarded as a welcome aid, not as an unnecessary requirement. See Fig. 4.12.

Make guide lines for finished pencil lettering *so lightly that they need not be erased*, as indeed they cannot be after the lettering has been completed. Guide lines should be barely visible at arm's length. Use a relatively hard pencil, such as a 4H to 6H, with a long, sharp, conical point, Fig. 2.11 (c).

### 4.14 Guide Lines for Capital Letters
Guide lines for vertical capital letters are shown in Fig. 4.13. On working drawings, capital letters are commonly made $\frac{1}{8}''$ (3.2 mm) high, with the space between lines of lettering from three-fifths to the full height of the letters. See Table 16.1 for ANSI-

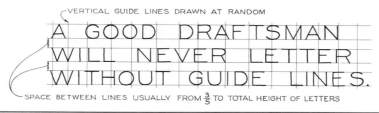

*Fig. 4.13* Guide Lines for Vertical Capital Letters.

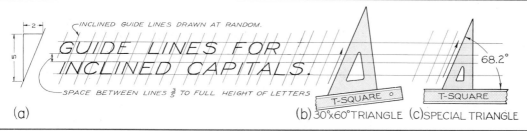

*Fig. 4.14* Guide Lines for Inclined Capital Letters.

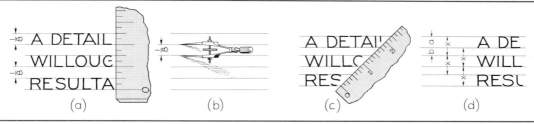

*Fig. 4.15* Spacing of Guide Lines.

recommended minimum letter heights on drawings. The vertical guide lines are not used to space the letters (as this should always be done by eye while lettering), but only to keep the letters uniformly vertical, and they should accordingly be drawn at random. Where several lines of letters are to be made, these vertical guide lines should be continuous from top to bottom of the lettered area, as shown.

Guide lines for inclined capital letters are shown in Fig. 4.14. The spacing of horizontal guide lines is the same as for vertical capital lettering. The American National Standard slope of 2 in 5 (or 68.2° with horizontal) may be established by drawing a "slope triangle," as shown at (a), and drawing the guide lines at random with the T-square and triangle, as shown at (b). Special triangles for the purpose may be used, as shown at (c), or the lines may be drawn with the Braddock-Rowe Lettering Triangle, Fig. 4.17, or the Ames Lettering Guide, Fig. 4.18.

A simple method of spacing horizontal guide lines is to use the scale, as shown in Fig. 4.15 (a), and merely set off a series of $\frac{1}{8}''$ spaces, making both the letters and the spaces between lines of letters $\frac{1}{8}''$ high. Another method of setting off equal spaces, $\frac{1}{8}''$ or otherwise, is to use the bow dividers, as shown at (b).

If it is desired to make the spaces between lines of letters less than the height of the letters, the methods shown at (c) and (d) will be convenient. At (c) the scale is placed diagonally, the letters in this case being four units high and the spaces between lines of lettering being three units. If the scale is rotated clockwise about the zero mark as a pivot, the height of the letters and the spaces between lines of letters diminish but remain proportional. If the scale is moved counterclockwise, the spaces are increased. The same unequal spacing may be accomplished with the bow dividers, as shown at (d). Let distance $x = a + b$, and set off $x$-distances, as shown.

When large and small capitals are used in combination, the small capitals should be three-fifths to

*Fig. 4.16*   Large and Small Capital Letters.

two-thirds as high as the large capitals, Fig. 4.16. This is conformity with the guide-line devices described below, §§4.15 and 4.16.

## 4.15 Lettering Triangles

Lettering triangles, which are available in a variety of shapes and sizes, are provided with sets of holes in which the pencil is inserted and the guide lines produced by moving the triangle with the pencil point along the T-square. The Braddock-Rowe lettering triangle, Fig. 4.17, is convenient for drawing guide lines for lettering and dimension figures and also for drawing section lines. In addition, the triangle is used as a utility 45° triangle. The numbers at the bottom of the triangle indicate heights of letters in thirty-seconds of an inch. Thus, to draw guide lines for $\frac{1}{8}''$ (3.2 mm) capitals, use the No. 4 set of holes. For lower-case letters, draw guidelines from every hole; for capitals, omit the second hole in each group. The spacing of holes is such that the lower portions of lower case letters are two-thirds as high as the capitals, and the spacing between lines of lettering is also two-thirds as high as the capitals.

The column of holes at the extreme left is used to draw guide lines for dimension figures $\frac{1}{8}''$ (3.2 mm) high and fractions $\frac{1}{4}''$ (6.4 mm) high, and also for section lines $\frac{1}{16}''$ (1.6 mm) apart.

## 4.16 Ames Lettering Guide

The Ames Lettering Guide, Fig. 4.18, is an ingenious transparent plastic device composed of a frame holding a disk with three columns of holes. The vertical distances between the holes may be adjusted quickly to the desired spacing for guide lines or section lines by simply turning the disk to one of the settings indicated at the bottom of the disk. These numbers indicate heights of letters in thirty-seconds of an inch. Thus, for $\frac{1}{8}''$ high letters, the No. 4 setting would be used. The center column of holes is used primarily to draw guide lines for numerals and fractions, the height of the whole number being two units and the height of the fraction four units. The No. 4 setting of the disc will provide guide lines for $\frac{1}{8}''$ whole numbers, with fractions twice as high, or $\frac{1}{4}''$, as shown at (a). Since the spaces are equal, these holes can also be used to draw equally spaced guide lines for lettering or to draw section lines. The Ames Lettering Guide is also available with metric graduations for desired metric spacing.

The two outer columns of holes are used to draw guide lines for capitals or lowercase letters, the column marked three-fifths being used where it is desired to make the lower portions of lowercase letters three-fifths the total height of the letters and the column marked two-thirds being used where the lower portion is to be two-thirds the total height of the letters. In each case, for capitals, the middle hole of each set is not used. The two-thirds and three-fifths also indicate the spaces between lines of letters.

The sides of the guide are used to draw inclined or vertical guide lines, as shown at (b) and (c).

## 4.17 Vertical Capital Letters and Numerals Fig. 4.19.

For convenience in learning the proportions of the letters and numerals, each character is shown in a grid 6 units high. Numbered arrows indicate the order and direction of strokes. The widths of the

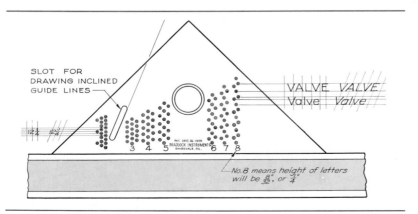

*Fig. 4.17*   Braddock-Rowe Lettering Triangle.

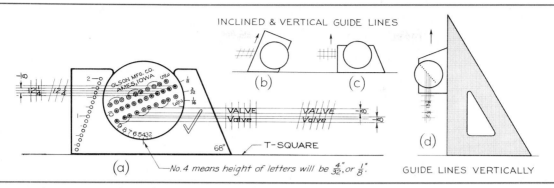

**Fig. 4.18** Ames Lettering Guide.

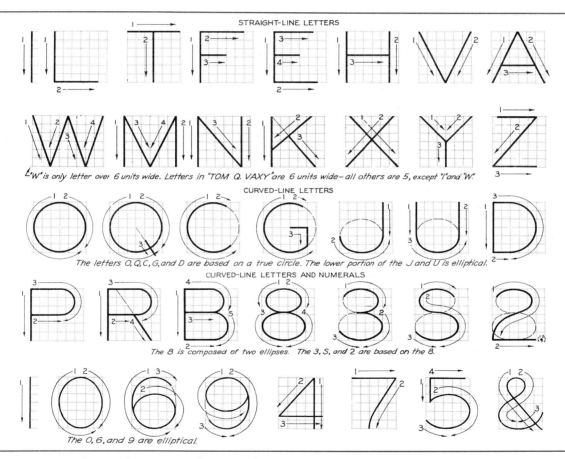

**Fig. 4.19** Vertical Capital Letters and Numerals.

letters can be easily remembered. The letter I, or the numeral 1, has no width. The W is 8 units wide ($1\frac{1}{3}$ times the height) and is the widest letter in the alphabet. All the other letters or numerals are either 5 or 6 units wide, and it is easy to remember the 6-unit letters because when assembled they spell TOM Q. VAXY. All numerals, except the 1, are 5 units wide.

All horizontal strokes are drawn to the right, and all vertical, inclined, and curved strokes are drawn downward, Fig. 4.9.

As shown in Fig. 4.19, the letters are classified

**Fig. 4.20** Inclined Capital Letters and Numerals.

as *straight-line letters* or *curved-line letters*. On the third row, the letters O, Q, C, and G are all based on the circle. The lower portions of the J and U are semiellipses, and the right sides of the D, P, R, and B are semicircular. The 8, 3, S, and 2 are all based on the figure 8, which is composed of a small ellipse over a larger ellipse. The 6 and 9 are based on the elliptical zero. The lower part of the 5 is also elliptical in shape.

## 4.18 Inclined Capital Letters and Numerals Fig. 4.20

The order and direction of the strokes and the proportions of the inclined capital letters and numerals are the same as those for the vertical characters. The methods of drawing guide lines for inclined capital letters are given in §4.14, and for numerals in §4.19. Inclined letters also are classified as straight-line letters or curved-line letters, most of the curves being elliptical in shape.

## 4.19 Guide Lines for Whole Numbers and Fractions

Complete guide lines should be drawn for whole numbers and fractions, especially by beginners. This means that both horizontal and vertical guide lines, or horizontal and inclined guide lines, should be drawn.

Draw five equally spaced guide lines for whole numbers and fractions, Fig. 4.21. Thus, fractions are twice the height of the corresponding whole numbers. Make the numerator and the denominator each about three-fourths as high as the whole number, to allow ample clear space between them and the fraction bar. For dimensioning, the most commonly used height for whole numbers is $\frac{1}{8}''$ (3.2 mm), and for fractions $\frac{1}{4}''$ (6.4 mm), as shown.

If the Braddock-Rowe triangle is used, the column of holes at the left produces five guide lines, $\frac{1}{16}''$ (1.6 mm) apart, Fig. 4.22.

If the Ames Lettering Guide, Fig. 4.18 is used with the No. 4 setting of the disk, the same five

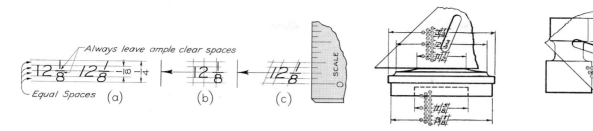

Fig. 4.21  Guide Lines for Dimension Figures.

Fig. 4.22  Use of Braddock-Rowe Triangle.

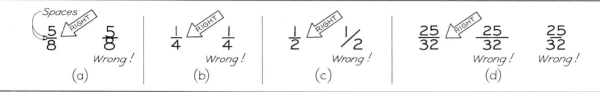

Fig. 4.23  Common Errors.

guide lines, $\frac{1}{16}''$ (1.6 mm) apart, may be drawn from the central column of holes.

Some of the most common errors in lettering fractions are illustrated in Fig. 4.23. Never let numerals touch the fraction bar, (a). Center the denominator under the numerator, (b). Never use an inclined fraction bar, (c), except when lettering in a narrow space, as in a parts list. Make the fraction bar slightly longer than the widest part of the fraction, (d).

## 4.20 Guide Lines for Lowercase Letters

Lowercase letters have four horizontal guide lines, called the cap line, waist line, base line, and drop line, Fig. 4.24. Strokes of letters that extend up to the cap line are called ascenders, and those that extend down to the drop line, descenders. Since there are only five letters that have descenders, the

drop line is little needed and is usually omitted. In spacing horizontal guide lines, space $a$ may vary from three-fifths to two-thirds of space $b$. Spaces $c$ are equal, as shown.

If it is desired to set off guide lines for letters $\frac{3}{16}''$ (4.8 mm) high with the scale (using two-thirds ratio), it is only necessary to set off equal spaces each $\frac{1}{16}''$ (1.6 mm), Fig. 4.25 (a). The lower portion of the letter thus would be $\frac{1}{8}''$ (3.2 mm), and the space between lines of letters would also be $\frac{1}{8}''$ (3.2 mm). If the scale is placed at an angle, the spaces will di-

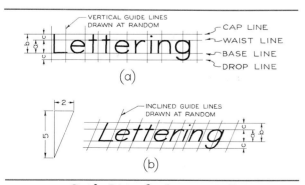

Fig. 4.24  Guide Lines for Lowercase Letters.

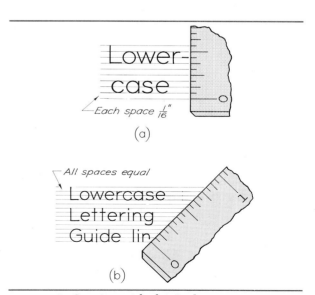

Fig. 4.25  Spacing with the Scale.

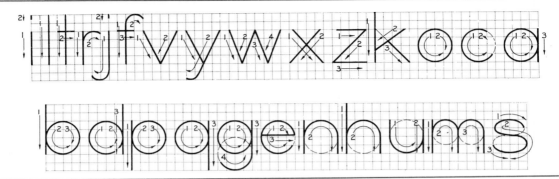

**Fig. 4.26**  Vertical Lowercase Letters.

**Fig. 4.27**  Inclined Lowercase Letters.

minish but remain equal, (b). Thus, this method may be easily used for various heights of lettering.

The Braddock-Rowe triangle, Fig. 4.17, and the Ames Lettering Guide, Fig. 4.18, produce guide lines for lowercase letters as described here and are highly recommended.

### 4.21 Vertical Lowercase Letters Fig. 4.26
Vertical lowercase letters are used largely on map drawings and very seldom on machine drawings. The shapes are based upon a repetition of the circle or circular arc and the straight line, with some variations. The lower part of the letter is usually two-thirds the height of the capital letter.

### 4.22 Inclined Lowercase Letters Fig. 4.27
The order and direction of the strokes and the proportions of inclined lowercase letters are the same as those of vertical lowercase letters. The slope of the letters is the same as for inclined capitals, or 68.2° with horizontal. The slope may be determined by drawing a "slope triangle" of 2 in 5, as shown in Fig. 4.24 (b), or with the aid of the inclined slot in the Braddock-Rowe Triangle, Fig. 4.17, or with the Ames Lettering Guide, Fig. 4.18 (b).

### 4.23 Spacing of Letters and Words
Uniformity in spacing of letters is a matter of equalizing spaces by eye. *The background areas between letters, not the distances between them, should be approximately equal.* In Fig. 4.28 (a) the actual distances are equal, but the letters do not appear equally spaced. At (b) the distances are intentionally unequal, but the background areas between letters are approximately equal, and the result is an even and pleasing spacing.

Some combinations, such as LT and VA, may even have to be slightly overlapped to secure good spacing. In some cases the width of a letter may be decreased. For example, the lower stroke of the L may be shortened when followed by A.

*Space words well apart, but space letters closely within words.* Make each word a compact unit well-separated from adjacent words. For either uppercase or lowercase lettering, make the spaces between words approximately equal to a capital O, Fig. 4.29. Avoid spacing letters too far apart and words too close together, as shown at (b). Samples of good spacing are shown in Fig. 4.11.

When it is necessary to letter to a stop line as in Fig. 4.30 (a), space each letter from *right to left*, as

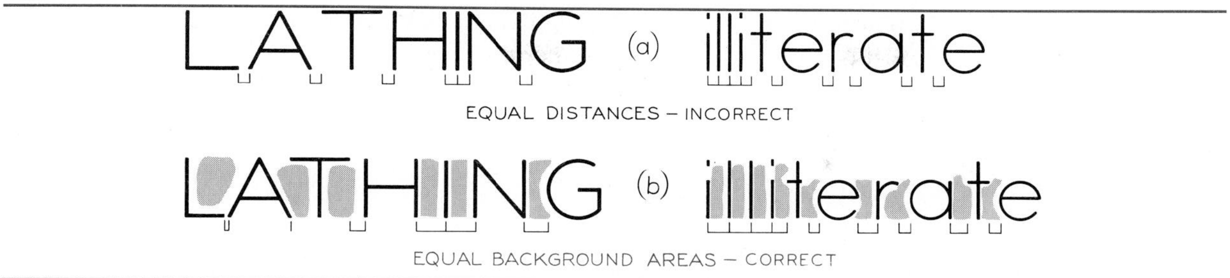

Fig. 4.28 Spacing Between Letters.

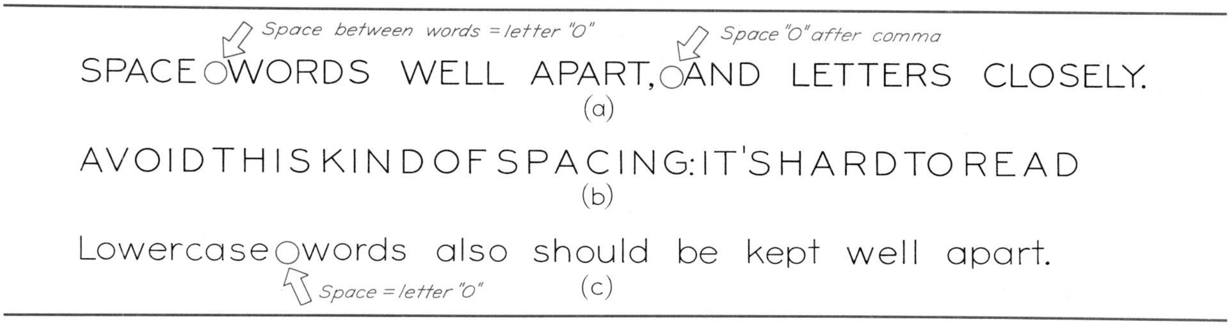

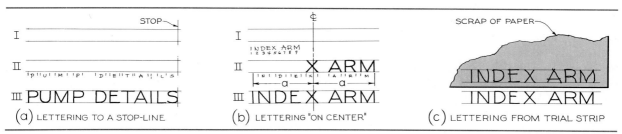

Fig. 4.29 Spacing Words.

Fig. 4.30 Spacing to a Stop Line and "on Center."

shown in step II, estimating the widths of the letters by eye. Then letter from *left to right*, as shown at III, and finally erase the spacing marks.

When it is necessary to space letters symmetrically about a center line, Fig. 4.30 (b), which is frequently the case in titles, Figs. 4.37 to 4.39, number the letters as shown, with the space between words considered as one letter. Then place the middle letter on center, making allowance for narrow letters (I's) or wide letters (W's) on either side. The X in Fig. 4.30 (b) is placed slightly to the left of center to compensate for the letter I, which has no width. Check with the dividers to make sure that distances *a* are exactly equal.

Another method is to letter roughly a trial line of lettering along the bottom edge of a scrap of paper, place it in position immediately above, as

shown at (c), and then letter the line in place. Be sure to use guide lines for the trial lettering.

## 4.24 Lettering Devices

The Leroy Standard Lettering Instrument, Fig. 4.31, is perhaps the most widely used lettering device. A guide pin follows grooved letters in a template, and the inking point moves on the paper. By adjusting the arm on the scribner, the letters may be made vertical or inclined. A number of templates, for letters and symbols, and sizes of pens are available, including templates for a wide variety of "built-up" letters similar to those made by the Varigraph and Letterguide, described shortly. Inside each pen is a cleaning pin used to keep the small tube open. These pins are easily broken, especially the small ones, when the pen is not promptly

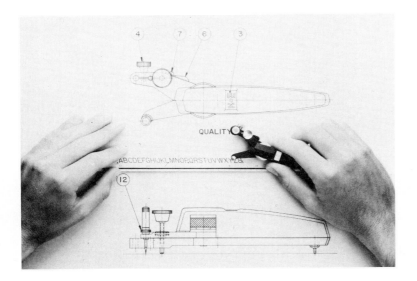

**Fig. 4.31** Leroy Standard Lettering Instrument. *Courtesy of Keuffel & Esser Co./Kratos*

cleaned. To clean a pen, draw it across a blotter until all ink has been absorbed; then insert the pin and remove it and wipe it with a cloth. Repeat this until the pin remains clean. If the ink has dried, the pens may be cleaned with Leroy pen-cleaning fluid, available at dealers. Leroy lettering sets, Fig. 4.32, are available as standard or metric sets. Both have the same style and features except for pen size designations. Leroy pens are also available in various sizes in standard or reservoir types.

The Wrico Lettering Guide, Fig. 4.33, consists of a scriber and templates similar to the Leroy system. Wrico letters more closely resemble American National Standard letters than do those of other sets.

The Varigraph is a more elaborate device for making a wide variety of either single-stroke letters or "built-up" letters. As shown in Fig. 4.34, a guide

pin is moved along the grooves in a template, and the pen forms the letters.

The Letterguide scriber, Fig. 4.35 is a much simpler instrument, which also makes a large variety of styles and sizes of letters when used with the various templates available. It also operates with a guide pin moving in the grooved letters of the template, while the pen, which is mounted on an adjustable arm, makes the letters in outline.

The Kroy Lettering Machine, Fig. 4.36, is a unique lettering machine that creates type on tape, which can then be applied on drawings, art work, posters, and so on. Lettering is produced by dialing the type disk to the desired character and pressing the print button. Spacing is automatic and adjustable. The machine is available in electric or manual models that use either 61- or 80-character type disks. A variety of type styles and sizes are available.

Various forms of press-on lettering and special lettering devices (typewriters, etc.) are available. In addition, the various computer-aided drafting systems have the capability to produce letters of dif-

**Fig. 4.32** Leroy Standard Lettering Set. *Courtesy of Keuffel & Esser Co./Kratos*

**Fig. 4.33** Wrico Lettering Guide. *Courtesy of Wood-Regan Instrument Co., Inc.*

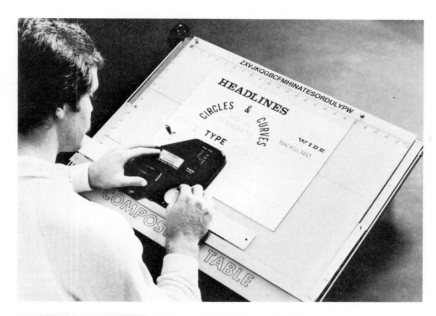

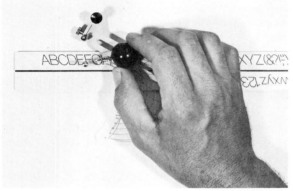

Fig. 4.34 The Varigraph Machine and Table. *Courtesy of Varigraph, Inc.*

Fig. 4.35 Letterguide. *Courtesy of Letterguide Co.*

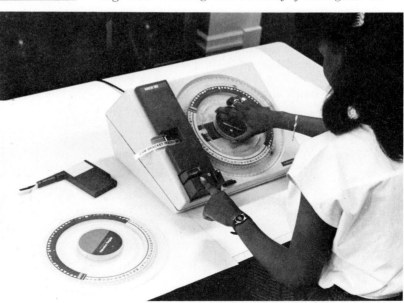

Fig. 4.36 Kroy Lettering Machine.

Equal
DRAWING TITLE
Scrap of paper underneath

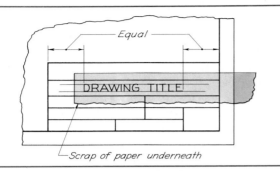

*Fig. 4.37* Centering Title in Title Box.

ferent heights and styles and to make changes as required. In whatever way the lettering is applied to the drawing and whatever style of lettering is used, the lettering must meet the requirements for legibility and microfilm reproduction.

## 4.25 Titles

The composition of titles on machine drawings is relatively simple. In most cases, the title and related information are lettered in "title boxes" or "title strips," which are printed directly on the drawing paper, tracing paper, or cloth; see for example, Figs. 16.24, 16.25, and 16.26. The main drawing title is usually centered in a rectangular space. This may be done by the method shown in Fig. 4.30 (b); or if the lettering is being done on tracing paper or cloth, the title may be lettered first on scrap paper and then placed underneath the tracing, as shown in Fig. 4.37, and then lettered directly over.

If a title box is not used, the title of a machine drawing may be lettered in the lower right corner of the sheet as a "balanced title," Fig. 4.38. A balanced title is simply one that is arranged symmetrically about an imaginary center line. These titles take such forms as the rectangle, the oval, the inverted pyramid, or any other simple symmetrical form.

Lettering
TOOL GRINDING MACHINE
TOOL REST SLIDE
SCALE : FULL SIZE
AMERICAN MACHINE COMPANY
NEW YORK CITY
DRAWN BY ____ CHECKED BY ____

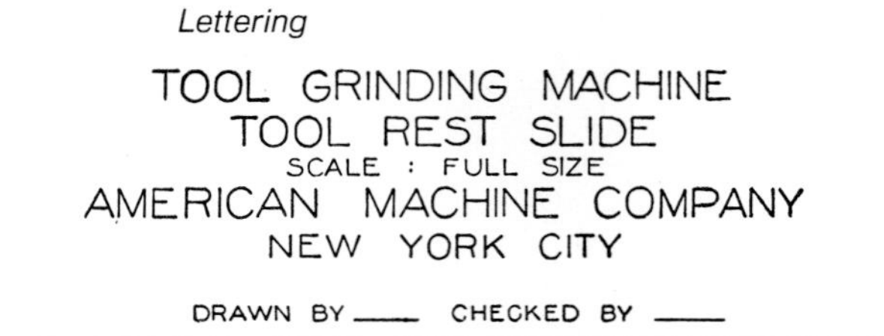

*Fig. 4.38* Balanced Machine-Drawing Title.

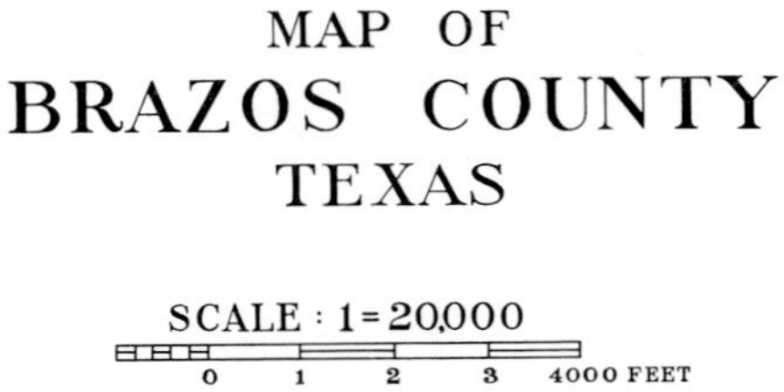

MAP OF
BRAZOS COUNTY
TEXAS
SCALE : 1=20,000
0 1 2 3 4000 FEET

*Fig. 4.39* Balanced Map Title.

On display drawings, or on highly finished maps or architectural drawings, titles may be composed of filled-in letters, usually Gothic or Roman, Fig. 4.39.

In any kind of title, the most important words are given most prominence by making the lettering larger, heavier, or both. Other data, such as scale and date may be displayed smaller.

## 4.26 Gothic Letters Fig. 4.40

Among the many forms of Gothic styles, including Old English and German Gothic, the so-called sans-serif Gothic letter is the only one of interest to engineers. It is from this style that the modern single-stroke engineering letters, discussed in the early part of this chapter, are derived. While they are

ABCDEFGHIJKLM
NOPQRSTUVWXYZ
&1234567890MAR

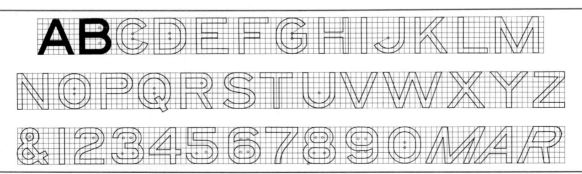

*Fig. 4.40* Gothic Capital Letters.

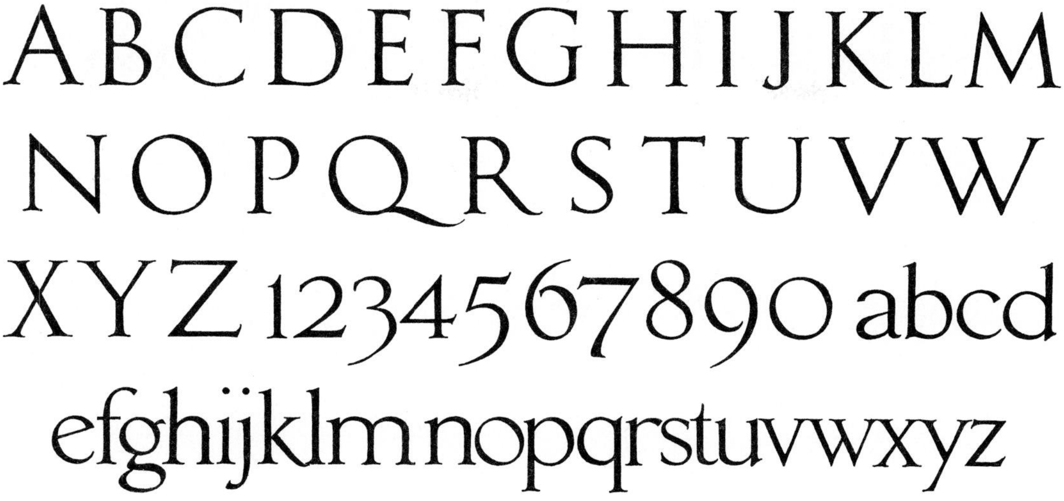

ABCDEFGHIJKLM
NOPQRSTUVW
XYZ 1234567890 abcd
efghijklmnopqrstuvwxyz

*Fig. 4.41* Old Roman Capitals, with Numerals and Lowercase of Similar Design.

ABCDEFGHIJKLM
NOPQRSTUVWXYZ
&1234567890ABLK

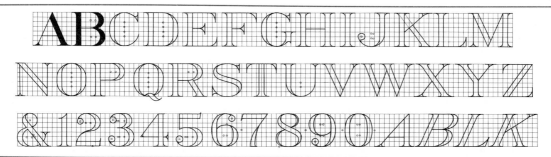

*Fig. 4.42* Modern Roman Capitals and Numerals.

admittedly not as beautiful as many other styles, they are very legible and comparatively easy to make.

### 4.27 Old Roman Letters Fig. 4.41

The Old Roman letter is the basis of all of our letters and is still regarded as the most beautiful. This letter is employed mostly by architects. Because of its great beauty, it is used almost exclusively on buildings and for inscriptions on bronze or stone.

### 4.28 Modern Roman Letters Figs. 4.42 and 4.43

The Modern Roman, or simply "Roman," letters were evolved during the eighteenth century by the type founders; the letters used in most modern newspapers, magazines, and books are of this style. The text of this book is set in Modern Roman capital and lowercase letters. These letters are often used on maps, especially for titles. They may be drawn in outline and then filled in, as shown in Fig. 4.42, or they may be produced with one of the broad-nib pens shown in Fig. 4.7.

A typical example of the use of Modern Roman in titles is shown in Fig. 4.39. Their use on maps is discussed in the next section.

### 4.29 Lettering on Maps

Modern Roman letters are generally used on maps, as follows.

1. *Vertical capitals.* Names of states, countries, townships, capital cities, large cities, and titles of maps.
2. *Vertical lowercase.* (First letter of each word a capital.) Names of small towns, villages, post offices, and so forth.
3. *Inclined capitals.* Names of oceans, bays, gulfs, sounds, large lakes, and rivers.

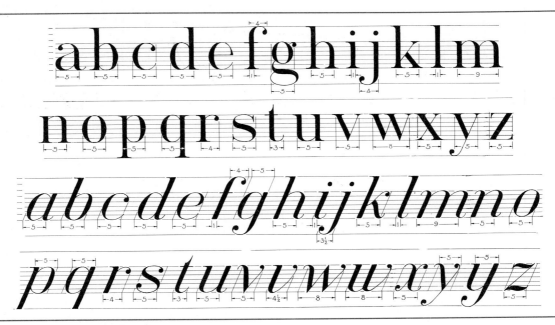

**Fig. 4.43**  Lowercase Modern Roman Letters.

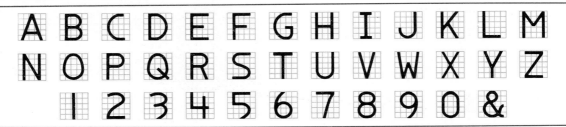

**Fig. 4.44**  Microfont Alphabet.

4. *Inclined lowercase, or "stump" letters.* (First letter of each word a capital.) Names of rivers, creeks, small lakes, ponds, marshes, brooks, and springs.

Prominent land features, such as mountains, plateaus, and canyons, are lettered in vertical Gothic, while the names of small land features, such as small valleys, islands, and ridges, are lettered in vertical lowercase Gothic. Names of railroads, tunnels, highways, bridges, and other public structures are lettered in inclined Gothic capitals.

### 4.30 Microfont Alphabet
The microfont alphabet, Fig. 4.44, is a recent adaptation of the single-stroke Gothic characters developed by the National Microfilm Association. It is designed for general usage and increased legibility in reproduction. Only the vertical style is shown.

### 4.31 Greek Alphabet
Greek letters are often used as symbols in both mathematics and technical drawing by the engineer. A Greek alphabet, showing both uppercase and lowercase letters, is given for reference purposes in Fig. 4.45.

| | | | | | | | | |
|---|---|---|---|---|---|---|---|---|
| A | $\alpha$ | alpha | I | $\iota$ | iota | P | $\rho$ | rho |
| B | $\beta$ | beta | K | $\kappa$ | kappa | $\Sigma$ | $s$ | sigma |
| $\Gamma$ | $\gamma$ | gamma | $\Lambda$ | $\lambda$ | lambda | T | $\tau$ | tau |
| $\Delta$ | $\delta$ | delta | M | $\mu$ | mu | $\Upsilon$ | $\upsilon$ | upsilon |
| E | $\epsilon$ | epsilon | N | $\nu$ | nu | $\Phi$ | $\phi$ | phi |
| Z | $\zeta$ | zeta | $\Xi$ | $\xi$ | xi | X | $\chi$ | chi |
| H | $\eta$ | eta | O | $o$ | omicron | $\Psi$ | $\psi$ | psi |
| $\Theta$ | $\theta$ | theta | $\Pi$ | $\pi$ | pi | $\Omega$ | $\omega$ | omega |

**Fig. 4.45**  Greek Alphabet.

# Lettering Exercises

Layouts for lettering practice are given in Figs. 4.46 to 4.49. Draw complete horizontal and vertical or inclined guide lines *very lightly*. Draw the vertical or inclined guide lines through the full height of the lettered area of the sheet. For practice in ink lettering, the last two lines and the title strip on each sheet may be lettered in ink, if assigned by the instructor. Omit all dimensions.

Sheets in convenient form for lettering practice may be found in *Principles of Engineering Graphics Problems* by Spencer, Hill, Loving, and Dygdon, a workbook designed to accompany this text that is also published by Macmillan Publishing Company.

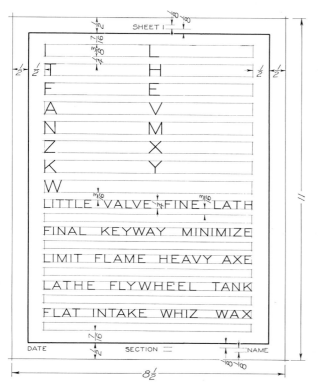

**Fig. 4.46** Lay out sheet, add vertical or inclined guide lines, and fill in vertical or inclined capital letters as assigned. For decimal-inch and millimeter equivalents of given dimensions, see table inside of front cover.

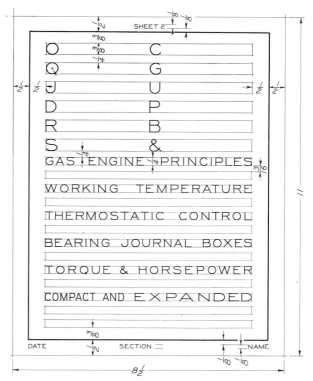

**Fig. 4.47** Lay out sheet, add vertical or inclined guide lines, and fill in vertical or inclined capital letters as assigned. For decimal-inch and millimeter equivalents of given dimensions, see table inside of front cover.

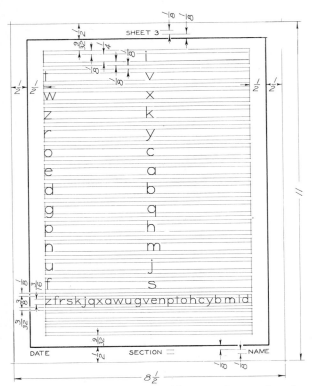

**Fig. 4.48** Lay out sheet, add vertical or inclined guide lines, and fill in vertical or inclined lower-case letters as assigned. For decimal-inch and millimeter equivalents of given dimensions, see table inside of front cover.

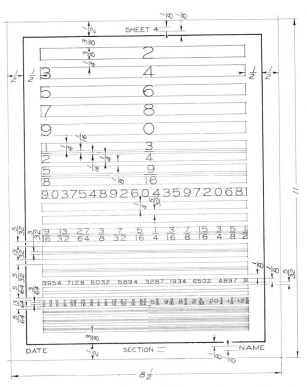

**Fig. 4.49** Lay out sheet, add vertical or inclined guide lines, and fill in vertical or inclined numerals as assigned. For decimal-inch and millimeter equivalents of given dimensions, see table inside of front cover.

# CHAPTER
# 5

# Geometric Constructions

Many of the constructions used in technical design drawings are based upon plane geometry, and every drafter, technician, or engineer should be sufficiently familiar with them to be able to apply them to the solutions of problems. Pure geometry problems may be solved only with the compass and a straightedge, and in some cases these methods may be used to advantage in technical drawing. However, the drafter or designer has available the T-square,* triangles, dividers, and other equipment, such as drafting machines, that in many cases can yield accurate results more quickly by what we may term "preferred methods." Therefore, many of the solutions in this chapter are practical adaptations of the principles of pure geometry.

This chapter is designed to present definitions of terms and geometric constructions of importance in technical drawing, suggest simplified methods of construction, point out practical applications, and afford opportunity for practice in accurate instrumental drawing. The problems at the end of this chapter may be regarded as a continuation of those at the end of Chapter 2.

In drawing these constructions, accuracy is most important. Use a sharp medium-hard lead (H to 3H) in your pencil and compasses. Draw construction lines extremely light—so light that they can hardly be seen when your drawing is held at arm's length. Draw all final and required lines medium to thin but dark.

*Hereafter, reference to the T-square could also refer to the parallel straightedge or drafting machine.

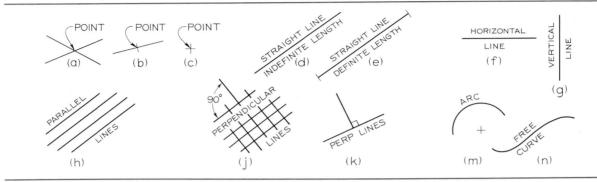

*Fig. 5.1*  Points and Lines.

## 5.1 **Points and Lines** Fig. 5.1

A *point* represents a location in space or on a drawing, and has no width, height, or depth. A point is represented by the intersection of two lines, (a), by a short crossbar on a line, (b), or by a small cross, (c). Never represent a point by a simple dot on the paper.

A line is defined by Euclid as "that which has length without breadth." A *straight line* is the shortest distance between two points and is commonly referred to simply as a "line." If the line is indefinite in extent, the length is a matter of convenience, and the endpoints are not fixed, (d). If the endpoints of the line are significant, they must be marked by means of small mechanically drawn crossbars, (e). Other common terms are illustrated from (f) to (h). Either straight lines or curved lines are parallel if the shortest distance between them remains constant. The common symbol for parallel lines is ∥, and for perpendicular lines it is ⊥ (singular) or ⊥s (plural). Two perpendicular lines may be marked with a "box" to indicate perpendicularity, as shown at (k). Such symbols may be used on sketches, but not in production drawings.

## 5.2 **Angles** Fig. 5.2

An angle is formed by two intersecting lines. A common symbol for angle is ∠ (singular) or ∠s (plural). There are 360 degrees (360°) in a full circle, as shown at (a). A degree is divided into 60 minutes (60′), and a minute is divided into 60 seconds (60″). Thus, 37° 26′ 10″ is read: 37 degrees, 26 minutes, and 10 seconds. When minutes alone are indicated, the number of minutes should be preceded by 0°, as 0° 20′.

The different kinds of angles are illustrated in (b) to (e). Two angles are *complementary*, (f), if they total 90°, and are *supplementary,* (g), if they total 180°. Most angles used in technical drawing can be drawn easily with the T-square or straightedge and triangles, Fig. 2.23. To draw odd angles, use the protractor, Fig. 2.24. For considerable accuracy, use a *vernier protractor*, or the tangent, sine, or chord methods, §5.20.

## 5.3 **Triangles** Fig. 5.3

A triangle is a plane figure bounded by three straight sides, and the sum of the interior angles is always 180°. A right triangle, (d), has one 90° angle, and the square of the hypotenuse is equal to the

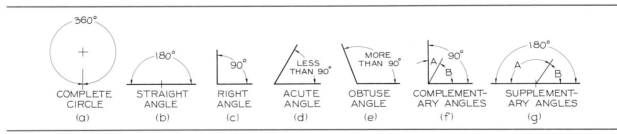

*Fig. 5.2*  Angles.

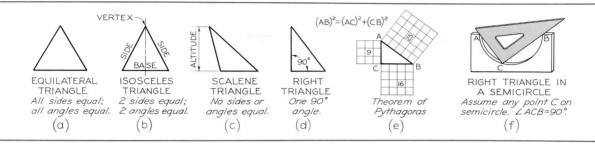

Fig. 5.3 Triangles.

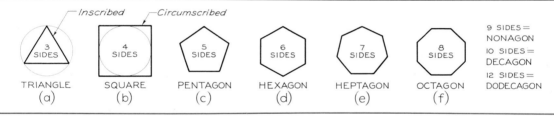

Fig. 5.4 Quadrilaterals.

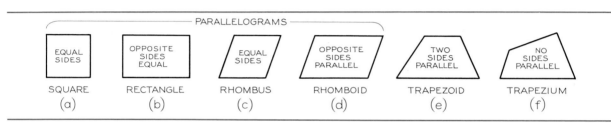

Fig. 5.5 Regular Polygons.

sum of the squares of the two sides, (e). As shown at (f), any triangle inscribed in a semicircle is a right triangle if the hypotenuse coincides with the diameter.

## 5.4 Quadrilaterals Fig. 5.4
A quadrilateral is a plane figure bounded by four straight sides. If the opposite sides are parallel, the quadrilateral is also a parallelogram.

## 5.5 Polygons Fig. 5.5
A polygon is any plane figure bounded by straight lines. If the polygon has equal angles and equal sides, it can be inscribed in or circumscribed around a circle and is called a *regular polygon*.

## 5.6 Circles and Arcs Fig. 5.6
A circle, (a), is a closed curve all points of which are the same distance from a point called the center. *Circumference* refers to the circle or to the distance

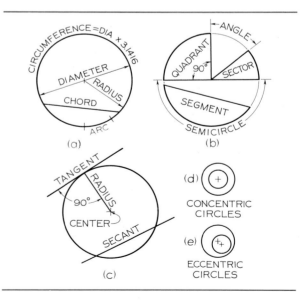

Fig. 5.6 The Circle.

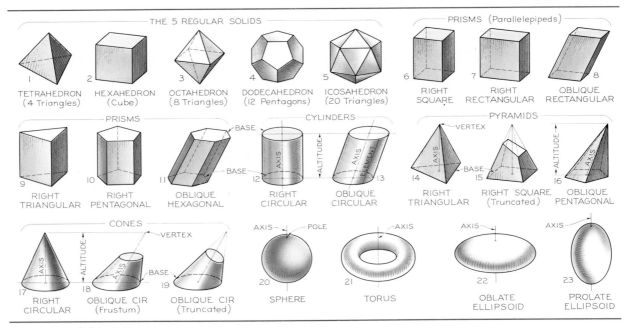

**Fig. 5.7** Solids.

around the circle. This distance equals the diameter multiplied by $\pi$ (called pi) or 3.1416. Other definitions are illustrated in the figure.

### 5.7 Solids Fig. 5.7

Solids bounded by plane surfaces are *polyhedra*. The surfaces are called faces, and if these are equal regular polygons, the solids are regular polyhedra.

A *prism* has two bases, which are parallel equal polygons, and three or more lateral faces, which are parallelograms. A triangular prism has a triangular base; a rectangular prism has rectangular bases; and so on. If the bases are parallelograms, the prism is a parallelepiped. A right prism has faces and lateral edges perpendicular to the bases; an oblique prism has faces and lateral edges oblique to the bases. If one end is cut off to form an end not parallel to the bases, the prism is said to be truncated.

A *pyramid* has a polygon for a base and triangular lateral faces intersecting at a common point called the vertex. The center line from the center of the base to the vertex is the axis. If the axis is perpendicular to the base, the pyramid is a right pyramid; otherwise it is an oblique pyramid. A triangular pyramid has a triangular base, a square pyramid has a square base, and so on. If a portion near the vertex has been cut off, the pyramid is truncated, or referred to as a frustum.

A *cylinder* is generated by a straight line, called the generatrix, moving in contact with a curved line and always remaining parallel to its previous position or to the axis. Each position of the generatrix is called an element of the cylinder.

A *cone* is generated by a straight line moving in contact with a curved line and passing through a fixed point, the vertex of the cone. Each position of the generatrix is an element of the cone.

A *sphere* is generated by a circle revolving about one of its diameters. This diameter becomes the axis of the sphere, and the ends of the axis are poles of the sphere.

A *torus* is generated by a circle (or other curve) revolving about an axis that is eccentric to the curve.

### 5.8 To Bisect a Line or a Circular Arc
Fig. 5.8

Given line or arc **AB**, as shown at (a), to be bisected:

I. From **A** and **B** draw equal arcs with radius greater than half **AB**.

II. and III. Join intersections **D** and **E** with a straight line to locate center **C**.

### 5.9 To Bisect a Line with Triangle and T-square Fig. 5.9

From endpoints **A** and **B**, draw construction lines at 30°, 45°, or 60° with the given line; then through their intersection, **C**, draw a line perpendicular to the given line to locate the center **D**, as shown.

To divide a line with the dividers, see §2.40.

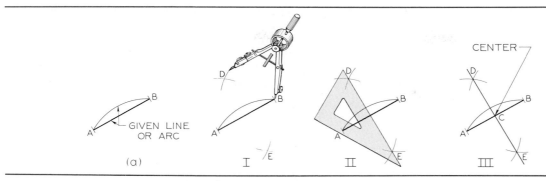

*Fig. 5.8*  Bisecting a Line or a Circular Arc (§5.8).

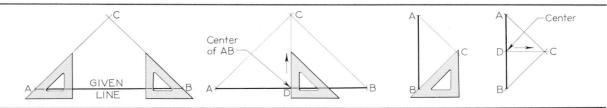

*Fig. 5.9*  Bisecting a Line with Triangle and T-square (§5.9).

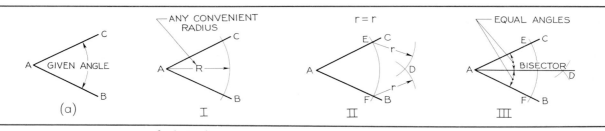

*Fig. 5.10*  Bisecting an Angle (§5.10).

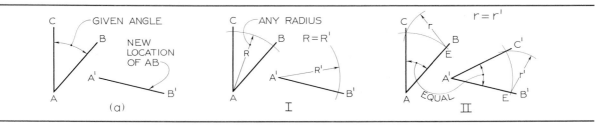

*Fig. 5.11*  Transferring an Angle (§5.11).

## 5.10 To Bisect an Angle Fig. 5.10

Given angle **BAC**, as shown at (a), to be bisected:

I.  Strike large arc **R**.

II.  Strike equal arcs **r** with radius slightly larger than half **BC**, to intersect at **D**.

III.  Draw line **AD**, which bisects angle.

## 5.11 To Transfer an Angle Fig. 5.11

Given angle **BAC**, as shown at (a), to be transferred to the new position at A′B′:

I.  Use any convenient radius **R**, and strike arcs from centers **A** and **A′**.

II.  Strike equal arcs **r**, and draw side **A′C′**.

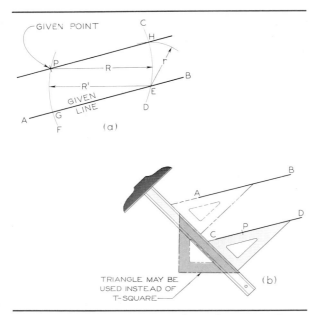

**Fig. 5.12** Drawing a Line Through a Point Parallel to a Line (§5.12).

## 5.12 To Draw a Line Through a Point and Parallel to a Line

***Fig. 5.12 (a)*** With given point P as center, and any convenient radius R, strike arc CD to intersect the given line AB at E. With E as center and the same radius, strike arc R′ to intersect the given line at G. With PG as radius, and E as center, strike arc r to locate point H. The line PH is the required line.

***Fig. 5.12 (b) Preferred Method*** Move the triangle and T-square as a unit until the triangle lines up with given line AB; then slide the triangle until its

edge passes through the given point P. Draw CD, the required parallel line. See also §2.21.

## 5.13 To Draw a Line Parallel to a Line and at a Given Distance
Let AB be the line and CD the given distance.

***Fig. 5.13 (a)*** With points E and F near A and B, respectively, as centers, and CD as radius, draw two arcs. The line GH, tangent to the arcs, is the required line.

***Fig. 5.13 (b) Preferred Method*** With any point E of the line as center and CD as radius, strike an arc JK. Move the triangle and T-square as a unit until the triangle lines up with the given line AB; then slide the triangle until its edge is tangent to the arc JK, and draw the required line GH.

***Fig. 5.13 (c)*** With centers selected at random on the curved line AB, and with CD as radius, draw a series of arcs; then draw the required line tangent to these arcs as explained in §2.54.

## 5.14 To Divide a Line into Equal Parts
Fig. 5.14

I. Draw a light construction line at any convenient angle from one end of line.

II. With dividers or scale, set off from intersection of lines as many equal divisions as needed, in this case, three.

III. Connect last division point to other end of line, using triangle and T-square, as shown.

IV. Slide triangle along T-square and draw parallel lines through other division points, as shown.

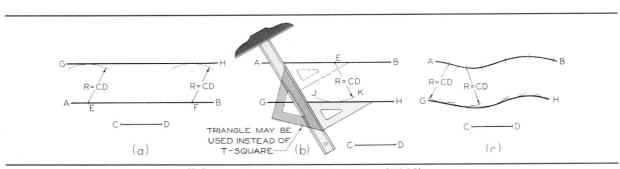

**Fig. 5.13** Drawing a Line Parallel to a Line at a Given Distance (§5.13).

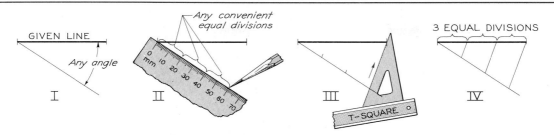

*Fig. 5.14*   Dividing a Line into Equal Parts (§5.14).

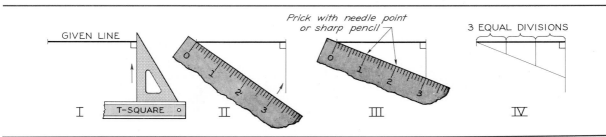

*Fig. 5.15*   Dividing a Line into Equal Parts (§5.15).

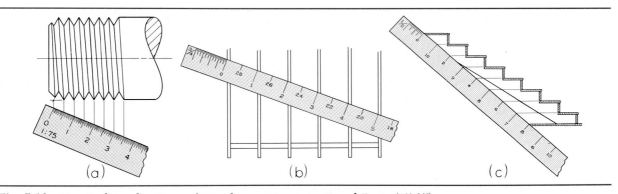

*Fig. 5.16*   Practical Applications of Dividing a Line into Equal Parts (§5.15).

## 5.15 To Divide a Line into Equal Parts
Fig. 5.15

I.   Draw vertical construction line at one end of given line.

II.   Set zero of scale at other end of line.

III.   Swing scale up until third unit falls on vertical line, and make tiny dots at each point, or prick points with dividers.

IV.   Draw vertical construction lines through each point.

Some practical applications of this method are shown in Fig. 5.16.

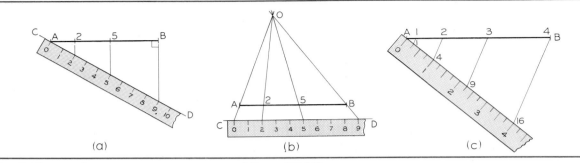

**Fig. 5.17**    Dividing a Line into Proportional Parts (§5.16).

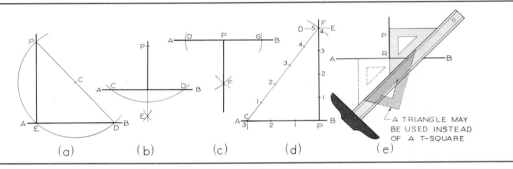

**Fig. 5.18**    Drawing a Line Through a Point and Perpendicular to a Line (§5.17).

### 5.16 To Divide a Line into Proportional Parts Fig. 5.17 (a) and (b)

Let it be required to divide the line AB into three parts proportional to 2, 3, and 4.

**Fig. 5.17 (a)   Preferred Method**   Draw a vertical line from point B. Select a scale of convenient size for a total of nine units and set the zero of the scale at A. Swing the scale up until the ninth unit falls on the vertical line. Along the scale, set off points for 2, 3, and 4 units, as shown. Draw vertical lines through these points.

**Fig. 5.17 (b)**   Draw a line CD parallel to AB and at any convenient distance. On this line, set off 2, 3, and 4 units, as shown. Draw lines through the ends of the two lines to intersect at the point O. Draw lines through O and the points 2 and 5 to divide AB into the required proportional parts.

Constructions of this type are useful in the preparation of graphs (Chapter 23).

**Fig. 5.17 (c)**   Given AB, to divide into proportional parts, in this case proportional to the square of $x$, where $x = 1, 2, 3, \ldots$ . Set zero of scale at end of line and set off divisions 4, 9, 16, . . . . Join the last division to the other end of the line, and draw parallel lines as shown. This method may be used for any power of $x$.

### 5.17 To Draw a Line Through a Point and Perpendicular to a Line Fig. 5.18

Given the line AB and a point P:

#### When the Point Is Not on the Line

Fig. 5.18 (a)

From P draw any convenient inclined line, as PD. Find center C of line PD, and draw arc with radius CP. The line EP is the required perpendicular.

**Fig. 5.18 (b)**   With P as center, strike an arc to intersect AB at C and D. With C and D as centers, and radius slightly greater than half CD, strike arcs to intersect at E. The line PE is the required perpendicular.

#### When the Point Is on the Line Fig. 5.18 (c)

With P as center and any radius, strike arcs to intersect AB at D and G. With D and G as centers, and radius slightly greater than half DG, strike equal arcs to intersect at F. The line PF is the required perpendicular.

***Fig. 5.18 (d)*** Select any convenient unit of length, for example, 6 mm or $\frac{1}{4}''$. With P as center, and 3 units as radius, strike an arc to intersect given line at C. With P as center, and 4 units as radius, strike arc DE. With C as center, and 5 units as radius, strike an arc to intersect DE at F. The line PF is the required perpendicular.

This method makes use of the 3–4–5 right triangle and is frequently used in laying off rectangular foundations of large machines, buildings, or other structures. For this purpose a steel tape may be used and distances of 30, 40, and 50 feet measured as the three sides of the right triangle.

***Fig. 5.18 (e)*** *Preferred Method* Move the triangle and T-square as a unit until the triangle lines up with AB; then slide the triangle until its edge passes through the point P (whether P is on or off the line), and draw the required perpendicular.

## 5.18 To Draw a Triangle with Sides Given Fig. 5.19

Given the sides A, B, and C, as shown at (a):

  I.   Draw one side, as C, in desired position, and strike arc with radius equal to given side A.

 II.   Strike arc with radius equal to given side B.

III.   Draw sides A and B from intersection of arcs, as shown.

## 5.19 To Draw a Right Triangle with Hypotenuse and One Side Given Fig. 5.20

Given sides S and R. With AB as a diameter equal to S, draw semicircle. With A as center, and R as radius, draw an arc intersecting the semicircle at C. Draw AC and CB to complete the right triangle.

## 5.20 To Lay Out an Angle Fig. 5.21

Many angles can be laid out directly with the triangle, Fig. 2.23, or they may be laid out with the protractor, Fig. 2.24. Other methods, where considerable accuracy is required, are as follows:

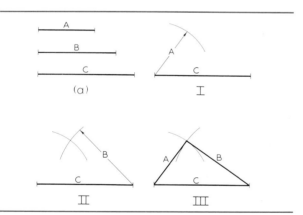

***Fig. 5.19*** Drawing a Triangle with Sides Given (§5.18).

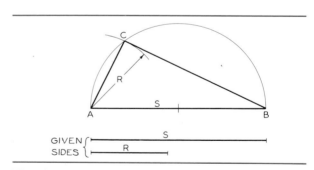

***Fig. 5.20*** Drawing a Right Triangle (§5.19).

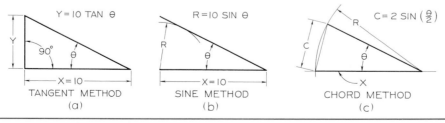

***Fig. 5.21*** Laying Out Angles (§5.20).

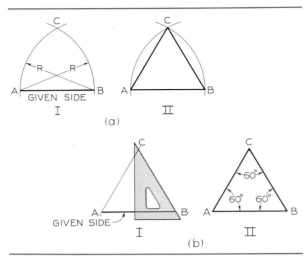

*Fig. 5.22*  Drawing an Equilateral Triangle (§5.21).

### Tangent Method Fig. 5.21 (a)

The tangent of angle $\theta$ is $\frac{y}{x}$, and $y = x \tan \theta$. To construct the angle, assume a convenient value for $x$, preferably 10 units of convenient length, as shown. (The larger the unit, the more accurate will be the construction.) Find the tangent of angle $\theta$ in a table of natural tangents, multiply by 10, and set off $y = 10 \tan \theta$.

EXAMPLE  To set off $31\frac{1}{2}°$, find the natural tangent of $31\frac{1}{2}°$, which is 0.6128. Then

$$y = 10 \text{ units} \times 0.6128 = 6.128 \text{ units}$$

### Sine Method Fig. 5.21 (b)

Draw line $x$ to any convenient length, preferably 10 units as shown. Find the sine of angle $\theta$ in a table of natural sines, multiply by 10, and strike arc $R = 10 \sin \theta$. Draw the other side of the angle tangent to the arc, as shown.

EXAMPLE  To set off $25\frac{1}{2}°$, find the natural sine of $25\frac{1}{2}°$, which is 0.4305. Then

$$R = 10 \text{ units} \times 0.4305 = 4.305 \text{ units}$$

### Chord Method Fig. 5.21 (c)

Draw line $x$ to any convenient length, draw arc with any convenient radius $R$, say, 10 units. Find the chordal length $C$ in a table of chords (see a machinists' handbook), and multiply the value by 10, since the table is made for a radius of 1 unit.

EXAMPLE  To set off 43° 20′, the chordal length $C$ for 1 unit radius, as given in a table of chords = 0.7384, and if $R = 10$ units, then $C = 7.384$ units.

If a table is not available, the chord $C$ may be calculated by the formula $C = 2 \sin \frac{\theta}{2}$.

EXAMPLE  Half of 43° 20′ = 21° 40′. The sine of 21° 40′ = 0.3692. $C = 2 \times 0.3692 = 0.7384$ for a 1 unit radius. For a 10 unit radius, $C = 7.384$ units.

### 5.21 To Draw an Equilateral Triangle
Given side AB.

*Fig. 5.22 (a)*  With A and B as centers and AB as radius, strike arcs to intersect at C. Draw lines AC and BC to complete the triangle.

*Fig. 5.22 (b)*  *Preferred Method*  Draw lines through points A and B making angles of 60° with the given line and intersecting at C, as shown.

### 5.22 To Draw a Square
*Fig. 5.23 (a)*  Given one side AB. Through point A, draw a perpendicular, Fig. 5.18 (c). With A as center, and AB as radius, draw the arc to intersect the

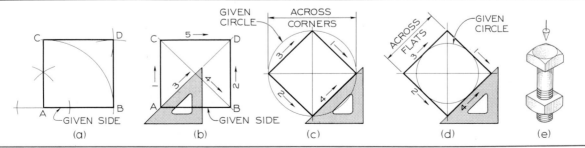

*Fig. 5.23*  Drawing a Square (§5.22).

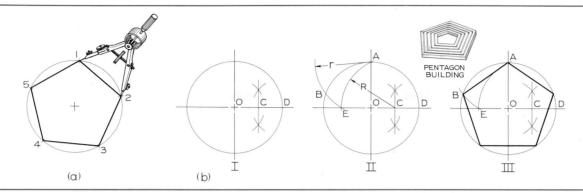

*Fig. 5.24*   Drawing a Pentagon (§5.23).

perpendicular at C. With B and C as centers, and AB as radius, strike arcs to intersect at D. Draw lines CD and BD.

*Fig. 5.23 (b)   Preferred Method*   Given one side AB. Using the T-square or parallel straightedge and 45° triangle, draw lines AC and BD perpendicular to AB and the lines AD and BC at 45° with AB. Draw line CD.

*Fig. 5.23 (c)   Preferred Method*   Given the circumscribed circle (distance "across corners"), draw two diameters at right angles to each other. The intersections of these diameters with the circle are vertexes of an inscribed square.

*Fig. 5.23 (d)   Preferred Method*   Given the inscribed circle (distance "across flats," as in drawing bolt heads), use the T-square (or parallel straightedge) and 45° triangle and draw the four sides tangent to the circle.

## 5.23 To Draw a Regular Pentagon
Given the circumscribed circle.

*Fig. 5.24 (a)   Preferred Method*   Divide the circumference of the circle into five equal parts with the dividers, and join the points with straight lines.

### *Geometrical Method*  Fig. 5.24 (b)
   I.   Bisect radius OD at C.

   II.   With C as center, and CA as radius, strike arc AE. With A as center, and AE as radius, strike arc EB.

   III.   Draw line AB; then set off distances AB around the circumference of the circle, and draw the sides through these points.

## 5.24 To Draw a Hexagon
Given the circumscribed circle.

*Fig. 5.25 (a)*   Each side of a hexagon is equal to the radius of the circumscribed circle. Therefore, using the compass or dividers and the radius of the circle, set off the six sides of the hexagon around the circle, and connect the points with straight lines. As a check on the accuracy of the construction, make sure that opposite sides of the hexagon are parallel.

*Fig. 5.25 (b)   Preferred Method*   This construction is a variation of the one shown at (a). Draw vertical and horizontal center lines. With A and B as centers

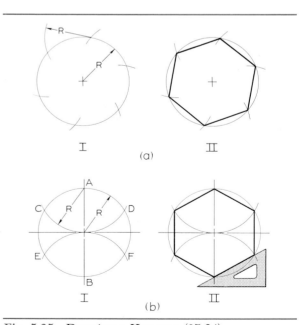

*Fig. 5.25*   Drawing a Hexagon (§5.24).

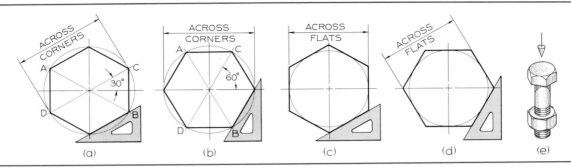

**Fig. 5.26**   Drawing a Hexagon (§5.25).

and radius equal to that of the circle, draw arcs to intersect the circle at **C**, **D**, **E**, and **F**, and complete the hexagon as shown.

### 5.25 To Draw a Hexagon
Given the circumscribed or inscribed circle. *Both Preferred Methods.*

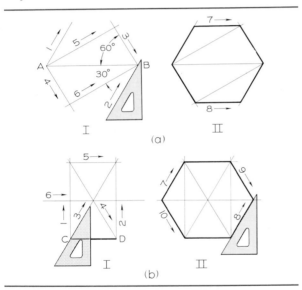

**Fig. 5.27**   Drawing a Hexagon (§5.26).

**Fig. 5.26 (a) and (b)**   Given the circumscribed circle (distance "across corners"). Draw vertical and horizontal center lines, and then diagonals **AB** and **CD** at 30° or 60° with horizontal; then with the 30° × 60° triangle and the T-square, draw the six sides as shown.

**Fig. 5.26 (c) and (d)**   Given the inscribed circle (distance "across flats"). Draw vertical and horizontal center lines; then with the 30° × 60° triangle and the T-square or straightedge draw the six sides tangent to the circle. This method is used in drawing bolt heads and nuts. For maximum accuracy, diagonals may be added as at (a) and (b).

### 5.26 To Draw a Hexagon Fig. 5.27
Using the 30° × 60° triangle and the T-square or straightedge, draw lines in the order shown at (a) where the distance **AB** ("across corners") is given, or as shown at (b) where a side **CD** is given.

### 5.27 To Draw an Octagon
***Fig. 5.28 (a)***   *Preferred Method*   Given inscribed circle, or distance "across flats." Using the T-square or straightedge and 45° triangle, draw the eight sides tangent to the circle, as shown.

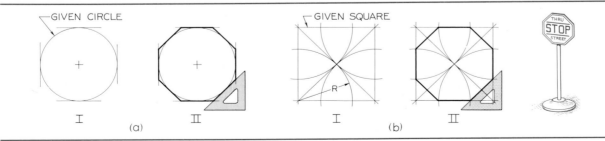

**Fig. 5.28**   Drawing an Octagon (§5.27).

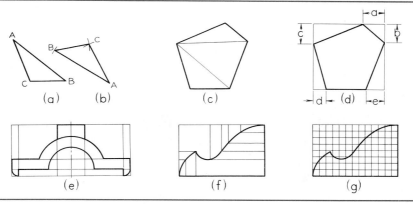

**Fig. 5.29**   Transferring a Plane Figure (§5.28).

**Fig. 5.28 (b)**   Given circumscribed square, or distance "across flats." Draw diagonals of square; then with the corners of the given square as centers, and with half the diagonal as radius, draw arcs cutting the sides as shown at I. Using the T-square and 45° triangle, draw the eight sides as shown at II.

## 5.28 To Transfer Plane Figures by Geometric Methods

### To Transfer a Triangle to a New Location
Fig. 5.29 (a) and (b)
Set off any side, as AB, in the new location, (b). With the ends of the line as centers and the lengths of the other sides of the given triangle, (a), as radii, strike two arcs to intersect at C. Join C to A and B to complete the triangle.

### To Transfer a Polygon by the Triangle Method  Fig. 5.29 (c)
Divide the polygon into triangles as shown, and transfer each triangle as explained previously.

### To Transfer a Polygon by the Rectangle Method  Fig. 5.29 (d)
Circumscribe a rectangle about the given polygon. Draw a congruent rectangle in the new location and locate the vertexes of the polygon by transferring location measurements a, b, c, and so on, along the sides of the rectangle to the new rectangle. Join the points thus found to complete the figure.

### To Transfer Irregular Figures  Fig. 5.29 (e)
Figures composed of rectangular and circular forms are readily transferred by enclosing the elementary features in rectangles and determining centers of arcs and circles. These may then be transferred to the new location.

### To Transfer Figures by Offset Measurements  Fig. 5.29 (f)
*Offset location measurements* are frequently useful in transferring figures composed of free curves. When the figure has been enclosed by a rectangle, the sides of the rectangle are used as reference lines for the location of points along the curve.

### To Transfer Figures by a System of Squares  Fig. 5.29 (g)
Figures involving free curves are easily copied, enlarged, or reduced by the use of a system of squares. For example, to enlarge a figure to double size, draw the containing rectangle and all small squares double their original size. Then draw the lines through the corresponding points in the new set of squares. See also Fig. 6.18.

## 5.29 To Transfer Drawings by Tracing-Paper Methods
To transfer a drawing to an opaque sheet, the following procedures may be used.

### Prick-Point Method
Lay tracing paper over the drawing to be transferred. With a sharp pencil, make a small dot directly over each important point on the drawing. Encircle each dot so as not to lose it. Remove the tracing paper, place it over the paper to receive the transferred drawing, and maneuver the tracing paper into the desired position. With a needle point (such as a point of the dividers), prick through each

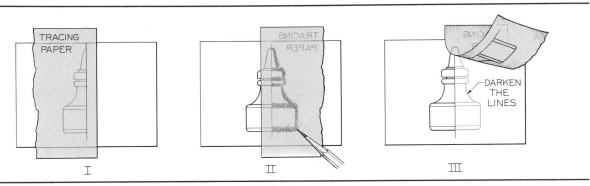

***Fig. 5.30***   Transferring a Symmetrical Half (§5.29).

dot. Remove the tracing paper and connect the prick-points to produce the lines as on the original drawing.

To transfer arcs or circles, it is only necessary to transfer the center and one point on the circumference. To transfer a free curve, transfer as many prick-points on the curve as desired.

### Tracing Method

Lay tracing paper over the drawing to be transferred, and make a pencil tracing of it. Turn the tracing paper over and mark over the lines with short strokes of a soft pencil so as to provide a coating of graphite over every line. Turn tracing face up and fasten in position where drawing is to be transferred. Trace over all lines of the tracing, using a hard pencil. The graphite on the back acts as a carbon paper and will produce dim but definite lines. Heavy in the dim lines to complete the transfer.

***Fig. 5.30***   If one-half of a symmetrical object has been drawn, as for the ink bottle at I, the other half

may be easily drawn with the aid of tracing paper as follows.

I.   Trace the half already drawn.

II.   Turn tracing paper over and maneuver to the position for the right half. Then trace over the lines freehand or mark over the lines with short strokes as shown.

III.   Remove the tracing paper, revealing the dim imprinted lines for the right half. Heavy in these lines to complete the drawing.

### 5.30 To Enlarge or Reduce a Drawing

***Fig. 5.31*** *(a)*   The construction shown is an adaptation of the parallel-line method, Figs. 5.14 and 5.15, and may be used whenever it is desired to enlarge or reduce any group of dimensions to the same ratio. Thus if full-size dimensions are laid off along the vertical line, the enlarged dimensions would appear along the horizontal line, as shown.

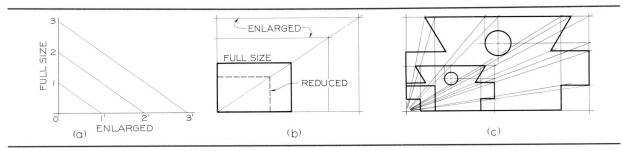

***Fig. 5.31***   Enlarging or Reducing (§5.30).

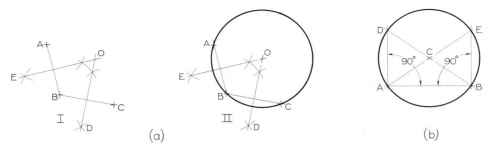

**Fig. 5.32**  Finding Center of Circle (§§5.31 and 5.32).

**Fig. 5.31 (b)**  To enlarge or reduce a rectangle (say, a sheet of drawing paper), a simple method is to use the diagonal, as shown.

**Fig. 5.31 (c)**  A simple method of enlarging or reducing a drawing is to make use of radial lines, as shown. The original drawing is placed underneath a sheet of tracing paper, and the enlarged or reduced drawing is made directly on the tracing paper.

## 5.31 To Draw a Circle Through Three Points Fig. 5.32 (a)

I.  Let A, B, and C be the three given points not in a straight line. Draw lines AB and BC, which will be chords of the circle. Draw perpendicular bisectors EO and DO, Fig. 5.8, intersecting at O.

II.  With center at O, draw required circle through the points.

## 5.32 To Find the Center of a Circle
Fig. 5.32 (b)
Draw any chord AB, preferably horizontal as shown. Draw perpendiculars from A and B, cutting circle at D and E. Draw diagonals DB and EA whose inter-

section C will be the center of the circle. This method uses the principle that any right triangle inscribed in a circle cuts off a semicircle, as was shown earlier, in Fig. 5.3 (f).

Another method, slightly longer, is to reverse the procedure of Fig. 5.32 (a). Draw any two non-parallel chords and draw perpendicular bisectors. The intersection of the bisectors will be the center of the circle.

## 5.33 To Draw a Circle Tangent to a Line at a Given Point Fig. 5.33
Given a line AB and a point P on the line, as shown at (a).

I.  At P erect a perpendicular to the line.

II.  Set off the radius of the required circle on the perpendicular.

III.  Draw circle with radius CP.

## 5.34 To Draw a Tangent to a Circle Through a Point
**Fig. 5.34 (a)  Preferred Method**  Given point P on the circle. Move the T-square and triangle as a unit until one side of the triangle passes through the

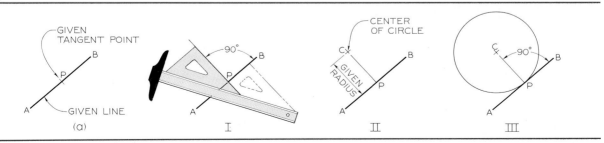

**Fig. 5.33**  Drawing a Circle Tangent to a Line (§5.33).

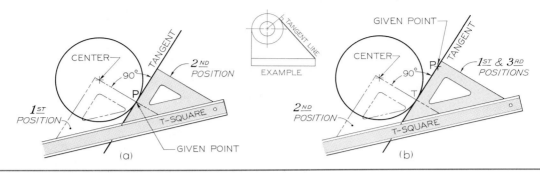

**Fig. 5.34**   Drawing a Tangent to a Circle Through a Point (§5.34).

point P and the center of the circle; then slide the triangle until the other side passes through point P, and draw the required tangent.

**Fig. 5.34 (b)**   Given point P outside the circle. Move the T-square and triangle as a unit until one side of the triangle passes through point P and, by inspection, is tangent to the circle; then slide the triangle until the other side passes through the center of the circle, and lightly mark the point of tangency T. Finally move the triangle back to its starting position, and draw the required tangent.

In both constructions either triangle may be used. Also, a second triangle may be used in place of the T-square.

## 5.35 To Draw Tangents to Two Circles
Fig. 5.35 (a) and (b)

Move the triangle and T-square as a unit until one side of the triangle is tangent, by inspection, to the two circles; then slide the triangle until the other side passes through the center of one circle, and lightly mark the point of tangency. Then slide the triangle until the side passes through the center of the other circle, and mark the point of tangency. Finally, slide the triangle back to the tangent position, and draw the tangent lines between the two points of tangency. Draw the second tangent line in a similar manner.

## 5.36 To Draw an Arc Tangent to a Line or Arc and Through a Point

**Fig. 5.36 (a)**   Given line AB, point P, and radius R. Draw line DE parallel to given line and distance R

from it. From P draw arc with radius R, cutting line DE at C, the center of the required tangent arc.

**Fig. 5.36 (b)**   Given line AB, with tangent point Q on the line, and point P. Draw PQ, which will be a chord of the required arc. Draw perpendicular bisector DE, and at Q erect a perpendicular to the line to intersect DE at C, the center of the required tangent arc.

**Fig. 5.36 (c)**   Given arc with center Q, point P, and radius R. From P strike arc with radius R. From Q strike arc with radius equal to that of the given arc plus R. The intersection C of the arcs is the center of the required tangent arc.

## 5.37 To Draw an Arc Tangent to Two Lines at Right Angles Fig. 5.37 (a)

I.   Given two lines at right angles to each other.

II.   With given radius R, strike arc intersecting given lines at tangent points T.

III.   With given radius R again, and with points T as centers, strike arcs intersecting at C.

IV.   With C as center and given radius R, draw required tangent arc.

## *For Small Radii* Fig. 5.37 (b)

For small radii, such as $\frac{1}{8}$R for fillets and rounds, it is not practicable to draw complete tangency constructions. Instead, draw a 45° bisector of the angle and locate the center of the arc by trial along this line, as shown.

Note that the center C can be located by intersecting lines parallel to the given lines, as shown in

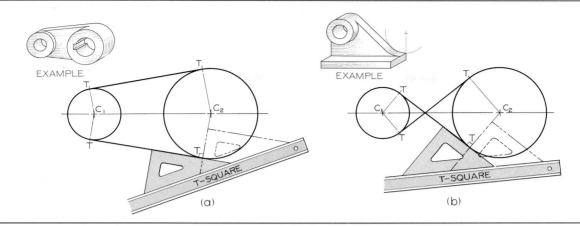

Fig. 5.35   Drawing Tangents to Two Circles (§5.35).

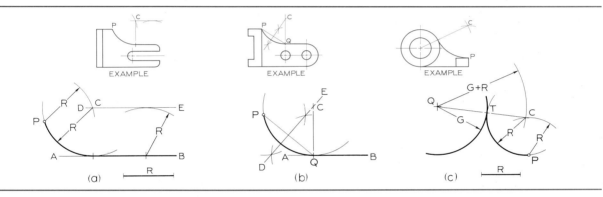

Fig. 5.36   Tangents (§5.36).

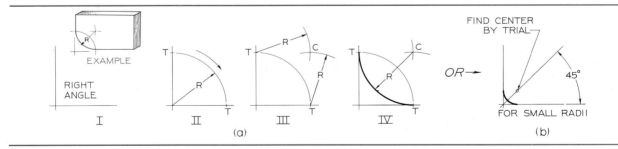

Fig. 5.37   Drawing a Tangent Arc in a Right Angle (§5.37).

Fig. 5.13 (b). However, the circle template can also be used to draw the arcs **R** for the parallel line method of Fig. 5.13 (b). While the circle template is very convenient to use for small radii up to about $\frac{5}{8}$″ or 16 mm, it is necessary that the diameter of a circle on the template precisely equals twice the required radius.

## 5.38 To Draw an Arc Tangent to Two Lines at Acute or Obtuse Angles Fig. 5.38 (a) or (b)

I.   Given two lines not making 90° with each other.

II.   Draw lines parallel to given lines, at distance **R** from them, to intersect at **C**, the required center.

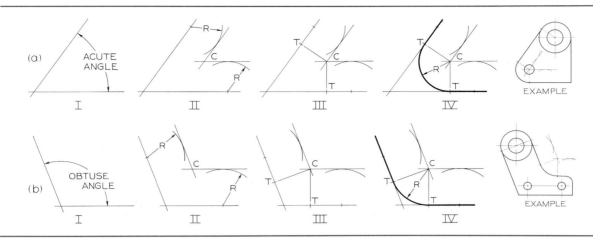

*Fig. 5.38*    Drawing Tangent Arcs (§5.38).

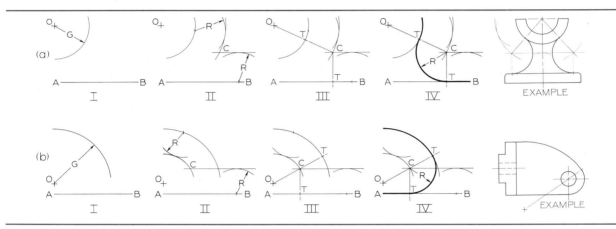

*Fig. 5.39*    Drawing Tangent Arcs (§5.39).

III.    From C drop perpendiculars to the given lines respectively to locate tangent points T.

IV.    With C as center and with given radius R, draw required tangent arc between the points of tangency.

### 5.39 To Draw an Arc Tangent to an Arc and a Straight Line Fig. 5.39 (a) or (b)

I.    Given arc with radius G and straight line AB.

II.    Draw straight line and an arc parallel, respectively, to the given straight line and arc at the required radius distance R from them, to intersect at C, the required center.

III.    From C drop a perpendicular to the given straight line to obtain one point of tangency T. Join the centers C and O with a straight line to locate the other point of tangency T.

IV.    With center C and given radius R, draw required tangent arc between the points of tangency.

### 5.40 To Draw an Arc Tangent to Two Arcs Fig. 5.40 (a) or (b)

I.    Given arcs with centers A and B, and required radius R.

II.    With A and B as centers, draw arcs parallel to the given arcs and at a distance R from them; their

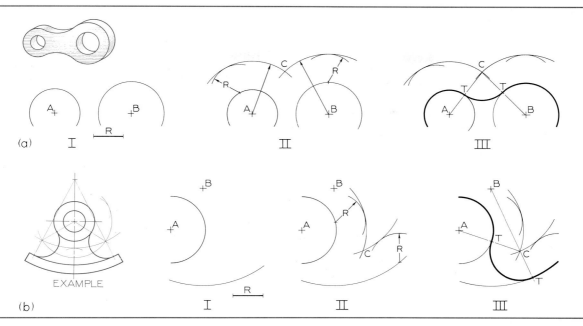

**Fig. 5.40** Drawing an Arc Tangent to Two Arcs (§5.39).

intersection **C** is the center of the required tangent arc.

III.   Draw lines of centers **AC** and **BC** to locate points of tangency **T**, and draw required tangent arc between the points of tangency, as shown.

## 5.41 To Draw an Arc Tangent to Two Arcs and Enclosing One or Both
### *The Required Arc Encloses Both Given Arcs* Fig. 5.41 (a)

With **A** and **B** as centers, strike arcs **HK − r** (given radius minus radius of small circle) and **HK − R** (given radius minus radius of large circle) intersecting at **G**, the center of the required tangent arc. Lines of centers **GA** and **GB** (extended) determine points of tangency **T**.

### *The Required Arc Encloses One Given Arc* Fig. 5.41 (b)

With **C** and **D** as centers, strike arcs **HK + r** (given radius plus radius of small circle) and **HK − R** (given radius minus radius of large circle) intersecting at **G**, the center of the required tangent arc. Lines of centers **GC** and **GD** (extended) determine points of tangency **T**.

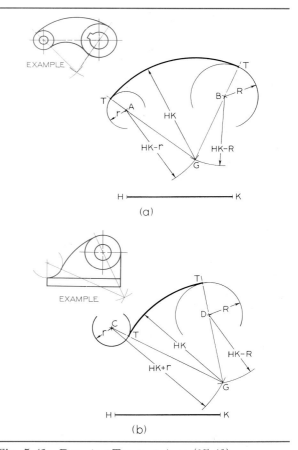

**Fig. 5.41** Drawing Tangent Arcs (§5.41).

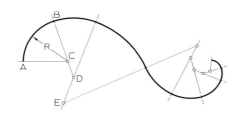

*Fig. 5.42*   A Series of Tangent Arcs (§5.42).

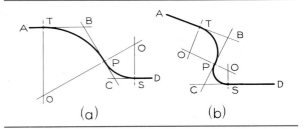

*Fig. 5.44*   Tangent Curves (§5.44).

## 5.42 To Draw a Series of Tangent Arcs Conforming to a Curve  Fig. 5.42

First sketch lightly a smooth curve as desired. By trial, find a radius R and a center C, producing an arc AB that closely follows that portion of the curve. The successive centers D, E, and so on, will be on lines joining the centers with the points of tangency, as shown.

## 5.43 To Draw an Ogee Curve

*Connecting Two Parallel Lines*  Fig. 5.43 (a)

Let NA and BM be the two parallel lines. Draw AB, and assume inflection point T (at midpoint if two equal arcs are desired). At A and B erect perpendiculars AF and BC. Draw perpendicular bisectors of AT and BT. The intersections F and C of these bisectors and the perpendiculars, respectively, are the centers of the required tangent arcs.

*Fig. 5.43 (b)*   Let AB and CD be the two parallel lines, with point B as one end of the curve and R the given radii. At B erect perpendicular to AB, make BG = R, and draw arc as shown. Draw line SP parallel to CD at distance R from CD. With center G, draw arc of radius 2R, intersecting line SP at O. Draw perpendicular OJ to locate tangent point J, and join centers G and O to locate point of tangency

T. Using centers G and O and radius R, draw the two tangent arcs as shown.

### Connecting Two Nonparallel Lines
Fig. 5.43 (c)

Let AB and CD be the two nonparallel lines. Erect perpendicular to AB at B. Select point G on the perpendicular so that BG equals any desired radius, and draw arc as shown. Erect perpendicular to CD at C and make CE = BG. Join G to E and bisect it. The intersection F of the bisector and the perpendicular CE, extended, is the center of the second arc. Join centers of the two arcs to locate tangent point T, the inflection point of the curve.

## 5.44 To Draw a Curve Tangent to Three Intersecting Lines

*Fig. 5.44 (a) and (b)*   Let AB, BC, and CD be the given lines. Select point of tangency P at any point on line BC. Make BT equal to BP, and CS equal to CP, and erect perpendiculars at the points P, T, and S. Their intersections O and Q are the centers of the required tangent arcs.

## 5.45 To Rectify a Circular Arc

To *rectify* an arc is to lay out its true length along a straight line. The constructions are approximate, but

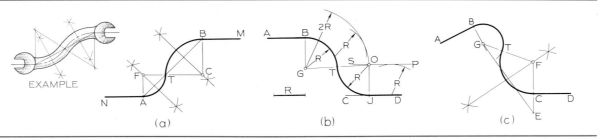

*Fig. 5.43*   Drawing an Ogee Curve (§5.43).

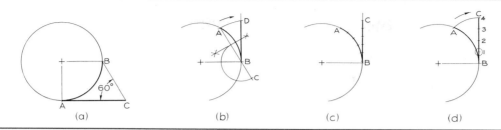

**Fig. 5.45** Rectifying Circular Arcs (§§5.45 and 5.46).

well within the range of accuracy of drawing instruments.

### To Rectify a Quadrant of a Circle, AB
Fig. 5.45 (a)
Draw AC tangent to the circle and BC at 60° to AC, as shown. The line AC is almost equal to the arc AB, the difference in length being about 1 in 240.

### To Rectify Arc AB Fig. 5.45 (b)
Draw tangent at B. Draw chord AB and extend it to C, making BC equal to half AB. With C as center and radius CA, strike the arc AD. The tangent BD is slightly shorter than the given arc AB. For an angle of 45° the difference in length is about 1 in 2866.

**Fig. 5.45 (c)** Use the bow dividers, and beginning at A, set off equal distances until the division point nearest to B is reached. At this point, reverse the direction and set off an equal number of distances along the tangent to determine point C. The tangent BC is slightly shorter than the given arc AB. If the angle subtended by each division is 10°, the error is approximately 1 in 830. ❧

NOTE   If the angle $\theta$ subtending an arc of radius $R$ is known, the length of the arc is $2\pi R \dfrac{\theta}{360°} = 0.01745R\theta$.

### 5.46 To Set Off a Given Length Along a Given Arc
**Fig. 5.45 (c)**   Reverse the preceding method so as to transfer distances from the tangent line to the arc.

**Fig. 5.45 (d)**   To set off the length BC along the arc BA, draw BC tangent to the arc at B. Divide BC into four equal parts. With center at 1, the first division point, and radius 1–C, draw the arc CA. The arc BA is practically equal to BC for angles less than 30°. For 45° the difference is approximately 1 in 3232, and for 60° it is about 1 in 835.

### 5.47 The Conic Sections Fig. 5.46
The conic sections are curves produced by planes intersecting a right circular cone. Four types of

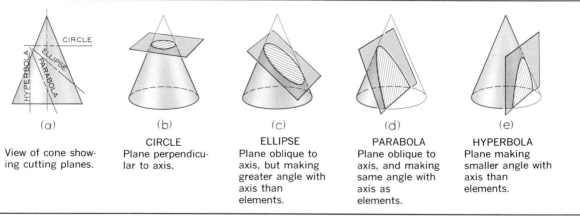

**Fig. 5.46** Conic Sections (§5.47).

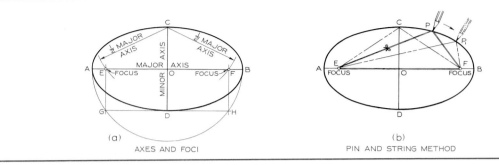

**Fig. 5.47** Ellipse Constructions (§5.48).

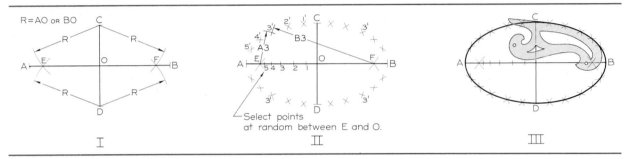

**Fig. 5.48** Drawing a Foci Ellipse (§5.49).

curves are produced: the *circle, ellipse, parabola,* and *hyperbola,* according to the position of the planes, as shown. These curves were studied in detail by the ancient Greeks, and are of great interest in mathematics, as well as in technical drawing. For equations, see any text on analytic geometry.

## 5.48 Ellipse Construction

The long axis of an ellipse is the major axis, and the short axis is the minor axis, Fig. 5.47 (a). The foci **E** and **F** are found by striking arcs with radius equal to half the major axis and with center at the end of the minor axis. Another method is to draw a semicircle with the major axis as diameter, then to draw **GH** parallel to the major axis and **GE** and **HF** parallel to the minor axis, as shown.

*An ellipse is generated by a point moving so that the sum of its distances from two points (the foci) is constant and equal to the major axis.* For example, Fig. 5.47 (b), an ellipse may be constructed by placing a looped string around the foci **E** and **F**, and around **C**, one end of the minor axis, and moving the pencil point **P** along its maximum orbit while the string is kept taut.

## 5.49 To Draw a Foci Ellipse Fig. 5.48

Let **AB** be the major axis and **CD** the minor axis. This method is the geometrical counterpart of the pin-and-string method. Keep the construction very light, as follows.

I.   To find foci **E** and **F**, strike arcs **R** with radius equal to half the major axis and with centers at the ends of the minor axis.

II.   Between **E** and **O** on the major axis, mark at random a number of points (spacing those on the left more closely), equal to the number of points desired in each quadrant of the ellipse. In this figure, five points were deemed sufficient. For large ellipses, more points should be used—enough to insure a smooth, accurate curve. Begin construction with any one of these points, such as **3**. With **E** and **F** as centers and radii **A–3** and **B–3**, respectively (from the ends of the major axis to point **3**), strike arcs to intersect at four points **3'**, as shown. Using the remaining points **1, 2, 4,** and **5,** for each find four additional points on the ellipse in the same manner.

III.   Sketch the ellipse lightly through the points;

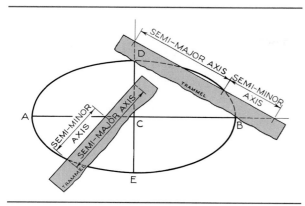

**Fig. 5.49**  Drawing a Trammel Ellipse (§5.50).

### 5.51 To Draw a Concentric-Circle Ellipse
Fig. 5.50

If a circle is viewed so that the line of sight is perpendicular to the plane of the circle, as shown for the silver dollar at (a), the circle will appear as a circle, in true size and shape. If the circle is viewed at an angle, as shown at (b), it will appear as an ellipse. If the circle is viewed edgewise, it appears as a straight line, as shown at (c). The case shown at (b) is the basis for the construction of an ellipse by the concentric-circle method, as follows (keep the construction very light).

I.   Draw circles on the major and minor axes using them as diameters and draw any diagonal XX through center O. From the points X, in which the diagonal intersects the large circle, draw lines XE parallel to the minor axes, and from points H, in which it intersects the small circle, draw lines HE parallel to the major axis. The intersections E are points on the ellipse. Two additional points, S and R, can be found by extending lines XE and HE, giving a total of four points from the one diagonal XX.

II.   Draw as many additional diagonals as needed to provide a sufficient number of points for a smooth and symmetrical ellipse, each diagonal accounting for four points on the ellipse. Notice that where the curve is sharpest (near the ends of the ellipse), the points are constructed closer together to better determine the curve.

III.   Sketch the ellipse lightly through the points, then heavy in the final ellipse with the aid of the irregular curve.

NOTE   It is evident at I, Fig. 5.50, that the ordinate EZ of the ellipse is to the corresponding ordinate XZ of the circle as $b$ is to $a$, where $b$ represents the semiminor axis and $a$ the semimajor axis.

then heavy in the final ellipse with the aid of the irregular curve. Fig. 2.66.

### 5.50 To Draw a Trammel Ellipse  Fig. 5.49

A "long trammel" or a "short-trammel" may be prepared from a small strip of stiff paper or thin cardboard, as shown. In both cases, set off on the edge of the trammel distances equal to the semimajor and semiminor axes. In one case these distances overlap; in the other they are end to end. To use either method, place the trammel so that the two of the points are on the respective axes, as shown; the third point will then be on the curve and can be marked with a small dot. Find additional points by moving the trammel to other positions, always keeping the two points exactly on the respective axes. Extend the axes to use the long trammel. Find enough points to insure a smooth and symmetrical ellipse. Sketch the ellipse lightly through the points; then heavy in the ellipse with the aid of the irregular curve, Fig. 2.66.

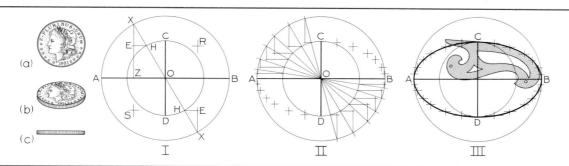

**Fig. 5.50**  Drawing a Concentric-Circle Ellipse (§5.51).

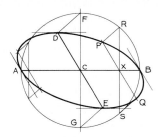

**Fig. 5.51** Oblique-Circle Ellipse (§5.52).

Thus, the area of the ellipse is equal to the area of the circumscribed circle multiplied by $\dfrac{b}{a}$; hence, it is equal to $\pi ab$.

### 5.52 To Draw an Ellipse on Conjugate Diameters—Oblique-Circle Method
Fig. 5.51

Let AB and DE be the given conjugate diameters. *Two diameters are conjugate when each is parallel to the tangents at the extremities of the other.* With center at C and radius CA, draw a circle; draw the diameter GF perpendicular to AB, and draw lines joining points D and F and points G and E.

Assume that the required ellipse is an oblique projection of the circle just drawn; the points D and E of the ellipse are the oblique projections of the points F and G of the circle, respectively; similarly, the points P and Q are the oblique projections of the points R and S, respectively. The points P and Q are determined by assuming the point X at any point on AB and drawing the lines RS and PQ, and RP and SQ, parallel, respectively, to GF and DE and FD and GE.

Determine at least five points in each quadrant (more for larger ellipses) by assuming additional points on the major axis and proceeding as explained for point X. Sketch the ellipse lightly through the points; then heavy in the final ellipse with the aid of the irregular curve, Fig. 2.66.

### 5.53 To Draw a Parallelogram Ellipse
Fig. 5.52 (a) and (b)

Given the major and minor axes, or the conjugate diameters AB and CD, draw a rectangle or parallelogram with sides parallel to the axes, respectively. Divide AO and AJ into the same number of equal parts, and draw *light* lines through these points from

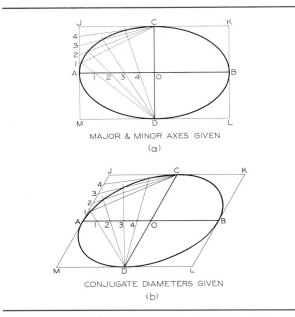

MAJOR & MINOR AXES GIVEN
(a)

CONJUGATE DIAMETERS GIVEN
(b)

**Fig. 5.52** Parallelogram Ellipse (§5.53).

the ends of the minor axis, as shown. The intersection of like-numbered lines will be points on the ellipse. Locate points in the remaining three quadrants in a similar manner. Sketch the ellipse lightly through the points; then heavy in the final ellipse with the aid of the irregular curve, Fig. 2.66.

### 5.54 To Find the Axes of an Ellipse, with Conjugate Diameters Given

*Fig. 5.53 (a)* Conjugate diameters AB and CD and the ellipse are given. With intersection O of the conjugate diameters (center of ellipse) as center, and any convenient radius, draw a circle to intersect the ellipse in four points. Join these points with straight lines, as shown; the resulting quadrilateral will be a rectangle whose sides are parallel, respectively, to the required major and minor axes. Draw the axes EF and GH parallel to the sides of the rectangle.

*Fig. 5.53 (b)* Ellipse only is given. To find the center of the ellipse, draw a circumscribing rectangle or parallelogram about the ellipse; then, draw diagonals to intersect at center O as shown. The axes are then found as shown at (a).

*Fig. 5.53 (c)* Conjugate diameters AB and CD only are given. With O as center and CD as diameter, draw a circle. Through center O and perpendicular

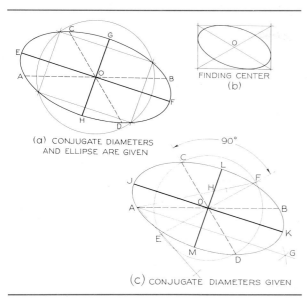

(b) FINDING CENTER

(a) CONJUGATE DIAMETERS AND ELLIPSE ARE GIVEN

(c) CONJUGATE DIAMETERS GIVEN

*Fig. 5.53*   Finding the Axes of an Ellipse (§5.54).

to CD, draw line EF. From points E and F, where this perpendicular intersects the circle, draw lines FA and EA to form angle FAE. Draw the bisector AG of this angle. The major axis JK will be parallel to this bisector, and the minor axis LM will be perpendicular to it. The length AH will be one half the

major axis, and HF one half the minor axis. The resulting major and minor axes are JK and LM, respectively.

### 5.55 To Draw a Tangent to an Ellipse

*Concentric Circle Construction* Fig. 5.54 (a) To draw a tangent at any point on the ellipse, as E, draw the ordinate at E to intersect the circle at V. Draw a tangent to the circle at V, §5.34, and extend it to intersect the major axis extended at G. The line GE is the required tangent.

To draw a tangent from a point outside the ellipse, as P, draw the ordinate PY and extend it. Draw DP, intersecting the major axis at X. Draw FX and extend it to intersect the ordinate through P at Q. Then, from similar triangles, QY:PY = OF:OD. Draw tangent to the circle from Q, §5.34, find the point of tangency R, and draw the ordinate at R to intersect the ellipse at Z. The line ZP is the required tangent. As a check on the drawing, the tangents RQ and ZP should intersect at a point on the major axis extended. Two tangents to the ellipse can be drawn from point P.

*Foci Construction* Fig. 5.54 (b)

To draw a tangent at any point on the ellipse, such as point 3, draw the focal radii E–3 and F–3, extend one, and bisect the exterior angle, as shown. The bisector is the required tangent.

To draw a tangent from any point outside the ellipse, such as point P, with center at P and radius PF, strike an arc as shown. With center at E and radius AB, strike an arc to intersect the first arc at points U. Draw the lines EU to intersect the ellipse at the points Z. The lines PZ are the required tangents.

### 5.56 Ellipse Templates

To save time in drawing ellipses, and to insure uniform results, ellipse templates, Fig. 5.55 (a), are often used. These are plastic sheets with elliptical openings in a wide variety of sizes, and usually come in sets of six or more sheets.

Ellipse guides are usually designated by the ellipse angle, the angle at which a circle is viewed to appear as an ellipse. In Fig. 5.55 (b) the angle between the line of sight and the edge view of the plane of the circle is found to be about 49°; hence, the 50° ellipse template is indicated. Ellipse templates are generally available in ellipse angles at 5° intervals, as 15°, 20°, 25°, and so on. On this 50°

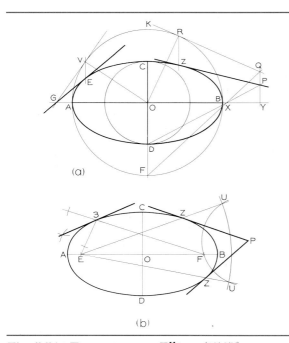

(a)

(b)

*Fig. 5.54*   Tangents to an Ellipse (§5.55).

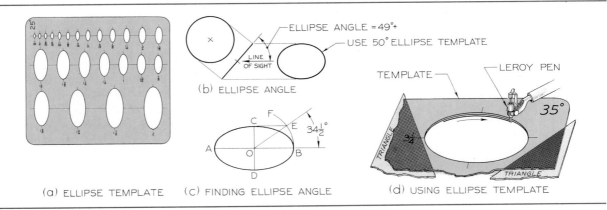

**Fig. 5.55**  Using the Ellipse Template (§5.56).

template a variety of sizes of 50° ellipses is provided, and it is only necessary to select the one that fits. If the ellipse angle is not easily determined, you can always look for the ellipse that is approximately as long and as "fat" as the ellipse to be drawn.

A simple construction for finding the ellipse angle when the views are not available is shown at (c). Using center O, strike arc BF; then draw CE parallel to the major axis. Draw diagonal OE, and measure angle EOB with the protractor, §2.19. Use the ellipse template nearest to this angle; in this case a 35° template is selected.

Since it is not feasible to have ellipse openings for every exact size that may be required, it is often necessary to use the template somewhat in the manner of an irregular curve. For example, if the opening is too long and too "fat" for the required ellipse, one end may be drawn and then the template shifted slightly to draw the other end. Similarly, one long side may be drawn and then the template shifted slightly to draw the opposite side. In such cases, leave gaps between the four segments, to be

filled in freehand or with the aid of an irregular curve. When the differences between the ellipse openings and the required ellipse are small, it is only necessary to lean the pencil slightly outward or inward from the guiding edge to offset the differences.

For inking the ellipses, the Leroy, Rapidograph, or Wrico pens are recommended. The Leroy pen is shown in Fig. 5.55 (d).

### 5.57 To Draw an Approximate Ellipse
Fig. 5.56

For many purposes, particularly where a small ellipse is required, the approximate circular-arc method is perfectly satisfactory. Such an ellipse is sure to be symmetrical and may be quickly drawn.

Given axes AB and CD.

I.   Draw line AC. With O as center and OA as radius, strike the arc AE. With C as center and CE as radius, strike the arc EF.

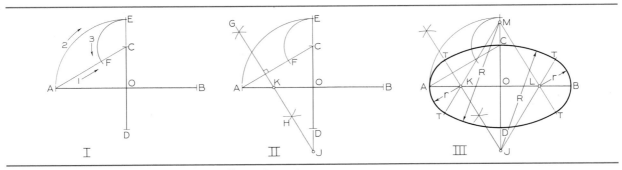

**Fig. 5.56**  Drawing an Approximate Ellipse (§5.57)

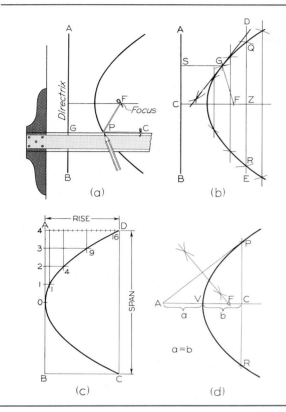

**Fig. 5.57** Drawing a Parabola (§5.58).

II. Draw perpendicular bisector **GH** of the line **AF**; the points **K** and **J**, where it intersects the axes, are centers of the required arcs.

III. Find centers **M** and **L** by setting off **OL = OK** and **OM = OJ**. Using centers **K**, **L**, **M**, and **J**, draw circular arcs as shown. The points of tangency **T** are at the junctures of the arcs on the lines joining the centers.

## 5.58 To Draw a Parabola

The curve of intersection between a right circular cone and a plane parallel to one of its elements, Fig. 5.46 (d), is a parabola. *A parabola is generated by a point moving so that its distances from a fixed point, the focus, and from a fixed line, the directrix, remain equal.* For example:

***Fig. 5.57 (a)*** Given focus **F** and directrix **AB**: A parabola may be generated by a pencil guided by a string, as shown. Fasten the string at **F** and **C**; its length is **GC**. The point **C** is selected at random, its

distance from **G** depending on the desired extent of the curve. Keep the string taut and the pencil against the T-square, as shown.

***Fig. 5.57 (b)*** Given focus **F** and directrix **AB**: Draw a line **DE** parallel to the directrix and at any distance **CZ** from it. With center at **F** and radius **CZ**, strike arcs to intersect the line **DE** in the points **Q** and **R**, which are points on the parabola. Determine as many additional points as are necessary to draw the parabola accurately, by drawing additional lines parallel to line **AB** and proceeding in the same manner.

A tangent to the parabola at any point **G** bisects the angle formed by the focal line **FG** and the line **SG** perpendicular to the directrix.

***Fig. 5.57 (c)*** Given the rise and span of the parabola: Divide **AO** into any number of equal parts, and divide **AD** into a number of equal parts amounting to the square of that number. From line **AB**, each point on the parabola is offset by a number of units equal to the square of the number of units from point **O**. For example, point **3** projects **9** units (the square of **3**). This method is generally used for drawing parabolic arches.

***Fig. 5.57 (d)*** Given points **P**, **R**, and **V** of a parabola, to find the focus **F**: Draw tangent at **P**, making **a = b**. Draw perpendicular bisector of **AP**, which intersects the axis at **F**, the focus of the parabola.

***Fig. 5.58 (a) or (b)*** Given rectangle or parallelogram **ABCD**: Divide **BC** into any even number of equal parts, and divide the sides **AB** and **DC** each into half as many parts, and draw lines as shown. The intersections of like-numbered lines are points on the parabola.

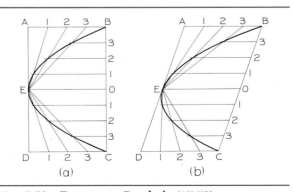

**Fig. 5.58** Drawing a Parabola (§5.58).

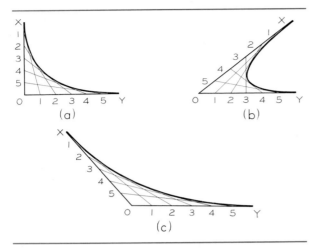

**Fig. 5.59**  Parabolic Curves (§5.59).

## Practical Applications

The parabola is used for reflecting surfaces for light and sound, for vertical curves in highways, for forms of arches, and approximately for forms of the curves of cables for suspension bridges. It is also used to show the bending moment at any point on a uniformly loaded beam or girder.

### 5.59 To Join Two Points by a Parabolic Curve Fig. 5.59

Let X and Y be the given points. Assume any point O, and draw tangents XO and YO. Divide XO and YO into the same number of equal parts, number the division points as shown, and connect corresponding points. These lines are tangents of the required parabola and form its envelope. Sketch a light smooth curve, and then heavy in the curve with the aid of the irregular curve, §2.54.

These parabolic curves are more pleasing in appearance than circular arcs and are useful in machine design. If the tangents OX and OY are equal, the axis of the parabola will bisect the angle between them.

### 5.60 To Draw a Hyperbola

The curve of intersection between a right circular cone and a plane making an angle with the axis smaller than that made by the elements, Fig. 5.46 (e), is a hyperbola. *A hyperbola is generated by a point moving so that the difference of its distances from two fixed points, the foci, is constant and equal to the transverse axis of the hyperbola.*

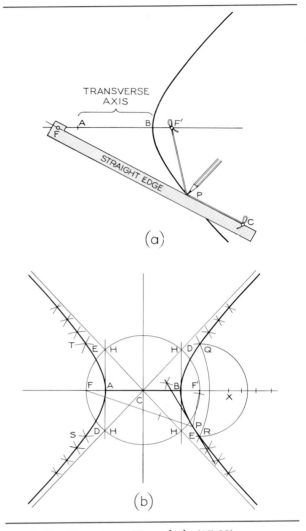

**Fig. 5.60**  Drawing a Hyperbola (§5.60).

**Fig. 5.60 (a)**  Let F and F′ be the foci and AB the transverse axis. The curve may be generated by a pencil guided by a string, as shown. Fasten a string at F′ and C; its length is FC minus AB. The point C is chosen at random; its distance from F depends on the desired extent of the curve.

Fasten the straightedge at F. If it is revolved about F, with the pencil point moving against it and with the string taut, the hyperbola may be drawn as shown.

**Fig. 5.60 (b)**  To construct the curve geometrically, select any point X on the transverse axis produced. With centers at F and F′ and BX as radius, strike the arcs DE. With the same centers, F and F′, and AX as radius, strike arcs to intersect the arcs first drawn in the points Q, R, S, and T, which are points of the required hyperbola. Find as many additional points as necessary to draw the curves accurately by selecting other points similar to point X along the transverse axis, and proceeding as described for point X.

To draw the tangent to a hyperbola at a given point P, bisect the angle between the focal radii FP and F'P. The bisector is the required tangent.

To draw the asymptotes HCH of the hyperbola, draw a circle with the diameter FF' and erect perpendiculars to the transverse axis at the points A and B to intersect the circle in the points H. The lines HCH are the required asymptotes.

## 5.61 To Draw an Equilateral Hyperbola
Fig. 5.61

Let the asymptotes OB and OA, at right angles to each other, and the point P on the curve be given.

***Fig. 5.61 (a)***   In an equilateral hyperbola the asymptotes, at right angles to each other, may be used as the axes to which the curve is referred. If a chord of the hyperbola is extended to intersect the axes, the intercepts between the curve and the axes are equal. For example, a chord through given point P intersects the axes at points 1 and 2, intercepts P–1 and 2–3 are equal, and point 3 is a point on the hyperbola. Likewise, another chord through P provides equal intercepts P–1' and 3'–2', and point 3' is a point on the curve. All chords need not be drawn through given point P, but as new points are established on the curve, chords may be drawn through them to obtain more points. After enough points are found to insure an accurate curve, the hyperbola is drawn with the aid of the irregular curve, §2.54.

***Fig. 5.61 (b)***   In an equilateral hyperbola, the coordinates are related so that their products remain constant. Through given point P, draw lines 1–P–Y and 2–P–Z parallel, respectively, to the axes. From the origin of coordinates O, draw any diagonal intersecting these two lines at points 3 and X. At these points draw lines parallel to the axes, intersecting at point 4, a point on the curve. Likewise, another diagonal from O intersects the two lines through P at points 8 and Y, and lines through these points parallel to the axes intersect at point 9, another point on the curve. A third diagonal similarly produces point 10 on the curve, and so on. Find as many

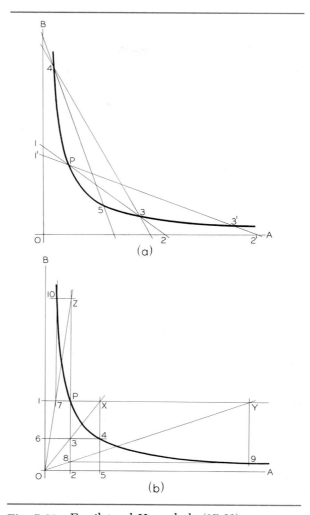

*Fig. 5.61*   Equilateral Hyperbola (§5.61).

points as necessary for a smooth curve, and draw the parabola with the aid of the irregular curve, §2.54. It is evident from the similar triangles O–X–5 and O–3–2 that lines P–1 × P–2 = 4–5 × 4–6.

The equilateral hyperbola can be used to represent varying pressure of a gas as the volume varies, since the pressure varies inversely as the volume; that is, pressure × volume is constant.

## 5.62 To Draw a Spiral of Archimedes
Fig. 5.62

To find points on the curve, draw lines through the pole C, making equal angles with each other, such as 30° angles, and beginning with any one line, set off any distance, such as 2 mm or $\frac{1}{16}''$; set off twice that distance on the next line, three times on the

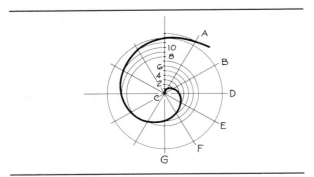

*Fig. 5.62*   Spiral of Archimedes (§5.62).

third, and so on. Through the points thus determined, draw a smooth curve, using the irregular curve, §2.54.

### 5.63 To Draw a Helix Fig. 5.63

*A helix is generated by a point moving around and along the surface of a cylinder or cone with a uniform angular velocity about the axis, and with a uniform linear velocity in the direction of the axis.* A cylindrical helix is generally known simply as a helix. The distance measured parallel to the axis traversed by the point in one revolution is called the lead.

If the cylindrical surface upon which a helix is generated is rolled out onto a plane, the helix becomes a straight line as shown in Fig. 5.63 (a), and

the portion below the helix becomes a right triangle, the altitude of which is equal to the lead of the helix and the length of the base equal to the circumference of the cylinder. Such a helix can, therefore, be defined as the shortest line that can be drawn on the surface of a cylinder connecting two points not on the same element.

To draw the helix, draw two views of the cylinder upon which the helix is generated, (b), and divide the circle of the base into any number of equal parts. On the rectangular view of the cylinder, set off the lead and divide it into the same number of equal parts as the base. Number the divisions as shown, in this case 16. When the generating point has moved one-sixteenth of the distance around the cylinder, it will have risen one-sixteenth of the lead; when it has moved halfway around the cylinder, it will have risen half the lead; and so on. Points on the helix are found by projecting up from point 1 in the circular view to line 1 in the rectangular view, from point 2 in the circular view to line 2 in the rectangular view, and so on.

The helix shown at (b) is a right-hand helix. In a left-hand helix, (c), the visible portions of the curve are inclined in the opposite direction, that is, downward to the right. The helix shown at (b) can be converted into a left-hand helix by interchanging the visible and hidden lines.

The helix finds many applications in industry, as in screw threads, worm gears, conveyors, spiral stairways, and so on. The stripes of a barber pole are helical in form.

The construction for a right-hand conical helix is shown at (d).

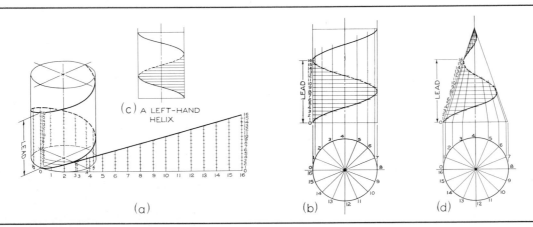

*Fig. 5.63*   Helix (§5.63).

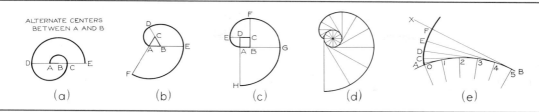

*Fig. 5.64*  Involutes (§5.64).

## 5.64 To Draw an Involute Fig. 5.64

The path of a point on a string, as the string unwinds from a line, a polygon, or a circle, is an involute.

### To Draw an Involute of a Line Fig. 5.64 (a)

Let AB be the given line. With AB as radius and B as center, draw the semicircle AC. With AC as radius and A as center, draw the semicircle CD. With BD as radius and B as center, draw the semicircle DE. Continue similarly, alternating centers between A and B, until a figure of the required size is completed.

### To Draw an Involute of a Triangle
Fig. 5.64 (b)

Let ABC be the given triangle. With CA as radius and C as center, strike the arc AD. With BD as radius and B as center, strike the arc DE. With AE as radius and A as center, strike the arc EF. Continue similarly until a figure of the required size is completed.

### To Draw an Involute of a Square
Fig. 5.64 (c)

Let ABCD be the given square. With DA as radius and D as center, draw the 90° arc AE. Proceed as for the involute of a triangle until a figure of the required size is completed.

### To Draw an Involute of a Circle Fig. 5.64 (d)

A circle may be regarded as a polygon with an infinite number of sides. The involute is constructed by dividing the circumference into a number of equal parts, drawing a tangent at each division point, setting off along each tangent the length of the corresponding circular arc, Fig. 5.45 (c), and drawing the required curve through the points set off on the several tangents.

*Fig. 5.64 (e)*  The involute may be generated by a point on a straight line that is rolled on a fixed circle. Points on the required curve may be determined by setting off equal distances 0–1, 1–2, 2–3, and so forth, along the circumference, drawing a tangent at each division point, and proceeding as explained for (d).

The involute of a circle is used in the construction of involute gear teeth. In this system, the involute forms the face and a part of the flank of the teeth of gear wheels; the outlines of the teeth of racks are straight lines.

## 5.65 To Draw a Cycloid Fig. 5.65

*A cycloid may be generated by a point P in the circumference of a circle that rolls along a straight line.*

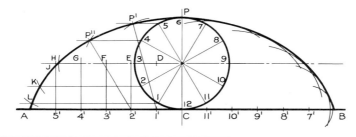

*Fig. 5.65*  Cycloid (§5.65).

Given the generating circle and the straight line AB tangent to it, make the distances CA and CB each equal to the semicircumference of the circle, Fig. 5.45 (c). Divide these distances and the semicircumference into the same number of equal parts, six, for instance, and number them consecutively as shown. Suppose the circle to roll to the left; when point 1 of the circle reaches point 1' of the line, the center of the circle will be at D, point 7 will be the highest point of the circle, and the generating point 6 will be at the same distance from the line AB as point 5 is when the circle is in its central position. Hence, to find the point P', draw a line through point 5 parallel to AB and intersect it with an arc drawn from the center D with a radius equal to that of the circle. To find point P'', draw a line through point 4 parallel to AB, and intersect it with an arc drawn from the center E, with a radius equal to that of the circle. Points J, K, and L are found in a similar manner.

Another method that may be employed is shown in the right half of the figure. With center at 11' and the chord 11–6 as radius, strike an arc. With 10' as center and the chord 10–6 as radius, strike an arc. Continue similarly with centers 9', 8', and 7'. Draw the required cycloid tangent to these arcs.

Either method may be used; however, the second is the shorter one and is preferred. It is evident, from the tangent arcs drawn in the manner just described, that the line joining the generating point and the point of contact for the generating circle is a normal of the cycloid; the lines 1'–P'' and 2'–P', for instance, are normals; this property makes the cycloid suitable for the outlines of gear teeth.

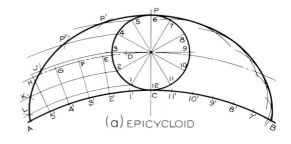

(a) EPICYCLOID

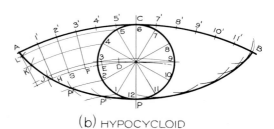

(b) HYPOCYCLOID

*Fig. 5.66*    Epicycloid and Hypocycloid (§5.66).

## 5.66 **To Draw an Epicycloid or a Hypocycloid** Fig. 5.66

If the generating point P is on the circumference of a circle that rolls along the convex side of a larger circle, (a), the curve generated is an epicycloid. If the circle rolls along the concave side of a larger circle, (b), the curve generated is a hypocycloid. These curves are drawn in a manner similar to the cycloid, Fig. 5.65. These curves, like the cycloid, are used to form the outlines of certain gear teeth and are, therefore, of practical importance in machine design.

# Geometric Construction Problems

Geometric constructions should be made very accurately, with a hard pencil (2H to 4H) having a long, sharp, conical point. Draw given and required lines dark and medium in thickness, and draw construction lines *very light*. Do not erase construction lines. Indicate points and lines as described in §5.2.

The instructor will make assignments from the many problems that follow. Use Layout A–2 (see inside of back cover) divided into four parts, as shown in Fig. 5.67, or Layout A4–2 (adjusted). Additional sheets with other problems selected from Figs. 5.68 to 5.79 and drawn on the same sheet layout, may be assigned by the instructor.

Many problems are dimensioned in the metric system. The instructor may assign the student to convert the remaining problems to metric measure. See inside front cover for decimal and millimeter equivalents.

The student should exercise care in setting up each problem so as to make the best use of the space available, to present the problem to best advantage, and to produce a pleasing appearance. Letter the principal points of all constructions in a manner similar to the various illustrations in this chapter.

Since many of the problems in this chapter are of a general nature, they can also be solved on most computer graphics systems. If a system is available, the instructor may choose to assign specific problems to be completed by this method.

Geometric construction problems in convenient form for solution may be found in *Principles of Engineering Graphics Problems* by Spencer, Hill, Loving, and Dygdon, a workbook designed to accompany this text that is also published by Macmillan Publishing Company.

---

*The first four problems are shown in Fig. 5.67.*

**Prob. 5.1**   Draw an inclined line **AB** 65 mm long and bisect it, Fig. 5.8.

**Prob. 5.2**   Draw any angle with vertex at **C**. Bisect it, Fig. 5.10, and transfer one half in new position at **D**, Fig. 5.11.

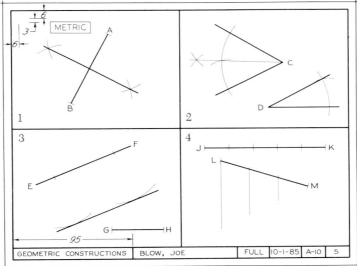

*Fig. 5.67*   Geometric Constructions.
Layout A–2 or A4–2 (adjusted). (Probs. 5.1 to 5.4)

**Prob. 5.35**   Draw major axis 102 mm long (horizontally) and minor axis 64 mm long, with their intersection at the center of the space. Draw ellipse by foci method with at least five points in each quadrant, Fig. 5.48.

**Prob. 5.36**   Draw axes as in Prob. 5.35, but draw ellipse by trammel method, Fig. 5.49.

**Prob. 5.37**   Draw axes as in Prob. 5.35, but draw ellipse by concentric-circle method, Fig. 5.50.

**Prob. 5.38**   Draw axes as in Prob. 5.35, but draw ellipse by parallelogram method, Fig. 5.52 (a).

**Prob. 5.39**   Draw conjugate diameters intersecting at center of space. Draw 88 mm diameter horizontally, and 70 mm diameter at 60° with horizontal. Draw oblique-circle ellipse, Fig. 5.51. Find at least 5 points in each quadrant.

**Prob. 5.40**   Draw conjugate diameters as in Prob. 5.39, but draw ellipse by parallelogram method, Fig. 5.52 (b).

**Prob. 5.41**   Draw axes as in Prob. 5.35, but draw approximate ellipse, Fig. 5.56.

**Prob. 5.42**   Draw a parabola with a vertical axis, and the focus 12 mm from the directrix, Fig. 5.57 (b). Find at least 9 points on the curve.

**Prob. 5.43**   Draw a hyperbola with a horizontal transverse axis 25 mm long and the foci 38 mm apart, Fig. 5.60 (b). Draw the asymptotes.

**Prob. 5.44**   Draw horizontal line near bottom of space, and vertical line near left side of space. Assume point P 16 mm to right of vertical line and 38 mm above horizontal line. Draw equilateral hyperbola through P and with reference to the two lines as asymptotes. Use either method of Fig. 5.61.

**Prob. 5.45**   Using the center of the space as the pole, draw a spiral of Archimedes with the generating point moving in a counterclockwise direction and away from the pole at the rate of 25 mm in each convolution, Fig. 5.62.

**Prob. 5.46**   Through center of space, draw horizontal center line, and on it construct a right-hand helix 50 mm diameter, 64 mm long, and with a lead of 25 mm, Fig. 5.63. Draw only a half-circular end view.

**Prob. 5.47**   Draw the involute of an equilateral triangle with 15 mm sides, Fig. 5.64 (b).

**Prob. 5.48**   Draw the involute of a 20 mm diameter circle, Fig. 5.64 (d).

**Prob. 5.49**   Draw a cycloid generated by a 30 mm diameter circle rolling along a horizontal straight line, Fig. 5.65.

**Prob. 5.50**   Draw an epicycloid generated by a 38 mm diameter circle rolling along a circular arc having a radius of 64 mm, Fig. 5.66 (a).

**Prob. 5.51**   Draw a hypocycloid generated by a 38 mm diameter circle rolling along a circular arc having a radius of 64 mm, Fig. 5.66 (b).

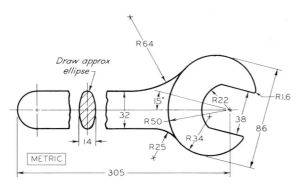

**Fig. 5.68**  Spanner.*

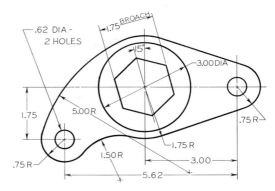

**Fig. 5.69**  Rocker Arm.*

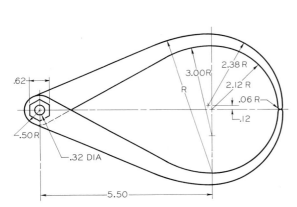

**Fig. 5.70**  Outside Caliper.*

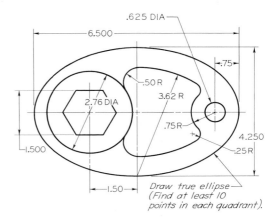

**Fig. 5.71**  Special Cam.*

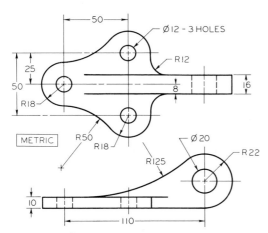

**Fig. 5.72**  Boiler Stay.*

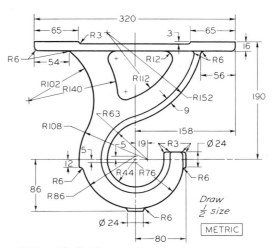

**Fig. 5.73**  Shaft Hanger Casting.*

*Using Layout A–2 or A4–2 (adjusted), draw assigned problem with instruments. Omit dimensions and notes unless assigned by instructor.

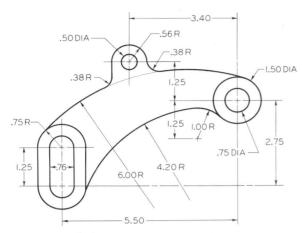

**Fig. 5.74**   Shift Lever.*

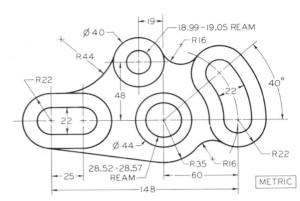

**Fig. 5.75**   Gear Arm.*

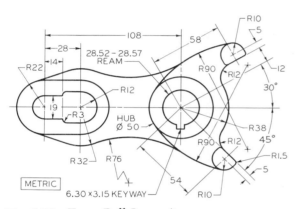

**Fig. 5.76**   Form Roll Lever.*

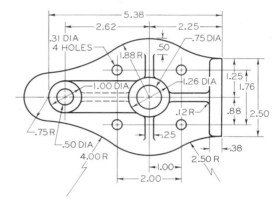

**Fig. 5.77**   Press Base.*

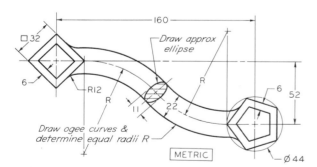

**Fig. 5.78**   Special S-Wrench.*

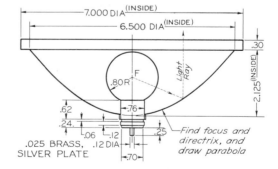

**Fig. 5.79**   Auto Headlight Reflector.*

*Using Layout A–2 or A4–2 (adjusted), draw assigned problem with instruments. Omit dimensions and notes unless assigned by instructor.

# CHAPTER

# 6

# Sketching and Shape Description

The importance of freehand drawing or sketching in engineering, design, and technical communications cannot be overestimated. To the person who possesses a complete knowledge of drawing as a language, the ability to execute quick, accurate, and clear sketches of ideas and designs constitutes a valuable means of expression. The old Chinese saying that "one picture is worth a thousand words" is not without foundation.

Most original design ideas find their first expression through the medium of a freehand sketch, §16.2. Freehand sketching is a valuable means of amplifying and clarifying, as well as recording, verbal explanations. Executives sketch freehand daily to explain their ideas to subordinates. Engineers often prepare their designs and turn them over to their detailers or designers in this convenient form as shown in the well-executed sketch of details for a steam locomotive, Fig. 6.1.

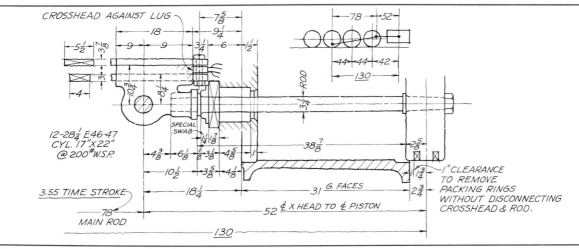

*Fig. 6.1* Typical Design Sketch.

## 6.1 Technical Sketching

Freehand sketches are of great assistance to designers in organizing their thoughts and recording their ideas. They are an effective and economical means of formulating various solutions to a given problem so that a choice can be made between them at the outset. Often much time can be lost if the designer starts his or her scaled layout before adequate preliminary study with the aid of sketches. Information concerning changes in design or covering replacement of broken parts or lost drawings is usually conveyed through sketches. Many engineers consider the ability to render serviceable sketches of greater value to them than skill in instrument drawing. The designer, technician, or engineer will find daily use for this valuable means of formulating, expressing, and recording ideas.

The degree of perfection required in a given sketch depends upon its use. Sketches hurriedly made to supplement oral description may be rough and incomplete. On the other hand, if a sketch is the medium of conveying important and precise information to engineers, technicians or skilled workers, it should be executed as carefully as possible under the circumstances.

The term "freehand sketch" is too often understood to mean a crude or sloppy freehand drawing in which no particular effort has been made. On the contrary, a freehand sketch should be made with care and with attention to proportion, clarity, and correct line widths.

## 6.2 Sketching Materials

One of the advantages of freehand sketching is that it requires only pencil, paper, and eraser—items that anyone has for ready use.

When sketches are made in the field, where an accurate record is required, a small notebook or sketching pad is frequently used. Often clipboards are employed to hold the paper. Graph paper is helpful to the sketcher, especially to the person who cannot sketch reasonably well without guide lines. Paper with 4, 5, 8, or 10 squares per inch is recommended. Such paper is convenient for sketching to scale since values can be assigned to the squares, and the squares counted to secure proportional distances, as shown in Fig. 6.2.

Sketching pads of plain tracing paper are available, accompanied by a master cross-section sheet. The drafter places a blank sheet over the master grid and can see the grid through the transparent sheet.

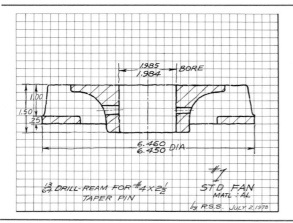

*Fig. 6.2* Sketch on Graph Paper.

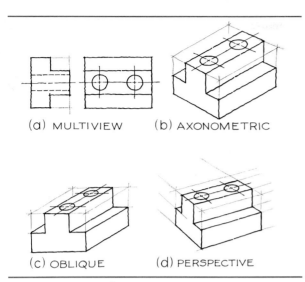

(a) MULTIVIEW          (b) AXONOMETRIC

(c) OBLIQUE            (d) PERSPECTIVE

*Fig. 6.3*  Types of Projection.

An alternate procedure is to draw, with instruments, a master cross-section sheet, using ink, making the squares either 10 mm or .25″ as desired. Ordinary bond typewriter paper is then placed over the master sheet and the sketch made thereon. Such a sketch is not only more uniform and "true" but shows up better because the cross-section lines are absent.

For isometric sketching, a specially ruled isometric paper is available, Fig. 6.26.

Soft pencils, such as HB or F, should be used for freehand sketching. For carefully made sketches, two soft erasers are recommended, a Pink Pearl or ParaPink and a Mars-Plastic, Fig. 2.16.

### 6.3 Types of Sketches

Since technical sketches are made of three-dimensional objects, the form of the sketch conforms approximately to one of the four standard types of projection, as shown in Fig. 6.3. In *multiview* projection, (a), the object is described by its necessary views, as discussed in §§6.11–6.24. Or the object may be shown pictorially in a single view, by *axonometric, oblique,* or *perspective sketches,* (b) to (d), as discussed in §§6.25–6.31.

### 6.4 Scale

Sketches usually are *not made to any scale.* Objects should be sketched in their correct proportions as accurately as possible, by eye. However, cross-section paper provides a ready scale (by counting squares) that may be used to assist in sketching to correct proportions. The size of the sketch is purely

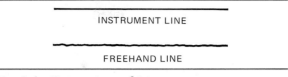

INSTRUMENT LINE

FREEHAND LINE

*Fig. 6.4*  Comparison of Lines.

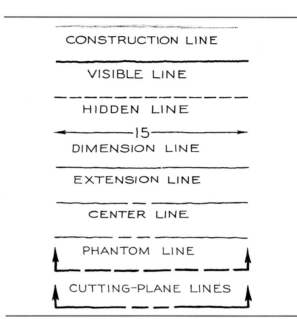

CONSTRUCTION LINE

VISIBLE LINE

HIDDEN LINE

15

DIMENSION LINE

EXTENSION LINE

CENTER LINE

PHANTOM LINE

CUTTING-PLANE LINES

*Fig. 6.5*  Sketch Lines.

optional, depending upon the complexity of the object and the size of paper available. Small objects are often sketched oversize so as to show the necessary details clearly.

### 6.5 Technique of Lines

The chief difference between an instrument drawing and a freehand sketch lies in the character or *technique* of the lines. A good freehand line is not expected to be rigidly straight or exactly uniform, as an instrument line. While the effectiveness of an instrument line lies in exacting uniformity, the quality of freehand line lies in its *freedom* and *variety,* Figs. 6.4 and 6.7.

Conventional lines, drawn instrumentally, are shown in Fig. 2.14, and the corresponding freehand renderings are shown in Fig. 6.5. The freehand construction line is a very light rough line in which some strokes may overlap. All other lines should be dark and clean-cut. Accent the ends of all dashes, and maintain a sharp contrast between the line

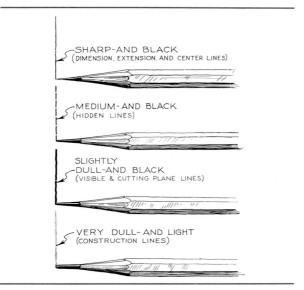

**Fig. 6.6**  Pencil Points.

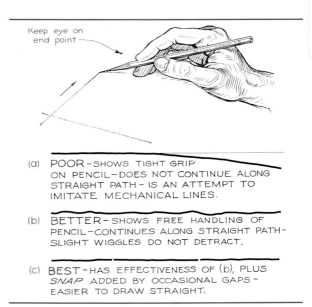

**Fig. 6.7**  Drawing Horizontal Lines.

thicknesses. Especially, make visible lines **heavy** so the outline will stand out clearly, and make hidden lines, center lines, dimension lines, and extension lines *thin*.

## 6.6  Sharpening Sketching Pencils

For sketching, use a mechanical pencil with a soft lead, such as HB or F, and sharpen it to a conical point, as shown in Fig. 2.11 (c). Use this sharp point for center lines, dimension lines, and extension lines. For visible lines, hidden lines, and cutting-plane lines, round off the point slightly to produce the desired thickness of line, Fig. 6.6. Make all lines dark, with the exception of construction lines, which should be very light.

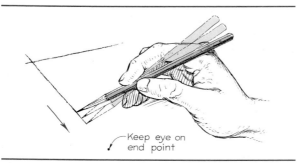

**Fig. 6.8**  Drawing Vertical Lines.

The use of thin-lead mechanical pencils with suitable diameters and grades of leads minimizes the need for sharpening and point dressing.

## 6.7  Straight Lines

Since the majority of lines on the average sketch are straight lines, it is necessary to learn to make them well. Hold the pencil naturally about $1\frac{1}{2}''$ back from the point, and approximately at right angles to the line to be drawn. Draw horizontal lines from left to right with a free and easy wrist-and-arm movement, Fig. 6.7. Draw vertical lines downward with finger-and-wrist movements, Fig. 6.8.

Inclined lines may be made to conform in direction to horizontal or vertical lines by shifting position with respect to the paper or by turning the paper slightly; hence, they may be drawn with the same general movements, Fig. 6.9.

In sketching long lines, mark the ends of the line with light dots, then move the pencil back and forth between the dots in long sweeps, keeping the eye always on the dot toward which the pencil is moving, the point of the pencil touching the paper lightly, and each successive stroke correcting the defects of the preceding strokes. When the path of the line has been established sufficiently, apply a little more pressure, replacing the trial series with a distinct line. Then, dim the line with a soft eraser and draw the final line clean-cut and dark, keeping the eye now on the point of the pencil.

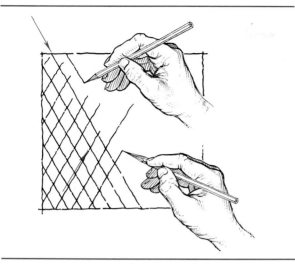

*Fig. 6.9* Drawing Inclined Lines.

An easy method of blocking in horizontal or vertical lines is to hold the hand and pencil rigidly and glide the fingertips along the edge of the pad or board, as shown in Fig. 6.10.

Another method, (b), is to mark the distance on the edge of a card or a strip of paper and transfer this distance at intervals, as shown; then draw the final line through these points. Or the pencil may

be held as shown at the lower part of (b), and distance marks made on the paper at intervals by tilting the lead down to the paper. It will be seen that both methods of transferring distances are substitutes for the dividers and will have many uses in sketching.

A common method of finding the midpoint of a line AB, shown at (c), is to hold the pencil in the left hand with the thumb gaging the estimated half-distance. Try this distance on the left and then on the right until the center is located by trial, and mark the center C, as shown. Another method is to mark the total distance AB on the edge of a strip of paper and then to fold the paper to bring points A and B together, thus locating center C at the crease. To find quarter points, the folded strip can be folded once more.

## 6.8 Circles and Arcs

Small circles and arcs can be easily sketched in one or two strokes, as for the circular portions of letters, without any preliminary blocking in.

One method of sketching a larger circle, Fig. 6.11, is first to sketch lightly the enclosing square, mark the midpoints of the sides, draw light arcs tangent to the sides of the square, and then heavy in the final circle.

Another method, Fig. 6.12, is to sketch the two

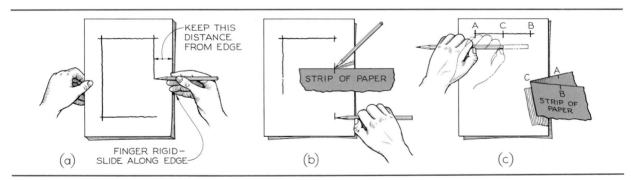

*Fig. 6.10* Blocking in Horizontal and Vertical Lines.

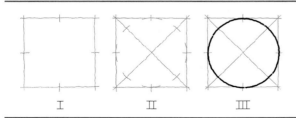

*Fig. 6.11* Sketching a Circle.

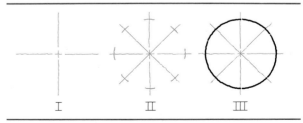

*Fig. 6.12* Sketching a Circle.

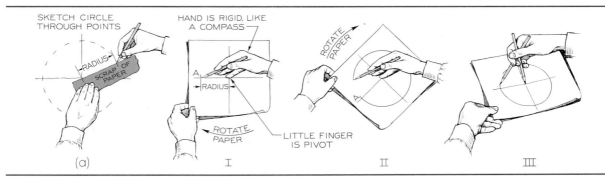

*Fig. 6.13*   Sketching Circles.

center lines, add light 45° radial lines, sketch light arcs across the lines at the estimated radius distance from the center, and finally sketch the required circle heavily. Dim all construction lines with a soft eraser before heavying in the final circle.

An excellent method, particularly for large circles, Fig. 6.13 (a), is to mark the estimated radius on the edge of a card or scrap of paper, to set off from the center as many points as desired, and to sketch the final heavy circle through these points.

The clever drafter will prefer the method at I and II, in which the hand is used as a compass. Place the tip of the little finger, or the knuckle joint of the little finger, at the center; "feed" the pencil out to the desired radius, hold this position rigidly, and carefully revolve the paper with the other hand, as shown. If you are using a sketching pad, place the pad on your knee and revolve the entire pad on the knee as a pivot.

At III, two pencils are held rigidly like a compass and the paper is slowly revolved.

Methods of sketching arcs, Fig. 6.14, are adaptations of those used for sketching circles. In general, it is easier to sketch arcs with the hand and pencil on the concave side of the curve. In sketching tangent arcs, always keep in mind the actual geometric constructions, carefully approximating all points of tangency.

## 6.9 Ellipses

If a circle is viewed obliquely, Fig. 5.50 (b), it appears as an ellipse. With a little practice, you can learn to sketch small ellipses with a free arm movement, Fig. 6.15 (a). Hold the pencil naturally, rest the weight on the upper part of the forearm, and move the pencil rapidly above the paper in the elliptical path desired; then lower the pencil so as to describe several light overlapping ellipses, as shown

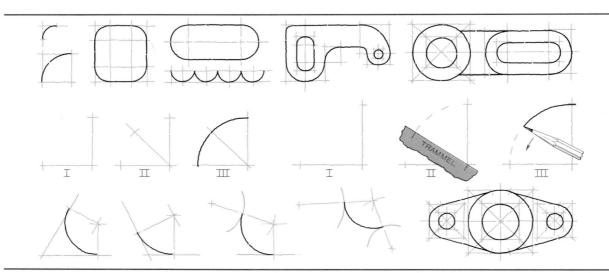

*Fig. 6.14*   Sketching Arcs.

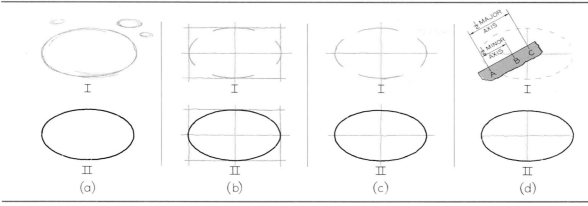

*Fig. 6.15*  Sketching Ellipses.

at I. Dim all lines with a soft eraser and heavy in the final ellipse, II.

Another method, (b), is to sketch lightly the enclosing rectangle, I, mark the midpoints of the sides, and sketch light tangent arcs, as shown. Then, II, complete the ellipse lightly, dim all lines with a soft eraser, and heavy in the final ellipse.

The same general procedure shown at (b) may be used in sketching the ellipse upon the given axes, as shown at (c).

The trammel method, (d), is excellent for sketching large ellipses. Prepare a "trammel" on the edge of a card or strip of paper, move it to different positions, and mark points on the ellipse at **A**. The trammel method is explained in §5.50. Sketch the final ellipse through the points, as shown. For sketching isometric ellipses, see §6.13.

## 6.10 **Proportions**

The most important rule in freehand sketching is *keep the sketch in proportion*. No matter how brilliant the technique or how well the small details are drawn, if the proportions—especially the large overall proportions—are bad, the sketch will be bad. First, the relative proportions of the height to the width must be carefully established; then as you proceed to the medium-sized areas and the small details, constantly compare each new estimated distance with already established distances.

If you are working from a given picture, such as the utility cabinet in Fig. 6.16 (a), it is first necessary to establish the relative width compared to the height. One way is to use the pencil as a measuring stick, as shown. In this case, the height is about $1\frac{3}{4}$ times the width. Then

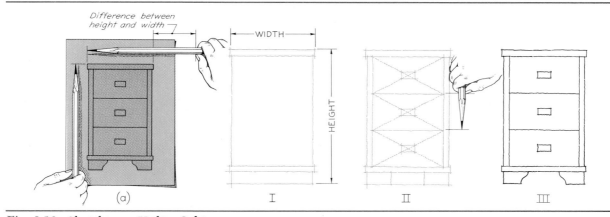

*Fig. 6.16*  Sketching a Utility Cabinet.

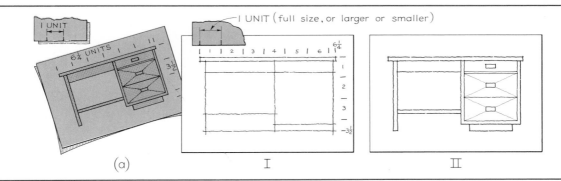

**Fig. 6.17**   Sketching a Desk.

I.   Sketch the enclosing rectangle in the correct proportion. In this case, the sketch is to be slightly larger than the given picture.

II.   Divide the available drawer space into three parts with the pencil by trial, as shown. Sketch light diagonals to locate centers of drawers, and block in drawer handles. Sketch all remaining details.

III.   Dim all construction with a soft eraser, and heavy in all final lines.

Another method of estimating distances is illustrated in Fig. 6.17. On the edge of a card or strip of paper, mark an arbitrary unit. Then see how many units wide and how many units high the desk is. If you are working from the actual object, you could use a scale, a piece of paper, or the pencil itself as a unit to determine the proportions.

To sketch an object composed of many curves to the same scale or to a larger or smaller scale, the method of "squares" is recommended, Fig. 6.18. On

the given picture, rule accurate grid lines to form squares of any convenient size. It is best to use a scale and some convenient spacing, such as either 50″ or 10 mm. On the new sheet rule a similar grid, making the spacing of the lines proportional to the original, but reduced or enlarged as desired. Make the final sketch by drawing the lines in and across the grid lines as in the original, as near as you can estimate by eye.

In sketching from an actual object, you can easily compare various distances on the object by using the pencil to compare measurements, as shown in Fig. 6.19. While doing this, do not change your position, and always hold your pencil at arm's length. The length sighted can then be compared in similar manner with any other dimension of the object. If the object is small, such as a machine part, you can compare distances in the manner of Fig. 6.16, by actually placing the pencil against the object itself.

In establishing proportions, the blocking-in method is recommended, especially for irregular

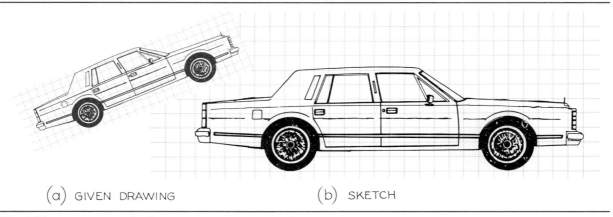

(a) GIVEN DRAWING                (b)   SKETCH

**Fig. 6.18**   Squares Method.

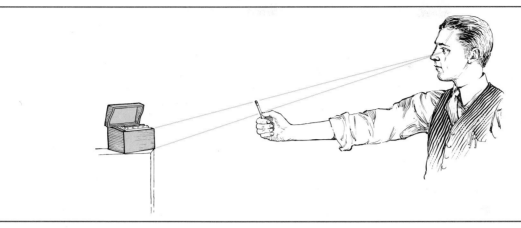

*Fig. 6.19*  Estimating Dimensions.

shapes. The steps for blocking in and completing the sketch of a Shaft Hanger are shown in Fig. 6.20. As always, first give attention to the main proportions, next to the general sizes and direction of flow of curved shapes, and finally to the snappy lines of the completed sketch.

In making sketches from actual machine parts, it is necessary to use the measuring tools used in the shop, especially those needed to determine dimensions that must be relatively accurate. For a discussion of these methods, see Chapters 12.

## 6.11 Pictorial Sketching

We shall now examine several simple methods of preparing pictorial sketches that will be of great assistance in learning the principles of multiview projection. A detailed and more technical treatment of pictorial drawing is given in Chapters 17 and 18.

## 6.12 Isometric Sketching

To make an isometric sketch from an actual object, hold the object in your hand and tilt it toward you, as shown in Fig. 6.21 (a). In this position, the front

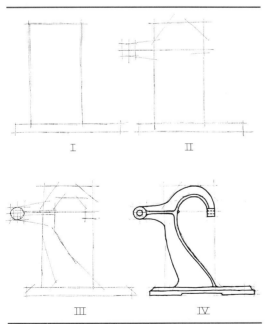

*Fig. 6.20*  Blocking in an Irregular Object (Shaft Hanger).

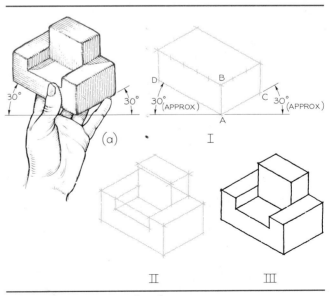

*Fig. 6.21*  Isometric Sketching.

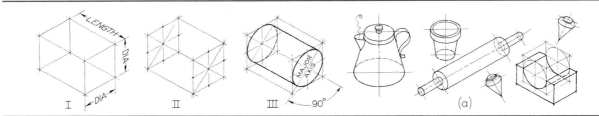

*Fig. 6.22*   Isometric Ellipses.

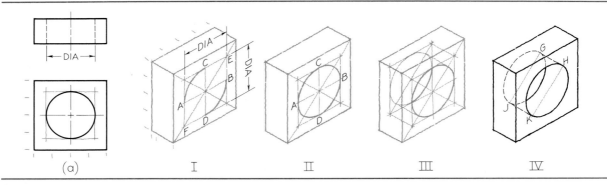

*Fig. 6.23*   Isometric Ellipses.

corner will appear vertical, and the two receding bottom edges and those parallel to them, respectively, will appear at about 30° with horizontal, as shown. The steps in sketching are

I.   Sketch the enclosing box lightly, making **AB** vertical and **AC** and **AD** approximately 30° with horizontal. These three lines are the *isometric axes*. Make **AB**, **AC**, and **AD** approximately proportional in length to the actual corresponding edges on the object. Sketch the remaining lines parallel, respectively, to these three lines.

II.   Block in the recess and the projecting block.

III.   Dim all construction lines with a soft eraser, and heavy in all final lines.

NOTE   The angle of the receding lines may be less than 30°, say, 20° or 15°. Although the result will not be an isometric sketch, the sketch may be more pleasing and effective in many cases.

## 6.13 Isometric Ellipses

As shown in Fig. 5.50 (b), a circle viewed at an angle appears as an ellipse. When objects having cylindrical or conical shapes are placed in the isometric or other oblique positions, the circles will be viewed at an angle and will appear as ellipses, Fig. 6.22.

The most important consideration in sketching isometric ellipses is: *The major axis of the ellipse is always at right angles to the center line of the cylinder, and the minor axis is at right angles to the major axis and coincides with the center line.*

Two views of a block with a large cylindrical hole are shown in Fig. 6.23 (a). The steps in sketching the object are

I.   Sketch the block and the enclosing parallelogram for the ellipse, making the sides of the parallelogram parallel to the edges of the block and equal in length to the diameter of the hole. Draw diagonals to locate the center of the hole, and then draw center lines **AB** and **CD**. Points A, B, C, and D will be midpoints of the sides of the parallelogram, and the ellipse will be tangent to the sides at those points. The major axis will be on the diagonal **EF**, which is at right angles to the center line of the hole, and the minor axis will fall along the short diagonal. Sketch long flat elliptical sides **CA** and **BD**, as shown.

II.   Sketch short small-radius arcs **CB** and **AD** to complete the ellipse. Avoid making the ends of the ellipse "squared off" or pointed like a football.

III.   Sketch lightly the parallelogram for the ellipse that lies in the back plane of the object, and sketch the ellipse in the same manner as the front ellipse.

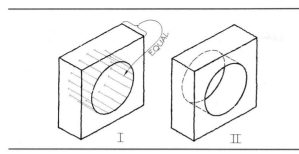

*Fig. 6.24*  Isometric Ellipses.

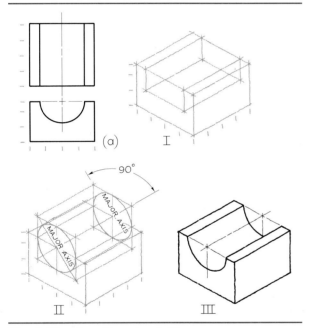

*Fig. 6.25*  Sketching Semiellipses.

IV.   Draw lines **GH** and **JK** tangent to the two ellipses. Dim all construction with a soft eraser, and heavy in all final lines.

Another method for determining the back ellipse is shown in Fig. 6.24.

I.   Select points at random on the front ellipse and sketch "depth lines" equal in length to the depth of the block.

II.   Sketch the ellipse through the ends of the lines, as shown.

Two views of a bearing with a semicylindrical opening are shown in Fig. 6.25 (a). The steps in sketching are

I.   Block in the object, including the rectangular space for the semicylinder.

II.   Block in the box enclosing the complete cylinder. Sketch the entire cylinder lightly.

III.   Dim all construction lines, and heavy in all final lines, showing only the lower half of the cylinder.

## 6.14 Sketching on Isometric Paper

Two views of a guide block are shown in Fig. 6.26 (a). The steps in sketching illustrate not only the use of isometric paper but also the sketching of individual planes or faces in order to build up a pictorial visualization from the given views.

I.   Sketch isometric of enclosing box, counting off the isometric grid spaces to equal the corresponding squares on the given views. Sketch surface **A**, as shown.

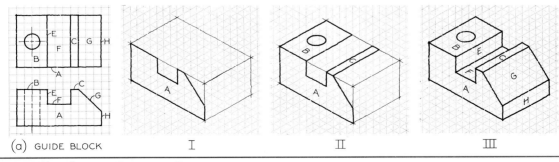

*Fig. 6.26*  Sketching on Isometric Paper.

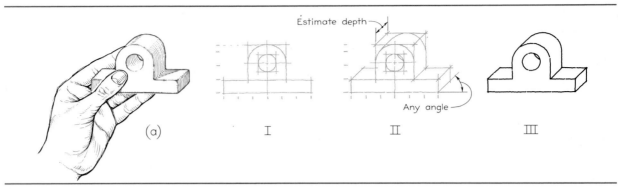

*Fig. 6.27*  Sketching in Oblique.

II.   Sketch additional surfaces B and C and the small ellipse.

III.   Sketch additional surfaces E, F, G, and H to complete the sketch.

## 6.15 Oblique Sketching
Another simple method for sketching pictorially is oblique sketching, Fig. 6.27. Hold the object in your hand, as shown at (a).

I.   Block in the front face of the bearing, as if you were sketching a front view.

II.   Sketch receding lines parallel to each other and at any convenient angle, say, 30° or 45° with horizontal, approximately. Cut off receding lines so that the depth appears correct. These lines may be full length, but a more natural appearance results if they are cut to three-quarters or one-half size, approximately. If they are full length, the sketch is a *cavalier* sketch. If half size, the sketch is a *cabinet* sketch. See Chapter 18.

III.   Dim all construction lines with a soft eraser and heavy in the final lines.

NOTE   Oblique sketching is a less suitable method for any object having circular shapes in or parallel to more than one plane of the object, because ellipses result when circular shapes are viewed obliquely. Therefore, place the object with most or all of the circular shapes toward you, so that they will appear as true circles and arcs in oblique sketching, as in Fig. 6.27.

## 6.16 Oblique Sketching on Graph Paper
Ordinary graph paper is suitable and convenient for oblique sketching. Two views of a bearing bracket are shown in Fig. 6.28 (a). The dimensions are determined simply by counting the squares.

I.   Sketch lightly the enclosing box construction. Sketch the receding lines at 45° diagonally through the squares. To establish the depth at a reduced scale, sketch the receding lines diagonally through

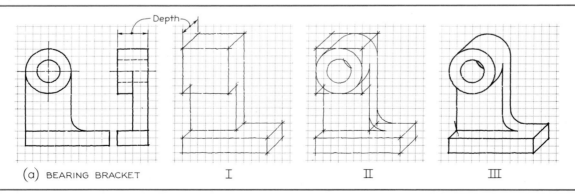

(a) BEARING BRACKET          I          II          III

*Fig. 6.28*  Oblique Sketching on Cross-section Paper.

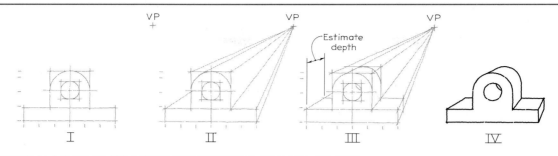

**Fig. 6.29** Sketching in One-Point Perspective.

half as many squares as the given number shown at (a).

II.   Sketch all arcs and circles.

III.   Heavy in all final lines.

## 6.17 Perspective Sketching

The bearing sketched in oblique in Fig. 6.27 can easily be sketched in *one-point perspective* (one vanishing point), as shown in Fig. 6.29.

I.   Sketch the true front face of the object, just as in oblique sketching. Select the vanishing point (VP) for the receding lines. In most cases, it is desirable to place VP above and to the right of the picture, as shown, although it can be placed anywhere in the vicinity of the picture. But if it is placed too close to the center, the lines will converge too sharply, and the picture will be distorted.

II.   Sketch the receding lines toward VP.

III.   Estimate the depth to look well, and sketch in the back portion of the object. Note that the back circle and arc will be slightly smaller than the front circle and arc.

IV.   Dim all construction lines with a soft eraser, and heavy in all final lines. Note the similarity between the perspective sketch and the oblique sketch in Fig. 6.27.

*Two-point perspective* (two vanishing points) is the most true to life of all pictorial methods, but it requires some natural sketching ability or considerable practice for best results. A simple method is shown in Fig. 6.30 that can be used successfully by the nonartistic student.

I.   Sketch front corner of desk in true height, and locate two *vanishing points* (VPL and VPR) on a *horizon* line (eye level). The distance CA may vary—

the greater it is, the higher the eye level will be and the more we will be looking down on top of the object. A good rule of thumb is to make C–VPL one-third to one-fourth of C–VPR.

II.   Estimate depth and width, and sketch enclosing box.

III.   Block in all details. Note that all parallel lines converge toward the same vanishing point.

IV.   Dim the construction lines with a soft eraser as necessary, and heavy in all final lines. Make the outlines thicker and the inside lines thinner, especially where they are close together.

## 6.18 Views of Objects

A pictorial drawing or a photograph shows an object as it *appears* to the observer, but not as it *is*. Such a picture cannot describe the object fully, no matter from which direction it is viewed, because it does not show the exact shapes and sizes of the several parts.

In industry, a complete and clear description of the shape and size of an object to be made is necessary, to make certain that the object will be manufactured exactly as intended by the designer. In order to provide this information clearly and accurately, a number of *views*, systematically arranged, are used. This system of views is called *multiview projection*. Each view provides certain definite information if the view is taken in a direction perpendicular to a principal face or side of the object. For example, as shown in Fig. 6.31 (a), an observer looking perpendicularly toward one face of the object obtains a true view of the shape and size of that side. This view as seen by the observer is shown at (b). (The observer is theoretically at an infinite distance from the object.)

An object has three principal dimensions: *width*, *height*, and *depth*, as shown at (a). In technical draw-

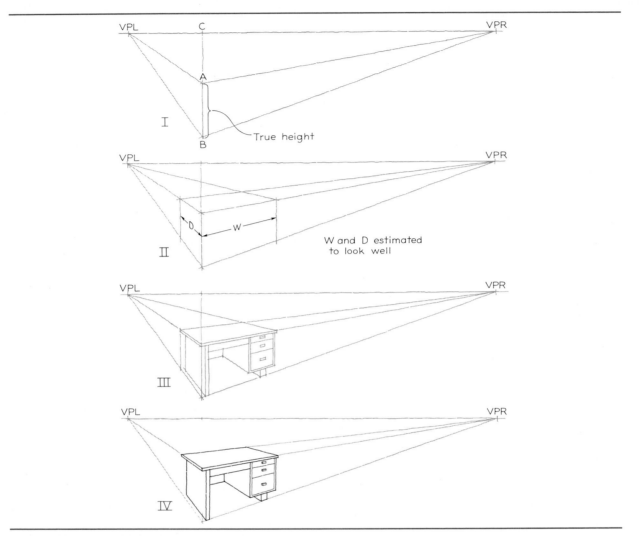

*Fig. 6.30*  Two-Point Perspective.

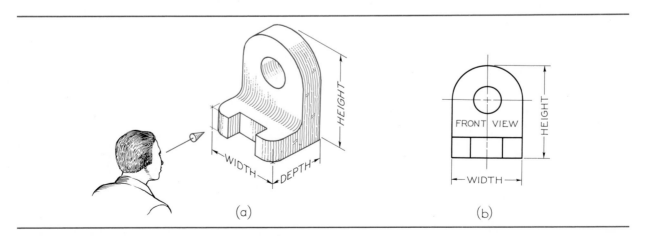

*Fig. 6.31*  Front View of an Object.

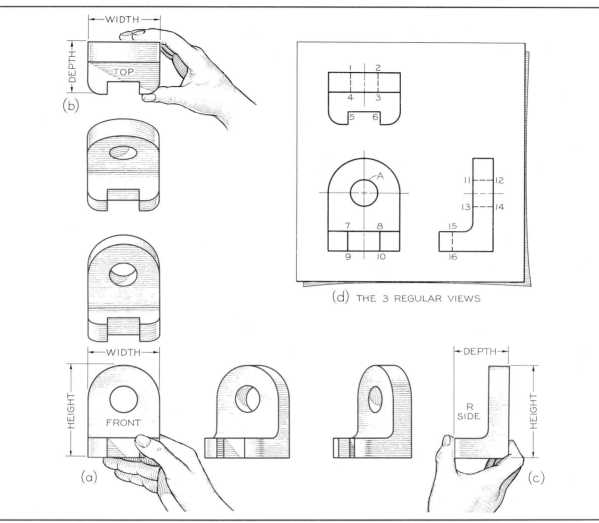

**Fig. 6.32** The Three Regular Views.

ing, these fixed terms are used for dimensions taken in these directions, regardless of the shape of the object. The terms "length" and "thickness" are not used because they cannot be applied in all cases. Note at (b) that the front view shows only the height and width of the object and not the depth. In fact, *any one view of a three-dimensional object can show only two dimensions; the third dimension will be found in an adjacent view.*

## 6.19 Revolving the Object

To obtain additional views, revolve the object as shown in Fig. 6.32. First, hold the object in the front-view position, as shown at (a).

To get the *top view*, (b), revolve the object so as to bring the *top of the object up and toward you.* To get the *right-side view*, (c), revolve the object so as to bring the *right side to the right and toward you.* To obtain views of any of the other sides, merely turn the object so as to bring those sides toward you.

The top, front, and right-side views, arranged closer together, are shown at (d). These are called the *three regular views* because they are the views most frequently used.

At this stage we can consider spacing between views as purely a matter of appearance. The views should be spaced well apart and yet close enough to appear related to each other. The space between

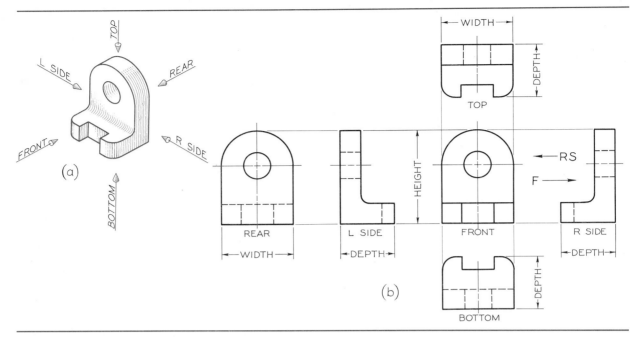

*Fig. 6.33*   The Six Views.

the front and top views may or may not be equal to the space between the front and side views. If dimensions (Chapter 13) are to be added to the sketch, sufficient space for them between views will have to be allowed.

An important advantage that a view has over a photograph of an object is that hidden features can be clearly shown by means of *hidden lines*, Fig. 2.14. In Fig. 6.32 (d), surface 7–8–9–10 in the front view appears as a visible line 5–6 in the top view and as a hidden line 15–16 in the side view. Also, hole A, which appears as a circle in the front view, shows as hidden lines 1–4 and 2–3 in the top view, and 11–12 and 13–14 in the side view. For a complete discussion of hidden lines, see §6.25.

Note, too, the use of center lines for the hole in Fig. 6.32 (d). See §6.26.

## 6.20 The Six Views

Any object can be viewed from six mutually perpendicular directions, as shown in Fig. 6.33 (a). Thus, six views may be drawn if necessary, as shown at (b). Except as explained in §7.8, these six views are always arranged as shown, which is the American National Standard arrangement of views. The *top*, *front*, and *bottom views* line up vertically, while the *rear*, *left-side*, *front*, and *right-side views* line

up horizontally. To draw a view out of place is a serious error, generally regarded as one of the worst mistakes one can make in this subject. See Fig. 6.47.

Note that the height is shown in the rear, left-side, front, and right-side views; the width is shown in the top, front, and bottom views; and the depth is shown in the four views that surround the front view, namely, the left-side, top, right-side, and bottom views. In each view, two of the principal dimensions are shown, and the third is not shown. Observe also that in the four views that surround the front view, the front of the object is faced toward the front view.

*Adjacent views are reciprocal.* If the front view, Fig. 6.33, is imagined to be the object itself, the right-side view is obtained by looking toward the right side of the front view, as shown by the arrow RS. Likewise, if the right-side view is imagined to be the object, the front view is obtained by looking toward the left side of the right-side view, as shown by the arrow F. The same relation exists between any two adjacent views.

Obviously, the six views may be obtained either by shifting the object with respect to the observer, as we have seen, Fig. 6.32, or by shifting the observer with respect to the object, Fig. 6.33. Another illustration of the second method is given in Fig. 6.34, showing six views of a house. The observer can

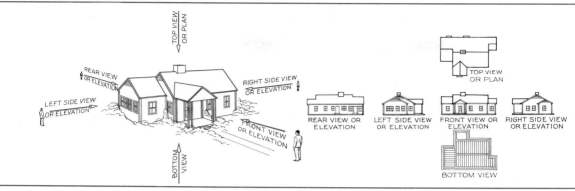

*Fig. 6.34*   Six Views of a House.

walk around the house and view its front, sides, and rear and can imagine the top view as seen from an airplane and the bottom or "worms's-eye view" as seen from underneath.* Notice the use of the terms *plan*, for the top view, and *elevation*, for all views showing the height of the building. These terms are regularly used in architectural drawing and, occasionally, with reference to drawings in other fields.

### 6.21 Orientation of Front View

Six views of a compact automobile are shown in Fig. 6.35. The view chosen for the front view in this case is the side, not the front of the automobile. In general, the front view should show the object in its operating position, particularly of familiar objects

*Architects frequently draw the views of a building on separate sheets because of the large sizes of the drawings.

such as the house shown above and the automobile. A machine part is often drawn in the position it occupies in the assembly. However, in most cases this is not important, and the drafter may assume the object to be in any convenient position. For example, an automobile connecting rod is usually drawn horizontally on the sheet, Fig. 16.37. Also, it is customary to draw screws, bolts, shafts, tubes, and other elongated parts in a horizontal position, not only because they are usually manufactured in this position but also because they can be presented more satisfactorily on paper in this position.

### 6.22 Choice of Views

A drawing for use in production should contain *only those views* needed for a clear and complete shape description of the object. These minimum required views are referred to as the *necessary views*. In se-

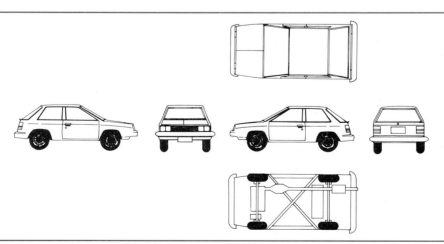

*Fig. 6.35*   Six Views of a Compact Automobile.

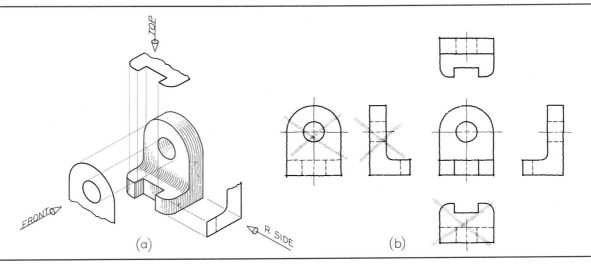

*Fig. 6.36*   Choice of Views.

lecting views, the drafter should choose those that show best the essential contours or shapes and should give preference to those with the least number of hidden lines.

As shown in Fig. 6.36 (a), three distinctive features of this object need to be shown on the drawing.

1.   Rounded top and hole, seen from the front.
2.   Rectangular notch and rounded corners, seen from the top.
3.   Right angle with filleted corner, seen from the side.

Another way to choose necessary views is to eliminate unnecessary views. At (b) a "thumbnail sketch" of the six views is shown. Both the front and rear views show the true shapes of the hole and the rounded top, but the front view is preferred because it has no hidden lines. Therefore, the rear view (which is seldom needed) is crossed out.

Both the top and bottom views show the rectangular notch and rounded corners, but the top view is preferred because it has fewer hidden lines.

Both the right-side and left-side views show the right angle with the filleted corner. In fact, in this case the side views are identical, except reversed. In such instances, it is customary to choose the right-side view.

The necessary views, then, are the three remaining views: the top, front, and right-side views. These are the three regular views referred to in connection with Fig. 6.32.

More complicated objects may require more than three views, or in many cases special views such as partial views, §7.9; sectional views, Chapter 9; auxiliary views, Chapter 10.

## 6.23 **Two-View Drawings**

Often only two views are needed to describe clearly the shape of an object. In Fig. 6.37 (a), the right-side view shows no significant contours of the object,

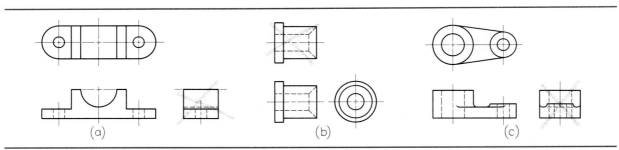

*Fig. 6.37*   Two Necessary Views.

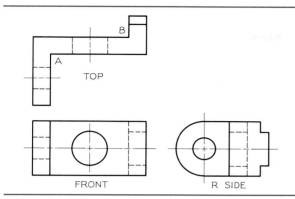

*Fig. 6.38*   Three Views.

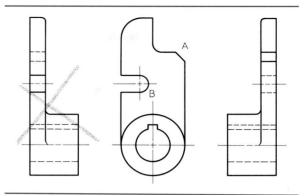

*Fig. 6.39*   Choice of Right-Side View.

and is crossed out. At (b) the top and front views are identical, so the top view is eliminated. At (c), no additional information not already given in the front and top views is shown in the side view, so the side view is unnecessary.

The question often arises: What are the absolute minimum views required? For example, in Fig.

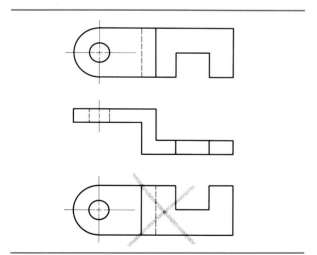

*Fig. 6.40*   Choice of Top View.

6.38, the top view might be omitted, leaving only the front and right-side views. However, it is more difficult to "read" the two views or visualize the object, because the characteristic "Z" shape of the top view is omitted. In addition, one must assume that corners A and B (top view) are square and not filleted. In this example, all three views are necessary.

If the object requires only two views, and the left-side and right-side views are equally descriptive, the right-side view is customarily chosen, Fig. 6.39. If contour A were omitted, then the presence of slot B would make it necessary to choose the left-side view in preference to the right-side view.

If the object requires only two views, and the top and bottom views are equally descriptive, the top view is customarily chosen, Fig. 6.40.

If only two views are necessary, and the top view and right-side view are equally descriptive, the combination chosen is that which spaces best on the paper, Fig. 6.41.

## 6.24 One-View Drawings

Frequently a single view supplemented by a note or lettered symbols is sufficient to describe clearly the shape of a relatively simple object. In Fig. 6.42

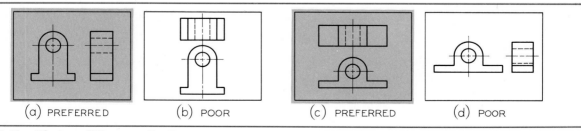

(a) PREFERRED        (b) POOR        (c) PREFERRED        (d) POOR

*Fig. 6.41*   Choice of Views to Fit Paper.

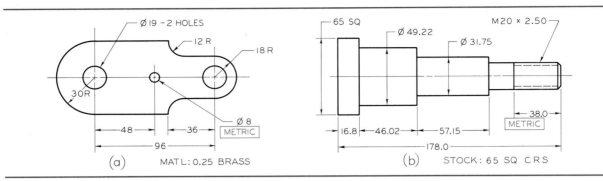

**Fig. 6.42**   One-View Drawings.

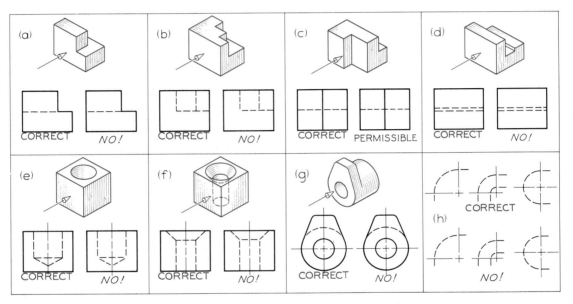

**Fig. 6.43**   Hidden-Line Practices.

(a), one view of the Shim, plus a note indicating the thickness as 0.25 mm, is sufficient. At (b), the left end is 65 mm square, the next portion is 49.22 mm diameter, the next is 31.75 mm diameter, and the portion with the thread is 20 mm diameter, as indicated in the note. Nearly all shafts, bolts, screws, and similar parts should be represented by single views in this manner.

## 6.25  Hidden Lines

Correct and incorrect practices in drawing hidden lines are illustrated in Fig. 6.43. In general, a hidden line should join a visible line except when it causes the visible line to extend too far, as shown at (a). In other words, *leave a gap whenever a hidden-line dash forms a continuation of a visible line*. Hidden lines should intersect to form L and T corners, as shown at (b). A hidden line preferably should "jump" a visible line when possible, (c). Parallel hidden lines should be drawn so that the dashes are staggered, in a manner similar to bricklaying, as at (d). When two or three hidden lines meet at a point, the dashes should join, as shown for the bottom of the drilled hole at (e), and for the top of a countersunk hole, (f). The example at (g) is similar to (a) in that hidden lines should not join visible lines when it makes the visible line extend too far. Correct and incorrect methods of drawing hidden arcs are shown at (h).

Poorly drawn hidden lines can easily spoil a drawing. Each dash should be carefully drawn about 5 mm long and spaced only about 1 mm apart, by eye. Accent the beginning and end of each dash by

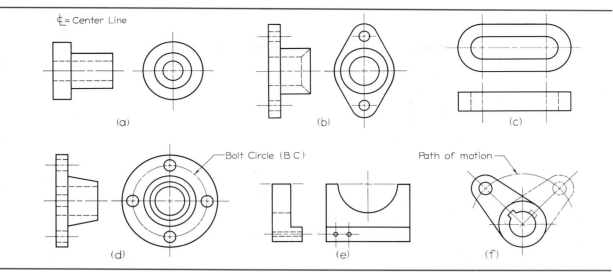

**Fig. 6.44** Center-Line Applications.

pressing down on the pencil, whether drawn free-hand or mechanically.

In general, views should be chosen that show features with visible lines, so far as possible. After this has been done, hidden lines should be used wherever necessary to make the drawing clear. Where they are not needed for clearness, hidden lines should be omitted, so as not to clutter the drawing any more than necessary and in order to save time. The beginner, however, would do well to be cautious about leaving out hidden lines until experience shows when they can be safely omitted.

## 6.26 Center Lines

Center lines (symbol: ₵) are used to indicate axes of symmetrical objects or features, bolt circles, and paths of motion. Typical applications are shown in Fig. 6.44. As shown at (a), a single center line is drawn in the longitudinal view and crossed center lines in the circular view. The small dashes should cross at the intersections of center lines. Center lines should extend uniformly about 8 mm outside the feature for which they are drawn.

The long dashes of center lines may vary from 20 to 40 mm or more in length, depending upon the size of the drawing. The short dashes should be about 5 mm long, with spaces about 2 mm. Center lines should always start and end with long dashes. Short center lines, especially for small holes, as at (e), may be made solid as shown. Always leave a gap as at (e) when a center line forms a continuation of

a visible or hidden line. Center lines should be thin enough to contrast well with the visible and hidden lines but dark enough to reproduce well.

Center lines are useful mainly in dimensioning and should be omitted from unimportant rounded or filleted corners and other shapes that are self-locating.

## 6.27 Sketching Two Views

The Support Block in Fig. 6.45 (a) requires only two views. The steps in sketching are

I. Block in lightly the enclosing rectangles for the two views. Sketch horizontal lines **1** and **2** to establish the height of the object, while making spaces **A** approximately equal. Sketch vertical lines **3**, **4**, **5**, and **6** to establish the width and depth in correct proportion to the already established height, while making spaces **B** approximately equal, and space **C** equal to or slightly less than space **B**.

II. Block in smaller details, using diagonals to locate the center, as shown. Sketch lightly the circle and arcs.

III. Dim all construction lines with a soft eraser, and heavy in all final lines.

## 6.28 Sketching Three Views

A Lever Bracket requiring three views is shown in Fig. 6.46 (a). The steps in sketching the three views are as follows.

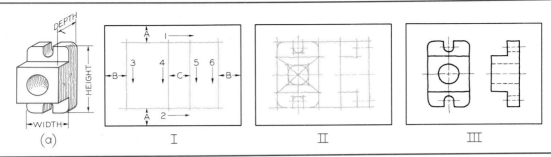

**Fig. 6.45** Sketching Two Views of a Support Block.

I.   Block in the enclosing rectangles for the three views. Sketch horizontal lines 1, 2, 3, and 4 to establish the height of the front view and the depth of the top view, while making spaces A approximately equal and space C equal to or slightly less than space A. Sketch vertical lines 5, 6, 7, and 8 to establish the width of the top and front views, and the depth of the side view. Make sure that this is in correct proportion to the height, while making spaces B approximately equal and space D equal to or slightly less than one space B. Note that spaces C and D are not necessarily equal, but are independent of each other. Similarly, spaces A and B are not necessarily equal. To transfer the depth dimension from the top view to the side view, use the edge of a card or strip of paper, as shown, or transfer the distance by using the pencil as a measuring stick, as shown in Fig. 6.10 (b) and (c). Note that *the depth in the top and side views must always be equal.*

II.   Block in all details lightly.

III.   Sketch all arcs and circles lightly.

IV.   Dim all construction lines with a soft eraser.

V.   Heavy in all final lines so that the views will stand out clearly.

## 6.29 Alignment of Views

Errors in arranging the views are so commonly made by students that it is necessary to repeat: The views must be drawn in accordance with the American National Standard arrangement, Fig. 6.33. In

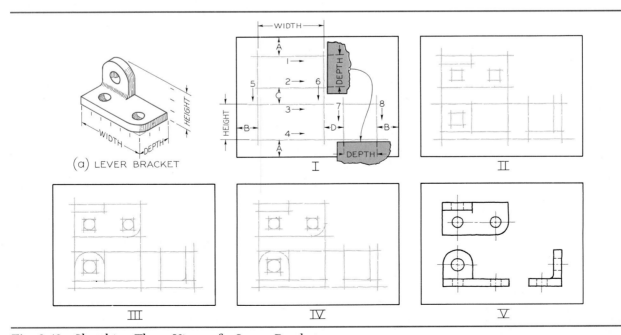

**Fig. 6.46** Sketching Three Views of a Lever Bracket.

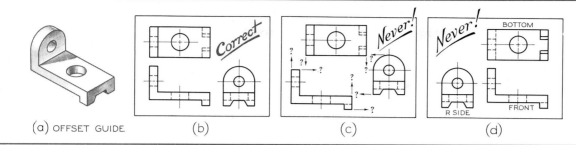

(a) OFFSET GUIDE    (b)    (c)    (d)

*Fig. 6.47*   Position of Views.

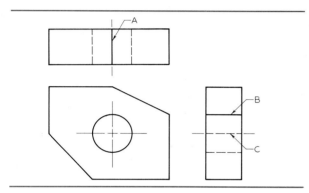

*Fig. 6.48*   Meaning of Lines.

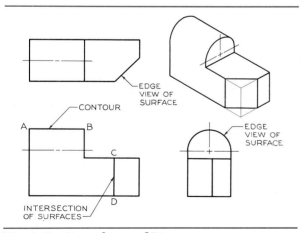

*Fig. 6.49*   Precedence of Lines.

Fig. 6.47 (a) an Offset Guide is shown that requires three views. These three views, correctly arranged, are shown at (b). The top view must be directly above the front view, and the right-side view di-

rectly to the right of the front view—not out of alignment, as at (c). Also, never draw the views in reversed positions, with the bottom over the front view, or the right-side to the left of the front view, as shown at (d), even though the views do line up with the front view.

## 6.30  Meaning of Lines

A visible line or a hidden line has three possible meanings, Fig. 6.48: (1) intersection of two surfaces, (2) edge view of a surface, and (3) contour view of a curved surface. Since *no shading is used on a working drawing,* it is necessary to examine all the views to determine the meaning of the lines. For example, the line **AB** at the top of the front view might be regarded as the edge view of a flat surface if we look at only the front and top views and do not observe the curved surface on top of the object as shown in the right-side view. Similarly, the vertical line **CD** in the front view might be regarded as the edge view of a plane surface if we look at only the front and side views. However, the top view shows that the line represents the intersection of an inclined surface.

## 6.31  Precedence of Lines

Visible lines, hidden lines, and center lines often coincide on a drawing, and it is necessary for the drafter to know which line to show. A visible line always takes precedence over (covers up) a center line or a hidden line, as shown at **A** and **B** in Fig. 6.49. A hidden line always takes precedence over a center line, as at **C**. Note that at **A** and **C** the ends of the center line are shown, but are separated from the view by short gaps.

# Sketching Problems

Figures 6.51 and 6.52 present a variety of objects from which the student is to sketch the necessary views. Using 8.5″ × 11.0″ graph paper, sketch a border and title strip and divide the sheet into two parts as shown in Fig. 6.50. Sketch two assigned problems per sheet, as shown. On the problems in Fig. 6.51, "ticks" are given that indicate .50″ or .25″ spaces. Thus, measurements may be easily spaced off on graph paper having .12″ or .25″ grid spacings.

If desired, the "ticks" on the problems in Fig. 6.51 may be used to indicate 10 mm and 5 mm spaces. Thus, metric measurements may be easily utilized on appropriate metric-grid graph paper.

On the problems in Fig. 6.52 no indications of size are given. The student is to sketch the necessary views of assigned problems to fit the spaces comfortably, as shown in Fig. 6.50. It is suggested that the student prepare a small paper scale, making the divisions equal to those on the paper scale in Prob. 1. This scale can be used to determine the approximate sizes. Let each division equal either .50″ or 10 mm on your sketch.

Missing-line and missing-view problems are given in Figs. 6.53 and 6.54, respectively. These are to be sketched, two problems per sheet, in the arrangement shown in Fig. 6.50. If the instructor so assigns, the missing lines or views may be sketched with a colored pencil. The problems given in Figs. 6.53 and 6.54 may be sketched in isometric on isometric paper or in oblique on graph paper.

Since many of the problems in this chapter are of a general nature, they can also be solved on most computer graphics systems. If a system is available, the instructor may choose to assign specific problems to be completed by this method.

Sketching problems in convenient form for solution are available in *Principles of Engineering Graphics Problems* by Spencer, Hill, Loving, and Dygdon, a workbook designed to accompany this text that is also published by Macmillan Publishing Company.

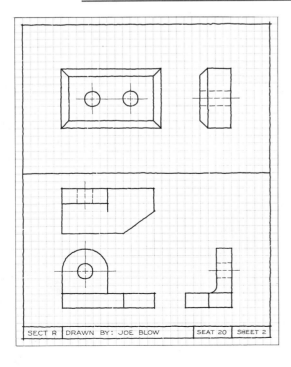

*Fig. 6.50*   Multiview Sketch (Layout A–1).

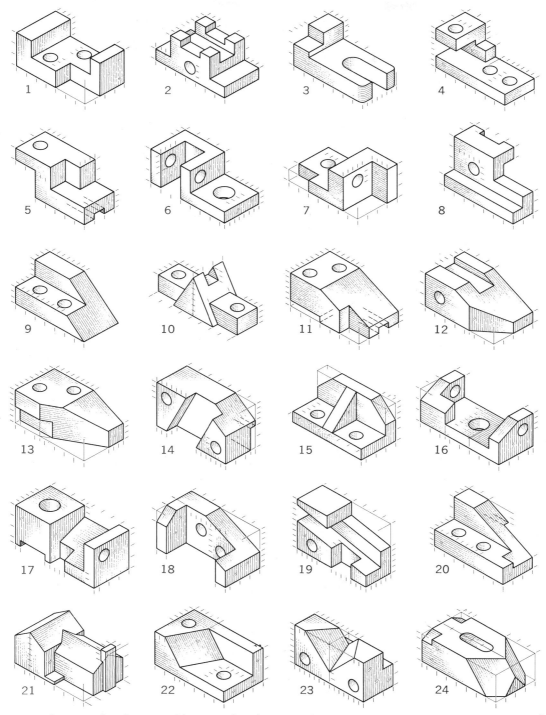

**Fig. 6.51**  Multiview Sketching Problems.   Sketch necessary views, using Layout A–1 or A4–1 adjusted (freehand), on graph paper or plain paper, two problems per sheet as in Fig. 6.50. The units shown may be either .50″ and .25″ or 10 mm and 5 mm. See instructions on page 150. All holes are through holes.

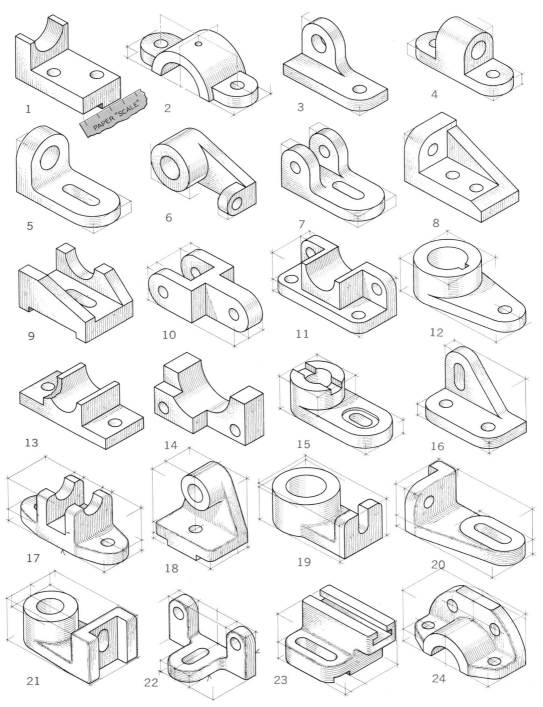

*Fig. 6.52*  Multiview Sketching Problems.   Sketch necessary views, using Layout A–1 or 4A–1 adjusted (freehand), on graph paper or plain paper, two problems per sheet as in Fig. 6.50. Prepare paper scale with divisions equal to those in Prob. 1, and apply to problems to obtain approximate sizes. Let each division equal either .50″ or 10 mm on your sketch. See instructions on page 150. For Probs. 17 to 24, study §§7.34 to 7.36.

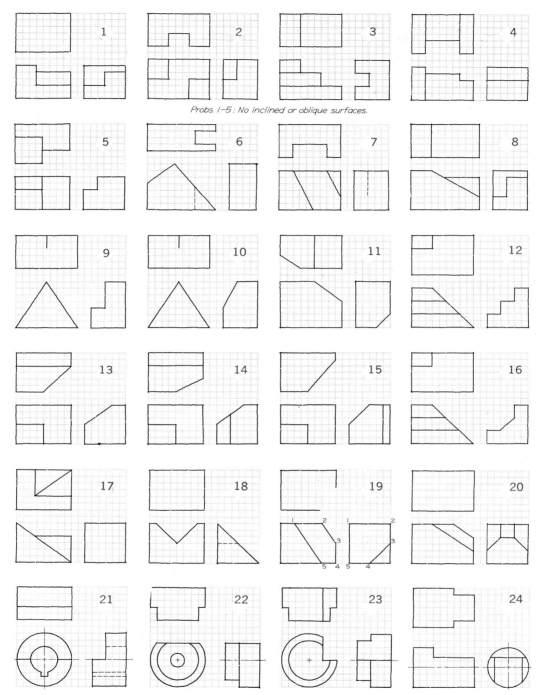

Probs. 1–5: No inclined or oblique surfaces.

**Fig. 6.53** Missing-Line Sketching Problems. (1) Sketch given views, using Layout A–1 or A4–1 adjusted (freehand), on graph paper or plain paper, two problems per sheet as in Fig. 6.50. Add missing lines. The squares may be either .25″ or 5 mm. See instructions on page 150. (2) Sketch in isometric on isometric paper or in oblique on cross-section paper.

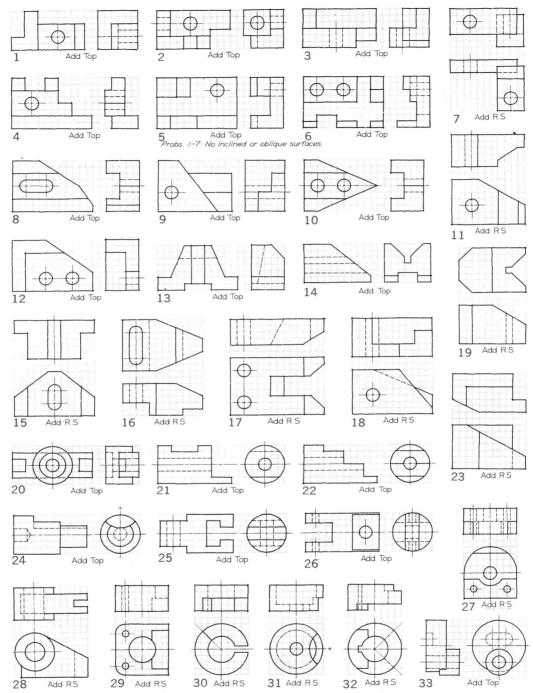

*Probs. 1–7: No inclined or oblique surfaces.*

**Fig. 6.54** Third-View Sketching Problems. (1) Using Layout A–1 or A4–1 adjusted (freehand), on graph paper or plain paper, two problems per sheet as in Fig. 6.50, sketch the two given views and add the missing views, as indicated. The squares may be either .25″ or 5 mm. See instructions on page 150. The given views are either front and right-side views or front and top views. Hidden holes with center lines are drilled holes. (2) Sketch in isometric paper or in oblique on cross-section paper.

# CHAPTER

# 7

# Multiview Projection

A view of a part for a design is known technically as a *projection*. A projection is a view conceived to be drawn or projected onto a plane known as the *plane of projection*. A system of views of an object formed by projectors from the object perpendicular to the desired planes of projection is known as orthographic or multiview projection. See ANSI Y14.3–1975 (R1980). This system of required views provides for the shape description of the object.

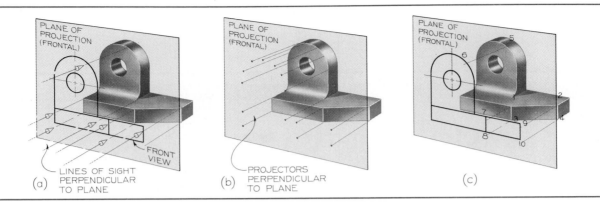

*Fig. 7.1*  Projection of an Object.

## 7.1 Projection Method

The method of viewing the part to obtain a *multi-view projection* is illustrated for a front view in Fig. 7.1 (a). Between the observer and the part, a transparent plane or pane of glass representing a plane of projection is located parallel to the front surfaces of the part. Shown on the plane of projection in outline is how the design appears to the observer. Theoretically, the observer is at an infinite distance from the part or object, so that the *lines of sight* are parallel.

In more precise terms, this view is obtained by drawing perpendiculars, called *projectors*, from all points on the edges or contours of the part or object to the plane of projection, (b). The collective piercing points of these projectors, being infinite in number, form lines on the pane of glass, as shown at (c).

Thus, as shown at (c), a projector from point **1** on the object pierces the plane of projection at point **7**, which is a view or projection of the point. The same procedure applies to point **2**, whose projection is point **9**. Since **1** and **2** are endpoints of a straight line on the object, the projections **7** and **9** are joined to give the projection of the line **7–9**. Similarly, if the projections of the four corners **1**, **2**, **3**, and **4** are found, the projections **7**, **9**, **10**, and **8** may be joined by straight lines to form the projection of the rectangular surface.

The same procedure can be applied to curved lines—for example, the top curved contour of the object. A point, **5**, on the curve is projected to the plane at **6**. The projection of an infinite number of such points, a few of which are shown at (b), on the plane of projection results in the projection of the

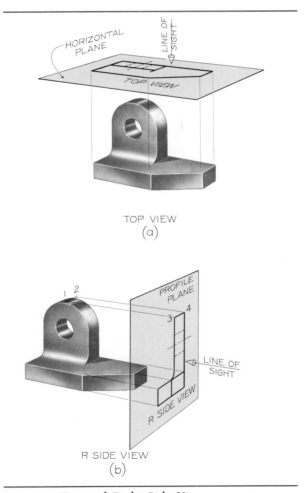

*Fig. 7.2*  Top and Right-Side Views.

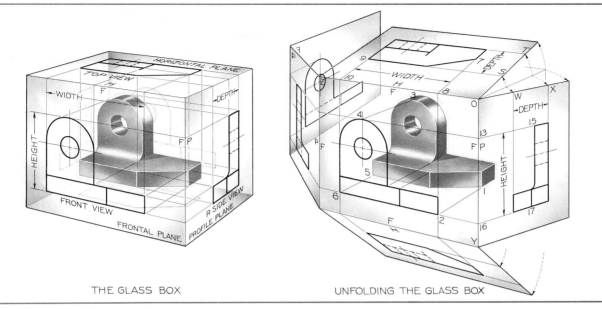

THE GLASS BOX          UNFOLDING THE GLASS BOX

*Fig. 7.3*   The Glass Box.

curve. If this procedure of projecting points is applied to all edges and contours of the object, a complete view or projection of the object results. This view is necessary in the shape description because it shows the true curvature of the top and the true shape of the hole.

A similar procedure may be used to obtain the top view, Fig. 7.2 (a). This view is necessary in the shape description because it shows the true angle of the inclined surface. In this view, the hole is invisible and its extreme contours are represented by hidden lines, as shown.

The right-side view, (b), is necessary because it shows the right-angled characteristic shape of the object and shows the true shape of the curved intersection. Note how the cylindrical contour on top of the object appears when viewed from the side. The extreme or contour element **1–2** on the object is projected to give the line **3–4** on the view. The hidden hole is also represented by projecting the extreme elements.

The plane of projection upon which the front view is projected is called the *frontal plane*, that upon which the top view is projected, the *horizontal plane*, and that upon which the side view is projected, the *profile plane*.

## 7.2 The Glass Box

If planes of projection are placed parallel to the principal faces of the object, they form a "glass box," as shown in Fig. 7.3 (a). Notice that the observer is always *on the outside looking in*, so that the object is seen through the planes of projection. Since the glass box has six sides, six views of the object can be obtained.

Note that the object has three principal dimensions: *width, height,* and *depth*. These are fixed terms used for dimensions in these directions, regardless of the shape of the object. See §6.18.

Since it is required to show the views of a solid or three-dimensional object on a flat sheet of paper, it is necessary to unfold the planes so that they will all lie in the same plane, Fig. 7.3 (b). All planes except the rear plane are hinged upon the frontal plane, the rear plane being hinged to the left-side plane, except as explained in §7.8. Each plane revolves outwardly from the original box position until it lies in the frontal plane, which remains stationary. The hinge lines of the glass box are known as *folding lines*.

The positions of these six planes, after they have been revolved, are shown in Fig. 7.4. Carefully identify each of these planes and corresponding

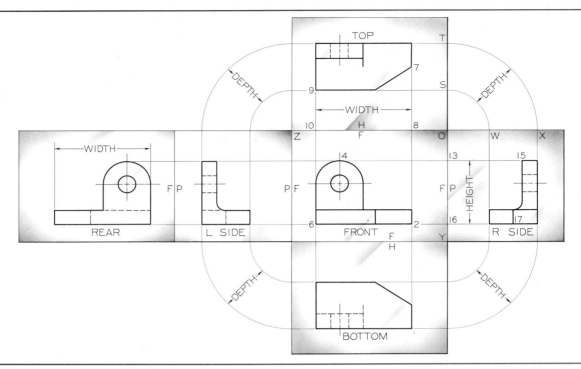

*Fig. 7.4* The Glass Box Unfolded.

views with its original position in the glass box, and repeat this mental procedure, if necessary, until the revolutions are thoroughly understood.

In Fig. 7.3 (b), observe that lines extend around the glass box from one view to another upon the planes of projection. These are the *projections of the projectors* from points on the object to the views. For example, the projector 1–2 is projected on the horizontal plane at 7–8 and on the profile plane at 16–17. When the top plane is folded up, lines 9–10 and 7–8 will become vertical and line up with 10–6 and 8–2, respectively. Thus, 9–10 and 10–6 form a single straight line 9–6, and 7–8 and 8–2 form a single straight line 7–2, as shown in Fig. 7.4. This explains why the top view is the same width as the front view and why it is placed directly above the front view. The same relation exists between the front and bottom views. Therefore, *the front, top, and bottom views all line up vertically and are the same width.*

In Fig. 7.3 (b), when the profile plane is folded out, lines 4–13 and 13–15 become a single straight line 4–15, and lines 2–16 and 16–17 become a single

straight line 2–17 as shown in Fig. 7.4. The same relation exists between the front, left-side, and rear views. Therefore, *the rear, left-side, front, and right-side views all line up horizontally and are the same height.*

In Fig. 7.3 (b), note that lines **OS** and **OW** and lines **ST** and **WX** are respectively equal. These lines of equal length are shown in the unfolded position in Fig. 7.4. Thus, it is seen that the top view must be the same distance from the folding line **OZ** as the right-side view is from the folding line **OY**. Similarly, the bottom view and the left-side view are the same distance from their respective folding lines as are the right-side view and the top view. Therefore, *the top, right-side, bottom, and left-side views are all equidistant from the respective folding lines, and are the same depth.* Note that in these four views that surround the front view, the front surfaces of the object are faced inward, or toward the front view. Observe also that the left-side and right-side views and the top and bottom views are the reverse of each other in outline shape. Similarly, the rear and front views are the reverse of each other.

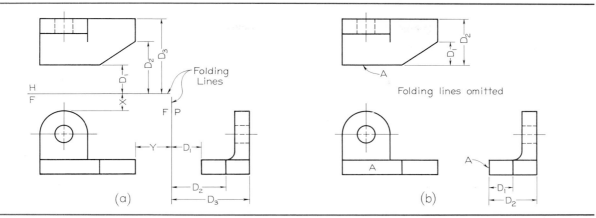

*Fig. 7.5*    Folding Lines.

## 7.3 Folding Lines

The three views of the object just discussed are shown in Fig. 7.5 (a), with folding lines between the views. These folding lines correspond to the hinge lines of the glass box, as we have seen. The H/F folding line, between the top and front views, is the intersection of the horizontal and frontal planes. The F/P folding line, between the front and side views, is the intersection of the frontal and profile planes. See Figs. 7.3 and 7.4.

The distances X and Y, from the front view to the respective folding lines, are not necessarily equal, since they depend upon the relative distances of the object from the horizontal and profile planes. However, as explained in §7.2, distances $D_1$, from the top and side views to the respective folding lines, must always be equal. Therefore, the views may be any desired distance apart, and the folding lines may be drawn anywhere between them, so long as distances $D_1$ are kept equal and the folding lines are at right angles to the projection lines between the views.

It will be seen that distances $D_2$ and $D_3$, respectively, are also equal and that the folding lines H/F and F/P are in reality reference lines for making equal *depth* measurements in the top and side views. Thus, any point in the top view is the same distance from H/F as the corresponding point in the side view is from F/P.

While it is necessary to understand the folding lines, particularly because they are useful in solving graphical problems in descriptive geometry, they are as a rule omitted in industrial drafting. The three views, with the folding lines omitted, are shown in Fig. 7.5 (b). Again, the distances between the top and front views and between the side and front views are not necessarily equal. Instead of using the folding lines as reference lines for setting off depth measurements in the top and side views, we may use the front surface A of the object as a reference line. In this way, $D_1$, $D_2$, and all other depth measurements are made to correspond in the two views in the same manner as if folding lines were used.

## 7.4 Two-View Instrumental Drawing

Let it be required to draw, full size with instruments on Layout A–2 (inside back cover), the necessary views of the Operating Arm shown in Fig. 7.6 (a). In this case, as shown by the arrows, only the front and top views are needed.

I.    Determine the spacing of the views. The width of the front and top views is approximately 152 mm (6″; 25.4 mm = 1″) and the width of the working space is approximately 266 mm (10½″). As shown at (b), subtract 152 mm from 266 mm and divide the result by 2 to get the value of space A. To set off the spaces, place the scale horizontally along the bottom of the sheet and make short vertical marks.

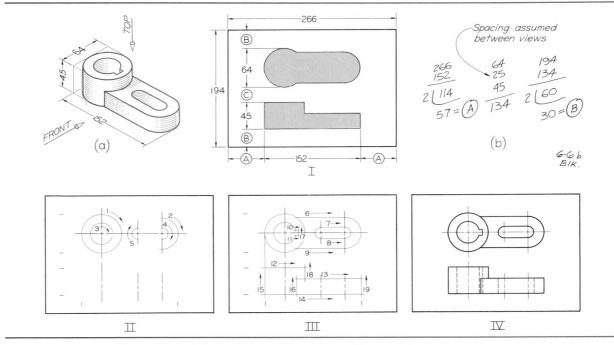

**Fig. 7.6** Two-View Instrumental Drawing (dimensions in millimeters).

The depth of the top view is approximately 64 mm (2½″) and the height of the front view is 45 mm (1¾″), while the height of the working space is 194 mm (7⅝″). Assume a space C, say 25 mm (1″), between views that will look well and that will provide sufficient space for dimensions, if any.

As shown at (b), add 64 mm, 25 mm, and 45 mm, subtract the total from 194 mm, and divide the result by 2 to get the value of space B. To set off the spaces, place the scale vertically along the left side of the sheet with the full-size scale on the left, and make short marks perpendicular to the scale. See Fig. 2.54 (III).

II. Locate center lines from spacing marks. Construct arcs and circles lightly.

III. Draw horizontal and then vertical construction lines in the order shown. Allow construction lines to cross at corners.

IV. Add hidden lines and heavy in all final lines, clean-cut and dark. The visible lines should be heavy enough to make the views stand out. The hid-

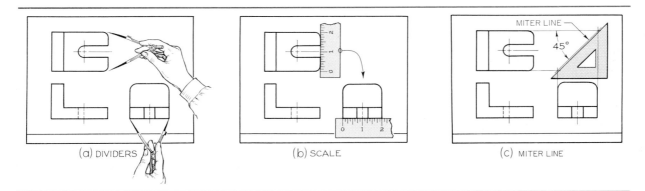

**Fig. 7.7** Transferring Depth Dimensions.

den lines and center lines should be sharp in contrast to the visible lines, but dark enough to reproduce well. See §2.46 for technique of pencil drawing. Construction lines need not be erased if drawn lightly. If you are working on tracing paper, hold the sheet up to the light to see if the density of your lines is sufficient to reproduce well.

## 7.5 Transferring Depth Dimensions
Since all depth dimensions in the top and side views must correspond point for point, accurate methods of transferring these distances, such as $D_1$ and $D_2$, Fig. 7.5 (b), must be used.

Professional drafters transfer dimensions between the top and side views either by dividers or scale, as shown in Fig. 7.7 (a) and (b). The scale method is especially convenient when the drafting machine, Fig. 2.70, is used because both vertical and horizontal scales are readily available. Beginners might find it convenient to use a 45° miter line to project dimensions between top and side views, as in Fig. 7.7 (c). Note that the right-side view may be moved to the right or left, or the top view may be moved upward or downward, by shifting the 45° line accordingly. It is not necessary to draw continuous lines between the top and side views via the miter line. Instead, make short dashes across the miter line and project from these.

The 45° miter-line method, Fig. 7.7 (c) is also convenient for transferring a large number of points, as when plotting a curve, Fig. 7.35.

## 7.6 Projecting a Third View
In Fig. 7.8 (top) is a pictorial drawing of a given object, three views of which are required. Each corner of the object is given a number, as shown. At I, the top and front views are shown, with each corner properly numbered in both views. Each number appears twice, once in the top view and once again in the front view.

If a point is *visible* in a given view, the number is placed *outside* the corner, but if the point is hidden, the numeral is placed *inside* the corner. For example, at I point 1 is visible in both views, and is therefore placed outside the corners in both views. However, point 2 is visible in the top view and the number is placed outside, while in the front view it is hidden and is placed inside.

This numbering system, in which points are identified by the same numbers in all views, is use-

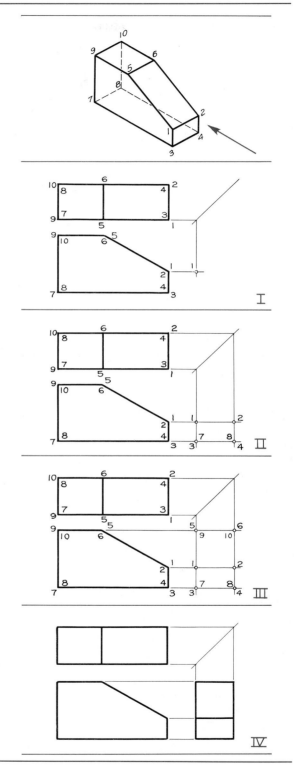

*Fig. 7.8*   Use of Numbers.

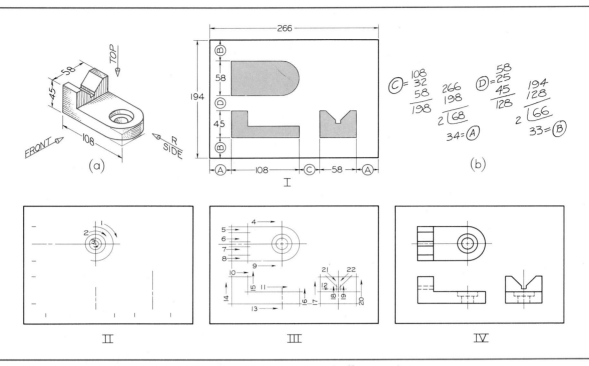

**Fig. 7.9** Three-View Instrumental Drawing (dimensions in millimeters).

ful in projecting known points in two views to unknown positions in a third view. Note that in this numbering system a given point has the same number in all views, and should not be confused with the numbering system used in Fig. 7.23 and others, in which a point has different numbers in each view.

Before starting to project the right-side view in Fig. 7.8, try to visualize the view as seen in the direction of the arrow (see pictorial drawing). Then construct the right-side view point by point, using a hard pencil and very light lines.

As shown at I, locate point 1 in the side view by projecting from point 1 in the top view and point 1 in the front view. In space II, project points 2, 3, and 4 in a similar manner to complete the vertical end surface of the object. In space III, project points 5 and 6 to complete the side view of the inclined surface 5–6–2–1. This completes the right-side view, since invisible points 9, 10, 8, and 7 are directly behind visible corners 5, 6, 4, and 3, respectively. Note that in the side view also, the invisible points are lettered *inside*, and the visible points *outside*.

As shown in space IV, the drawing is completed by heavying in the lines in the right-side view.

## 7.7 Three-View Instrumental Drawing

Let it be required to draw, full size with instruments on Layout A–2, the necessary views of the V-Block in Fig. 7.9 (a). In this case, as shown by the arrows, three views are needed.

I. Determine the spacing of the views. The width of the front view is 108 mm, and the depth of the side view is 58 mm, while the width of the working space is 266 mm. Assume a space C between views, say, 32 mm, that will look well and will allow sufficient space for dimensions, if any.

As shown at (b), add 108 mm, 32 mm, and 58 mm, subtract the total from 266 mm, and divide the result by 2 to get the value of space A. To set off these horizontal spacing measurements, place the scale along the bottom of the sheet and make short vertical marks.

The depth of the top view is 58 mm, and the height of the front view is 45 mm, while the height of the working space is 194 mm. Assume a space D between views, say, 25 mm. As shown in §7.3, space D need not be the same as space C. As shown at (b), add 58 mm, 25 mm, and 45 mm, subtract the total from 194 mm, and divide the result by 2 to get the

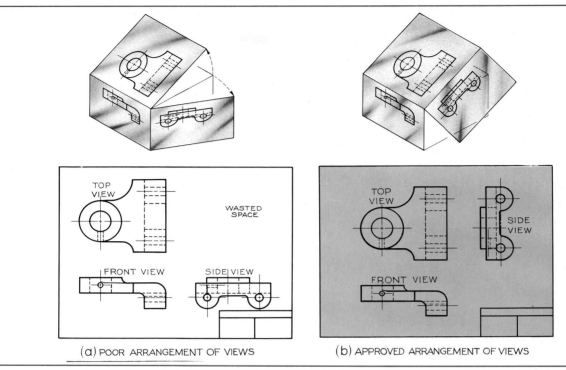

Fig. 7.10   Position of Side View.

value of space **B**. To set off these vertical spacing measurements, place the scale along the left side of the sheet with the scale used on the left, and make short marks perpendicular to the scale. Allow for dimensions, if any.

II.   Locate the center lines from the spacing marks. Construct lightly the arcs and circles.

III.   Draw horizontal, then vertical, then inclined construction lines, in the order shown. Allow construction lines to cross at the corners. Do not complete one view at a time; construct the views simultaneously.

IV.   Add hidden lines and heavy in all final lines, clean-cut and dark. A convenient method of transferring a hole diameter from the top view to the side view is to use the compass with the same setting used for drawing the hole. The visible lines should be heavy enough to make the views stand out. The hidden lines and center lines should be sharp in contrast to the visible lines, but dark enough to reproduce well. Construction lines need not be erased if they are drawn lightly. If you are working on tracing paper, hold the sheet up to the light to see if the density of your lines is sufficient to reproduce well. See §2.46.

## 7.8 Alternate Positions of Views

If three views of a wide flat object are drawn, using the conventional arrangement of views, Fig. 7.10 (a), a large wasted space is left on the paper, as shown. In such cases, the profile plane may be considered hinged to the horizontal plane instead of the frontal plane, as shown at (b). This places the side view beside the top view, which results in better spacing and in some cases makes the use of a reduced scale unnecessary.

It is also permissible in extreme cases to place the side view across horizontally from the bottom view, in which case the profile plane is considered hinged to the bottom plane of projection. Similarly, the rear view may be placed directly above the top view or under the bottom view, if necessary, in which case the rear plane is considered hinged to the horizontal or bottom plane, as the case may be, and then rotated into coincidence with the frontal plane.

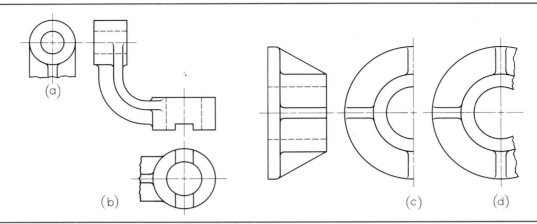

*Fig. 7.11*   Partial Views.

## 7.9 Partial Views

A view may not need to be complete but may show only what is necessary in the clear description of the object. Such a view is a partial view, Fig. 7.11. A break line, (a), may be used to limit the partial view; the contour of the part shown may limit the view, (b); or if symmetrical, a half-view may be drawn on one side of the center line, (c), or a partial view, "broken out," may be drawn as at (d). The half shown at (c) and (d) should be the near side, as shown. For half-views in connection with sections, see Fig. 9.31.

Do not place a break line so as to coincide with a visible or hidden line.

Occasionally the distinctive features of an object are on opposite sides, so that in either complete side view there would be a considerable overlapping of shapes, resulting in an unintelligible view. In such cases two side views are often the best solution, Fig. 7.12. Observe that the views are partial views, in both of which certain visible and invisible lines have been omitted for clearness.

## 7.10 Revolution Conventions

In some cases, regular multiview projections are awkward, confusing, or actually misleading. For example, Fig. 7.13 (a) shows an object that has three triangular ribs, three holes equally spaced in the base, and a keyway. The right-side view at (b) is a regular projection and is not recommended. The lower ribs appear in a foreshortened position, the holes do not appear in their true relation to the rim of the base, and the keyway is projected as a confusion of hidden lines.

The conventional method shown at (c) is preferred, not only because it is simpler to read, but also requires less drafting time. Each of the features mentioned has been revolved in the front view to lie along the vertical center line from where it is projected to the correct side view at (c).

At (d) and (e) are shown regular views of a flange with many small holes. The hidden holes at (e) are confusing and take unnecessary time to draw. The preferred representation at (f) shows the holes revolved, and the drawing is clear.

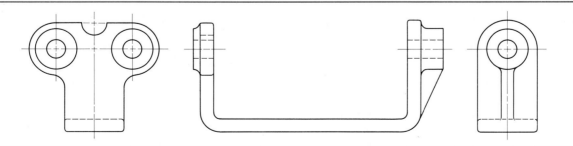

*Fig. 7.12*   Incomplete Side Views.

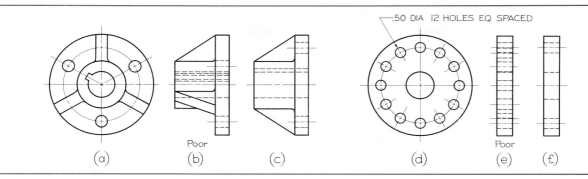

*Fig. 7.13*   Revolution Conventions.

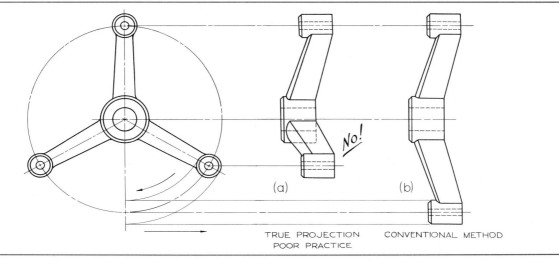

*Fig. 7.14*   Revolution Conventions.

Another example is shown in Fig. 7.14. As shown at (a), a regular projection results in a confusing foreshortening of an inclined arm. In order to preserve the appearance of symmetry about the common center, the lower arm is revolved to line up vertically in the front view so that it projects true length in the side view at (b).

Revolutions of the type discussed here are frequently used in connection with sectioning. Such sectional views are called *aligned sections*, §9.13.

## 7.11 Removed Views

A removed view, Fig. 7.15, is a complete or partial view removed to another place on the sheet so that it no longer is in direct projection with any other view. Such a view may be used to show some feature of the object more clearly, possibly to a larger scale,

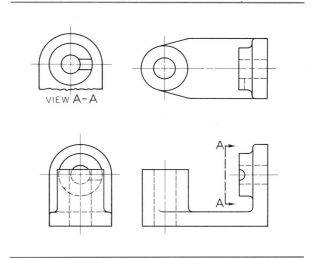

*Fig. 7.15*   Removed View.

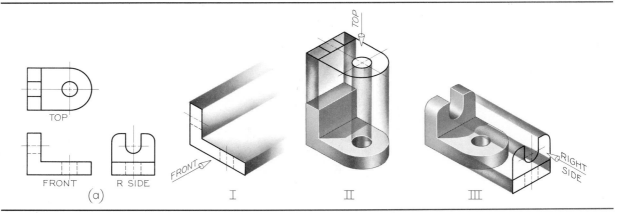

***Fig. 7.16*** Visualizing from Given Views.

### 7.12 Visualization

As stated in §1.9, the ability to *visualize* or *think in three dimensions* is one of the most important requisites of the successful engineer or scientist. In practice, this means the ability to study the views of an object and to form a mental picture of it—to *visualize* its three-dimensional shape. To the designer it means the ability to *synthesize* or form a mental picture before the object even exists and the ability to express this image in terms of views. The engineer is the master planner in the construction of new machines, structures, or processes. The ability to visualize, and to use the language of drawing as a means of communication or recording of mental images, is indispensable.

Even the experienced engineer or designer cannot look at a multiview drawing and instantly visualize the object represented (except for the simplest shapes) any more than we can grasp the ideas on a book page merely at a glance. It is necessary to *study* the drawing, to read the lines in a logical way, to piece together the little things until a clear idea of the whole emerges. How this is done is described in §§7.13–7.32.

The viewing-plane line is used to indicate the part being viewed, the arrows at the corners showing the direction of sight. See §9.5. The removed view should be labeled VIEW A–A or VIEW B–B and so on, the letters referring to those placed at the corners of the viewing-plane line.

### 7.13 Visualizing the Views

A method of reading drawings that is essentially the reverse mental process to that of obtaining the views by projection is illustrated in Fig. 7.16. The given views of an Angle Bracket are shown at (a).

I.  The front view shows that the object is L-shaped, the height and width of the object, and the thickness of the members. The meaning of the hidden and center lines is not yet clear, nor do we yet know the depth of the object.

II.  The top view tells us that the horizontal member is rounded on the end and has a round hole. Some kind of slot is indicated at the left end. The depth and width of the object are shown.

III.  The right-side view tells us that the left end of the object has rounded corners at the top and has an open-end slot in a vertical position. The height and depth of the object are shown.

Thus, each view provides certain definite information regarding the shape of the object. All views must be considered in order to visualize the object completely.

### 7.14 Models

One of the best aids to visualization is an actual model of the object. Such a model need not be made accurately to scale and may be made of any convenient material, such as modeling clay, soap, wood, styrofoam, or any material that can be easily carved or cut.

A typical example of the use of soap or clay

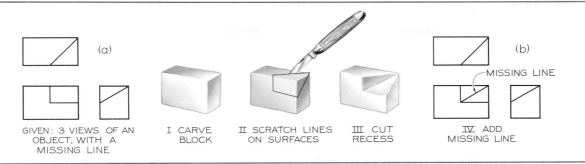

**Fig. 7.17** Use of Model to Aid Visualization.

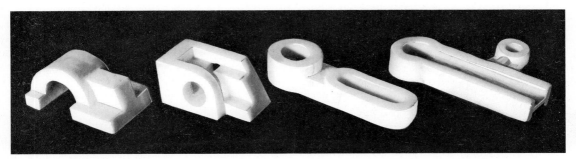

**Fig. 7.18** Soap Models.

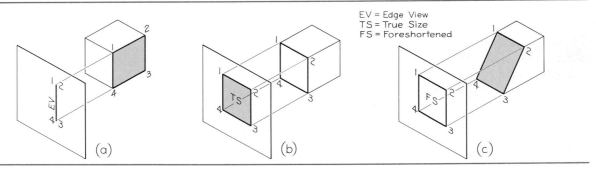

**Fig. 7.19** Projections of Surfaces.

models is shown in Fig. 7.17, in which three views of an object are given, (a), and the student is to supply a missing line. The model is carved as shown in I, II, and III, and the "missing" line, discovered in the process, is added to the drawing as shown at (b).

Some typical examples of soap models are shown in Fig. 7.18.

## 7.15 Surfaces, Edges, and Corners

In order to analyze and synthesize multiview projections, it is necessary to consider the component elements that make up most solids. A *surface* (plane)

may be bounded by straight lines or curves, or a combination of them. A surface may be *frontal, horizontal,* or *profile,* according to the plane of projection to which it is parallel. See §7.1.

If a plane surface is perpendicular to a plane of projection, it appears as a line, *edge view* (EV), Fig. 7.19 (a). If it is parallel, it appears as a surface, *true size* (TS), (b). If it is situated at an angle, it appears as a surface, *foreshortened* (FS), (c). Thus, *a plane surface always projects as a line or a surface.*

The intersection of two plane surfaces produces an *edge,* or a straight line. Such a line is common to both surfaces and forms a boundary line for each. If an edge is perpendicular to a plane of projection, it

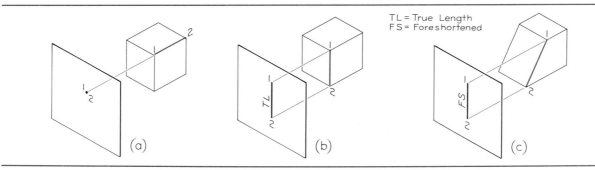

*Fig. 7.20*    Projections of Lines.

appears as a point, Fig. 7.20 (a); otherwise it appears as a line, (b) and (c). If it is parallel to the plane of projection, it shows true length, (b); if not parallel, it appears foreshortened, (c). Thus, a *straight line always projects as a straight line or as a point*. A line may be *frontal, horizontal,* or *profile,* according to the plane of projection to which it is parallel.

A *corner,* or point, is the common intersection of three or more surfaces or edges. A point always appears as a point in every view.

## 7.16 Adjacent Areas

Consider a given top view, as shown at Fig. 7.21 (a). Lines divide the view into three areas. Each of these must represent a surface *at a different level*. Surface A may be high and surfaces B and C lower, as shown at (b). Or B may be lower than C, as shown at (c). Or B may be highest, with C and A each lower, (d). Or one or more surfaces may be inclined, as at (e). Or one or more surfaces may be cylindrical, as at (f), and so on. Hence the rule: *No two adjacent areas can lie in the same plane.*

The same reasoning can apply, of course, to the adjacent areas in any given view. Since an area (surface) on a view can be interpreted in several different ways, it is necessary to observe other views also in order to determine which interpretation is correct.

## 7.17 Similar Shapes of Surfaces

If a surface is viewed from several different positions, it will in each case be seen to have a certain number of sides and to have a certain characteristic shape. An L-shaped surface, Fig. 7.22 (a), will appear as an L-shaped figure in every view in which it does not appear as a line. A T-shaped surface, (b) a U-shaped surface, (c), or a hexagonal surface, (d), will in each case have the same number of sides and the same characteristic shape in every view in which it appears as a surface.

This repetition of shapes is one of our best means for analyzing views.

## 7.18 Reading a Drawing

Let it be required to read or visualize the object shown by three views in Fig. 7.23. Since no lines are curved, the object is made up of plane surfaces.

Surface 2–3–10–9–6–5 in the top view is an L-

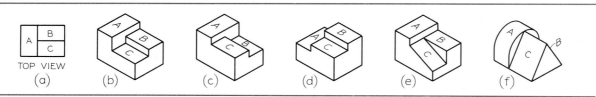

*Fig. 7.21*    Adjacent Areas.

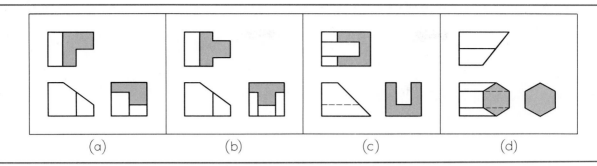

*Fig. 7.22* Similar Shapes.

shaped surface of six sides. It appears in the side view at 16–17–21–20–18–19 and is L-shaped and six-sided. No such shape appears in the front view, but we note that points 2 and 5 line up with 11 in the front view, points 6 and 9 line up with 13, and points 3 and 10 line up with 15. Evidently line 11–15 in the front view is the edge view of the L-shaped surface.

Surface 11–13–12 in the front view is triangular in shape, but no corresponding triangles appear in either the top or the side view. We note that point 12 lines up with 8 and 4 and that point 13 lines up with 6 and 9. However, surface 11–13–12 of the front view cannot be the same as surface 4–6–9–8 in the top view because the former has three sides and the latter has four. Obviously, the triangular surface appears as a line 4–6 in the top view and as a line 16–19 in the side view.

Surface 12–13–15–14 in the front view is trapezoidal in shape. But there are no trapezoids in the top and side views, so the surface evidently appears

in the top view as line 7–10 and in the side view as line 18–20.

The remaining surfaces can be identified in the same manner, whence it will be seen that the object is bounded by seven plane surfaces, two of which are rectangular, two triangular, two L-shaped, and one trapezoidal.

Note that the numbering system used in Fig. 7.23 is different from that in Fig. 7.8 in that different numbers are used for all points and there is no significance in a point being inside or outside a corner.

## 7.19 Normal Surfaces

A normal surface is *a plane surface that is parallel to a plane of projection.* It appears in true size and shape on the plane to which it is parallel, and as a vertical or a horizontal line on adjacent planes of projection.

In Fig. 7.24 four stages in machining a block of steel to produce the final Tool Block in space IV are shown. All surfaces are normal surfaces. In space I, normal surface A is parallel to the horizontal plane and appears true size in the top view at 2–3–7–6, as line 9–10 in the front view, and as line 17–18 in the side view. Normal surface B is parallel to the profile plane and appears true size in the side view at 17–18–20–19, as line 3–7 in the top view, and as line 10–13 in the front view. Normal surface C, an inverted T-shaped surface, is parallel to the frontal plane and appears true size in the front view at 9–10–13–14–16–15–11–12, as line 5–8 in the top view, and as line 17–21 in the side view.

All other surfaces of the object may be visualized in a similar manner. In the four stages of Fig. 7.24, observe carefully the changes in the views produced by the machining operations, including the introduction of new surfaces, new visible edges and hid-

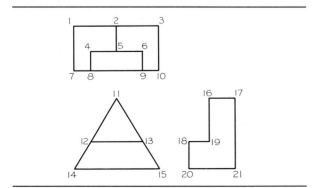

*Fig. 7.23* Reading a Drawing.

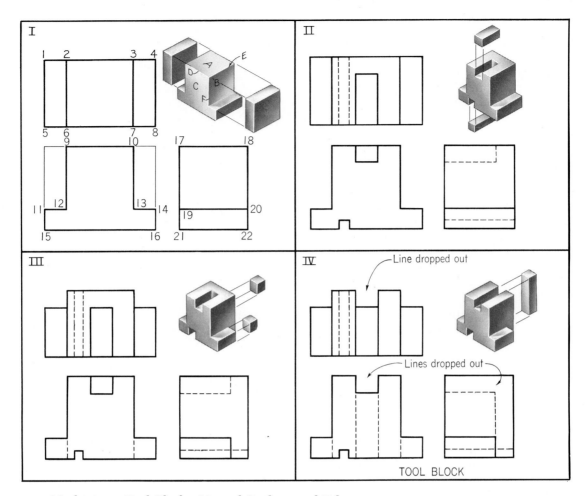

*Fig. 7.24*   Machining a Tool Block—Normal Surfaces and Edges.

den edges, and the dropping out of certain lines as the result of a new cut.

The top view in space I is cut by lines 2–6 and 3–7, which means that there are three surfaces, 1–2–6–5, 2–3–7–6, and 3–4–8–7. In the front view, surface 9–10 is seen to be the highest, and surfaces 11–12 and 13–14 are at the same lower level. In the side view both of these latter surfaces appear as one line 19–20. Surface 11–12 might appear as a hidden line in the side view, but surface 13–14 appears as a visible line 19–20, which covers up the hidden line and takes precedence over it. See §6.31.

## 7.20 Normal Edges
A normal edge is *a line that is perpendicular to a plane of projection*. It will appear as a point on the

plane of projection to which it is perpendicular and as a line in true length on adjacent planes of projection. In space I of Fig. 7.24, edge D is perpendicular to the profile plane of projection and appears as point 17 in the side view. It is parallel to the frontal and horizontal planes of projection, and is shown true length at 9–10 in the front view and 6–7 in the top view. Edges E and F are perpendicular, respectively, to the frontal and horizontal planes of projection, and their views may be similarly analyzed.

## 7.21 Inclined Surfaces
An inclined surface is *a plane surface that is perpendicular to one plane of projection but inclined to adjacent planes*. An inclined surface will project

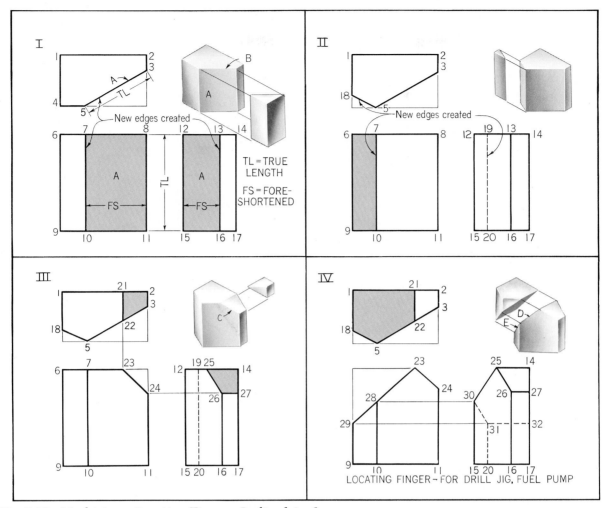

*Fig. 7.25*  Machining a Locating Finger—Inclined Surfaces.

as a straight line on the plane to which it is perpendicular, and it will appear foreshortened (FS) on planes to which it is inclined, the degree of foreshortening being proportional to the angle of inclination.

In Fig. 7.25 four stages in machining a Locating Finger are shown, producing several inclined surfaces. In space I, inclined surface A is perpendicular to the horizontal plane of projection and appears as line 5–3 in the top view. It is shown as a foreshortened surface in the front view at 7–8–11–10 and in the side view at 12–13–16–15. Note that the surface is more foreshortened in the side view than in the front view because the plane makes a greater angle

with the profile plane of projection than with the frontal plane of projection.

In space III, edge 23–24 in the front view is the edge view of an inclined surface that appears in the top view as 21–2–3–22 and in the side view as 25–14–27–26. Note that 25–14 is equal in length to 21–22 and that the surface has the same number of sides (four) in both views in which it appears as a surface.

In space IV, edge 29–23 in the front view is the edge view of an inclined surface that appears in the top view as visible surface 1–21–22–5–18 and in the side view as invisible surface 25–14–32–31–30. While the surface does not appear true size in any

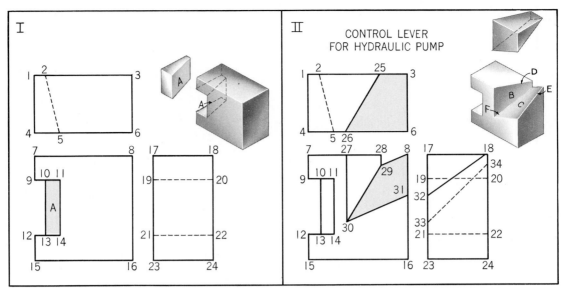

**Fig. 7.26**    Machining a Control Lever—Inclined and Oblique Surfaces.

view, it does have the same characteristic shape and the same number of sides (five) in the views in which it appears as a surface.

In order to obtain the true size of an inclined surface, it is necessary to construct an auxiliary view (Chapter 10) or to revolve the surface until it is parallel to a plane of projection (Chapter 11).

## 7.22 Inclined Edges

An inclined edge is *a line that is parallel to a plane of projection but inclined to adjacent planes.* It will appear true length on the plane to which it is parallel and foreshortened on adjacent planes, the degree of foreshortening being proportional to the angle of inclination. The true-length view of an inclined line is always inclined, while the foreshortened views are either vertical or horizontal lines.

In space I of Fig. 7.25, inclined edge B is parallel to the horizontal plane of projection, and appears true length in the top view at 5–3. It is foreshortened in the front view at 7–8 and in the side view at 12–13. Note that plane A produces two normal edges and two inclined edges.

In spaces III and IV, some of the sloping lines are not inclined lines. In space III, the edge appears in the top view at 21–22, in the front view at 23–24, and in the side view at 14–27 is an inclined line. However, the edge that appears in the top view at 22–23, in the front view at 23–24, and in the side

view at 25–26 is not an inclined line by the definition given here. Actually, it is an oblique line, §7.24.

## 7.23 Oblique Surfaces

An oblique surface is *a plane that is oblique to all planes of projection.* Since it is not perpendicular to any plane, it cannot appear as a line in any view. Since it is not parallel to any plane, it cannot appear true size in any view. Thus, an oblique surface always appears as a foreshortened surface in all three views.

In space II of Fig. 7.26, oblique surface C appears in the top view at 25–3–6–26 and in the front view at 29–8–31–30. What are its numbers in the side view? Note that any surface appearing as a line in any view cannot be an oblique surface. How many inclined surfaces are there? How many normal surfaces?

To obtain the true size of an oblique surface, it is necessary to construct a secondary auxiliary view, §§10.21 and 10.22, or to revolve the surface until it is parallel to a plane of projection, §11.11.

## 7.24 Oblique Edges

An oblique edge is *a line that is oblique to all planes of projection.* Since it is not perpendicular to any plane, it cannot appear as a point in any view. Since it is not parallel to any plane, it cannot appear true length in any view. An oblique edge appears foreshortened and in an inclined position in every view.

In space II of Fig. 7.26, oblique edge F appears in the top view at 26–25, in the front view at 30–29, and in the side view at 33–34.

## 7.25 Parallel Edges

If a series of parallel planes is intersected by another plane, the resulting lines of intersection will be parallel, Fig. 7.27 (a). At (b) the top plane of the object intersects the front and rear planes, producing the parallel edges 1–2 and 3–4. If two lines are parallel in space, their projections in any view are parallel. The example in (b) is a special case in which the two lines appear as points in one view and coincide as a single line in another and should not be regarded as an exception to the rule. Note that even in the pictorial drawings the lines are shown parallel.

Parallel inclined lines are shown in (c), and parallel oblique lines in (d).

In Fig. 7.28 it is required to draw three views of the object after a plane has been passed through the points A, B, and C. As shown at (b), only points that lie in the same plane are joined. In the front view, join points A and C, which are in the same plane, extending the line to P on the vertical front edge of the block extended. In the side view, join P to B, and in the top view, join B to A. Complete the draw-

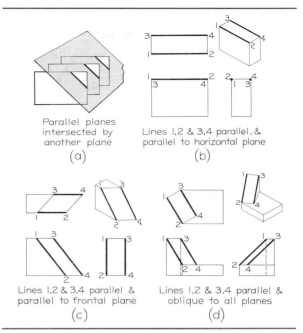

Fig. 7.27  Parallel Lines.

ing by applying the rule: *Parallel lines in space will be projected as parallel lines in any view*. The remaining lines are thus drawn parallel to lines AP, PB, and BA.

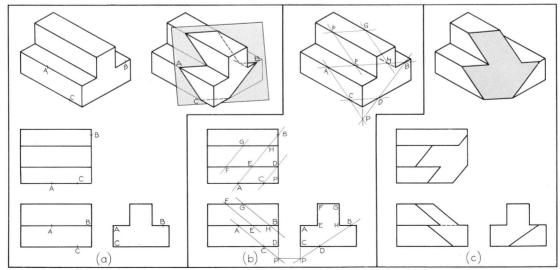

*Fig. 7.28*  Oblique Surface.

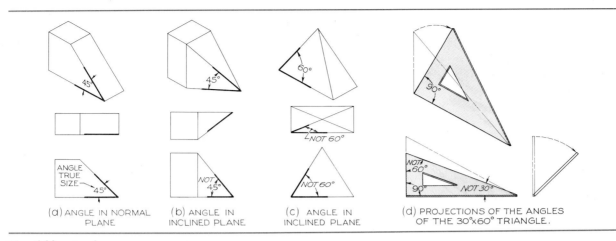

*Fig. 7.29*   Angles.

## 7.26 Angles

If an angle is in a normal plane—that is, parallel to a plane of projection—the angle will be shown true size on the plane of projection to which it is parallel, Fig. 7.29 (a).

If the angle is in an inclined plane, (b) and (c), the angle may be projected either larger or smaller than the true angle, depending upon its position. At (b) the 45° angle is shown *oversize* in the front view, and at (c) the 60° angle is shown *undersize* in both views.

A 90° angle will be projected true size, even though it is in an inclined plane, provided one leg

of the angle is a normal line, as shown at (d). In this figure, the 60° angle is projected *oversize* and the 30° angle *undersize*. Study these relations, using your own 30° × 60° triangle as a model.

## 7.27 Curved Surfaces

Rounded surfaces are common in engineering practice because they are easily formed on the lathe, the drill press, and other machines using the principle of rotation either of the "work" or of the cutting tool. The most common are the cylinder, cone, and

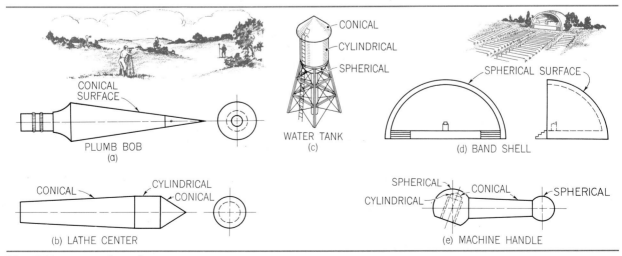

*Fig. 7.30*   Curved Surfaces.

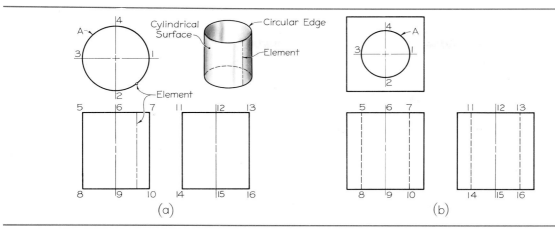

*Fig. 7.31* Cylindrical Surfaces.

sphere, a few of whose applications are shown in Fig. 7.30. For other geometric solids, see Fig. 5.7.

### 7.28 Cylindrical Surfaces

Three views of a *right-circular cylinder*, the most common type, are shown in Fig. 7.31 (a). The single cylindrical surface is intersected by two plane (normal) surfaces, forming two curved lines of intersection or *circular edges* (the bases of the cylinder). These circular edges are the only actual edges on the cylinder. Fig. 7.31 (b) shows a cylindrical hole in a right square prism.

The cylinder is represented on a drawing by its circular edges and the contour elements. An *element* is a straight line on the cylindrical surface, parallel to the axis, as shown in the pictorial view of the cylinder at (a). In this figure, at both (a) and (b), the circular edges appear in the top views as circles A, in the front views as horizontal lines 5–7 and 8–10, and in the side views as horizontal lines 11–13 and 14–16.

The contour elements 5–8 and 7–10 in the front views appear as points 3 and 1 in the top views. The contour elements 11–14 and 13–16 in the side views appear as points 2 and 4 in the top views.

In Fig. 7.32 four possible stages in machining a Cap are shown, producing several cylindrical surfaces. In space I, the removal of the two upper corners forms cylindrical surface A which appears in the top view as surface 1–2–4–3, in the front view as arc 5, and in the side view as surface 8–9–Y–X.

In space II, a large reamed hole shows in the front view as circle 16, in the top view as cylindrical surface 12–13–15–14, and in the side view as cylindrical surface 17–18–20–19.

In space III, two drilled and counterbored holes are added, producing four more cylindrical surfaces and two normal surfaces. The two normal surfaces are those at the bottoms of the counterbores.

In space IV, a cylindrical cut is added, producing two cylindrical surfaces that appear edgewise in the front view as arcs 30 and 33, in the top view as surfaces 21–22–26–25 and 23–24–28–27 and in the side view as surfaces 36–37–40–38 and 41–42–44–43.

### 7.29 Deformities of Cylinders

In shop practice, cylinders are usually machined or formed so as to introduce other surfaces, usually plane surfaces. In Fig. 7.33 (a) is shown a cut that introduces two normal surfaces. One surface appears as line 3–4 in the top view, as surface 6–7–10–9 in the front view, and as line 13–16 in the side view. The other appears as line 15–16 in the side view, as line 9–10 in the front view, and as surface 3–4, arc 2 in the top view.

All elements touching arc 2, between 3 and 4 in the top view, become shorter as a result of the cut. For example, element A, which shows as a point in the top view, now becomes CD in the front view, and 15–17 in the side view. As a result of the cut, the front half of the cylindrical surface has changed

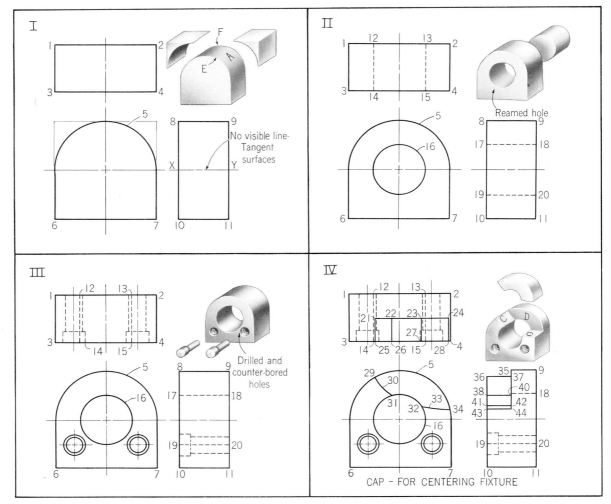

**Fig. 7.32**　Machining a Cap—Cylindrical Surfaces.

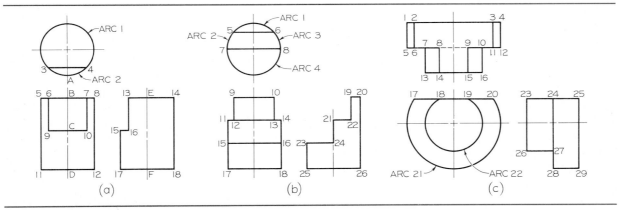

**Fig. 7.33**　Deformities of Cylinders.

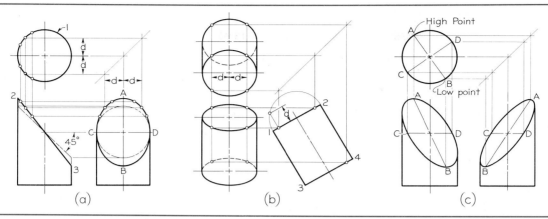

*Fig. 7.34*  Cylinders and Ellipses.

from 5–8–12–11 to 5–6–9–10–7–8–12–11 (front view). The back half remains unchanged.

At (b) two cuts introduce four normal surfaces. Note that surface 7–8 (top view) is through the center of the cylinder, producing in the side view line 21–24 and in the front view surface 11–14–16–15 equal in width to the diameter of the cylinder. Surface 15–16 (front view) is read in the top view as 7–8–arc 4. Surface 11–14 (front view) is read in the top view as 5–6–arc 3–8–7–arc 2.

At (c) two cylinders on the same axis are shown, intersected by a normal surface parallel to the axis. Surface 17–20 (front view) is 23–25 in the side view, and 2–3–11–9–15–14–8–6 in the top view. A common error is to draw a visible line in the top view between 8 and 9. However, this would produce two surfaces 2–3–11–6 and 8–9–15–14 not in the same plane. In the front view, the larger surface appears as line 17–20 and the smaller, as line 18–19. These lines coincide; hence, they are all one surface, and there can be no visible line joining 8 and 9 in the top view.

The vertical surface that appears in the front view at 17–18–arc 22–19–20–arc 21 appears as a line in the top view at 5–12, which explains the hidden line 8–9 in the top view.

## 7.30  Cylinders and Ellipses

If a cylinder is cut by an inclined plane, as in Fig. 7.34 (a), the inclined surface is bounded by an ellipse. The ellipse appears as circle 1 in the top view, as straight line 2–3 in the front view, and as ellipse ADBC in the side view. Note that circle 1 in the top view would remain a circle regardless of the angle of the cut. If the cut is 45° with horizontal, the ellipse will appear as a circle in the side view (see phantom lines) since the major and minor axes in that view would be equal. To find the true size and shape of the ellipse, an auxiliary view will be required, with the line of sight perpendicular to surface 2–3 in the front view, §10.12.

Since the major and minor axes AB and CD are known, the ellipse can be drawn by any of the methods in Figs. 5.48 to 5.50 and 5.52 (a) (true ellipses) or by the aid of an ellipse template, Fig. 5.55.

If the cylinder is tilted forward, (b), the bases or circular edges 1–2 and 3–4 (side view) become ellipses in the front and top views. Points on the ellipses can be plotted from the semicircular end view of the cylinder, as shown, distances d being equal. Since the major and minor axes for each ellipse are known, the ellipses can be drawn with the aid of an ellipse template, or by any of the true ellipse methods, or by the approximate method.

If the cylinder is cut by an oblique plane, (c), the elliptical surface appears as an ellipse in two views. In the top view, points A and B are selected, diametrically opposite, as the high and low points in the ellipse, and CD is drawn perpendicular to AB. These are the projections of the major and minor axes, respectively, of the actual ellipse in space. In the front and side views, points A and B are assumed at the desired altitudes. Since CD appears true length in the top view, it will appear horizontal in the front and side views, as shown. These axes in the front and side views are the conjugate axes of the ellipses. The ellipses may be drawn upon these axes by the method of Fig. 5.51 or 5.52 (b) or by trial with the aid of an ellipse template, Fig. 5.55.

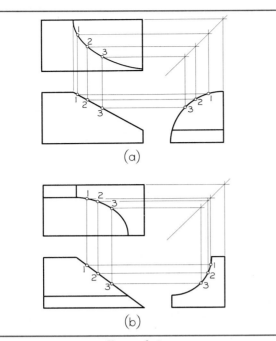

Fig. 7.35 Plotting Elliptical Curves.

In Fig. 7.35, the intersection of a plane and a quarterround molding is shown at (a), and with a cove molding at (b). In both figures, assume points 1, 2, 3, . . ., at random in the side views in which the cylindrical surfaces appear as curved lines, and project the points to the front and top views, as shown. A sufficient number of points should be used to insure smooth curves. Draw the final curves through the points with the aid of the irregular curve, §2.54.

### 7.31 Space Curves

The views of a space curve are established by the projections of points along the curve, Fig. 7.36. In this figure any points 1, 2, 3, . . ., are selected along the curve in the top view and then projected to the side view (or the reverse), and points are located in the front view by projecting downward from the top view and across from the side view. The resulting curve in the front view is drawn with the aid of the irregular curve, §2.54.

### 7.32 Intersections and Tangencies

No line should be drawn where a curved surface is tangent to a plane surface, Fig. 7.37 (a), but when a curved surface *intersects* a plane surface, a definite

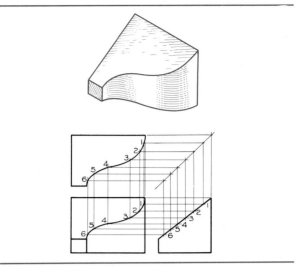

Fig. 7.36 Space Curve.

edge is formed, (b). If curved surfaces are arranged as at (c), no lines appear in the top view, as shown. If the surfaces are arranged as at (d), a vertical surface in the front view produces a line in the top view. Other typical intersections and tangencies of surfaces are shown from (e) to (h). To locate the point of tangency A in (g), refer to Fig. 5.34 (b).

The intersection of a small cylinder with a large cylinder is shown in Fig. 7.38 (a). The intersection is so small that it is not plotted, a straight line being used instead. At (b) the intersection is larger, but still not large enough to justify plotting the curve. The curve is approximated by drawing an arc whose radius r is the same as radius R of the large cylinder.

The intersection at (c) is significant enough to justify constructing the true curve. Points are selected at random in the circle in the side or top view, and these are then projected to the other two views to locate points on the curve in the front view, as shown. A sufficient number of points should be used, depending upon the size of the intersection, to insure a smooth and accurate curve. Draw the final curve with the aid of the irregular curve, §2.54.

At (d), the cylinders are the same diameter. The figure of intersection consists of two semiellipses that appear as straight lines in the front view.

If the intersecting cylinders are holes, the intersections would be similar to those for the external cylinders in Fig. 7.38. See also Fig. 9.34 (d).

In Fig. 7.39 (a), a narrow prism intersects a cylinder, but the intersection is insignificant and is ignored. At (b) the prism is larger and the intersection

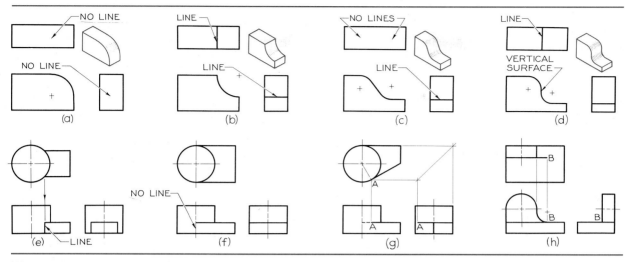

**Fig. 7.37** Intersections and Tangencies.

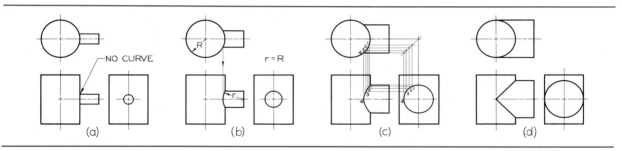

**Fig. 7.38** Intersections of Cylinders.

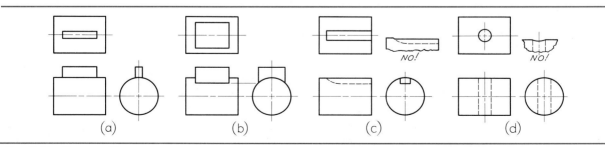

**Fig. 7.39** Intersections.

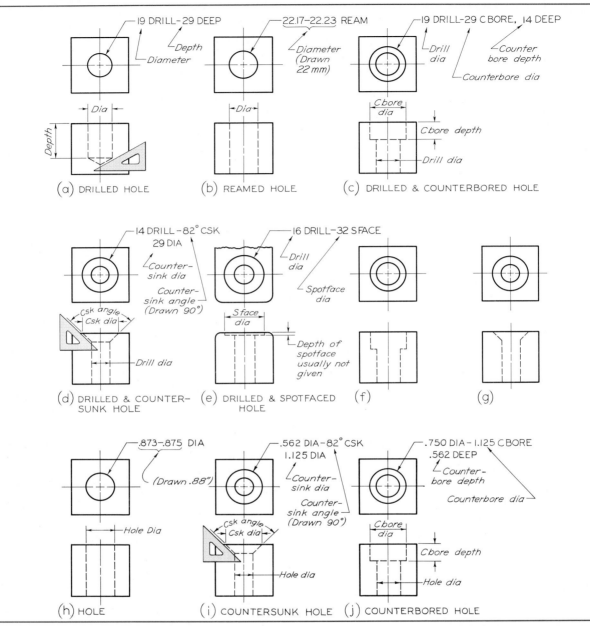

**Fig. 7.40** How to Represent Holes.    Dimensions for (a)–(e) in metric. For threaded holes, see §15.24.

is noticeable enough to warrant construction, as shown. At (c) and (d) a keyseat and a small drilled hole, respectively, are shown; in both cases the intersection is not important enough to construct.

## 7.33 How to Represent Holes

The correct methods of representing most common types of machined holes are shown in Fig. 7.40. Instructions to the machinist are given in the form of notes, and the drafter represents the holes in con-

formity with these specifications. In general, the notes tell the machine operator what to do and in which order it is to be done. Hole sizes are always specified by diameter—never by radius. For each operation specified, the diameter is given first, followed by the method such as drill, ream, and so on, as shown in (a) and (b).

The size of the hole may be specified as a diameter without the specific method such as drill, ream, and so on, since the selection of the method

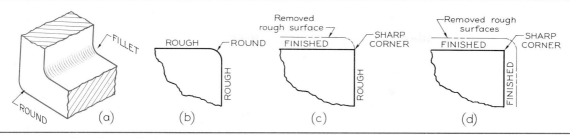

*Fig. 7.41*  Rough and Finished Surfaces.

will depend upon available production facilities. See (h) to (j).

A drilled hole is a *through* hole if it goes through a member. If the hole has a specified depth, as shown at (a), the hole is called a *blind* hole. The depth includes the cylindrical portion of the hole only. The point of the drill leaves a conical bottom in the hole, drawn approximately with the 30° × 60° triangle, as shown. For drill sizes, see Appendix 16 (Twist Drill Sizes). For abbreviations, see Appendix 4.

A through-drilled or reamed hole is drawn as shown at (b). The note tells how the hole is to be produced—in this case by reaming. Note that tolerances are ignored in actually laying out the diameter of a hole.

At (c) a hole is drilled and then the upper part is enlarged cylindrically to a specified diameter and depth.

At (d) a hole is drilled and then the upper part is enlarged conically to a specified angle and diameter. The angle is commonly 82° but is drawn 90° for simplicity.

At (e) a hole is drilled and then the upper part is enlarged cylindrically to a specified diameter. The depth usually is not specified, but is left to the shop to determine. For average cases, the depth is drawn 1.5 mm ($\frac{1}{16}''$).

For complete information about how holes are

made in the shop, see §12.16. For further information on notes, see §13.24.

## 7.34 Fillets and Rounds

A rounded interior corner is called a fillet, and a rounded exterior corner a round, Fig. 7.41 (a). Sharp corners should be avoided in designing parts to be cast or forged not only because they are difficult to produce but also because, in the case of interior corners, they are a source of weakness and failure. See §12.5 for shop processes involved.

Two intersecting rough surfaces produce a rounded corner, (b). If one of these surfaces is machined, (c), or if both surfaces are machined, (d), the corner becomes sharp. Therefore, on drawing a rounded corner means that both intersecting surfaces are rough, and a sharp corner means that one or both surfaces are machined. On working drawings, fillets and rounds are never shaded. The presence of the curved surfaces is indicated only where they appear as arcs, except as shown in Fig. 7.45.

Fillets and rounds should be drawn with the filleted corners of the triangle, a special fillets and rounds template, or a circle template.

## 7.35 Runouts

The correct method of representing fillets in connection with plane surfaces tangent to cylinders is shown in Fig. 7.42. These small curves are called

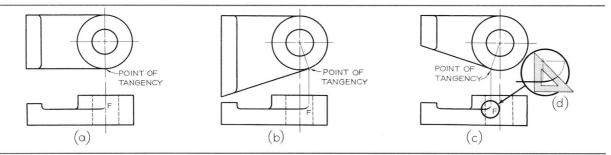

*Fig. 7.42*  Runouts.

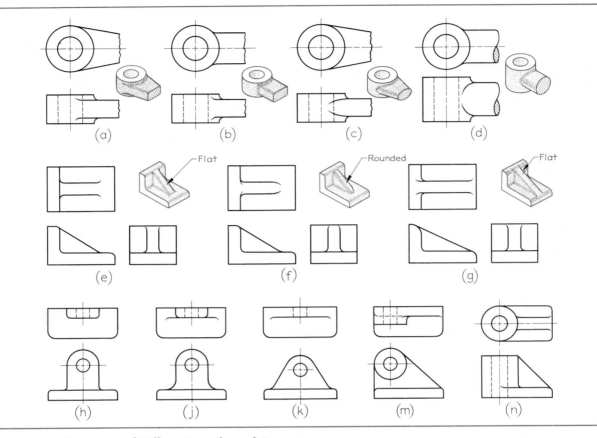

*Fig. 7.43*    Conventional Fillets, Rounds, and Runouts.

runouts. Note that the runouts F have a radius equal to that of the fillet and a curvature of about one-eighth of a circle, (d).

Typical filleted intersections are shown in Fig. 7.43. The runouts from (a) to (d) differ because of the different shapes of the horizontal intersecting members. At (e) and (f) the runouts differ because the top surface of the web at (e) is flat, with only slight rounds along the edge, while the top surface of the web at (f) is considerably rounded. When two different sizes of fillets intersect, as at (g) and (j), the direction of the runout is dictated by the larger fillet, as shown.

### 7.36 Conventional Edges
Rounded and filleted intersections eliminate sharp edges and sometimes make it difficult to present a clear shape description. In fact, true projection in some cases may be actually misleading, as in Fig. 7.44 (a), in which the side view of the railroad rail

is quite blank. A more clear representation results if lines are added for rounded and filleted edges, as shown at (b) and (c). The added lines are projected from the actual intersections of the surfaces as if the fillets and rounds were not present.

In Fig. 7.45, two top views are shown for each given front view. The upper top views are nearly devoid of lines that contribute to the shape descriptions, while the lower top views, in which lines are used to represent the rounded and filleted edges, are quite clear. Note, in the lower top views at (a) and (c), the use of small Y's where rounded or filleted edges meet a rough surface. If such an edge intersects a finished surface, no Y is shown.

### 7.37 Right-Hand and Left-Hand Parts
In industry many individual parts are located symmetrically so as to function in pairs. These opposite parts are often exactly alike, as for example, the hub caps used on the left and right sides of the auto-

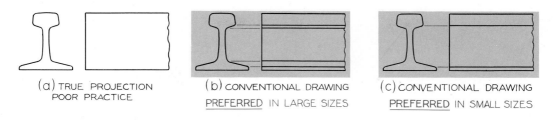

**Fig. 7.44** Conventional Repression of a Rail.

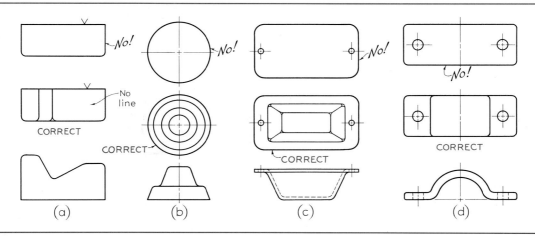

**Fig. 7.45** Conventional Edges.

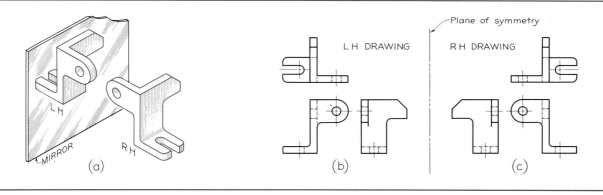

**Fig. 7.46** Right-Hand and Left-Hand Parts.

mobile. In fact, whenever possible, for economy's sake the designer will design identical parts for use on both the right and left. But opposite parts often cannot be exactly alike, such as a pair of gloves or a pair of shoes. Similarly, the right-front fender of an automobile cannot be the same shape as the left-front fender. Therefore, a left-hand part is not sim-

ply a right-hand part turned around; the two parts will be opposite and not interchangeable.

A left-hand part is referred to as an LH part, and a right-hand part as an RH part. In Fig. 7.46 (a), the image in the mirror is the "other hand" of the part shown. If the part in front of the mirror is an RH part, the image shows the LH part. No matter how

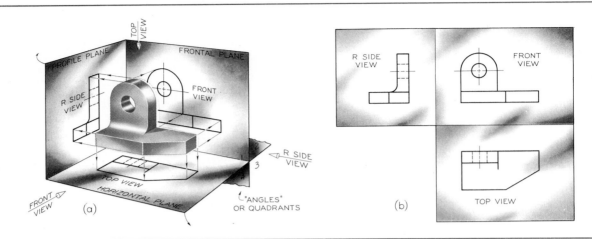

***Fig. 7.47*** First-Angle Projection.

the object is turned, the image will show the LH part. At (b) and (c) are shown LH and RH drawings of the same object, and it will be seen that the drawings are also symmetrical with respect to a reference-plane line between them.

If you hold a drawing faced against a window-pane or a light table so that the lines can be seen through the paper, you can trace the reverse image of the part on the back or on tracing paper, which will be a drawing of the opposite part.

It is customary in most cases to draw only one of two opposite parts and to label the one that is drawn with a note, such as **LH PART SHOWN, RH OPPOSITE.** If the opposite-hand shape is not clear, a separate drawing must be made for it and properly identified.

## 7.38 First-Angle Projection

If the vertical and horizontal planes of projection are considered indefinite in extent and intersecting at 90° with each other, the four dihedral angles produced are the *first, second, third,* and *fourth* angles, Fig. 7.47 (a). The profile plane intersects these two planes and may extend into all angles. If the object is placed below the horizontal plane and behind the vertical plane as in the glass box, Fig. 7.3, the object is said to be in the third angle. In this case, as we have seen, the observer is always "outside, looking in," so that for all views the lines of sight proceed from the eye *through the planes of projection and to the object.*

If the object is placed above the horizontal plane and in front of the vertical plane, the object is in the

first angle. In this case, the observer always looks *through the object and to the planes of projection.* Thus, the right-side view is still obtained by looking toward the right side of the object, the front by looking toward the front, and the top by looking down toward the top; but the views are projected from the object onto a plane in each case. When the planes are unfolded, as at (b), the right-side view falls at the left of the front view, and the top view falls below the front view, as shown. A comparison between first-angle orthographic projection and third-angle orthographic projection is shown in Fig. 7.48. The front, top, and right-side views shown in Fig. 7.47 (b) for first-angle projection are repeated in Fig. 7.48 (a). The front, top, and right-side views for third-angle projection of Fig. 7.4 are repeated at (b). Ultimately, the only difference between third-angle and first-angle projection is in the arrangement of the views. Still, confusion and possibly manufacturing errors may result when the user reading a first-angle drawing thinks it is a third-angle drawing, or vice versa. To avoid misunderstanding, international projection symbols, shown in Fig. 7.48, have been developed to distinguish between first-angle and third-angle projections on drawings. On drawings where the possibility of confusion is anticipated, these symbols may appear in or near the title box.

In the United States and Canada, and to some extent in England, third-angle projection is standard, while in most of the rest of the world, first-angle projection is used. First-angle projection was originally used all over the world, including the United States, but in this country it was abandoned around 1890.

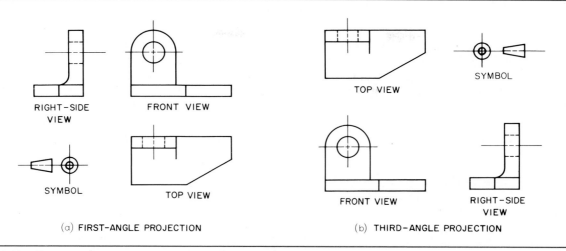

RIGHT−SIDE VIEW

FRONT VIEW

SYMBOL

TOP VIEW

(a) FIRST−ANGLE PROJECTION

TOP VIEW

SYMBOL

FRONT VIEW

RIGHT−SIDE VIEW

(b) THIRD−ANGLE PROJECTION

*Fig. 7.48*   First-Angle Projection Compared to Third-Angle Projection.

# Multiview Projection Problems

The following problems are intended primarily to afford practice in instrumental drawing, but any of them may be sketched freehand on graph paper or plain paper. Sheet layouts, Figs. 7.49 and 7.50, or inside back cover, are suggested, but the instructor may prefer a different sheet size or arrangement.

Dimensions may or may not be required by the instructor. If they are assigned, the student should study §§13.1 to 13.25. In the given problems, whether in multiview or in pictorial form, it is often not possible to give dimensions in the preferred places or, occasionally, in the standard manner. The student is expected to move dimensions to the preferred locations and otherwise to conform to the dimensioning practices recommended in Chapter 13.

For the problems in Figs. 7.54 to 7.89, it is suggested that the student make a thumbnail sketch of the necessary views, in each case, and obtain his instructor's approval before starting the mechanical drawing.

For additional problems, see Fig. 10.29. Draw top views instead of auxiliary views.

Since many of the problems in this chapter are of a general nature, they can also be solved on most computer graphics systems. If a system is available, the instructor may choose to assign specific problems to be completed by this method.

Problems in convenient form for solution may be found in *Principles of Engineering Graphics Problems* by Spencer, Hill, Loving, and Dygdon, a workbook designed to accompany this text that is also published by Macmillan Publishing Company.

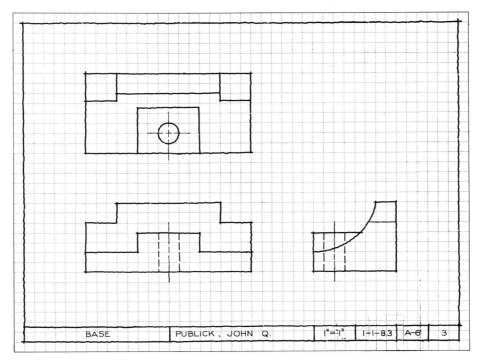

*Fig. 7.49* Freehand Sketch (Layout A–2 or A4–2 adjusted).

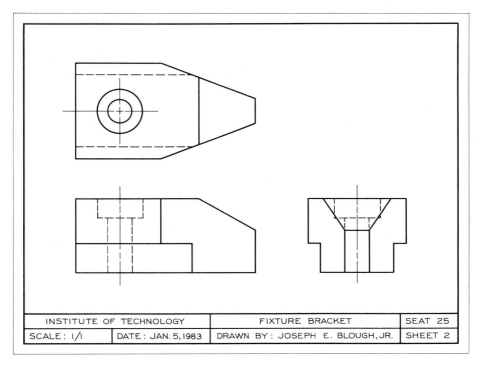

*Fig. 7.50* Mechanical Drawing (Layout A–3 or A4–3 adjusted).

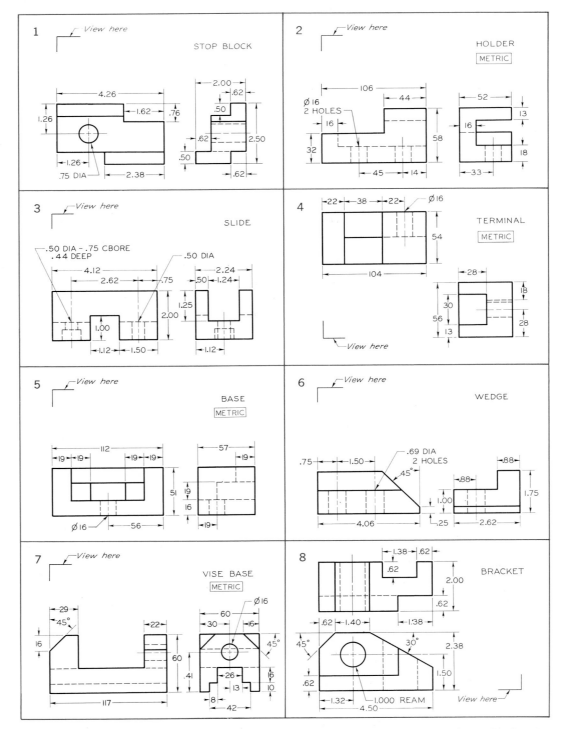

**Fig. 7.51** Missing-View Problems. Using Layout A–2 or 3 or Layout A4–2 or 3 (adjusted), sketch or draw with instruments the given views, and add the missing view, as shown in Figs. 7.49 and 7.50. If dimensions are required, study §§13.1 to 13.25. Use metric or decimal-inch dimensions as assigned by the instructor. Move dimensions to better locations where possible. In Probs. 1 to 5, all surfaces are normal surfaces.

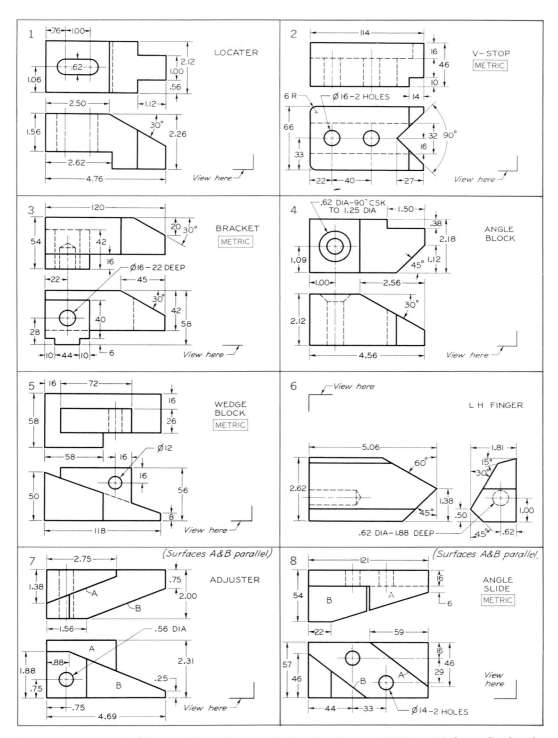

**Fig. 7.52  Missing-View Problems.**   Using Layout A–2 or 3 or Layout A4–2 or 3 (adjusted), sketch or draw with instruments the given views, and add the missing view, as shown in Figs. 7.49 and 7.50. If dimensions are required, study §§13.1 to 13.25. Use metric or decimal-inch dimensions as assigned by the instructor. Move dimensions to better locations where possible.

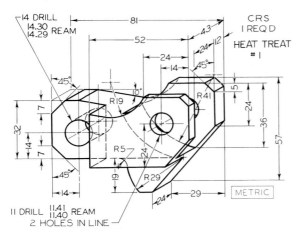

**Fig. 7.60  Roller Lever.** Using Layout A–3 or A4–3 (adjusted), draw or sketch necessary views.*

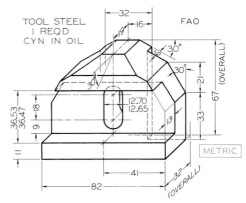

**Fig. 7.61  Support.** Using Layout A–3 or A4–3 (adjusted), draw or sketch necessary views.*

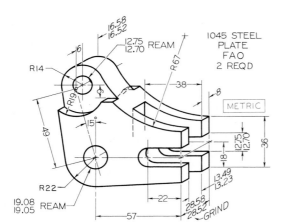

**Fig. 7.62  Toggle Lever.** Using Layout A–3 or A4–3 (adjusted), draw or sketch necessary views.*

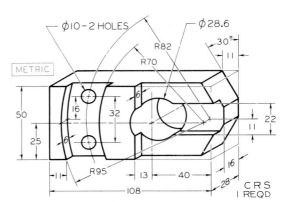

**Fig. 7.63  Index Slide.** Using Layout A–3 or A4–3 (adjusted), draw or sketch necessary views.*

*If dimensions are required, study §§13.1 to 13.25. Use metric or decimal-inch dimensions as assigned by the instructor.

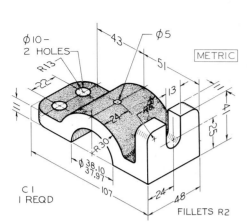

**Fig. 7.64** Frame Guide. Using Layout A–3 or A4–3 (adjusted), draw or sketch necessary views.*

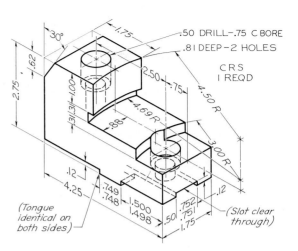

**Fig. 7.65** Chuck Jaw. Using Layout A–3 or A4–3 (adjusted), draw or sketch necessary views.*

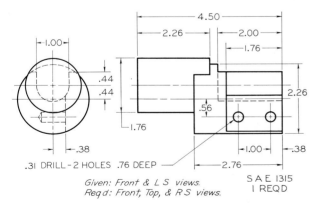

Given: Front & L S views.
Req'd: Front, Top, & R S views.

**Fig. 7.66** Tool Holder. Using Layout A–3 or A4–3 (adjusted), draw or sketch necessary views.*

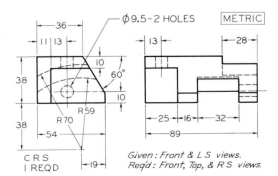

Given: Front & L S views.
Req'd: Front, Top, & R S views.

**Fig. 7.67** Shifter Block. Using Layout A–3 or A4–3 (adjusted), draw or sketch necessary views.*

*If dimensions are required, study §§13.1 to 13.25. Use metric or decimal-inch dimensions as assigned by the instructor.

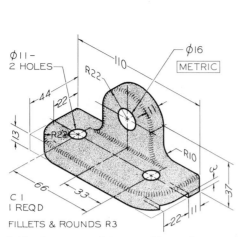

**Fig. 7.68** Cross-Feed Stop.   Using Layout A–3 or A4–3 (adjusted), draw or sketch necessary views.*

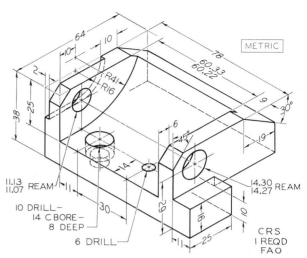

**Fig. 7.69** Hinge Block.   Using Layout A–3 or A4–3 (adjusted), draw or sketch necessary views.*

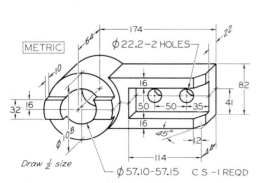

**Fig. 7.70.** Lever Hub.   Using Layout A–3 or A4–3 (adjusted), draw or sketch necessary views.*

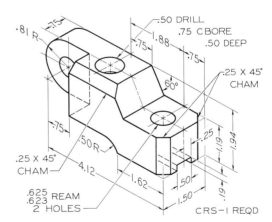

**Fig. 7.71** Vibrator Arm.   Using Layout A–3 or A4–3 (adjusted), draw or sketch necessary views.*

*If dimensions are required, study §§13.1 to 13.25. Use metric or decimal-inch dimensions as assigned by the instructor.

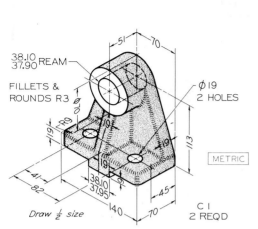

**Fig. 7.72** Counter Bearing Bracket. Using Layout A–3 or A4–3 (adjusted), draw or sketch necessary views.*

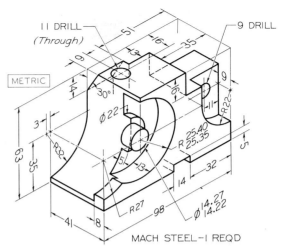

**Fig. 7.73** Tool Holder. Using Layout A–3 or A4–3 (adjusted), draw or sketch necessary views.*

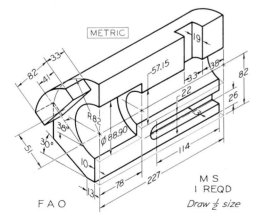

**Fig. 7.74** Control Block. Using Layout A–3 or A4–3 (adjusted), draw or sketch necessary views.*

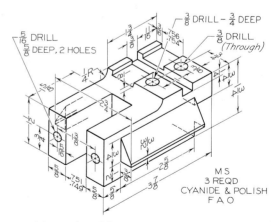

**Fig. 7.75** Tool Holder. Using Layout A–3 or A4–3 (adjusted), draw or sketch necessary views.*

*If dimensions are required, study §§13.1 to 13.25. Use metric or decimal-inch dimensions as assigned by the instructor.

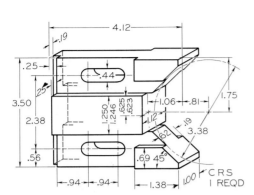

**Fig. 7.76** Locating V-Block. Using Layout A–3 or A4–3 (adjusted), draw or sketch necessary views.*

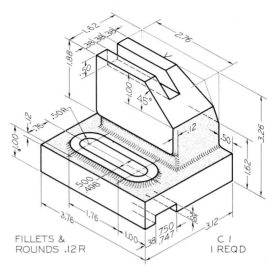

**Fig. 7.77** Door Bearing. Using Layout B–3 or A4–3 (adjusted), draw or sketch necessary views.*

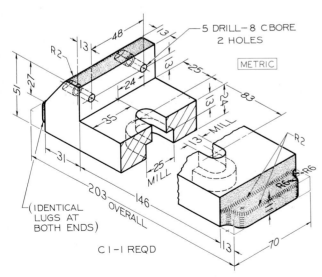

**Fig. 7.78** Vise Base. Using Layout B–3 or A3–3 (adjusted), draw or sketch necessary views.*

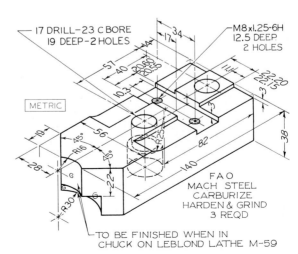

**Fig. 7.79** Chuck Jaw. Using Layout B–3 or A3–3 (adjusted), draw or sketch necessary views.*

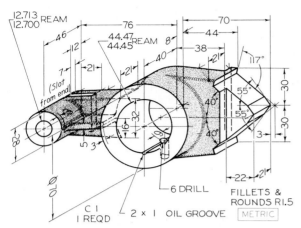

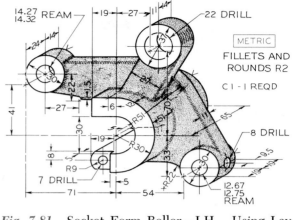

**Fig. 7.80** Motor Switch Lever. Using Layout B–3 or A3–3 (adjusted), draw or sketch necessary views.*

**Fig. 7.81** Socket Form Roller—LH. Using Layout B–4 or A3–3 (adjusted), draw or sketch necessary views.*

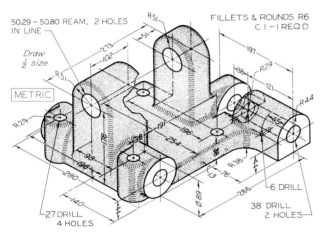

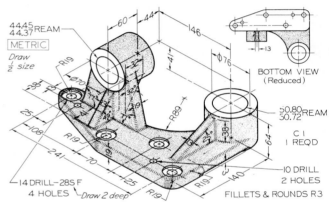

**Fig. 7.82** Automatic Stop Base. Using Layout C–3 or A2–3, draw or sketch necessary views.*

**Fig. 7.83** Lever Bracket. Using Layout C–3 or A2–3, draw or sketch necessary views.*

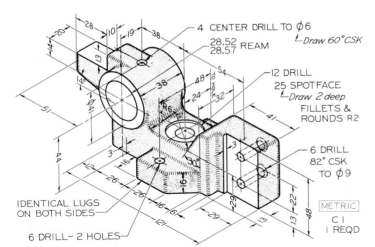

**Fig. 7.84** Gripper Rod Center. Using Layout B–3 or A3–3, draw or sketch necessary views.*

*If dimensions are required, study §§13.1 to 13.25. Use metric or decimal-inch dimensions as assigned by the instructor.

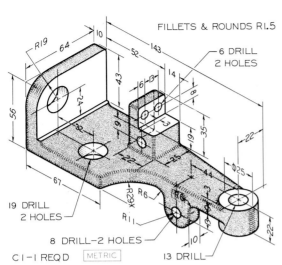

**Fig. 7.85** Mounting Bracket. Using Layout B–3 or A3–3, draw or sketch necessary views.*

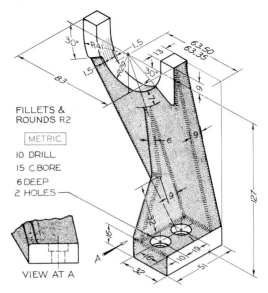

**Fig. 7.86** LH Shifter Fork. Using Layout B–3 or A3–3, draw or sketch necessary views.*

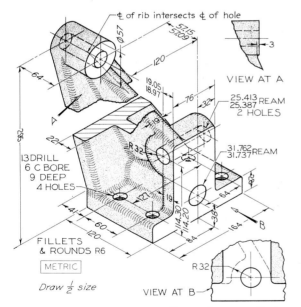

**Fig. 7.87** Ejector Base. Using Layout C–4 or A2–4, draw or sketch necessary views.*

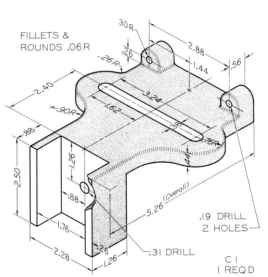

**Fig. 7.88** Tension Bracket. Using Layout C–4 or A2–4, draw or sketch necessary views.*

*If dimensions are required, study §§13.1 to 13.25. Use metric or decimal-inch dimensions as assigned by the instructor.

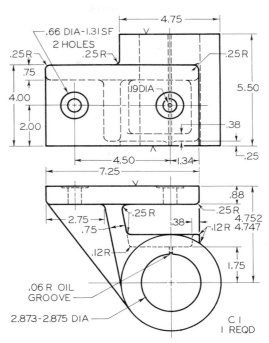

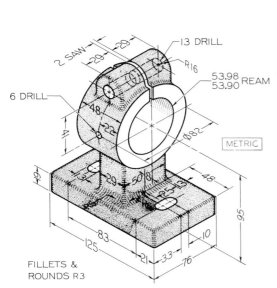

**Fig. 7.89** Feed Guide. Using Layout C–4 or A2–4, draw or sketch necessary views.*

**Fig. 7.90** Feed Shaft Bracket.
Given: Front and top views.
Required: Front, top, and right-side views, half size (Layout B–3 or A3–3).†

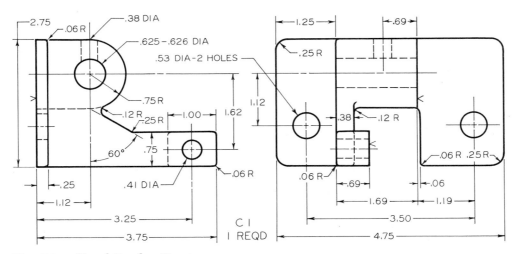

**Fig. 7.91** Knurl Bracket Bearing.
Given: Front and left-side views.
Required: Take front as top view on new drawing, and add front and right-side views (Layout B–3 or A3–3).†

*If dimensions are required, study §§13.1 to 13.25. Use metric or decimal-inch dimensions as assigned by the instructor.

†Draw or sketch necessary views. If dimensions are required, study §§13.1 to 13.25. Use metric or decimal-inch dimensions as assigned by the instructor.

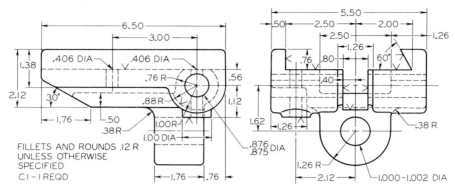

**Fig. 7.92** Sliding Nut for Mortiser.
Given: Top and right-side views.
Required: Front, top, and left-side views, full size (Layout C–4 or A2–4).*

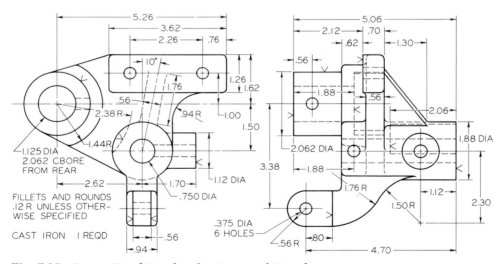

**Fig. 7.93** Power Feed Bracket for Universal Grinder.
Given: Front and right-side views.
Required: Front, top, and left-side views, full size (Layout C–4 or A2–4).

*Draw or sketch necessary views. If dimensions are required, study §§13.1 to 13.25. Use metric or decimal-inch dimensions as assigned by the instructor.

# Computer Graphics: CAD System Operation

by Gary R. Bertoline*

The operation of a CAD system takes time to learn because of the numbers of commands and options available to the user. This chapter is a generic introduction to the use of CAD for creating engineering drawings. Every CAD system is a little different in the geometry that is created and edited. However, there are many similarities among systems. For example, to erase a line from a drawing using CAD, the user would select the proper command and pick the line on the screen with the input device. The differences among systems are in the exact steps for executing a command, but the general procedures are similar.

This chapter will describe many of the most common CAD procedures for creating and editing a drawing. Each command is described by its function and how it is executed. The generic description serves as an introduction to CAD system operation.

*Assistant Professor Engineering Graphics, The Ohio State University, Columbus, OH.

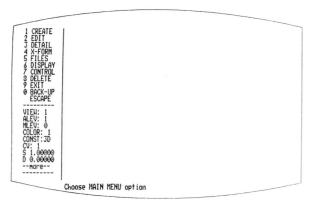

*Fig. 8.1*  Screen Menu.

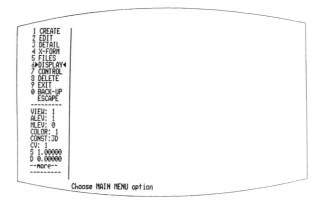

*Fig. 8.2*  Highlighted Menu Option Ready for Selection by the Mouse Pick Button.

## 8.1 Basic System Operation

The process of loading the software into a memory device, such as the hard disk drive, is called *installing*. After the software is installed it must be configured. *Configuring* is the process used to match the software to the hardware devices used on a particular workstation. The display device, cursor control device, output devices, and other variables are specified during configuration.

To begin using a CAD system, the hardware devices must be turned on and the software loaded from the storage device into the computer. This is referred to as the "start-up" sequence and varies widely among CAD software programs. After the software is loaded, the user is given a choice of starting a new drawing or loading an old one. The user interacts with the computer and software through the input devices and by responding to software messages or prompts. A *prompt* is a message that appears on the screen to assist the user of the software. For example, when drawing a circle, the computer might display the following message on screen: `Input DIAMETER then press RETURN.`

Most CAD systems use menus to display the options necessary to create engineering drawings. Typically a *menu*, Fig. 8.1, is a list of software commands displayed on screen and/or tablet to assist the user in making drawings. Selection of some menu items only reveals other menu choices to further define your selection. For example, if you select menu command CIRCLE, another menu might be displayed on screen that offers six different methods of drawing circles, such as center point and radius, center point and diameter, two points, or three

points. Some menu commands will result in some immediate action by the computer.

Menu items are selected by moving the screen cursor onto the menu item on the screen. This usually results in the menu item being highlighted on screen. After the menu item is highlighted, Fig. 8.2, the pick button on the input device is pressed. Tablet menu items are selected by moving the puck or stylus over the menu item and pressing the pick button, Fig. 8.3. Some systems allow menu selection from the keyboard by typing in the command name and pressing RETURN.

A menu usually is displayed vertically along the left or right side of the display screen. Some CAD systems use pulldown menus that disappear after the menu item is selected.

The *cursor* is a large plus sign (+) that is moved around the screen by movement of the cursor con-

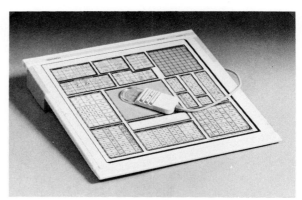

*Fig. 8.3*  Selecting Menu Items from a Tablet. *Courtesy of Hewlett Packard.*

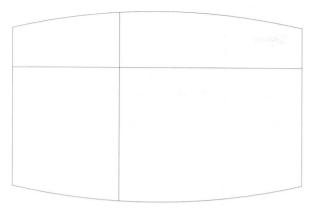

**Fig. 8.4** Screen Cursor Movement.

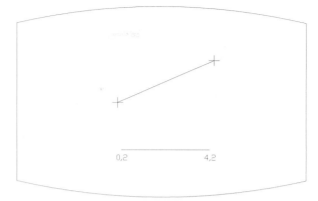

**Fig. 8.6** Lines Drawn by Cursor Picking and X–Y Coordinate Points.

trol device, the mouse, stylus, or puck. Movement of the cursor control device causes a corresponding movement of the screen cursor, Fig. 8.4.

## 8.2 Screen Display

Figure 8.5 shows a typical CAD screen display featuring the important parts. The menu area on this software is arranged vertically on the left side of the screen. The cursor tracking area displays the current X–Y coordinate position of the cursor on screen. The prompt line is located on the bottom of the screen.

## 8.3 Drawing Geometry

Geometry created with CAD is usually referred to as entities. *Entities* are the basic geometric elements

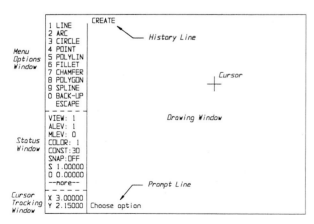

**Fig. 8.5** Typical Screen Display. *Courtesy of CAD-KEY.*

or groups of elements used to create drawings. For example, lines, circles, arcs, points, rectangles, polygons, dimensions, cross hatching, and splines are considered entities and are the basic building blocks for a drawing. Most entities are drawn by selecting the appropriate menu item and then picking points on the screen with the cursor or entering coordinate values from the keyboard.

### Drawing a Line

Generally the user creates a line by picking the two locations of the end points of the line with the cursor or by entering the X–Y coordinate points, Fig. 8.6. Often five or more different types of lines are available. Different line types are used on a CAD drawing by changing the default setting. A *default* is the original setting of the CAD software. For example, the default setting for the line type usually is an object line, the default setting for the text style might be Gothic. Default settings are changed through the software. Changing the line type to hidden will result in dashed lines instead of solid lines. Hidden lines are drawn until the setting is changed back to solid or to some other line style.

Construction lines can be used with a CAD system to lay out a drawing just as they are used with traditional tools. There are a number of different ways that construction lines can be used with CAD. One method is to draw construction lines in a different line style, such as dot or phantom, or in a different color. The construction lines would then be erased, turned off, or placed on a different layer after the drawing is complete and before plotting. These techniques are explained later in this chapter.

### Drawing Circles

Most CAD systems provide a variety of methods to draw circles, for example, locating the center point and then inputting the radius or diameter, picking two points, or picking three points. Points are located by picking them with the cursor or by entering coordinate values. The point and diameter option requires that the location of the center point be picked by the cursor or input from the keyboard. A prompt requests the user to input the diameter. The user responds by typing the numeric value of the diameter followed by RETURN or ENTER. The circle is drawn on screen as specified, Fig. 8.7

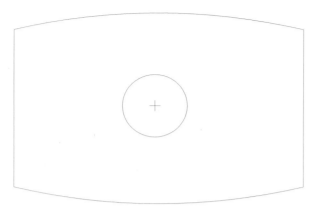

**Fig. 8.7**    Circle Drawn by Picking the Center Point and Entering the Diameter.

### Drawing Arcs

Arcs are drawn like circles except the start and end points must be specified. The radius for the arc is entered by keyboard. The end points are specified from the keyboard or through cursor picks. For example, the arc shown in Fig. 8.8 was drawn by entering the radius from the keyboard and picking the two end points with the cursor.

### Points

Points are typically used for construction purposes with CAD. A point is located and displayed on a screen in a number of different forms, such as a small plus (+) or a small x. The point locations are then used for construction points or handles, Fig. 8.9.

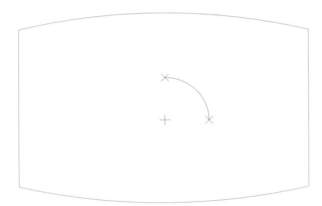

**Fig. 8.8**    Drawing an Arc.

### Splines

*Splines* are a series of smooth curves drawn through a string of points. Splines are used when a smooth curve must be fitted between a series of points, such as the layout of a truncated cone. Spline is a drawing command that creates an irregular curve. Splines are drawn by locating a series of construction points. The spline is added to the drawing by picking the construction points. The CAD software will automatically draw a smooth curve from the first construction point to the last that contains all the points between, Fig. 8.10.

### Polygons

The polygon command is used to create any type of regular polygon by specifying the number of sides, the distance across the corners or flats of the polygon, and its location on the drawing. The hexagon

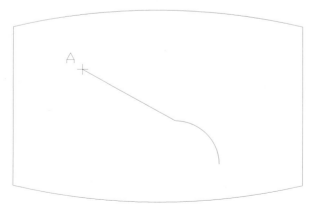

**Fig. 8.9**    Drawing a Line from a Point.

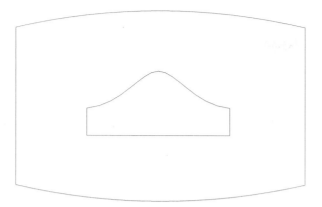

*Fig. 8.10* Spline Used for the Layout of a Truncated Cylinder.

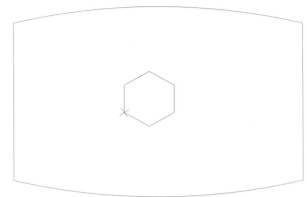

*Fig. 8.11* Polygon Command Used to Create a Hexagon.

shown in Fig. 8.11 is created by entering six for the number of sides and picking a point on the corner for its size.

## 8.4 The Location Menu

Ends of lines, arcs, and other entities are located on the points for accurate placement of geometry. Intersections of construction lines are also used for accurate placement of entities.

The use of construction entities, such as points and construction lines, and existing entities for accurate placement of geometry involves a special set of commands. These commands snap new entities to the location of current geometry. A line that was drawn between a point labeled *A* and the end of an arc is shown in Fig. 8.9. The set of commands used to place an entity by specifying its relationship to existing geometry is called a *location menu* or *snap menu*. This menu has commands such as end point, center, midpoint, perpendicular, parallel, grid, and intersection.

## 8.5 Files

Drawings created on a CAD system are saved on floppy or hard disk for storage and retrieval for editing. While creating a drawing it is good practice to save it occasionally to prevent loss due to human or computer error. After the file is saved it can be retrieved for further work or editing, copied for backup, or erased.

Most CAD systems allow the user to combine or merge drawing files. Various methods are used, but basically all systems are similar. One method of merging files is to have one drawing file on screen, then retrieve the other one and place it with the current drawing on screen. Another method is to create a special file called a pattern, block, or symbol file. A *symbol* file is any defined group of entities that can be stored, retrieved, and placed onto an existing drawing. Symbol files are similar to templates that are used with traditional tools. In fact, some CAD systems call these special files templates.

Most systems allow the symbol to be scaled and rotated before being placed on the drawing. Common applications of pattern or symbol files are to add a title block and border or to add commonly used symbols to a drawing, such as electronic schematic components or fasteners, Fig. 8.4. Virtually any drawing created on a CAD system can become a symbol file. The frequent use of symbol files is a great time saver.

## 8.6 Display Options

One of the most powerful features of CAD systems is the ability to make full scale drawings of anything from miniaturized electronic circuit boards to maps of the world. This is done by use of the display commands.

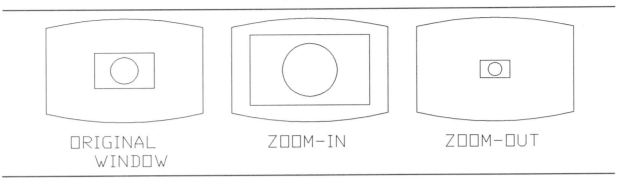

Fig. 8.12   Examples of Zoom.

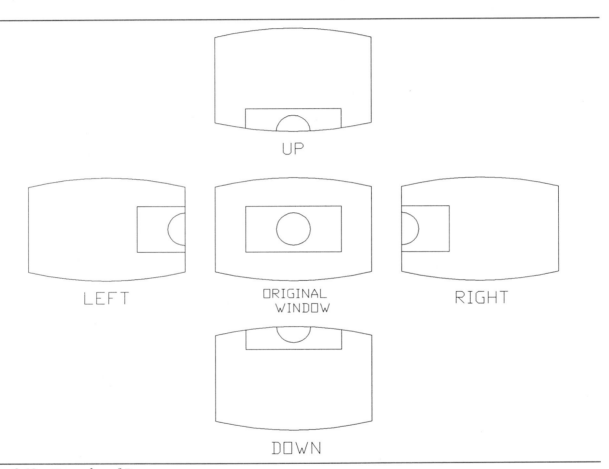

Fig. 8.13   Examples of Pan.

### Zoom Window

To use the display commands it is helpful to think of the screen display as your window or camera view of the drawing. By moving closer to the window or putting a zoom lens on the camera, you can see the drawing close up. This display command is called *zoom* or *window*, Fig. 8.12, and it allows the user to view any part of the drawing at any size. For example, designing a small circuit board would be impossible without the zoom feature, which allows the user to get in close and see the circuit much larger on the screen.

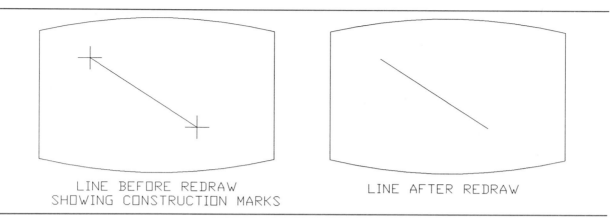

*Fig. 8.14*   Redraw Used to Remove Construction Markers.

## Pan

Another display command, called *pan*, Fig. 8.13, allows the user to move the window left or right or up and down. For example, if you were working on a map of the world and wanted to focus on North America, pan and zoom could be used to display only that part of the map.

## Redraw

The *redraw* command Fig. 8.14, is used to refresh the screen image to remove temporary markers. Most CAD systems use various types of markers, in the form of ×, ○, triangles, squares, to indicate start and end points of entities and other construction points. These are temporary markers that are removed by using the redraw command. It may be helpful to think of this as removing the erasure crumbs from a drawing that is produced with hand tools.

## Grid

Snap and grid are two common display options used as an aid in placement of entities on a drawing. A *grid*, Fig. 8.15, is a series of small marks displayed on screen similar to grid paper. Grids are of equal or unequal horizontal and vertical spacing. Some CAD programs can display isometric grids or grids rotated at any user-specified angle. Grids are considered as construction points and will not plot.

## Snap

The *snap* command is used to make the screen cursor snap or jump in specified increments. For example, a snap spacing of 0.50″ causes the screen cursor to jump at 0.50″ increments horizontally and

*Fig. 8.15*   Screen Grid.

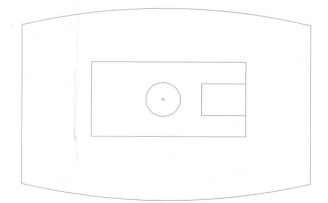

*Fig. 8.16*   Drawing Produced Using Grid and Snap.

vertically. When snap is on, it is impossible to pick a point between the snap setting with the cursor. Snap can be set with or independent of the grid if desired. A drawing created using a grid and snap setting of 0.50″ is shown in Fig. 8.16.

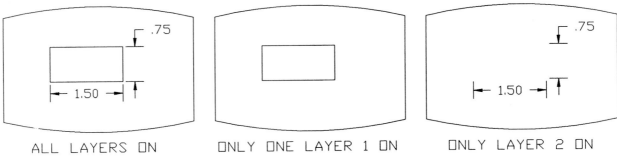

*Fig. 8.17*   Entities Separated by Layer.

### Layers

CAD systems allow the user to separate different entities or parts of a drawing onto different layers or levels. Use of the *layers* display option separates parts of a drawing in the same way as use of overlays or separate drawing sheets prepared with hand tools. One, some, or all different layers can be displayed, as desired, Fig. 8.17, and any part or parts of the drawing can be plotted. For example, an architectural floor plan might have the layout of the building on one layer, the dimensions on another, and the plumbing details on a third.

### 8.7 Erasing Entities

CAD systems provide many different methods of erasing entities from a drawing. Entities are erased singly, by group (all dimensions or hidden lines), by layer or by window. A single entity is erased by picking it with the cursor and removing it from the screen and the data base. The window option allows the user to define a rectangular window. All entities fully within the window are deleted, Fig. 8.18. Most CAD systems provide a method of retrieving erased

entities. The *recall* command is used to return erased entities to the screen and the drawing data base. Usually there is some limit to the number of erased entities that a CAD system stores for retrieval with the recall command.

### 8.8 Adding Text to a Drawing

One of the great features of CAD is the ease of adding text to a drawing. Most CAD systems provide the user with many text options, such as different text styles, user-specified height, aspect ratio, slant, and rotation. Text is added to a drawing by specifying the variables, picking the text location point on screen with the cursor, then keying the text string into the computer with the keyboard.

### 8.9 Adding Dimensions

Automatic dimensioning commands are used for linear, circles, arc entities, and angular measurements. To place a horizontal, vertical, or aligned dimension, the end points of the line are picked with the cursor, then the position of the dimension text

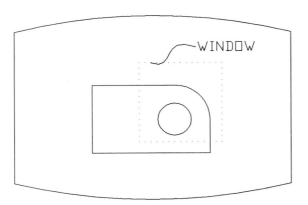
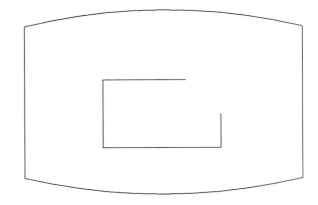

*Fig. 8.18*   Window Delete.

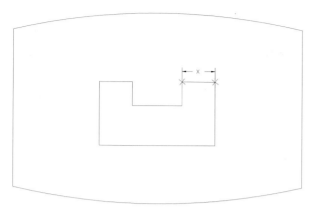

*Fig. 8.19*  Adding a Horizontal Dimension.

is specified. All the components of a dimension are automatically drawn on screen, Fig. 8.19. The dimension line, extension lines, arrows, text, extension line space from the object, and other variables all are controlled by the user.

Arcs are automatically dimensioned by selecting the arc option and picking the arc with the cursor. The leader, arrow, and text are automatically added to the drawing. Dimensioning circles is the same as the arc except a phi symbol ($\Phi$) precedes the text.

## 8.10 Adding Cross Hatching for Section Lines

Sectional drawings require the use of section lines or cross hatching. Many CAD systems provide the ANSI standard cross hatching symbols. Section lines are added to a drawing by defining the area to be cross hatched. The cross hatching boundaries are defined by picking the entities that surround the area or by using the window option to define an area to be cross hatched. The user selects a standard pat-

tern or can create a unique pattern by specifying the angle, spacing of lines, and other variables.

## 8.11 Editing a Drawing

Existing drawings are changed by using automated editing commands. One of the greatest advantages of CAD over traditional tools is the speed with which it can make changes to existing drawings. *Edit* commands are used to change existing drawing entities, such as line type, color, layer and other variables of entities, and text parameters.

### *Fillets and Chamfers*

Square corners are edited to fillets and rounds using the fillet command. The *fillet* command automatically adds fillets or rounds to corners at any specified radius when the corner is selected. The square corner is trimmed and a tangent arc is added to make the fillet or round. The *chamfer* command automatically places a chamfer at a square corner and works the same as fillet.

### *Trim and Extend*

*Trim* or *break* is an editing command that is used to erase parts of lines, arcs, circles, and other entities by picking the entity and the start and end points of the part to be erased. The trim command is used like an erasing shield. The opposite of trim would be extend. *Extend* is used to lengthen an entity such as a line from the current end point to a new point, such as the end point of another line.

### *Move*

At times it is desirable to move entities, such as a side view of a three-view drawing, to another place on the sheet. The *move* command, Fig. 8.20, is used

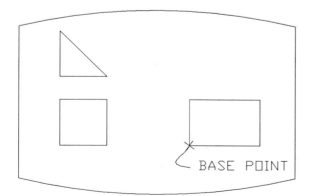
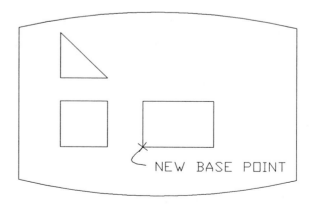

*Fig. 8.20*  The Move Command.

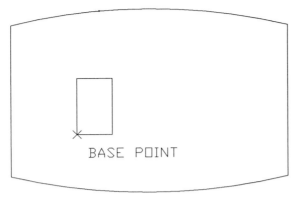
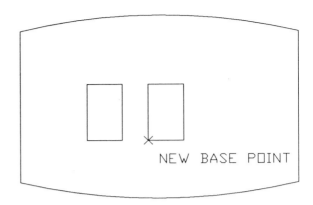

*Fig. 8.21*    The Copy Command.

to change the position of entities on a drawing. The entities to be moved are defined by picking them with the cursor or by using a window or some other method. The base point or handle is defined. A *handle* is a defined point that is used to copy, move, or place entities or blocks on a drawing. Usually the handle is an easily remembered point such as the center of a hole or the corner of an object. After the handle point is defined, the new position for the handle is defined by picking with the cursor or entering X-Y coordinate values.

### Drag

Some CAD systems have a drag function to assist in the placement of copies. *Drag* is a function that dynamically shows the new position of entities and blocks being copied, moved, or scaled as the cursor changes position on the screen. It is a quick redraw function of the entities and is very useful when trying to locate copies accurately or to preview the new position before its final position is picked.

### Copy

The *copy* command, Fig. 8.21, is used to make a duplicate(s) of selected entities. After the entity to be copied and the handle is defined, a new base point is selected and the copy made. A copy can be scaled or rotated, or more than one can be created. When more than one copy is created in a series of rows and columns, it is called an *array*.

### Rotate

*Rotate*, Fig. 8.22, is used to position entities at a new angle and can be combined with the copy and move commands. The rotate command works the same as the copy or move command except the defined entities are rotated at a specified angle about

the base point or axis. An object rotated counterclockwise 45 degrees is shown in Fig. 8.22. The bolt circle drawn in Fig. 8.23 was created using the copy and rotate command.

### Scale

The *scale* command is used to enlarge or reduce the size of a drawing. Scale is useful for creating enlarged details of parts, for example. Entities to be scaled are picked, a handle is defined, then the new scale is entered. The new position for the base point is picked, and the entities are drawn to the new scale. Scale changes the geometric data base of the drawing and is very different from the display option, which changes only the size of the displayed image. For example, using the scale command to change a drawing to half size will cause a 1″ line to measure 0.50″ as well as change the displayed image.

### Mirror

One other editing command that can speed drawing time is the mirror command, Fig. 8.24. *Mirror* will make a mirrored copy of defined entities around a selected axis. This command is especially useful with symmetrical objects.

### 8.12  Plotting

After the drawing is complete with border lines and title block, some type of hard copy output is desired. Virtually anything drawn on a CAD system should be drawn full scale and then plotted to fit the selected paper size. The plot sequence is used to scale the size of the drawing to fit the selected paper.

Before plotting make sure that only those layers to be plotted are displayed and saved. Most CAD

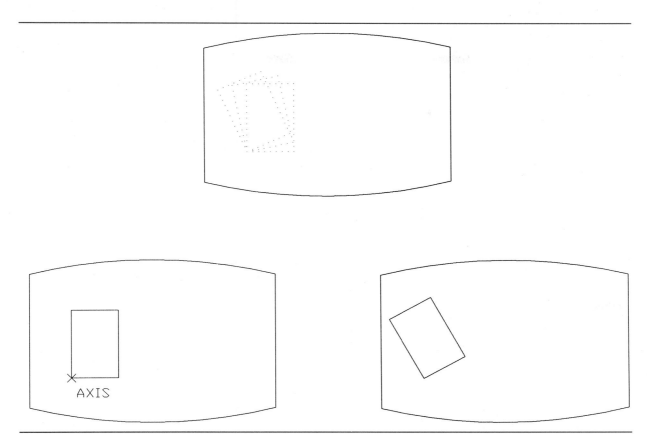

*Fig. 8.22*    The Rotate Command.

systems allow the user many different plotting options by specifying variables such as scale, rotation of the plot, and control of pens. A pen plotter draws ANSI standard line weights by automatically selecting the required pen widths. Many times pen plots will reveal errors that are not seen on screen, in which case the drawing should be edited. It is important to keep a clean data base because CAD drawings often are used as the data base for CAM. Plots are useful to check drawings for accuracy and lines drawn on top of each other, which cannot be seen on screen. These extra lines must be erased from the drawing.

## 8.13 Three-Dimensional Modeling

A *model* is a graphic representation of an object. Most designs are three dimensional and can be drawn on a CAD system as a 3D model. There are many advantages to creating a 3D model of a design on a CAD system. Three-dimensional models are less abstract and more easily recognized than ortho-

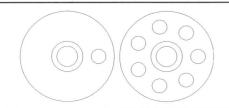

*Fig. 8.23*    The Rotate-Copy Command.

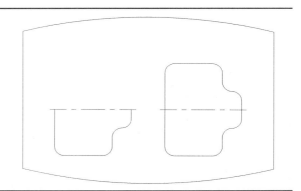

*Fig. 8.24*    The Mirror Command.

graphic views. The 3D model also is used as the input for CAM.

There are three types of 3D CAD models: wireframe, surface, and solid. The wireframe model is the most abstract and least realistic. The solid model is the least abstract and most realistic. The surface model is somewhere between the other two models.

## 8.14 Wireframe Modeling

A *wireframe* model, Fig. 8.25, is a 3D representation of an object in which lines represent the edges of surfaces. It is the most abstract type of model and sometimes is very difficult to visualize. Wireframe models are used as the input for CAM. Most CAD software programs allow a number of different methods of creating a wireframe model.

### Methods of Creating Wireframe Models

*Extrusion* is a technique for creating a 3D wireframe model by copying a 2D profile and extending it to a depth defined by the operator. The result is a 3D wireframe of the profile, Figs. 8.26 and 8.27.

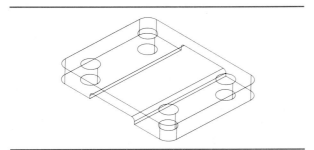

*Fig. 8.25*    Abstract Wireframe Model.

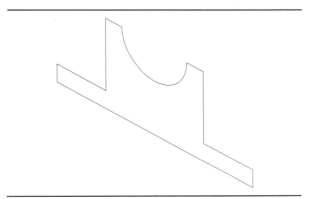

*Fig. 8.26*    Isometric View of a Profile to Be Extruded.

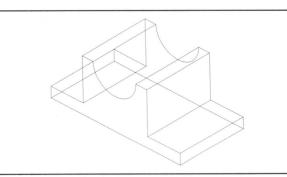

*Fig. 8.27*    Three-Dimensional Model Created from the Profile.

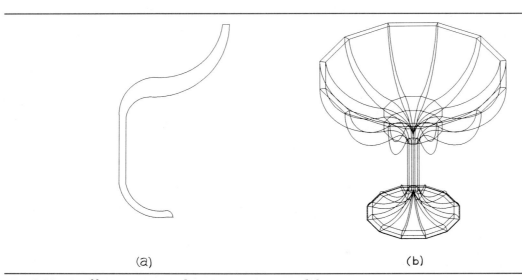

(a)                              (b)

*Fig. 8.28*    Profile View Rotated to Create a 3D Model.

*Rotation* produces wireframe models by rotating a cross-section or profile of the part about an axis. It is similar to extrusion except it is swept about an axis, Fig. 8.28. Other methods of creating wireframe models are: extrusion with scale, Fig. 8.29, and using primitive shapes to build models, Fig. 8.30.

### Editing the Model

Once the model is extruded, it can be edited and viewed from different points. Some CAD systems can automatically remove hidden lines using a command called *hide*, Fig. 8.31. After the model is completed, orthographic views are created by changing the line of sight so that it is perpendicular to the front, top, and profile faces of the model, Fig. 8.32. After the orthographic views are created, they must be edited to remove extraneous lines and to add hidden and center lines so the drawing conforms with the standards.

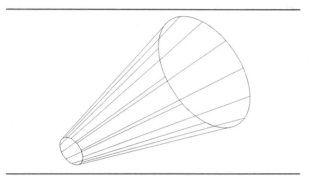

*Fig. 8.29*   Extrusion with Scale to Form a Truncated Cone.

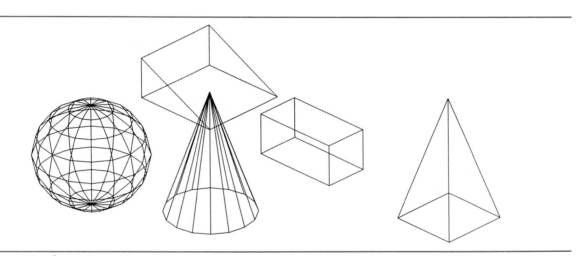

*Fig. 8.30*   Wireframe Primitives.

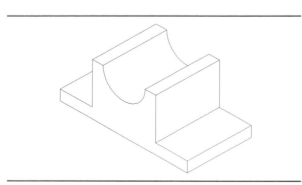

*Fig. 8.31*   Model with Hidden Lines Removed.

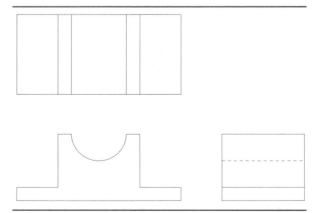

*Fig. 8.32*   Orthographic Views Created from Model.

## 8.15 Surface Models

A *surface model* is one of three models that is a graphic representation of an object created by defining surfaces. After the surfaces are defined, the model can be shaded with different colors and light sources. After the model is shaded, it looks like a solid model.

## 8.16 Solid Models

A solid model is a graphic representation of an object that is the most realistic and least abstract of the three types of models. Only solid models allow the user to check for interference of assemblies and accurately analyze the mass properties. There are two main types of solid models: CSG (constructive solid geometry), which uses primitive shapes to build the model, and B-rep (boundary representation), which combines lines, points, curves, and surfaces into a solid model. B-rep models are used for complex shapes, such as the hulls of ships. After the model is completed, it can be analyzed for its mass properties, edited, then shaded, Fig. 8.33.

**Fig. 8.33**  Solid Model. *Courtesy of CADKEY.*

## 8.17 Finite Element Analysis

After a model is created it can be tested by creating a FEM (finite element model). A FEM is a model represented by a mesh with values assigned to the intersection points of the lines of the mesh. The intersection points are called nodes and can be assigned loads, temperature, or other variables. The

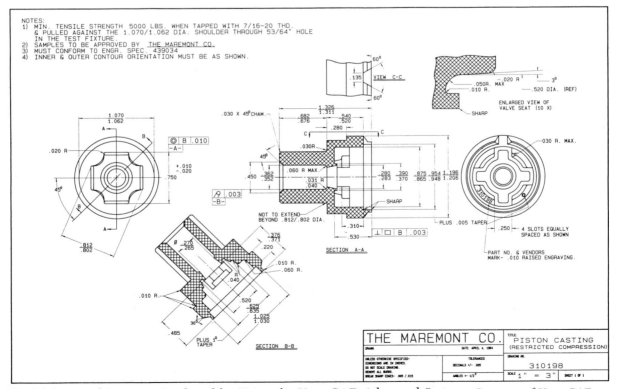

**Fig. 8.34**  Detail Drawing Produced by Using the VersaCAD Advanced System. *Courtesy of VersaCAD.*

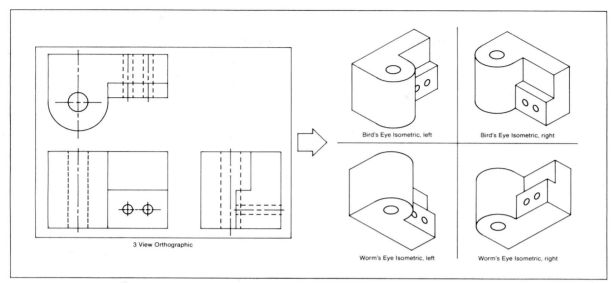

***Fig. 8.35*** Orthographic to Isometric Conversion. The Auto-trol Orthographic to Axonometric Package (OTAP) system can be used to convert an orthographic drawing to axonometric. *Courtesy of Auto-trol Technology Corporation.*

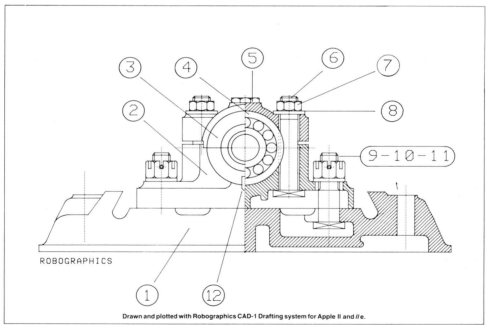

***Fig. 8.36*** Computer-Generated Assembly Drawing in Half Section. *Courtesy of Chessel-Robocom Corporation.*

computer will then analyze the model and display areas of high stress or temperature in red and areas of low stress or temperature in blue.

## 8.18 Summary

In the near future more design will be done on CAD systems. The 3D models created with CAD will be-come the input for many other operations. The fundamentals of engineering graphics will, however, remain the foundation necessary to create designs and drawings on a CAD system. This chapter has presented a generic introduction to CAD with explanations of the basic commands necessary to create and edit simple drawings. Examples of various CAD-produced drawings are shown in Figs. 8.34 to 8.39.

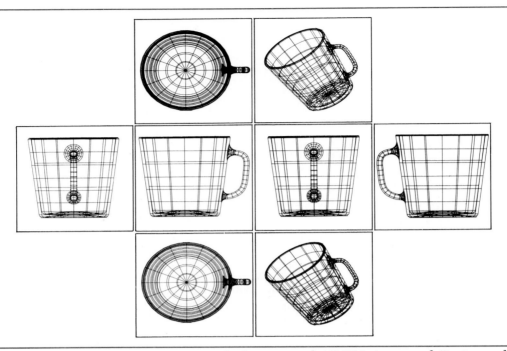

*Fig. 8.37*    Mold Design and Production.    With the Auto-trol AD/380 Automated Design and Drafting System, a part can be displayed in multiple views so the mold designer can see the part from all angles. *Courtesy of Auto-trol Technology Corporation.*

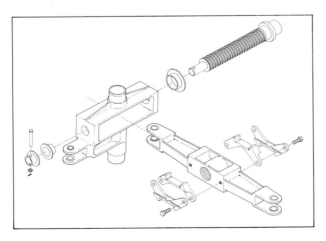

*Fig. 8.38*    Exploded Assembly Drawing Produced from Engineering Data by Using the Auto-trol Orthographic to Axonometric Package (OTAP) System. *Courtesy of Auto-trol Technology Corporation.*

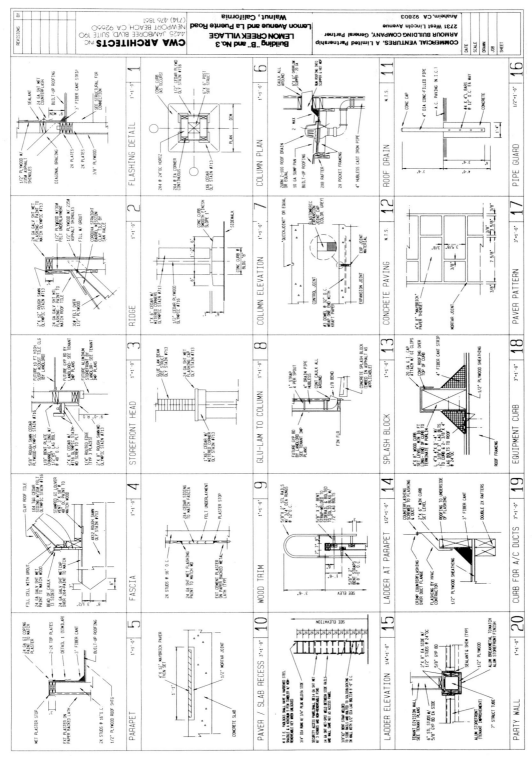

*Fig. 8.39* Architectural Details Produced by Using the VersaCAD System. *Courtesy of VersaCAD.*

## *Computer Graphics Problems*

The problems that follow are selected from other chapters in this text for solution by a student using a computer-aided drafting system rather than traditional drafting methods. The text of this chapter will serve as a reference when a CAD system is used to create the drawings assigned by the instructor.

Additional computer graphics problems in convenient form for solution are available in *Principles of Engineering Graphics Problems* by Spencer, Hill, Loving, and Dygdon, a workbook designed to accompany this text that is also published by Macmillan Publishing Company.

---

**Prob. 8.1**   Using CAD, draw those figures assigned by your instructor from Instrumental Drawing Problems in Chapter 2.

**Prob. 8.2**   Using CAD, draw those figures assigned by your instructor from Geometric Constructions Problems of Chapter 5.

**Prob. 8.3**   Using CAD, make a three-view drawing of the safety key, Fig. 7.54; finger guide, Fig. 7.55; roller lever, Fig. 7.60; index slide, Fig. 7.63; cross-feed stop, Fig. 7.68; tool holder, Fig. 7.73; and/or vise base, Fig. 7.78.

**Prob. 8.4**   Using CAD, make a sectional drawing of the truck wheel, Fig. 9.40; cup washer, Fig. 9.43; adjuster base, Fig. 9.49; bracket, Fig. 9.53; and/or oil retainer, Fig. 9.54.

**Prob. 8.5**   Using CAD, dimension the drawing(s) from Prob. 8.3.

**Prob. 8.6**   Using CAD, make a complete set of working drawings of the hand rail column, Fig. 16.65; drill jig, Fig. 16.66; tool post, Fig. 16.67; or belt tightener, Fig. 16.68.

**Prob. 8.7**   Using CAD, make a 3D model of one of the drawings from Prob. 8.3.

# Sectional Views

The basic method of representing parts for designs by views, or projections, has been explained in previous chapters. By means of a limited number of carefully selected views, the external features of the most complicated designs can be fully described.

However, we are frequently confronted with the necessity of showing more or less complicated interiors of parts that cannot be shown clearly by means of hidden lines. We accomplish this by slicing through the part much as one would cut through an apple or a melon. A cutaway view of the part is then drawn; it is called a *sectional view*, a *cross section*, or simply a *section*. See ANSI Y14.2M–1979 (R1987) and Y14.3–1975 (R1987) for complete standards for multiview and sectional-view drawings.

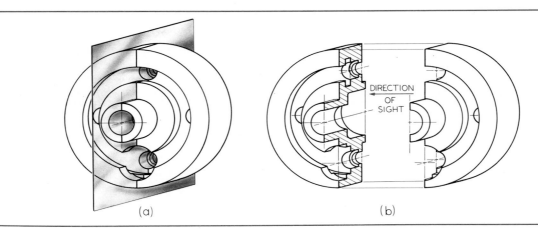

*Fig. 9.1*   A Section.

## 9.1 Sectioning

To produce a sectional view, a *cutting plane*, §9.5, is assumed to be passed through the part for the design, as shown in Fig. 9.1 (a). Then, at (b) the cutting plane is removed and the two halves drawn apart, exposing the interior construction. In this case, the direction of sight is toward the left half, as shown, and for purposes of the section the right half

is mentally discarded. The sectional view will be in the position of a right-side view.

## 9.2 Full Sections

The sectional view obtained by passing the cutting plane fully through the object is called a *full section*, Fig. 9.2 (c). A comparison of this sectional view with the left-side view, (a), emphasizes the advantage in

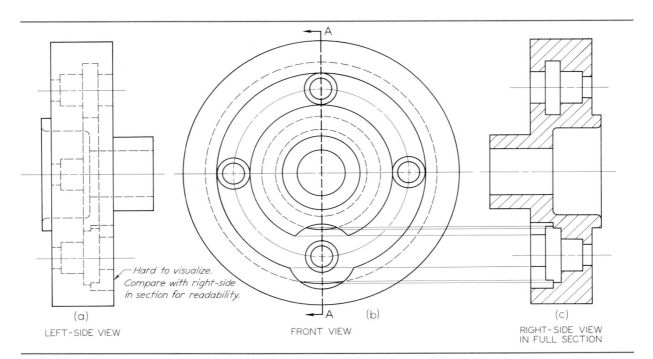

*Fig. 9.2*   Full Section.

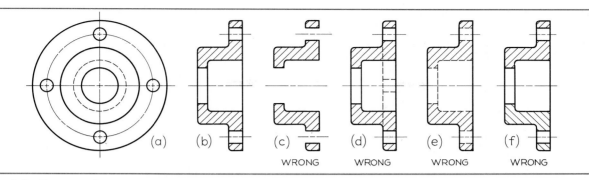

*Fig. 9.3*    Lines in Sectioning.

clearness of the former. The left-side view would naturally be omitted. In the front view, the cutting plane appears as a line, called a *cutting-plane line*, §9.5. The arrows at the ends of the cutting-plane line indicate the direction of sight for the sectional view.

Note that in order to obtain the sectional view, the right half is only *imagined* to be removed and not actually shown removed anywhere except in the sectional view itself. In the sectional view, the section-lined areas are those portions that have been in actual contact with the cutting plane. Those areas are *crosshatched* with thin parallel section lines spaced carefully by eye. In addition, the visible parts behind the cutting plane are shown but not crosshatched.

As a rule, the location of the cutting plane is obvious from the section itself, and, therefore, the cutting-plane line is omitted. It is shown in Fig. 9.2 for illustration only. Cutting-plane lines should, of course, be used wherever necessary for clearness, as in Figs. 9.21, 9.22, 9.24, and 9.25.

## 9.3 Lines in Sectioning

A correct front view and sectional view are shown in Fig. 9.3 (a) and (b). In general, *all visible edges and contours behind the cutting plane should be shown;* otherwise a section will appear to be made up of disconnected and unrelated parts, as shown in (c). Occasionally, however, visible lines behind the cutting plane are not necessary for clearness and should be omitted.

Sections are used primarily to replace hidden-line representation; hence, as a rule, *hidden lines should be omitted in sectional views.* As shown in Fig. 9.3 (d), the hidden lines do not clarify the drawing; they tend to confuse, and they take unnecessary time to draw. Sometimes hidden lines are necessary for clearness and should be used in such cases, especially if their use will make it possible to omit a view, Fig. 9.4.

A section-lined area is always completely bounded by a visible outline—never by a hidden line as in Fig. 9.3 (e), since in every case the cut surfaces and their boundary lines will be visible.

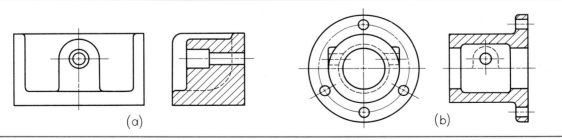

*Fig. 9.4*    Hidden Lines in Sections.

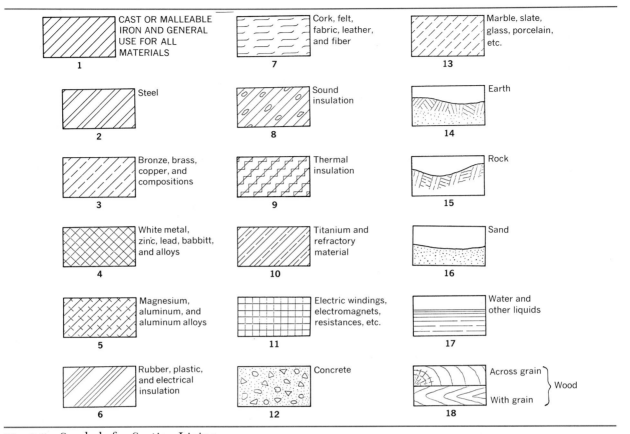

*Fig. 9.5*   Symbols for Section Lining.

Also, a visible line can never cut across a section-lined area.

In a sectional view of a part, alone or in assembly, the section lines in all sectioned areas must be parallel, not as shown in Fig. 9.3 (f). The use of section lining in opposite directions is an indication of different parts, as when two or more parts are adjacent in an assembly drawing, Fig. 16.42.

## 9.4 Section Lining

Symbolic section-lining symbols, Fig. 9.5, have been used to indicate the material to be used in producing the object. These symbols represented the general types only, such as cast-iron, brass, and steel. Now, however, there are so many different materials, and each general type has so many subtypes, that a general name or symbol is not enough. For example, there are hundreds of different kinds of steel alone. Since detailed specifications of material must be lettered in the form of a note or in the title strip, the general-purpose (cast-iron) section lining is used for all materials on detail drawings (single parts).

Symbolic section lining may be used in assembly drawings in cases where it is desirable to distinguish the different materials; otherwise, the general-purpose symbol is used for all parts. For assembly sections, see §16.22.

The correct method of drawing section lines is shown in Fig. 9.6 (a). Draw the section lines with a sharp medium-grade pencil (H or 2H) with a conical point as shown in Fig. 2.11 (c). Always draw the lines at 45° with horizontal as shown, unless there is some advantage in using a different angle. Space the section lines as uniformly as possible by eye from about approximately 1.5 mm ($\frac{1}{16}''$) to 3 mm ($\frac{1}{8}''$) or more apart, depending on the size of the drawing or of the sectioned area. *For average drawings, space the lines about 2.5 mm ($\frac{3}{32}''$) or more apart.* As a rule, space the lines as generously as possible and yet close enough to distinguish clearly the sectioned areas.

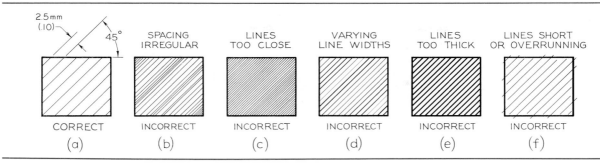

**Fig. 9.6** Section-Lining Technique.

After the first few lines have been drawn, look back repeatedly at the original spacing to avoid gradually increasing or decreasing the intervals, Fig. 9.6 (b). Beginners almost invariably draw section lines too close together, (c). This is very tedious because with small spacing the least inaccuracy in spacing is conspicuous.

Section lines should be uniformly thin, never varying in thickness, as in (d). There should be a marked contrast in thickness of the visible outlines and the section lines. Section lines should not be too thick, as in (e). Also avoid running section lines beyond the visible outlines or stopping the lines too short, as in (f).

If section lines drawn at 45° with horizontal would be parallel or perpendicular (or nearly so) to a prominent visible outline, the angle should be changed to 30°, 60°, or some other angle, Fig. 9.7.

Dimensions should be kept off sectioned areas, but when this is unavoidable the section lines should be omitted where the dimension figure is placed. See Fig. 13.13.

Section lines may be drawn adjacent to the boundaries of the sectioned areas (outline sectioning) providing that clarity is not sacrificed. See Fig. 16.6.

## 9.5 The Cutting Plane

The cutting plane is indicated in a view adjacent to the sectional view, Fig. 9.8. In this view, the cutting plane appears edgewise, as a line called the *cutting-plane line*. Alternate styles of cutting-plane lines are shown in Fig. 9.9. See also Fig. 2.14. The first form, Fig. 9.9 (a), composed of equal dashes each about 6 mm ($\frac{1}{4}''$) or more long plus the arrowheads, is the standard in the automotive industry. This form without the dashes between the ends is especially desirable on complicated drawings. The form shown in (b), composed of alternate long dashes and pairs of short dashes plus the arrowheads, has been in general use for a long time. Both lines are drawn the same thickness as visible lines. Arrowheads indicate the direction in which the cutaway object is viewed.

Capital letters are used at the ends of the cutting-plane line when necessary to identify the cutting-plane line with the indicated section. This most often occurs in the case of multiple sections, Fig. 9.25, or removed sections, Fig. 9.21.

As shown in Fig. 9.8, sectional views occupy normal projected positions in the standard arrangement of views. At (a) the cutting plane is a frontal plane, §7.15, and appears as a line in the top view. The

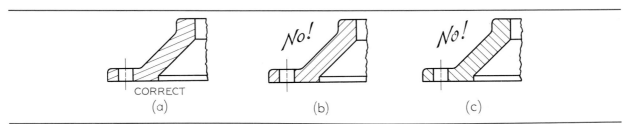

**Fig. 9.7** Direction of Section Lines.

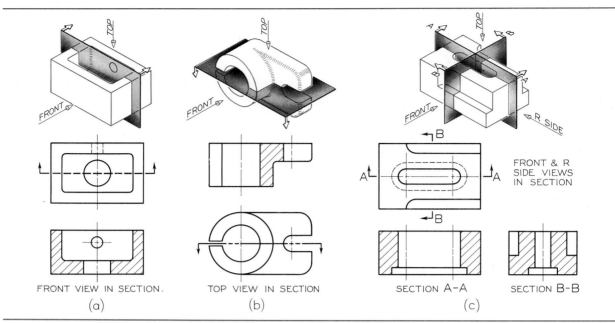

**Fig. 9.8** Cutting Planes and Sections.

front half of the object (lower half in the top view) is imagined removed. The arrows at the ends of the cutting-plane line point in the direction of sight for a front view, that is, away from the front view or section. Note that the arrows do not point in the direction of withdrawal of the removed portion. The resulting full section may be referred to as the "front view in section," since it occupies the front view position.

In Fig. 9.8 (b), the cutting plane is a horizontal plane, §7.15, and appears as a line in the front view. The upper half of the object is imagined removed. The arrows point toward the lower half in the same direction of sight as for a top view, and the resulting full section is a "top view in section."

In Fig. 9.8 (c), two cutting planes are shown, one a frontal plane and the other a profile plane, §7.15, both of which appear edgewise in the top view. Each section is completely independent of the other and drawn as if the other were not present. For section

A–A, the front half of the object is imagined removed. The back half is then viewed in the direction of the arrows for a front view, and the resulting section is a "front view in section." For section B–B, the right half of the object is imagined removed. The left half then is viewed in the direction of the arrows for a right-side view, and the resulting section is a "right-side view in section." The cutting-plane lines are preferably drawn through an exterior view, in this case the top view, as shown, instead of a sectional view.

The cutting-plane lines in Fig. 9.8 are shown for purposes of illustration only. They are generally omitted in cases such as these, in which the location of the cutting plane is obvious. When a cutting-plane line coincides with a center line, the cutting-plane line takes precedence.

Correct and incorrect relations between cutting-plane lines and corresponding sectional views are shown in Fig. 9.10.

**Fig. 9.9** Cutting-Plane Lines (Full Size).

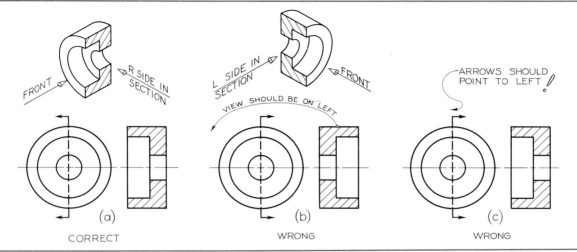

*Fig. 9.10*  Cutting Planes and Sections.

## 9.6 Visualizing a Section

Two views of an object to be sectioned, having a drilled and counterbored hole, are shown in Fig. 9.11 (a). The cutting plane is assumed along the horizontal center line in the top view, and the front half of the object (lower half of the top view) is imagined removed. A pictorial drawing of the remaining back half is shown at (b). The two cut surfaces produced by the cutting plane are 1–2–5–6–10–9 and 3–4–12–11–7–8. However, the corresponding section at (c) is incomplete because certain visible lines are missing.

If the section is viewed in the direction of sight, as shown at (b), arcs A, B, C, and D will be visible. As shown at (d), these arcs will appear as straight lines 2–3, 6–7, 5–8, and 10–11. These lines may also be accounted for in other ways. The top and bottom surfaces of the object appear in the section as lines 1–4 and 9–12. The bottom surface of the counterbore appears in the section as line 5–8. Also, the semicylindrical surfaces for the back half of the counterbore and of the drilled hole will appear as rectangles in the section at 2–3–8–5 and 6–7–11–10.

The front and top views of a Collar are shown in Fig. 9.12 (a), and a right-side view in full section is required. The cutting plane is understood to pass along the center lines AD and EL. If the cutting plane were drawn, the arrows would point to the left in conformity with the direction of sight (see arrow) for the right-side view. The right-side of the object is

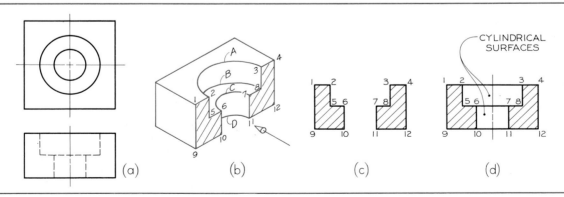

*Fig. 9.11*  Visualizing a Section.

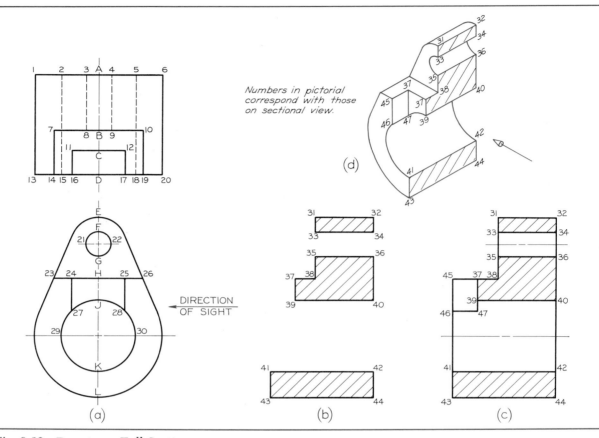

*Fig. 9.12*   Drawing a Full Section.

imagined removed and the left half will be viewed in the direction of the arrow, as shown pictorially at (d). The cut surfaces will appear edgewise in the top and front views along AD and EL; and since the direction of sight for the section is at right angles to them, they will appear in true size and shape in the sectional view. Each sectioned area will be completely enclosed by a boundary of visible lines. The sectional view will show, in addition to the cut surfaces, all visible parts behind the cutting plane. No hidden lines will be shown.

Whenever a surface of the object (plane or cylindrical) appears as a line and is intersected by a cutting plane that also appears as a line, a new edge (line of intersection) is created that will appear as a *point* in that view. Thus, in the front view, the cutting plane creates new edges appearing as points at E, F, G, H, J, K, and L. In the sectional view, (b), these are horizontal lines 31–32, 33–34, 35–36, 37–38, 39–40, 41–42, and 43–44.

Whenever a surface of the object appears as a surface (that is, not as a line) and is cut by a cutting

plane that appears as a line, a new edge is created that will appear as a line in the view, coinciding with the cutting-plane line, and as a line in the section.

In the top view, D is the *point view* of a vertical line KL in the front view and 41–43 in the section at (b). Point C is the point view of a vertical line HJ in the front view and 37–39 in the section. Point B is the point view of two vertical lines EF and GH in the front view, and 31–33 and 35–38 in the section. Point A is the point view of three vertical lines EF, GJ, and KL in the front view, and 32–34, 36–40, and 42–44 in the section. This completes the boundaries of three sectioned areas 31–32–34–33, 35–36–40–39–37–38, and 41–42–44–43. It is only necessary now to add the visible lines beyond the cutting plane.

The semicylindrical left half F–21–G of the small hole (front view) will be visible as a rectangle in the sections at 33–34–36–35, as shown at (c). The two semicircular arcs will appear as straight lines in the section at 33–35 and 34–36.

Surface 24–27, appearing as a line in the front

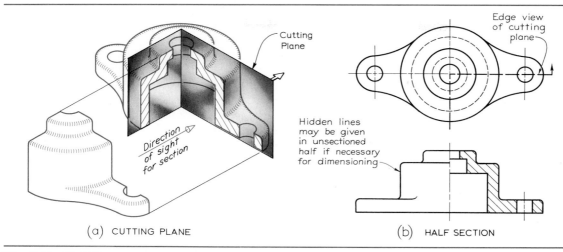

**Fig. 9.13**    Half Section.

view, appears as a line 11–16 in the top view and as surface 45–37–47–46, true size, in the section at (c).

Cylindrical surface J–29–K, appearing as an arc in the front view, appears in the top view as 2–A–C–11–16–15, and in the section as 46–47–39–40–42–41. Thus, arc 27–29–K (front view) appears in the section, (c), as straight lines 46–41; and arc J–29–K appears as straight line 40–42.

All cut surfaces here are part of the same object; hence, the section lines must all run in the same direction, as shown.

## 9.7 Half Sections

If the cutting plane passes halfway through the object, the result is a half section, Fig. 9.13. A half section has the advantage of exposing the interior of one half of the object and retaining the exterior of the other half. Its usefulness is, therefore, largely limited to symmetrical objects. It is not widely used in detail drawings (single parts) because of this limitation of symmetry and also because of difficulties in dimensioning internal shapes that are shown in part only in the sectioned half, Fig. 9.13 (b).

In general, hidden lines should be omitted from both halves of a half section. However, they may be used in the unsectioned half if necessary for dimensioning.

The greatest usefulness of the half section is in assembly drawing, Fig. 16.42, in which it is often necessary to show both internal and external construction on the same view, but without the necessity of dimensioning.

As shown in Fig. 9.13 (b), a center line is used to separate the halves of the half section. The American National Standards Institute recommends a center line for the division line between the sectioned half and the unsectioned half of a half-sectional view, although in some cases the same overlap of the exterior portion, as in a broken-out section, is preferred. See Fig. 9.33 (b). Either form is acceptable.

## 9.8 Broken-Out Sections

It often happens that only a partial section of a view is needed to expose interior shapes. Such a section, limited by a break line, Fig. 2.14, is called a broken-out section. In Fig. 9.14, a full or half section is not necessary, a small broken-out section being suffi-

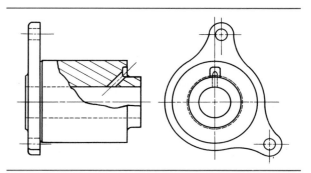

**Fig. 9.14**    Broken-Out Section.

cient to explain the construction. In Fig. 9.15, a half section would have caused the removal of half the keyway. The keyway is preserved by breaking out around it. Note that in this case the section is limited partly by a break line and partly by a center line.

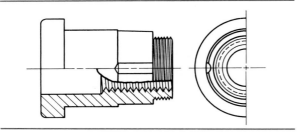

*Fig. 9.15*   Break Around Keyway.

## 9.9 **Revolved Sections**

The shape of the cross section of a bar, arm, spoke, or other elongated object may be shown in the longitudinal view by means of a revolved section, Fig. 9.16. Such sections are made by assuming a plane perpendicular to the center line or axis of the bar or other object, as shown in Fig. 9.17 (a), then revolving the plane through 90° about a center line at right angles to the axis, as at (b) and (c).

The visible lines adjacent to a revolved section may be broken out if desired, as shown in Figs. 9.16 (k) and 9.18.

The superimposition of the revolved section requires the removal of all original lines covered by it, Fig. 9.19. The true shape of a revolved section should be retained after the revolution of the cutting plane, regardless of the direction of the lines in the view, Fig. 9.20.

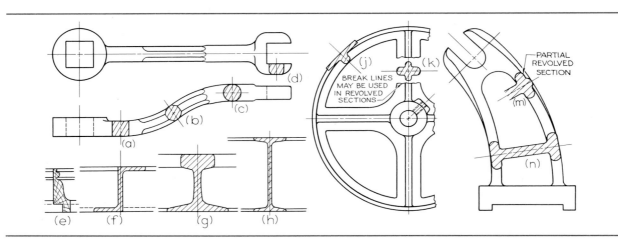

*Fig. 9.16*   Revolved Sections.

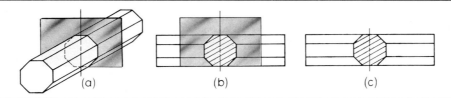

*Fig. 9.17*   Use of the Cutting Plane in Revolved Sections.

*Fig. 9.18*   Conventional Breaks Used with Revolved Sections.

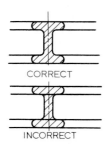

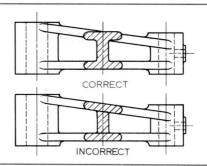

**Fig. 9.19**  A Common Error in Drawing Revolved Sections.

**Fig. 9.20**  A Common Error in Drawing Revolved Sections.

## 9.10 Removed Sections

A removed section is one not in direct projection from the view containing the cutting plane—that is, it is not positioned in agreement with the standard arrangement of views. This displacement from the normal projection position should be made without turning the section from its normal orientation.

Removed sections, Fig. 9.21, should be labeled, such as SECTION A–B and SECTION B–B, corresponding to the letters at the ends of the cutting-plane line.

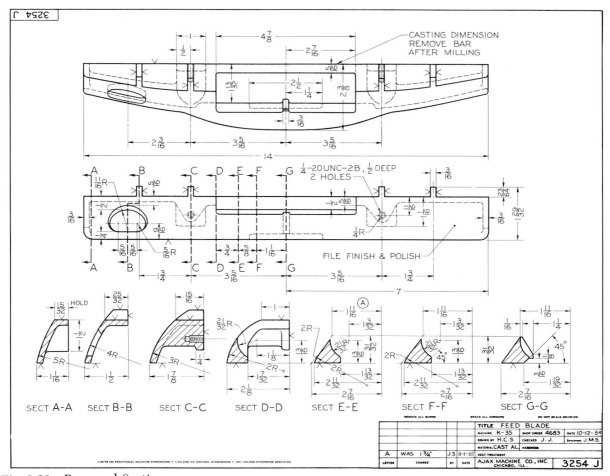

**Fig. 9.21**  Removed Sections.

They should be arranged in alphabetical order from left to right on the sheet. Section letters should be used in alphabetical order, but letters I, O, and Q should not be used because they are easily confused with the numeral 1 or the zero.

A removed section is often a partial section. Such a removed section, Fig. 9.22, is frequently drawn to an enlarged scale, as shown. This is often desirable in order to show clear delineation of some small detail and to provide sufficient space for dimensioning. In such case the enlarged scale should be indicated beneath the section title.

A removed section should be placed so that it no longer lines up in projection with any other view. It should be separated clearly from the standard arrangement of views. See Fig. 14.9.

Whenever possible removed sections should be on the same sheet with the regular views. If a section must be placed on a different sheet, cross-references should be given on the related sheets. A

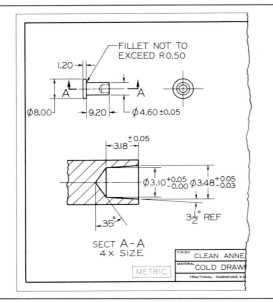

**Fig. 9.22**   Removed Section.

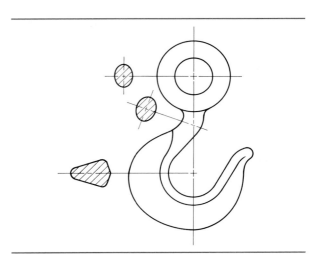

**Fig. 9.23**   Removed Sections.

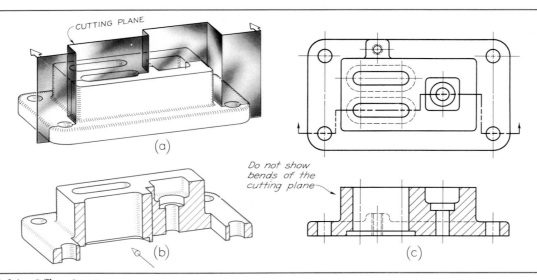

**Fig. 9.24**   Offset Section.

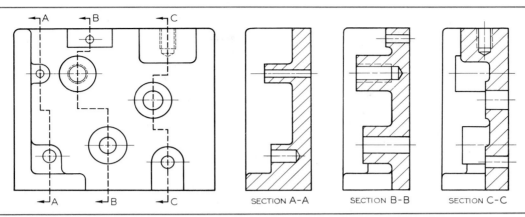

*Fig. 9.25* Three Offset Sections.

note should be given below the section title, such as

SECTION B–B ON SHEET 4, ZONE A3

A similar note should be placed on the sheet on which the cutting-plane line is shown, with a leader pointing to the cutting-plane line and referring to the sheet on which the section will be found.

Sometimes it is convenient to place removed sections on center lines extended from the section cuts, Fig. 9.23.

## 9.11 Offset Sections

In sectioning through irregular objects, it is often desirable to show several features that do not lie in a straight line, by "offsetting" or bending the cutting plane. Such a section is called an offset section. In Fig. 9.24 (a) the cutting plane is offset in several places in order to include the hole at the left end, one of the parallel slots, the rectangular recess, and one of the holes at the right end. The front portion of the object is then imagined to be removed, (b). The path of the cutting plane is shown by the cutting-plane line in the top view at (c), and the resulting offset section is shown in the front view. The offsets or bends in the cutting plane are all 90° and are *never shown in the sectional view*.

Figure 9.24 also illustrates how hidden lines in a section eliminate the need for an additional view. In this case, an extra view would be needed to show the small boss on the back if hidden lines were not shown.

An example of multiple offset sections is shown in Fig. 9.25. Notice that the visible background

shapes without hidden lines appear in each sectional view.

## 9.12 Ribs in Section

To avoid a false impression of thickness and solidity, ribs, webs, gear teeth, and other similar flat features are not sectioned even though the cutting plane passes along the center plane of the feature. For example, in Fig. 9.26, the cutting plane A–A passes flatwise through the vertical web, or rib, and the

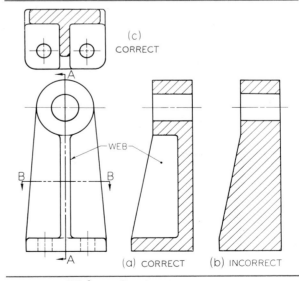

*Fig. 9.26* Webs in Section.

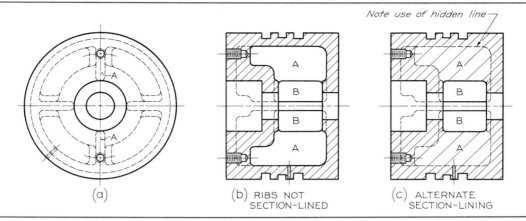

**Fig. 9.27** Alternate Section Lining.

web is not section-lined, (a). *Such thin features should not be section-lined, even though the cutting plane passes through them.* The incorrect section is shown at (b). Note the false impression of thickness or solidity resulting from section lining the rib.

If the cutting plane passes *crosswise* through a rib or any thin member, as does the plane **B–B** in Fig. 9.26, the member should be section-lined in the usual manner, as shown in the top view at (c).

In some cases, if a rib is not section-lined when the cutting plane passes through it flatwise, it is difficult to tell whether the rib is actually present, as for example, ribs **A** in Fig. 9.27 (a) and (b). It is difficult to distinguish spaces **B** as open spaces and spaces **A** as ribs. In such cases, double-spaced *section lining* of the ribs should be used, (c). This consists simply in continuing alternate section lines through the ribbed areas, as shown.

### 9.13 Aligned Sections

In order to include in a section certain angled elements, the cutting plane may be bent so as to pass through those features. The plane and feature are then imagined to be revolved into the original plane. For example, in Fig. 9.28, the cutting plane was bent to pass through the angled arm and then revolved to a vertical position (aligned), from where it was projected across to the sectional view.

In Fig. 9.29 the cutting plane is bent so as to include one of the drilled and counterbored holes in the sectional view. The correct section view at (b) gives a clearer and more complete description than does the section at (c), which was taken along the vertical center line of the front view—that is, without any bend in the cutting plane.

In such cases, the angle of revolution should always be less than 90°.

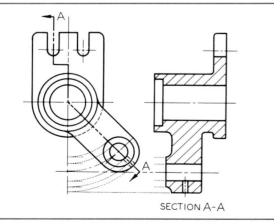

**Fig. 9.28** Aligned Section.

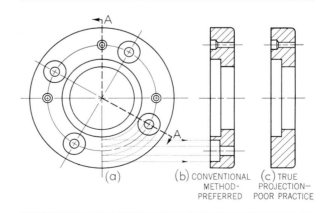

**Fig. 9.29** Aligned Section.

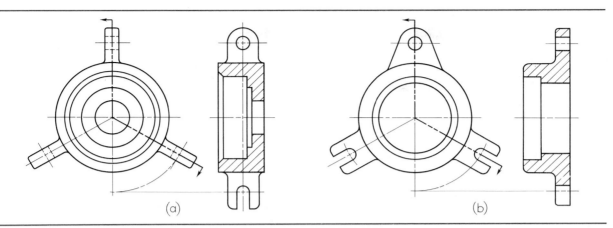

*Fig. 9.30* Aligned Sections.

The student is cautioned *not to revolve* features when clearness is not gained. In some cases the revolving features will result in a loss of clarity. Examples in which revolution should not be used are Fig. 9.39, Probs. 17 and 18.

In Fig. 9.30 (a) is an example in which the projecting lugs were not sectioned on the same basis that ribs are not sectioned. At (b) the projecting lugs are located so that the cutting plane would pass through them crosswise; hence, they are sectioned.

Another example involving rib sectioning and also aligned sectioning is shown in Fig. 9.31. In the circular view, the cutting plane is offset in circular-arc bends to include the upper hole and upper rib, the keyway and center hole, the lower rib, and one of the lower holes. These features are imagined to be revolved until they line up vertically and are then projected from that position to obtain the section at (b). Note that the ribs are not sectioned. If a regular full section of the object were drawn, without the use of conventions discussed here, the resulting section, (c), would be both incomplete and confusing and, in addition, would take more time to draw.

In sectioning a pulley or any spoked wheel, it is standard practice to revolve the spokes if necessary (if there is an odd number) and not to section-line

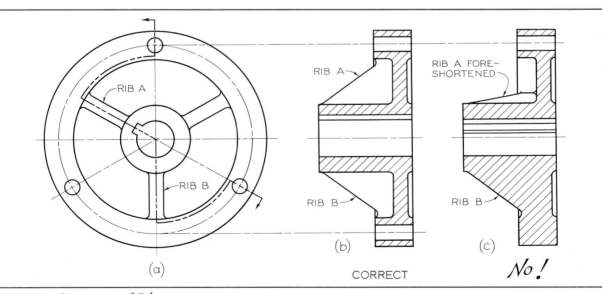

*Fig. 9.31* Symmetry of Ribs.

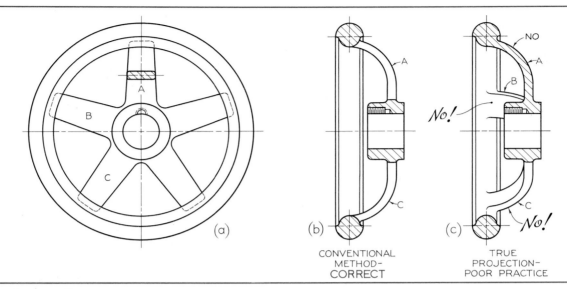

*Fig. 9.32*  Spokes in Section.

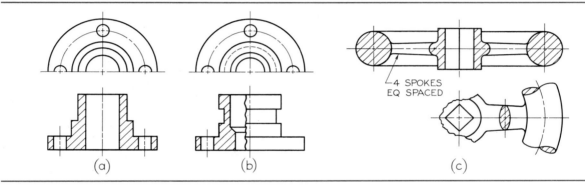

*Fig. 9.33*  Partial Views.

the spokes, Fig. 9.32 (b). If the spoke is sectioned, as shown at (c), the section gives a false impression of continuous metal. If the lower spoke is not revolved, it will be foreshortened in the sectional view in which it presents an "amputated" and wholly misleading appearance.

Figure 9.32 also illustrates correct practice in omitting visible lines in a sectional view. Notice that spoke **B** is omitted at (b). If it were included, (c), the spoke would be foreshortened, difficult and time consuming to draw, and confusing to the reader of the drawing.

### 9.14 Partial Views

If space is limited on the paper or if it is necessary to save drafting time, *partial views* may be used in connection with sectioning, Fig. 9.33. *Half views* are

shown at (a) and (b) in connection with a full section and a half section, respectively. Note that in each case the back half of the object in the circular view is shown, in conformity with the idea of removing the front portion of the object in order to expose the back portion for viewing in section. See also §7.9.

Another method of drawing a partial view is to break out much of the circular view, retaining only those features that are needed for minimum representation, Fig. 9.33 (c).

### 9.15 Intersections in Sectioning

Where an intersection is small or unimportant in a section, it is standard practice to disregard the true projection of the figure of intersection as shown in Fig. 9.34 (a) and (c). Larger figures of intersection

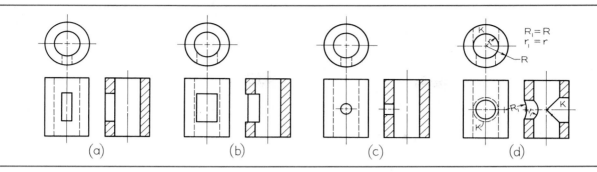

*Fig. 9.34*  Intersections.

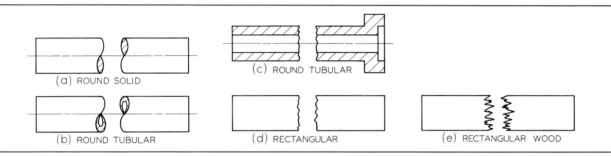

*Fig. 9.35*  Conventional Breaks.

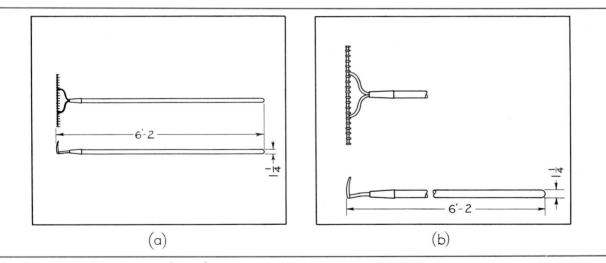

*Fig. 9.36*  Use of Conventional Breaks.

may be projected, as shown at (b), or approximated by circular arcs, as shown for the smaller hole at (d). Note that the larger hole K is the same diameter as the vertical hole. In such cases the curves of intersection (ellipses) appear as straight lines, as shown. See also Figs. 7.38 and 7.39.

## 9.16 Conventional Breaks

In order to shorten a view of an elongated object, conventional breaks are recommended, as shown in Fig. 9.35. For example, the two views of a garden rake are shown in Fig. 9.36 (a), drawn to a small scale to get it on the paper. At (b) the handle was

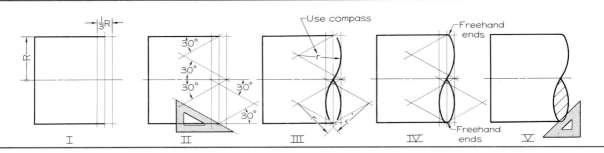

*Fig. 9.37*   Steps in Drawing S-Break for Solid Shaft.

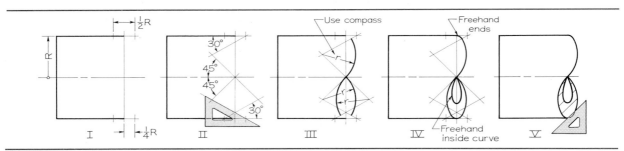

*Fig. 9.38*   Steps in Drawing S-Breaks for Tubing.

"broken," a long central portion removed, and the rake then drawn to a larger scale, producing a much more clear delineation.

Parts thus broken must have the same section throughout, or if tapered they must have a uniform taper. Note at (b) the full-length dimension is given, just as if the entire rake were shown.

The breaks used on cylindrical shafts or tubes are often referred to as "S-breaks" and in the industrial drafting room are usually drawn entirely freehand or partly freehand and partly with the ir-regular curve or the compass. By these methods, the result is often very crude, especially when attempted by beginners. Simple methods of construction for use by the student or the industrial drafter are shown in Figs. 9.37 and 9.38 and will always produce a professional result. Excellent S-breaks are also obtained with an S-break template.

Breaks for rectangular metal and wood sections are always drawn freehand, as shown in Fig. 9.35. See also Fig. 9.18 which illustrates the use of breaks in connection with the revolved sections.

# *Sectioning Problems*

Any of the following problems, Figs. 9.39 to 9.56 may be drawn freehand or with instruments, as assigned by the instructor. However, the problems in Fig. 9.39 are especially suitable for sketching on 8.5″ × 11.0″ graph paper with appropriate grid squares. Two problems can be drawn on one sheet, using Layout A–1 similar to Fig. 6.50, with borders drawn freehand. If desired, the problems may be sketched on plain drawing paper. Before making any sketches, the student should study carefully §§6.1 to 6.10.

The problems in Figs. 9.40 to 9.51 are intended to be drawn with instruments, but may be drawn freehand, if desired. If metric or decimal dimensions are required, the student should first study §§13.1 to 13.25. If an ink tracing is required, the student is referred to §§2.50 to 2.52.

Since many of the problems in this chapter are of a general nature, they can also be solved on most computer graphics systems. If a system is available, the instructor may choose to assign specific problems to be completed by this method.

Sketching problems in convenient form for solution may be found in *Principles of Engineering Graphics Problems* by Spencer, Hill, Loving, and Dygdon, a workbook designed to accompany this text that is also published by Macmillan Publishing Company.

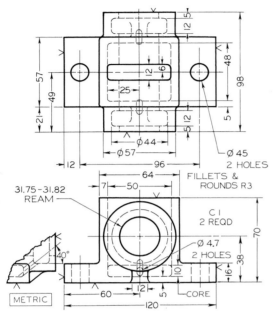

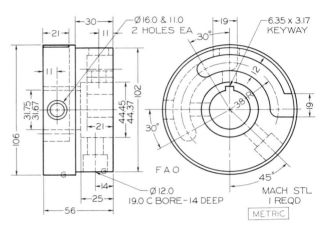

**Fig. 9.51**   Auxiliary Shaft Bearing.
Given: Front and top views.
Required: Front and top views and right-side view in full section [Layout B–4 or A3–4 (adjusted)].*

**Fig. 9.52**   Traverse Spider.
Given: Front and left-side views.
Required: Front and right-side views and top view in full section [Layout B–4 or A3–4 (adjusted)].*

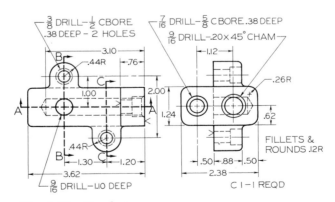

**Fig. 9.53**   Bracket.
Given: Front and right-side views.
Required: Take front as new top; then add right-side view, front view in full section A–A, and sections B–B and C–C [Layout B–4 or A3–4 (adjusted)].*

*If dimensions are required, study §§13.1 to 13.25. Use metric or decimal-inch dimensions as assigned by the instructor.

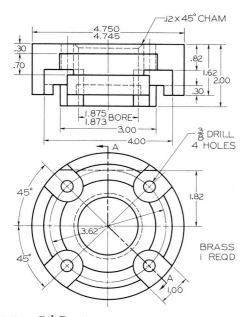

**Fig. 9.54** Oil Retainer.
Given: Front and top views.
Required: Front view and section A–A [Layout B–4 or A3–4 (adjusted)].*

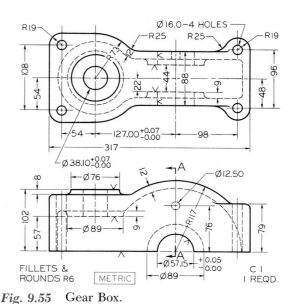

**Fig. 9.55** Gear Box.
Given: Front and top views.
Required: Front in full section, bottom view, and right-side section A–A; half size [Layout B–4 or A3–4 (adjusted)].*

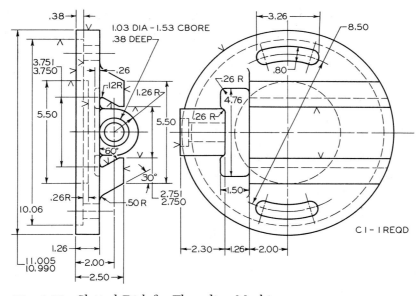

**Fig. 9.56** Slotted Disk for Threading Machine.
Given: Front and left-side views.
Required: Front and right-side views and top full-section view; half size [Layout B–4 or A3–4 (adjusted)].*

*If dimensions are required, study §§13.1 to 13.25. Use metric or decimal-inch dimensions as assigned by the instructor.

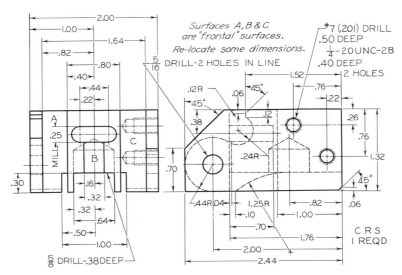

**Fig. 9.57** Cocking Block.

Given: Front and right-side views.

Required: Take front as new top view; then add new front view, and right-side view in full section. Draw double size on Layout C–4 or A2–4.*

*If dimensions are required, study §§13.1 to 13.25. Use metric or decimal-inch dimensions as assigned by the instructor.

# CHAPTER
# 10

# Auxiliary Views

Many objects are of such shape that their principal faces cannot always be assumed parallel to the regular planes of projection. For example, in Fig. 10.1 (a), the base of the design for the bearing is shown in its true size and shape, but the rounded upper portion is situated at an angle with the planes of projection and does not appear in its true size and shape in any of the three regular views.

In order to show the true circular shapes, it is necessary to assume a direction of sight perpendicular to the planes of those curves, as shown at (b). The resulting view is known as an *auxiliary view*. This view, together with the top view, completely describes the object. The front and right-side views are not necessary.

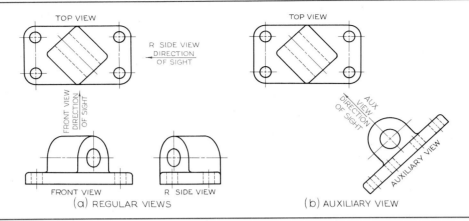

*Fig. 10.1*   Regular Views and Auxiliary Views.

## 10.1 Definitions

A view obtained by a projection on any plane other than the horizontal, frontal, and profile projection planes is called an *auxiliary view*. A *primary* auxiliary view is projected on a plane that is perpendicular to one of the principal planes of projection and is inclined to the other two. A *secondary* auxiliary view is projected from a primary auxiliary view and on a plane inclined to all three principal projection planes. See §10.19.

## 10.2 The Auxiliary Plane

In Fig. 10.2 (a), the object shown has an inclined surface that does not appear in its true size and shape in any regular view. The auxiliary plane is assumed parallel to the inclined surface P, that is,

perpendicular to the line of sight, which is at right angles to that surface. The auxiliary plane is then perpendicular to the frontal plane of projection and hinged to it.

When the horizontal and auxiliary planes are unfolded to appear in the plane of the front view, as shown at (b), the *folding lines* represent the hinge lines joining the planes. The drawing is simplified, as shown at (c), by retaining the folding lines, H/F and F/1, and omitting the planes. As will be shown later, the folding lines may themselves be omitted in the actual drawing. The inclined surface P is shown in its true size and shape in the auxiliary view, the long dimension of the surface being projected directly from the front view and the *depth* from the top view.

It should be observed that the positions of the

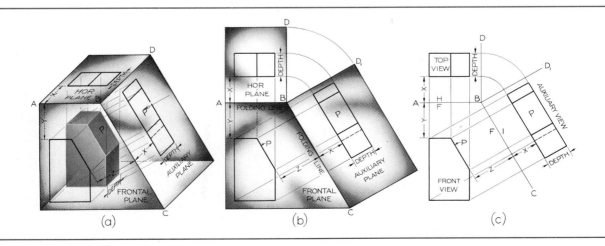

*Fig. 10.2*   An Auxiliary View.

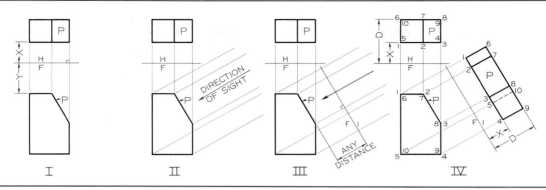

**Fig. 10.3** To Draw an Auxiliary View—Folding-Line Method.

folding lines depend upon the relative positions of the planes of the glass box at (a). If the horizontal plane is moved upward, the distance Y is increased. If the frontal plane is brought forward, the distances X are increased but remain *equal*. If the auxiliary plane is moved to the right, the distance Z is increased. Note that both the top and auxiliary views show the *depth* of the object.

## 10.3 To Draw an Auxiliary View— Folding-Line Method

As shown in Fig. 10.2 (c), the folding lines are the hinge lines of the glass box. Distances X must be equal, since they both represent the distance of the front surface of the object from the frontal plane of projection.

Although distances X must remain equal, distances Y and Z, from the front view to the respective folding lines, may or may not be equal.

The steps in drawing an auxiliary view with the aid of the folding lines, shown in Fig. 10.3, are described as follows:

I.    The front and top views are given. It is required to draw an auxiliary view showing the true size and shape of inclined surface P. Draw the folding line H/F between the views at right angles to the projection lines. Distances X and Y may or may not be equal, as desired.

NOTE    In the following steps, manipulate the triangle (either triangle) as shown in Fig. 10.4 to draw lines parallel or perpendicular to the inclined face.

II.    Assume arrow, indicating direction of sight,

perpendicular to surface P. Draw light projection lines from the front view parallel to the arrow, or perpendicular to surface P.

III.    Draw folding line F/1 for the auxiliary view at right angles to the projection lines and at any convenient distance from the front view.

IV.    Draw the auxiliary view, using the numbering system explained in §7.6. Locate all points the same distances from folding line F/1 as they are from folding line H/F in the top view. For example, points 1 to 5 are distance X from the folding lines in both the top and auxiliary views, and points 6 to 10 are distance D from the corresponding folding lines. Since the object is viewed in the direction of the arrow, it will be seen that edge 5–10 will be hidden in the auxiliary view.

## 10.4 Reference Planes

In drawing the auxiliary views shown in Figs. 10.2 (c) and 10.3, the folding lines represent the edge views of the front plane of projection. In effect, the

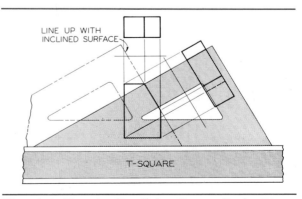

**Fig. 10.4** Drawing Parallel or Perpendicular Lines.

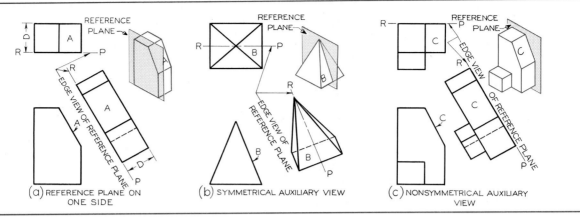

(a) REFERENCE PLANE ON ONE SIDE    (b) SYMMETRICAL AUXILIARY VIEW    (c) NONSYMMETRICAL AUXILIARY VIEW

*Fig. 10.5*   Position of the Reference Plane.

frontal plane is used as a *reference plane*, or *datum plane*, for transferring distances (*depth measurements*) from the top view to the auxiliary view.

Instead of using one of the planes of projection as a reference plane, it is often more convenient to assume a reference plane inside the glass box parallel to the plane of projection and touching or cutting through the object. For example, Fig. 10.5 (a), a reference plane is assumed to coincide with the front surface of the object. This plane appears edgewise in the top and auxiliary views, and the two reference lines are then used in the same manner as folding lines. Dimensions D, to the reference lines, are equal. The advantage of the reference-plane method is that fewer measurements are required, since some points of the object lie in the reference plane.

The reference plane may coincide with the front surface of the object as at (a), or it may cut through the object as at (b) if the object is symmetrical, or the reference plane may coincide with the back surface of the object as at (c), or through any intermediate point of the object.

The reference plane should be assumed in the position most convenient for transferring distances with respect to it. Remember the following:

1. Reference lines, like folding lines, are always at right angles to the projection lines between the views.

2. A reference plane appears as a line in two alternate views, never in adjacent views.

3. Measurements are always made at right angles to the reference lines or parallel to the projection lines.

4. In the auxiliary view, all points are at the same distances from the reference line as the corresponding points are from the reference line in the *second previous view*, or alternate view.

## 10.5 To Draw an Auxiliary View— Reference-Plane Method

The object shown in Fig. 10.6 (a) is numbered as explained in §7.6. To draw the auxiliary view, proceed as follows:

I.   Draw two views of the object, and assume an arrow indicating the direction of sight for the auxiliary view of surface A.

II.   Draw projection lines parallel to the arrow.

III.   Assume reference plane coinciding with back surface of object as shown at (a). Draw reference lines in the top and auxiliary views at right angles to the projection lines: *these are the edge views of the reference plane.*

IV.   Draw auxiliary view of surface A. It will be true size and shape because the direction of sight was taken perpendicular to that surface. Transfer depth measurements from the top view to the auxiliary view with dividers or scale. Each point in the auxiliary view will be on its projection line from the front view and the same distance from the reference line as it is in the top view to the corresponding reference line.

V.   Complete the auxiliary view by adding other visible edges and surfaces of the object. Each numbered point in the auxiliary view lies on its projection line from the front view and is the same dis-

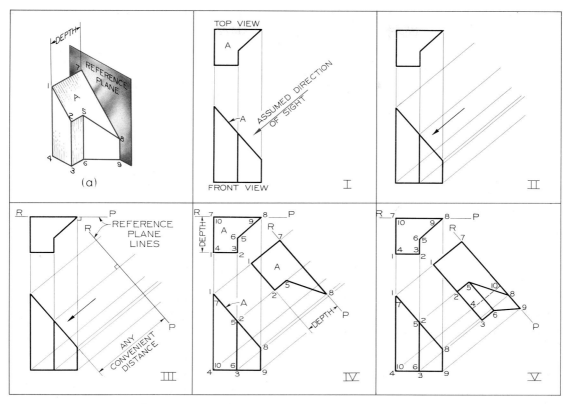

***Fig. 10.6*** To Draw an Auxiliary View—Reference-Plane Method.

tance from the reference line as it is in the top view. Note that two surfaces of the object appear as lines in the auxiliary view.

## 10.6 Classification of Auxiliary Views

Auxiliary views are classified and named according to the principal dimensions of the object shown in the auxiliary view. For example, the auxiliary view in Fig. 10.6 is a *depth auxiliary view* because it shows the principal dimension of the object, *depth*. Any auxiliary view projected from the front view, also known as front adjacent view, will show the depth of the object and is a depth auxiliary view.

Similarly, any auxiliary view projected from the top view, also known as top-adjacent view, is a height auxiliary view, and any auxiliary view projected from the side view (either side), also known as side adjacent view, is a width-auxiliary view. For examples of height auxiliary views, see Figs. 10.1 (b) and 10.13 (b). Depth auxiliary views are illustrated in Figs. 10.27 and 10.33.

## 10.7 Depth Auxiliary Views

An infinite number of auxiliary planes can be assumed perpendicular to, and hinged to, the frontal plane (**F**) of projection. Five such planes are shown in Fig. 10.7 (a), the horizontal plane being included to show that it is similar to the others. In all of these views the principal dimension, *depth*, is shown; hence all of the auxiliary views are depth auxiliary views.

The unfolded auxiliary planes are shown in (b), which also shows how the depth dimension may be projected from the top view to all auxiliary views. The arrows indicate the directions of sight for the several views, and the projection lines are respectively parallel to these arrows. The arrows may be assumed but need not be actually drawn, since the projection lines determine the direction of sight. The folding lines are perpendicular to the arrows and the corresponding projection lines. Since the auxiliary planes can be assumed at any distance from the object, it follows that the folding lines may be any distance from the front view.

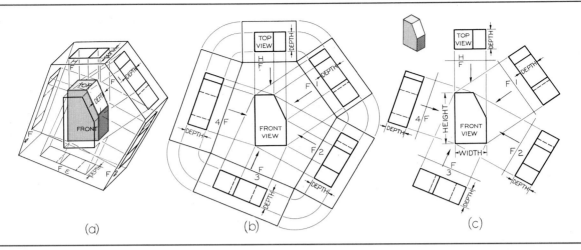

**Fig. 10.7** Depth Auxiliary Views.

The complete drawing, with the outlines of the planes of projection omitted, is shown at (c). This shows the drawing as it would appear on paper, in which use is made of reference planes as described in §10.4, all depth dimensions being measured perpendicular to the reference line in each view.

Note that the front view shows the *height* and the *width* of the object, *but not the depth*. The depth is shown in all views that are projected from the front view; hence, this rule: *The principal dimension shown in an auxiliary view is that one which is not shown in the adjacent view from which the auxiliary view was projected.*

## 10.8 Height Auxiliary Views

An infinite number of auxiliary planes can be assumed perpendicular to, and hinged to, the horizontal plane (H) of projection, several of which are shown in Fig. 10.8 (a). The front view and all of the auxiliary views show the principal dimension, *height*. Hence, all of the auxiliary views are height auxiliary views.

The unfolded projection planes are shown at (b), and the complete drawing, with the outlines of the planes of projection omitted, is shown at (c). All reference lines are perpendicular to the corresponding projection lines, and all height dimensions are

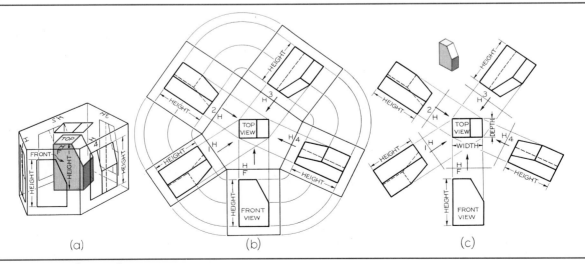

**Fig. 10.8** Height Auxiliary Views.

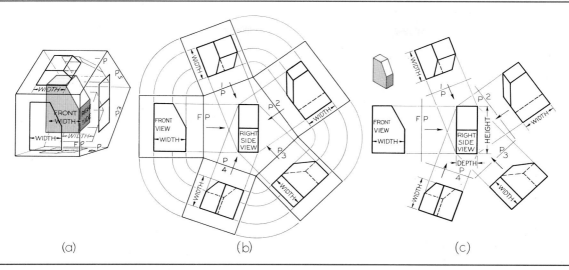

**Fig. 10.9**  Width Auxiliary Views.

measured parallel to the projection lines, or perpendicular to the reference lines, in each view. Note that in the view projected from, which is the top view, the only dimension *not shown* is height.

## 10.9  Width Auxiliary Views

An infinite number of auxiliary planes can be assumed perpendicular to, and hinged to, the profile plane (P) of projection, several of which are shown in Fig. 10.9 (a). The front view and all of the auxiliary views show the principal dimension, *width*. Hence, all of the auxiliary views are width auxiliary views.

The unfolded planes are shown at (b), and the complete drawing, with the outlines of the planes of projection omitted, is shown at (c). All reference lines are perpendicular to the corresponding projection lines, and all width dimensions are measured parallel to the projection lines, or perpendicular to the reference lines, in each view. Note that in the right-side view, from which the auxiliary views are projected, the only dimension *not shown* is width.

## 10.10  Revolving a Drawing

In Fig. 10.10 (a) is a drawing showing top, front, and auxiliary views. At (b) the drawing is shown revolved, as indicated by the arrows, until the auxiliary view and the front view line up horizontally. Although the views remain exactly the same, the

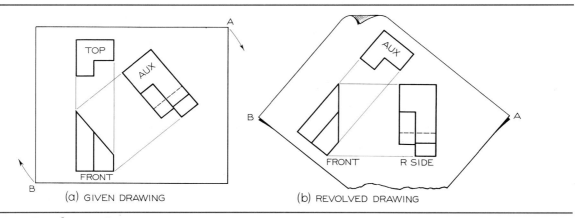

**Fig. 10.10**  Revolving a Drawing.

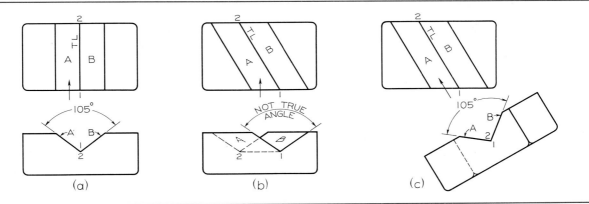

*Fig. 10.11*    Dihedral Angles.

names of the views are changed if drawn in this position. The auxiliary view now becomes a right-side view, and the top view becomes an auxiliary view. Some students find it easier to visualize and draw an auxiliary view when revolved to the position of a regular view in this manner. In any case, it should be understood that an auxiliary view basically is like any other view.

## 10.11 Dihedral Angles

The angle between two planes is a dihedral angle. One of the principal uses of auxiliary views is to show dihedral angles in true size, mainly for dimensioning purposes. In Fig. 10.11 (a) a block is shown with a V-groove situated so that the true dihedral angle between inclined surfaces A and B is shown in the front view.

Assume a line in a plane. For example, draw a straight line on a sheet of paper; then hold the paper so as to view the line as a point. You will observe that when the line appears as a point, the plane containing the line appears as a line. Hence, this rule: *To get the edge view of a plane, find the point view of any line in that plane.*

In Fig. 10.11 (a), line 1–2 is the line of intersection of planes A and B. Now, line 1–2 lies in both planes at the same time; therefore, a point view of this line will show both planes as lines, and the angle between them is the dihedral angle between the planes. Hence, this rule: *To get the true angle between two planes, find the point view of the line of intersection of the planes.*

At (b), the line of intersection 1–2 does not appear as a point in the front view; hence, planes A and B do not appear as lines, and the true dihedral

angle is not shown. Assuming that the actual angle is the same as at (a), does the angle show larger or smaller than at (a)? The drawing at (b) is unsatisfactory. The true angle does not appear because the direction of sight (see arrow) is not parallel to the line of intersection 1–2.

At (c) the direction of sight arrow is taken parallel to the line 1–2, producing an auxiliary view in which line 1–2 appears as a point, planes A and B appear as lines, and the true dihedral angle is shown. *To draw a view showing a true dihedral angle, assume the direction of sight parallel to the line of intersection between the planes of the angle.*

## 10.12 Plotted Curves

As shown in §7.30, if a cylinder is cut by an inclined plane, the inclined surface is elliptical in shape. In Fig. 7.34 (a), such a surface is produced, but the ellipse does not show true size and shape because the plane of the ellipse is not seen at right angles in any view.

In Fig. 10.12 (a), the line of sight is taken perpendicular to the edge view of the inclined surface, and the resulting ellipse is shown in true size and shape in the auxiliary view. The major axis is found by direct projection from the front view, and the minor axis is equal to the diameter of the cylinder. The left end of the cylinder (a circle) will appear as an ellipse in the auxiliary view, the major axis of which is equal to the diameter of the cylinder.

Since this is a symmetrical object, the reference plane is assumed to be located through the center, as shown. To plot points on the ellipses, select points on the circle of the side view, and project them across to the inclined surface or to the left-end sur-

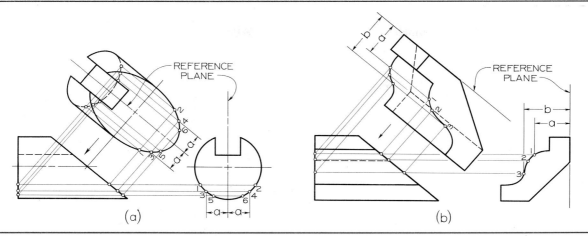

*Fig. 10.12* Plotted Curves.

face, and then upward to the auxiliary view. In this manner, two points can be projected each time, as shown for points 1–2, 3–4, and 5–6. Distances *a* are equal and are transferred from the side view to the auxiliary view with the aid of dividers. A sufficient number of points must be projected to establish the curves accurately. Use the irregular curve as described in §2.54.

Since the major and minor axes are known, any of the true ellipse methods of Figs. 5.48 to 5.50 and 5.52 (a) may be used. If an approximate ellipse is adequate for the job in hand, the method of Fig. 5.57 can be used. But the quickest and easiest method is to use an ellipse template, as explained in §5.56.

In Fig. 10.12 (b), the auxiliary view shows the true size and shape of the inclined cut through a piece of molding. The method of plotting points is similar to that explained for the ellipse in Fig. 10.12 (a).

## 10.13 Reverse Construction

In order to complete the regular views, it is often necessary to construct an auxiliary view first. For example, in Fig. 10.13 (a) the upper portion of the right-side view cannot be constructed until the auxiliary view is drawn and points established on the curves and then projected back to the front view as shown.

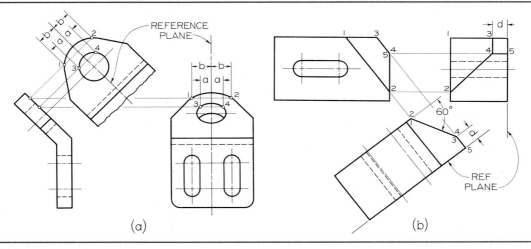

*Fig. 10.13* Reverse Construction.

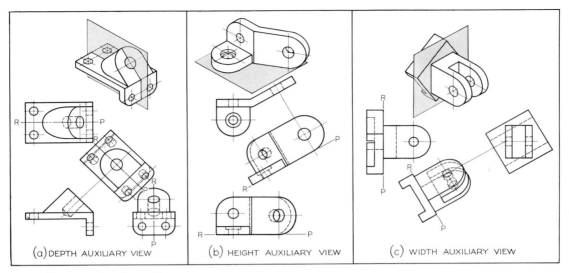

**Fig. 10.14**   Primary Auxiliary Views.

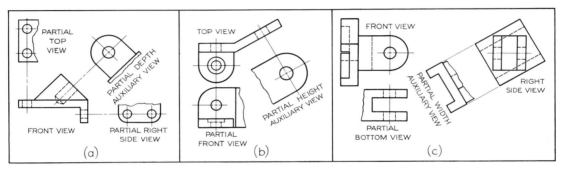

**Fig. 10.15**   Partial Views.

At (b), the 60° angle and the location of line 1–2 in the front view are given. In order to locate line 3–4 in the front view, the lines 2–4, 3–4, and 4–5 in the side view, it is necessary first to construct the 60° angle in the auxiliary view and project back to the front and side views, as shown.

## 10.14 Partial Auxiliary Views

The use of an auxiliary view often makes it possible to omit one or more regular views and thus to simplify the shape description, as shown in Fig. 10.1 (b). In Fig. 10.14 three complete auxiliary-view drawings are shown. Such drawings take a great deal of time to draw, particularly when ellipses are involved, as is so often the case, and the completeness of detail may add nothing to clearness or may even detract from it because of the clutter of lines. However, in these cases some portion of every view is needed—no view can be completely eliminated, as was done in Fig. 10.1 (b).

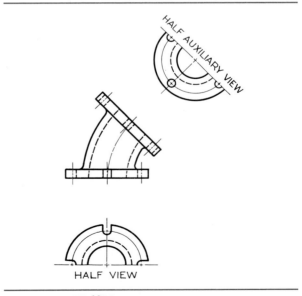

**Fig. 10.16**   Half Views.

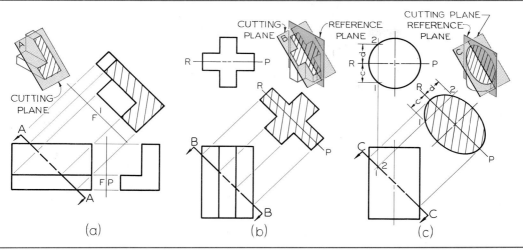

**Fig. 10.17** Auxiliary Sections.

As described in §7.9, *partial views* are often sufficient, and the resulting drawings are considerably simplified and easier to read. Similarly, as shown in Fig. 10.15, partial regular views and partial auxiliary views are used with the same result. Usually a break line is used to indicate the imaginary break in the views. *Do not draw a break line coinciding with a visible line or a hidden line.*

In order to clarify the relationship of views, the auxiliary views should be connected to the views from which they are projected, either with a center line or with one or two projection lines. This is particularly important with regard to partial views that often are small and easily appear to be "lost" and not related to any view.

### 10.15 Half Auxiliary Views

If an auxiliary view is symmetrical, and if it is necessary to save space on the drawing or to save time in drafting, only a half of the auxiliary view may be drawn, as shown in Fig. 10.16. In this case, a half of a regular view is also shown, since the bottom flange is also symmetrical. See §§7.9 and 9.14. Note that in each case the *near half* is shown.

### 10.16 Hidden Lines in Auxiliary Views

In practice, hidden lines should be omitted in auxiliary views, as in ordinary views, §6.25, unless they are needed for clearness. The beginner, however, should show all hidden lines, especially if the auxiliary view of the entire object is shown. Later, in advanced work, it will become clearer as to when hidden lines can be omitted.

### 10.17 Auxiliary Sections

An auxiliary section is simply an auxiliary view in section. In Fig. 10.17 (a), note the cutting-plane line and the terminating arrows that indicate the direction of sight for the auxiliary section. Observe that the section lines are drawn at approximately 45° with visible outlines. In an auxiliary section drawing, the entire portion of the object behind the cutting plane may be shown, as at (a), or the cut surface alone, as at (b) and (c).

An auxiliary section through a cone is shown in Fig. 10.18. This is one of the conic sections, §5.47, in this case a parabola. The parabola may be drawn by other methods, Figs. 5.57 and 5.58, but the

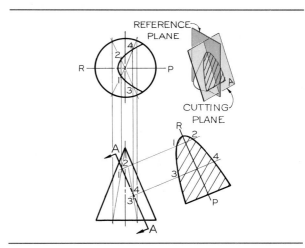

**Fig. 10.18** Auxiliary Section.

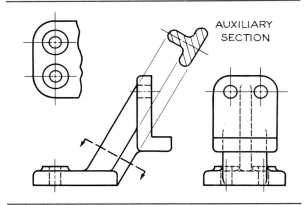

*Fig. 10.19*   Auxiliary Section.

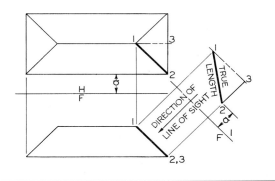

*Fig. 10.20*   True Length of a Line by Means of an Auxiliary View.

method shown here is by projection. In Fig. 10.18, elements of the cone are drawn in the front and top views. These intersect the cutting plane at points **1**, **2**, **3**, and so on. These points are established in the top view by projecting upward to the top views of the corresponding elements. In the auxiliary section, all points on the parabola are the same distance from the reference plane **RP** as they are in the top view.

A typical example of an auxiliary section in machine drawing is shown in Fig. 10.19. Here, there is not sufficient space for a revolved section, §9.9, although a removed section, §9.10, could have been used instead of an auxiliary section.

## 10.18 True Length of Line — Auxiliary-View Method

A line will show in true length when projected to a projection plane parallel to the line.

In Fig. 10.20, let it be required to find the true length of the hip rafter **1–2** by means of a depth auxiliary view.

I.   Assume an arrow perpendicular to **1–2** (front view) indicating the direction of sight, and place the **H/F** folding line as shown.

II.   Draw the **F/1** folding line perpendicular to the arrow and at any convenient distance from **1–2** (front view), and project the points **1** and **3** toward it.

III.   Set off the points **1** and **2** in the auxiliary view at the same distance from the folding line as they are in the top view. The triangle **1–2–3** in the auxiliary view shows the true size and shape of the roof section **1–2–3**, and the distance **1–2** in the auxiliary view is the true length of the hip rafter **1–2**.

To find the true length of a line by revolution, see §11.10.

## 10.19 Successive Auxiliary Views

Up to this point we have dealt with *primary* auxiliary views, that is, single auxiliary views projected from one of the regular views. In Fig. 10.21, auxiliary view **1** is a primary auxiliary view, projected from the top view.

From primary auxiliary view **1**, a *secondary* auxiliary view **2** can be drawn; then from it a third auxiliary view **3**, and so on. An infinite number of such successive auxiliary views may be drawn, a process that may be likened to the "chain reaction" in a nuclear explosion.

However, secondary auxiliary view **2** is not the only one that can be projected from primary auxiliary view **1** and thus start an independent "chain reaction." As shown by the arrows around view **1**, an infinite number of secondary auxiliary views, with different lines of sight, may be projected. *Any auxiliary view projected from a primary auxiliary view is a secondary auxiliary view.* Furthermore, any succeeding auxiliary view may be used to project an infinite number of "chains" of views from it.

In this example, folding lines are more convenient than reference-plane lines. In auxiliary view **1**, all numbered points of the object are the same distance from folding line **H/1** as they are in the front view from folding line **H/F**. These distances, such as distance **a**, are transferred from the front view to the auxiliary view with the aid of dividers.

To draw the secondary auxiliary view **2**, drop the front view from consideration, and center attention

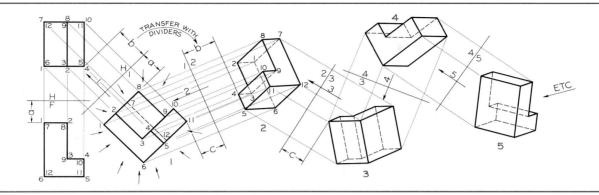

*Fig. 10.21*    Successive Auxiliary Views.

on the sequence of three views: the top view, view 1, and view 2. Draw arrow 2 toward view 1 in the direction desired for view 2, and draw light projection lines parallel to the arrow. Draw folding line 1/2 perpendicular to the projection lines and at any convenient distance from view 1. Locate all numbered points in view 2 from folding line 1/2 at the same distances they are in the top view from folding line H/1, using the dividers to transfer distances. For example, transfer distance **b** to locate points **4** and **5**. Connect points with straight lines, and determine visibility. The corner nearest the observer (**11**) for view 2 will be visible, and the one farthest away (**1**) will be hidden, as shown.

To draw views **3**, **4**, and so on, repeat this procedure, remembering that each time we will be concerned only with a sequence of three views. In drawing any auxiliary view, the paper may be revolved so as to make the last two views line up as regular views.

## 10.20 Uses of Auxiliary Views

Generally, auxiliary views are used to show the true shape or true angle of features that appear distorted in the regular views. Basically, auxiliary views have the following four uses.

1. True length of line (TL), §10.18.
2. Point view of line, §10.11.
3. Edge view of plane (EV), §10.21.
4. True size of plane (TS), §10.21.

## 10.21 True Size of an Oblique Surface— Folding-Line Method

A typical requirement of a secondary auxiliary view is to show the true size and shape of an oblique surface, such as surface 1–2–3–4 in Fig. 10.22. In this case folding lines are used, but the same results can be obtained with reference lines. Proceed as follows.

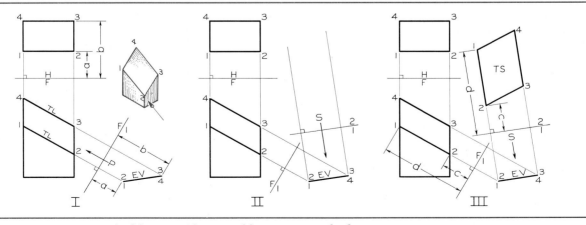

*Fig. 10.22*    True Size of Oblique Surface—Folding-Line Method.

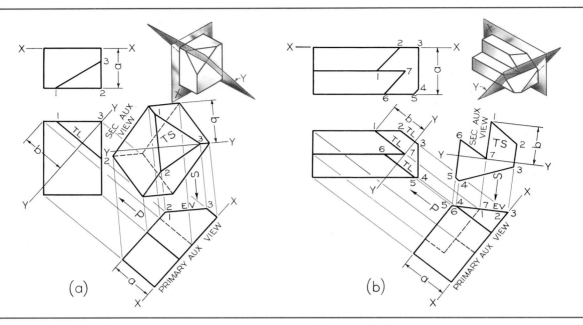

**Fig. 10.23** True Size of an Oblique Surface—Reference-Plane Method.

I.  Draw primary auxiliary view showing surface 1–2–3–4 as a line. As explained in §10.11, the edge view (EV) of a plane is found by getting the point view of a line in that plane. To get the point view of a line, the line of sight must be assumed parallel to the line. Therefore, draw arrow P parallel to lines 1–2 and 3–4, which are true length (TL) in the front view, and draw projection lines parallel to the arrow. Draw folding line H/F between the top and front views and F/1 between the front and auxiliary views, perpendicular to the respective projection lines. All points in the auxiliary view will be the same distance from the folding line F/1 as they are in the top view from folding line H/F. Lines 1–2 and 3–4 will appear as points in the auxiliary view, and plane 1–2–3–4 will therefore appear *edgewise*, that is, as a line.

II.  Draw arrow S perpendicular to the edge view of plane 1–2–3–4 in the primary auxiliary view, and draw projection lines parallel to the arrow. Draw folding line 1/2 perpendicular to these projection lines and at a convenient distance from the primary auxiliary view.

III.  Draw secondary auxiliary view. Locate each point (transfer with dividers) the same distance from the folding line 1/2 as it is in the front view to the folding line F/1, as for example, dimensions c and d. The true size (TS) of the surface 1–2–3–4 will be shown in the secondary auxiliary view, since the di-rection of sight, arrow S, was taken perpendicular to it.

## 10.22 True Size of an Oblique Surface—Reference-Plane Method

In Fig. 10.23 (a), it is required to draw an auxiliary view in which triangular surface 1–2–3 will appear in true size and shape. In order for the true size of the surface to appear in the secondary auxiliary view, arrow S must be assumed perpendicular to the edge view of that surface; so it is necessary to have the edge view of surface 1–2–3 in the primary auxiliary view first. In order to do this, the direction of sight, arrow P, must be parallel to a line in surface 1–2–3 that appears true length (TL) in the front view. Hence, arrow P is drawn parallel to line 1–2 of the front view, line 1–2 will appear as a point in the primary auxiliary view, and surface 1–2–3 must therefore appear edgewise in that view.

In this case it is convenient to use reference lines and to assume the reference plane X (for drawing the primary auxiliary view) coinciding with the back surface of the object, as shown. For the primary auxiliary view, all depth measurements, as a in the figure, are transferred with dividers from the top view with respect to the reference line X–X.

For the secondary auxiliary view, reference plane Y is assumed cutting through the object for

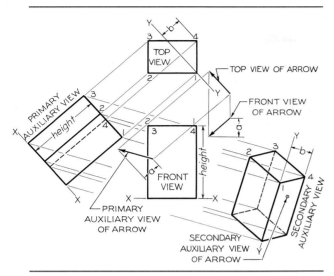

*Fig. 10.24* Secondary Auxiliary View with Oblique Direction of Sight Given.

convenience in transferring measurements. All measurements perpendicular to Y–Y in the secondary auxiliary view are the same as between the reference plane and the corresponding points in the front view. Note that corresponding measurements must be *inside* (toward the central view in the sequence of three views) or *outside* (away from the central view). For example, dimension b is on the side of Y–Y *away* from the primary auxiliary view in both places.

In Fig. 10.23 (b) it is required to find the true size and shape of surface 1–2–3–4–5–6–7 and not to draw the complete secondary auxiliary view. The method is similar to that just described.

## 10.23 Secondary Auxiliary View, Oblique Direction of Sight Given

In Fig. 10.24 two views of a block are given, with two views of an arrow indicating the direction in which it is desired to look at the object to obtain a view. Proceed as follows.

I.  *Draw primary auxiliary view of both the object and the assumed arrow,* which will show the true length *of the arrow.* In order to do this, assume a horizontal reference plane X–X in the front and auxiliary views, as shown. Then assume a direction of sight perpendicular to the given arrow. In the front view, the butt end of the arrow is a distance a higher than the arrow point, and this distance is

transferred to the primary auxiliary view as shown. All *height* measurements in the auxiliary view correspond to those in the front view.

II.  *Draw secondary auxiliary view,* which will show the arrow as a point. This can be done because the arrow shows in true length in the primary auxiliary view, and projection lines for the secondary auxiliary view are drawn parallel to it. Draw reference line Y–Y, for the secondary auxiliary view perpendicular to these projection lines. In the top view, draw Y–Y perpendicular to the projection lines to the primary auxiliary view. All measurements, such as b, with respect to Y–Y correspond in the secondary auxiliary view and the top view.

It will be observed that the secondary auxiliary views of Figs. 10.23 (a) and 10.24 have considerable pictorial value. These are trimetric projections, §17.31. However, the direction of sight could be assumed, in the manner of Fig. 10.24, to produce either isometric or dimetric projections. If the direction of sight is assumed parallel to the diagonal of a cube, the resulting view is an *isometric projection,* §17.3.

A typical application of a secondary auxiliary view in machine drawing is shown in Fig. 10.25. All

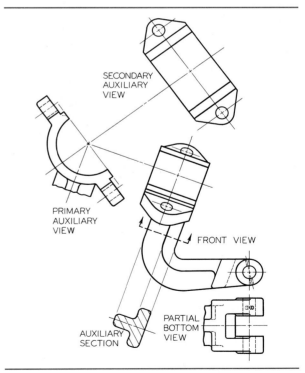

*Fig. 10.25* Secondary Auxiliary View—Partial Views.

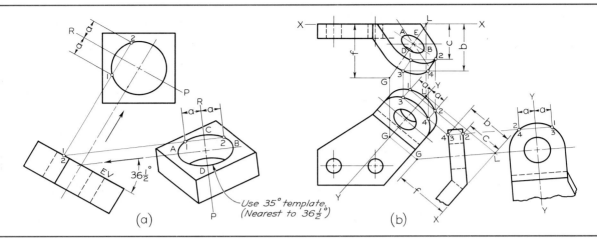

*Fig. 10.26*   Ellipses.

views are partial view, except the front view. The partial secondary auxiliary view illustrates a case in which break lines are not needed. Note the use of an auxiliary section to show the true shape of the arm.

## 10.24 Ellipses

As shown in §7.30, if a circle is viewed obliquely, the result is an ellipse. This often occurs in successive auxiliary views, because of the variety of directions of sight. In Fig. 10.26 (a) the hole appears as a true circle in the top view. The circles appear as straight lines in the primary auxiliary view, and as ellipses in the secondary auxiliary view. In the latter, the major axis **AB** of the ellipse is parallel to the projection lines and equal in length to the true diameter of the circle as shown in the top view. The minor axis **CD** is perpendicular to the major axis, and its foreshortened length is projected from the primary auxiliary view.

The ellipse can be completed by projecting points, such as **1** and **2**, symmetrically located about the reference plane **RP** coinciding with **CD** with distances **a** equal in the top and secondary auxiliary views as shown, and finally, after a sufficient number of points have been plotted, by applying the irregular curve, §2.54

Since the major and minor axes are easily found, any of the true-ellipse methods of Figs. 5.48 to 5.50 and 5.52 (a) may be used, or an approximate ellipse, Fig. 5.56, may be found sufficiently accurate for a particular drawing. Or the ellipses may be easily and rapidly drawn with the aid of an ellipse template, §5.57. The "angle" of ellipse to use is the one that most closely matches the angle between the direction of sight arrow and the plane (**EV**) containing the circle, as seen in this case in the primary auxiliary view. Here the angle is $36\frac{1}{2}°$, so a 35° ellipse is selected.

At (b) successive auxiliary views are shown in which the true circular shapes appear in the secondary auxiliary view, and the elliptical projections in the front and top views. It is necessary to construct the circular shapes in the secondary auxiliary view, then to project plotted points back to the primary auxiliary view, the front view, and finally to the top view, as shown in the figure for points **1**, **2**, **3**, and **4**. The final curves are then drawn with the aid of the irregular curve.

If the major and minor axes are found, any of the true-ellipse methods may be used; or better still, an ellipse template, §5.57, may be employed. The major and minor axes are easily established in the front view, but in the top view, they are more difficult to find. The major axis **AB** is at right angles to the center line **GL** of the hole, and equal in length to the true diameter of the hole. The minor axis **ED** is at right angles to the major axis. Its length is found by plotting several points in the vicinity of one end of the minor axis, or by using descriptive geometry to find the angle between the line of sight and the inclined surface, and by this angle selecting the ellipse guide required.

# Auxiliary View Problems

The problems in Figs. 10.27 to 10.49 are to be drawn with instruments or freehand. If partial auxiliary views are not assigned, the auxiliary views are to be complete views of the entire object, including all necessary hidden lines.

It is often difficult to space properly the views of an auxiliary view drawing. In some cases it may be necessary to make a trial blocking out on a preliminary sheet before starting the actual drawing. Allowances for dimensions must be made if metric or decimal dimensions are to be included. In such case, the student should study §§13.1 to 13.25.

Since many of the problems in this chapter are of a general nature, they can also be solved on most computer graphics systems. If a system is available, the instructor may choose to assign specific problems to be completed by this method.

Auxiliary view problems in convenient form for solution may be found in *Principles of Engineering Graphics Problems* by Spencer, Hill, Loving, and Dygdon, a workbook designed to accompany this text that is also published by Macmillan Publishing Company.

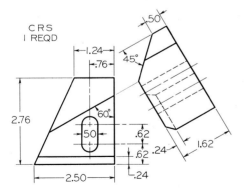

**Fig. 10.27   RH Finger.**
Given: Front and auxiliary views.
Required: Complete front, auxiliary, left-side, and top views [Layout A–3 or A4–3 (adjusted)].

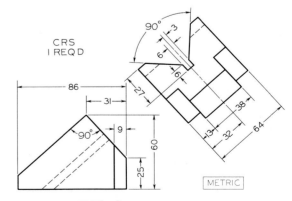

**Fig. 10.28   V-Block.**
Given: Front and auxiliary views.
Required: Complete front, top, and auxiliary views [Layout A–3 or A4–3 (adjusted)].

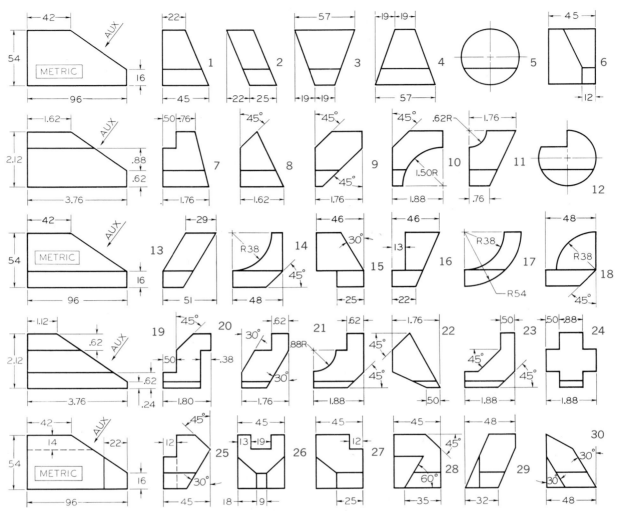

**Fig. 10.29  Auxiliary View Problems.**  Make freehand sketch or instrument drawing of selected problem as assigned by instructor. Draw given front and right-side views, and add incomplete auxiliary view, including all hidden lines [Layout A–3 or A4–3 (adjusted)]. If assigned, design your own right-side view consistent with given front view, and then add complete auxiliary view. Problems 1 to 6, 13 to 18, and 25 to 30 are given in metric dimensions. Problems 7 to 12 and 19 to 24 are given in decimal-inch dimensions.

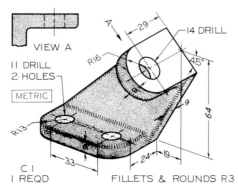

**Fig. 10.30  Anchor Bracket.**  Draw necessary views or partial views [Layout A–3 or A4–3 (adjusted)].*

*If dimensions are required, study §§13.1 to 13.25. Use metric or decimal-inch dimensions as assigned by the instructor.

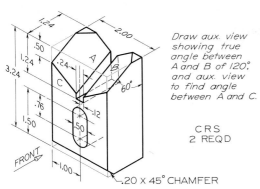

**Fig. 10.31** Centering Block. Draw complete front, top, and right-side views, plus indicated auxiliary views (Layout B–3 or A3–3).*

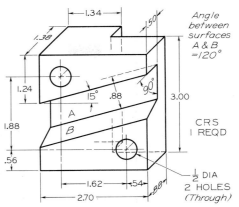

**Fig. 10.32** Clamp Slide. Draw necessary views completely (Layout B–3 or A3–3).*

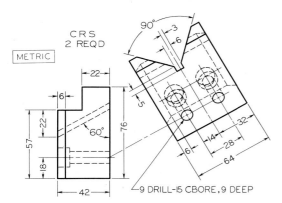

**Fig. 10.33** Guide Block.
Given: Right-side and auxiliary views.
Required: Right-side, auxiliary, plus front and top views—all complete (Layout B–3 or A3–3).*

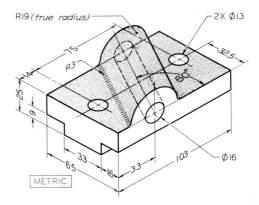

**Fig. 10.34** Angle Bearing. Draw necessary views, including a complete auxiliary view [Layout A–3 or A4–3 (adjusted)].*

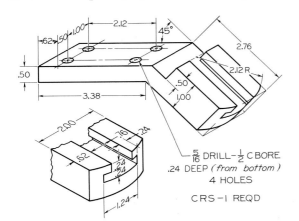

**Fig. 10.35** Guide Bracket. Draw necessary views or partial views (Layout B–3 or A3–3).*

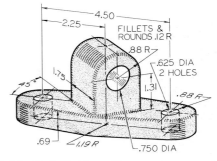

**Fig. 10.36** Rod Guide. Draw necessary views, including complete auxiliary view showing true shape of upper rounded portion [Layout B–4 or A3–4 (adjusted)].*

*If dimensions are required, study §§13.1 to 13.25. Use metric or decimal-inch dimensions as assigned by the instructor.

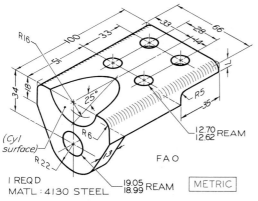

**Fig. 10.37** Angle Guide. Draw necessary views, including a partial auxiliary view of cylindrical recess [Layout B–4 or A3–4 (adjusted)].*

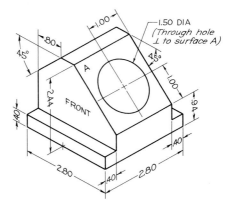

**Fig. 10.38** Holder Block. Draw front and right-side views (2.80″ apart) and complete auxiliary view of entire object showing true shape of surface **A** and all hidden lines [Layout A–3 or A4–3 (adjusted)].*

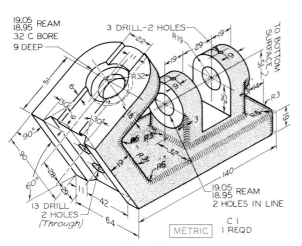

**Fig. 10.39** Control Bracket. Draw necessary views, including partial auxiliary views and regular views (Layout C–4 or A2–4).*

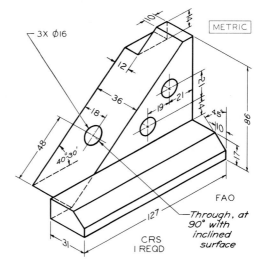

**Fig. 10.40** Adjuster Block. Draw necessary views, including complete auxiliary view showing true shape of inclined surface [Layout B–4 or A3–4 (adjusted)].*

*If dimensions are required, study §§13.1 to 13.25. Use metric or decimal-inch dimensions as assigned by the instructor.

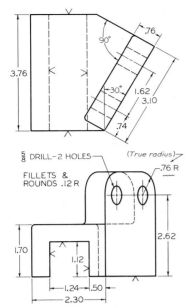

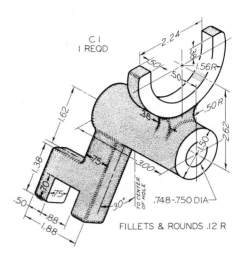

**Fig. 10.41** Drill Press Bracket. Draw given views and add complete auxiliary view showing true shape of inclined face [Layout B–4 or A3–4 (adjusted)].*

**Fig. 10.42** Shifter Fork. Draw necessary views, including partial auxiliary view showing true shape of inclined arm [Layout B–4 or A3–4 (adjusted)].*

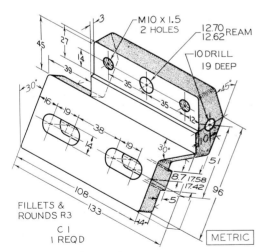

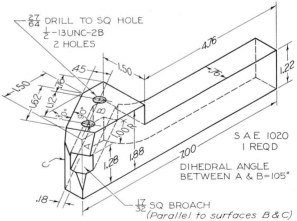

**Fig. 10.43** Cam Bracket. Draw necessary views or partial views as needed. For threads, see §§15.9 and 15.10 [Layout B–4 or A3–4 (adjusted)].*

**Fig. 10.44** RH Tool Holder. Draw necessary views, including partial auxiliary views showing 105° angle and square hole true size. For threads, see §§15.9 and 15.10 [Layout B–4 or A3–4 (adjusted)].*

*If dimensions are required, study §§13.1 to 13.25. Use metric or decimal-inch dimensions as assigned by the instructor.

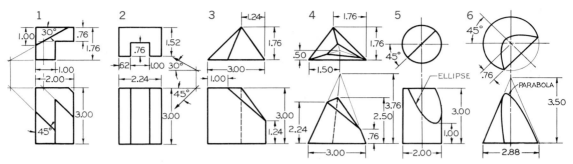

**Fig. 10.45** Draw secondary auxiliary views, complete, which (except Prob. 2) will show the true sizes of the inclined surfaces. In Prob. 2, draw secondary auxiliary view as seen in direction of arrow (Layout B–3 or A3–3).*

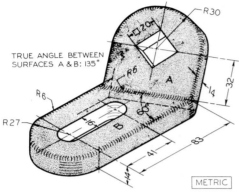

**Fig. 10.46** Control Bracket. Draw necessary views including primary and secondary auxiliary views so that the latter shows true shape of oblique surface A [(Layout B–4 or A3–4 (adjusted)].*

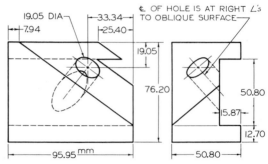

**Fig. 10.47** Holder Block. Draw given views and primary and secondary auxiliary views so that the latter shows true shape of oblique surface [Layout B–4 or A3–4 (adjusted)].*

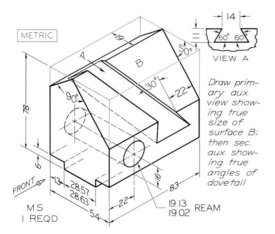

**Fig. 10.48** Adjustable Stop. Draw complete front and auxiliary views plus partial right-side view. Show all hidden lines (Layout C–4 or A2–4).*

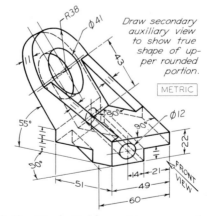

**Fig. 10.49** Tool Holder. Draw complete front view, and primary and secondary auxiliary views as indicated [Layout B–4 or A3–4 (adjusted)].*

*If dimensions are required, study §§13.1 to 13.25. Use metric or decimal-inch dimensions as assigned by the instructor.

# CHAPTER
# 11

# Revolutions

To obtain an auxiliary view, the observer changes position with respect to the object, as shown by the arrow in Fig. 11.1 (a). The auxiliary view shows the true size and shape of surface A. Exactly the same view of the object also can be obtained by moving the object with respect to the observer, as shown at (b). Here the object is revolved until surface A appears in its true size and shape in the right-side view.

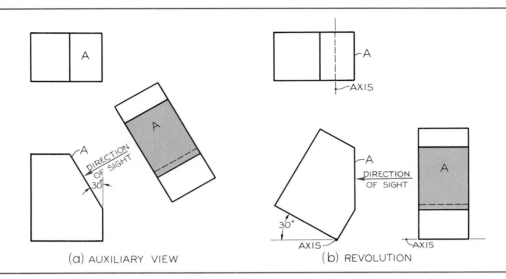

**Fig. 11.1**  Auxiliary View and Revolution Compared.

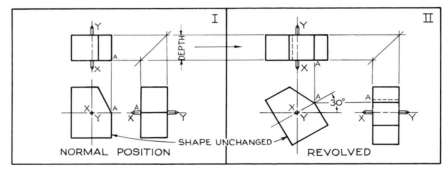

**Fig. 11.2**  Primary Revolution About an Axis Perpendicular to Frontal Plane.

## 11.1 Axis of Revolution

In Fig. 11.1 (b) the axis of revolution is assumed perpendicular to the frontal plane of projection. Note that the view in which the axis of revolution appears as a point (in this case the front view) *revolves but does not change shape* and that in the views in which the axis is shown as a line in true length the *dimensions of the object parallel to the axis do not change.*

To make a revolution drawing, the view on the plane of projection that is perpendicular to the axis of revolution is drawn first, since it is the only view that remains unchanged in size and shape. This view is drawn revolved either *clockwise* or *counterclockwise* about a point that is the end view, or point view, of the axis of revolution. This point may be assumed at any convenient point on or outside the view. The other views are then projected from this view.

The axis of revolution is usually considered perpendicular to one of the three principal planes of projection. Thus, an object may be revolved about an axis perpendicular to the horizontal, frontal, or profile planes of projection, and the views drawn in the new positions. Such a process is called a *primary revolution*. If this drawing is then used as a basis for another revolution, the operation is called *successive revolutions*. Obviously, this process may be continued indefinitely, which reminds us of the "chain reaction" in successive auxiliary views, Fig. 10.21.

## 11.2 Revolution About Axis Perpendicular to Frontal Plane

A primary revolution is illustrated in Fig. 11.2. An imaginary axis **XY** is assumed, about which the object is to revolve to the desired position. In this case the axis is selected perpendicular to the frontal

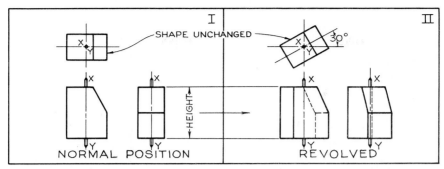

*Fig. 11.3*   Primary Revolution About an Axis Perpendicular to Horizontal Plane.

plane of projection, and during the revolution all points of the object describe circular arcs parallel to that plane. The axis may pierce the object at any point or may be exterior to it. In space II, the front view is drawn *revolved* (but not changed in shape) through the angle desired (30° in this case), and the top and side views are obtained by projecting from the front view. The *depth* of the top view and the side view is found by projecting from the top view of the first unrevolved position (space I) *because the depth, since it is parallel to the axis, remains unchanged.* If the front view of the revolved position is drawn directly without first drawing the normal unrevolved position, the depth of the object, as shown in the revolved top and side views, may be drawn to known dimensions. No difficulty should be encountered by the student who understands how to obtain projections of points and lines, §7.6.

Note the similarity between the top and side views in space II of Fig. 11.2 and some of the auxiliary views of Fig. 10.7 (c).

## 11.3 Revolution About Axis Perpendicular to Horizontal Plane

A revolution about an axis perpendicular to the horizontal plane of projection is shown in Fig. 11.3. An imaginary axis XY is assumed perpendicular to the top plane of projection, the top view is drawn revolved (but not changed in shape) to the desired position (30° in this case), and the other views are obtained by projecting from this view. During the revolution, all points of the object describe circular arcs parallel to the horizontal plane. The *heights* of all points in the front and side views in the revolved position remain unchanged, since they are measured parallel to the axis, and may be drawn by projecting from the initial front and side views of space I.

Note the similarity between the front and side views in space II of Fig. 11.3 and some of the auxiliary views of Fig. 10.8 (c).

## 11.4 Revolution About Axis Perpendicular to Profile Plane

A revolution about an axis XY perpendicular to the profile plane of projection is illustrated in Fig. 11.4. During the revolution, all points of the object describe circular arcs parallel to the profile plane of projection. The widths of the top and front views in the revolved position remain unchanged, since they

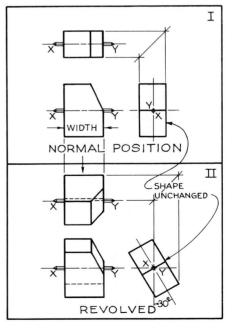

*Fig. 11.4*   Primary Revolution About an Axis Perpendicular to Profile Plane.

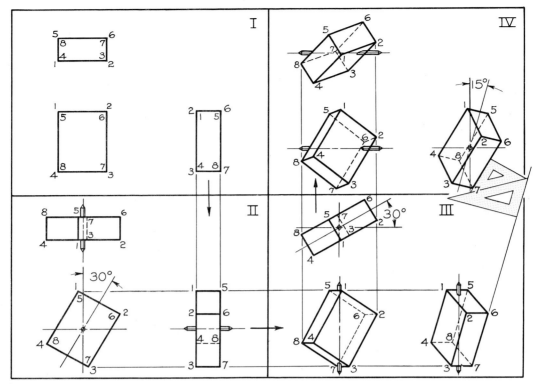

**Fig. 11.5**    Successive Revolutions of a Prism.

are measured parallel to the axis, and may be obtained by projection from the top and front views of space I, or may be set off by direct measurement.

Note the similarity between the top and front views in space II of Fig. 11.4 and some of the auxiliary views of Fig. 10.9 (c).

## 11.5 Successive Revolutions

It is possible to draw an object in an infinite number of revolved positions by making successive revolutions. Such a procedure, Fig. 11.5, limited to three or four stages, offers excellent practice in multiview projection. While it is possible to make several revolutions of a simple object without the aid of a system of numbers, it is absolutely necessary in successive revolutions to assign a number or a letter to every corner of the object. See §7.6.

The numbering or lettering must be consistent in the various views of the several stages of revolution. Figure 11.5 shows four sets of multiview drawings numbered I, II, III, and IV, respectively. These represent the same object in different positions with reference to the planes of projection.

In space I, the object is represented in its normal position, with its faces parallel to the planes of projection. In space II, the object is represented after it has been revolved clockwise through an angle of 30° about an axis perpendicular to the frontal plane. The drawing in space II is placed under space I so that the side view, whose width remains unchanged, can be projected from space I to space II as shown.

During the revolution, all points of the object describe circular arcs parallel to the frontal plane of projection and remain at the same distance from that plane. The side view, therefore, may be projected from the side view of space I and the front view of space II. The top view may be projected in the usual manner from the front and side views of space II.

In space III, the object is taken as represented in space II and is revolved counterclockwise through an angle of 30° about an axis perpendicular to the horizontal plane of projection. During the revolution, all points describe *horizontal* circular arcs and remain at the same distance from the horizontal plane of projection. The top view is copied from space II but is revolved through 30°. The front and side views are obtained by projecting from the front

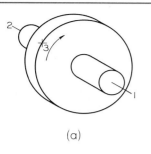

(a)

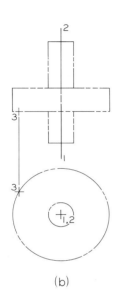

(b)

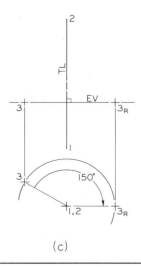

(c)

**Fig. 11.6**  Revolution of a Point About a Normal Axis.

and side views of space II and from the top view of space III.

In space IV, the object is taken as represented in space III and is revolved clockwise through 15° about an axis perpendicular to the profile plane of projection. During the revolution, all points of the object describe circular arcs parallel to the profile plane of projection and remain at the same distance from that plane. The side view is copied, §5.28, from the side view of space III but revolved through 15°. The front and top views are projected from the side view of space IV and from the top and front views of space III.

Another convenient method of copying a view in a new revolved position is to use tracing paper as described in §5.29. Either a tracing can be made and transferred by rubbing, or the prick points may be made and transferred, as shown.

In spaces III and IV of Fig. 11.5, each view is an axonometric projection, §17.2. An isometric projection can be obtained by revolution, as shown in Fig. 17.3, and a dimetric projection, §17.28, can be constructed in a similar manner. If neither an isometric nor a dimetric projection is specifically sought, the successive revolution will produce a trimetric projection, §17.31, as shown in Fig. 11.5.

## 11.6  Revolution of a Point—Normal Axis

Examples of the revolution of a point about a straight-line axis are often found in design problems that involve pulleys, gears, cranks, linkages, and so on. For example, in Fig. 11.6 (a), as the disk is revolved, point **3** moves in a circular path lying in a plane perpendicular to the axis **1–2**. This relationship is represented in the two views at (b). Note in this instance that the axis is normal or perpendicular to the frontal plane of projection, resulting in a front view that shows a point view of the axis and a true-size view of the circular path of revolution for point **3**. The top view shows the path of revolution in edge view and perpendicular to the true-length view of the axis. Similar two-view relationships would occur if the axis were perpendicular or normal to either the horizontal or profile planes of projection.

The clockwise revolution through 150° for point **3** is illustrated at (c).

## 11.7  Revolution of a Point—Inclined Axis

In Fig. 11.7 (a) the axis of revolution for point **3** is positioned parallel to the frontal plane and inclined to the horizontal and profile projection planes. Since the axis **1–2** is true length in the front view, the

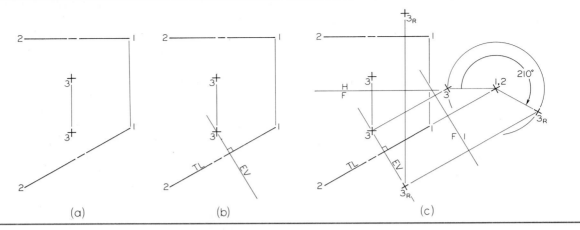

*Fig. 11.7*  Revolution of a Point About an Inclined Axis.

edge view of the path of revolution can be located as at (b). In order to establish the circular path of revolution for point **3**, an auxiliary view showing the axis in point view is required as at (c). The required revolution of point **3** (in this case, **210°**) is now performed in this circular view. The revolved position of the point is projected to the given front and top views, as shown.

Note the similarity of the relationships of the front view and auxiliary view and the constructions shown in Fig. 11.6 (c).

## 11.8 Revolution of a Point—Oblique Axis

In Fig. 11.8 (a), the axis of revolution for point **3** is oblique to all principal planes of projection and, therefore, is shown neither in true length nor as a point view in the top, front, or profile views. To establish the necessary true length and point view of the axis **1–2** in adjacent views, two successive auxiliary views are required, as shown at (b). The required revolved position of point **3** can now be located and then projected back to complete the given front and top views.

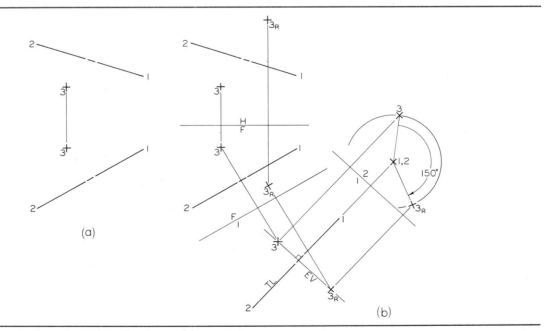

*Fig. 11.8*  Revolution of a Point About an Oblique Axis.

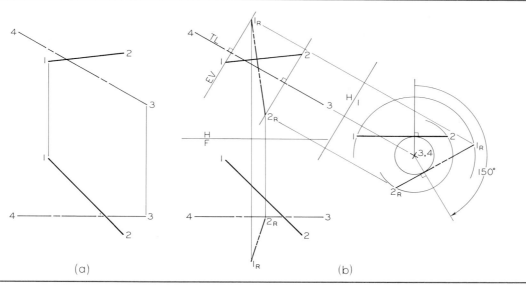

**Fig. 11.9** Revolution of a Line About an Inclined Axis.

## 11.9 Revolution of a Line

The procedure for the revolution of a line about an axis is very similar to that required for the revolution of a point, §11.6. All points on a line must revolve through the same angle, or the revolved line becomes altered.

In Fig. 11.9 (a), the line **1–2** is to be revolved through 150° about the inclined axis **3–4**.

Since the axis **3–4** is given in true length in the top view, an auxiliary view is required to provide a point view of the axis, as shown in Fig. 11.9 (b). The necessary revolution can then be made about point view **3–4**. In order to insure that all points on the line rotate through the same number of degrees, note that a construction circle tangent to line **1–2** is

drawn and that a perpendicular through the tangency point becomes the reference for measurement of the angle of rotation. The circular arc paths for points **1** and **2** locate the points **1$_R$** and **2$_R$**, as the revolved position of the line is drawn perpendicular to the radial line subtending the 150° arc of revolution, and tangent to the smaller circle. The alternate-position line is used to distinguish the revolved-position line from the original given line.

## 11.10 True Length of a Line — Revolution Method

If a line is parallel to one of the planes of projection, its projection on that plane is equal in length to the line, Fig. 7.20. In Fig. 11.10 (a), the element **AB** of

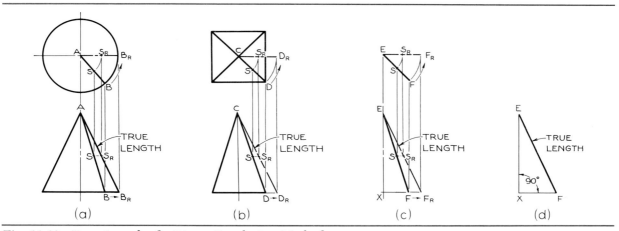

**Fig. 11.10** True Length of a Line—Revolution Method.

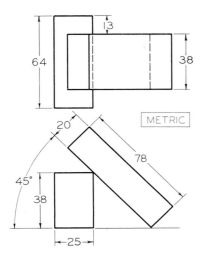

**Fig. 11.17** Using Layout A–2 or A–3 or Layout A4–2 or A4–3 (adjusted), draw three views of the blocks but revolved 30° clockwise about an axis perpendicular to the top plane of projection. Do not change the relative positions of the blocks.

**Fig. 11.18 opposite** Use Layout A–1 or A4–1 (adjusted), and divide the working area into four equal areas for four problems per sheet to be assigned by the instructor. Data for the layout of each problem are given by a coordinate system in metric dimensions. For example, in Prob. 1, point 1 is located by the scale coordinates (28 mm, 38 mm, 76 mm). The first coordinate locates the front view of the point from the left edge of the problem area. The second one locates the front view of the point from the bottom edge of the problem area. The third one locates either the top view of the point from the bottom edge of the problem area or the side view of the point from the left edge of the problem area. Inspection of the given problem layout will determine which application to use.

1. Revolve clockwise point 1(28, 38, 76) through 210° about the axis 2(51, 58, 94) – 3(51, 8, 94).
2. Revolve point 3(41, 38, 53) about the axis 1(28, 64, 74) – 2(28, 8, 74) until point 3 is at the farthest distance behind the axis.
3. Revolve point 3(20, 8, 84) about the axis 1(10, 18, 122) – 2(56, 18, 76) through 210° and to the rear of line 1–2.
4. Revolve point 3(5, 53, 53) about the axis 1(10, 13, 71) – 2(23, 66, 71) to its extreme position to the left in the front view.
5. Revolve point 3(15, 8, 99) about the axis 1(8, 10, 61) – 2(33, 25, 104) through 180°.
6. By revolution find the true length of line 1(8, 48, 64) – 2(79, 8, 119). Scale: 1:100.
7. Revolve line 3(30, 38, 81) – 4(76, 51, 114) about axis 1(51, 33, 69) – 2(51, 33, 122) until line 3–4 is shown true length and below the axis 1–2. Scale: 1:20.
8. Revolve line 3(53, 8, 97) – 4(94, 28, 91) about the axis 1(48, 23, 81) – 2(91, 23, 122) until line 3–4 is in true length and above the axis.
9. Revolve line 3(28, 15, 99) – 4(13, 30, 84) about the axis 1(20, 20, 97) – 2(43, 33, 58) until line 3–4 is level and above the axis.

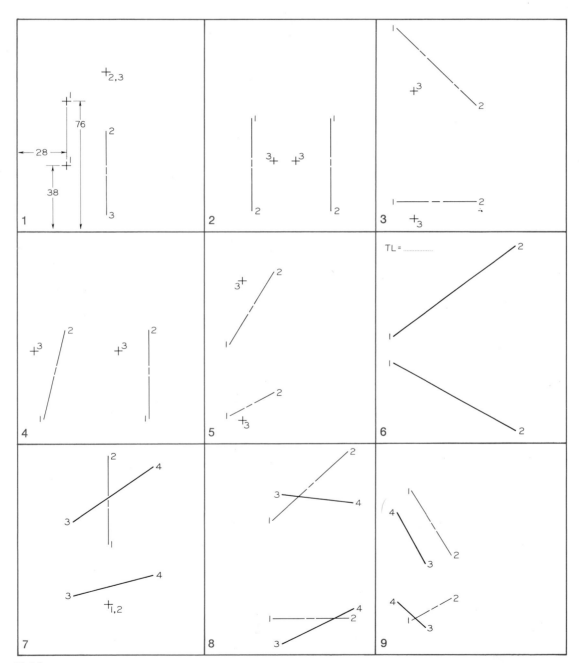

*Fig. 11.18*

# Manufacturing Processes

by J. George H. Thompson*
and John Gilbert McGuire[†]
Revised by Stephen A. Smith[‡]

The test of the usefulness of any working drawing is whether the object described can be satisfactorily produced without further information than that furnished on the drawing. The drawing must give information as to shape, size, material, and finish and, where necessary, indicate the manufacturing processes required to produce the desired object.

It is the purpose of this chapter to provide beginning engineers with some information about certain fundamental terms and processes and to assist them in using this information on their drawings. The drafter and engineer in any organization must have a thorough knowledge of manufacturing processes and methods before they can properly indicate on drawings the machining operations, heat-treatment, finish, and the accuracy desired on each part.

*Late Professor Emeritus of Mechanical Engineering, Texas A&M University.
†Assistant Dean of Engineering Emeritus, Texas A&M University.
‡Tool Development Engineer, Packaging Corporation of America, A Tenneco Company.

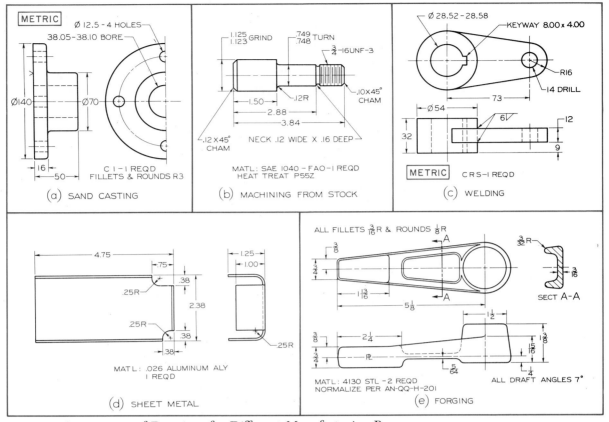

**Fig. 12.1**   Comparison of Drawings for Different Manufacturing Processes.

## 12.1 Production Processes

A manufacturing department starts with what might be called *raw stock* and modifies this until it agrees with the detail drawing. The shape of raw stock usually has to be altered.

Changing the shape and size of the material of which a part is being made requires one or more of the following processes: (1) removing part of the original material, (2) adding more material, and (3) redistributing original material. Cutting, such as turning on a lathe, punching holes by means of a power press, or cutting with a laser system, removes material. Welding, brazing, soldering, metal spraying, and electrochemical plating add material. Forging, pressing, drawing, extruding, spinning, and plastics processing redistribute material.

## 12.2 Manufacturing Methods and the Drawing

Before preparing a drawing for the production of a part, the drafter should consider what manufacturing processes are to be used. These processes will determine the representation of the detailed features of the part, the choice of dimensions, and the machining or processing accuracy. Principal types of metal forming are (1) casting, (2) machining from standard stock, (3) welding, (4) forming from sheet stock, and (5) forging. A knowledge of these processes, along with a thorough understanding of the intended use of the part, will help determine some basic manufacturing processes. Drawings that reflect these manufacturing methods are shown in Fig. 12.1.

In sand casting, Fig. 12.1 (a), all cast surfaces remain rough textured, with all corners filleted or rounded. Sharp corners indicate at least one of the surfaces is finished, §7.34, and finish marks are shown on the edge view of the finished surface. See §§13.16 and 13.17.

In drawings of parts machined from standard stock, Fig. 12.1 (b), most surfaces are represented as machined. In some cases, as on shafting, the surface existing on the raw stock is often accurate enough without further finishing. Corners are usually sharp, but fillets and rounds are machined when

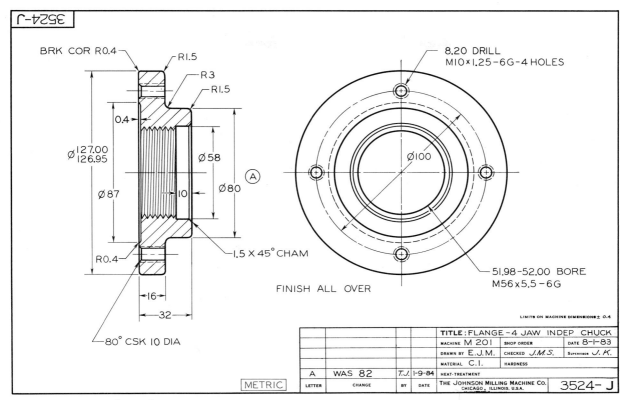

*Fig. 12.2* A Detail Working Drawing.

necessary. For example, an interior corner may be machined with a radius to provide greater strength.

On welding drawings, (c), the several pieces are cut to size, brought together, and then welded. Notice that lines are shown where the separate pieces are joined.

On sheet-metal drawings, (d), the thickness of material is uniform and is usually given in the material specification note rather than by a dimension on the drawing. Bend radii and bend reliefs at corners are specified according to standard practice.

For forged parts, §12.21, separate drawings are usually made for the diemaker and for the machinist. All corners are rounded and filleted and are so shown on the drawing. The draft is drawn to scale, and is usually specified by degrees in a note.

## 12.3 Sand Casting
Although a number of different casting processes are used, sand molds are the most common. Sand molds are made by ramming sand around a pattern, and then carefully removing it, leaving a cavity that ex-

actly matches the pattern to receive the molten metal. The pattern must be of such a shape that it will "pull away" from the sand. The plane of separation of the two mold halves is the *parting line* on the pattern.

Since shrinkage occurs when metal cools, patterns are made slightly oversize. The patternmaker accomplishes this by increasing the pattern size by predetermined amounts, dependent upon the kind of metal being used in the casting. Due to shrinkage and draft, small holes are better drilled in the casting, and large holes are better cored (cast-in) and then bored.

## 12.4 The Patternmaker and the Drawing
The patternmaker receives the working drawing showing the object in its completed state, including all dimensions and finish marks. Usually the same drawing is used by the patternmaker and the machinist; hence, it should contain all dimensions and notes needed by both, as shown in Fig. 12.2. Some companies follow the practice of dimensioning a

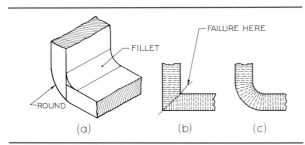

**Fig. 12.3**  Fillets and Rounds.

copy of the drawing in colored pencil with the patternmaker's information. Pattern dimensions, §13.27, typically need to be accurate only to within $\frac{1}{32}''$ or $\frac{1}{16}''$. More critical tolerances are required for machining however.

Finish marks, §13.17, are as important to the patternmaker as to the machinist because additional material must be provided on each surface that will eventually be machined. For small and medium-sized castings, 1.5 mm ($\frac{1}{16}''$ approximate) to 3 mm ($\frac{1}{8}''$ approximate) is usually sufficient; larger allowances are made if there is probability of distortion or warping. On the flange, Fig. 12.2, it is necessary for the patternmaker to provide material for finish on all surfaces, since the note indicates finish all over (FAO).

## 12.5 Fillets and Rounds

Fillets (inside rounded corners) and rounds (outside rounded corners) must be included in the pattern,

in order to provide for maximum strength as well as a pleasing appearance in the finished casting, Fig. 12.3. Crystals of cooling metal tend to arrange themselves perpendicular to the exterior surfaces as indicated at (b) and (c). If the corners of a casting are rounded as shown at (c), a much stronger casting results.

Rounds are constructed by rounding off the corners of the pattern by sanding, planing, or turning, while fillets are constructed of preformed leather, wax, or wood.

For some cylindrical patterns it is possible to form on the lathe the proper radius for the fillets and rounds as an integral part of the wood pattern.

All fillets and rounds should be shown on the drawing, either freehand or drawn to scale with the use of a compass or with a circle template. For a general discussion of fillets and rounds from the standpoint of representation on the drawing, see §§7.34 to 7.36. For dimensioning of fillets and rounds, see §13.16.

## 12.6 Machine Tools

Some of the more common machine tools are the engine lathe, drill press, milling machine, shaper, planer, grinding machine, and boring mill. Brief descriptions of these machines follow. See also §12.23 for examples of how these machine tools can be blended into automated processing.

## 12.7 Engine Lathe

The engine lathe, Fig. 12.4, is one of the most versatile machines used in the machine shop, and on it

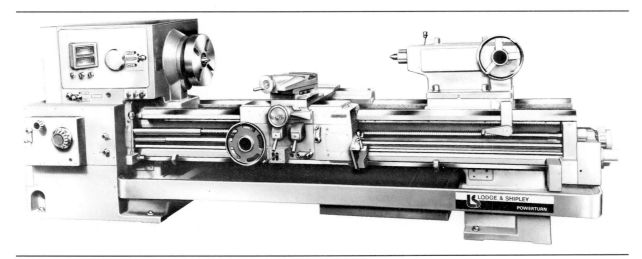

**Fig. 12.4**  Engine Lathe. *Courtesy of Lodge & Shipley*

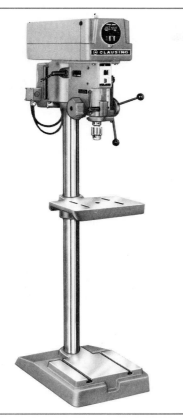

Fig. 12.6 Radial Drill Press. *Courtesy of Clausing Machine Tools*

Fig. 12.5 Drill Press. *Courtesy of Clausing Machine Tools*

are performed such operations as turning, boring, reaming, facing, threading, and knurling.

The workpiece is held in the lathe in a chuck (essentially a rotating vise). The cutting tool is fastened in the tool holder of the lathe and fed mechanically into the work as required.

## 12.8 Drill Press

The drill press, Fig. 12.5, is one of the most frequently used machine tools. Some of the operations that may be performed on this machine are drilling, reaming, boring, tapping, spot facing, counterboring, and countersinking.

The *radial* drill press, shown in Fig. 12.6, with its adjustable head and spindle, is very versatile and is especially suitable for large work.

## 12.9 Milling Machine

On the milling machine, Fig. 12.7, cutting is accomplished by feeding the work into a rotating cutter. Large plane milling machines, similar to the planer

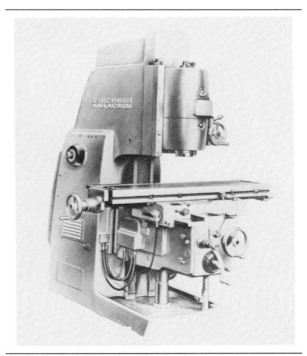

Fig. 12.7 Milling Machine. *Courtesy of Cincinnati Milacron*

**Fig. 12.8** Typical Milling Cutters. *Courtesy of Sharpaloy Division, Precision Industries, Inc.*

shown in Fig. 12.11, are also important production machines. In such plane mills, rotating milling cutters, Fig. 12.8, are used. In addition, it is practical to drill, ream, and bore on the milling machine.

The accuracy of many types of machine tools is dependent on the degree of accuracy to which the machine is set up and maintained. A laser-based measurement system, Fig. 12.9, is used in the calibration of the positioning accuracy of numerically controlled machine tools such as milling machines.

## 12.10 Shaper

On the shaper, Fig. 12.10, work is held in a vise while a single-pointed nonrotating cutting tool mounted in a reciprocating head is forced to move in a straight line past the stationary work. Between succeeding strokes of the tool, the vise that holds the work is fed mechanically into the path of the tool for the next cut alongside the one just completed. Shapers are often used to cut external and internal keyways, gear racks, dovetails, and T-slots.

## 12.11 Planer

On the planer, Fig. 12.11, work is fixed to a table that is moved mechanically so that the cutting takes place between stationary tools and moving work. This machine, with its reciprocating bed, and a tool head that is adjustable both horizontally and vertically, is used principally for machining large plane surfaces, or surfaces on a large number of pieces, as shown. In extremely large planers, the table is sta-

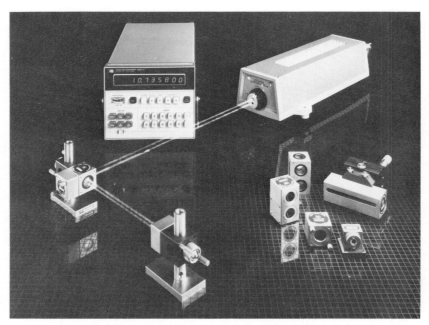

**Fig. 12.9** Laser Machine Tool Calibration System. *Courtesy of Hewlett Packard.*

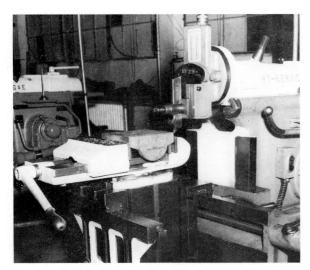

Fig. 12.10  Machining Plane Surface on Shaper.

tionary and the cutting tool is carried along a track past the work.

## 12.12 Boring Mill

The vertical boring mill, Fig. 12.12, is used for facing, turning, and boring heavy work weighing up to 20 tons. The vertical boring mill has a large rotating table and a nonrotating cutting tool that moves mechanically into the work.

The horizontal boring machine is suitable for

Fig. 12.12  Vertical Boring Mill.

accurate boring, reaming, facing, counterboring, and milling of pieces larger than could be handled on the typical milling machine.

The jig borer is a precision machine that somewhat resembles a drill press in its basic features of

Fig. 12.11  Planer.

**Fig. 12.13**   Vertical and Horizontal Turning Center. *Courtesy of Ingersoll Milling Machine Co.*

a rotating vertical spindle supporting a cutting tool and a stationary horizontal table for holding the work.

## 12.13 Vertical and Horizontal Turning Center

Vertical and horizontal turning centers, Fig. 12.13, eliminate the need for individual shaping, planing, and boring machines as previously described. These computer-driven machining centers not only replace planer mills, but they also do away with the need for moving a machined part from a milling machine to a drilling machine to a boring machine, and so on. The automatic tool changer on this machine stores 48 different cutters weighing up to 120 pounds each.

## 12.14 Grinding Machine

A grinding machine is used for removing a relatively small amount of material to bring the work to a very fine and accurate finish. In grinding, the work is fed mechanically against the rapidly rotating grinding wheel, and the depth of cut may be varied from 0.03 mm (.001″) to 0.0064 mm (.00025″).

Surface grinders, Fig. 12.14, reciprocate the work on a table that is simultaneously fed transverse to the grinding wheel. The unit shown lets the operator dial into the electronic memory the precise amount of metal to be removed during each of the roughing and finishing cuts and the amount of metal to be removed with each pass and even allows the

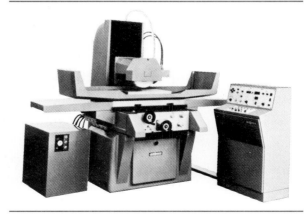

**Fig. 12.14**   Electronic Sequence-Controlled Surface Grinder. *Courtesy of Clausing/Jakobsen*

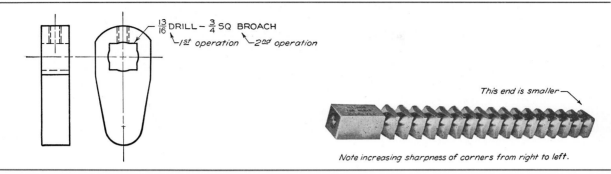

*Fig. 12.15*   Broaching.

operator to dial in the amount of wheel dressing to be done each cycle. The electronic memory automatically compensates for the loss of abrasive on the grinding wheel as the work proceeds.

## 12.15 Broaching

Broaching is similar to a single-stroke filing operation. As the broach is forced through the work, each succeeding tooth bites deeper and deeper into the metal, thus enlarging and forming the keyway as the broach passes through the hole.

A typical broach, along with the corresponding drawing calling for its use, is shown in Fig. 12.15.

## 12.16 Holes

Rough holes are produced in metal by coring, piercing (punching) and flame cutting. Drilling, Fig. 12.16(a), although superior, does not produce a hole

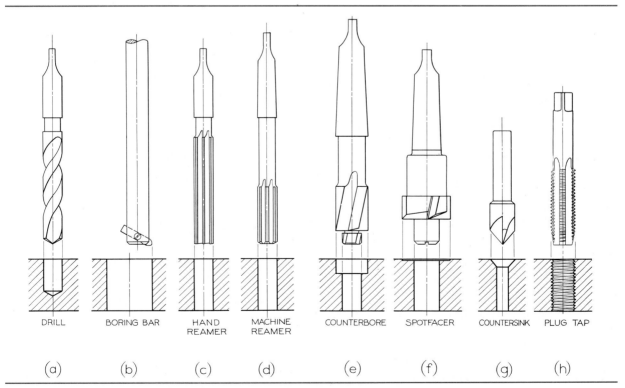

*Fig. 12.16*   Types of Machined Holes.

of extreme accuracy in roundness, straightness, or size.

Drills frequently cut holes slightly larger than their nominal size. A twist drill (Appendix 16) is somewhat flexible, which makes it tend to follow a path of least resistance. For work that demands greater accuracy, drilling is followed by boring, Fig. 12.16 (b), or by reaming, (c) or (d). When a drilled hole is to be finished by boring, it is drilled slightly undersize and the boring tool, which is supported by a relatively rigid bar, generates a hole that is round and straight. Reaming is also used for enlarging and improving the surface quality of a drilled or bored hole. Reamers are a finishing tool, and best results are achieved by limiting the material removed to .004″ to .012″ on the diameter.

Counterboring, Fig. 12.16 (e), is the cutting of an enlarged cylindrical portion of a previously pro-

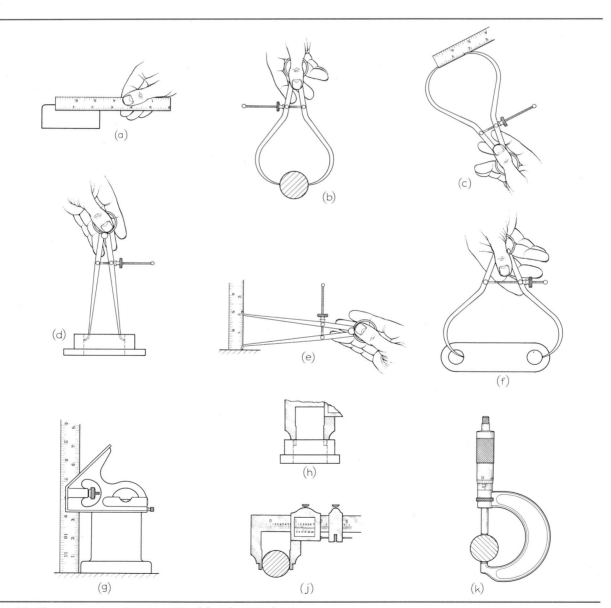

**Fig. 12.17**    Measuring Devices Used by the Machinist.

duced hole, usually to receive the head of a fillister-head or socket-head screw, Figs. 15.33 and 15.34.

Spotfacing is similar to counterboring, but is quite shallow, usually about 1.5 mm ($\frac{1}{16}''$) deep or just deep enough to clean a rough surface, Fig. 12.16 (f), or to finish the top of a boss to form a bearing surface. Although the depth of a spotface is commonly drawn 1.5 mm ($\frac{1}{16}''$) deep, the actual depth is usually left up to the machinist.

Countersinking, Fig. 12.16 (g), is the process of cutting a conical taper at one end of a hole, usually to receive the head of a flat-head screw, Figs. 15.32 and 15.33.

Tapping is the threading of previously drilled small holes by the use of one or more styles of taps, Fig. 12.16 (h).

## 12.17 Measuring Devices Used in Manufacturing

Although the machinist uses various measuring devices depending upon the kind of dimensions (fractional, decimal, or metric) shown on the drawing, it is evident that to dimension correctly, the engineering designer must have at least a working knowledge of the common measuring tools. The machinists' steel rule, or scale, is a commonly used measuring tool in the shop, Fig. 12.17 (a). The smallest division on one scale of this rule is $\frac{1}{64}''$, and such a scale is used for common fractional dimensions. Also, many machinists' rules have a decimal scale with the smallest division of .01″, which is used for dimensions given on the drawing by the decimal system, §13.10. For checking the nominal size of outside diameters, the outside spring caliper and steel scale are used as shown at (b) and (c). Likewise, the inside spring caliper is used for checking nominal dimensions, as shown at (d) and (e). Another use for the outside caliper, (f), is to check the nominal distance between holes (center-to-center). The combination square may be used for checking height, as shown at (g), and for a variety of other measurements. Measuring devices are also available that have metric scales.

For dimensions that require more precise measurements, the vernier caliper, (h) and (j), or the micrometer caliper, (k), may be used. It is common practice to check measurements to 0.025 mm (.001″) with these instruments and in some instances they are used to measure directly to 0.0025 mm (.0001″).

Computerized measuring devices have broadened the range of accuracy previously attainable.

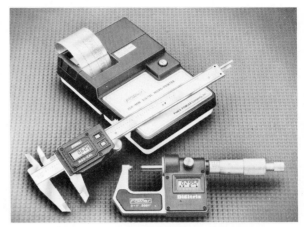

*Fig. 12.18* Computerized Measurement System. *Courtesy of Fred V. Fowler Co., Inc.*

Figure 12.18 illustrates an ultraprecision electronic digital readout micrometer and caliper that contain integral microprocessors. In addition to the hand-held printer/recorder providing a hard-copy output of measurements, the printer also calculates and lists statistical mean, minimum, and maximum values as well as standard deviation.

## 12.18 Chipless Machining

Several manufacturing processes do not employ the cutting action described in most of the previous sections.

Chemical milling removes material through chemical reactions at the surface of the work piece. An etchant, acidic or basic, is agitated over an immersed workpiece that is masked where no metal removal is desired. For a material with high homogeneity, the tolerances obtainable in such an operation are ±0.08 mm (±.003″).

Electrodischarge machining (EDM) utilizes high-energy electric sparks that build up high charge densities on the surface of the work material. Thermal stresses, exceeding the strength of the work material, cause minute particles to break away from the work. Though this process is slow, it allows intricate shapes to be cut in hard materials, such as tungsten carbides.

*Laser machine tools* are helping designers and manufacturing engineers increase productivity over a wide range of machine applications. Laser systems reduce costs and improve the quality of many manufactured products that require cutting, engraving,

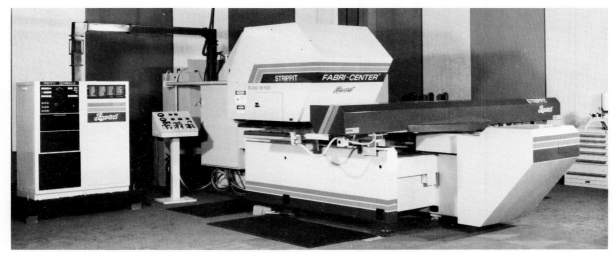

**Fig. 12.19** Numerically Controlled Hole Punching and Laser Cutting System. *Courtesy of Strippit/Di-Acro Houdaille*

scribing, drilling, welding, perforating, or heat treating.

Metal cutting is the single largest application for most lasers. Laser processing offers several advantages over typical cutting processes, including greater accuracy and flexibility, reduced costs, and higher material throughput.

The computerized holemaking system shown in Fig. 12.19 combines a precision punching machine to produce conventional round and shaped holes using punch and die tooling, with a laser beam. Linear cutting rates range from 200 IPM (inches per minute) for .125″ acrylic and 160 IPM for .039″ low-carbon steel to 20 IPM for .250″ low-carbon steel.

### 12.19 Welding

Welding is a process of joining metals by fusion. Although welding by lasers is becoming popular in sophisticated plants, arc welding, gas welding, resistance welding, and atomic hydrogen welding of stock plates, tubing, and angles are more commonly used.

### 12.20 Stock Forms

Many standardized structural shapes are available in stock sizes for the fabrication of parts or structures. Among these are bars of various shapes, flat stock, rolled structural shapes, and extrusions, Fig. 12.20. The tube shown at (e) is often round, square, or rectangular in shape.

### 12.21 Forging

Forging is the process of shaping metal to a desired form by means of pressing or hammering. Generally, forging is hot forging, Fig. 12.21, in which the metal is heated to a required temperature before forging. Some softer metals can be forged without

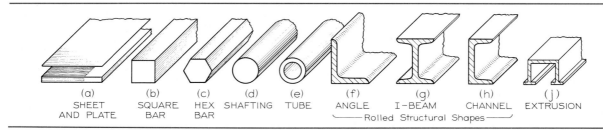

| (a) | (b) | (c) | (d) | (e) | (f) | (g) | (h) | (j) |
| --- | --- | --- | --- | --- | --- | --- | --- | --- |
| SHEET AND PLATE | SQUARE BAR | HEX BAR | SHAFTING | TUBE | ANGLE | I-BEAM | CHANNEL | EXTRUSION |

└─── Rolled Structural Shapes ───┘

**Fig. 12.20** Common Stock Forms.

Fig. 12.21   Drop Forging. *Courtesy of Jervis B. Webb Co.*

heating, and this is cold forging. *Draft*, or taper, must be provided on all forgings.

## 12.22 Heat-Treating

Heat-treating is when heat is used to alter the properties of metal. Different procedures affect metal in different ways. Annealing and normalizing involve heating to a critical temperature range, and then slowly cooling to soften the metal, thereby releasing internal stresses developed in previous manufacturing processes.

*Hardening* requires heating to above the critical temperature followed by rapid cooling—quenching in oil, water, brine, or in some instances in air. *Tempering* reduces internal stresses caused by hardening and also improves the toughness and ductility. *Surface hardening* is a way of hardening the surface of a steel part while leaving the inside of the piece soft. Surface hardening is accomplished by *carburizing*, followed by heat-treatment; by *cyaniding*, followed by heat-treatment; by *nitriding;* by *induction hardening;* and by *flame hardening.*

Lasers are widely used for transformation hardening of selected surface areas of metal parts. Laser hardening applies less heat to the part than do tra-

ditional methods, and thermal distortion is greatly reduced.

## 12.23 Automation

Automation is the term applied to systems of automatic machines and processes. These machines and processes are essentially the same as previously discussed, with the addition of mechanisms to control the sequence of operation, movement of tool, flow, and so forth. Little, if any, operator interaction is required once the equipment has been "set up."

Most contain built-in computers to control all of their functions, such as the surface grinder in Fig. 12.14. Some machines are used in combination with an industrial robot, Fig. 12.22, where the robot takes the place of an operator for the load and unload functions for the machine. Robots are also used extensively for spray painting and welding applications as well as for around-the-clock performance in environments where a human operator could not exist.

When a machine's cycle is relatively long and the robot is underutilized, two or more machines can be grouped together in such a fashion that the

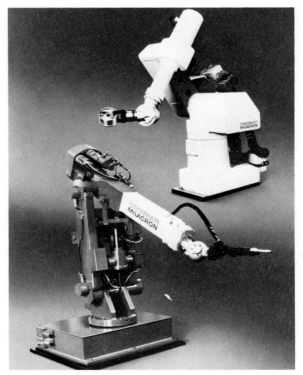

Fig. 12.22   Industrial Robots. *Courtesy of Cincinnati Milacron*

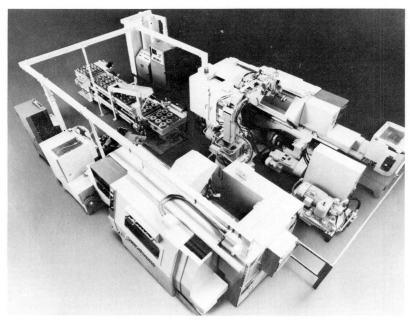

*Fig. 12.23*  Flexible Manufacturing System. *Courtesy of Cincinnati Milacron*

robot can service each. The flexible manufacturing system in Fig. 12.23 consists of a unique computer numerically controlled (CNC) turning system equipped with fully automated features, designed for reliable batch manufacturing of quality parts while virtually unattended. The two turning centers are rear loaded by a heavy-duty industrial robot. The system also features automatic tool changing from an 84-tool storage magazine as well as automatic postprocess gaging and a take-away parts conveyor.

The manufacturing system in Fig. 12.24 is based on a different concept. The parts being produced are the front- and rear-end housings for automotive air-conditioning compressors. A conveyor system circulates the housing parts throughout the 20-station system, where various facing, boring, drilling, chamfering, probing, reaming, gaging, wirebrushing, washing, blow drying, and unloading and reject-diverting operations are performed. The disadvantage of a system such as this is that it is usually custom-made for the specific application and, therefore, is not considered flexible enough for most smaller manufacturing operations.

Large-scale manufacturing operations such as automotive plants must be automated to remain competitive. Figure 12.25 shows Pontiac Fiero bod-

ies being carried on a conveyor into a one-of-a-kind milling and drilling machine.

## 12.24 Plastics Processing

The growth of the plastics industry has expanded in recent decades to the point that the domestic demand for plastics reached approximately 40 billion pounds in 1984. Plastic resins used in motor vehicle production alone accounted for about 1.6 billion pounds. As an example of this tremendous increase, the 3000-pound Chevrolet Corvette contains about 580 pounds of various plastic parts.

The two main families of plastics are known as *thermosetting* and *thermoplastic*. The thermosetting plastics will take a set and will not soften when reheated, whereas thermoplastics will soften whenever heat is applied.

Typical plastic processing operations include extrusion, blow molding, compression molding, transfer molding, injection molding, and thermoforming.

## 12.25 Extrusion Molding

An extrusion-molding machine transfers solid plastic particles, additives, colorants, and regrinds from a feed hopper into a heating chamber. The extruder screw conveys the material forward through the bar-

*Fig. 12.24* Automated Manufacturing System. *Courtesy of Cargill Detroit*

*Fig. 12.25* Automated Material Conveying. *Courtesy of Jervis B. Webb Co.*

rel toward a die. The contour of the die will determine the cross-sectional shape of the extruded member. Typical uses of the extrusion process are in compounding and pelletizing of bulk plastics, pipe and profile extrusion, blown film and sheet manufacturing, and blow molding.

## 12.26 Blow Molding

Blow-molding operations are classified as *extrusion, injection,* and *stretch* blow molding. In extrusion blow-molding, an open-ended parison (hollow tube) is extruded, and the mold is closed on the parison, thus closing the bottom of the tube. When compressed air is introduced through the blow pin or blow needles, the parison expands to take the shape of the mold. The part is cooled by being in contact with the cooler mold surface and then is removed from the mold. Automotive coolant reservoirs, seat backs, ductwork, wheels, toy parts, juice and handled bottles are common extrusion blow-molded shapes.

In injection blow molding, the parison is formed by injection molding a test-tube-shaped preform on a core pin and is then transferred to the blow-molding station. An advantage of the injection blow-molding process is that there is no flash to trim and any thread detail on the neck is of injection-molding quality.

In stretch blow molding, also called orientation blow molding, a parison is formed by either an extrusion or injection-molding operation and is subsequently stretched and blown in both an axial and radial direction. Advantages of this process are greater strength, better clarity, improved impact strength, and increased stiffness.

## 12.27 Compression Molding

In compression molding, a charge of plastic material, usually thermosetting, is placed into a mold, the mold is closed, and a plunger compresses the material so that heat and pressure modifies the plastic to take the shape of the mold cavity. The pressure is maintained during the curing cycle, after which the plunger is removed and the part is ejected. Preformed, preheated discs are often used for better control of the volume of material, thereby reducing variations in part thickness or part density.

## 12.28 Transfer Molding

In the transfer-molding process, an amount of thermosetting material is placed in a chamber (usually referred to as a pot) and then is forced out of the pot and into the closed cavities where polymerization takes place. The plastics formulations used are generally of a softer plastic than those that are used in compression molding. This enables the design of the part to include ribs and thin sections that are not formable in compression molding.

## 12.29 Injection Molding

Injection molding, Fig. 12.26, is primarily used to manufacture thermoplastic products.

*Fig. 12.26*   Injection-Molding Machine. *Courtesy of Reed Prentice*

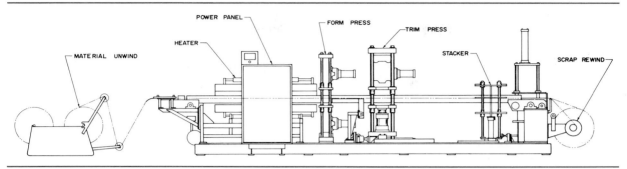

**Fig. 12.27**  Thermoforming Machine. *Courtesy of Packaging Industries*

Materials are melted and fed through the machine and injected into the mold where the plastic solidifies into the shape of the mold cavities. The press is opened after the plastic has sufficiently cooled and the parts are then ejected. Owing to the relatively high costs of tooling, the process is typically and only economically suited for mass production.

### 12.30 Thermoforming

Thermoforming is a process in which one or both sides of a thermoplastic sheet are heated so that it can be formed into various products. Figure 12.27 represents a thermoformer that feeds roll stock up to .050″ through the heater zone and into a form area. Once the heated sheet is clamped in place over the mold (some machines have a form area of up to 50″ wide by 50″ long), pressure is exerted on the outside of the sheet, and a vacuum is drawn to evacuate the air that is trapped between the sheet and the mold. The sheet is held in place on the mold until the plastic is cooled sufficiently to maintain the part shape, and compressed air is then introduced through the vacuum holes to lift the parts off the mold. The sheet is then indexed to the die cutting and stack stations.

Figure 12.28 shows a machine that is geared toward high-volume production runs. The parts being molded are high-impact foam polystyrene egg cartons and are being formed at about 200 cartons per minute, cut on a matched metal punch press, and then stacked automatically for packing and shipping.

**Fig. 12.28**  Thermoforming Machine. *Courtesy of Brown Machine, Plastics Machinery John Brown*

# CHAPTER

# 13

# Dimensioning

We have all heard of "rule of thumb." Actually, at one time an inch was defined as the width of a thumb, and a foot was simply the length of a man's foot. In old England, an inch used to be "three barley corns, round and dry." In the time of Noah and the Ark, the *cubit* was the length of a man's forearm, or about 18″.

In 1791, France adopted the *meter** (1 meter = 39.37″; 1″ = 25.4 mm), from which the decimalized metric system evolved. In the meantime England was setting up a more accurate determination of the *yard*, which was legally defined in 1824 by act of Parliament. A foot was $\frac{1}{3}$ yard, and an inch was $\frac{1}{36}$ yard. From these specifications, graduated rulers, scales, and many types of measuring devices have been developed to achieve even more accuracy of measurement and inspection.

Until this century, common fractions were considered adequate for dimensions, but as designs became more complicated, and as it became necessary to have interchangeable parts in order to support mass production, more accurate specifications were required and it became necessary to turn to the decimal-inch system or the SI system. See §§13.9 and 13.10.

*In the SI system the meter is now defined as a length equal to the distance traveled by light in a vacuum during a time interval of $\frac{1}{299\ 792\ 458}$ second.

## 13.1 Metric Units

The current rapid growth of worldwide science and commerce has fostered an international system of units (SI), based on the meter and suitable for measurements in physical science and engineering. The seven basic units of measure are the meter (length), kilogram (mass), second (time), ampere (electric current), kelvin (thermodynamic temperature), mole (amount of substance), and candela (luminous intensity).

The SI system is gradually coming into use in the United States, especially by the many multinational companies in the chemical, electronic, and mechanical industries. A tremendous effort is now under way to convert all standards of the American National Standards Institute (ANSI) to the SI units in conformity with the International Standards Organization (ISO) standards.

See inside the front cover of this book for a table for converting fractional inches to decimal inches or millimeters. For the International System of Units and their United States equivalents, see Appendix 31.

## 13.2 Size Description

In addition to a complete *shape description* of an object, as discussed in previous chapters, a drawing of the design must also give a complete *size description;* that is, it must be *dimensioned.* See ANSI Y14.5M–1982 (R1988).

The need for *interchangeability* of parts is the basis for the development of modern methods of size description. Drawings today must be dimensioned so that production personnel in widely separated places can make mating parts that will fit properly when brought together for final assembly or when used as repair or replacement parts by the customer, §13.26.

The increasing need for precision manufacturing and the necessity to control sizes for interchangeability has shifted responsibility for size control to the designing engineer and the drafter. The production worker no longer exercises judgment in engineering matters, but only on the proper execution of instructions given on the drawings. Therefore, it is necessary for engineers and designers to be familiar with materials and methods of construction and with production requirements. The engineering student or designer should seize every opportunity to become familiar with the fundamental manufacturing processes, especially *patternmaking, foundry, forging,* and *machine shop practice.*

The drawing should show the object in its completed condition and should contain all necessary information to bring it to that final state. Therefore, in dimensioning a drawing, the designer and the drafter should keep in mind the finished piece, the production processes required, and above all the function of the part in the total assembly. Whenever possible—that is, when there is no conflict with functional dimensioning, §13.4—dimensions should be given that are convenient for the individual worker or the production engineer. These dimensions should be given so that it will not be necessary to scale or assume any dimensions. Do not give dimensions to points or surfaces that are not accessible to the worker.

Dimensions should not be duplicated or superfluous, §13.30. Only those dimensions should be given that are needed to produce and inspect the part exactly as intended by the designer. The student often makes the mistake of giving the dimensions that are used *to make the drawing.* These are not necessarily the dimensions required. There is much more to the theory of dimensioning, as we shall see.

## 13.3 Scale of Drawing

Drawings should be made to scale, and the scale should be indicated in the title block even though the worker is never expected to scale the drawing or print for a needed dimension. See §2.31 for indications of scales.

A heavy straight line should be drawn under any dimension that is not to scale, Fig. 13.15, or the abbreviation *NTS* (not to scale) should be indicated. This procedure may be necessary when a change is made in the drawing that is not important enough to justify making an entirely new drawing.

## 13.4 Learning to Dimension

Dimensions are given in the form of linear distances, angles, or notes irrespective of the dimensioning units being used. The ability to dimension properly in millimeters, decimal inch, or fractional inch requires the following:

1. The student must learn the *technique of dimensioning:* the character of the lines, the spacing of dimensions, the making of arrowheads, and so forth. A typical dimensioned drawing is shown in Fig. 13.1. Note the strong contrast between the visible lines of the object and the thin lines used for the dimensions.

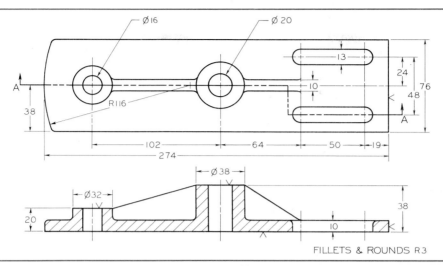

**Fig. 13.1** Dimensioning Technique.   Dimensions in millimeters.

2. The student must learn the rules of *placement of dimensions* on the drawing. These practices assure a logical and practical arrangement with maximum legibility.

3. The student should learn the *choice of dimensions*. Formerly, manufacturing processes were considered the governing factor in dimensioning. Now function is considered first and the manufacturing processes second. The proper procedure is to dimension tentatively for function and then review the dimensioning to see if any improvements from the standpoint of production can be made without adversely affecting the functional dimensioning. A "geometric breakdown," §13.20, will assist the beginner in selecting dimensions. In most cases dimensions thus determined will be functional, but this method should be accompanied by a logical analysis of the functional requirements.

## 13.5 Lines Used in Dimensioning

A dimension line, Fig. 13.2 (a), is a thin, dark, solid line terminated by arrowheads, which indicates the direction and extent of a dimension. In machine drawing, the dimension line is broken, usually near the middle, to provide an open space for the dimension figure. In structural and architectural drawing, it is customary to place the dimension figure above an unbroken dimension line.

As shown in Fig. 13.2 (b), the dimension line nearest the object outline should be spaced at least 10 mm ($\frac{3}{8}''$) away. All other parallel dimension lines should be at least 6 mm ($\frac{1}{4}''$) apart, and more if space is available. *The spacing of dimension lines should be uniform throughout the drawing.*

An *extension line*, (a), is a thin, dark, solid line that "extends" from a point on the drawing to which a dimension refers. The dimension line meets the extension lines at right angles except in special cases, as in Fig. 13.6 (a). A gap of about 1.5 mm

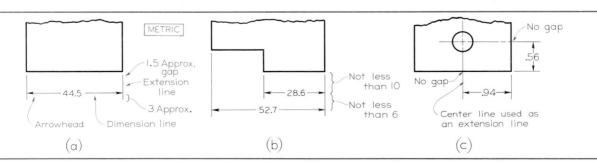

**Fig. 13.2** Dimensioning Technique.

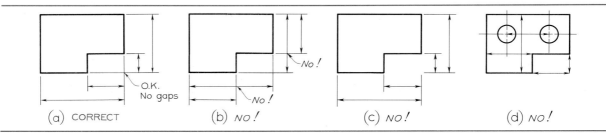

*Fig. 13.3*  Dimension and Extension Lines.

($\frac{1}{16}''$) should be left where the extension line would join the object outline. The extension line should extend about 3 mm ($\frac{1}{8}''$) beyond the outermost arrowhead, (a) and (b).

The foregoing dimensions for lettering height, spacing, and so on should be increased approximately 50 percent for drawings that are to be microfilmed and blown back to one-half size for the working print. Otherwise the lettering and dimensioning often are not legible.

A *center line* is a thin, dark line composed of alternate long and short dashes and is used to represent axes of symmetrical parts and to denote centers. As shown in Fig. 13.2 (c), center lines are commonly used as extension lines in locating holes and other features. When so used, the center line crosses over other lines of the drawing without gaps. A center line should always end in a long dash.

## 13.6 Placement of Dimension and Extension Lines

A correct example of the placement of dimension lines and extension lines is shown in Fig. 13.3 (a). The shorter dimensions are nearest to the object outline. Dimension lines should not cross extension lines as at (b), which results from placing the shorter dimensions outside. Note that it is perfectly satisfactory to cross extension lines, as shown at (a). They should never be shortened, as at (c). A dimension line should never coincide with, or form a continuation of, any line of the drawing, as shown at (d). Avoid crossing dimension lines wherever possible.

Dimensions should be lined up and grouped together as much as possible, as shown in Fig. 13.4 (a), and not as at (b).

In many cases, extension lines and center lines must cross visible lines of the object, Fig. 13.5 (a). When this occurs, gaps should not be left in the lines, as at (b).

Dimension lines are normally drawn at right angles to extension lines, but an exception may be made in the interest of clearness, as shown in Fig. 13.6 (a). In crowded conditions, gaps in extension lines near arrowheads may be left, in order to clarify the dimensions, as shown at (b). In general, avoid dimensioning to hidden lines, as shown at (c).

## 13.7 Arrowheads

Arrowheads, Fig. 13.7, indicate the extent of dimensions. They should be uniform in size and style throughout the drawing and not varied according to the size of the drawing or the length of dimensions. Arrowheads should be drawn freehand and the length and width should be in a ratio of 3:1. The length of the arrowhead should be equal to the height of the dimension whole numbers. For average use, make arrowheads about 3 mm ($\frac{1}{8}''$) long and

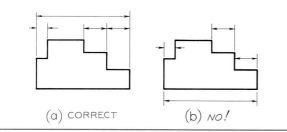

*Fig. 13.4*  Grouped Dimensions.

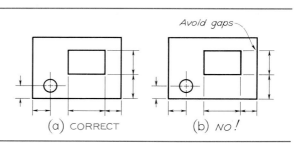

*Fig. 13.5*  Crossing Lines.

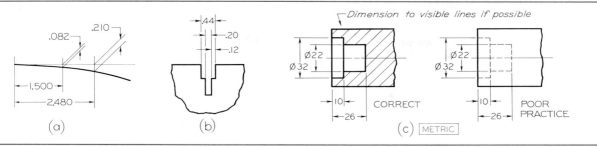

**Fig. 13.6** Placement of Dimensions.

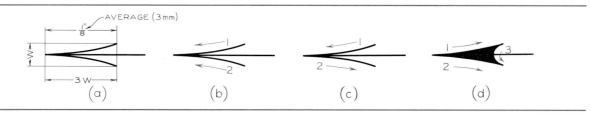

**Fig. 13.7** Arrowheads.

very narrow, (a). Use strokes toward the point or away from the point desired, (b) to (d). The method at (b) is easier when the strokes are drawn toward the drafter. For best appearance, fill in the arrowhead as at (d).

## 13.8 Leaders

A leader, Fig. 13.8, is a thin solid line leading from a note or dimension, and terminated by an arrowhead or a dot touching the part to which attention is directed. Arrowheads should always terminate on a line such as the edge of a hole; dots should be within the outline of the object. A leader should generally be an inclined straight line, if possible, except for the short horizontal shoulder (6 mm or $\frac{1}{4}''$, approx.) extending from midheight of the lettering *at the beginning or end of a note.*

A leader to a circle should be radial; that is, if extended, it would pass through the center. A draw-

ing presents a more pleasing appearance if leaders near each other are drawn parallel. Leaders should cross as few lines as possible and should never cross each other. They should not be drawn parallel to nearby lines of the drawing, allowed to pass through a corner of the view, drawn unnecessarily long, or drawn horizontally or vertically on the sheet. A leader should be drawn at a large angle and terminate with the appropriate arrowhead, or with a dot, as shown in Fig. 13.8 (f).

## 13.9 Fractional, Decimal, and Metric Dimensions

In the early days of machine manufacturing in this country, the worker would scale the undimensioned design drawing to obtain any needed dimensions, and it was the worker's responsibility to see to it that the parts fitted together properly. Workers were skilled, and it should not be thought that very

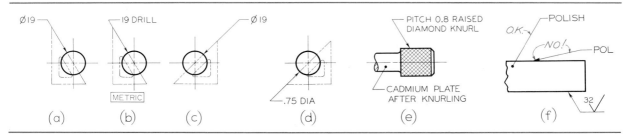

**Fig. 13.8** Leaders.

accurate and excellent fits were not obtained. Hand-built machines were often beautiful examples of precision craftsmanship.

The system of units and common fractions is still used in architectural and structural work where close accuracy is relatively unimportant and where the steel tape or framing square is used to set off measurements. Architectural and structural drawings are therefore often dimensioned in this manner. Also, certain commercial commodities, such as pipe and lumber, are identified by standard nominal designations that are close approximations of actual dimensions.

As industry has progressed, there has been greater and greater demand for more accurate specifications of the important functional dimensions—more accurate than the $\frac{1}{64}''$ permitted by the engineers', architects', and machinists' scale. Since it was cumbersome to use still smaller fractions, such as $\frac{1}{128}$ or $\frac{1}{256}$, it became the practice to give decimal dimensions, such as 4.2340 and 3.815, for the dimensions requiring accuracy. However, some dimensions, such as standard nominal sizes of materials, punched holes, drilled holes, threads, keyways, and other features produced by tools that are so designated are still expressed in whole numbers and common fractions.

Thus, drawings may be dimensioned entirely with whole numbers and common fractions, or entirely with decimals, or with a combination of the two. However, more recent practice adopted the decimal-inch system and current practice also utilizes the metric system as recommended by ANSI. Millimeters and inches in the decimal form can be added, subtracted, multiplied, and divided more easily than can fractions.

Also the decimal system is compatible with the calibrations of machine tool controls, the requirements for numerically controlled machine tools, and for computer-programmed digital plotting. For an example of a computer-made drawing, see Fig. 16.33; also see Chapter 8.

For inch–millimeter equivalents of decimal and common fractions, see inside the front cover of this book. For additional metric equivalents, see Appendix 31. For rounding off decimals, see §13.10.

## 13.10 Decimal Systems

A decimal system, based upon the decimal inch or the millimeter as a linear unit of measure, has many advantages and is compatible with most measuring devices and machine tools. Metric measurement is based upon the meter as a linear unit of measure, but the millimeter is used on most engineering drawings. To facilitate the changeover to metric dimensions during the transition, some drawings are dual-dimensioned in millimeters and decimal inches. See §13.11.

Total decimal dimensioning is *preferred* by the American National Standard Institute.

*Complete decimal dimensioning* employs decimals for all dimensions and designations except where certain commercial commodities, such as pipe and lumber, are identified by standardized nominal designations. *Combination dimensioning* employs decimals for all dimensions except the designations of nominal sizes of parts or features such as bolts, screw threads, keyseats, or other standardized fractional designations. [ANSI Y14.5M–1982 (R1988)].

In these systems, two-place inch or one-place millimeter decimals are used when a common fraction has been regarded as sufficiently accurate.

*In the combination dimensioning system, common fractions* may continue to be used to indicate nominal sizes of materials, drilled holes, punched holes, threads, keyways, and other standard features.

One-place millimeter decimals are used when tolerance limits of ±0.1 mm or more can be permitted. Two (or more)-place millimeter decimals are used for tolerance limits less than ±0.1 mm. Fractions are considered to have the same tolerance as two-place decimal-inch dimensions when determining the number of places to retain in the conversion to millimeters. Keep in mind that 0.1 mm is approximately equal to .004 inch.

Two-place inch decimals are used when tolerance limits of ±.01″, or more, can be permitted. Three or more decimal places are used for tolerance limits less than ±.01″. In two-place decimals, the second place preferably should be an even digit (for example, .02, .04, and .06 are preferred to .01, .03, or .05) so that when the dimension is divided by 2, as is necessary in determining the radius from a diameter, the result will be a decimal of two places. However, odd two-place decimals are used when required for design purposes, such as in dimensioning points on a smooth curve or when strength or clearance is a factor.

A typical example of the use of the complete decimal-inch system is shown in Fig. 13.9. The use of the preferred decimal-millimeter system is shown in Fig. 13.10.

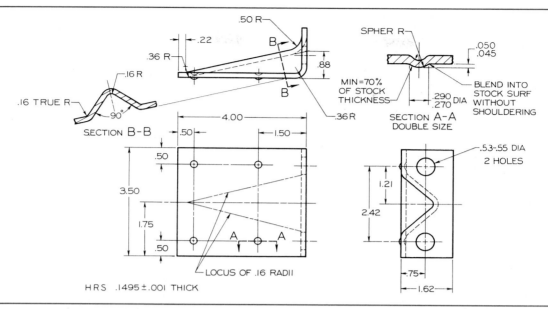

**Fig. 13.9** Complete Decimal Dimensioning.

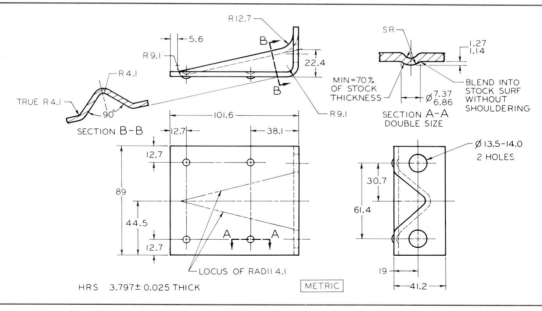

**Fig. 13.10** Complete Metric Dimensioning.

When a decimal value is to be rounded off to fewer places than the calculated number, irrespective of the unit of measurement involved, the method prescribed is as follows.

The last figure to be retained should not be changed when the figure beyond the last figure to be retained is less than 5.

**EXAMPLE**   3.46325, if rounded off to three places, should be 3.463.

The last figure to be retained should be increased by 1 when the figure beyond the last figure to be retained is greater than 5.

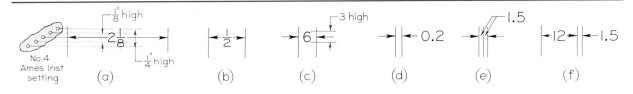

**Fig. 13.11**   Dimension Figures.   Metric dimensions (c) through (f).

EXAMPLE   8.37652, if rounded off to three places, should be 8.377.

The last figure to be retained should be unchanged if it is even, or increased by 1 if odd, when followed by exactly 5.

EXAMPLE   4.365 becomes 4.36 when rounded off to two places. Also 4.355 becomes 4.36 when cut off to two places.

The use of the metric system means not only a changeover of measuring equipment but also a changeover in thinking on the part of drafters and designers. They must stop thinking in terms of inches and common fractions and think in terms of millimeters and other SI units. Dimensioning practices remain essentially the same; only the units are changed. Compare Figs. 13.9 and 13.10.

Shop scales and drafting scales for use in the decimal systems are available in a variety of forms. The drafting scale is known as the *decimal scale*, and is discussed in §2.29. See inside the front cover for two-, three-, and four-place decimal equivalent table.

Once the metric system is installed, the advantages in computation, in checking, and in simplified dimensioning techniques are considerable.

## 13.11 Dimension Figures

The importance of good lettering of dimension figures cannot be overstated. The shop produces according to the directions on the drawing, and to save time and prevent costly mistakes, all lettering should be perfectly legible. A complete discussion of numerals is given in §§4.17 to 4.19. The standard height for whole numbers is $\frac{1}{8}''$ (3 mm), and for fractions double that, or $\frac{1}{4}''$ (6 mm). Beginners should use guide lines, as shown in Figs. 4.21 and 4.22. The numerator and denominator of a fraction should be clearly separated from the fraction bar, and the fraction bar should always be horizontal, as in Fig. 4.23 (c). An exception to this may be made in crowded places, such as parts lists, but never in dimensioning.

Legibility should never be sacrificed by crowding dimension figures into limited spaces. For every such case there is a practical and effective method, as shown in Fig. 13.11. At (a) and (b) there is enough space for the dimension line, the numeral, and the arrowheads. At (c) there is only enough room for the figure, and the arrowheads are placed outside. At (d) both the arrowheads and the figure are placed outside. Other methods are shown at (e) and (f).

If necessary, a removed partial view may be drawn to an enlarged scale to provide the space needed for clear dimensioning, Fig. 9.22.

Methods of lettering and displaying decimal dimension figures are shown in Fig. 13.12. All numerals are 3 mm ($\frac{1}{8}''$) high whether on one or two lines. The space between lines of numerals is 1.5 mm ($\frac{1}{16}''$) or 0.8 mm ($\frac{1}{32}''$) on each side of a dimension line. To draw guide lines with the Ames Lettering Guide, Fig. 4.18, use the No. 4 setting or appropriate metric setting and the center column of holes.

Make all decimal points bold, allowing ample space. Where the metric dimension is a whole number, neither a decimal point nor a zero is given, Fig. 13.11 (c) and (f). Where the metric dimension is less than 1 millimeter, a zero precedes the decimal point, as at (d). Where the dimension exceeds a whole number by a fraction of 1 millimeter, the last digit to the right of the decimal point is not followed by a zero, (e) and (f), except when expressing tolerances.

Where the decimal-inch dimension is used on drawings, a zero is not used before the decimal point of values less than 1 inch, Fig. 13.12 (f)–(j). The decimal-inch dimension is expressed to the same number of decimal places as its tolerance. Thus, zeros are added to the right of the decimal point as necessary, as at (e).

Never letter a dimension figure over any line on the drawing, but break the line if necessary. Place dimension figures outside a sectioned area if possible, Fig. 13.13 (a). When a dimension must be

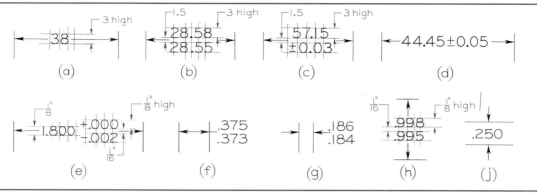

Fig. 13.12 Decimal Dimension Figures. Metric dimensions (a) through (d).

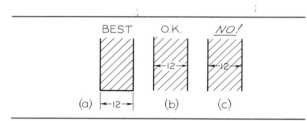

Fig. 13.13 Dimensions and Section Lines. Metric.

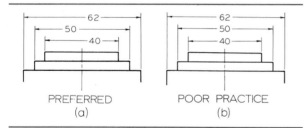

Fig. 13.14 Staggered Numerals. Metric

placed on a sectioned area, leave an opening in the section lining for the dimension figure, (b).

In a group of parallel dimension lines, the numerals should be staggered, as in Fig. 13.14 (a), and not stacked up one above the other, as at (b).

Dual dimensioning, though not recommended, is used to show metric and decimal-inch dimensions on the same drawing. Two methods of displaying the dual dimensions are as follows.

## Position Method

The millimeter value is placed above the inch dimension and is separated by a dimension line or an added line for some dimensions when the unidirectional system of dimensioning is used. An alternative arrangement in a single line places the millimeter dimension to the left of the inch dimension, separated by a slash line (virgule). Each drawing should illustrate the dimension identification as $\frac{\text{MILLIMETER}}{\text{INCH}}$ or MILLIMETER/INCH. (Placement of the inch dimension above or to the left of the millimeter is also acceptable.)

EXAMPLES

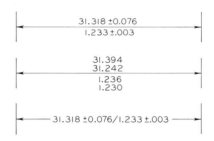

## Bracket Method

The millimeter dimension is enclosed in square brackets, [ ]. The location of this dimension is optional but should be uniform on any drawing, that is, above or below or to the left or right of the inch dimension. Each drawing should include a note to identify the dimension values as DIMENSIONS IN [ ] ARE MILLIMETERS.

EXAMPLES

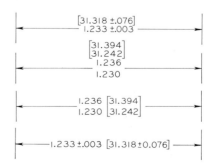

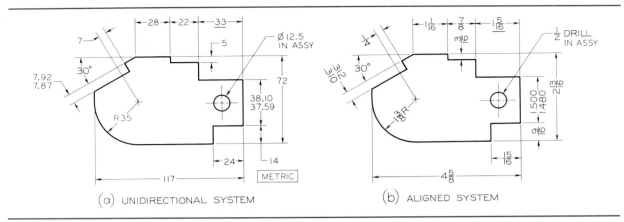

Fig. 13.15　Directions of Dimension Figures.

When converting a decimal-inch dimension to millimeters, multiply the inch dimension by 25.4 and round off to one less digit to the right of the decimal point than for the inch value (see §13.10). When converting a millimeter dimension to inches, divide the millimeter dimension by 25.4 and round off to one more digit to the right of the decimal point than for the millimeter value.

### 13.12 Direction of Dimension Figures

Two systems of reading direction for dimension figures are available. In the preferred *unidirectional system*, approved by ANSI, Fig. 13.15 (a), all dimension figures and notes are lettered horizontally on the sheet and are read from the bottom of the drawing. The unidirectional system has been extensively adopted in the aircraft, automotive, and other industries because it is easier to use and read, especially on large drawings. In the *aligned system*, Fig. 13.15 (b), all dimension figures are aligned with the

dimension lines so that they may be read from the right side of the sheet. Dimension lines in this system should not run in the directions included in the shaded area of Fig. 13.16, if avoidable.

In both systems, dimensions and notes shown with leaders are aligned with the bottom of the drawing. Notes without leaders should also be aligned with the bottom of the drawing.

### 13.13 Millimeters and Inches

*Millimeters* are indicated by the lowercase letters mm placed one space to the right of the numeral; thus, 12.5 mm. *Meters* are indicated by the lowercase letter m placed similarly; thus, 50.6 m. *Inches* are indicated by the symbol ″ placed slightly above and to the right of the numeral; thus, $2\frac{1}{2}$″. *Feet* are indicated by the symbol ′ similarly placed; thus, 3′–0, 5′–6, 10′–0$\frac{1}{4}$. It is customary in such expressions to omit the inch marks.

It is standard practice to omit mm designations and inch marks on a drawing except when there is a possibility of misunderstanding. For example, 1 VALVE should be 1″ VALVE, and 1 DRILL should be 1″ DRILL or 1 mm DRILL. Where some inch dimensions are shown on a millimeter-dimensioned drawing, the abbreviation IN. follows the inch values.

In some industries all dimensions, regardless of size, are given in inches; in others dimensions up to 72″ inclusive are given in inches, and those greater are given in feet and inches. In structural and architectural drafting, all dimensions of 1′ or over are usually expressed in feet and inches.

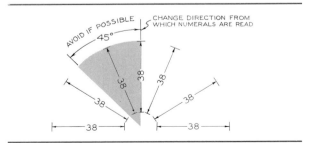

Fig. 13.16　Directions of Dimensions.

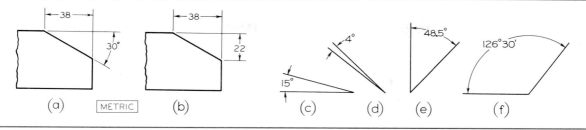

*Fig. 13.17*   Angles.

## 13.14 Dimensioning Angles

Angles are dimensioned, preferably, by means of an angle in degrees and a linear dimension, Fig. 13.17 (a), or by means of coordinate dimensions of the two legs of a right triangle, (b). The coordinate method is more suitable for work requiring a high degree of accuracy. Variations of angle (in degrees) are hard to control because the amount of variation increases with the distance from the vertex of the angle. Methods of indicating various angles are shown from (c) to (f). Tolerances of angles are discussed in §14.18.

When degrees alone are indicated, the symbol ° is used. When minutes alone are given, the number should be preceded by 0°.

EXAMPLE    0° 23′.

In all cases, whether in the unidirectional system or in the aligned system, the dimension figures for an-

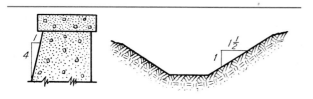

*Fig. 13.18*   Angles in Civil Engineering Projects.

gles are lettered on horizontal guide lines. For a general discussion of angles, see §5.2.

In civil engineering drawings, *slope* represents the angle with the horizontal, while *batter* is the angle referred to the vertical. Both are expressed by making one member of the ratio equal to 1, as shown in Fig. 13.18. *Grade,* as of a highway, is similar to slope but is expressed in percentage of rise per 100′ of run. Thus a 20′ rise in a 100′ run is a grade of 20 percent.

In structural drawings, angular measurements are made by giving the ratio of "run" to "rise," with the larger size being 12″. These right triangles are referred to as *bevels*.

## 13.15 Dimensioning Arcs

A circular arc is dimensioned in the view in which its true shape is shown by giving the numerical denoting its radius, preceded by the abbreviation R, as shown in Fig. 13.19. The centers may be indicated by small crosses to clarify the drawing but not for small or unimportant radii. Crosses should not be shown for undimensioned arcs. As shown at (a) and (b), when there is room enough, both the numeral and the arrowhead are placed inside the arc. At (c) the arrowhead is left inside, but the numeral had to be moved outside. At (d) both the arrowhead and the numeral had to be moved outside. At (e) is

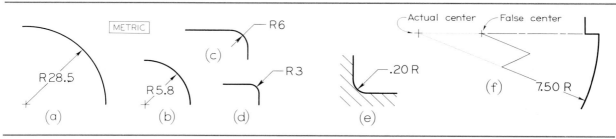

*Fig. 13.19*   Dimensioning Arcs.

shown an alternate method to (c) or (d) to be used when section lines or other lines are in the way. Note that in the unidirectional system, all of these numerals are lettered horizontally on the sheet.

For a long radius, as shown at (f), when the center falls outside the available space, the dimension line is drawn toward the actual center; but a false center may be indicated and dimension line "jogged" to it, as shown.

## 13.16 Fillets and Rounds

Individual fillets and rounds are dimensioned as any arc, as shown in Fig. 13.19 (c), (d), and (e). If there are only a few and they are obviously the same size, as in Fig. 13.43 (5), one typical radius is sufficient. However, fillets and rounds are often quite numerous on a drawing and most of them are likely to be some standard size, as R3 and R6 when dimensioning in metric or .125R and .250R when using the decimal-inch system.

In such cases it is customary to give a note in the lower portion of the drawing to cover all uniform fillets and rounds, thus,

FILLETS R6 AND ROUNDS R3 UNLESS
OTHERWISE SPECIFIED

or

ALL CASTING RADII R6 UNLESS NOTED

or simply

ALL FILLETS AND ROUNDS R6.

For a discussion of fillets and rounds in the shop, see §12.5.

## 13.17 Finish Marks

A *finish mark* is used to indicate that a surface is to be machined, or finished, as on a rough casting or forging. To the patternmaker or diemaker, a finish mark means that allowance of extra metal in the rough workpiece must be provided for the machining. See §12.4. On drawings of parts to be machined from rolled stock, finish marks are generally unnecessary, for it is obvious that the surfaces are finished. Similarly, it is not necessary to show finish marks when an operation is specified in a note that indicates machining, such as drilling, reaming, boring, countersinking, counterboring, and broaching, or when the dimension implies a finished surface, such as ∅6.22–6.35 (metric) or 2.45–2.50 DIA (decimal-inch).

Three styles of finish marks, the general ∨ symbol, the new basic √ symbol, and the traditional ⨍ symbol, are used to indicate an ordinary smooth machined surface. The ∨ symbol, Fig. 13.20 (a), is like a capital V, made about 3 mm ($\frac{1}{8}''$) high in conformity with the height of dimensioning lettering. The extended √ symbol, preferred by ANSI, Fig. 13.20 (b), is like a larger capital with the right leg extended. The short leg is made about 5 mm ($\frac{3}{16}''$) high and the height of the long leg is about 10 mm ($\frac{3}{8}''$). The basic symbol may be altered for more elaborate surface texture specifications; see §14.22.

For best results all finished marks should be drawn with the aid of a template or the 30° × 60° triangle. The point of the ∨ symbol should be directed inward toward the body of metal in a manner similar to that of a tool bit. The √ symbol is not shown upside down; see Figs. 13.35 and 14.45.

The preferred form and placement for the ⨍ symbol are described and given in Fig. 13.20 (e). The ⨍ symbol is in limited use and found mainly

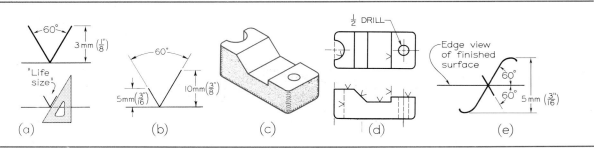

*Fig. 13.20*    Finish Marks.

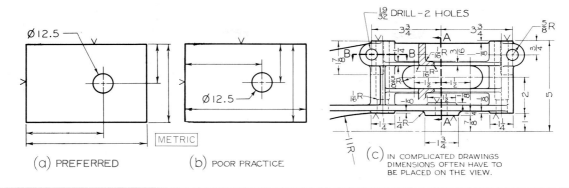

*Fig. 13.21*    Dimensions On or Off the Views.

on drawings made in accordance with earlier drafting standards.

At (c) is shown a simple casting having several finished surfaces, and at (d) are shown two views of the same casting, showing how the finish marks are indicated on a drawing. *The finish mark is shown only on the edge view of a finished surface and is repeated in any other view in which the surface appears as a line, even if the line is a hidden line.*

The several kinds of finishes are detailed in machine shop practice manuals. The following terms are commonly used: *finish all over, rough finish, file finish, sand blast, pickle, scrape, lap, hone, grind, polish, burnish, buff, chip, spotface, countersink, counterbore, core, drill, ream, bore, tap, broach, knurl,* and so on. When it is necessary to control the surface texture of finished surfaces beyond that of an ordinary machine finish, the new basic $\sqrt{}$ symbol is used as a base for the more elaborate surface quality symbols as discussed in §14.22.

If a part is to be finished all over, finish marks should be omitted, and a general note, such as FINISH ALL OVER or FAO, should be lettered on the lower portion of the sheet.

## 13.18 Dimensions On or Off Views

*Dimensions should not be placed upon a view unless doing so promotes the clearness of the drawing.* The ideal form is shown in Fig. 13.21 (a), in which all dimensions are placed outside the view. Compare this with the evidently poor practice shown at (b). This is not to say that a dimension should never be

placed on a view, for in many cases, particularly in complicated drawings, this is necessary, as shown at (c). Certain radii and other dimensions are given on the views, but in each case investigation will reveal a good reason for placing the dimension on the view. *Place dimensions outside of views, except where directness of application and clarity are gained by placing them on the views where they will be closer to the features dimensioned.* When a dimension must be placed in a sectioned area or on the view, leave an opening in the sectioned area or a break in the lines for the dimension figures, Figs. 13.13 (b) and 13.21 (c).

## 13.19 Contour Dimensioning

Views are drawn to describe the shapes of the various features of the object, and dimensions are given to define exact sizes and locations of those shapes. It follows that *dimensions should be given where the shapes are shown,* that is, in the views where the contours are delineated, as shown in Fig. 13.22 (a). Incorrect placement of the dimensions is shown at (b).

If individual dimensions are attached directly to the contours that show the shapes being dimensioned, this will automatically prevent the attachment of dimensions to hidden lines, as shown for the depth 10 of the slot at (b). It will also prevent the attachment of dimensions to a visible line, the meaning of which is not clear in a particular view, such as dimension 20 for the height of the base at (b).

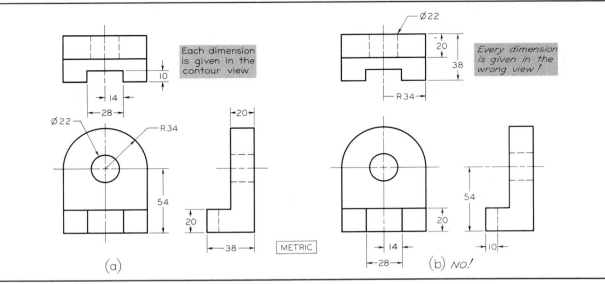

**Fig. 13.22**   Contour Dimensioning.

Although the placement of notes for holes follows the contour rule wherever possible, as shown at (a), the diameter of an external cylindrical shape is preferably given in the rectangular view where it can be readily found near the dimension for the length of the cylinder, as shown in Figs. 13.23 (b), 13.26, and 13.27.

## 13.20 Geometric Breakdown

Engineering structures are composed largely of simple geometric shapes, such as the prism, cylinder, pyramid, cone, and sphere, as shown in Fig. 13.23 (a). They may be exterior (positive) or interior (negative) forms. For example, a steel shaft is a positive cylinder, and a round hole is a negative cylinder.

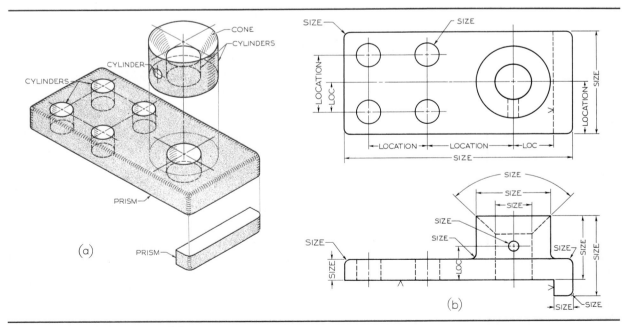

**Fig. 13.23**   Geometric Breakdown.

These shapes result directly from the design necessity to keep forms as simple as possible and from the requirements of the fundamental manufacturing operations. Forms having plane surfaces are produced by planing, shaping, milling, and so forth, while forms having cylindrical, conical, or spherical surfaces are produced by turning, drilling, reaming, boring, countersinking, and other rotary operations. See Chapter 12, "Manufacturing Processes."

The dimensioning of engineering structures involves two basic steps:

1. Give the dimensions showing the *sizes* of the simple geometric shapes, called *size dimensions*.

2. Give the dimensions *locating* these elements with respect to each other, called *location dimensions*.

The process of geometric analysis is very helpful in dimensioning any object, but must be modified when there is a conflict with either the function of the part in the assembly or with the manufacturing requirements in the shop.

Figure 13.23 (b) is a multiview drawing of the object shown in isometric at (a). Here it will be seen that each geometric shape is dimensioned with size dimensions and that these shapes are then located with respect to each other with location dimensions. Note that a *location dimension locates a three-dimensional geometric element* and not just a surface; otherwise, all dimensions would have to be classified as location dimensions.

## 13.21 Size Dimensions—Prisms

The right rectangular prism, Fig. 5.7, is probably the most common geometric shape. Front and top views are dimensioned as shown in Fig. 13.24 (a) or (b). The height and width are given in the front view and the depth in the top view. The vertical dimensions can be placed on the left or right provided both of them are placed in line. The horizontal dimension applies to both the front and top views and should be placed between them, as shown, and not above the top or below the front view.

Front and side views should be dimensioned as at (c) or (d). The horizontal dimensions can be placed above or below the views, provided both are placed in line. The dimension between views applies to both views and should not be placed elsewhere without a special reason.

An application of size dimensions to a machine

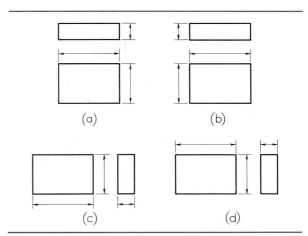

**Fig. 13.24** Dimensioning Rectangular Prisms.

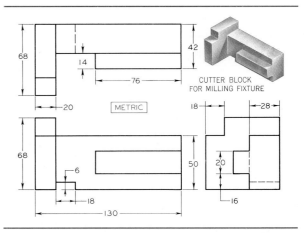

**Fig. 13.25** Dimensioning a Machine Part That is Composed of Prismatic Shapes.

part composed entirely of rectangular prisms is shown in Fig. 13.25.

## 13.22 Size Dimensions—Cylinders

The right circular cylinder is the next most common geometric shape and is commonly seen as a shaft or a hole. The general method of dimensioning a cylinder is to give both its diameter and its length in the rectangular view, Fig. 13.26. If the cylinder is drawn in a vertical position, the length or altitude of the cylinder may be given at the right as at (a), or on the left as at (b). If the cylinder is drawn in a

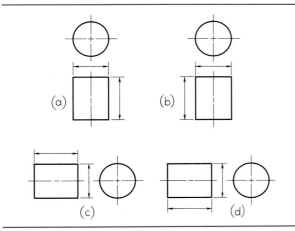

**Fig. 13.26** Dimensioning Cylinders.

horizontal position, the length may be given above the rectangular view as at (c) or below as at (d). An application showing the dimensioning of cylindrical shapes is shown in Fig. 13.27. The use of a diagonal diameter in the circular view, in addition to the method shown in Fig. 13.26, is not recommended except in special cases when clearness is gained thereby. The use of several diagonal diameters on the same center is definitely to be discouraged, since the result is usually confusing.

The radius of a cylinder should never be given, since measuring tools, such as the micrometer caliper, are designed to check diameters.

Small cylindrical holes, such as drilled, reamed, or bored holes, are usually dimensioned by means of notes specifying the diameter and the depth, with or without manufacturing operations, Figs. 13.27 and 13.32.

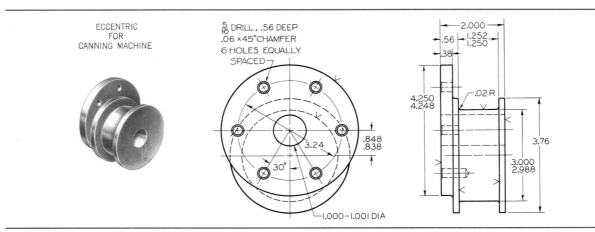

**Fig. 13.27** Dimensioning a Machine Part That Is Composed of Cylindrical Shapes.

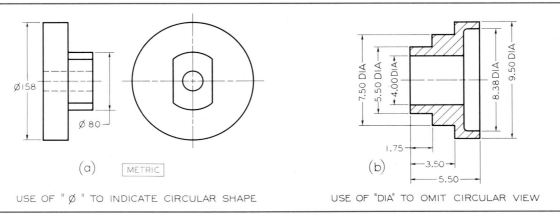

**Fig. 13.28** Use of ∅ or DIA in Dimensioning Cylinders.

The diameter symbol ⌀ should be given before all diametral dimensions on metric drawings rather than the abbreviation DIA following the dimension that is used on decimal-inch drawings, Fig. 13.28 (a) [ANSI Y14.5M–1982 (R1988)]. In some cases, the symbol on metric drawings or the abbreviation on decimal-inch drawings may be used to eliminate the circular view, as in (b).

## 13.23 Symbols and Size Dimensions—Miscellaneous Shapes

Traditional terms and abbreviations used to describe various shapes and manufacturing processes, in addition to size specifications, are employed throughout this text. A variety of dimensioning symbols, introduced by ANSI [Y14.5M–1982 (R1988)] to replace traditional terms or abbreviations, are given with construction details in Fig. 13.29. Traditional terms and abbreviations are suitable for use where the symbols are not desired. Typical applications of some of these symbols are given in Fig. 13.30.

A triangular prism is dimensioned, Fig. 13.31 (a) by giving the height, width, and displacement of the top edge in the front view and the depth in the top view.

A rectangular pyramid is dimensioned, (b), by giving the heights in the front view, and the dimen-

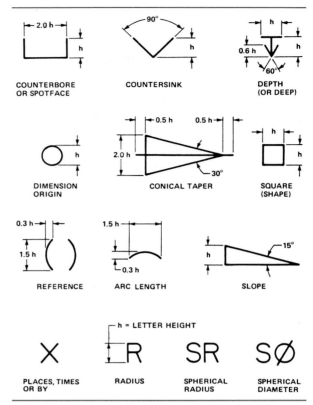

*Fig. 13.29*  Form and Proportion of Dimensioning Symbols   [ANSI Y14.5M–1982 (R1988)].

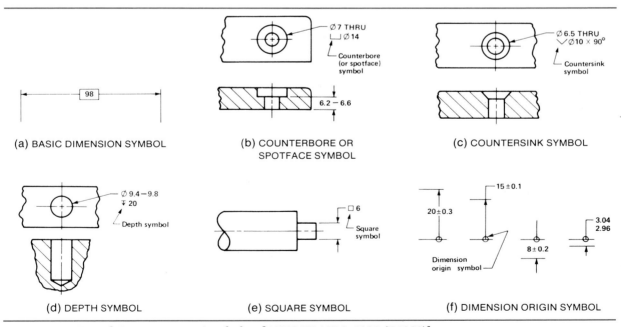

*Fig. 13.30*   Use of Dimensioning Symbols   [ANSI Y14.5M–1982 (R1988)].

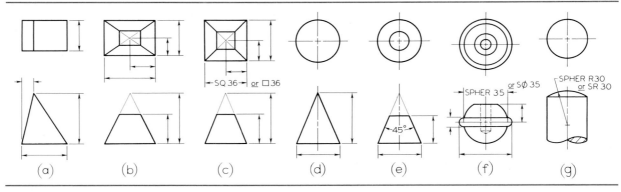

*Fig. 13.31*   Dimensioning Various Shapes.

sions of the base and the centering of the vertex in the top view. If the base is square, (c), it is necessary to give the dimensions for only one side of the base, provided it is labeled **SQ** as shown or preceded by the square symbol □. See Fig. 13.29 for form and proportions of dimensioning symbols.

A cone is dimensioned, (d), by giving its altitude and diameter of the base in the triangular view. A frustum of a cone may be dimensioned, (e), by giving the vertical angle and the diameter of one of the bases. Another method is to give the length and the diameters of both ends in the front view. Still another is to give the diameter at one end and the amount of taper per foot in a note, §13.33.

In Figure 13.31 (f) is shown a two-view drawing of a plastic knob. The main body is spherical and is dimensioned by giving its diameter preceded by the abbreviation and symbol for spherical diameter **S∅** or followed by the abbreviation **SPHER**. A bead around the knob is in the shape of a torus, Fig. 5.7, and it is dimensioned by giving the thickness of the ring and the outside diameter, as shown. At (g) a

spherical end is dimensioned by a radius preceded by the abbreviation SR or followed by **SPHER R**.

Internal shapes corresponding to the external shapes in Fig. 13.31 would be dimensioned in a similar manner.

## 13.24 Size Dimensioning of Holes

Holes that are to be drilled, bored, reamed, punched, cored, and so on are usually specified by standard notes, as shown in Figs. 7.40, 13.32 (a), and 13.44. The order of items in a note corresponds to the order of procedure in the shop in producing the hole. Two or more holes are dimensioned by a single note, the leader pointing to one of the holes, as shown at the top of Fig. 13.32.

As illustrated in Figs. 7.40 and 13.32, the leader of a note should, as a rule, point to the circular view of the hole. It should point to the rectangular view only when clearness is promoted thereby. When the circular view of the hole has two or more concentric circles, as for counterbored, countersunk, or tapped

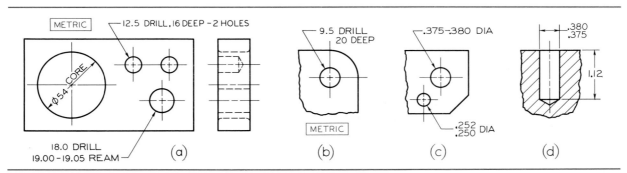

*Fig. 13.32*   Dimensioning Holes.

holes, the arrowhead should touch the outer circle, Fig. 13.44 (b), (c), and (e)–(j).

*Notes should always be lettered horizontally on the drawing paper, and guide lines should always be used.*

The use of decimal fractions to designate metric or inch drill sizes has gained wide acceptance,* Fig. 13.32 (b). For numbered or letter-size drills, Appendix 16, it is recommended that the decimal size be given in this manner, or given in parentheses; thus, #28 (.1405) DRILL, or "P" (.3230) DRILL. Metric drills are all decimal size and are not designated by number or letter.

On drawings of parts to be produced in large quantity for interchangeable assembly, dimensions and notes may be given without specification of the manufacturing process to be used. Only the dimensions of the holes are given, without reference to whether the holes are to be drilled, reamed, or punched, as in Fig. 13.32 (c) and (d). It should be realized that even though manufacturing operations are omitted from a note, the tolerances indicated would tend to dictate the manufacturing processes required.

## 13.25 Location Dimensions

After the geometric shapes composing a structure have been dimensioned for *size*, as discussed, *location dimensions* must be given to show the relative positions of these geometric shapes, as shown in Fig. 13.23. Fig. 13.33 (a) shows that rectangular shapes, whether in the form of solids or of recesses, are located with reference to their faces. At (b), cylindrical or conical holes or bosses, or other symmetical shapes, are located with reference to their center lines.

As shown in Fig. 13.34, location dimensions for holes are preferably given in the circular view of the holes.

Location dimensions should lead to finished surfaces wherever possible, Fig. 13.35, because rough castings and forgings vary in size, and unfinished surfaces cannot be relied upon for accurate measurements. Of course, the *starting dimension*, used in locating the first machined surface on a rough casting or forging, must necessarily lead from a rough

*Although drills are still listed fractionally in manufacturers' catalogs, many companies have supplemented drill and wire sizes with a decimal value. In many cases the number, letter, or common fraction has been replaced by the decimal-inch size. Metric drills are usually listed separately with a decimal-millimeter value.

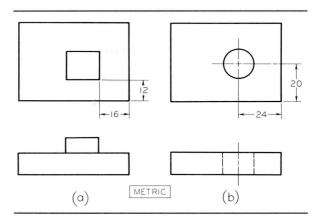

***Fig. 13.33*** Location Dimensions.

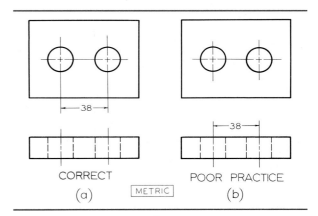

***Fig. 13.34*** Locating Holes.

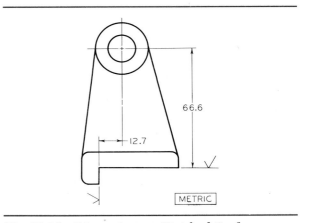

***Fig. 13.35*** Dimensions to Finished Surfaces.

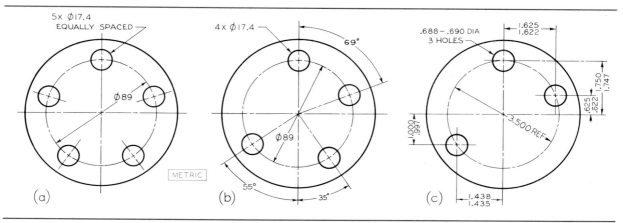

**Fig. 13.36**   Locating Holes About a Center.

surface, or from a center or a center line of the rough piece. See Figs. 16.74 and 16.76.

In general, location dimensions should be built from a finished surface as a datum plane, or from an important center or center line.

When several cylindrical surfaces have the same center line, as in Fig. 13.28 (b), it is not necessary to locate them with respect to each other.

Holes equally spaced about a common center may be dimensioned, Fig. 13.36 (a), by giving the diameter (diagonally) of the *circle of centers*, or *bolt circle*, and specifying "equally spaced" in the note. Repetitive features or dimensions may be specified by the use of an **X** preceded with a numeral to indicate the number of times or places the feature is required. Allow a space between the letter **X** and the dimension as at (a) and (b).

Holes unequally spaced, (b), are located by means of the bolt circle diameter plus angular measurements with reference to *only one* of the center lines, as shown.

Where greater accuracy is required, coordinate dimensions should be given, as at (c). In this case, the diameter of the bolt circle is marked **REF**, or enclosed in parentheses, ( ), Fig. 13.29, to indicate that it is to be used only as a *reference dimension*. Reference dimensions are given for information only. They are not intended to be measured and do not govern the manufacturing operations. They represent calculated dimensions and are often useful in showing the intended design sizes. See Fig. 13.36 (c).

When several nonprecision holes are located on a common arc, they are dimensioned, Fig. 13.37 (a), by giving the radius and the angular measurements from a *base line*, as shown. In this case, the base line is the horizontal center line.

At (b) the three holes are on a common center line. One dimension locates one small hole from the center; the other gives the distance between the small holes. Note the omission of a dimension at **X**. This method is used when (as is usually the case) the distance between the small holes is the important consideration. If the relation between the center hole and each of the small holes is more important, then include the distance at **X**, and mark the overall dimension **REF**.

At (c) is another example of coordinate dimensioning. The three small holes are on a bolt circle whose diameter is marked **REF** for reference purposes only. From the main center, the small holes are located in two mutually perpendicular directions.

Another example of locating holes by means of linear measurements is shown at (d). In this case, one such measurement is made at an angle to the coordinate dimensions because of the direct functional relationship of the two holes.

At (e) the holes are located from two *base lines* or *datums*. When all holes are located from a common datum, the sequence of measuring and machining operations is controlled, overall tolerance accumulations are avoided, and proper functioning of the finished part is assured as intended by the designer. The datum surfaces selected must be more accurate than any measurement made from them, must be accessible during manufacture, and must be arranged so as to facilitate tool and fixture design. Thus it may be necessary to specify accuracy of the datum surfaces in terms of straightness, roundness, flatness, and so forth. See §14.16.

At (f) is shown a method of giving, in a single line, all of the dimensions from a common datum.

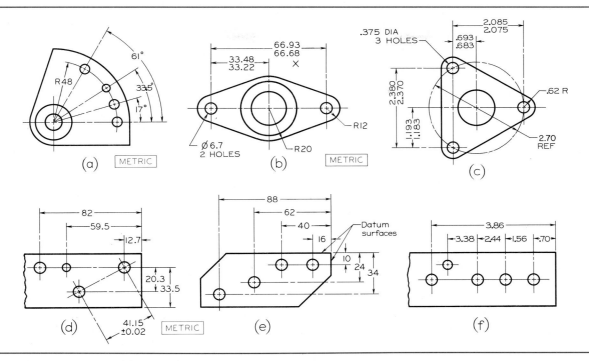

**Fig. 13.37** Locating Holes.

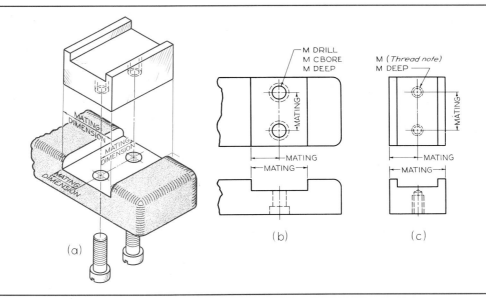

**Fig. 13.38** Mating Dimensions.

Each dimension except the first has a single arrowhead and is accumulative in value. The final and longest dimension is separate and complete.

These methods of locating holes are equally applicable to locating pins or other symmetrical features.

## 13.26 Mating Dimensions

In dimensioning a single part, its relation to mating parts must be taken into consideration. For example, in Fig. 13.38 (a), a guide block fits into a slot in a base. Those dimensions common to both parts are *mating dimensions*, as indicated.

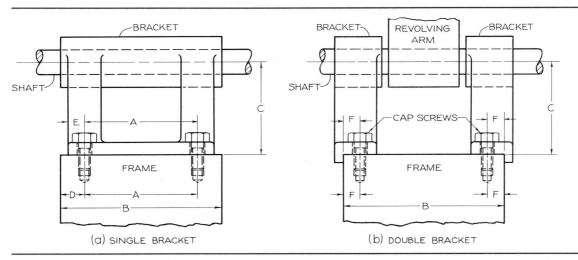

**Fig. 13.39**   Bracket Assembly.

These mating dimensions should be given on the multiview drawings in the corresponding locations, as shown at (b) and (c). Other dimensions are not mating dimensions since they do not control the accurate fitting together of two parts. The actual *values* of two corresponding mating dimensions may not be exactly the same. For example, the width of the slot at (b) may be dimensioned $\frac{1}{32}''$ (0.8 mm) or several thousandths of an inch larger than the width of the block at (c), but these are mating dimensions figured from a single basic width. It will be seen that the mating dimensions shown might have been arrived at from a geometric breakdown, §13.20. However, the mating dimensions need to be identified so that they can be specified in the corresponding locations on the two parts and so that they

can be given with the degree of accuracy commensurate with the proper fitting of the parts.

In Fig. 13.39 (a) the dimension A should appear on both the drawings of the bracket and of the frame and, therefore, is a necessary mating dimension. At (b), which shows a redesign of the bracket into two parts, dimension A is not used on either part, as it is not necessary to control closely the distance between the cap screws. But dimensions F are now essential mating dimensions and should appear correspondingly on the drawings of both parts. The remaining dimensions E, D, B, and C, at (a) are not considered to be mating dimensions, since they do not directly affect the mating of the parts.

## 13.27 Machine, Pattern, and Forging Dimensions

In Fig. 13.38 (a), the base is machined from a rough casting; the patternmaker needs certain dimensions to make the pattern, and the machinist needs certain dimensions for the machining. In some cases one dimension will be used by both. Again, in most cases, these dimensions will be the same as those resulting from a geometric breakdown, §13.20, but it is important to identify them in order to assign values to them.

The same part is shown in Fig. 13.40, with the machine dimensions and pattern dimensions identified by the letters M and P. The patternmaker is interested only in the dimensions required to make the pattern, and the machinist, in general, is concerned only with the dimensions needed to machine

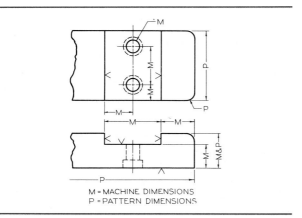

M = MACHINE DIMENSIONS
P = PATTERN DIMENSIONS

**Fig. 13.40**   Machine and Pattern Dimensions.

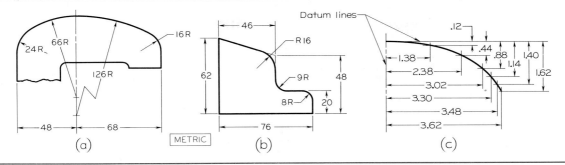

*Fig. 13.41*    Dimensioning Curves.

the part. Frequently, a dimension that is convenient for the machinist is not convenient for the pattern-maker, or vice versa. Since the patternmaker uses the drawing only once, while making the pattern, and the machinist refers to it continuously, the dimensions should be given primarily for the convenience of the machinist.

If the part is large and complicated, two separate drawings are sometimes made, one showing the pattern dimensions and the other the machine dimensions. The usual practice, however, is to prepare one drawing for both the patternmaker and the machinist. See §12.4.

For forgings, it is common practice to make separate forging drawings and machining drawings. A forging drawing of a connecting rod, showing only the dimensions needed in the forge shop, is shown in Fig. 16.36. A machining drawing of the same part, but containing only the dimensions needed in the machine shop, is shown in Fig. 16.37.

Unless a decimal system is used, §13.10, the pattern dimensions are nominal, usually to the nearest $\frac{1}{16}''$, and given in whole numbers and common fractions. If a machine dimension is given in whole numbers and common fractions, the machinist is usually allowed a tolerance (permissible in variation in size) of $\pm\frac{1}{64}''$. Some companies specify a tolerance of $\pm.010''$ on all common fractions. If greater accuracy is required, the dimensions are given in decimal form. Metric dimensions are given to one or more places and decimal-inch dimensions are given to three or more places, §§13.10 and 14.1. Remember that 0.1 mm is approximately .004 inch.

## 13.28 Dimensioning of Curves

Curved shapes may be dimensioned by giving a group of radii, as shown in Fig. 13.41 (a). Note that in dimensioning the **R126** arc whose center is in-

accessible, the center may be moved inward along a center line, and a jog made in the dimension line. See also Fig. 13.19 (f). Another method is to dimension the outline envelope of a curved shape so that the various radii are self-locating from "floating centers," as at (b). Either a circular or a noncircular curve may be dimensioned by means of coordinate dimensions referred to datums, as shown at (c). See also Fig. 13.6 (a).

## 13.29 Dimensioning of Rounded-End Shapes

The method used for dimensioning rounded-end shapes depends upon the degree of accuracy required, Fig. 13.42. When precision is not necessary, the methods used are those that are convenient for manufacturing, as at (a), (b), and (c).

At (a) the link, to be cast or to be cut from sheet metal or plate, is dimensioned as it would be laid out for manufacture, by giving the center-to-center distance and the radii of the ends. Note that only one such radius dimension is necessary, but that the number of places may be included with the size dimension.

At (b) the pad on a casting, with a milled slot, is dimensioned from center to center for the convenience of both the patternmaker and the machinist in layout. An additional reason for the center-to-center distance is that it gives the total travel of the milling cutter, which can be easily controlled by the machinist. The width dimension indicates the diameter of the milling cutter; hence, it is incorrect to give the radius of a machined slot. On the other hand, a cored slot (see §12.3) should be dimensioned by radius in conformity with the patternmaker's layout procedure.

At (c) the semicircular pad is laid out in a similar

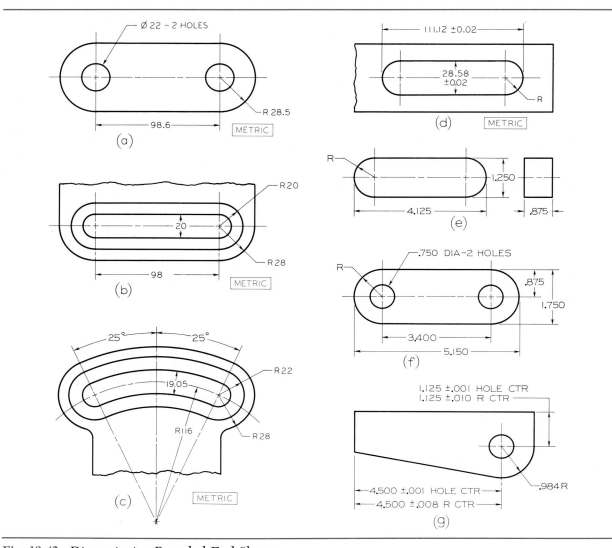

**Fig. 13.42**  Dimensioning Rounded-End Shapes.

manner to the pad at (b), except that angular dimensions are used. Angular tolerances, §14.18, can be used if necessary.

When accuracy is required, the methods shown at (d) to (g) are recommended. Overall lengths of rounded-end shapes are given in each case, and radii are indicated, but without specific values. In the example at (f), the center-to-center distance is required for accurate location of the holes.

At (g) the hole location is more critical than the location of the radius; hence, the two are located independently, as shown.

## 13.30 Superfluous Dimensions

Though it is necessary to give all dimensions, the designer should avoid giving unnecessary or superfluous dimensions, Fig. 13.43. Dimensions should not be repeated on the same view or on different views, nor should the same information be given in two different ways.

Figure 13.43 (2) illustrates a type of superfluous dimensioning that should generally be avoided, especially in machine drawing where accuracy is important. The production personnel should not be allowed a choice between two dimensions. *Avoid*

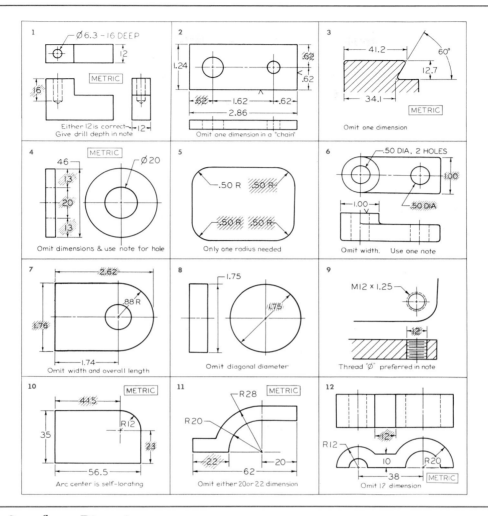

*Fig. 13.43* Superfluous Dimensions.

*"chain" dimensioning*, in which a complete series of detail dimensions is given, together with an overall dimension. In such cases, one dimension of the chain should be omitted, as shown, so that the machinist is obliged to work from one surface only. This is particularly important in tolerance dimensioning, §14.1, where an accumulation of tolerances can cause serious difficulties. See also §14.9.

Some inexperienced detailers have the habit of omitting both dimensions, such as those at the right at (2), on the theory that the holes are symmetrically located and will be understood to be centered. One of the two location dimensions should be given.

As shown at (5), when one dimension clearly applies to several identical features, it need not be repeated, but the number of places should be indicated. Dimensions for fillets and rounds and other noncritical features need not be repeated nor number of places specified, as at (5).

For example, the radii of the rounded ends in Fig. 13.42 (a) to (f) need not be repeated, and in Fig. 13.1 both ribs are obviously the same thickness so it is unnecessary to repeat the 10 mm dimension.

### 13.31 Notes

It is usually necessary to supplement the direct dimensions with notes, Fig. 13.44. They should be brief and should be carefully worded so as to be capable of only one interpretation. *Notes should al-*

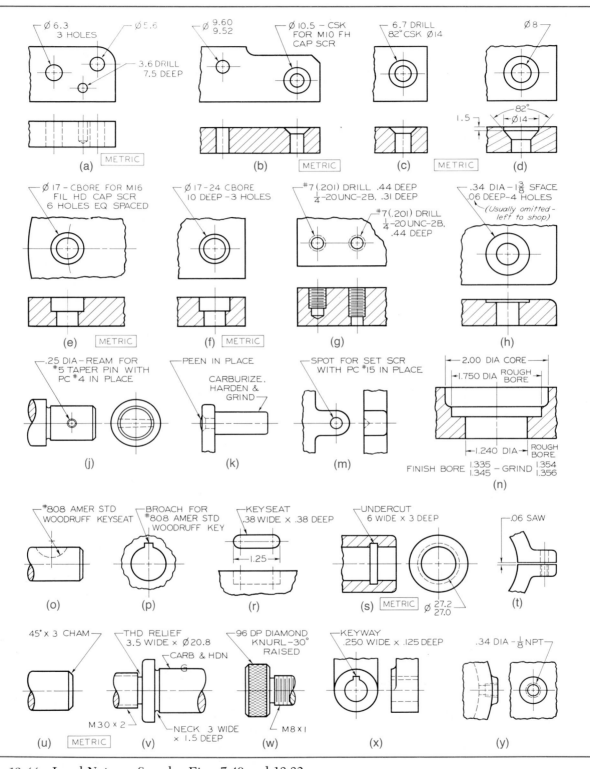

**Fig. 13.44**    Local Notes.    See also Figs. 7.40 and 13.32.

ways be lettered horizontally on the sheet, with guide lines, and arranged in a systematic manner. They should not be lettered in crowded places. Avoid placing notes between views, if possible. They should not be lettered closely enough to each other to confuse the reader or close enough to another view or detail to suggest application to the wrong view. Leaders should be as short as possible and cross as few lines as possible. They should never run through a corner of a view or through any specific points or intersections.

Notes are classified as *general notes* when they apply to an entire drawing and as *local notes* when they apply to specific items.

## General Notes

General notes should be lettered in the lower right-hand corner of the drawing, above or to the left of the title block, or in a central position below the view to which they apply.

EXAMPLES

FINISH ALL OVER (FAO)

BREAK SHARP EDGES TO R0.8

G33106 ALLOY STEEL–BRINELL 340–380

ALL DRAFT ANGLES 3° UNLESS OTHERWISE SPECIFIED

DIMENSIONS APPLY AFTER PLATING

In machine drawings, the title strip or title block will carry many general notes, including material, general tolerances, heat treatment, and pattern information. See Fig. 16.25.

## Local Notes

Local notes apply to specific operations only and are connected by a leader to the point at which such operations are performed.

EXAMPLES

6.30 DRILL–4 HOLES

45°x1.6 CHAMFER

96 DP DIAMOND KNURL, RAISED

The leader should be attached at the front of the first word of a note, or just after the last word, and not at any intermediate place.

For information on notes applied to holes, see §13.24.

Certain commonly used abbreviations may be used freely in notes, such as THD, DIA, MAX. The less common abbreviations should be avoided as much as possible. All abbreviations should conform to ANSI Y1.1–1972 (R1984). See Appendix 4 for American National Standard abbreviations. See Figs. 13.29 and 13.30 for form and use of alternative dimensioning symbols.

In general, leaders and notes should not be placed on the drawing until the dimensioning is substantially completed. If notes are lettered first, almost invariably they will be in the way of necessary dimensions and will have to be moved.

## 13.32 Dimensioning of Threads

Local notes are used to specify dimensions of threads. For tapped holes the notes should, if possible, be attached to the circular views of the holes, as shown in Fig. 13.44 (g). For external threads, the notes are usually placed in the longitudinal views where the threads are more easily recognized, as at (v) and (w). For a detailed discussion of thread notes, see §15.21.

## 13.33 Dimensioning of Tapers

A *taper* is a conical surface on a shaft or in a hole. The usual method of dimensioning a taper is to give the amount of taper in a note such as TAPER 0.167 ON DIA (often TO GAGE added), and then give the diameter at one end, plus the length, or give the diameter at both ends and omit the length. *Taper on diameter means the difference in diameter per unit of length.*

*Standard machine tapers* are used on machine spindles, shanks of tools, pins, for example, and are described in "Machine Tapers," ANSI B5.10–1981 (R1987). Such standard tapers are dimensioned on a drawing by giving the diameter, usually at the large end, the length, and a note, such as NO. 4 AMERICAN NATIONAL STANDARD TAPER. See Fig. 13.45 (a).

For not-too-critical requirements, a taper may be dimensioned by giving the diameter at the large end, the length, and the included angle, all with proper tolerances, (b). Or the diameters of both ends, plus the length, may be given with necessary tolerances.

For close-fitting tapers, the amount of *taper per unit on diameter* is indicated as shown at (c) and (d). A gage line is selected and located by a comparatively generous tolerance, while other dimensions are given appropriate tolerances as required.

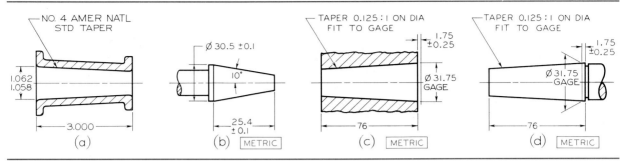

**Fig. 13.45**    Dimensioning Tapers.

## 13.34 Dimensioning of Chamfers

A *chamfer* is a beveled or sloping edge, and it is dimensioned by giving the length of the offset and the angle, as in Fig. 13.46 (a). A 45° chamfer also may be dimensioned in a manner similar to that

shown at (a), but usually it is dimensioned by note without or with the word **CHAMFER** as at (b).

## 13.35 Shaft Centers

Shaft centers are required on shafts, spindles, and other conical or cylindrical parts for turning, grinding, and other operations. Such a center may be dimensioned as shown in Fig. 13.47. Normally the centers are produced by a combined drill and countersink.

## 13.36 Dimensioning Keyways

Methods of dimensioning keyways for Woodruff keys and stock keys are shown in Fig. 13.48. Note in both cases, the use of a dimension to center the keyway in the shaft or collar. The preferred method of dimensioning the depth of a keyway is to give the dimension from the bottom of the keyway to the opposite side of the shaft or hole, as shown. The method of computing such a dimension is shown at (d). Values for **A** may be found in machinists' handbooks.

     For general information about keys and keyways, see §15.34.

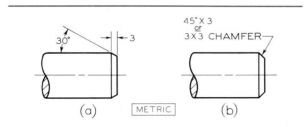

**Fig. 13.46**    Dimensioning Chamfers.

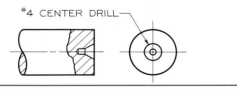

**Fig. 13.47**    Shaft Center.

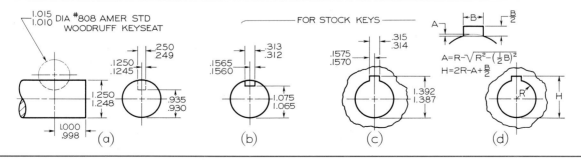

**Fig. 13.48**    Dimensioning Keyways.

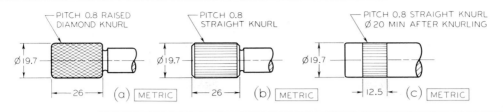

Fig. 13.49 Dimensioning Knurls.

## 13.37 Dimensioning of Knurls

A *knurl* is a roughened surface to provide a better handgrip or to be used for a press fit between two parts. For handgripping purposes, it is necessary only to give the pitch of the knurl, the type of knurling, and the length of the knurled area, Fig. 13.49 (a) and (b). To dimension a knurl for a press fit, the toleranced diameter before knurling should be given, (c). A note should be added giving the pitch and type of knurl and the minimum diameter after knurling. See ANSI B94.6–1984.

## 13.38 Dimensioning Along Curved Surfaces

When angular measurements are unsatisfactory, chordal dimensions, Fig. 13.50 (a), or linear dimensions upon the curved surfaces, as shown at (b), may be given.

## 13.39 Sheet-Metal Bends

In sheet-metal dimensioning, allowance must be made for bends. The intersection of the plane surfaces adjacent to a bend is called the *mold line*, and this line, rather than the center of the arc, is used to determine dimensions, Fig. 13.51. The following procedure for calculating bends is typical. If the two inner plane surfaces of an angle are extended, their

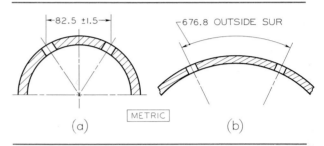

Fig. 13.50 Dimensioning Along Curved Surfaces.

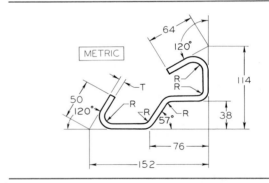

Fig. 13.51 Profile Dimensioning.

line of intersection is called the IML or *inside mold line*, Fig. 13.52 (a) to (c). Similarly, if the two outer plane surfaces are extended, they produce the OML or *outside mold line*. The *center line of bend* ($\mathcal{C}$B)

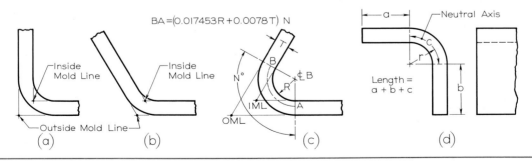

$$BA = (0.017453R + 0.0078T) N$$

Fig. 13.52 Bends.

refers primarily to the machine on which the bend is made and is at the center of the bend radius.

The length, or *stretchout*, of the pattern equals the sum of the flat sides of the angle plus the distance around the bend measured along the *neutral axis*. The distance around the bend is called the *bend allowance*. When metal bends, it compresses on the inside and stretches on the outside. At a certain zone in between, the metal is neither compressed nor stretched. This is called the neutral axis. See Fig. 13.52 (d). The neutral axis is usually assumed to be 0.44 of the thickness from the inside surface of the metal.

The developed length of material, or bend allowance (BA), to make the bend is computed from the empirical formula

$$BA = (0.017453R + 0.0078T)N$$

where $R$ = radius of bend, $T$ = metal thickness, and $N$ = number of degrees of bend. See Fig. 13.52 (c).

## 13.40 Tabular Dimensions
A series of objects having like features but varying in dimensions may be represented by one drawing, Fig. 13.53. Letters are substituted for dimension figures on the drawing, and the varying dimensions are given in tabular form. The dimensions of many standard parts are given in this manner in the various catalogs and handbooks.

## 13.41 Standards
Dimensions should be given, wherever possible, to make use of readily available materials, tools, parts, and gages. The dimensions for many commonly used machine elements, such as bolts, screws, nails, keys,

tapers, wire, pipes, sheet metal, chains, belts, ropes, pins, and rolled metal shapes, have been standardized, and the drafter must obtain these sizes from company standards manuals, from published handbooks, from American National Standards, or from manufacturers' catalogs. Tables of some of the more common items are given in the Appendix of this text.

Such standard parts are not delineated on detail drawings unless they are to be altered for use, but are drawn conventionally on assembly drawings and are listed in parts lists, §16.14. Common fractions are often used to indicate the nominal sizes of standard parts or tools. If the complete decimal-inch system is used, all such sizes ordinarily are expressed by decimals; for example, .250 DRILL instead of $\frac{1}{4}$ DRILL. If the all-metric system of dimensioning is used, then the *preferred* metric drill of the approximate same size (.2480″) would be indicated as a 6.30 DRILL.

## 13.42 Coordinate Dimensioning
In general, the basic coordinate dimensioning practices are compatible with the data requirements for tape or computer-controlled automatic production machines. However, to design for automated production, the designer and/or drafter should first consult the manufacturing machine manuals before making the drawings for production. Certain considerations should be noted:

1. A set of three mutually perpendicular datum or reference planes is usually required for coordinate dimensioning. These planes either must be obvious or clearly identified. See Fig. 13.54.

2. The designer selects as origins for dimensions those surfaces or other features most important to the functioning of the part. Enough of these

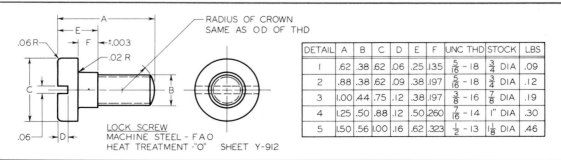

| DETAIL | A | B | C | D | E | F | UNC THD | STOCK | LBS |
|---|---|---|---|---|---|---|---|---|---|
| 1 | .62 | .38 | .62 | .06 | .25 | .135 | $\frac{5}{16}$ - 18 | $\frac{3}{4}$ DIA | .09 |
| 2 | .88 | .38 | .62 | .09 | .38 | .197 | $\frac{5}{16}$ - 18 | $\frac{3}{4}$ DIA | .12 |
| 3 | 1.00 | .44 | .75 | .12 | .38 | .197 | $\frac{3}{8}$ - 16 | $\frac{7}{8}$ DIA | .19 |
| 4 | 1.25 | .50 | .88 | .12 | .50 | .260 | $\frac{7}{16}$ - 14 | 1″ DIA | .30 |
| 5 | 1.50 | .56 | 1.00 | .16 | .62 | .323 | $\frac{1}{2}$ - 13 | $1\frac{1}{8}$ DIA | .46 |

*Fig. 13.53* Tabular Dimensioning.

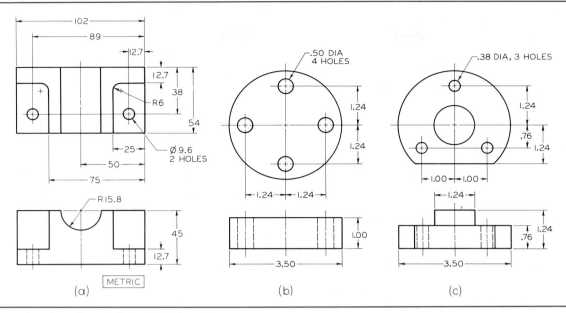

**Fig. 13.54** Coordinate Dimensioning.

features are selected to position the part in relation to the set of mutually perpendicular planes. All related dimensions on the part are then made from these planes.

3. All dimensions should be in decimals.

4. Angles should be given, where possible, in degrees and decimal parts of degrees.

5. Standard tools, such as drills, reamers, and taps, should be specifed wherever possible.

6. All tolerances should be determined by the design requirements of the part, not by the capability of the manufacturing machine.

## 13.43 Do's and Don'ts of Dimensioning

The following checklist summarizes briefly most of the situations in which a beginning designer is likely to make a mistake in dimensioning. The student should check the drawing by this list before submitting it to the instructor.

1. Each dimension should be given clearly, so that it can be interpreted in only one way.

2. Dimensions should not be duplicated or the same information be given in two different ways—dual dimensioning excluded—and no dimensions should be given except those needed to produce or inspect the part.

3. Dimensions should be given between points or surfaces that have a functional relation to each other or that control the location of mating parts.

4. Dimensions should be given to finished surfaces or important center lines in preference to rough surfaces wherever possible.

5. Dimensions should be so given that it will not be necessary for the machinist to calculate, scale, or assume any dimension.

6. Dimensions should be attached to the view where the shape is best shown (contour rule).

7. Dimensions should be placed in the views where the features dimensioned are shown true shape.

8. Avoid dimensioning to hidden lines wherever possible.

9. Dimensions should not be placed upon a view unless clearness is promoted and long extension lines are avoided.

10. Dimensions applying to two adjacent views should be placed between views, unless clearness is promoted by placing some of them outside.

11. The longer dimensions should be placed outside all intermediate dimensions, so that dimension lines will not cross extension lines.

12. In machine drawing, omit all unit marks, except when necessary for clearness; for example, 1″ VALVE or 1 mm DRILL.

13. Do not expect production personnel to assume that a feature is centered (as a hole on a plate), but give a location dimension from one side. However, if a hole is to be centered on a symmetrical rough casting, mark the center line and omit the locating dimension from the center line.

14. A dimension should be attached to only one view, not to extension lines connecting two views.

15. Detail dimensions should "line up" in chain fashion.

16. Avoid a complete chain of detail dimensions; better omit one, otherwise add REF (reference) to one detail dimension or the overall dimension or enclose the dimension within parentheses, ( ).

17. A dimension line should never be drawn through a dimension figure. A figure should never be lettered over any line of the drawing. Break the line if necessary.

18. Dimension lines should be spaced uniformly throughout the drawing. They should be at least 10 mm ($\frac{3}{8}$″) from the object outline and 6 mm ($\frac{1}{4}$″) apart.

19. No line of the drawing should be used as a dimension line or coincide with a dimension line.

20. A dimension line should never be joined end to end (chain fashion) with any line of the drawing.

21. Dimension lines should not cross, if avoidable.

22. Dimension lines and extension lines should not cross, if avoidable (extension lines may cross each other).

23. When extension lines cross extension lines or visible lines, no break in either line should be made.

24. A center line may be extended and used as an extension line, in which case it is still drawn like a center line.

25. Center lines should generally not extend from view to view.

26. Leaders for notes should be straight, not curved, and pointing to the center of circular views of holes wherever possible.

27. Leaders should slope at 45°, or 30°, or 60° with horizontal but may be made at any convenient angle except vertical or horizontal.

28. Leaders should extend from the beginning or from the end of a note, the horizontal "shoulder" extending from midheight of the lettering.

29. Dimension figures should be approximately centered between the arrowheads, except that in a "stack" of dimensions, the figures should be "staggered."

30. Dimension figures should be about 3 mm ($\frac{1}{8}$″) high for whole numbers and 6 mm ($\frac{1}{4}$″) high for fractions.

31. Dimension figures should never be crowded or in any way made difficult to read.

32. Dimension figures should not be lettered over lines or sectioned areas unless necessary, in which case a clear space should be reserved for the dimension figures.

33. Dimension figures for angles should generally be lettered horizontally.

34. Fraction bars should never be inclined except in confined areas, such as in tables.

35. The numerator and denominator of a fraction should never touch the fraction bar.

36. Notes should always be lettered horizontally on the sheet.

37. Notes should be brief and clear, and the wording should be standard in form, Fig. 13.44.

38. Finish marks should be placed on the edge views of all finished surfaces, including hidden edges and the contour and circular views of cylindrical surfaces.

39. Finish marks should be omitted on holes or other features where a note specifies a machining operation.

40. Finish marks should be omitted on parts made from rolled stock.

41. If a part is finished all over, omit all finish marks, and use the general note FINISH ALL OVER, or FAO.

42. A cylinder is dimensioned by giving both its diameter and length in the rectangular view, except when notes are used for holes. A diagonal diameter in the circular view may be used in cases where clearness is gained thereby.

43. Holes to be bored, drilled, reamed, and so on are size-dimensioned by notes in which the leaders preferably point toward the center of the circular views of the holes. Indications of manufacturing processes may be omitted from notes.

44. Drill sizes are preferably expressed in decimals. For drills designated by number or letter, the decimal size must also be given.

45. In general, a circle is dimensioned by its diameter, an arc by its radius.

46. Avoid diagonal diameters, except for very large holes and for circles of centers. They may be used on positive cylinders when clearness is gained thereby.

47. A metric diameter dimension value should always be preceded by the symbol ∅ and the diameter dimension value in inches should be followed by DIA.

48. A metric radius dimension should always be preceded by the letter R and the radius value in inches should be followed by the letter R. The radial dimension line should have only one arrowhead, and it should pass through or point through the arc center and touch the arc.

49. Cylinders should be located by their center lines.

50. Cylinders should be located in the circular views, if possible.

51. Cylinders should be located by coordinate dimensions in preference to angular dimensions where accuracy is important.

52. When there are several rough noncritical features obviously the same size (fillets, rounds, ribs, etc.), it is necessary to give only typical (abbreviation TYP) dimensions, or to use a note.

53. When a dimension is not to scale, it should be underscored with a heavy straight line or marked NTS or NOT TO SCALE.

54. Mating dimensions should be given correspondingly on drawings of mating parts.

55. Pattern dimensions should be given in two-place decimals or in common whole numbers and fractions to the nearest $\frac{1}{16}$″.

56. Decimal dimensions should be used for all machining dimensions.

57. Avoid cumulative tolerances, especially in limit dimensioning, described in §14.9.

## *Dimensioning Problems*

It is expected that most of the student's practice in dimensioning will be in connection with working drawings assigned from other chapters. However, a limited number of special dimensioning problems are available here in Figs. 13.55 and 13.56. The problems are designed for Layout A–3 (8.5″ × 11.0″) and are to be drawn with instruments and dimensioned to a full-size scale. Layout A4–3 (297 mm × 420 mm) may be used with appropriate adjustments in the title strip layout.

Since many of the problems in this and other chapters are of a general nature, they can also be solved on most computer graphics systems. If a system is available, the instructor may choose to assign specific problems to be completed by this method.

Dimensioning problems in convenient form for solution may be found in *Principles of Engineering Graphics Problems* by Spencer, Hill, Loving, and Dygdon, a workbook designed to accompany this text that is also published by Macmillan Publishing Company.

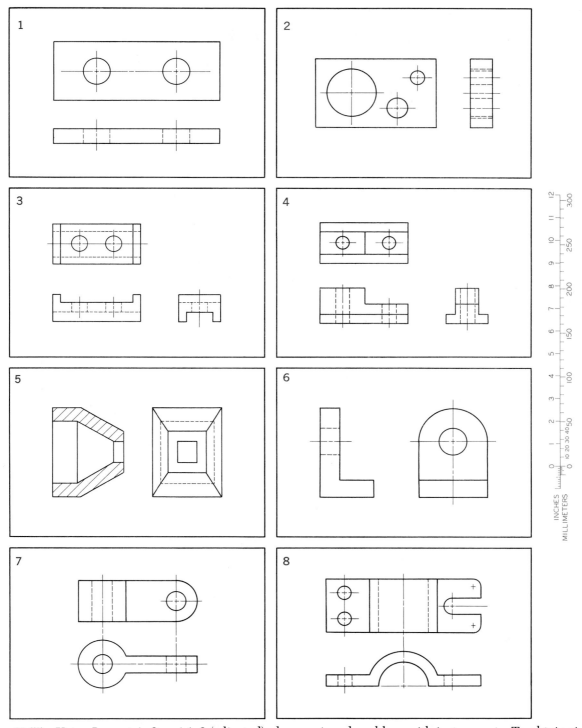

**Fig. 13.55** Using Layout A–3 or A4–3 (adjusted), draw assigned problem with instruments. To obtain sizes, place bow dividers on the views on this page and transfer to scale at the side to obtain values. Dimension drawing completely in one-place millimeters or two-place inches as assigned, full size. See inside front cover for decimal-inch and millimeter equivalents.

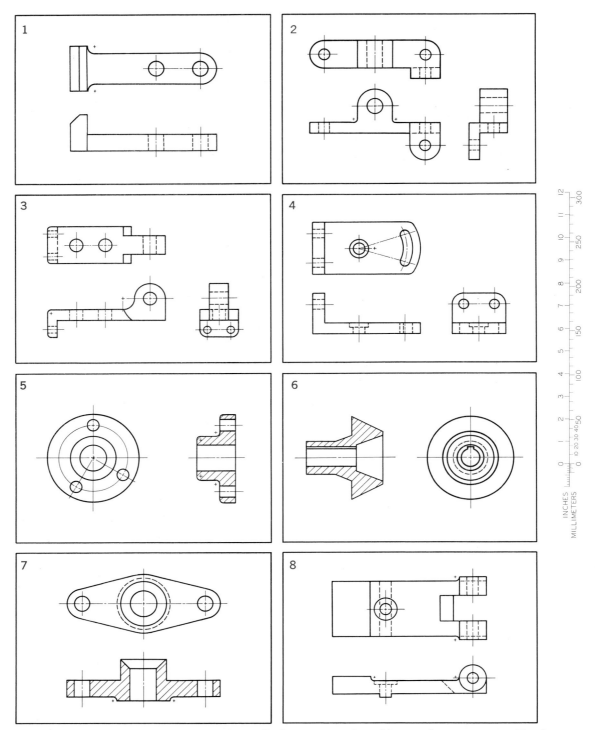

*Fig. 13.56* Using Layout A–3 or A4–3 (adjusted), draw assigned problem with instruments. To obtain sizes, place bow dividers on the views on this page and transfer to scale at the side to obtain values. Dimension drawing completely in one-place millimeters or two-place inches as assigned, full size. See inside front cover for decimal-inch and millimeter equivalents.

# CHAPTER

# 14

# Tolerancing

Interchangeable manufacturing, by means of which parts can be made in widely separated localities and then be brought together for assembly, where the parts will all fit together properly, is an essential element of mass production. Without interchangeable manufacturing, modern industry could not exist, and without effective size control by the engineer, interchangeable manufacturing could not be achieved.

For example, an automobile manufacturer not only subcontracts the manufacture of many parts of a design to other companies, but must be concerned with parts for replacement. All parts in each category must be near enough alike so that any one of them will fit properly in any assembly. Unfortunately, *it is impossible to make anything to exact size*. Parts can be made to very close dimensions, even to a few millionths of an inch or thousandths of a millimeter (e.g., gage blocks), but such accuracy is extremely expensive.

However, exact sizes are not needed, only varying degrees of accuracy according to functional requirements. A manufacturer of children's tricycles would soon go out of business if the parts were made with jet-engine accuracy, as no one would be willing to pay the price. So what is needed is a means of specifying dimensions with whatever degree of accuracy may be required. The answer to the problem is the specification of a *tolerance* on each dimension.

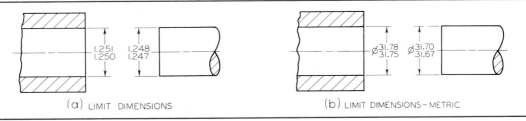

**Fig. 14.1**  Fits Between Mating Parts.

## 14.1 Tolerance Dimensioning

*Tolerance* is the total amount a specific dimension is permitted to vary, which is the difference between the maximum and the minimum limits [ANSI Y14.5M–1982 (R1988)]. For example, a dimension given as **1.625 ± .002** means that it may be (on the manufactured part) **1.627"** or **1.623"**, or anywhere between these *limit dimensions*. The tolerance, or total amount of variation "tolerated," is .004". Thus, it becomes the function of the detailer or designer to specify the allowable error that may be tolerated for a given dimension and still permit the satisfactory functioning of the part. Since greater accuracy costs more money, the detailer or designer will not specify the closest tolerance, but instead will specify as generous a tolerance as possible.

In order to control the dimensions of quantities of the two parts so that any two mating parts will be interchangeable, it is necessary to assign tolerances to the dimensions of the parts, as shown in Fig. 14.1 (a). The diameter of the hole may be machined not less than **1.250"** and not more than **1.251"**, these two figures representing the *limits* and the difference between them, .001", being the *tolerance*. Likewise, the shaft must be produced between the limits of **1.248"** and **1.247"**, the tolerance on the shaft being the difference between these, or **.001"**. The metric

versions for these limit dimensions for the hole and shaft are shown at (b). The difference in the dimensions for either the hole or shaft is 0.03 mm, the total *tolerance*.

A pictorial illustration of the dimensions in Fig. 14.1 (a) is shown in Fig. 14.2 (a). The maximum shaft is shown solid, and the minimum shaft is shown in phantom. The difference in diameters, .001", is the tolerance on the shaft. Similarly, the tolerance on the hole is the difference between the two limits shown, or .001". The loosest fit, or maximum clearance, occurs when the smallest shaft is in the largest hole, as shown at (b). The tightest fit, or minimum clearance, occurs when the largest shaft is in the smallest hole, as shown at (c). The difference between these, .002", is the *allowance*. The average clearance is .003", which is the same difference as allowed in the example of Fig. 14.1 (a); thus, any shaft will fit any hole interchangeably.

When expressed in metric dimensions, the limits for the hole are 31.75 mm and 31.78 mm, the difference between them, 0.03 mm, being the tolerance. Similarly, the limits for the shaft are 31.70 mm and 31.67 mm, the tolerance on the shaft being the difference between them, or 0.03 mm.

When parts are required to fit properly in assembly but not to be interchangeable, the size of

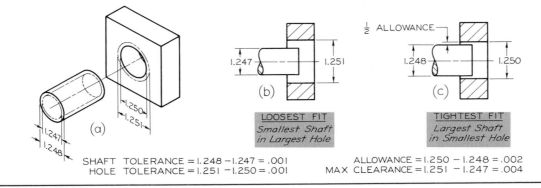

**Fig. 14.2**  Limit Dimensions.

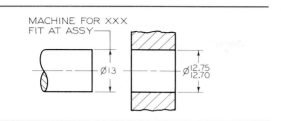

**Fig. 14.3** Noninterchangeable Fit.

one part need not be toleranced, but indicated to be made to fit at assembly, Fig. 14.3.

## 14.2 Definitions of Terms

At this point, it is well to fix in mind the definitions of certain terms [ANSI Y14.5M–1982 (R1988)].

*Nominal size* The designation that is used for the purpose of general identification is usually expressed in common fractions. In Fig. 14.1, the nominal size of both hole and shaft, which is $1\frac{1}{4}''$, would be **1.25"** or **31.75 mm** in a decimal system of dimensioning.

*Basic size or dimension* The theoretical size from which limits of size are derived by the application of allowances and tolerances. It is the size from which limits are determined for the size, shape, or location of a feature. In Fig. 14.1 (a), the basic size is the decimal equivalent of the nominal size $1\frac{1}{4}''$, or **1.250"** or **31.75 mm** at (b).

*Actual size* The measured size of the finished part.

*Tolerance* The total amount by which a given dimension may vary, or the difference between the limits. In Fig. 14.2 (a) the tolerance on either the shaft or hole is the difference between the limits, or **.001"**.

*Limits* The maximum and minimum sizes indicated by a toleranced dimension. In Fig. 14.2 (a) the limits for the hole are **1.250"** and **1.251"**, and for the shaft are **1.248"** and **1.247"**.

*Allowance* The minimum clearance space (or maximum interference) intended between the maximum material condition (MMC) of mating parts. In Fig. 14.2 (c) the allowance is the difference between the smallest hole, **1.250"**, and the largest shaft, **1.248"**, or **.002"**. Allowance, then, represents the tightest permissible fit and is simply the smallest hole minus the largest shaft. For clearance fits, this difference will be positive, while for interference fits it will be negative.

## 14.3 Fits Between Mating Parts

"Fit is the general term used to signify the range of tightness or looseness that may result from the application of a specific combination of allowances and tolerances in mating parts." [ANSI Y14.5M–1982 (R1988)]. There are four general types of fits between parts.

*Clearance fit* In which an internal member fits in an external member (as a shaft in a hole), and always leaves a space or clearance between the parts. In Fig. 14.2 (c) the largest shaft is **1.248"** and the smallest hole is **1.250"**, which permits a minimum air space of **.002"** between the parts. This space is the allowance, and in a clearance fit it is always positive.

*Interference fit* In which the internal member is larger than the external member such that there is always an actual interference of metal. In Fig. 14.4 (a) the smallest shaft is **1.2513"**, and the largest hole is **1.2506"**, so that there is an actual interference of metal amounting to at least **.0007"**. Under maximum material conditions the interference would be **.0019"**. This interference is the allowance, and in an interference fit it is always negative.

*Transition fit* In which the fit might result in either a clearance or interference condition. In Fig. 14.4 (b) the smallest shaft, **1.2503"**, will fit in the largest hole, **1.2506"**, with **.0003"** to spare. But the largest

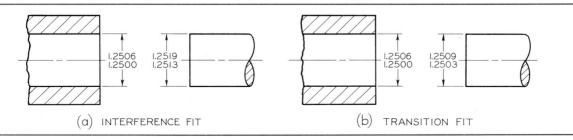

**Fig. 14.4** Fits Between Parts.

shaft, 1.2509″, will have to be forced into the smallest hole, 1.2500″, with an interference of metal (negative allowance) of .0009″.

*Line fit*   In which limits of size are so specified that a clearance or surface contact may result when mating parts are assembled.

## 14.4 Selective Assembly

If allowances and tolerances are properly given, mating parts can be completely interchangeable. But for close fits, it is necessary to specify very small allowances and tolerances, and the cost may be very high. In order to avoid this expense, either manual or computer controlled *selective assembly* is often used. In selective assembly, all parts are inspected and classified into several grades according to actual sizes, so that "small" shafts can be matched with "small" holes, "medium" shafts with "medium" holes, and so on. In this way, very satisfactory fits may be obtained at much less expense than by machining all mating parts to very accurate dimensions. Since a transition fit may or may not represent an interference of metal, interchangeable assembly generally is not as satisfactory as selective assembly.

## 14.5 Basic Hole System

Standard reamers, broaches, and other standard tools are often used to produce holes, and standard plug gages are used to check the actual sizes. On the other hand, shafting can easily be machined to any size desired. Therefore, toleranced dimensions are commonly figured on the so-called *basic hole system*. In this system, the *minimum hole is taken as the basic size*, an allowance is assigned, and tolerances are applied on both sides of, and away from, this allowance.

In Fig. 14.5 (a) the minimum size of the hole, .500″, is taken as the basic size. An allowance of .002″, is decided upon and subtracted from the basic hole size, giving the maximum shaft, .498″. Tolerances of .002″ and .003″, respectively, are applied to the hole and shaft to obtain the maximum hole of .502″ and the minimum shaft of .495″. Thus the minimum clearance between the parts becomes .500″ − .498″ = .002″ (smallest hole minus largest shaft), and the maximum clearance is .502″ − .495″ = .007″ (largest hole minus smallest shaft).

In the case of an interference fit, the maximum shaft size would be found by *adding the desired allowance* (maximum interference) to the basic hole

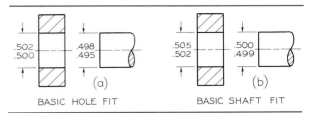

*Fig. 14.5*   Basic Hole and Basic Shaft Systems.

size. In Fig. 14.4 (a), the basic size is 1.2500″. The maximum interference decided upon was .0019″, which added to the basic size gives 1.2519″, the largest shaft size.

The basic hole size can be changed to the basic shaft size by subtracting the allowance for a clearance fit, or adding it for an interference fit. The result is the largest shaft size, which is the new basic size.

## 14.6 Basic Shaft System

In some branches of industry, such as textile machinery manufacturing, in which use is made of a great deal of cold-finished shafting, the *basic shaft system* is often used. This system should be used only when there is a reason for it. For example, it is advantageous when several parts having different fits, but one nominal size, are required on a single shaft. In this system, *the maximum shaft is taken as the basic size*, an allowance for each mating part is assigned, and tolerances are applied on both sides of, and away from, this allowance.

In Fig. 14.5 (b) the maximum size of the shaft, .500″, is taken as the basic size. An allowance of .002″ is decided upon and added to the basic shaft size, giving the minimum hole, .502″. Tolerances of .003″ and .001″, respectively, are applied to the hole and shaft to obtain the maximum hole .505″ and the minimum shaft .499″. Thus the minimum clearance between the parts is .502″ − .500″ = .002″ (smallest hole minus largest shaft), and the maximum clearance is .505″ − .499″ = .006″ (largest hole minus smallest shaft).

In the case of an interference fit, the minimum hole size would be found by *subtracting the desired allowance from the basic shaft size*.

The basic shaft size may be changed to the basic hole size by adding the allowance for a clearance fit or by subtracting it for an interference fit. The result is the smallest hole size, which is the new basic size.

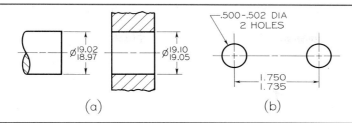

*Fig. 14.6*    Method of Giving Limits.

## 14.7 Specification of Tolerances

A tolerance of a decimal dimension must be given in decimal form to the same number of places. See Fig. 14.8.

*General tolerances* on decimal dimensions in which tolerances are not given may also be covered in a printed note, such as

DECIMAL DIMENSIONS TO BE HELD TO ± .001.

Thus if a dimension 3.250 is given, the worker machines between the limits 3.249 and 3.251. See Fig. 14.9.

Tolerances for metric dimensions may be covered in a note, such as the commonly used

METRIC DIMENSIONS TO BE HELD TO ± 0.08.

Thus, when the given dimension of 3.250″ is converted to millimeters, the worker machines between the limits of 82.63 mm and 82.47 mm.

Every dimension on a drawing should have a tolerance, either direct or by general tolerance note, except that commercial material is often assumed to have the tolerances set by commercial standards.

It is customary to indicate an overall general tolerance for all common fraction dimensions by means of a printed note in or just above the title block.

EXAMPLES

ALL FRACTIONAL DIMENSIONS $\pm \frac{1}{64}″$
UNLESS OTHERWISE SPECIFIED.

HOLD FRACTIONAL DIMENSIONS TO $\pm \frac{1}{64}″$
UNLESS OTHERWISE NOTED.

See Fig. 14.9. General angular tolerances also may be given, as

ANGULAR TOLERANCE ± 1°.

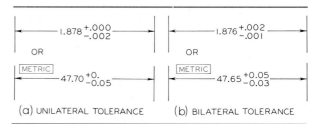

(a) UNILATERAL TOLERANCE    (b) BILATERAL TOLERANCE

*Fig. 14.7*    Tolerance Expression.

Several methods of expressing tolerances in dimensions are approved by ANSI [Y14.5M–1982 (R1988)] as follows.

1. *Limit Dimensioning.* In this preferred method, the maximum and minimum limits of size and location are specified, as shown in Fig. 14.6. The high limit (maximum value) is placed above the low limit (minimum value). See Fig. 14.6 (a). In single-line note form, the low limit precedes the high limit separated by a dash, Fig. 14.6 (b).

2. *Plus and Minus Dimensioning.* In this method the basic size is followed by a plus and minus expression of tolerance resulting in either a unilateral or bilateral tolerance as in Fig. 14.7. If two unequal tolerance numbers are given, one plus and one minus, the plus is placed above the minus. One of the numbers may be zero, if desired. If a single tolerance value is given, it is preceded by the plus-or-minus symbol ( ± ), Fig. 14.8. This method should be used when the plus and minus values are equal.

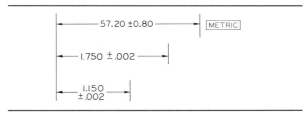

*Fig. 14.8*    Bilateral Tolerances.

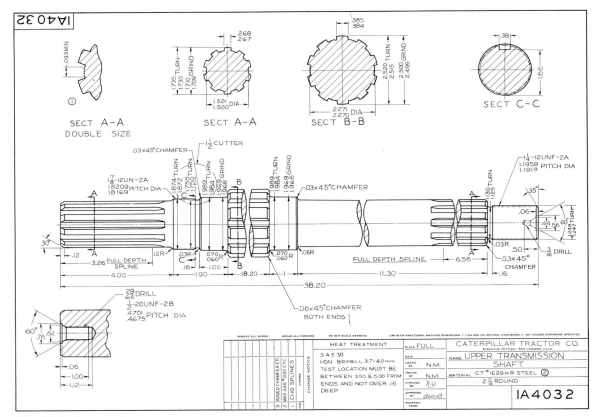

*Fig. 14.9*   Limit Dimensions.

The *unilateral system* of tolerances allows variations in only one direction from the basic size. This method is advantageous when a critical size is approached as material is removed during manufacture, as in the case of close-fitting holes and shafts. In Fig. 14.7 (a) the basic size is 1.878″ (47.70 mm). The tolerance .002″ (0.05 mm) is all in one direction—toward the smaller size. If this is a shaft diameter, the basic size 1.878″ (47.70 mm) is the size nearest the critical size because it is nearest to the tolerance zone; hence, the tolerance is taken *away* from the critical size. A unilateral tolerance is always all plus or all minus; that is, either the plus or the minus value must be zero. However, the zeros should be given as shown at (a).

The *bilateral system* of tolerances allows variations in both directions from the basic size. Bilateral tolerances are usually given with location dimensions or with any dimensions that can be allowed to vary in either direction. In Fig. 14.7 (b) the basic size is 1.876″ (47.65 mm), and the actual size may be larger by .002″ (0.05

mm) or smaller by .001″ (0.03 mm). If it is desired to specify an equal variation in both directions, the combined plus or minus symbol (±) is used with a single value, as shown in Fig. 14.8.

A typical example of limit dimensioning is given in Fig. 14.9.

3. *Single-Limit Dimensioning.* It is not always necessary to specify both limits. MIN or MAX is often placed after a number to indicate minimum or maximum dimensions desired where other elements of design determine the other unspecified limit. For example, a thread length may be dimensioned thus: |←——1.500——→| MIN FULL THD or a radius dimensioned: .05 R MAX——◁ Other applications include depths of holes, chamfers, and so on.

4. *Angular tolerances* are usually bilateral and in terms of degrees, minutes, and seconds.

EXAMPLES   25° ± 1°, 25° 0′ ± 0° 15′, or 25° ± 0.25°. See also §14.18.

# 14.8 American National Standard Limits and Fits

The American National Standards Institute has issued the ANSI B4.1–1967 (R1987), "Preferred Limits and Fits for Cylindrical Parts," defining terms and recommending preferred standard sizes, allowances, tolerances, and fits in terms of the decimal inch. This standard gives

a series of standard types and classes of fits on a unilateral hole basis such that the fit produced by mating parts in any one class will produce approximately similar performance throughout the range of sizes. These tables prescribe the fit for any given size, or type of fit; they also prescribe the standard limits for the mating parts which will produce the fit.

The tables are designed for the basic hole system, §14.5. See Appendices 5 to 9. For coverage of the metric system of tolerances and fits see §§14.11 to 14.13, and Appendices 11 to 14.

Letter symbols to identify the five types of fits are

RC   Running or Sliding Clearance Fits
LC   Locational Clearance Fits
LT   Transition Clearance or Interference Fits
LN   Locational Interference Fits
FN   Force or Shrink Fits

These letter symbols, plus a number indicating the class of fit within each type, are used to indicate a complete fit. Thus, FN 4 means a Class 4 Force Fit. The fits are described [ANSI B4.1–1967 (R1987)] as follows.

### RUNNING AND SLIDING FITS

Running and sliding fits, for which description of classes of fits and limits of clearance are given [Appendix 5], are intended to provide a similar running performance, with suitable lubrication allowance, throughout the range of sizes. The clearances for the first two classes, used chiefly as slide fits, increase more slowly with diameter than the other classes, so that accurate location is maintained even at the expense of free relative motion.

### LOCATIONAL FITS

Locational fits [Appendices 6 to 8] are fits intended to determine only the location of the mating parts; they may provide rigid or accurate

location, as with interference fits, or provide some freedom of location, as with clearance fits. Accordingly they are divided into three groups: clearance fits, transition fits, and interference fits.

### FORCE FITS

Force or shrink fits [Appendix 9] constitute a special type of interference fit, normally characterized by maintenance of constant bore pressures throughout the range of sizes. The interference therefore varies almost directly with diameter, and the difference between its minimum and maximum value is small, to maintain the resulting pressures within reasonable limits.

In the tables for each class of fit, the range of nominal sizes of shafts or holes is given in inches. To simplify the tables and reduce the space required to present them, the other values are given in thousandths of an inch. Minimum and maximum limits of clearance are given, the top number being the least clearance, or the allowance, and the lower number the maximum clearance, or the greatest looseness of fit. Then, under the heading "Standard Limits" are given the limits for the hole and for the shaft that are to be applied algebraically to the basic size to obtain the limits of size for the parts, using the basic hole system.

For example, take a 2.0000″ basic diameter with a Class RC 1 fit. This fit is given in Appendix 5. In the column headed "Nominal Size Range, Inches," find 1.97–3.15, which embraces the 2.0000″ basic size. Reading to the right we find under "Limits of Clearance" the values 0.4 and 1.2, representing the maximum and minimum clearance between the parts *in thousandths of an inch*. To get these values in inches, simply multiply by one thousandth; thus, $\frac{4}{10} \times \frac{1}{1000} = .0004″$. To convert 0.4 thousandths to inches, simply move the decimal point three places to the left; thus: .0004″. Therefore, for this 2.0000″ diameter, with a Class RC 1 fit, the minimum clearance, or allowance, is .0004″, and the maximum clearance, representing the greatest looseness, is .0012″.

Reading farther to the right, we find under "Standard Limits" the value +0.5, which converted to inches is .0005″. Add this to the basic size thus: 2.0000″ + .0005″ = 2.0005″, the upper limit of the hole. Since the other value given for the hole is zero, the lower limit of the hole is the basic size of the hole, or 2.0000″. The hole would then be dimensioned as

$$\begin{array}{cc} 2.0005 \\ 2.0000 \end{array} \quad or \quad 2.0000 \begin{array}{c} +.0005 \\ -.0000 \end{array}$$

The limits for the shaft are read as $-.0004''$ and $-.0007''$. To get the limits of the shaft, subtract these values from the basic size; thus,

$$2.0000'' - .0004'' = 1.9996'' \text{ (upper limit)}$$
$$2.0000'' - .0007'' = 1.9993'' \text{ (lower limit)}$$

The shaft would then be dimensioned in inches as follows.

$$\begin{array}{cc} 1.9996 \\ 1.9993 \end{array} \quad or \quad 1.9996 \begin{array}{c} +.0000 \\ -.0003 \end{array}$$

## 14.9 Accumulation of Tolerances

In tolerance dimensioning, it is very important to consider the effect of one tolerance on another. When the location of a surface in a given direction is affected by more than one tolerance figure, the tolerances are *cumulative*. For example, in Fig. 14.10 (a), if dimension Z is omitted, surface A would be controlled by both dimensions X and Y, and there could be a total variation of .010'' instead of the variation of .005'' permitted by dimension Y, which is the dimension directly applied to surface A. Further, if the part is made to all the minimum tolerances of X, Y, and Z, the total variation in the length of the part will be .015'', and the part can be as short as 2.985''. However, the tolerance on the overall dimension W is only .005'', permitting the part to be only as short as 2.995''. The part is superfluously dimensioned.

In some cases, for functional reasons, it may be desired to hold all three small dimensions X, Y, and Z closely without regard to the overall length. In

such a case the overall dimension is just a *reference dimension* and should be marked REF. In other cases it may be desired to hold two small dimensions X and Y and the overall closely without regard to dimension Z. In that case, dimension Z should be omitted, or marked REF.

As a rule, it is best to dimension each surface so that it is affected by only one dimension. This can be done by referring all dimensions to a single datum surface, such as B, as shown at (b). See also Fig. 13.37 (d), (e), and (f).

## 14.10 Tolerances and Machining Processes

As has been repeatedly stated in this chapter, tolerances should be as coarse as possible and still permit satisfactory use of the part. If this is done, great savings can be effected as a result of the use of less expensive tools, lower labor and inspection costs, and reduced scrapping of material.

Figure 14.11 shows a chart of tolerance grades obtainable in relation to the accuracy of machining processes that may be used as a guide by the designer. Metric values may be ascertained by multiplying the given decimal-inch values by 25.4 and rounding off the product to one less place to the right of the decimal point than given for the decimal-inch value. See §13.10. For detailed information on manufacturing processes and measuring devices see Chapter 12.

## 14.11 Metric System of Tolerances and Fits

The preceding material on limits and fits between mating parts is suitable, without need of conversion, for the decimal-inch system of measurement. A sys-

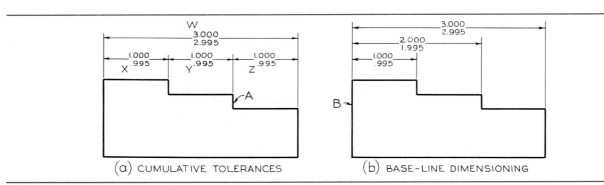

*Fig. 14.10*   Cumulative Tolerances.

| Range of Sizes | | Tolerances | | | | | | | | |
|---|---|---|---|---|---|---|---|---|---|---|
| From | To & Incl. | | | | | | | | | |
| .000 | .599 | .00015 | .0002 | .0003 | .0005 | .0008 | .0012 | .002 | .003 | .005 |
| .600 | .999 | .00015 | .00025 | .0004 | .0006 | .001 | .0015 | .0025 | .004 | .006 |
| 1.000 | 1.499 | .0002 | .0003 | .0005 | .0008 | .0012 | .002 | .003 | .005 | .008 |
| 1.500 | 2.799 | .00025 | .0004 | .0006 | .001 | .0015 | .0025 | .004 | .006 | .010 |
| 2.800 | 4.499 | .0003 | .0005 | .0008 | .0012 | .002 | .003 | .005 | .008 | .012 |
| 4.500 | 7.799 | .0004 | .0006 | .001 | .0015 | .0025 | .004 | .006 | .010 | .015 |
| 7.800 | 13.599 | .0005 | .0008 | .0012 | .002 | .003 | .005 | .008 | .012 | .020 |
| 13.600 | 20.999 | .0006 | .001 | .0015 | .0025 | .004 | .006 | .010 | .015 | .025 |

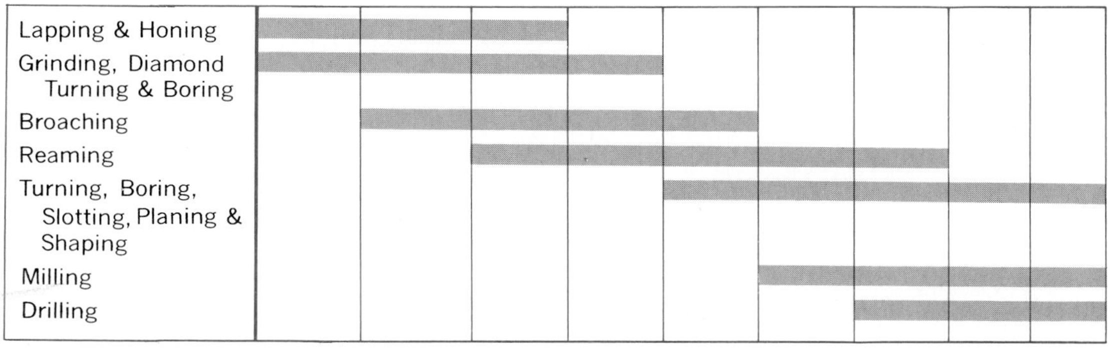

Lapping & Honing

Grinding, Diamond
  Turning & Boring

Broaching

Reaming

Turning, Boring,
  Slotting, Planing &
  Shaping

Milling

Drilling

*Fig. 14.11* Tolerances Related to Machining Processes.

tem of preferred metric limits and fits by the International Organization for Standardization (ISO) is in the ANSI B4.2 standard. The system is specified for holes, cylinders, and shafts, but it is also adaptable to fits between parallel surfaces of such features as keys and slots. The following terms for metric fits, although somewhat similar to those for decimal-inch fits, are illustrated in Fig. 14.12.

*Basic size*   The size from which limits or deviations are assigned. Basic sizes, usually diameters, should be selected from a table of preferred sizes. See Fig. 14.17.

*Deviation*   The difference between the basic size and the hole or shaft size. (This is equivalent to the tolerance in the decimal-inch system.)

*Upper deviation*   The difference between the basic size and the permitted maximum size of the part. (This compares with the maximum tolerance in the decimal-inch system.)

*Lower deviation*   The difference between the basic size and the minimum permitted size of the part. (This compares with the minimum tolerance in the decimal-inch system.)

*Fundamental deviation*   The deviation closest to the basic size. (This compares with the minimum allowance in the decimal-inch system.)

*Tolerance*   The difference between the permitted minimum and maximum size of a part.

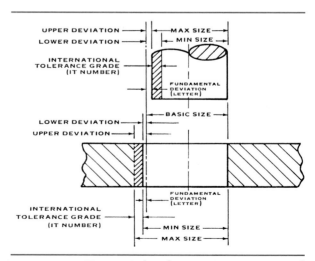

*Fig. 14.12*   Terms Related to Metric Limits and Fits   [ANSI B4.2–1978 (R1984)].

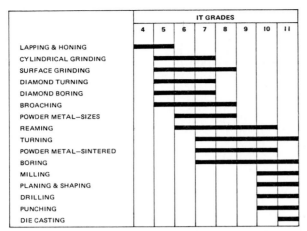

**Fig. 14.13** International Tolerance Grades Related to Machining Processes [ANSI B4.2–1978 (R1984)].

*International tolerance grade (IT)* A set of tolerances that varies according to the basic size and provides a uniform level of accuracy within the grade. For example, in the dimension 50 H8 for a close-running fit, the IT grade is indicated by the numeral 8. (The letter H indicates the tolerance is on the hole for the 50 mm dimension.) In all, there are 18 IT grades—IT01, IT0, and IT1 through IT16. See Figs. 14.13 and 14.14 for IT grades related to machining processes and the practical use of the IT grades. See also Appendix 10.

*Tolerance zone* The tolerance and its position in relation to basic size. It is established by a combination of the fundamental deviation indicated by a letter and the IT grade number. In the dimension of 50 H8, for the close running fit, the H8 specifies the tolerance zone. See Fig. 14.14.

*Hole-basis system of preferred fits* A system based upon the basic diameter as the minimum size. For the generally preferred hole-basis system, the fundamental deviation is specified by the uppercase letter H. See Fig. 14.15 (a).

*Shaft-basis system of preferred fits* A system based upon the basic diameter as the maximum size of the shaft. The fundamental deviation is given by the lowercase letter f. See Fig. 14.15 (b).

*Interference fit* A fit that results in an interference fit between two mating parts under *all* tolerance conditions.

*Transition fit* A fit that results in either a clearance or an interference condition between two assembled parts.

*Tolerance symbols* Symbols used to specify the tolerances and fits for mating parts. See Fig. 14.15 (c). For the hole-basis system, the **50** indicates the diameter in millimeters; the fundamental deviation for the hole is indicated by the capital letter H and for the shaft it is indicated by the lowercase letter f. The numbers following the letters indicate this IT grade. Note that the symbols for the hole and shaft are separated by the slash (slanting

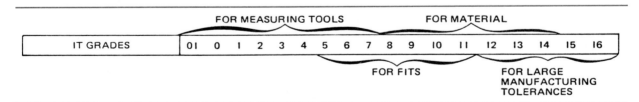

**Fig. 14.14** Practical Use of International Tolerance Grades.

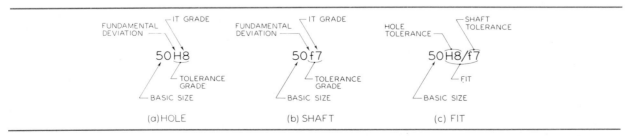

**Fig. 14.15** Application of Definitions and Symbols to Holes and Shafts [ANSI B4.2–1978 (R1984)].

| 50 H8 | 50H8$\left(\dfrac{50.039}{50.000}\right)$ | $\dfrac{50.039}{50.000}$(50H8) |
|:---:|:---:|:---:|
| (a) PREFERRED | (b) | (c) |

*Fig. 14.16* Acceptable Methods of Giving Tolerance Symbols [ANSI Y14.5M–1982 (R1988)].

line). Tolerance symbols may be given in several acceptable forms as in Fig. 14.16 for a 50 mm diameter hole. The values in parentheses are for reference only and may be omitted. The upper and lower limit values may be found in Appendix 11.

## 14.12 Preferred Sizes

The preferred basic sizes for computing tolerances are given in Table 14.1. Basic diameters should be selected from the first choice column since these are readily available stock sizes for round, square, and hexagonal products.

## 14.13 Preferred Fits

The symbols for either the hole-basis or shaft-basis preferred fits (clearance, transition, and interference) are given in Table 14.2. Fits should be selected from this table for mating parts where possible.

The values corresponding to the fits are found in Appendices 11 to 14. Although second and third choice basic size diameters are possible, they must be calculated from tables not included in this text. For the generally preferred hole-basis system note that the ISO symbols range from H11/c11 (loose running) to H7/u6 (force fit). For the shaft-basis system, the preferred symbols range from C11/h11 (loose fit) to U7/h6 (force fit).

Assume that it is desired to use the symbols to specify the dimensions for a free-running fit (hole basis) for a proposed diameter of 48 mm. Since 48 mm is not listed as a preferred size in Table 14.1, the design is altered to use the acceptable 50 mm diameter. From the preferred fits descriptions in Table 14.2, the free-running fit (hole-basis) is H9/d9. To determine the upper and lower deviation

**Table 14.1** *Preferred Sizes* [ANSI B4.2–1978 (R1984)]

| Basic Size, mm | | Basic Size, mm | | Basic Size, mm | |
|:---:|:---:|:---:|:---:|:---:|:---:|
| First Choice | Second Choice | First Choice | Second Choice | First Choice | Second Choice |
| 1 | | 10 | | 100 | |
| | 1.1 | | 11 | | 110 |
| 1.2 | | 12 | | 120 | |
| | 1.4 | | 14 | | 140 |
| 1.6 | | 16 | | 160 | |
| | 1.8 | | 18 | | 180 |
| 2 | | 20 | | 200 | |
| | 2.2 | | 22 | | 220 |
| 2.5 | | 25 | | 250 | |
| | 2.8 | | 28 | | 280 |
| 3 | | 30 | | 300 | |
| | 3.5 | | 35 | | 350 |
| 4 | | 40 | | 400 | |
| | 4.5 | | 45 | | 450 |
| 5 | | 50 | | 500 | |
| | 5.5 | | 55 | | 550 |
| 6 | | 60 | | 600 | |
| | 7 | | 70 | | 700 |
| 8 | | 80 | | 800 | |
| | 9 | | 90 | | 900 |
| | | | | 1000 | |

**Table 14.2**   *Preferred Fits* [ANSI B4.2–1978 (R1984)]

| ISO Symbol | | Description | |
| --- | --- | --- | --- |
| **Hole Basis** | **Shaft[a] Basis** | | |
| **Clearance Fits** H11/c11 | C11/h11 | ***Loose running***   fit for wide commercial tolerances or allowances on external members. | ↑ |
| H9/d9 | D9/h9 | ***Free running***   fit not for use where accuracy is essential, but good for large temperature variations, high running speeds, or heavy journal pressures. | More clearance |
| H8/f7 | F8/h7 | ***Close running***   fit for running on accurate machines and for accurate location at moderate speeds and journal pressures. | |
| H7/g6 | G7/h6 | ***Sliding fit***   not intended to run freely, but to move and turn freely and locate accurately. | |
| **Transition Fits** H7/h6 | H7/h6 | ***Locational clearance***   fit provides snug fit for locating stationary parts; but can be freely assembled and disassembled. | |
| H7/k6 | K7/h6 | ***Locational transition***   fit for accurate location, a compromise between clearance and interference. | |
| H7/n6 | N7/h6 | ***Locational transition***   fit for more accurate location where greater interference is permissible. | More interference |
| **Interference Fits** H7/p6 | P7/h6 | ***Locational interference***   fit for parts requiring rigidity and alignment with prime accuracy of location but without special bore pressure requirements. | |
| H7/s6 | S7/h6 | ***Medium drive***   fit for ordinary steel parts or shrink fits on light sections, the tightest fit usable with cast iron. | |
| H7/u6 | U7/h6 | ***Force***   fit suitable for parts which can be highly stressed or for shrink fits where the heavy pressing forces required are impractical. | ↓ |

[a]The transition and interference shaft basis fits shown do not convert to exactly the same hole basis fit conditions for basic sizes in range from Q through 3 mm. Interference fit P7/h6 converts to a transition fit H7/p6 in the above size range.

limits of the hole as given in the preferred hole-basis table, Appendix 11, follow across from the basic size of 50 to H9 under "Free running." The limits for the hole are 50.000 and 50.062 mm. Then, the upper and lower limits of deviation for the shaft are found in the d9 column under "Free running." They are 49.920 and 49.858 mm, respectively. Limits for other fits are established in a similar manner.

The limits for the shaft basis dimensioning are determined similarly from the preferred shaft basis table in Appendix 13. See Figs. 14.16 and 14.17 for acceptable methods of specifying tolerances by symbols on drawings. A single note for the mating parts (free-running fit, hole basis) would be ⌀50 H9/d9, Fig. 14.17.

## 14.14 Geometric Tolerancing

Geometric tolerances state the maximum allowable variations of a form or its position from the perfect geometry implied on the drawing. The term "geo-metric" refers to various forms, such as a plane, a cylinder, a cone, a square, a hexagon. Theoretically these are perfect forms, but, because it is impossible to produce perfect forms, it may be necessary to specify the amount of variation permitted. These tolerances specify either the diameter or the width of a tolerance zone within which a surface or the axis of a cylinder or a hole must be if the part is to meet the required accuracy for proper function and fit. When tolerances of form are not given on a drawing, it is customary to assume that, regardless of form variations, the part will fit and function satisfactorily.

Tolerances of form and position or location control such characteristics as straightness, flatness, parallelism, perpendicularity (squareness), concentricity, roundness, angular displacement, and so on.

Methods of indicating geometric tolerances by means of *geometric characteristic symbols*, as recommended by ANSI, rather than by traditional notes, are discussed and illustrated subsequently.

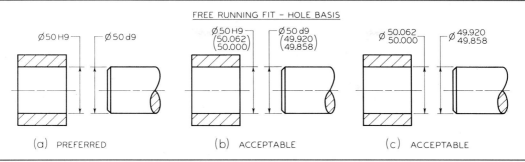

Fig. 14.17  Methods of Specifying Tolerances with Symbols for Mating Parts.

See the latest Dimensioning and Tolerancing Standard [Y14.5M–1982 (R1988)] for more complete coverage.

## 14.15 Symbols for Tolerances of Position and Form

Since traditional narrative notes for specifying tolerances of *position* (location) and *form* (shape) may be confusing or not clear, may require much of the space available on the drawing, and often may not be understood internationally, most multinational companies have adopted symbols for such specifications [ANSI Y14.5M–1982 (R1988)]. These ANSI symbols provide an accurate and concise means of specifying geometric characteristics and tolerances in a minimum of space, Table 14.3. The symbols may be supplemented by notes if the precise geometric requirements cannot be conveyed by the symbols. For construction details of the geometric tolerancing symbols, see Appendix 32.

Combinations of the various symbols and their meanings are given in Fig. 14.18. Application of the symbols to a drawing are illustrated in Fig. 14.43.

The geometric characteristic symbols plus the

## Table 14.3  *Geometric Characteristics and Modifying Symbols* [ANSI Y14.5M–1982 (R1988)]

Geometric characteristic symbols

| | Type of Tolerance | Characteristic | Symbol |
|---|---|---|---|
| For individual features | Form | Straightness | — |
| | | Flatness | ▱ |
| | | Circularity (roundness) | ○ |
| | | Cylindricity | /○/ |
| For individual or related features | Profile | Profile of a line | ⌒ |
| | | Profile of a surface | ⌓ |
| For related features | Orientation | Angularity | ∠ |
| | | Perpendicularity | ⊥ |
| | | Parallelism | // |
| | Location | Position | ⊕ |
| | | Concentricity | ◎ |
| | Runout | Circular runout | ↗ |
| | | Total runout | ↗↗ a |

Modifying symbols

| Term | Symbol |
|---|---|
| At maximum material condition | Ⓜ |
| Regardless of feature size | Ⓢ |
| At least material condition | Ⓛ |
| Projected tolerance zone | Ⓟ |
| Diameter | ∅ |
| Spherical diameter | **S∅** |
| Radius | **R** |
| Spherical radius | **SR** |
| Reference | ( ) |
| Arc length | ⌒ |

a Arrowhead(s) may be filled in.

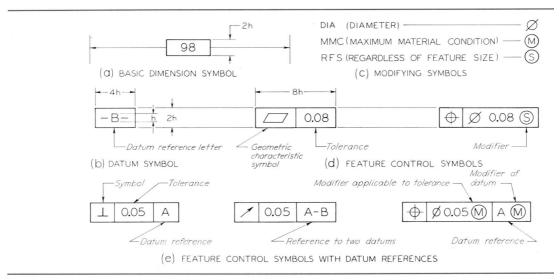

**Fig. 14.18**   Use of Symbols for Tolerance of Position and Form    [ANSI Y14.5M–1982 (R1988)].

supplementary symbols are further explained and illustrated with material adapted from ANSI Y14.5M–1982 (R1988), as follows.

### Basic Dimension Symbol

The basic dimension is identified by the enclosing frame symbol, Fig. 14.18 (a). The basic dimension (size) is the value used to describe the theoretically exact size, shape, or location of a feature. It is the basis from which permissible variations are established by tolerances on other dimensions in notes, or in feature control frames.

### Datum Identifying Symbol

The datum identifying symbol consists of frame containing a reference letter preceded and followed by a dash, Fig. 14.18 (b). A point, line, plane, cylinder, or other geometric form assumed to be exact for purposes of computation may serve as a datum from which the location or geometric relationship of features of a part may be established.

### Supplementary Symbols

The symbols for MMC (maximum material condition, i.e., minimum hole diameter, maximum shaft diameter) and RFS (regardless of feature size—the tolerance applies to any size of the feature within its size tolerance and or the actual size of a datum feature) are illustrated in Fig. 14.18 (c). The abbreviations MMC and RFS are also used in notes. See also Table 14.3.

The symbol for diameter is used instead of the abbreviation DIA to indicate a diameter, and it precedes the specified tolerance in a feature control symbol, Fig. 14.18 (d). This symbol for diameter instead of the abbreviation DIA may be used on a drawing, and it should precede the dimension. For narrative notes, the abbreviation DIA is preferred.

### Combined Symbols

Individual symbols, datum reference letters, needed tolerances, and so on may be combined in a single frame, Fig. 14.18 (e).

A position of form tolerance is given by a feature control symbol made up of a frame about the appropriate geometric characteristic symbol plus the allowable tolerance. A vertical line separates the symbol and the tolerance, Fig. 14.18 (d). Where needed, the tolerance should be preceded by the symbol for diameter and followed by the symbol for MMC or RFS.

A tolerance of position or form related to a datum is so indicated in the feature control symbol by placing the datum reference letter following either the geometric characteristic symbol or the tolerance. Vertical lines separate the entries, and, where applicable, the datum reference letter entry includes the symbol for MMC or RFS. See Fig. 14.18.

## 14.16 Positional Tolerances

In §13.25 are shown a number of examples of the traditional methods of locating holes, that is, by means of rectangular coordinates or angular dimen-

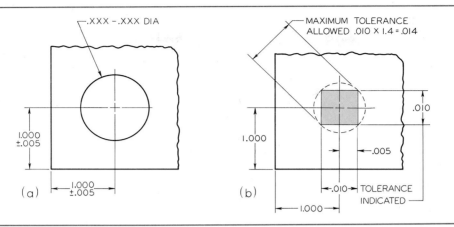

*Fig. 14.19*   Tolerance Zones.

sions. Each dimension has a tolerance, either given directly or indicated on the completed drawing by a general note.

For example, in Fig. 14.19 (a) is shown a hole located from two surfaces at right angles to each other. As shown at (b), the center may lie anywhere within a square tolerance zone the sides of which are equal to the tolerances. Thus, the total variations along either diagonal of the square by the coordinate method of dimensioning will be 1.4 times greater than the indicated tolerance. Hence, a .014 diameter tolerance zone would increase the square tolerance zone area 57 percent without exceeding the tolerance permitted along the diagonal of the square tolerance zone.

Features located by toleranced angular and radial dimensions will have a wedge-shaped tolerance zone. See Fig. 14.28.

If four holes are dimensioned with rectangular coordinates as in Fig. 14.20 (a), acceptable patterns for the square tolerance zones for the holes are shown at (b) and (c). The locational tolerances are actually greater than indicated by the dimensions.

Feature control symbols are related to the feature by one of several methods illustrated in Fig. 14.43. The following methods are preferred:

1. Adding the symbol to a note or dimension pertaining to the feature.

2. Running a leader from the symbol to the feature.

3. Attaching the side, end, or corner of the symbol frame to an extension line from the feature.

4. Attaching a side or end of the symbol frame to the dimension line pertaining to the feature.

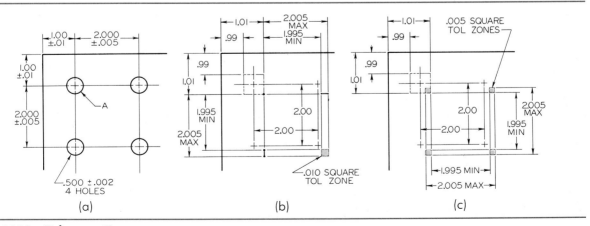

*Fig. 14.20*   Tolerance Zones.

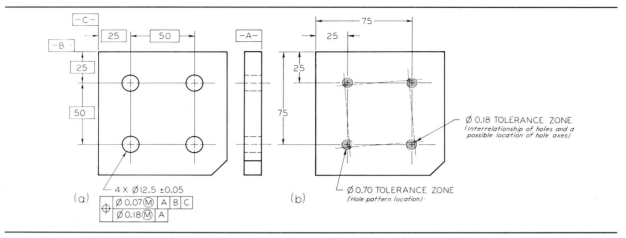

**Fig. 14.21**    True-Position Dimensioning    [ANSI Y14.5M–1982 (R1988)].

In Fig. 14.20 (a), hole **A** is selected as a datum, and the other three are located from it. The square tolerance zone for hole **A** results from the tolerances on the two rectangular coordinate dimensions locating hole **A**. The sizes of the tolerance zones for the other three holes result from the tolerances between the holes, while their locations will vary according to the actual location of the datum hole **A**. Two of the many possible zone patterns are shown at (b) and (c).

Thus, with the dimensions shown at (a), it is difficult to say whether the resulting parts will actually fit the mating parts satisfactorily even though they conform to the tolerances shown on the drawing.

These disadvantages are overcome by giving exact theoretical locations by untoleranced dimensions and then specifying by a note how far actual positions may be displaced from these locations. This is called *true-position dimensioning*. It will be seen that the tolerance zone for each hole will be a circle, the size of the circle depending upon the amount of variation permitted from "true position."

A true-position dimension denotes the theoretically exact position of a feature. The location of each feature such as a hole, slot, stud, and so on, is given by untoleranced basic dimensions identified by the enclosing frame or symbol. To prevent misunderstandings, true position should be established with respect to a datum.

In simple arrangements, the choice of a datum may be obvious and not require identification.

Positional tolerancing is identified by a characteristic symbol directed to a feature, which establishes a circular tolerance zone, Fig. 14.21.

Actually, the "circular tolerance zone" is a cylindrical tolerance zone (the diameter of which is equal

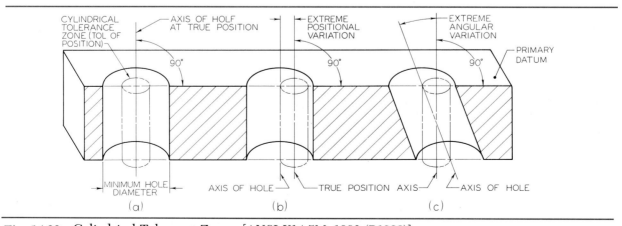

**Fig. 14.22**    Cylindrical Tolerance Zone    [ANSI Y14.5M–1982 (R1988)].

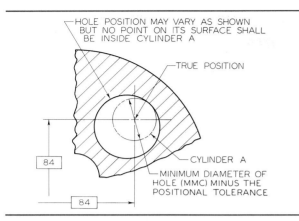

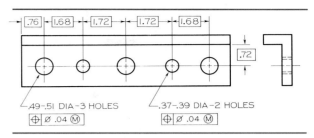

Fig. 14.24   No Tolerance Accumulation.

*Fig. 14.23*   True Position Interpretation [ANSI Y14.5M–1982 (R1988)].

to the positional tolerance and its length is equal to the length of the feature unless otherwise specified), and its axis must be within this cylinder, Fig. 14.22.

The center line of the hole may coincide with the center line of the cylindrical tolerance zone, (a), or it may be parallel to it but displaced so as to remain within the tolerance cylinder, (b), or it may be inclined while remaining within the tolerance cylinder, (c). In this last case we see that the positional tolerance also defines the limits of squareness variation.

In terms of the cylindrical surface of the hole, the positional tolerance specification indicates that all elements on the hole surface must be on or outside a cylinder whose diameter is equal to the minimum diameter (MMC; §14.17) or the maximum diameter of the hole minus the positional tolerance (diameter, or twice the radius), with the center line of the cylinder located at true position, Fig. 14.23.

The use of basic untoleranced dimensions to locate features at true position avoids one of the chief difficulties in tolerancing—the accumulation of tol-

erances, §14.9, even in a chain of dimensions. Fig. 14.24.

While features, such as holes and bosses, may vary in any direction from the true-position axis, other features, such as slots, may vary on either side of a true-position plane, Fig. 14.25.

Since the exact locations of the true positions are given by untoleranced dimensions, it is important to prevent the application of general tolerances to these. A note should be added to the drawing such as

**GENERAL TOLERANCES DO NOT APPLY TO BASIC TRUE–POSITION DIMENSIONS.**

## 14.17  Maximum Material Condition

*Maximum material condition,* usually abbreviated to MMC, means that a feature of a finished product contains the maximum amount of material permitted by the toleranced size dimensions shown for that feature. Thus, we have MMC when holes, slots, or other internal features are at minimum size, or when shafts, pads, bosses, and other external features are at their maximum size. We have MMC for both mating parts when the largest shaft is in the smallest hole and there is the least clearance between the parts.

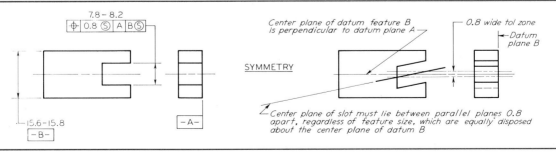

*Fig. 14.25*   Positional Tolerancing for Symmetry   [ANSI Y14.5M–1982 (R1988)].

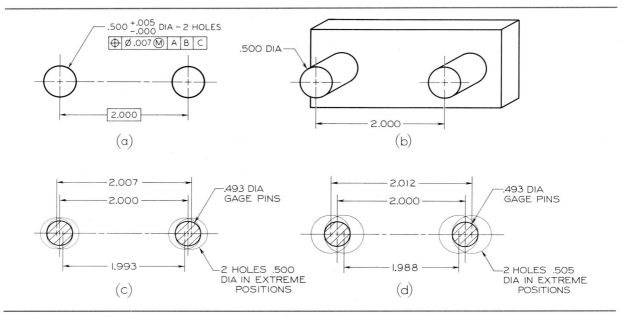

**Fig. 14.26**    Maximum and Minimum Material Conditions—Two-Hole Pattern    [ANSI Y14.5M–1982 (R1988)].

In assigning positional tolerance to a hole, it is necessary to consider the size limits of the hole. If the hole is at MMC (smallest size), the positional tolerance is not affected, but if the hole is larger, the available positional tolerance is greater. In Fig. 14.26 (a) two half-inch holes are shown. If they are exactly .500″ in diameter (MMC, or smallest size) and are exactly 2.000″ apart, they should receive a gage, (b), made of two round pins .500″ in diameter fixed in a plate 2.000″ apart. However, the center-to-center distance between the holes may vary from 1.993″ to 2.007″.

If the .500″ diameter holes are at their extreme positions, (c), the pins in the gage would have to be .007″ smaller, or .493″ diameter, to enter the holes. Thus, if the .500″ diameter holes are located at the maximum distance apart, the .493″ diameter gage pins would contact the inner sides of the holes; and if the holes are located at the minimum distance apart, the .493″ diameter pins would contact the outer surfaces of the holes, as shown. If gagemakers' tolerances are not considered, the gage pins would have to be .493″ diameter and exactly 2.000 apart if the holes are .500 diameter, or MMC.

If the holes are .505″ diameter—that is, at maximum size, as at (d)—they will be accepted by the same .493″ diameter gage pins at 2.000″ apart if the inner sides of the holes contact the inner sides of the gage pins and the outer sides of the holes contact the outer sides of the gage pins, as shown. Thus the holes may be 2.012″ apart, which is beyond the tolerance permitted for the center-to-center distance between the holes. Similarly, the holes may be as close together as 1.988″ from center to center, which again is outside the specified positional tolerance.

Thus, when the holes are not at MMC—that is, when they are at maximum size—a greater positional tolerance becomes available. Since all features may vary in size, it is necessary to make clear on the drawing at what basic dimension the true position applies. In all but a few exceptional cases, the additional positional tolerance available, when holes are larger than minimum size, is acceptable and desirable. Parts thus accepted can be freely assembled whether the holes or other features are within the specified positional tolerance or not. This practice has been recognized and used in manufacturing for years, as is evident from the use of fixed-pin gages, which have been commonly used to inspect parts and control the least favorable condition of assembly. Thus it has become common practice for both manufacturing and inspection to assume that positional tolerance applies to MMC and that greater positional tolerance becomes permissible when the part is not at MMC.

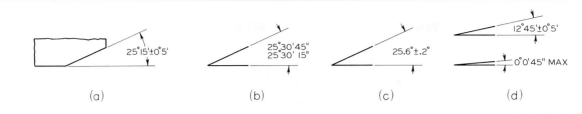

**Fig. 14.27**   Tolerances of Angles.

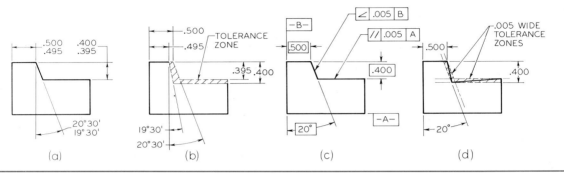

**Fig. 14.28**   Angular Tolerance Zones    [ANSI Y14.5M–1982 (R1988)].

To avoid possible misinterpretation as to whether maximum material condition (MMC) or regardless of feature size (RFS) applies, it should be clearly stated on the drawing by the addition of MMC or RFS symbols to each applicable tolerance or by suitable coverage in a document referenced on the drawing.

When MMC or RFS is not specified on the drawing with respect to an individual tolerance, datum reference or both, the following rules shall apply:

1.  True-position tolerances and related datum references apply at MMC.

2.  Angularity, parallelism, perpendicularity, concentricity, and symmetry tolerances, including related datum references, apply RFS. No element of the actual feature shall extend beyond the envelope of the perfect form at MMC.

## 14.18 Tolerances of Angles

Bilateral tolerances have been given traditionally on angles as illustrated in Fig. 14.27. Consequently, the wedge-shaped tolerance zone increases as the distance from the vertex of the angle increases. Thus,

the tolerance had to be figured after considering the total displacement at the point farthest from the vertex of the angle before a tolerance could be specified that would not exceed the allowable displacement. The use of angular tolerances may be avoided by using gages. Taper turning is often handled by machining to fit a gage or by fitting to the mating part.

If an angular surface is located by a linear and an angular dimension, Fig. 14.28 (a), the surface must lie within a tolerance zone as shown at (b). The angular zone will be wider as the distance from the vertex increases. In order to avoid the accumulation of tolerances, that is, to decrease the tolerance zone, the *basic angle* tolerancing method of (c) is recommended [ANSI Y14.5M–1982 (R1988)]. The angle is indicated as basic with the proper symbol and no angular tolerance is specified. The tolerance zone is now defined by two parallel planes, resulting in improved angular control, (d).

## 14.19 Form Tolerances for Single Features

Straightness, flatness, roundness, cylindricity, and, in some instances, profile are form tolerances applicable to single features.

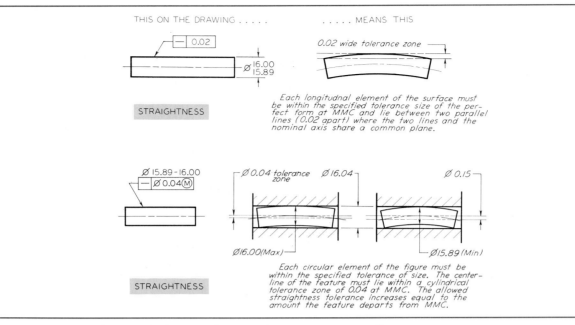

**Fig. 14.29**  Specifying Straightness   [ANSI Y14.5M–1982 (R1988)].

### Straightness Tolerance

A straightness tolerance specifies a tolerance zone within which an axis or all points of the considered element must lie, Fig. 14.29. Straightness is a condition where an element of a surface or an axis is a straight line.

### Flatness Tolerance

A flatness tolerance specifies a tolerance zone defined by two parallel planes within which the surface must lie, Fig. 14.30. Flatness is the condition of a surface having all elements in one plane.

### Roundness (Circularity) Tolerance

A roundness tolerance specifies a tolerance zone bounded by two concentric circles within which each circular element of the surface must lie, Fig.

14.31. Roundness is a condition of a surface of revolution where, for a cone or cylinder, all points of the surface intersected by any plane perpendicular to a common axis are equidistant from that axis. For a sphere, all points of the surface intersected by any plane passing through a common center are equidistant from that center.

### Cylindricity Tolerance

A cylindricity tolerance specifies a tolerance zone bounded by two concentric cylinders within which the surface must lie, Fig. 14.32. This tolerance applies to both circular and longitudinal elements of the entire surface. Cylindricity is a condition of a surface of revolution in which all points of the surface are equidistant from a common axis. When no tolerance of form is given, many possible shapes may

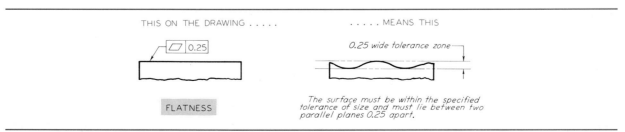

**Fig. 14.30**  Specifying Flatness   [ANSI Y14.5M–1982 (R1988)].

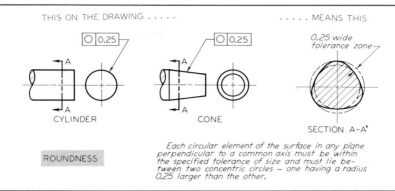

THIS ON THE DRAWING . . . . .          . . . . MEANS THIS

CYLINDER          CONE

ROUNDNESS

SECTION A-A'

Each circular element of the surface in any plane perpendicular to a common axis must be within the specified tolerance of size and must lie between two concentric circles — one having a radius 0.25 larger than the other.

**Fig. 14.31** Specifying Roundness for a Cylinder or Cone [ANSI Y14.5M–1982 (R1988)].

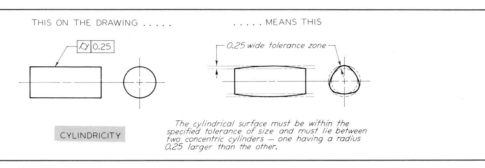

THIS ON THE DRAWING . . . . .          . . . . . MEANS THIS

0.25 wide tolerance zone

CYLINDRICITY

The cylindrical surface must be within the specified tolerance of size and must lie between two concentric cylinders — one having a radius 0.25 larger than the other.

**Fig. 14.32** Specifying Cylindricity [ANSI Y14.5M–1982 (R1988)].

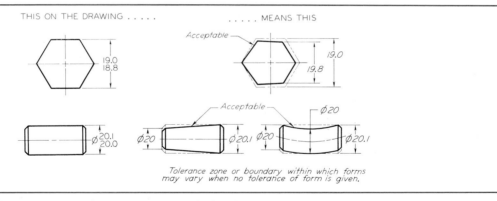

THIS ON THE DRAWING . . . . .          . . . . . MEANS THIS

Acceptable

Acceptable

Tolerance zone or boundary within which forms may vary when no tolerance of form is given.

**Fig. 14.33** Acceptable Variations of Form—No Specified Tolerance of Form.

exist within a tolerance zone, as illustrated in Fig. 14.33.

### Profile Tolerance

A profile tolerance specifies a uniform boundary or zone along the true profile within which all elements of the surface must lie, Figs. 14.34 and 14.35. A profile is the outline of an object in a given plane (two-dimensional) figure. Profiles are formed by projecting a three-dimensional figure onto a plane or by taking cross sections through the figure with

the resulting profile composed of such elements as straight lines, arcs, or other curved lines.

## 14.20 Form Tolerances for Related Features

Angularity, parallelism, perpendicularity, and, in some instances, profile are form tolerances applicable to related features. These tolerances control the attitude of features to one another [ANSI Y14.5M–1982 (R1988)].

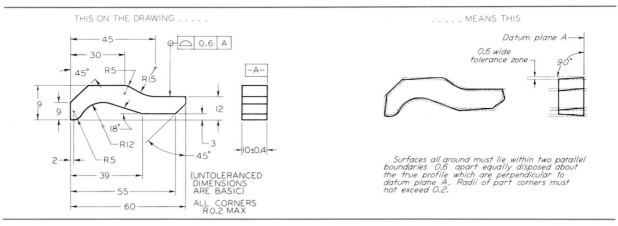

**Fig. 14.34** Specifying Profile of a Surface All Around  [ANSI Y14.5M–1982 (R1988)].

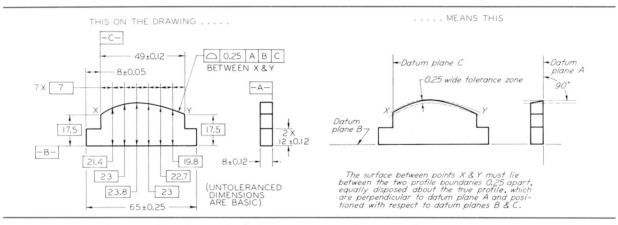

**Fig. 14.35** Specifying Profile of a Surface Between Points  [ANSI Y14.5M–1982 (R1988)].

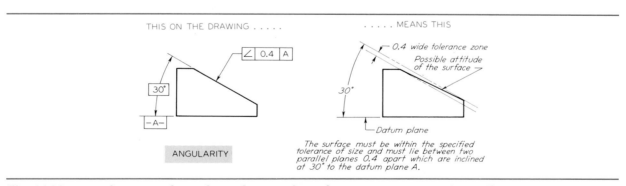

**Fig. 14.36** Specifying Angularity for a Plane Surface  [ANSI Y14.5M–1982 (R1988)].

### Angularity Tolerance

An angularity tolerance specifies a tolerance zone defined by two parallel planes at the specified basic angle (other than 90°) from a datum plane or axis within which the surface or the axis of the feature must lie, Fig. 14.36.

### Parallelism Tolerance

A parallelism tolerance specifies a tolerance zone defined by two parallel planes or lines parallel to a datum plane or axis within which the surface or axis of the feature must lie or the parallelism tolerance may specify a cylindrical tolerance zone parallel to

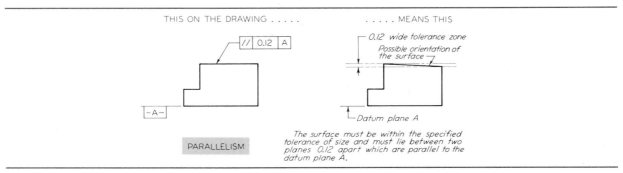

**Fig. 14.37**   Specifying Parallelism for a Plane Surface   [ANSI Y14.5M–1982 (R1988)].

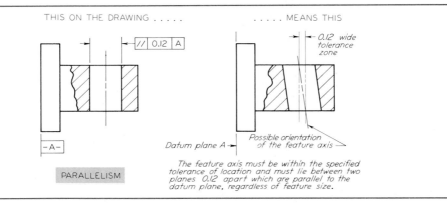

**Fig. 14.38**   Specifying Parallelism for an Axis Feature RFS   [ANSI Y14.5M–1982 (R1988)].

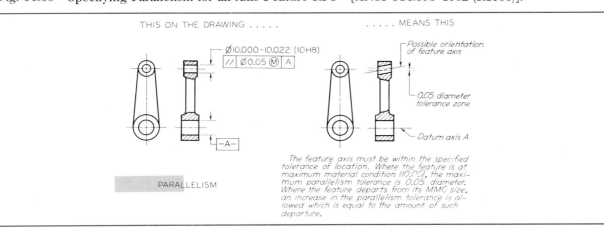

**Fig. 14.39**   Specifying Parallelism for an Axis Feature at MMC   [ANSI Y14.5M–1982 (R1988)].

a datum axis within which the axis of the feature must lie, Figs. 14.37 to 14.39.

### Perpendicularity Tolerance

Perpendicularity is a condition of a surface, median plane, or axis at 90° to a datum plane or axis. A perpendicularity tolerance specifies one of the following:

1. A tolerance zone defined by two parallel planes perpendicular to a datum plane, datum axis, or axis within which the surface of the feature must lie, Fig. 14.40.

2. A cylindrical tolerance zone perpendicular to a datum plane within which the axis of the feature must lie, Fig. 14.41.

THIS ON THE DRAWING . . . . .                    . . . . . MEANS THIS

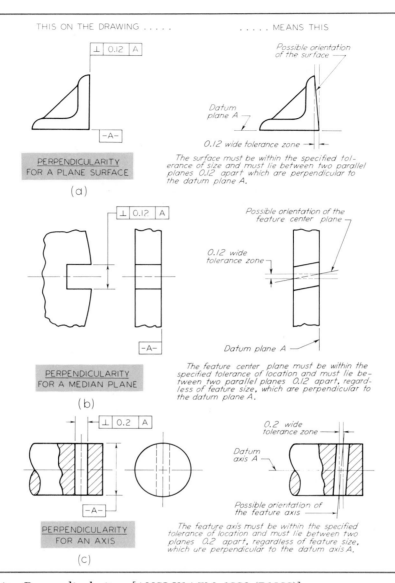

**Fig. 14.40**   Specifying Perpendicularity   [ANSI Y14.5M–1982 (R1988)].

THIS ON THE DRAWING . . . . .                    . . . . . MEANS THIS

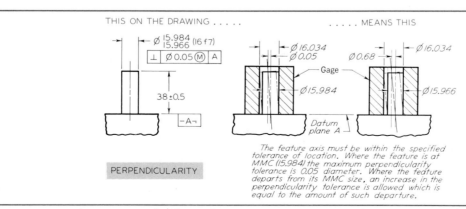

**Fig. 14.41**   Specifying Perpendicularity for an Axis, Pin, or Boss   [ANSI Y14.5M–1982 (R1988)].

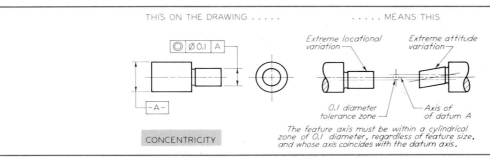

**Fig. 14.42**   Specifying Concentricity   [ANSI Y14.5M–1982 (R1988)].

### *Concentricity Tolerance*

Concentricity is the condition where the axes of all cross-sectional elements of a feature's surface of revolution are common to the axis of a datum feature. A concentricity tolerance specifies a cylindrical tolerance zone whose axis coincides with a datum axis and within which all cross sectional axes of the feature being controlled must lie, Fig. 14.42.

### 14.21 Application of Geometric Tolerancing

The use of various feature control symbols in lieu of notes for position and form tolerance dimensions as abstracted from ANSI Y14.5M–1982 (R1988) is il-

lustrated in Fig. 14.43. For a more detailed treatment of geometric tolerancing, consult the latest ANSI Y14.5 Dimensioning and Tolerancing standard.

### 14.22 Surface Roughness, Waviness, and Lay

The modern demands of the automobile, the airplane, and other machines that can stand heavier loads and higher speeds with less friction and wear have increased the need for accurate control of surface quality by the designer irrespective of the size of the feature. Simple finish marks are not adequate to specify surface finish on such parts.

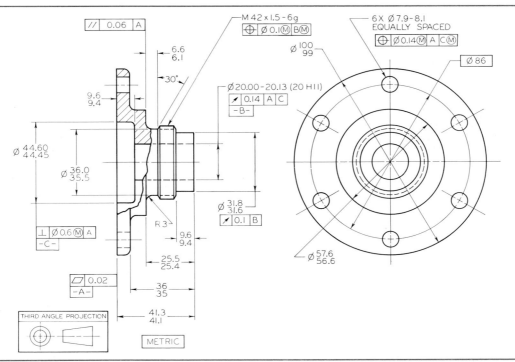

**Fig. 14.43**   Application of Symbols to Position and Form Tolerance Dimensions   [ANSI Y14.5M–1982 (R1988)].

| | Symbol | Meaning |
|---|---|---|
| (a)  | | Basic Surface Texture Symbol. Surface may be produced by any method except when the bar or circle, (b) or (d), is specified. |
| (b)  | | Material Removal By Machining Is Required. The horizontal bar indicates that material removal by machining is required to produce the surface and that material must be provided for that purpose. |
| (c) | 3.5  | Material Removal Allowance. The number indicates the amount of stock to be removed by machining in millimeters (or inches). Tolerances may be added to the basic value shown or in a general note. |
| (d)  | | Material Removal Prohibited. The circle in the vee indicates that the surface must be produced by processes such as casting, forging, hot finishing, cold finishing, die casting, powder metallurgy or injection molding without subsequent removal of material. |
| (e)  | | Surface Texture Symbol. To be used when any surface characteristics are specified above the horizontal line or to the right of the symbol. Surface may be produced by any method except when the bar or circle, (b) or (d), is specified. |
| (f) | 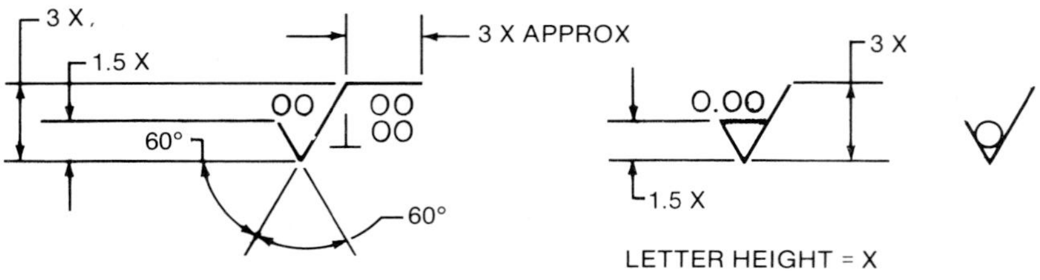 3 X, 1.5 X, 60°, 3 X APPROX, OO OO, ⊥ OO, 60°, 0.00, 3 X, 1.5 X, LETTER HEIGHT = X | |

**Fig. 14.44**   Surface Texture Symbols and Construction    [ANSI Y14.36–1978 (R1987)].

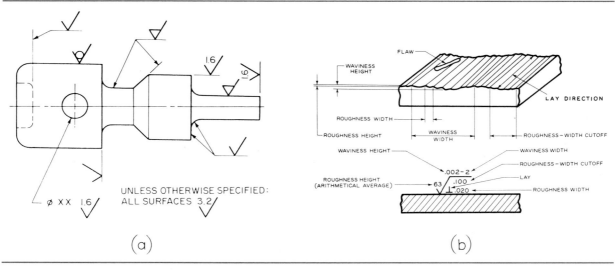

**Fig. 14.45**   Application of Surface Texture Symbols and Surface Characteristics    [ANSI Y14.36–1978 (R1987)].

**Table 14.4**  *Preferred Series Roughness Average Values*
$(R_a)$ [ANSI Y14.36–1978 (R1987)]
Recommended values are in color.

| Micrometers[a] ($\mu$m) | Microinches ($\mu$in.) | Micrometers[a] ($\mu$m) | Microinches ($\mu$in.) |
|---|---|---|---|
| 0.012 | 0.5 | 1.25 | 50 |
| 0.025 | 1 | 1.60 | 63 |
| 0.050 | 2 | 2.0 | 80 |
| 0.075 | 3 | 2.5 | 100 |
| 0.10 | 4 | 3.2 | 125 |
| 0.125 | 5 | 4.0 | 180 |
| 0.15 | 6 | 5.0 | 200 |
| 0.20 | 8 | 6.3 | 250 |
| 0.25 | 10 | 8.0 | 320 |
| 0.32 | 13 | 10.0 | 400 |
| 0.40 | 16 | 12.5 | 500 |
| 0.50 | 20 | 15 | 600 |
| 0.63 | 25 | 20 | 800 |
| 0.80 | 32 | 25 | 1000 |
| 1.00 | 40 | | |

[a]Micrometers are the same as thousandths of a millimeter (1 $\mu$m = 0.001 mm).

Surface finish is intimately related to the functioning of a surface, and proper specification of finish of such surfaces as bearings and seals is necessary. Surface-quality specifications should be used only where needed, since the cost of producing a finished surface becomes greater as the quality of the surface called for is increased. Generally, the ideal surface finish is the roughest one that will do the job satisfactorily.

The system of surface texture symbols recommended by ANSI [Y14.36–1978 (R1987)] for use on drawings, regardless of the system of measurement used, is now broadly accepted by American industry. These symbols are used to define *surface texture, roughness*, and *lay*. See Fig. 14.44 for meaning and construction of these symbols. The basic surface texture symbol in Fig. 14.45 (a) indicates a finished or machined surface by any method just as does the general V symbol, Fig. 13.20 (a). Modifications to the basic surface texture symbol, (b) through (d), define restrictions on material removal for the finished surface. Where surface texture values other than roughness average $(R_a)$ are specified, the symbol must be drawn with the horizontal extension as shown in (e). Construction details for the symbols are given in (f).

Applications of the surface texture symbols are given in Fig. 14.45 (a). Note that the symbols read from the bottom and/or the right side of the drawing and that they are not drawn at any angle or upside down.

Measurements for roughness and waviness, unless otherwise specified, apply in the direction that gives the maximum reading, usually across the lay. See Fig. 14.45 (b). The recommended roughness height values are given in Table 14.4.

When it is necessary to indicate the roughness-width cutoff values, the standard values to be used are listed in Table 14.5. If no value is specified, the 0.80 value is assumed.

**Table 14.5**  *Standard Roughness Sampling Length (Cutoff) Values* [ANSI Y14.36–1978 (R1987)]

| Millimeters (mm) | Inches (in.) | Millimeters (mm) | Inches (in.) |
|---|---|---|---|
| 0.08 | .003 | 2.5 | .1 |
| 0.25 | .010 | 8.0 | .3 |
| 0.80 | .030 | 25.0 | 1.0 |

When maximum waviness height values are required, the recommended values to be used are as given in Table 14.6.

When it is desired to indicate lay, the lay symbols in Fig. 14.46 are added to the surface texture symbol as per the examples given. Selected applications of the surface texture values to the symbol are given and explained in Fig. 14.47.

A typical range of surface roughness values that may be obtained from various production methods is shown in Fig. 14.48. Preferred roughness-height values are shown at the top of the chart.

**Table 14.6** *Preferred Series Maximum Waviness Height Values* [ANSI Y14.36–1978 (R1987)]

| Millimeters (mm) | Inches (in.) | Millimeters (mm) | Inches (in.) |
|---|---|---|---|
| 0.0005 | .00002 | 0.025 | .001 |
| 0.0008 | .00003 | 0.05 | .002 |
| 0.0012 | .00005 | 0.08 | .003 |
| 0.0020 | .00008 | 0.12 | .005 |
| 0.0025 | .0001 | 0.20 | .008 |
| 0.005 | .0002 | 0.25 | .010 |
| 0.008 | .0003 | 0.38 | .015 |
| 0.012 | .0005 | 0.50 | .020 |
| 0.020 | .0008 | 0.80 | .030 |

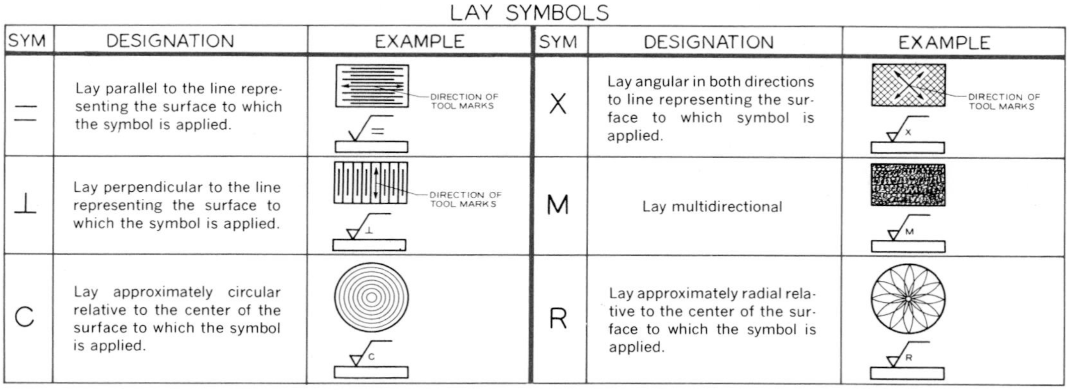

LAY SYMBOLS

| SYM | DESIGNATION | EXAMPLE | SYM | DESIGNATION | EXAMPLE |
|---|---|---|---|---|---|
| — | Lay parallel to the line representing the surface to which the symbol is applied. | DIRECTION OF TOOL MARKS | X | Lay angular in both directions to line representing the surface to which symbol is applied. | DIRECTION OF TOOL MARKS |
| ⊥ | Lay perpendicular to the line representing the surface to which the symbol is applied. | DIRECTION OF TOOL MARKS | M | Lay multidirectional | |
| C | Lay approximately circular relative to the center of the surface to which the symbol is applied. | | R | Lay approximately radial relative to the center of the surface to which the symbol is applied. | |

**Fig. 14.46** Lay Symbols [ANSI Y14.36–1978 (R1987)].

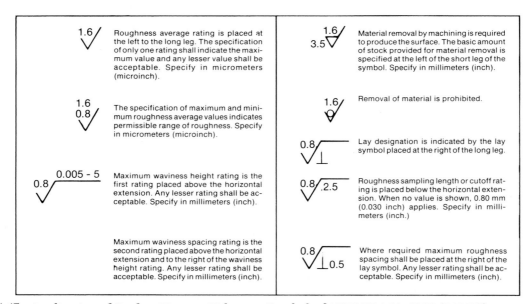

**Fig. 14.47** Application of Surface Texture Values to Symbol [ANSI Y14.36–1978 (R1987)].

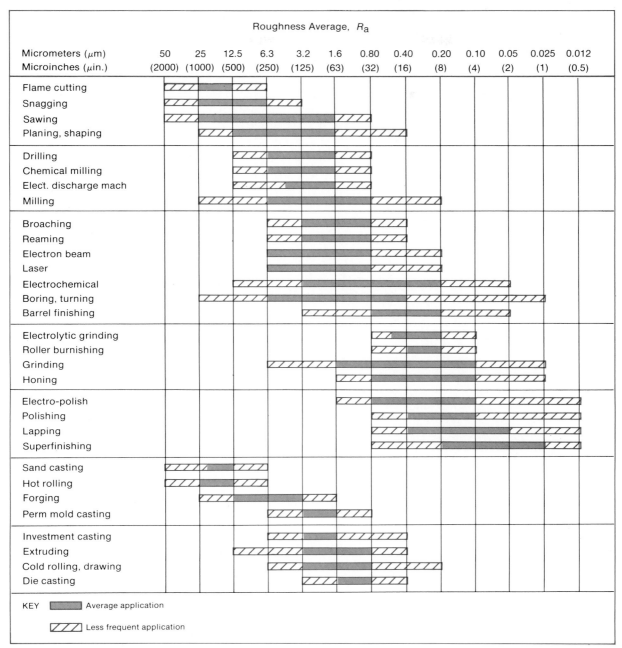

Fig. 14.48 Surface Roughness Produced by Common Production Methods [ANSI/ASME B46.1–1985]. The ranges shown are typical of the processes listed. Higher or lower values may be obtained under special conditions.

# CHAPTER
# 15

# Threads, Fasteners, and Springs*

The concept of the screw thread seems to have occurred first to Archimedes, the third-century B.C. mathematician, who wrote briefly on spirals and invented or designed several simple devices applying the screw principle. By the first century B.C. the screw was a familiar element, but was crudely cut from wood or filed by hand on a metal shaft. Nothing more was heard of the screw thread in Europe until the fifteenth century.

Leonardo da Vinci understood the screw principle, and he has left sketches showing how to cut screw threads by machine. In the sixteenth century, screws appeared in German watches, and screws were used to fasten suits of armour. In 1569 the screw-cutting lathe was invented by the Frenchman Besson, but the method did not take hold for another century and a half; nuts and bolts continued to be made largely by hand. In the eighteenth century, screw manufacturing got started in England during the Industrial Revolution.

*For a listing of American National Standards Institute (ANSI) standards for threads, fasteners, and springs, see Appendix 1.

## 15.1 Standardized Screw Threads

In early times, there was no such thing as standardization. Nuts made by one manufacturer would not fit the bolts of another. In 1841 Sir Joseph Whitworth started crusading for a standard screw thread, and soon the Whitworth thread was accepted throughout England.

In 1864, the United States adopted a thread proposed by William Sellers of Philadelphia, but the Sellers' nuts would not screw on a Whitworth bolt, or vice versa. In 1935 the American Standard thread, with the same 60° V form of the old Sellers' thread, was adopted in the United States. Still there was no standardization among countries. In peacetime it was a nuisance; in World War I it was a serious inconvenience; and in World War II the obstacle was so great that the Allies decided to do something about it. Talks began among the Americans, British, and the Canadians, and in 1948 an agreement was reached on the unification of American and British screw threads. The new thread was called the *Unified screw thread*, and it represents a compromise between the American Standard and Whitworth systems, allowing complete interchangeability of threads in the three countries.

In 1946 an International Organization for Standardization (ISO) committee was formed to establish a single international system of metric screw threads. Consequently, through the cooperative efforts of the Industrial Fasteners Institute (IFI), several committees of the American National Standards Institute, and the ISO representatives, a metric fastener standard (IFI-500–1975) was prepared.

Today screw threads are vital to our industrial life. They are designed for hundreds of different purposes, the three basic applications being (1) to *hold parts* together, (2) to *adjust parts* with reference to each other, and (3) to *transmit power*.

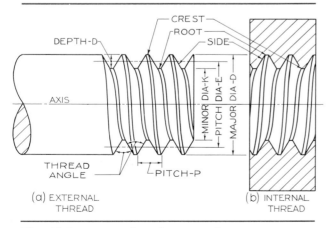

(a) EXTERNAL THREAD   (b) INTERNAL THREAD

*Fig. 15.1*   Screw-Thread Nomenclature.

## 15.2 Definitions of Terms

The following definitions apply to screw threads in general, Fig. 15.1.

*Screw thread*   A ridge of uniform section in the form of a helix, §5.63, on the external or internal surface of a cylinder.

*External thread*   A thread on the outside of a member, as on a shaft.

*Internal thread*   A thread on the inside of a member, as in a hole.

*Major diameter*   The largest diameter of a screw thread (applies to both internal and external threads).

*Minor diameter*   The smallest diameter of a screw thread (applies to both internal and external threads).

*Pitch*   The distance from a point on a screw thread to a corresponding point on the next thread measured parallel to the axis. The pitch $P$ is equal to 1 divided by the number of threads per inch.

*Pitch diameter*   The diameter of an imaginary cylinder passing through the threads so as to make equal the widths of the threads and the widths of the spaces cut by the cylinder.

*Lead*   The distance a screw thread advances axially in one turn.

*Angle of thread*   The angle included between the sides of the thread measured in a plane through the axis of the screw.

*Crest*   The top surface joining the two sides of a thread.

*Root*   The bottom surface joining the sides of two adjacent threads.

*Side*   The surface of the thread that connects the crest with the root.

*Axis of screw*   The longitudinal center line through the screw.

*Depth of thread*   The distance between the crest and the root of the thread measured normal to the axis.

*Form of thread*   The cross section of thread cut by a plane containing the axis.

*Series of thread*   Standard number of threads per inch for various diameters.

## 15.3 Screw-Thread Forms

Various forms of threads, Fig. 15.2, are used to hold parts together, to adjust parts with reference to each other, or to transmit power. The 60° *Sharp-V thread*

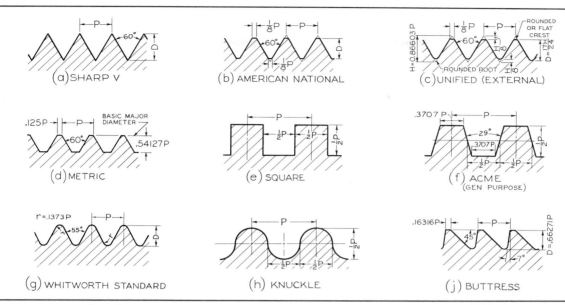

**Fig. 15.2**   Screw-Thread Forms.

was originally called the United States Standard thread, or the Sellers' thread. For purposes of certain adjustments, the Sharp-V thread is useful with the increased friction resulting from the full thread face. It is also used on brass pipe work.

The *American National thread* with flattened roots and crests is a stronger thread. This form replaced the Sharp-V thread for general use.

The *Unified thread* is the standard thread agreed upon by the United States, Canada, and Great Britain in 1948, and is gradually replacing the American National form. The crest of the external thread may be flat or rounded, and the root is rounded; otherwise, the thread form is essentially the same as the American National.

The *metric thread* is the standard screw thread agreed upon for international screw thread fasteners. The crest and root are flat, but the external thread is often rounded if formed by a rolling process. The form is similar to the American National and the Unified threads but with less depth of thread.

The *square thread* is theoretically the ideal thread for power transmission, since its face is nearly at right angles to the axis, but due to the difficulty of cutting it with dies and because of other inherent disadvantages, such as the fact that split nuts will not readily disengage, the square thread has been displaced to a large extent by the Acme thread. The square thread is not standardized.

The *Acme thread* is a modification of the square

thread and has largely replaced it. It is stronger than the square thread, is easier to cut, and has the advantage of easy disengagement from a split nut, as on the lead screw of a lathe.

The *standard worm thread* (not shown) is similar to the Acme thread, but is deeper. It is used on shafts to carry power to worm wheels.

The *Whitworth thread* has been the British standard and is being replaced by the Unified thread. Its uses correspond to those of the American National thread.

The *knuckle thread* is usually rolled from sheet metal but is sometimes cast, and is used in modified forms in electric bulbs and sockets, bottle tops, and the like.

The *buttress thread* is designed to transmit power in one direction only and is used in large guns, in jacks, and in other mechanisms of similar high strength requirements.

## 15.4  Thread Pitch

The *pitch* of any thread form is the distance parallel to the axis between corresponding points on adjacent threads, Figs. 15.1 (a) and 15.2. For metric threads, this distance is specified in millimeters.

The pitch for a metric thread that is included with the major diameter in the thread designation determines the size of the thread, as shown in Fig. 15.3 (b). For example, M10 × 1.5. See also §15.21, or Appendix 15, for more information.

For threads dimensioned in inches, the pitch is

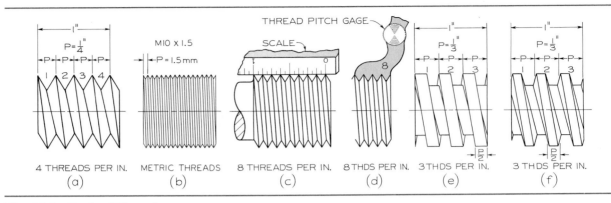

*Fig. 15.3*   Pitch of Threads.

equal to 1 divided by the number of threads per inch. The thread tables give the number of threads per inch for each standard diameter. Thus, a Unified coarse thread, Appendix 15, of 1″ diameter, has eight threads per inch, and the pitch $P$ equals $\frac{1}{8}$″.

As shown in Fig. 15.3 (a), if a thread has only four threads per inch, the pitch and the threads themselves are quite large. If there are, say, sixteen threads per inch, the pitch is only $\frac{1}{16}$″, and the threads are relatively small, similar to the threads shown at (b).

The pitch or the number of threads per inch can easily be measured with a scale, (c), or with a *thread-pitch gage*, (d).

## 15.5 Right-Hand and Left-Hand Threads

A right-hand thread is one that advances into a nut when turned clockwise, and a left-hand thread is one that advances into a nut when turned counterclockwise, Fig. 15.4. A thread is always considered to be right-hand (RH) unless otherwise specified. A left-hand thread is always labeled **LH** on a drawing. See Fig. 15.19 (a).

## 15.6 Single and Multiple Threads

A *single* thread, as the name implies, is composed of one ridge, and the lead is therefore equal to the pitch. *Multiple* threads are composed of two or more ridges running side by side. As shown in Fig. 15.5 (a) to (c), the *slope line* is the hypotenuse of a right triangle whose short side equals $\frac{1}{2}P$ for single threads, P for double threads, $1\frac{1}{2}P$ for triple threads, and so on. This applies to all forms of threads. In *double* threads, the lead is twice the pitch; in *triple* threads the lead is three times the pitch, and so on. On a drawing of a single or triple thread, a root is opposite a crest; in the case of a double or quadruple thread, a root is drawn opposite a root. Therefore, in one turn, a double thread advances twice as far as a single thread, and a triple thread advances three times as far.

RH double square and RH triple Acme threads are shown at (d) and (e), respectively.

Multiple threads are used wherever quick motion, but not great power, is desired, as on fountain pens, toothpaste caps, valve stems, and the like. The threads on a valve stem are frequently multiple threads, to impart quick action in opening and clos-

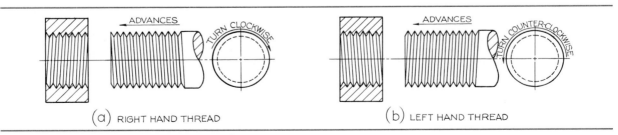

*Fig. 15.4*   Right-Hand and Left-Hand Threads.

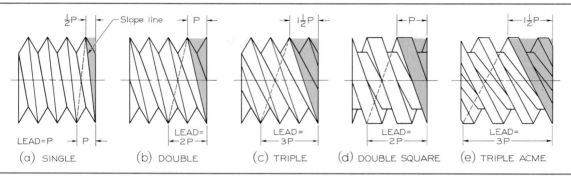

*Fig. 15.5* Multiple Threads.

ing the valve. Multiple threads on a shaft can be recognized and counted by observing the number of thread endings on the end of the screw.

## 15.7 Thread Symbols

There are three methods of representation for showing screw threads on drawings—the *schematic,* the *simplified,* and the *detailed.* For clarity of representation and where good judgment dictates, schematic, simplified, and detailed thread symbols may be combined on a single drawing.

Two sets of thread symbols, the schematic and the more common simplified, are used to represent threads of small diameter, under approximately 1″ or 25 mm diameter on the drawing. The symbols are the same for all forms of threads, such as metric, Unified, square, and Acme.

The detailed representation is a close approximation of the exact appearance of a screw thread. The true projection of the helical curves of a screw thread, Fig. 15.1, presents a pleasing appearance, but this does not compensate for the laborious task of plotting the helices, §5.63. Consequently, the true projection is used rarely in practice.

When the diameter of the thread on the drawing is over approximately 1″ or 25 mm, a pleasing drawing may be made by the *detailed representation* method, in which the true profiles of the threads (any form of thread) are drawn; but the helical curves are replaced by straight lines, as shown in Fig. 15.6.* Whether the crests or roots are flat or rounded, they are represented by single lines and not double lines as in Fig. 15.1; consequently, American National and Unified threads are drawn in exactly the same way.

## 15.8 External Thread Symbols

Simplified external thread symbols are shown in Fig. 15.7 (a) and (b). The threaded portions are indicated by hidden lines parallel to the axis at the approximate depth of the thread, whether in section or in elevation. Use the schematic depth of thread

*A thread 42 mm or 1⅝″ diameter, if drawn half size, would be less than 25 mm or 1″ diameter on the drawing and hence would be too small for this method of representation.

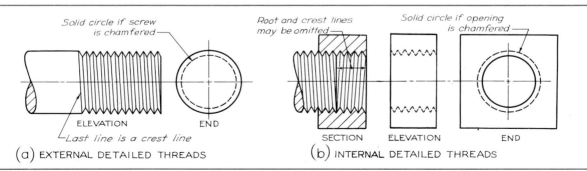

*Fig. 15.6* Detailed Metric, American National, and Unified Threads.

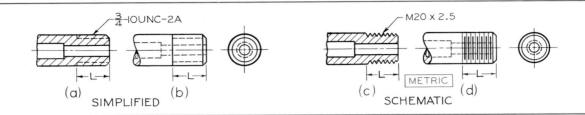

*Fig. 15.7*   External Thread Symbols.

as given in the table in Fig. 15.9 (a), to draw these lines, (d).

When the schematic form is shown in section, (c), it is necessary to show the V's; otherwise no threads would be evident. However, it is not necessary to show the V's to scale or according to the actual slope of the crest lines. To draw the V's, use the schematic thread depth, Fig. 15.9 (a), and let the pitch be determined by the 60° V's.

Schematic threads in elevation, Fig. 15.7 (d), are indicated by alternate long and short lines at right angles to the center line, the root lines being preferably thicker than the crest lines. Although theoretically the crest lines would be spaced according to actual pitch, the lines would often be very crowded and tedious to draw, thus defeating the purpose of the symbol, to save drafting time. In practice, the experienced drafter spaces the crest lines carefully by eye, and then adds the heavy root lines spaced by eye halfway between the crest lines. In general, the spacing should be proportionate for all diameters. For convenience in drawing, proportions for the schematic symbol are given in Fig. 15.9.

## 15.9 Internal Thread Symbols

Internal thread symbols are shown in Fig. 15.8. Note that the only differences between the schematic and simplified internal thread symbols occur in the sectional views. The representation of the schematic thread in section, Fig. 15.8 (m), (o), and (r), is exactly the same as the external symbol in Fig. 15.7 (d). Hidden threads, by either method, are represented by pairs of hidden lines. The hidden dashes should be staggered, as shown.

In the case of blind tapped holes, the drill depth normally is drawn at least three schematic pitches beyond the thread length, as shown in Fig. 15.8 (d), (e), (n), and (o). The representations at (f) and (p) are used to represent the use of a bottoming tap, when the length of thread is the same as the depth of drill. See also §15.24.

## 15.10 To Draw Thread Symbols

Fig. 15.9 (a) shows a table of values of depth and pitch to use in drawing thread symbols. These values are selected to produce a well-proportioned

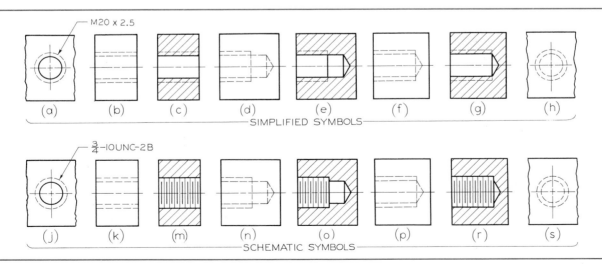

*Fig. 15.8*   Internal Thread Symbols.

| MAJOR DIAMETER | #5 (.125) TO #12 (.216) | $\frac{1}{4}$ | $\frac{5}{16}$ | $\frac{3}{8}$ | $\frac{7}{16}$ | $\frac{1}{2}$ | $\frac{9}{16}$ | $\frac{5}{8}$ | $\frac{11}{16}$ | $\frac{3}{4}$ | $\frac{13}{16}$ | $\frac{7}{8}$ | $\frac{15}{16}$ | 1 |
|---|---|---|---|---|---|---|---|---|---|---|---|---|---|---|
| DEPTH, D | $\frac{1}{32}$ | $\frac{1}{32}$ | $\frac{1}{32}$ | $\frac{1}{32}$ | $\frac{3}{64}$ | $\frac{3}{64}$ | $\frac{1}{16}$ | $\frac{1}{16}$ | $\frac{1}{16}$ | $\frac{1}{16}$ | $\frac{5}{64}$ | $\frac{3}{32}$ | $\frac{3}{32}$ | $\frac{3}{32}$ |
| PITCH, P | $\frac{3}{64}$ | $\frac{1}{16}$ | $\frac{1}{16}$ | $\frac{1}{16}$ | $\frac{1}{16}$ | $\frac{3}{32}$ | $\frac{3}{32}$ | $\frac{3}{32}$ | $\frac{3}{32}$ | $\frac{1}{8}$ | $\frac{1}{8}$ | $\frac{1}{8}$ | $\frac{1}{8}$ | $\frac{1}{8}$ |

(a)

*(For metric values: 1" = 25.4 mm or see inside front cover.)*

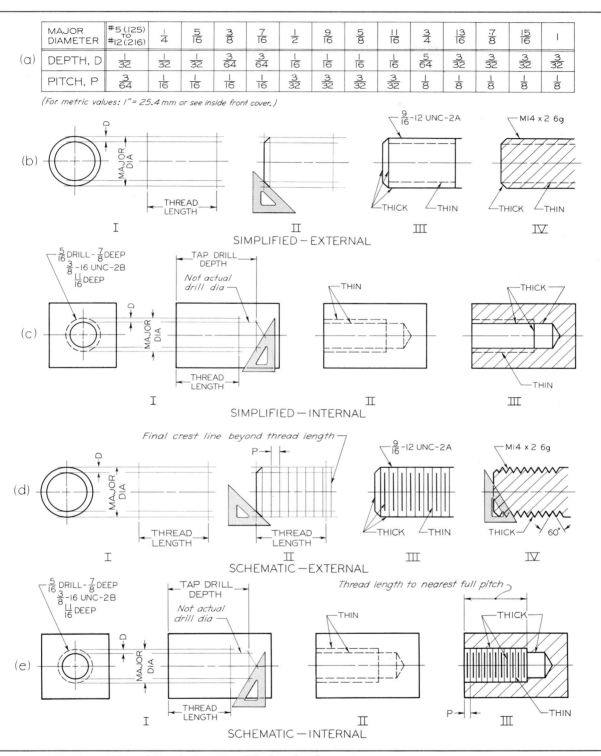

**Fig. 15.9** To Draw Thread Symbols—Simplified and Schematic.

symbol and to be convenient to set off with the scale. The experienced drafter will carefully space the lines by eye, but the student should use the scale. Note that the values of D and P are for the diameter *on the drawing.* Thus a $1\frac{1}{2}''$ diameter thread at half-scale would be $\frac{3}{4}''$ diameter on the drawing, and values of D and P for a $\frac{3}{4}''$ major diameter would be used. Nominal diameters for metric threads are treated in a similar manner.

### Simplified Symbols

The steps for drawing the simplified symbols for an external thread in elevation and in section are shown in Fig. 15.9 (b). The thread depth from the table at (a) is used for establishing the pairs of hidden lines that represent the threads in elevation and in section. No pitch measurement is needed. The completed symbols are shown at III and IV.

The steps for drawing the simplified symbol for an internal thread in section are shown in Fig. 15.9 (c). The simplified representation for the internal thread in elevation is identical to that used for schematic representation, as the threads are indicated by pairs of hidden lines as shown at II. The simplified symbol for an internal thread in section is shown at III. The major diameter of the thread is represented by hidden lines across the sectioned area.

### Schematic Symbols

The steps for drawing the schematic symbols for an external thread in elevation and in section are shown in Fig. 15.9 (d). Note that when the pitches P are set off in II, the final crest line for a full pitch may fall beyond the actual thread length as shown. The completed schematic symbol for an external thread in elevation is shown at III. The completed schematic symbol for an external thread in section is shown at IV. The schematic thread depth is used for drawing the V's, and the pitch is established by the 60° V's.

The steps for drawing the schematic symbols for an internal thread in elevation and in section are shown at (e). Here again the symbol thread length may be slightly longer than the actual given thread length. If the tap drill depth is known or given, the drill is drawn to that depth, as shown. If the thread note omits this information, as is often done in practice, the drafter merely draws the hole three thread pitches (schematic) beyond the thread length. The tap drill diameter is represented approximately, as shown, and not to actual size. The completed schematic symbol for an internal thread in elevation is

shown at II. Pairs of hidden lines represent the threads, and the hidden-line dashes are staggered. The completed schematic symbol for an internal thread in section is shown at III. The schematic internal thread in section is represented in the same manner as for the schematic external thread.

## 15.11 Detailed Representation—Metric, Unified, and American National Threads

The detailed representation for metric, unified, and American National threads is the same, since the flats, if any, are disregarded. The steps in drawing these threads are shown in Fig. 15.10.

I.  Draw center line and lay out length and major diameter.

II.  Find the number of threads per inch in Appendix 15 for American National and Unified threads. This number depends upon the major diameter of the thread, whether the thread is internal or external. Find P (pitch) by dividing 1 by the number of threads per inch, §15.4. The pitch for metric threads is given directly in the thread designation. For example the M14 × 2 thread has a pitch of 2 mm. See Appendix 15. Establish the slope of the thread by offsetting the slope line $\frac{1}{2}$P for single threads, P for double threads, $1\frac{1}{2}$P for triple threads, and so on.* For right-hand external threads, the slope line slopes upward to the left; for left-hand external threads, the slope line slopes upward to the right. If the numbers of threads per inch conforms to the scale, the pitch can be set off directly. For example, eight threads per inch can easily be set off with the architects' scale, and ten threads per inch with the engineers' scale. Otherwise, use the bow dividers or use the parallel-line method shown in Fig. 15.10 (II).

III.  From the pitch points, draw crest lines parallel to slope line. These should be dark thin lines. Slide triangle on T-square (or another triangle) to make lines parallel. Draw two V's to establish depth of thread, and draw guide lines for root of thread, as shown.

IV.  Draw 60° V's finished weight. These V's should stand vertically; that is, they should not "lean" with the thread.

*These offsets are the same in terms of P for any form of thread.

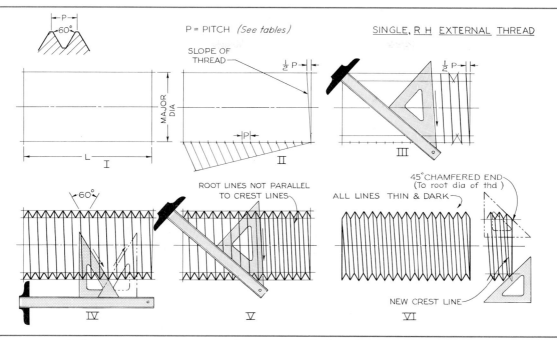

Fig. 15.10 Detailed Representation—External Metric, Unified, and American National Threads.

V. Draw root lines dark at once. Root lines will *not* be parallel to crest lines. Slide triangle on straightedge to make root lines parallel.

VI. When the end is chamfered (usually 45° with end of shaft, sometimes 30°), the chamfer extends to the thread depth. The chamfer creates a new crest line, which is then drawn between the two new crest points. It is not parallel to the other crest lines. In the final drawing, all thread lines should be approximately the same weight—thin, but dark.

The corresponding internal detailed threads, in section, are drawn as shown in Fig. 15.11. Notice that for LH threads the lines slope upward to the left, as shown at (a), while for RH threads the lines slope upward to the right, as at (b). Make all final thread lines medium-thin but dark.

## 15.12 Detailed Representation of Square Threads

The steps in drawing the detailed representation of an external square thread when the major diameter is over 1″ or 25 mm (approx.) on the drawing are shown in Fig. 15.12.

I. Draw center line, and lay out length and major diameter of thread. Determine P by dividing 1 by

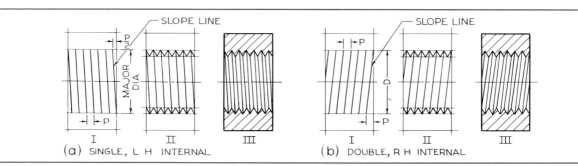

Fig. 15.11 Detailed Representation—Internal Metric, Unified, and American National Threads.

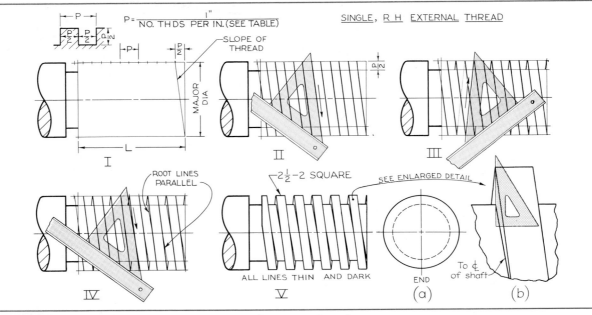

**Fig. 15.12** Detailed Representation—External Square Threads.

the number of threads per inch. See Appendix 22. For a single RH thread, the lines slope upward to the left, and the slope line is offset *as for all single threads of any form.* On the upper line, set off spaces equal to $\frac{P}{2}$, as shown, using a scale if possible; otherwise, use the bow dividers or the parallel-line method to space the points.

II.   From the $\frac{P}{2}$ points on the upper line, draw fairly thin lines. Draw guide lines for root of thread, making the depth $\frac{P}{2}$ as shown.

III.   Draw parallel visible back edges of threads.

IV.   Draw parallel visible root lines. Note enlarged detail at (b).

V.   Accent the lines. All lines should be thin and dark.

Note the end view of the shaft at (a). The root circle is hidden; no attempt is made to show the true projection. If the end is chamfered, a solid circle would be drawn instead of the hidden circle.

An assembly drawing, showing an external square thread partly screwed into the nut, is shown in Fig. 15.15. The detail of the square thread at A is the same as shown in Fig. 15.12. But when the external and internal threads are assembled, the

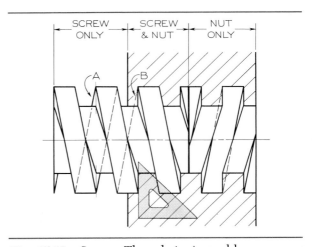

**Fig. 15.13** Square Threads in Assembly.

thread in the nut overlaps and covers up half of the V, as shown at B.

The internal thread construction is the same as in Fig. 15.14. Note that the thread lines representing the back half of the internal threads (since the thread is in section) slope in the opposite direction from those on the front side of the screw.

Steps in drawing a single internal square thread in section are shown in Fig. 15.14. Note in step II that a crest is drawn opposite a root. This is the case

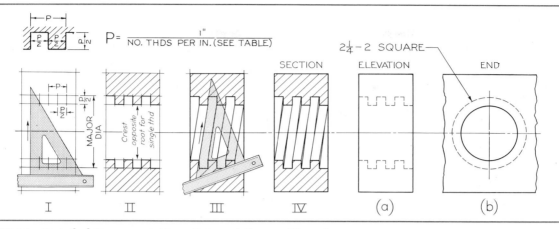

**Fig. 15.14** Detailed Representation—Internal Square Threads.

for both single and triple threads. For double or quadruple threads, a crest is opposite a crest. Thus, the construction in steps I and II is the same for any multiple of thread. The differences are developed in step III where the threads and spaces are distinguished and outlined.

The same internal thread is shown in elevation (external view) in Fig. 15.14 (a). The profiles of the threads are drawn in their normal position, but with hidden lines, and the sloping lines are omitted for simplicity. The end view of the same internal thread

is shown at (b). Note that the hidden and solid circles are opposite those for the end view of the shaft. See Fig. 15.12 (a).

## 15.13 Detailed Representation of Acme Thread

The steps in drawing the detailed representation of Acme threads when the major diameter is over 1″ or 25 mm (approx.) on the drawing are shown in Fig. 15.15.

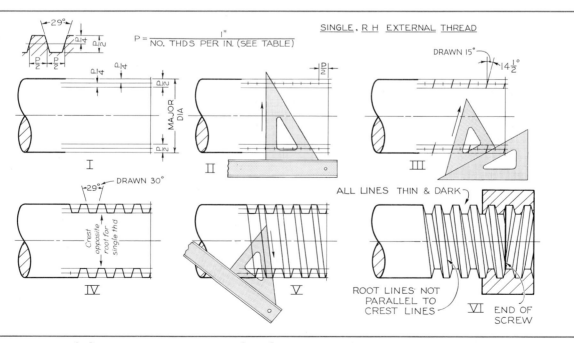

**Fig. 15.15** Detailed Representation—Acme Threads.

I.  Draw center line, and lay out length and major diameter of thread. Determine P by dividing 1 by the number of threads per inch. See Appendix 22. Draw construction lines for the root diameter, making the thread depth $\frac{P}{2}$. Draw construction lines halfway between crest and root guide lines.

II.  On the intermediate construction lines, lay off $\frac{P}{2}$ spaces, as shown. Setting off spaces directly with a scale is possible (for example, if $\frac{P}{2} = \frac{1}{10}''$, use the engineers' scale); otherwise, use bow dividers or parallel-line method.

III.  Through alternate points, draw construction lines for sides of threads (draw 15° instead of $14\frac{1}{2}°$).

IV.  Draw construction lines for other sides of threads. Note that for single and triple threads, a crest is opposite a root, while for double and quadruple threads, a crest is opposite a crest. Heavy in tops and bottoms of threads.

V.  Draw parallel crest lines, final weight at once.

VI.  Draw parallel root lines, final weight at once, and heavy in the thread profiles. All lines should be thin and dark. Note that the internal threads in the back of the nut slope in the opposite direction to the external threads on the front side of the screw.

End views of Acme threaded shafts and holes are drawn exactly like those for the square thread, Figs. 15.12 and 15.14.

## 15.14  Use of Phantom Lines

In representing objects having a series of identical features, phantom lines, Fig. 15.16, may be used to save time. Threaded shafts and springs thus represented may be shortened without the use of conventional breaks, but must be correctly dimensioned. The use of phantom lines is limited almost entirely to detailed drawings.

## 15.15  Threads in Section

Detailed representations of large threads in section are shown in Figs. 15.6 and 15.10 to 15.15. As indicated by the note in Fig. 15.6 (b), the root lines and crest lines may be omitted in internal sectional views, if desired.

External thread symbols are shown in section in Fig. 15.7. Note that in the schematic symbol, the V's must be drawn. Internal thread symbols in section are shown in Fig. 15.8.

Threads in an assembly drawing are shown in Fig. 15.17. It is customary not to section a stud or a nut, or any solid part, unless necessary to show some internal shapes. See §15.22. Note that when external and internal threads are sectioned in assembly, the V's are required to show the threaded connection.

## 15.16  American National Thread

The old American National thread was adopted in 1935. The *form*, or profile, Fig. 15.2 (b), is the same as the old Sellers' profile, or U.S. Standard, and is known as the *National form*. The methods of representation are the same as for the Unified and metric threads. American National threads are being replaced by the Unified and metric threads. However, they may still be encountered on earlier drawings. Five *series* of threads were embraced in the old ANSI standards.

1.  *Coarse thread*—A general-purpose thread for holding purposes. Designated NC (National Coarse).
2.  *Fine thread*—A greater number of threads per inch; used extensively in automotive and aircraft construction. Designated NF (National Fine).
3.  *8-pitch thread*—All diameters have 8 threads per inch. Used on bolts for high-pressure pipe flanges, cylinder-head studs, and similar fasteners. Designated 8N (National form, 8 threads per inch).

*Fig. 15.16*   Use of Phantom Lines.

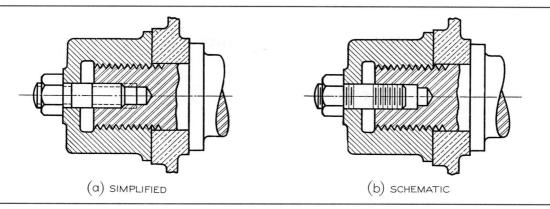

(a) SIMPLIFIED        (b) SCHEMATIC

*Fig. 15.17*  Threads in Assembly.

4. *12-pitch thread*—All diameters have 12 threads per inch; used in boiler work and for thin nuts on shafts and sleeves in machine construction. Designated 12N (National form, 12 threads per inch).

5. *16-pitch thread*—All diameters have 16 threads per inch; used where necessary to have a fine thread regardless of diameter, as on adjusting collars and bearing retaining nuts. Designated 16N (National form, 16 threads per inch).

## 15.17 Unified Extra Fine Threads

*The Unified extra fine thread series* has many more threads per inch for given diameters than any series of the American National or Unified.

The form of thread is the same as the American National. These small threads are used in thin metal where the length of thread engagement is small, in cases where close adjustment is required, and where vibration is great. It is designated UNEF (extra fine).

## 15.18 American National Thread Fits

For general use, three classes of screw thread fits between mating threads (as between bolt and nut) have been established by ANSI.

These fits are produced by the application of tolerances listed in the standard and are described as follows:

*Class 1 fit*—Recommended only for screw-thread work where clearance between mating parts is essential for rapid assembly and where shake or play is not objectionable.

*Class 2 fit*—Represents a high quality of commercial thread product and is recommended for the great bulk of interchangeable screw thread work.

*Class 3 fit*—Represents an exceptionally high quality of commercially threaded product and is recommended only in cases where the high cost of precision tools and continual checking are warranted.

The class of fit desired on a thread is indicated in the thread note, as shown in §15.21.

## 15.19 Metric and Unified Threads

The preferred metric thread for commercial purposes conforms to the International Organization for Standardization publication [ISO 68] basic profile M for metric threads. This M profile design is comparable to the Unified inch profile, but they are not interchangeable. For commercial purposes, two series of metric threads are preferred—coarse (general purpose) and fine—thus drastically reducing the number of previously used thread series. See Appendix 15.

The Unified thread constitutes the present American National Standard. Some of the earlier American National threads are still included in the new standard. See Appendix 15. The standard lists eleven different series of numbers of threads per inch for the various standard diameters, together with selected combinations of special diameters and pitches.

The eleven series are the *coarse thread series* (UNC or NC) recommended for general use corresponding to the old National coarse thread; the *fine*

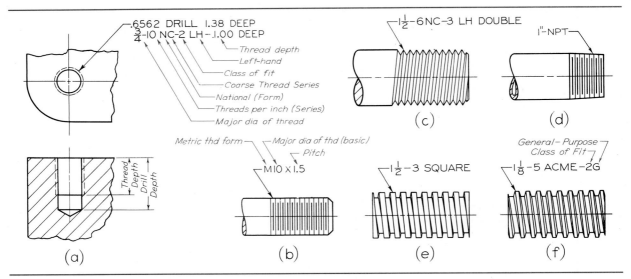

*Fig. 15.18*   Thread Notes.

*thread series* (UNF or NF), for general use in automotive and aircraft work and in applications where a finer thread is required; the *extra fine series* (UNEF or NEF), which is the same as the SAE extra fine series, used particularly in aircraft and aeronautical equipment and generally for threads in thin walls; and the eight series of 4, 6, 8, 12, 16, 20, 28, and 32 threads with constant pitch. The 8UN or 8N, 12UN or 12N, and 16UN or 16N series are recommended for the uses corresponding to the old 8-, 12-, and 16-pitch American National threads. In addition, there are three special thread series—UNS, NS, and UN—that involve special combinations of diameter, pitch, and length of engagement.

## 15.20 Thread Fits, Metric and Unified

For some specialized metric thread applications, the tolerances and deviations are specified by tolerance grade, tolerance position, class, and length of engagement.* Two classes of metric thread fits are generally recognized. The first class of fits is for general-purpose applications and has a tolerance class of **6H** for internal threads and a class **6g** for external threads. The second class of fits is used where closer fits are necessary and has a tolerance of class of **6H** for internal threads and a class of **5g6g** for external threads. Metric thread tolerance classes of **6H/6g** are generally assumed if not otherwise designated and

*ISO Standards Handbook No. 18, 1984.

are used in applications comparable to the **2A/2B** inch classes of fits.

The single-tolerance designation of **6H** refers to both the tolerance grade and position for the pitch diameter and the minor diameter for an internal thread. The single tolerance of designation of **6g** refers to both the tolerance grade and position for the pitch diameter and the major diameter of the external thread. A double designation **5g6g** indicates a separate tolerance grade for the pitch diameter and for the major diameter of the external thread.

The standard for Unified screw threads specifies tolerances and allowances defining the several classes of fit (degree of looseness or tightness) between mating threads. In the symbols for fit, the letter A refers to external threads and B to internal threads. There are three classes of fit each for external threads (1A, 2A, 3A) and internal threads (1B, 2B, 3B). Classes 1A and 1B have generous tolerances, facilitating rapid assembly and disassembly. Classes 2A and 2B are used in the normal production of screws, bolts, and nuts, as well as a variety of general applications. Classes 3A and 3B provide for applications needing highly accurate and close-fitting threads.

## 15.21 Thread Notes

Thread notes for metric, Unified, and American National screw threads are shown in Figs. 15.18 and 15.19. These same notes or symbols are used in correspondence, on shop and storeroom records, and

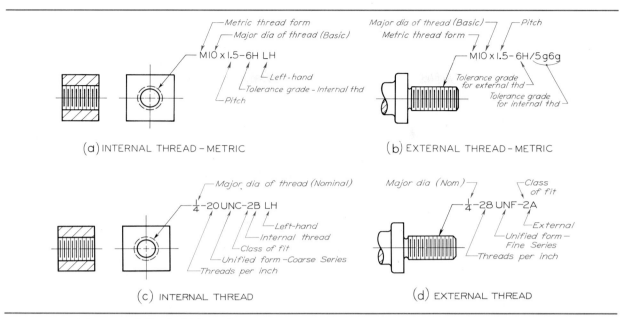

**Fig. 15.19** Metric and Unified Thread Notes.

in specifications for parts, taps, dies, tools, and gages.

Metric screw threads are designated basically by the letter **M** for metric profile followed by the nominal size (basic major diameter) and the pitch, both in millimeters and separated by the symbol ×. For example, the basic thread note M10 × 1.5 is adequate for most commercial purposes, Fig. 15.18 (b). If the generally understood tolerances need to be specified, the tolerance class of **6H** for the internal thread or the tolerance class of **6g** for the external thread is added to the basic note, Fig. 15.19 (a). Where closer-mating threads are desired, a tolerance of **6H** for the internal thread and tolerance classes of **5g6g** for the external thread are added to the basic note. When the thread note refers to mating parts, a single basic note is adequate with the addition of the internal and external thread tolerance classes separated by the slash. The basic note for mating threads now becomes M10 × 1.5–6H/6g for the general-purpose thread or M10 × 1.5–6H/5g6g for the close-fitting thread, Fig. 15.19 (b). Should the thread be left-hand, LH is added to the thread note. (Absence of LH indicates an RH thread.) If it is necessary to indicate the length of the thread engagement, the letter **S** (short), **N** (normal), or **L** (long) is added to the thread note. For example, the single note M10 × 1.5–6H/6g–N–LH combines the specifications for internal and external mating

left-hand metric threads of 10 mm diameter and 1.5 mm pitch with general-purpose tolerances and normal length of engagement.

A thread note for a blind tapped hole is shown in Fig. 15.18 (a). In a complete note, the tap drill and depth should be given, though in practice they are often omitted and left to the shop. For tap drill sizes, see Appendix 15. If the LH symbol is omitted, the thread is understood to be RH. If the thread is a multiple thread, the word **DOUBLE**, **TRIPLE**, or **QUADRUPLE** should precede the thread depth; otherwise, the thread is understood to be single. Thread notes for holes are perferably attached to the circular views of the holes, as shown.

Thread notes for external threads are preferably given in the longitudinal view of the threaded shaft, as shown in Fig. 15.18 (b)–(f). Examples of 8-, 12-, and 16-pitch threads, not shown in the figure, are 2–8N–2, 2–12N–2, and 2–16N–2. A sample special thread designation is 1½–7N–LH.

General-purpose Acme threads are indicated by the letter **G**, and centralizing Acme threads by the letter **C**. Typical thread notes are 1¾–4 ACME–2G or 1¾–6 ACME–4C.

Thread notes for Unified threads are shown in Fig. 15.19 (c) and (d). Unified threads are distinguished from American National threads by the insertion of the letter **U** before the series letters, and by the letters **A** or **B** (for external or internal, respectively) after the numeral designating the class of fit. If the letters LH are omitted, the thread is understood to be RH. Some typical thread notes are

$$\tfrac{1}{4}\text{--}20\ \text{UNC--2A TRIPLE}$$

$$\tfrac{9}{16}\text{--}18\ \text{UNF--2B}$$

$$1\tfrac{3}{4}\text{--}16\ \text{UN--2A}$$

## 15.22 American National Standard Pipe Threads

The American National Standard for pipe threads, originally known as the Briggs standard, was formulated by Robert Briggs in 1882. Two general types of pipe threads have been approved as American National Standard: *taper pipe threads* and *straight pipe threads.*

The profile of the taper pipe thread is illustrated in Fig. 15.20. The taper of the standard tapered pipe thread is 1 in 16 or .75″ per foot measured on the diameter and along the axis. The angle between the sides of the thread is 60°. The depth of the sharp V is .8660$p$, and the basic maximum depth of the truncated thread is .800$p$, where $p$ = pitch. The basic pitch diameters, $E_0$ and $E_1$, and the basic length of the effective external taper thread, $L_2$, are determined by the formulas

$$E_0 = D - (.050D + 1.1)\frac{1}{n}$$

$$E_1 = E + .0625L_1$$

$$L_2 = (.80D + 6.8)\frac{1}{n}$$

where $D$ = basic O.D. of pipe, $E_0$ = pitch diameter of thread at end of pipe, $E_1$ = pitch diameter of

thread at large end of internal thread, $L_1$ = normal engagement by hand, and $n$ = number of threads per inch.

The ANSI also recommended two modified taper pipe threads for (1) dryseal pressure-tight joints (.88″ per foot taper) and (2) rail fitting joints. The former is used to provide a metal-to-metal joint, eliminating the need for a sealer, and is used in refrigeration, marine, automotive, aircraft, and ordnance work. The latter is used to provide a rigid mechanical thread joint as required in rail fitting joints.

While taper pipe threads are recommended for general use, there are certain types of joints where straight pipe threads are used to advantage. The number of threads per inch, the angle, and the depth of thread are the same as on the taper pipe thread, but the threads are cut parallel to the axis. Straight pipe threads are used for pressure-tight joints for pipe couplings, fuel and oil line fittings, drain plugs, free-fitting mechanical joints for fixtures, loose-fitting mechanical joints for locknuts, and loose-fitting mechanical joints for hose couplings.

Pipe threads are represented by detailed or symbolic methods in a manner similar to the representation of Unified and American National threads. The symbolic representation (schematic or simplified) is recommended for general use regardless of diameter, Fig. 15.21. The detailed method is recommended only when the threads are large and when it is desired to show the profile of the thread as, for example, in a sectional view of an assembly.

As shown in Fig. 15.21, it is not necessary to draw the taper on the threads unless there is some reason to emphasize it, since the thread note indicates whether the thread is straight or tapered. If it

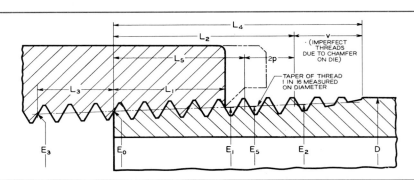

*Fig. 15.20*  American National Standard Taper Pipe Thread  [ANSI B2.1–1968].

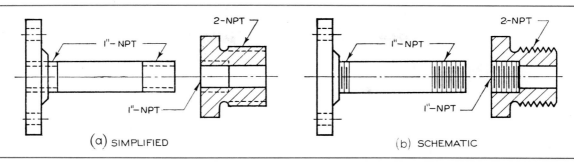

Fig. 15.21   Conventional Representation of Pipe Threads.

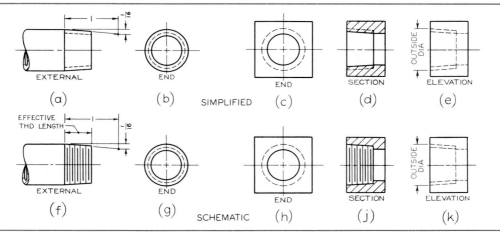

Fig. 15.22   Conventional Pipe Thread Representation.

is desired to show the taper, it should be exaggerated, as shown in Fig. 15.22, where the taper is drawn $\frac{1}{16}''$ per $1''$ *on radius* (or $6.75''$ per $1'$ *on diameter*) instead of the actual taper of $\frac{1}{16}''$ *on diameter*. American National Standard taper pipe threads are indicated by a note giving the nominal diameter followed by the letters **NPT** (National pipe taper), as shown in Fig. 15.21. When straight pipe threads are specified, the letters **NPS** (National pipe straight) are used. In practice, the tap drill size is normally not given in the thread note.

## 15.23 Bolts, Studs, and Screws

The term *bolt* is generally used to denote a "through bolt" that has a head on one end and is passed through clearance holes in two or more aligned parts and is threaded on the other end to receive a nut to tighten and hold the parts together, Fig. 15.23 (a). See also §§15.25 and 15.26.

A hexagon head *cap screw*, (b), is similar to a bolt, except that it generally has greater length of

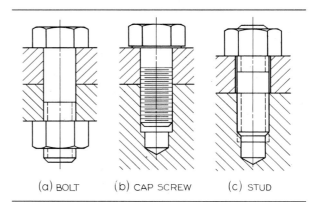

Fig. 15.23   Bolt, Cap Screw, and Stud.

thread, for when it is used without a nut, one of the members being held together is threaded to act as a nut. It is screwed on with a wrench. Cap screws are not screwed into thin materials if strength is desired. See §15.29.

A *stud*, (c), is a steel rod threaded on both ends. It is screwed into place with a pipe wrench or, pref-

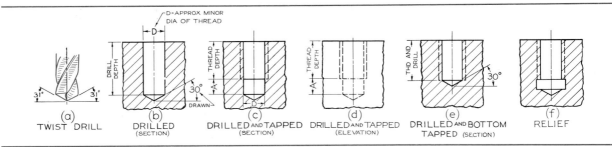

*Fig. 15.24*   Drilled and Tapped Holes.

erably, with a stud driver. As a rule, a stud is passed through a clearance hole in one member and screwed into another member, and a nut is used on the free end, as shown.

A *machine screw*, Fig. 15.33, is similar to the slotted-head cap screws but, in general, is smaller. It may be used with or without a nut.

A *set screw*, Fig. 15.34, is a screw with or without a head that is screwed through one member and whose special point is forced against another member to prevent relative motion between the two parts.

It is not customary to section bolts, nuts, screws, and similar parts when drawn in assembly, as shown in Figs. 15.23 and 15.32, because they do not themselves require sectioning for clearness. See §16.22.

## 15.24 Tapped Holes

The bottom of a drilled hole is conical in shape, as formed by the point of the twist drill, Fig. 15.24 (a) and (b). When an ordinary drill is used in connection with tapping, it is referred to as a *tap drill*. On drawings, an angle of 30° is used to approximate the actual 31°.

The thread length is the length of full or perfect threads. The tap drill depth is the depth of the cylindrical portion of the hole and does not include the cone point, (b). The portion A of the drill depth shown beyond the threads at (c) and (d) includes the several imperfect threads produced by the chamfered end of the tap. This distance A varies according to drill size and whether a plug tap, Fig. 12.16 (h), or a bottoming tap is used to finish the hole. For drawing purposes, when the tap drill depth is not specified, the distance A may be drawn equal to three schematic thread pitches, Fig. 15.9.

A tapped hole finished with a bottoming tap is drawn as shown in Fig. 15.24 (e). Blind bottoming holes should be avoided wherever possible. A better procedure is to cut a relief with its diameter slightly

greater than the major diameter of the thread, Fig. 15.24 (f).

One of the chief causes of tap breakage is insufficient tap drill depth, in which the tap is forced against a bed of chips in the bottom of the hole. Therefore, the drafter should never draw a blind hole when a through hole of not much greater length can be used. When a blind hole is necessary however, the tap drill depth should be generous. Tap drill sizes for Unified, American National, and metric threads are given in Appendix 15. It is good practice to give the tap drill size in the thread note, §15.21.

The thread length in a tapped hole depends upon the major diameter and the material being tapped. In Fig. 15.25, the minimum engagement length X, when both parts are steel, is equal to the diameter D of the thread. When a steel screw is screwed into cast iron, brass, or bronze, X = 1½D; when screwed into aluminum, zinc, or plastic, X = 2D.

Since the tapped thread length contains only full threads, it is necessary to make this length only one or two pitches beyond the end of the engaging screw. In simplified or schematic representation, the threads are omitted in the bottoms of tapped

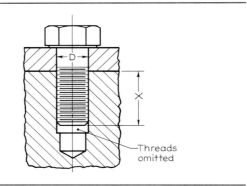

*Fig. 15.25*   Tapped Holes.

holes so as to show the ends of the screws clearly, Fig. 15.25.

When a bolt or a screw is passed through a clearance hole in one member, the hole may be drilled 0.8 mm ($\frac{1}{32}''$) larger than the screw up to $\frac{3}{8}''$ or 10 mm diameter and 1.5 mm ($\frac{1}{16}''$) larger for larger diameters. For more precise work, the clearance hole may be only 0.4 mm ($\frac{1}{64}''$) larger than the screw up to $\frac{3}{8}''$ or 10 mm diameter and 0.8 mm ($\frac{1}{32}''$) larger for larger diameters. Closer fits may be specified for special conditions.

The clearance spaces on each side of a screw or bolt need not be shown on a drawing unless it is necessary to show that there is no thread engagement, in which case the clearance spaces are drawn about 1.2 mm ($\frac{3}{64}''$) wide for clarity.

## 15.25 Standard Bolts and Nuts

*American National Standard* bolts and nuts,* metric and inch series, are produced in the hexagon form, and the square form is only produced in the inch series, Fig. 15.26. Square heads and nuts are chamfered at 30°, and hexagon heads and nuts are chamfered at 15° to 30°. Both are drawn at 30° for simplicity.

### Bolt Types

Bolts are grouped according to use: *regular* bolts for general service, and *heavy* bolts for heavier service or easier wrenching. Square bolts come only in the regular type; hexagon bolts, screws, and nuts and square nuts are standard in both types.

### Finish

Square bolts and nuts, hexagon bolts, and hexagon flat nuts are *unfinished*. Unfinished bolts and nuts are not machined on any surface except for the threads. Traditionally, hexagon bolts and hexagon nuts have been available as unfinished, semifinished, or finished. According to the latest standards, hexagon cap screws and finished hexagon bolts have been consolidated into a single product—*hex cap screws*—thus eliminating the regular semifinished hexagon bolt classification. Heavy semifinished hexagon bolts and heavy finished hexagon bolts also have been combined into a single product called *heavy hex screws*. Hexagon cap screws, heavy hexagon screws, and all hexagon nuts, except hexagon flat nuts, are considered *finished* to some degree and are characterized by a "washer face" machined or

*The ANSI standards cover several bolts and nuts. For complete details, see the standards.

otherwise formed on the bearing surface. The washer face is $\frac{1}{64}''$ thick (drawn $\frac{1}{32}''$), and its diameter is equal to $1\frac{1}{2}$ times the body diameter $D$ for the inch series.

For nuts the bearing surface also may be a circular surface produced by chamfering. Hexagon screws and hexagon nuts have closer tolerances and may have a more finished appearance, but are not completely machined. There is no difference in the drawing for the degree of finish on finished screws and nuts.

*Metric* bolts, cap screws, and nuts are produced in the hexagon form, Fig. 15.26 (a). The hexagon heads and nuts are chamfered 15° to 30°. Both are drawn at 30° for simplicity.

### Bolt Types

Metric hexagon bolts are grouped according to use: *regular* and *heavy* bolts and nuts for general service and *high-strength* bolts and nuts for structural bolting.

### Finish

Hexagon cap screws, hexagon bolts, and hexagon nuts (including metric) are considered finished to some degree and are usually characterized by a "washer face" machined or otherwise formed on the bearing surface. The washer face is approximately 0.5 mm thick (drawn 1 mm) and its diameter is equal to $1\frac{1}{2}$ times the body diameter $D$.

For hexagon nuts the circular bearing surface may be produced by chamfering. Hexagon head bolts (usually the larger sizes) are available without the washer-face bearing surface.

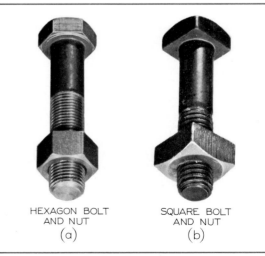

HEXAGON BOLT AND NUT
(a)

SQUARE BOLT AND NUT
(b)

*Fig. 15.26*  Standard Bolts and Nuts.

## Proportions

Sizes based on diameter $D$ of the bolt body (including metric), Fig. 15.27, which are either exact formula proportions or close approximations for drawing purposes, are as follows.

*Regular hexagon and square bolts and nuts:*

$$W = 1\tfrac{1}{2}D \qquad H = \tfrac{2}{3}D \qquad T = \tfrac{7}{8}D$$

where $W$ = width across flats, $H$ = head height, and $T$ = nut height.

*Heavy hexagon bolts and nuts and square nuts:*

$$W = 1\tfrac{1}{2}D + \tfrac{1}{8}'' \text{ (or } + 3 \text{ mm)}$$
$$H = \tfrac{2}{3}D \qquad T = D$$

The washer face is always included in the head or nut height for finished hexagon screw heads and nuts.

## Threads

Square and hex bolts, hex cap screws, and finished nuts in the inch series are usually Class 2 and may have coarse, fine, or 8-pitch threads. Unfinished nuts have coarse threads, Class 2B. For diameter and pitch specifications for metric threads, see Appendix 18.

## Thread Lengths

*For bolts or screws up to 6" (150 mm) in length:*

$$\text{Thread length} = 2D + \tfrac{1}{4}'' \text{ (or } + 6 \text{ mm)}$$

*For bolts or screws over 6" in length,*

$$\text{Thread length} = 2D + \tfrac{1}{2}'' \text{ (or } + 12 \text{ mm)}$$

Fasteners too short for these formulas are threaded as close to the head as practicable. For drawing purposes, this may be taken as three pitches, approximately. The threaded end may be rounded or chamfered, but is usually drawn with a 45° chamfer from the thread depth, Fig. 15.27.

## Bolt Lengths

Lengths of bolts have not been standardized because of the endless variety required by industry. Short bolts are typically available in standard length

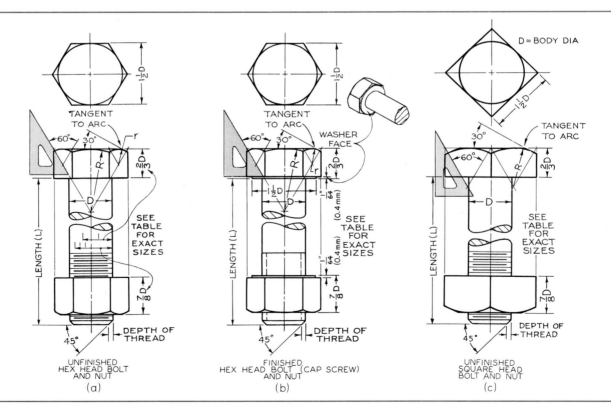

**Fig. 15.27**   Bolt Proportions (Regular).

increments of $\frac{1}{4}''$ (6 mm), while long bolts come in increments of $\frac{1}{2}$ to $1''$ (12 to 25 mm). For dimensions of standard bolts and nuts, see Appendix 18.

## 15.26 To Draw Standard Bolts

In practice, standard bolts and nuts are not shown on detail drawings unless they are to be altered, but they appear so frequently on assembly drawings that

a suitable but rapid method of drawing them must be used. Time-saving templates are available, or they may be drawn from exact dimensions taken from tables (see Appendix 18) if accuracy is important, as in figuring clearances, but in the great majority of cases the conventional representation, in which proportions based upon the body diameter are used, will be sufficient, and a considerable amount of time may be saved. Three typical bolts illustrating the use of these proportions for the regular bolts are shown in Fig. 15.27.

Although the curves produced by the chamfer on the bolt heads and nuts are hyperbolas, in practice these curves are always represented approximately by means of circular arcs, as shown in Fig. 15.27.

Generally, bolt heads and nuts should be drawn "across corners" in all views, regardless of projection. This conventional violation of projection is used to prevent confusion between the square and hexagon heads and nuts and to show actual clearances. Only when there is a special reason should bolt heads and nuts be drawn across flats. In such cases, the conventional proportions shown in Fig. 15.28 are used.

Steps in drawing hexagon bolts, cap screws, and nuts are illustrated in Fig. 15.29, and those for square bolts and nuts in Fig. 15.30. Before drawing a bolt, the diameter of the bolt, the length (from the

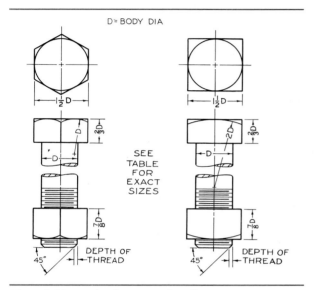

*Fig. 15.28*    Bolts "Across Flats."

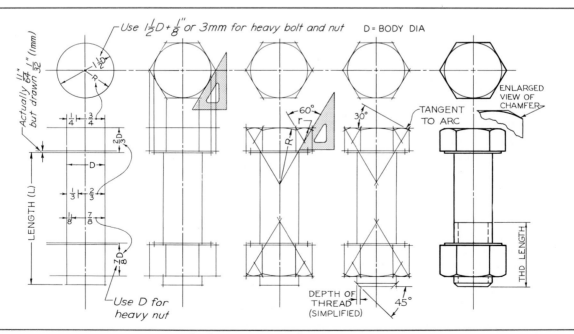

*Fig. 15.29*    Steps in Drawing a Finished Hexagon Head Bolt (Cap Screw) and Hexagon Nut.

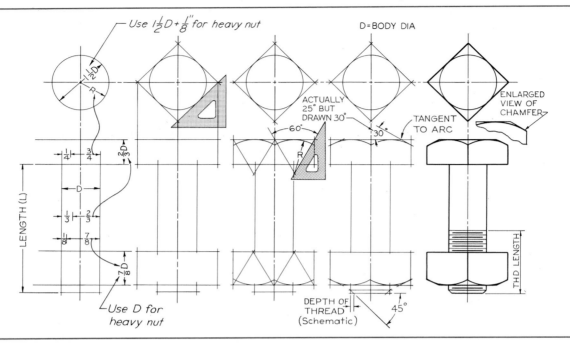

*Fig. 15.30*  Steps in Drawing Square-Head Bolt and Square Nut.

underside of the bearing surface to the tip), the style of head (square or hexagon), and the type (regular or heavy), as well as the finish, must be known.

If only the longitudinal view of a bolt is needed, it is necessary to draw only the lower half of the top views in Figs. 15.29 and 15.30 *with light construction lines* in order to project the corners of the hexagon or square to the front view. These construction lines may then be erased if desired.

The head and nut heights can be spaced off with the dividers on the shaft diameter and then transferred as shown in both figures, or the scale may be used as in Fig. 5.15. The heights should not be determined by arithmetic.

The $\frac{1}{64}''$ (0.4 mm) washer face has a diameter equal to the distance across flats of the bolt head or nut. It appears only on the metric and finished hexagon screws or nuts, the washer face thickness being drawn at $\frac{1}{32}''$ (1 mm) for clearness. The $\frac{1}{32}''$ (1 mm) is included in the head or nut height.

Threads should be drawn in simplified or schematic form for body diameters of 1" (25 mm) or less on the drawing, Fig. 15.9 (b) or (d), and by detailed representation for larger diameters, §§15.7 and 15.8. The threaded end of the screw should be chamfered at 45° from the schematic thread depth, Fig. 15.9 (a).

On drawings of small bolts or nuts under approximately $\frac{1}{2}''$ diameter (12 mm), where the cham-

fer is hardly noticeable, the chamfer on the head or nut may be omitted in the longitudinal view.

Many styles of templates are available for saving time in drawing bolt heads and nuts. One of these, the Draftsquare, is illustrated in Fig. 2.68 (b).

## 15.27 Specifications for Bolts and Nuts

In specifying bolts in parts lists, in correspondence, or elsewhere, the following information must be covered in order:

1.  Nominal size of bolt body
2.  Thread specification or thread note (see §15.21)
3.  Length of bolt
4.  Finish of bolt
5.  Style of head
6.  Name

**EXAMPLE (Complete)**

$$\frac{3}{4}\text{--}10 \text{ UNC--2A} \times 2\frac{1}{2} \text{ HEXAGON CAP SCREW}$$

**EXAMPLE (Abbreviated)**

$$\frac{3}{4} \times 2\frac{1}{2} \text{ HEX CAP SCR}$$

**EXAMPLE (Metric)**

  M8 × 1.25–40, HEX CAP SCR

Nuts may be specified as follows:

**EXAMPLE (Complete)**

  $\frac{5}{8}$–11 UNC–2B SQUARE NUT

**EXAMPLE (Abbreviated)** $\frac{5}{8}$ SQ NUT

**EXAMPLE (Metric)** M8 × 1.25 HEX NUT

For either bolts or nuts, the words REGULAR or GENERAL PURPOSE are assumed if omitted from the specification. If the heavy series is intended, the word HEAVY should appear as the first word in the name of the fastener. Likewise, HIGH STRENGTH STRUCTURAL should be indicated for such metric fasteners. However, the number of the specific ISO standard is often included in the metric specifications; for example, HEXAGON NUT ISO 4032 M12

× 1.75. Similarly, finish need not be mentioned if the fastener or nut is correctly named. See §15.25.

### 15.28 Locknuts and Locking Devices

Many types of special nuts and devices to prevent nuts from unscrewing are available, some of the most common of which are illustrated in Fig. 15.31. The American National Standard *jam nuts,* (a) and (b), are the same as the hexagon or hexagon flat nuts, except that they are thinner. The application at (b), where the larger nut is on top and is screwed on more tightly, is recommended. They are the same distance across flats as the corresponding hexagon nuts ($1\frac{1}{2}D$ or $1\frac{1}{2}D + \frac{1}{8}''$). They are slightly over $\frac{1}{2}D$ in thickness, but are drawn $\frac{1}{2}D$ for simplicity. They are available with or without the washer face in the regular and heavy types. The tops of all are flat and chamfered at 30°, and the finished forms have either a washer face or a chamfered bearing surface.

The lock washer, shown at (c), and the cotter pin, (e), (g), and (h), are very common. See Appendixes 27 and 30. The set screw, (f), is often made to press against a plug of softer material, such as brass, which in turn presses against the threads without deforming them.

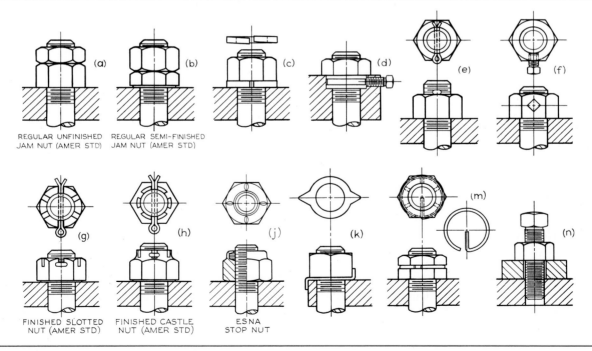

**Fig. 15.31** Locknuts and Locking Devices.

For use with cotter pins (see Appendix 30), a hex slotted nut, (g), and a hex castle nut, (h), as well as a hex thick slotted nut and a heavy hex thick slotted nut are recommended.

Similar metric locknuts and locking devices are available. See fastener catalogs for details.

## 15.29 Standard Cap Screws

Five types of American National Standard cap screws are shown in Fig. 15.32. The first four of these have standard heads, while the socket head cap screws, (e), have several different shapes of round heads and sockets. Cap screws are regularly produced in finished form and are used on machine tools and other machines, for which accuracy and appearance are important. The ranges of sizes and exact dimensions are given in Appendixes 18 and 19. The hexagon head cap screw and hex socket head

cap screw in several forms are available in metric. See Appendix 18.

Cap screws ordinarily pass through a clearance hole in one member, as explained in §15.24, and screw into another. The clearance hole need not be shown on the drawing when the presence of the unthreaded clearance hole is obvious.

Cap screws are inferior to studs if frequent removal is necessary; hence, they are used on machines requiring few adjustments. The slotted or socket-type heads are best under crowded conditions.

The actual standard dimensions may be used in drawing the cap screws whenever exact sizes are necessary, but this is seldom the case. In Fig. 15.32 the dimensions are given in terms of body diameter *D*, and they closely conform to the actual dimensions. The resulting drawings are almost exact reproductions and are easy to draw. The hexagonal

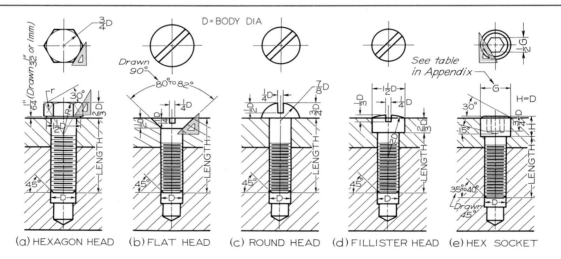

(a) HEXAGON HEAD     (b) FLAT HEAD     (c) ROUND HEAD     (d) FILLISTER HEAD     (e) HEX SOCKET

***Hexagon Head Screws:*** Coarse, Fine, or 8-Thread Series, 2A. Thread length $= 2D + \frac{1}{4}''$ up to 6″ long and $2D + \frac{1}{2}''$ if over 6″ long. For screws too short for formula, threads extend to within $2\frac{1}{2}$ threads of the head for diameters up to 1″. Screw lengths not standardized. For suggested lengths for metric Hexagon Head Screws, see Appendix 15.

***Slotted Head Screws:*** Coarse, Fine, or 8-Thread Series, 2A. Thread length $= 2D + \frac{1}{4}''$. Screw lengths not standardized. For screws too short for formula, threads extend to within $2\frac{1}{2}$ threads of the head.

***Hexagon Socket Screws:*** Coarse or Fine Threads, 3A. Coarse thread length $= 2D + \frac{1}{2}''$ where this would be over $\frac{1}{2}L$; otherwise thread length $= \frac{1}{2}L$. Fine thread length $= 1\frac{1}{2}D + \frac{1}{2}''$ where this would be over $\frac{3}{8}L$; otherwise thread length $= \frac{3}{8}L$. Increments in screw lengths $= \frac{1}{8}''$ for screws $\frac{1}{4}''$ to 1″ long, $\frac{1}{4}''$ for screws 1″ to 3″ long, and $\frac{1}{2}''$ for screws $3\frac{1}{2}''$ to 6″ long.

*Fig. 15.32* Standard Cap Screws. See Appendixes 18 and 19.

head cap screw is drawn in the manner shown in Fig. 15.29. The points are drawn chamfered at 45° from the schematic thread depth.

For correct representation of tapped holes, see §15.24. For information on drilled, countersunk, or counterbored holes, see §§7.33 and 12.16.

In an assembly section, it is customary not to section screws, bolts, shafts, or other solid parts whose center lines lie in the cutting plane. Such parts in themselves do not require sectioning and are, therefore, shown "in the round," Fig. 15.32 and §16.22.

Note that screwdriver slots are drawn at 45° in the circular views of the heads, without regard to true projection, and that threads in the bottom of the tapped holes are omitted so that the ends of the screws may be clearly seen. A typical cap screw note is as follows:

**EXAMPLE (Complete)**

$\frac{3}{8}$–16 UNC–2A × 2$\frac{1}{2}$ HEXAGON HEAD CAP SCREW

**EXAMPLE (Abbreviated)**

$\frac{3}{8}$ × 2$\frac{1}{2}$ HEX HD CAP SCR

**EXAMPLE (Metric)**

M20 × 2.5 × 80 HEX HD CAP SCR

## 15.30 Standard Machine Screws

Machine screws are similar to cap screws but are in general smaller (.060″ to .750″ dia). There are eight ANSI-approved forms of heads shown in Appendix 20. The hexagonal head may be slotted if desired. All others are available in either slotted or recessed-head forms. Standard machine screws are regularly produced with a naturally bright finish, not heat treated, and are regularly supplied with plain-sheared ends, not chamfered. For similar metric machine screw forms and specifications, see Appendix 20.

Machine screws are particularly adapted to screwing into thin materials, and all the smaller-numbered screws are threaded nearly to the head. They are used extensively in firearms, jigs, fixtures, and dies. Machine screw nuts are used mainly on the round head, pan head, and flat head types, and are usually hexagonal in form.

Exact dimensions of machine screws are given in Appendix 20, but they are seldom needed for drawing purposes. The four most common types of machine screws are shown in Fig. 15.33, where proportions based on diameter $D$ conform closely to the actual dimensions and produce almost exact drawings. Clearance holes and counterbores should be made slightly larger than the screws, as explained in §15.24.

Note that the threads in the bottom of the tapped holes are omitted so that the ends of the screws will be clearly seen. Observe also that it is conventional practice to draw the screwdriver slots at 45° in the circular view without regard to true projection.

A typical machine screw note is as follows:

**EXAMPLE (Complete)**

No. 10 (.1900)–32 NF–3 × $\frac{5}{8}$ FILLISTER HEAD

MACHINE SCREW

**EXAMPLE (Abbreviated)**

No. 10 (.1900) × $\frac{5}{8}$ FILL HD MACH SCR

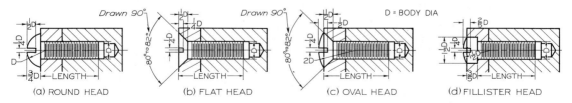

(a) ROUND HEAD    (b) FLAT HEAD    (c) OVAL HEAD    (d) FILLISTER HEAD

*Threads:* National Coarse or Fine, Class 2 fit. On Screws 2″ long or less, threads extend to within 2 threads of head; on longer screws thread length = 1$\frac{3}{4}$″. Screw lengths not standardized.

*Fig. 15.33* Standard Machine Screws. See Appendix 20.

**EXAMPLE (Metric)**

$$M8 \times 1.25 \times 30 \text{ SLOTTED PAN HEAD}$$
$$\text{MACHINE SCREW}$$

## 15.31 Standard Set Screws

The function of set screws, Fig. 15.34 (a), is to prevent relative motion, usually rotary, between two parts, such as the movement of the hub of a pulley on a shaft. A set screw is screwed into one part so that its point bears firmly against another part. If the point of the set screw is cupped, (e), or if a flat is milled on the shaft, (a), the screw will hold much more firmly. Obviously, set screws are not efficient when the load is heavy or is suddenly applied. Usually they are manufactured of steel and case hardened.

The American National Standard square head set screw and slotted headless set screw are shown in Fig. 15.34 (a) and (b). Two American National Standard socket set screws are illustrated at (c) and (d). American National Standard set screw points are shown from (e) to (k). The headless set screws have come into greater use because the projecting head of headed set screws has caused many industrial casualties; this has resulted in legislation prohibiting their use in many states.

Most of the dimensions in Fig. 15.34 are American National Standard formula dimensions, and the resulting drawings are almost exact representations.

Metric hexagon socket headless set screws with the full range of points are available and are represented in the same manner as given in Fig. 15.34. Nominal diameters of metric hex socket set screws are 1.6, 2, 2.5, 3, 4, 5, 6, 8, 10, 12, 16, 20, and 24 mm.

Square head set screws have coarse, fine, or 8-pitch threads, Class 2A, but are usually furnished with coarse threads, since the square head set screw is generally used on the rougher grades of work. Slotted headless and socket set screws have coarse or fine threads, Class 3A.

Nominal diameters of set screws range from number 0 up through 2″, set screw lengths are standardized in increments of $\frac{1}{32}$″ to 1″ depending upon the overall length of the set screw.

Metric set screw length increments range from 0.5 to 4 mm, again depending upon overall screw length.

Set screws are specified as follows:

**EXAMPLE (Complete)**

$$\frac{3}{8}\text{--16 UNC--2a} \times \frac{3}{4} \text{ SQUARE HEAD}$$
$$\text{FLAT POINT SET SCREW}$$

**EXAMPLES (Abbreviated)**

$$\frac{3}{8} \times 1\frac{1}{4} \text{ SQ HD FL PT SS}$$

$$\frac{7}{16} \times \frac{3}{4} \text{ HEX SOC CUP PT SS}$$

$$\frac{1}{4}\text{--20 UNC 2A} \times \frac{5}{8} \text{ SLOT. HDLS CONE PT SS}$$

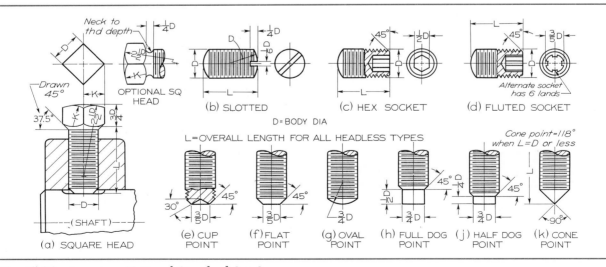

**Fig. 15.34** American National Standard Set Screws.

EXAMPLE (Metric)

M10 × 1.5 × 12 HEX SOCKET HEAD SET SCREW

## 15.32 American National Standard Wood Screws

Wood screws with three types of heads—flat, round, and oval—have been standardized, Fig. 15.35. The dimensions shown closely approximate the actual dimensions and are more than sufficiently accurate for use on drawings.

The Phillips style recessed head is also available on several types of fasteners as well as wood screws. Three styles of cross recesses have been standardized by the ANSI. Many examples may be seen on the automobile. A special screwdriver is used, as shown in Fig. 15.36 (q), and results in rapid assembly without damage to the head.

## 15.33 Miscellaneous Fasteners

Many other types of fasteners have been devised for specialized uses. Some of the more common types are shown in Fig. 15.36. A number of these are American National Standard round head bolts including carriage, button head, step, and countersunk bolts.

Aero-thread inserts, or heli-coil inserts, as shown at (p), are shaped like a spring except that the cross section of the wire conforms to threads on the screw and in the hole. These are made of phosphor bronze or stainless steel, and they provide a hard, smooth protective lining for tapped threads in soft metals and in plastics.

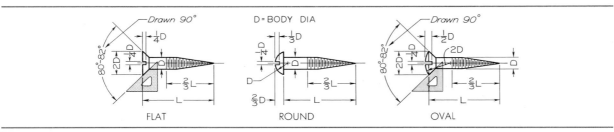

*Fig. 15.35* American National Standard Wood Screws.

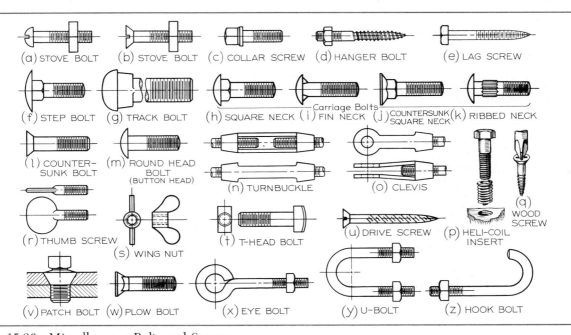

*Fig. 15.36* Miscellaneous Bolts and Screws.

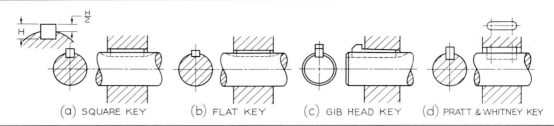

*Fig. 15.37*   Square and Flat Keys.

## 15.34 Keys

*Keys* are used to prevent relative movement between shafts and wheels, couplings, cranks, and similar machine parts attached to or supported by shafts, Fig. 15.37.

For heavy duty, rectangular keys (flat or square) are suitable, and sometimes two rectangular keys are necessary for one connection. For even stronger connections, interlocking *splines* may be machined on the shaft and in the hole. See Fig. 14.9.

A *square key* is shown in Fig. 15.37 (a) and a *flat key* in (b). The widths of keys generally used are about one-fourth the shaft diameter. In either case, one-half the key is sunk into the shaft. The depth of the keyway or the keyseat is measured on the side — not the center, (a). Square and flat keys may have the top surface tapered $\frac{1}{8}''$ per foot, in which case they become square taper or flat taper keys.

A rectangular key that prevents rotary motion but permits relative longitudinal motion is a *feather key,* and is usually provided with *gib heads,* or otherwise fastened so it cannot slip out of the keyway. A *gib head key* is shown at (c). It is exactly the same as the square taper or flat taper key, except that a gib head, which provides for easy removal, is

added. Square and flat keys are made from cold-finished stock and are not machined. For dimensions, see Appendix 21.

The *Pratt & Whitney key,* (d), is rectangular in shape, with semicylindrical ends. Two-thirds of the height of the P & W key is sunk into the shaft keyseat. See Appendix 25.

The *Woodruff key* is semicircular in shape, Fig. 15.38. The key fits into a semicircular key slot cut with a Woodruff cutter, as shown, and the top of the key fits into a plain rectangular keyway. Sizes of keys for given shaft diameters are not standardized, but for average conditions it will be found satisfactory to select a key whose diameter is approximately equal to the shaft diameter. For dimensions, see Appendix 23.

A *keyseat* is in a shaft; a *keyway* is in the hub or surrounding part.

Typical specifications for keys are

$$\frac{1}{4} \times 1\frac{1}{2} \text{ SQ KEY}$$

$$\text{No. 204 WOODRUFF KEY}$$

$$\frac{1}{4} \times \frac{1}{16} \times 1\frac{1}{2} \text{ FLAT KEY}$$

$$\text{No. 10 P \& W KEY}$$

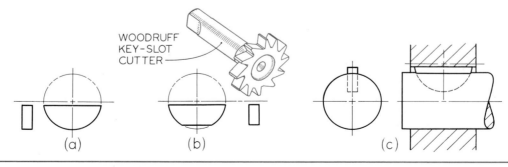

*Fig. 15.38*   Woodruff Keys and Key-Slot Cutter.

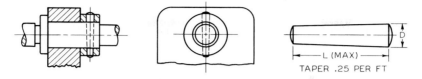

*Fig. 15.39* Taper Pin.

Notes for nominal specifications of keyways and keyseats are shown in Fig. 13.44 (o), (p), (r) and (x). For production work, keyways and keyseats should be dimensioned as shown in Fig. 13.48. See manufacturers' catalogs for specifications for metric counterparts.

## 15.35 Machine Pins
Machine pins include taper pins, straight pins, dowel pins, clevis pins, and cotter pins. For light work, the taper pin is effective for fastening hubs or collars to shafts, as shown in Fig. 15.39, in which the hole through the collar and shaft is drilled and reamed when the parts are assembled. For slightly heavier duty, the taper pin may be used parallel to the shaft as for square keys. See Appendix 29.

Dowel pins are cylindrical or conical in shape and are used for a variety of purposes, chief of which is to keep two parts in a fixed position or to preserve alignment. The dowel pin is most commonly used and is recommended where accurate alignment is essential. Dowel pins are usually made of steel and are hardened and ground in a centerless grinder.

## 15.36 Rivets
*Rivets* are regarded as permanent fastenings as distinguished from removable fastenings, such as bolts and screws. They are generally used to hold sheet metal or rolled steel shapes together and are made of wrought iron, carbon steel, copper, or occasionally other metals.

## 15.37 Springs
A *spring* is a mechanical device designed to store energy when deflected and to return the equivalent amount of energy when released. Springs are classified as *helical springs*, Fig. 15.40, or *flat springs*, Fig. 15.44.

There are three types of helical springs: *compression springs,* which offer resistance to a compressive force, Fig. 15.40 (a) to (e), *extension*

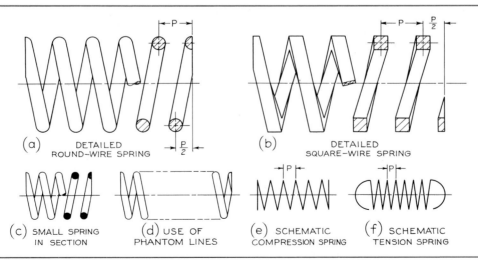

*Fig. 15.40* Helical Springs.

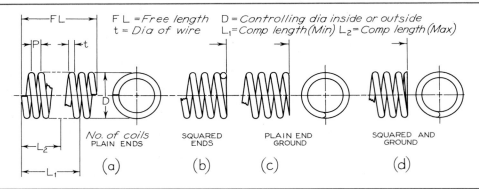

FL = Free length   D = Controlling dia inside or outside
t = Dia of wire    L₁ = Comp length (Min)   L₂ = Comp length (Max)

No. of coils
PLAIN ENDS

SQUARED
ENDS

PLAIN END
GROUND

SQUARED AND
GROUND

(a)        (b)        (c)        (d)

*Fig. 15.41*  Compression Springs.

*springs*, which offer resistance to a pulling force, Fig. 15.42, and *torsion springs*, which offer resistance to a torque load or twisting force, Fig. 15.43.

On working drawings, true projections of helical springs are never drawn because of the labor involved. As in the drawing of screw threads, the detailed and schematic methods, employing straight lines in place of helical curves, are used as shown in Fig. 15.40.

The elevation view of the square-wire spring is similar to the square thread with the core of the shaft removed, Fig. 15.12. Standard section lining is used if the areas in section are large, as shown in Fig. 15.40 (a) and (b). If these areas are small, the sectioned areas may be made solid black, (c). In cases where a complete picture of the spring is not necessary, phantom lines may be used to save time

in drawing the coils, (d). If the drawing of the spring is too small to be represented by the outlines of the wire, it may be drawn by the schematic method, in which single lines are used, (e) and (f).

Compression springs have *plain ends*, Fig. 15.41 (a), or *squared (closed) ends*, (b). The ends may be *ground* as at (c), or both *squared and ground* as at (d). Required dimensions are indicated in the figure. When required, RH or LH is specified.

An extension spring may have any one of many types of ends, and it is therefore necessary to draw the spring or at least the ends and a few adjacent coils, Fig. 15.42. Note the use of phantom lines to show the continuity of coils.

A typical torsion spring drawing is shown in Fig.

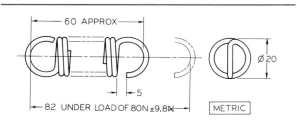

60 APPROX

Ø 20

5

82 UNDER LOAD OF 80N ±9.8N

METRIC

MATERIAL:  2.00 OIL TEMPERED SPRING STEEL WIRE
          14.5 COILS RIGHT HAND
          MACHINE LOOP AND HOOK IN LINE
          SPRING MUST EXTEND TO 110 WITHOUT SET
FINISH:  BLACK JAPAN

*Fig. 15.42*  Extension Spring Drawing.

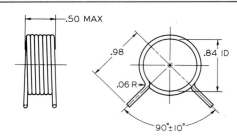

.50 MAX

.98

.84 ID

.06 R

90° ±10°

MATERIAL : .059 MUSIC WIRE
          6.75 COILS RIGHT HAND NO INITIAL TENSION
TORQUE : 2.50 INCH LB AT 155° DEFLECTION SPRING MUST
          DEFLECT 180° WITHOUT PERMANENT SET AND
          MUST OPERATE FREELY ON .75 DIAMETER SHAFT
FINISH : CADMIUM OR ZINC PLATE

*Fig. 15.43*  Torsion Spring Drawing.

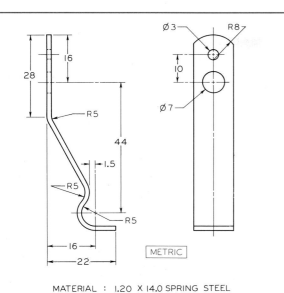

MATERIAL : 1.20 X 14.0 SPRING STEEL
HEAT TREAT : 44-48 C ROCKWELL
FINISH : BLACK OXIDE AND OIL

**Fig. 15.44** Flat Spring Drawing.

15.43, and a typical flat spring drawing is shown in Fig. 15.44. Other types of flat springs are *power springs* (or flat coil springs), *Belleville springs* (like spring washers), and *leaf springs* (commonly used in automobiles).

### 15.38 To Draw Helical Springs

The construction for a schematic elevation view of a compression spring having six total coils is shown in Fig. 15.45 (a). Since the ends are closed, or squared, two of the six coils are "dead" coils, leaving only four full pitches to be set off along the top of the spring, as shown.

If there are $6\frac{1}{2}$ total coils, as at (b), the $\frac{P}{2}$ spacings will be on opposite sides of the spring. The construction of an extension spring with six active coils and loop ends is shown at (c).

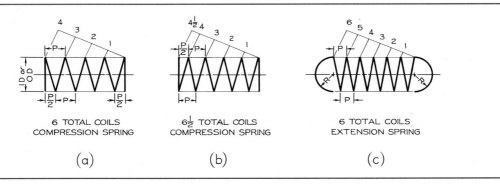

**Fig. 15.45** Schematic Spring Representation.

# *Thread and Fastener Problems*

It is expected that the student will make use of the information in this chapter and in various manufacturers' catalogs in connection with the working drawings at the end of the next chapter, where many different kinds of threads and fasteners are required. However, several problems are included here for specific assignment in this area, Figs. 15.46 to 15.49. All are to be drawn on tracing paper or detail paper, size B or A3 sheet (see inside back cover).

Thread and fastener problems in convenient form for solution may be found in *Principles of Engineering Graphics Problems* by Spencer, Hill, Loving, and Dygdon, a workbook designed to accompany this text that is also published by Macmillan Publishing Company.

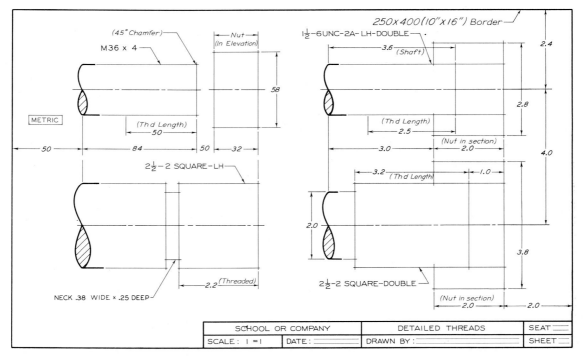

**Fig. 15.46** Draw specified detailed threads arranged as shown (Layout B–3 or A3–3). Omit all dimensions and notes given in inclined letters. Letter only the thread notes and the title strip.

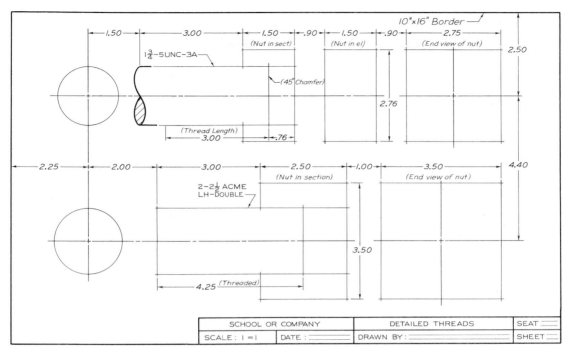

**Fig. 15.47** Draw specified detailed threads, arranged as shown (Layout B–3 or A3–3). Omit all dimensions and notes given in inclined letters. Letter only the thread notes and the title strip.

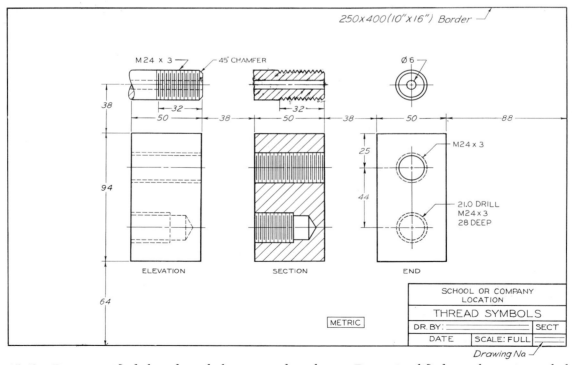

**Fig. 15.48** Draw specified thread symbols, arranged as shown. Draw simplified or schematic symbols, as assigned by instructor (Layout B–5 or A3–5). Omit all dimensions and notes given in inclined letters. Letter only the drill and thread notes, the titles of the views, and the title strip.

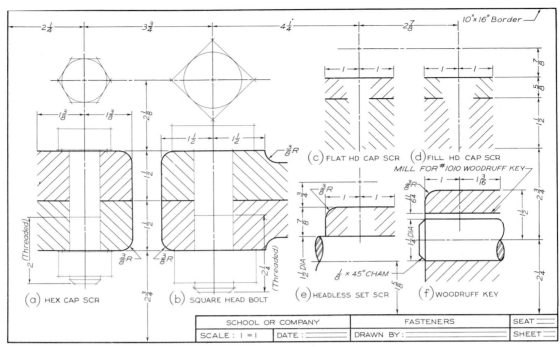

**Fig. 15.49** Draw fasteners, arranged as shown (Layout B–3 or A3–3). At (a) draw $\frac{7}{8}$–9 UNC–2A × 4 Hex Cap Screw. At (b) draw $1\frac{1}{8}$–7 UNC–2A × $4\frac{1}{4}$ Sq Hd Bolt. At (c) draw $\frac{3}{8}$–16 UNC–2A × $1\frac{1}{2}$ Flat Hd Cap Screw. At (d) draw $\frac{7}{16}$–14 UNC–2A × 1 Fill Hd Cap Screw. At (e) draw $\frac{1}{2}$ × 1 Headless Slotted Set Screw. At (f) draw front view of No. 1010 Woodruff Key. Draw simplified or schematic thread symbols as assigned. Letter titles under each figure as shown.

# CHAPTER
# 16

# Design and Working Drawings

The many products, systems, and services that enrich our standard of living are largely the result of the design activities of engineers. It is principally this design activity that distinguishes engineering from science and research; the engineer is a designer, a creator, or a "builder."

The design process is an exciting and challenging effort and the engineer-designer relies heavily upon graphics as a means to create, record, analyze, and communicate to others design concepts or ideas. The ability to communicate verbally, symbolically, and graphically is essential.

## 16.1 Design Sources

There are two general types of design: *empirical design*, sometimes referred to as conceptual design, and *scientific design*. In scientific design, use is made of the principles of physics, mathematics, chemistry, mechanics, and other sciences in the new or revised design of devices, structures, or systems intended to function under specific conditions. In empirical design, much use is made of the information in handbooks, which in turn has been learned by experience. Nearly all technical design is a combination of scientific and empirical design. Therefore, a competent designer has both adequate engineering and scientific knowledge and access to the many handbooks related to the field.

You may not yet have acquired the necessary educational background and experience to undertake a sophisticated design. And you probably do not have access to the variety of experts on materials, production methods, business economics, legal considerations, or other vital areas that probably would be available in a large company. Nevertheless, through conscientious application of your design ability and the sources of information that are available, you can go far toward the creation or improvement of some device, system, or service that *works* and is uniquely your own. The working models of the Postal Scale and the Educational Toys in Fig. 16.1 are examples of student design efforts.

## 16.2 Design Concepts

New ideas or design concepts must exist initially in the mind of the designer. In order to capture, preserve, and develop these ideas, the designer makes liberal use of freehand sketches of views and pictorials, Chapter 6. These sketches are revised or redrawn as the concept is developed. All sketches should be preserved for reference and dated as a record of the development of the design.

At some point in the development of the idea, you will probably find it to your advantage to pool your ideas with those of others and begin working in a team effort; such a team may include others familiar with problems of materials, production, marketing, and so on. In industry, the project becomes a team effort long before the product is produced and marketed. Obviously, the design process is not a haphazard operation of an inventor working in a garage or basement, although it might well begin in that manner. Industry could not long survive if its products were determined in a haphazard manner. Hence, nearly all successful companies support a well-organized design effort, and the vitality of the company depends to a large extent on the *planned* output of its designers.

Since it is important for you to be able to work effectively with others in a group or team, you must be able to express yourself clearly and concisely. Do not underestimate the importance of your communication skills, your ability to express your ideas verbally (written and spoken), symbolically (equations, formulas, etc.), and *graphically*. See page 430 for suggested format of a student design report.

The graphical skills include the ability to present information and ideas clearly and effectively in the form of sketches, drawings, graphs, and so on. This textbook is dedicated to helping you develop your communication skills in graphics.

(a) POSTAL SCALE                (b) EDUCATIONAL TOYS

*Fig. 16.1*  Working Models of Student Designs.

## 16.3 The Design Process

Design is the ability to combine ideas, scientific principles, resources, and often existing products into a solution of a problem. This ability to solve problems in design is the result of an organized and orderly approach to the problem known as the *design process*.

The design process leading to manufacturing, assembly, marketing, service, and the many activities necessary for a successful product is composed of several easily recognized phases. Although many industrial groups may identify them in their own particular way, a convenient procedure for the design of a new or improved product is in five stages as follows:

1. Identification of problem.
2. Concepts and ideas.
3. Compromise solutions.
4. Models and/or prototypes.
5. Production and/or working drawings.

Ideally, the design moves through the stages as shown in Fig. 16.2, but if a particular stage proves unsatisfactory, it may be necessary to return to a previous stage and repeat the procedure as indicated by the dashed-line paths. This repetitive procedure is often referred to as *looping*.

## 16.4 Stage 1—Identification of Problem

The design activity begins with the recognition and determination of a need or want for a product, service, or system and the economic feasibility of fulfilling this need. Engineering design problems may range from the need for a simple and inexpensive container opener such as the pull tab commonly used on beverage cans, Fig. 16.3, to the more complex problems associated with the needs of air and ground travel, space exploration, environmental control, and so forth. Although the product may be very simple, such as the pull tab on a beverage can,

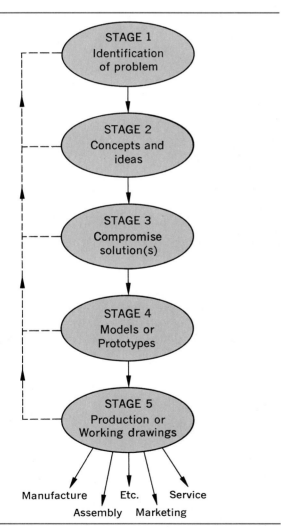

*Fig. 16.2*   Stages of the Design Process.

*Fig. 16.3*   Pull-Tab Can Opener. *John Schultz— PAR/NYC*

*Fig. 16.4*   Airport Transit System.
*Courtesy of Westinghouse Electric Corp.*

*Fig. 16.5*   Lunar Roving Vehicle.
*Courtesy of The Boeing Co.*

the production tools and dies require considerable engineering and design effort. The airport automated transit system design, Fig. 16.4, meets the need of moving people efficiently between the terminal areas. The system is capable of moving 3300 people every 10 minutes.

The Lunar Roving Vehicle, Fig. 16.5, is a solution to a need in the space program to explore larger areas of the lunar surface. This vehicle is the end result of a great deal of design work associated with the support systems and the related hardware.

At the problem identification stage, either the designer recognizes that there does exist a need requiring a design solution or, perhaps more often, a directive is received to that effect from management. No attempt is made at this time to set goals or criteria for the solution.

Information concerning the identified problem becomes the basis for a problem proposal, which may be a paragraph or a multipage report presented for formal consideration. A proposal is a plan for action that will be followed to solve the problem. The proposal, if approved, becomes an agreement to follow the plan. In the classroom, the agreement is made between you and your instructor on the identification of the problem and your proposed plan of action.

Following approval of the proposal, further aspects of the problem are explored. Available information related to the problem is collected and parameters or guide lines for time, cost, function, and so on are defined within which you will work. For example: What is the design expected to do? What is the estimated cost limit? What is the market potential? What can it be sold for? When will the prototype be ready for testing? When must production drawings be ready? When will production begin? When will the product be available on the market?

The parameters of a design problem, including the time schedule, are established at this stage. Nearly all designs represent a compromise, and the amount of time budgeted to a project is no exception.

## 16.5 Stage 2—Concepts and Ideas

At this stage, many ideas are collected, "wild" or otherwise, for possible solutions to the problem. The ideas are broad and unrestricted to permit the possibility of new and unique solutions. This compilation of ideas may be from individuals or they may come from group or team "brainstorming" sessions wherein a suggested idea often generates many more ideas from the group. As the ideas are elicited, they are written down and/or recorded in graphic form (multiview or pictorial sketches) for future consideration and refinement.

The larger the collection of ideas, the greater are the chances of finding one or more ideas suitable for further refinement. All sources of ideas such as technical literature, reports, design and trade journals, patents, and existing products are explored. The Greenfield Village Museum in Dearborn, Michigan, the Museum of Science and Industry in Chicago, trade exhibitions, large hardware and supply stores, mail order catalogs, and so on are all excellent sources for ideas. Another source is the user of an existing product who, often has suggestions for improvement. The potential user may be helpful with specific reactions to the proposed solution.

No attempt is made to evaluate ideas at this stage. All notes and sketches are signed, dated, and retained for possible patent proof.

## 16.6 Stage 3—Compromise Solutions

Various features of the many conceptual ideas generated in the preceding stages are selected after careful consideration and combined into one or more promising compromise solutions. At this point the best of the solutions is evaluated in detail, and attempts are made to simplify it and thereby make manufacture and performance more efficient.

The sketches of the design are often followed by a study of suitable materials and of motion problems that may be involved. What source of power is to be used—manual, electric motor, or what? What type of motion is needed? Is it necessary to translate rotary motion to linear motion or vice versa? Many of these problems are solved graphically by means of a schematic drawing in which various parts are shown in skeleton form. A pulley is represented by a circle, meshing gears by tangent pitch circles, an arm by a single line, paths of motion by center lines, and so on. At this time, too, certain basic calculations, such as those related to velocity and acceleration, may be made, if required.

These preliminary studies are followed by a *design layout*, made with instruments, or a *layout sketch* from which a drafter makes an accurate to-scale instrumental drawing so that actual sizes and proportions can be clearly visualized, Fig. 16.6. At this time all parts are carefully designed for strength and function. Costs are constantly kept in mind, for no matter how well the device performs, it must be built to sell for a profit or the time and development costs will have been a loss.

During the layout process, great reliance is placed upon what has gone on before. Experience provides a sense of proportion and size that permits the noncritical or more standard features to be designed by eye or with the aid of empirical data. Stress analysis and detailed computation may be necessary in connection with high speeds, heavy loads, or special requirements or conditions.

As shown in Fig. 16.6, the layout is an assembly showing how parts fit together and the basic proportions of the various parts. Auxiliary views or sections are used, if necessary. Section lining may be used sparingly to save time. All lines should be sharp and the drawing made as accurately as possible, since most dimensions are omitted except for a few key ones that will be used in the detail or working drawings for production. Any notes or other information related to the detail drawing should be given on the layout.

Special attention is given to clearances of moving parts, ease of assembly, and serviceability. Standard parts are used wherever possible for it is less costly to use stock items. Most companies maintain some form of an *engineering standards manual*, which contains much of the empirical data and detailed information that is regarded as "company standard." Materials and costs are carefully considered. Although functional considerations must come first, manufacturing problems must be kept constantly in mind.

A great many design problems are concerned with the improvement of an existing product or with the redesign of a device from a different approach

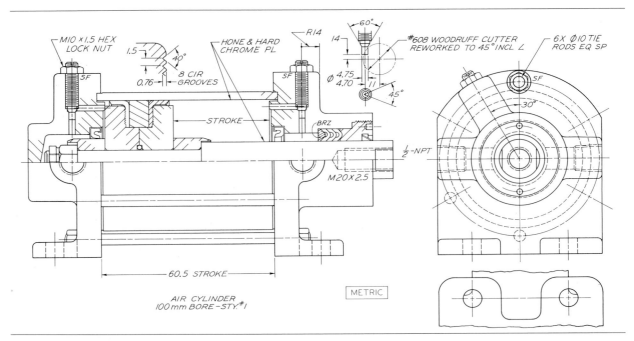

*Fig. 16.6*   Design Layout.

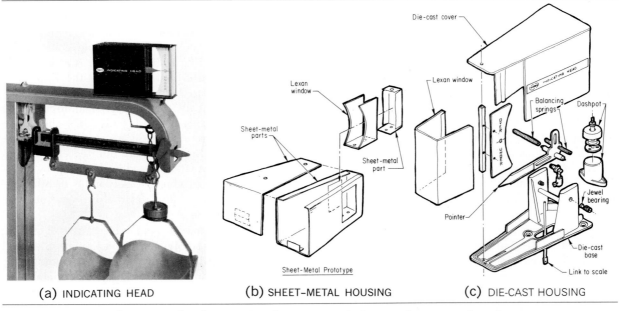

(a) INDICATING HEAD          (b) SHEET–METAL HOUSING          (c) DIE-CAST HOUSING

*Fig. 16.7*   Improved Design of Indicating Head. *Courtesy of Ohaus Scale Corp. and* Machine Design

in which many of the details will be similar to others previously used. For example, in Fig. 16.7, the Indicating Head is attached to a portable beam scale to add damping, sensitivity, and improved visibility to the weight readout. The original design of the housing was made of three sheet-metal parts with a plastic window, (b). The new two-piece design of a die-cast housing and larger plastic window, (c) provides more resistance to abuse and a drop in the unit cost of the housing after the first 2400 units, to less than one-third of the cost of the original sheet-metal design. Very often a change in material or a

Fig. 16.8  Redesigned Totalizer Wheel for Adding Machine. *Courtesy of Addmaster Corp. and E. I. du Pont de Nemours & Co., Inc.*

slight change in the shape of some part may be made without any loss of effectiveness and yet may save hundreds or thousands of dollars. The *ideal design* is the one that will do the job required at the lowest possible cost.

The Totalizer Wheel, Fig. 16.8, represents a cost-reducing redesign of an assembly in an adding machine. The redesigned wheel replaces an assembly of the 23 parts shown and continues to act as an indexing gear, integral bearing, integral spring, position stop, and print wheel.

An example of design from a different approach is the Electric Wheel used in heavy duty four-wheel drive earth-moving equipment, Fig. 16.9. This self-contained wheel design eliminates the usual restrictive drive train components, such as the drive shaft, universal joints, differential gears, and transmission, and makes possible nonslip traction at each wheel. The motor in the wheel is powered by a heavy-duty diesel electric generating system aboard the equipment.

The designs for a large vacation-area complex include unique and unusual approaches or systems to the problem of housing and transportation. The 14-story A-frame hotel, as shown by the model, Fig. 16.10 (a), is serviced by water craft, surface vehicles, and special monorail trains that pass through the

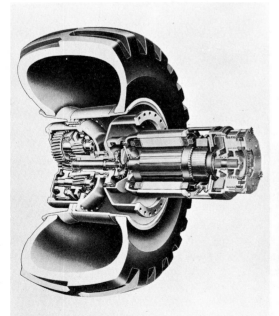

Fig. 16.9  Electric Wheel. Model L-700, LeTro-Loader, manufactured by Marathon LeTourneau Co., Equipment Division, Longview, Texas. *Courtesy of Marathon LeTourneau Co.*

(a) MODEL

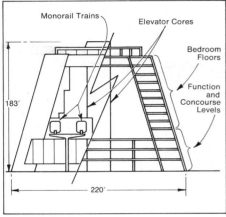

Monorail Trains

Elevator Cores

Bedroom Floors

Function and Concourse Levels

183'

220'

(b) ELEVATION

(c) MONORAIL TRAIN

**Fig. 16.10**   Recreation Park. *(a, b) © Walt Disney Productions; (c) courtesy of* Engineering New-Record

structure as indicated in (b). A pictorial of the train is shown in (c).

In a revised electronic organ design, the room-filling pipes and bellows have been replaced by a digital musical computer composed of aerospace microelectronics that requires about 1 cubic foot, Fig. 16.11. The computer contains some 48,000 transistors and enables the organ to be played in a virtually unlimited number of voices.

An improved design of a freight handling system for the transporting of subcompact automobiles includes the unique design of a special railroad car called Vert-A-Pack, Fig. 16.12. The sides of the five compartments on each side, which hold three cars each, are hinged at the bottom for use as ramps for efficient drive-on loading and drive-away unloading. The autos lock into place as the ramps are raised, and when the compartments are closed the autos are protected from vandalism and the weather.

## 16.7 Stage 4—Models and Prototypes

A model to scale is often constructed to study, analyze, and refine a design, Figs. 16.1, 16.11, and 16.17. The model of the Carveyor, Fig. 16.13, shows how it works and how people may be moved in such areas as airports, shopping centers, and college campuses. It is a loop system and is designed to carry up to 22,000 people per hour.

To instruct the model shop craftsperson in the construction of the prototype or model, dimensioned sketches and/or rudimentary working drawings are required. A full-size working model made to final specifications, except possibly for materials, is known as a *prototype*. The prototype is tested, modified where necessary, and the results noted in the revision of the sketches and working drawings.

If the prototype proves to be unsatisfactory, it may be necessary to return to a previous stage in the design process and repeat the procedures. It

Fig. 16.11 Electronic Organ. *Courtesy of Allen Organ Co.*

Fig. 16.12 Railway Auto Transport System. *Courtesy of American Iron and Steel Institute*

Fig. 16.13 Carveyor. *Courtesy of Goodyear Tire and Rubber Co.*

must be remembered that time and expense ceilings always limit the duration of this "looping." Eventually, a decision must be reached for the production model.

## 16.8 Stage 5—Working Drawings

To produce or manufacture a product, a final set of production or working drawings is made, checked, and approved.

In industry the approved production design layouts are turned over to the engineering department for the production drawings. The drafter, or detailers, "pick off" the details from the layouts with the aid of the scale or dividers. The necessary views §6.22, are drawn for each part to be made and complete dimensions and notes, Chapter 13, are added

so that the drawings will describe these parts completely. These working drawings of the individual parts are also known as *detail drawings*, §16.10.

Unaltered standard parts, §13.41, do not require a detail drawing but are shown conventionally on the assembly drawing and listed with specifications in the parts list, §16.14.

A detail drawing of one of the parts from the design layout of Fig. 16.4 is shown in Fig. 16.14. For

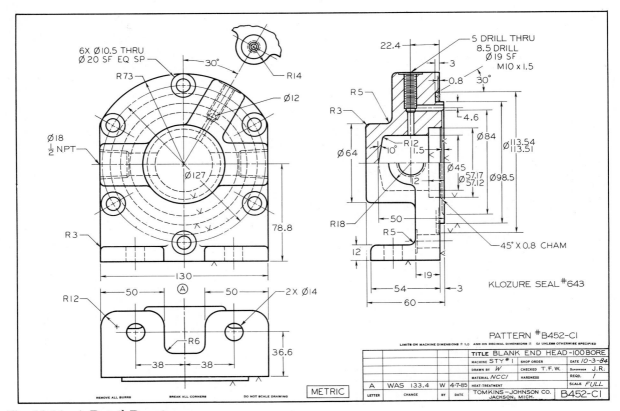

*Fig. 16.14*  A Detail Drawing.

details concerning working drawings, see §§16.10 to 16.19.

After the parts have been detailed, an *assembly drawing* is made, showing how all the parts go together in the complete product. The assembly may be made by tracing the various details in place directly from the detail drawings, or the assembly may be traced from the original design layout, but if either is done, the value of the assembly for checking purposes, §16.25, will be largely lost. The various types of assemblies are discussed in §§16.20 to 16.25.

Finally, in order to protect the manufacturer, a *patent drawing*, which is often a form of assembly, is prepared and filed with the U.S. Patent Office. Patent drawings are line shaded, often lettered in script, and otherwise follow rules of the Patent Office, §16.26.

## 16.9 Design of a New Product

An example of the design and development of a new product is that of the Cordless Electric Eraser shown in Fig. 16.15.

*Fig. 16.15*  Cordless Eraser with Recharging Console. *Courtesy of Pierce Business Products, Inc.*

## Stage 1—Identification of the Problem

In order to determine the feasibility of the *first* cordless eraser, opinions and ideas were solicited from many sources, including engineers, drafters, drafting teachers, drafting supply house managers, drafting supply store owners, and others. Price ranges and estimated sales were also carefully explored. This extensive survey indicated that there was a need and a potential market for a cordless eraser, provided it was convenient to use, durable, lightweight, versatile, maintenance free, safe, and competitively priced.

## Stage 2—Concepts and Ideas

Various cord erasers on the market were examined, tested, and analyzed for possible improvements. Several cordless devices on the market were also studied. Power and speed requirements for efficient erasing of pencil and/or ink lines were determined. Several methods of holding the eraser were reviewed. Various eraser refills for electric erasers were tested. With this collection of information as a background, several questions now needed to be considered. Should a direct drive and a large motor or a reduction drive and a small motor be used? What power source should be used—replaceable or rechargeable batteries? What recharging arrangements are necessary for rechargeable battery power? What about safety precautions with 110 volt power for the charger? What materials are suitable? What bearings are available? What is a suitable way to chuck the eraser? Should long or short eraser plugs or only short ones be used? What standard parts are available for such items as the motor, bearings, batteries, recharging unit, and power cord?

How many ways can these components be arranged for the solution?

## Stage 3—Compromise Solution

The cordless version of the eraser with a charging unit in a separate stand or console was selected as the preferred goal. The power train of a small battery-driven motor with a pinion gear meshed with a larger spur gear on the main shaft provided adequate power and speed for the eraser chuck. The batteries provided power long enough for normal usage. Recharging would occur while the eraser was at rest on the charging console. Careful selection of components and materials could lead to a durable and lightweight unit.

The few simple components of the system could be arranged in several ways. Thus, some flexibility in the final design was possible. Pictorials of several concepts, Fig. 16.16, were made and evaluated for balance, handling qualities, and appearance.

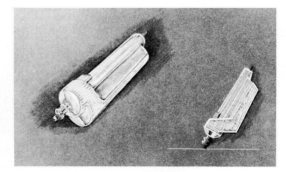
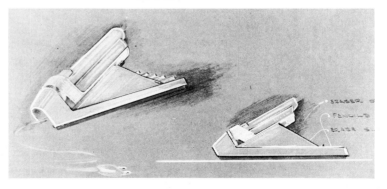

**Fig. 16.16** Preliminary Design Pictorials of Cordless Eraser. *Courtesy of Pierce Business Products, Inc.*

(a) WORKING MODEL            (b) PRODUCTION MODEL

**Fig. 16.17**    Models of Cordless Eraser. *Courtesy of Pierce Business Products, Inc.*

### Stage 4—Prototype

A Prototype or working model, Fig. 16.17 (a), was built, tested, refined, and restyled into the final production design, (b). The section of the production model, Fig. 16.18, shows the selected arrangements of components; note that all the components are held in place in one half of the molded shell without additional parts or fasteners. The Main Shaft assembly is shown in Fig. 16.19. To achieve the goals of lightweight durability and a competitive selling price, the engineers selected a molding of reinforced nylon resin rather than a metal tubing. A manufac-

turing cost saving of over 50 percent for the main shaft alone was made possible by the elimination of several secondary operations. For example, in Fig. 16.19 note the following features.

1. No thread cutting required—threads are molded.

2. No groove cutting—molded hubs locate the bearings and also eliminate the need for retaining rings.

3. No slot cutting for chuck removal slot—slot is molded.

**Fig. 16.18**    Section of Cordless Eraser Production Model. *Courtesy of Pierce Business Products, Inc.*

**Fig. 16.19**    Main Shaft Assembly. *Courtesy of Pierce Business Products, Inc.*

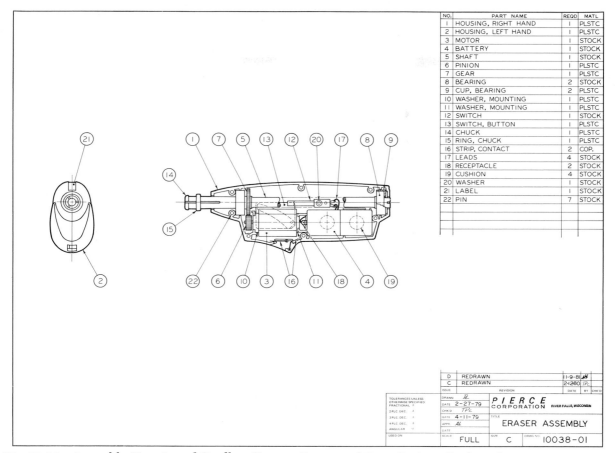

| NO. | PART NAME | REQD | MATL |
|---|---|---|---|
| 1 | HOUSING, RIGHT HAND | 1 | PLSTC |
| 2 | HOUSING, LEFT HAND | 1 | PLSTC |
| 3 | MOTOR | 1 | STOCK |
| 4 | BATTERY | 1 | STOCK |
| 5 | SHAFT | 1 | STOCK |
| 6 | PINION | 1 | PLSTC |
| 7 | GEAR | 1 | PLSTC |
| 8 | BEARING | 2 | STOCK |
| 9 | CUP, BEARING | 2 | PLSTC |
| 10 | WASHER, MOUNTING | 1 | PLSTC |
| 11 | WASHER, MOUNTING | 1 | PLSTC |
| 12 | SWITCH | 1 | STOCK |
| 13 | SWITCH, BUTTON | 1 | PLSTC |
| 14 | CHUCK | 1 | PLSTC |
| 15 | RING, CHUCK | 1 | PLSTC |
| 16 | STRIP, CONTACT | 2 | COP. |
| 17 | LEADS | 4 | STOCK |
| 18 | RECEPTACLE | 2 | STOCK |
| 19 | CUSHION | 4 | STOCK |
| 20 | WASHER | 1 | STOCK |
| 21 | LABEL | 1 | STOCK |
| 22 | PIN | 7 | STOCK |

| D | REDRAWN | | 11-9-81 | JH | |
| C | REDRAWN | | 2-1280 | TPc | |
| ISSUE | | REVISION | | DATE | BY | CHK'D |

TOLERANCES UNLESS OTHERWISE SPECIFIED
FRACTIONAL ±
2 PLC. DEC. ±
3 PLC. DEC. ±
4 PLC. DEC. ±
ANGULAR

DRAWN R. DATE 2-27-79
CHK'D TPc
DATE 4-11-79
APPR AL
DATE
USED ON
SCALE FULL SIZE C DRWG NO. 10038-01

PIERCE CORPORATION RIVER FALLS, WISCONSIN
TITLE ERASER ASSEMBLY

**Fig. 16.20** Assembly Drawing of Cordless Eraser. *Courtesy of Pierce Business Products, Inc.*

4. No external grinding of shaft is necessary—precision molding gives a diameter suitable for installation of the high-speed bearings.

5. No keyseat required—molded-in key eliminates need for a key.

6. No keyway required on gear—molded-in keyway. No key or retaining ring installations facilitate more simplified assembly.

The foregoing considerations are typical of the attention given all components in the design.

### Stage 5—Production Drawings

Complete sets of detail and assembly drawings were made for the eraser and the charging console. The assembly drawing for the eraser is shown in Fig. 16.20 and the assembly drawing of the charger base is shown in Fig. 16.21. A detail of the shaft is given in Fig. 16.22. Standard parts were specified on separate sheets or in a parts list. (Space limitations do not permit including more of the drawings necessary for the product.)

## 16.10 Working Drawings

Working drawings, which normally include assembly and details, are the specifications for the manufacture of a design. Therefore, they must be neatly made and carefully checked. The working drawings of the individual parts are also referred to as *detail drawings*. See §§16.11 to 16.18.

## 16.11 Number of Details per Sheet

Two general methods are followed in industry regarding the grouping of details on sheets. If the machine or structure is small or composed of few parts, all the details may be shown on one large sheet, Fig. 16.23.

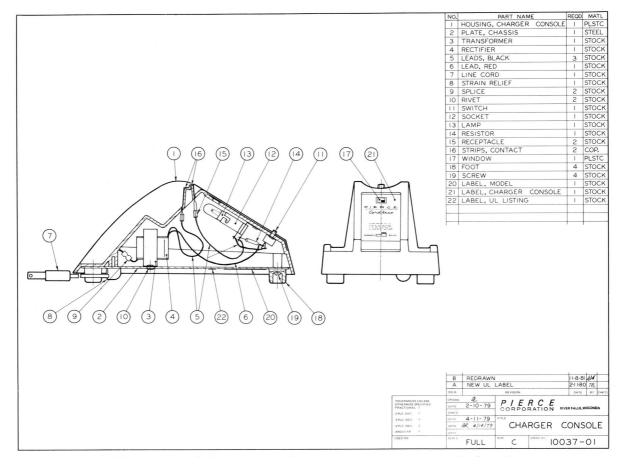

| NO. | PART NAME | REQD | MATL |
|---|---|---|---|
| 1 | HOUSING, CHARGER CONSOLE | 1 | PLSTC |
| 2 | PLATE, CHASSIS | 1 | STEEL |
| 3 | TRANSFORMER | 1 | STOCK |
| 4 | RECTIFIER | 1 | STOCK |
| 5 | LEADS, BLACK | 3 | STOCK |
| 6 | LEAD, RED | 1 | STOCK |
| 7 | LINE CORD | 1 | STOCK |
| 8 | STRAIN RELIEF | 1 | STOCK |
| 9 | SPLICE | 2 | STOCK |
| 10 | RIVET | 2 | STOCK |
| 11 | SWITCH | 1 | STOCK |
| 12 | SOCKET | 1 | STOCK |
| 13 | LAMP | 1 | STOCK |
| 14 | RESISTOR | 1 | STOCK |
| 15 | RECEPTACLE | 2 | STOCK |
| 16 | STRIPS, CONTACT | 2 | COP. |
| 17 | WINDOW | 1 | PLSTC |
| 18 | FOOT | 4 | STOCK |
| 19 | SCREW | 4 | STOCK |
| 20 | LABEL, MODEL | 1 | STOCK |
| 21 | LABEL, CHARGER CONSOLE | 1 | STOCK |
| 22 | LABEL, UL LISTING | 1 | STOCK |

| | | | | |
|---|---|---|---|---|
| B | REDRAWN | | 11-8-81 | |
| A | NEW UL LABEL | | 2-1-80 | |
| ISSUE | REVISION | | DATE | BY CHK'D |

TOLERANCES UNLESS OTHERWISE SPECIFIED FRACTIONAL ±

2 PLC. DEC. ±
3 PLC. DEC. ±
4 PLC. DEC. ±
ANGUL AR ±

USED ON

DRAWN   2-10-79
CHK'D
DATE 4-11-79
APPR AR 4/4/79
DATE

**PIERCE CORPORATION** RIVER FALLS, WISCONSIN

TITLE   CHARGER CONSOLE

SCALE FULL   SIZE C   DRWG N° 10037-01

*Fig. 16.21*   Assembly Drawing of Charger Console. *Courtesy of Pierce Business Products, Inc.*

When larger or more complicated mechanisms are represented, the details may be drawn on several large sheets, several details to the sheet, and the assembly drawn on a separate sheet. Most companies have now adopted the practice of drawing only one detail per sheet, however simple or small, Fig. 16.22. The basic 8.5″ × 11.0″ or 210 mm × 297 mm sheet is most commonly used for details, multiples of these sizes being used for larger details or the assembly. For standard sheet sizes, see §2.63.

When several details are drawn on one sheet, careful consideration must be given to spacing. The drafter should determine the necessary views for each detail, and *block in all views lightly before beginning to draw any view*, as shown in Fig. 16.23. Ample space should be allowed for dimensions and notes. A simple method to space the views is to cut out rectangular scraps of paper roughly equal to the sizes of the views and to move these around on the sheet until a suitable spacing is determined. The corner locations are then marked on the sheet, and the scraps of paper are discarded.

The same scale should be used for all details on a single sheet, if possible. When this is not possible, the scales for the dissimilar details should be clearly noted under each.

## 16.12 Title and Record Strips

The function of the title and record strip is to show, in an organized manner, all necessary information not given directly on the drawing with its dimensions and notes. Obviously, the type of title used depends upon the filing system in use, the processes of manufacture, and the requirements of the product. The following information should generally be given in the title form:

1. Name of the object represented.
2. Name and address of the manufacturer.

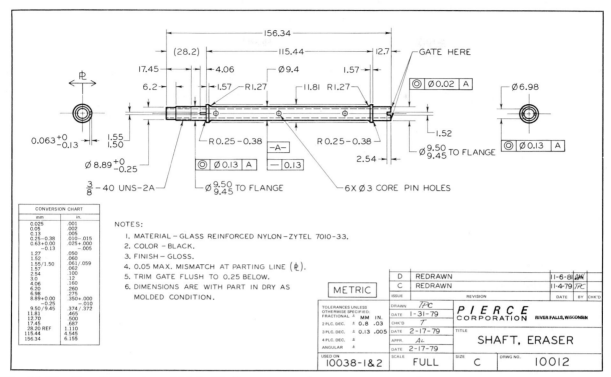

**Fig. 16.22** Detail Drawing of Main Shaft. *Courtesy of Pierce Business Products, Inc.*

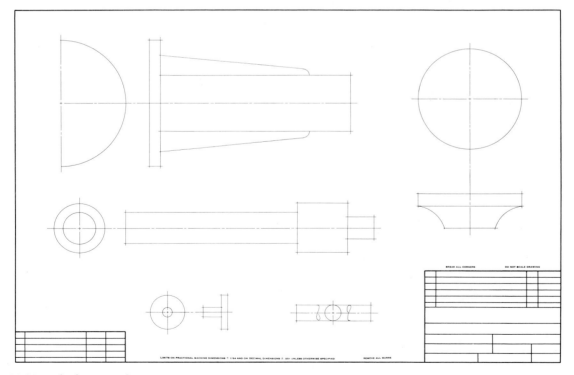

**Fig. 16.23** Blocking In the Views

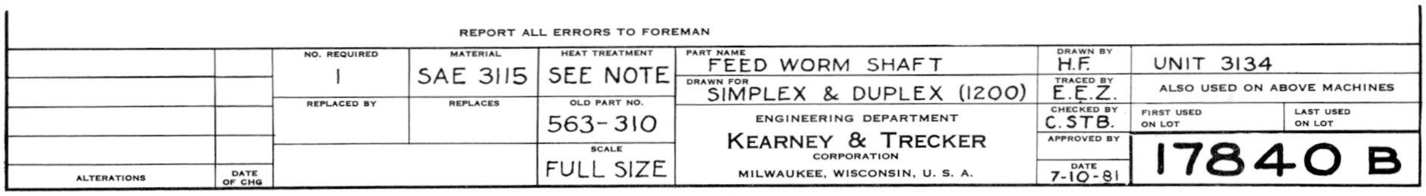

| | | NO. REQUIRED | MATERIAL | HEAT TREATMENT | PART NAME FEED WORM SHAFT | DRAWN BY H.F. | UNIT 3134 | |
|---|---|---|---|---|---|---|---|---|

REPORT ALL ERRORS TO FOREMAN

| | | NO. REQUIRED 1 | MATERIAL SAE 3115 | HEAT TREATMENT SEE NOTE | PART NAME FEED WORM SHAFT | DRAWN BY H.F. | UNIT 3134 | |
| REPLACED BY | REPLACES | | | OLD PART NO. 563-310 | DRAWN FOR SIMPLEX & DUPLEX (1200) ENGINEERING DEPARTMENT KEARNEY & TRECKER CORPORATION | TRACED BY E.E.Z. CHECKED BY C.STB. | ALSO USED ON ABOVE MACHINES FIRST USED ON LOT / LAST USED ON LOT | |
| ALTERATIONS | DATE OF CHG | | | SCALE FULL SIZE | MILWAUKEE, WISCONSIN, U. S. A. | APPROVED BY DATE 7-10-81 | 17840 B | |

Fig. 16.24  Title Strip.

3.  Name and address of the purchasing company, if any.

4.  Signature of the drafter who made the drawing, and the date of completion.

5.  Signature of the checker, and the date of completion.

6.  Signature of the chief drafter, chief engineer, or other official, and the date of approval.

7.  Scale of the drawing.

8.  Number of the drawing.

Other information may be given, such as material, quantity, heat treatment, finish, hardness, pattern number, estimated weight, superseding and superseded drawing numbers, symbol of machine, and many other items, depending upon the plant organization and the peculiarities of the product. Some typical commercial titles are shown in Figs. 16.24, 16.25, and 16.26. See inside back cover for traditional title forms and ANSI-approved sheet sizes.

The title form is usually placed along the bottom of the sheet, Fig. 16.24, or in the lower right-hand corner of the sheet, Fig. 16.26, because drawings are often filed in flat, horizontal drawers, and the title must be easily found. However, many filing systems are in use, and the location of the title form is governed by the system employed.

Lettering should be single-stroke vertical or inclined Gothic capitals, Figs. 4.20 and 4.21. The items in the title form should be lettered in accord-

ance with their relative importance. The drawing number should receive greatest emphasis, closely followed by the name of the object and the name of the company. The date, scale, and drafter's and checker's names are important, but they do not deserve prominence. Greater importance of items is indicated by heavier lettering, larger lettering, wider spacing of letters, or by a combination of these methods. See Table 16.1 for recommended letter heights.

Many companies have adopted their own title forms or those preferred by ANSI and have them printed on standard-size sheets, so that the drafters need merely fill in the blank spaces.

Drawings constitute important and valuable information regarding the products of a manufacturer. Hence, carefully designed, well-kept, systematic files are generally maintained for the filing of drawings.

## 16.13 Drawing Numbers

Every drawing should be numbered. Some companies use serial numbers, such as 60412, or a number with a prefix or suffix letter to indicate the sheet size, as A60412 or 60412–A. The size A sheet would probably be the standard 8.5″ × 11.0″ or 9.0″ × 12.0″, and the B size a multiple thereof. Many different numbering schemes are in use in which various parts of the drawing number indicate different things, such as model number of the machine and the general nature or use of the part. In general, it

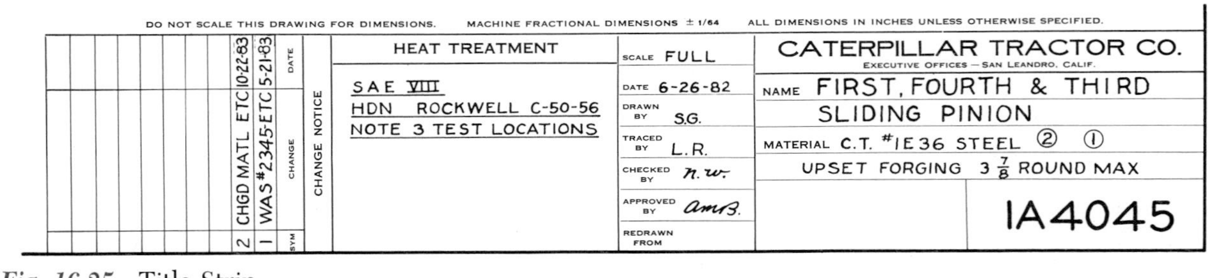

DO NOT SCALE THIS DRAWING FOR DIMENSIONS.     MACHINE FRACTIONAL DIMENSIONS ± 1/64     ALL DIMENSIONS IN INCHES UNLESS OTHERWISE SPECIFIED.

| | | | | | | | 2 CHGD MATL ETC 10-22-83 / 1 WAS #2345 ETC 5-21-83 | DATE | CHANGE NOTICE | HEAT TREATMENT SAE VIII HDN ROCKWELL C-50-56 NOTE 3 TEST LOCATIONS | SCALE FULL / DATE 6-26-82 / DRAWN BY S.G. / TRACED BY L.R. / CHECKED BY n.w. / APPROVED BY amrß. / REDRAWN FROM | CATERPILLAR TRACTOR CO. EXECUTIVE OFFICES — SAN LEANDRO, CALIF. NAME FIRST, FOURTH & THIRD SLIDING PINION MATERIAL C.T. #1E36 STEEL ② ① UPSET FORGING 3 7/8 ROUND MAX |
|---|---|---|---|---|---|---|---|---|---|---|---|---|

1A4045

Fig. 16.25  Title Strip.

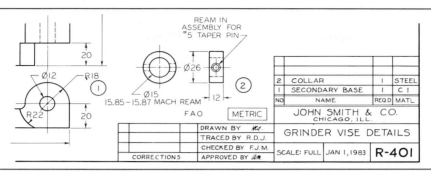

**Fig. 16.26**   Identification of Details with Parts List.

is best to use a simple numbering system and not to load the number with too many indications.

The drawing number should be lettered 7 mm (.250″) high in the lower-right and upper-left corners of the sheet, Fig. 16.40.

## 16.14 Parts Lists

A bill of material, or *parts list*, consists of an itemized list of the several parts of a structure shown on a detail drawing or an assembly drawing [ANSI Y14.34M–1982 (R1988)]. This list is often given on a separate sheet, but is frequently lettered directly on the drawing, Fig. 16.39. The title strip alone is

sufficient on detail drawings of only one part, Fig. 16.22, but a parts list is necessary on detail drawings of several parts, Fig. 16.26.

Parts lists on machine drawings contain the part numbers or symbols, a descriptive title of each part, the number required, the material specified, and frequently other information, such as pattern numbers, stock sizes of materials, and weights of parts.

Parts are listed in general order of size or importance. The main castings or forgings are listed first, parts cut from cold-rolled stock second, and standard parts such as fasteners, bushings, and roller bearings third. If the parts list rests on top of the title box or strip, the order of items should be from

**Table 16.1**   *Recommended[a] Minimum Letter Heights*

| Use | Minimum Letter Heights | | Drawing Size |
|---|---|---|---|
| | Freehand | Instrumental | |
| Drawing number in title block | .312″ ($\frac{5}{16}$)   7 mm | .290″   7 mm | Larger than 17″ × 22″ |
| | .250″ ($\frac{1}{4}$)   7 mm | .240″   7 mm | Up to and including 17″ × 22″ |
| Drawing title | .250″ ($\frac{1}{4}$)   7 mm | .240″   7 mm | All |
| Section and tabulation letters | .250″ ($\frac{1}{4}$)   7 mm | .240″   7 mm | |
| Zone letters and numerals in borders | .188″ ($\frac{3}{16}$)   5 mm | .175″   5 mm | |
| Dimensions, tolerances, limits, notes, subtitles for special views, tables, revisions, and zone letters for the body of the drawing | .125″ ($\frac{1}{8}$)   3.5 mm | .120″   3.5 mm | Up to and including 17″ × 22″ |
| | .156″ ($\frac{5}{32}$)   5 mm | .140″   5 mm | Larger than 17″ × 22″ |

[a]ANSI Y14.2M–1979 (R1987).

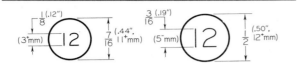

*Fig. 16.27*  Identification Numbers.

the bottom upward, Figs. 16.26 and 16.39, so that new items can be added later, if necessary. If the parts list is placed in the upper-right corner, the items should read downward.

Each detail on the drawing may be identified with the parts list by the use of a small circle containing the part number, placed adjacent to the detail, as in Fig. 16.26. One of the sizes in Fig. 16.27 will be found suitable, depending on the size of the drawing.

Standard parts §13.41 purchased or company produced, are not drawn, but are listed in the parts list. Bolts, screws, bearings, pins, keys, and so on are identified by the part number from the assembly drawing and are specified by name and size or number.

## 16.15 Zoning
To facilitate locating an item on a large or complex drawing, regular ruled intervals are labeled along the margins, often the right and lower margins only. The intervals on the horizontal margin are labeled from right to left with numerals and the intervals on the vertical margin are labeled from bottom to top with letters. See Fig. 16.45.

## 16.16 Checking
The importance of accuracy in technical drawing cannot be overestimated. In commercial offices, errors sometimes cause tremendous unnecessary expenditures. *The drafter's signature on a drawing identifies who is responsible for the accuracy of the work.*

In small offices, checking is usually done by the designer or by one of the drafters. In large offices, experienced engineers are employed who devote a major part of their time to checking drawings.

The pencil drawing, upon completion, is carefully checked and signed by the drafter who made it. The drawing is then checked by the designer for function, economy, practicability, and so on. Corrections, if any, are then made by the original drafter.

The final checker should be able to discover all remaining errors, and, to be effective, the work must be done in a systematic way, studying the drawing with particular attention to the following points.

1. Soundness of design, with reference to function, strength, materials, economy, manufacturability, serviceability, ease of assembly and repair, lubrication, and so on.

2. Choice of views, partial views, auxiliary views, sections, line work, lettering, and so on.

3. Dimensions, with special reference to repetition, ambiguity, legibility, omissions, errors, and finish marks. Special attention should be given to tolerances.

4. Standard parts. In the interest of economy, as many parts as possible should be standard.

5. Notes, with special reference to clear wording and legibility.

6. Clearances. Moving parts should be checked in all possible positions to assure freedom of movement.

7. Title form information.

## 16.17 Drawing Revisions
Changes on drawings are necessitated by changes in design, changes in tools, desires of customers, or by errors in design or in production. In order that the sources of all changes of information on drawings may be understood, verified, and accessible, an accurate record of all changes should be made on the drawings. The record should show the character of the change, by whom, when, and why made.

The changes are made by erasures directly on the original drawing or by means of erasure fluid on a reproduction print. Additions are simply drawn in on the original. The removal of information by crossing out is not recommended. If a dimension is not noticeably affected by a change, it may be underlined with a heavy line as shown in Fig. 13.15 to indicate that it is not to scale. In any case, prints of each issue or microfilms are kept on file to show how the drawing appeared before the revision. New prints are issued to supersede old ones each time a change is made.

If considerable change on a drawing is necessary, a new drawing may be made and the old one then stamped OBSOLETE and placed in the "obsolete" file. In the title block of the old drawing, the words

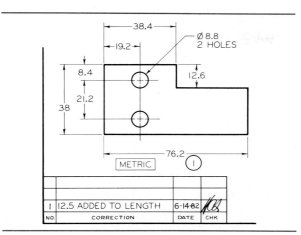

**Fig. 16.28**  Revisions.

"SUPERSEDED BY . . ." or "REPLACED BY . . ." are entered followed by the number of the new drawing. On the new drawing, under "SUPERSEDES . . ." or "REPLACES . . .," the number of the old drawing is entered.

Various methods are used to reference the area on a drawing where the change is made, with the entry in the revision block. The most common is to place numbers or letters in small circles near the places where the changes were made and to use the same numbers or letters in the revision block, Fig. 16.28. On zoned drawings, §16.15, the zone of the correction would be shown in the revision block. In addition, a brief description of the change should be made, and the date and the initials of the person making the change should be given.

## 16.18 Simplified Drafting

Drafting time is a considerable element of the total cost of a product. Consequently, industry attempts to reduce drawing costs by simplifying its drafting practices, but without loss of clarity to the user.

The American National Standard Drafting Manual, published by the American National Standards Institute, incorporates the best and most representative practices in this country, and the authors are in full accord with them. These standards advocate simplification in many ways, for example, partial views, half views, thread symbols, piping symbols, and single-line spring drawings. Any line or lettering on a drawing that is not needed for clarity should be omitted.

A summary of practices to simplify drafting is as follows:

1. Use word description in place of drawing wherever practicable.

2. Never draw an unnecessary view. Often a view can be eliminated by using abbreviations or symbols such as HEX, SQ, DIA, ∅, □, and ℄.

3. Draw partial views instead of full views wherever possible. Draw half views of symmetrical parts.

4. Avoid elaborate, pictorial, or repetitive detail as much as possible. Use phantom lines to avoid drawing repeated features, §15.14.

5. List rather than draw, when possible, standard parts such as bolts, nuts, keys, and pins.

6. Omit unnecessary hidden lines. See §6.25.

7. Use outline section lining in large sectioned areas wherever it can be done without loss of clarity.

8. Omit unnecessary duplication of notes and lettering.

9. Use symbolic representation wherever possible, such as piping symbols and thread symbols.

10. Draw freehand, or mechanically plus freehand, wherever practicable.

11. Avoid hand lettering as much as possible. For example, parts lists should be typed on a separate sheet.

12. Use laborsaving devices wherever feasible, such as templates and plastic overlays.

13. Use electronic devices or computer graphics systems wherever feasible for design, drawing, and repetitive work.

Some industries have attempted to simplify their drafting practices even more. Until these practices are accepted generally by industry and in time find their way into the ANSI standards, the students should follow the ANSI standards as exemplified throughout this book. Fundamentals should come first—shortcuts perhaps later.

## 16.19 Computer Graphics

A computer is an electronic machine that is capable of performing specific tasks at incredibly high speeds. The use of the computer today in various business and industrial operations is very well

***Fig. 16.29*** Bausch & Lomb Producer Electronic/CAD System. *Courtesy of Bausch & Lomb*

***Fig. 16.30*** Bruning SPECTRA Computer-Aided Design System. *Courtesy of Bruning*

known. Engineers and scientists have used computers for many years to perform the mathematical calculations required in their work. In recent years, new and innovative technological developments in computer applications have expanded the versatility of the computer into a still more useful engineering and design tool.

*Computer graphics* is a general term used to define any procedure that uses computers to generate, process, and display graphic images. *Computer-aided design* or *computer-aided drafting (CAD)* and *computer-aided design and drafting (CADD)*, or other comparable terms, are used synonymously and refer to a specific process that uses a computer

**Fig. 16.31**   Prime CAD Workstation. *Courtesy of Prime Computer, Inc.*

system to assist in the creation, modification, and display of a drawing or design.

All CAD systems have similar hardware components that include input devices, a central processing unit, data storage devices, and output devices. Two typical CAD systems are shown in Figs. 16.29 and 16.30. All CAD systems must also have software, which are the programs and instructions that permit the computer system to operate.

The designer or drafter can create a drawing by first entering the appropriate data into the system at a CAD workstation, Fig. 16.31, using any one of several different input devices. The immediate result of the input is a graphic display on the screen of a cathode-ray tube (CRT). The designer can then view and analyze the drawing, make changes, delete sections, and "think out" the design on the tube much the same way as would be done on a drawing board. The final drawing displayed on the tube may then be stored in memory, or a permanent hard-copy drawing made using an output device such as the plotter shown in Fig. 16.32. Thus, by means of

**Fig. 16.32**   Plotter. *Courtesy of Versatec, Inc.*

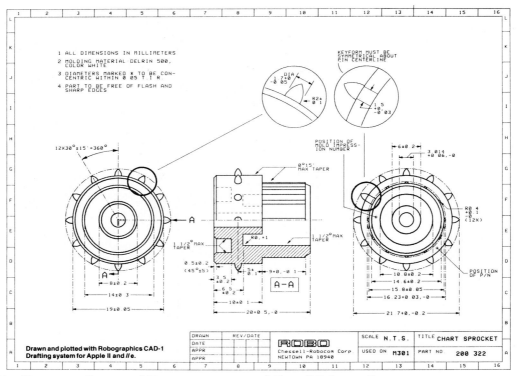

**Fig. 16.33**  Computer-Generated Drawing. *Courtesy of Chessell-Robocom Corp.*

programmed data supplied to the computer, the designer or drafter is able to secure from a plotter a drawing complete with all required lines, lettering, dimensions, and so forth, Fig. 16.33.

*Computer-aided design/computer aided manufacturing (CAD/CAM)* refers to the integration of computers into the entire design-to-manufacturing process of a product or plant. In this method, the data are transmitted directly from the CAD system to the manufacturing plant to manage and control operations. The CAD/CAM system shown in Fig. 16.34 is used to cut seat patterns for Chrysler cars and trucks. A computer-controlled trim cutter in Chrysler's trim plant, Fig. 16.34 (a), makes precision cuts in bolts of material according to directions given by the CAD terminal operator, Fig. 16.34 (b), located in the company's corporate engineering complex. Before CAD/CAM the cutting of seat patterns was a painstaking manual job, requiring a worker to climb on top of layers of cloth and jigsaw through them following paper patterns the same way a seamstress follows a dress pattern.

The procedure for the design and preparation of engineering drawings and the production of machine parts by computer for the development of a new aircraft is shown in the work flow diagram, Fig. 16.35. Each of the operations from design to manufacture is systematically monitored and controlled by a CAD/CAM system.

The engineers and drafters must have a thorough understanding of the graphic language in order to prepare the correct input data for the computer and to evaluate the output of the plotter. The computer will do only what it is programmed to do. Regardless of how complex or automated the method of making a drawing becomes, the time-proven mode of graphic expression is indispensable for purposes of communication, specifications, records, and so on.

For additional information about computer-aided design and drafting, see Chapters 3 and 8.

## 16.20 Assembly Drawings

An assembly drawing shows the assembled machine or structure, with all detail parts in their functional positions. Assembly drawings vary in character ac-

(a)

(b)

*Fig. 16.34* CAD/CAM System. *Courtesy of Control Data Corporation*

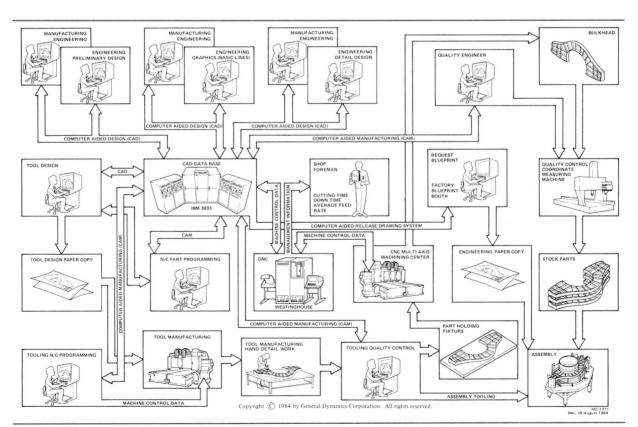

MC-1371
Rev. 28 August 1984

*Fig. 16.35* Work Flow Diagram for Producing Detail and Assembly Drawings, and Machine Parts by Numerical Control. *Courtesy of Fort Worth Division of General Dynamics Corporation*

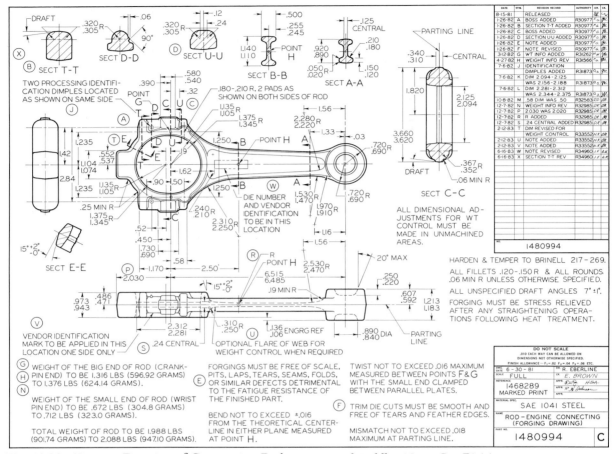

**Fig. 16.36**   Forging Drawing of Connecting Rod. *Courtesy of Cadillac Motor Car Division*

cording to use, as follows: (1) design assemblies, or layouts, discussed in §16.6, (2) general assemblies, (3) working drawing assemblies, (4) outline or installation assemblies, and (5) check assemblies.

## 16.21 General Assemblies

A set of working drawings includes the *detail drawings* of the individual parts and the *assembly drawing* of the assembled unit. The detail drawings of an automobile connecting rod are shown in Figs. 16.36 and 16.37, and the corresponding assembly drawing is shown in Fig. 16.38. Such an assembly, showing only one unit of a larger machine, is often referred to as a *subassembly*.

An example of a complete general assembly appears in Fig. 16.39, which shows the assembly of a

hand grinder. Another example of a subassembly is shown in Fig. 16.40.

### 1. Views

In selecting the views for an assembly drawing, the purpose of the drawing must be kept in mind: to show how the parts fit together in the assembly and to suggest the function of the entire unit, not to describe the shapes of the individual parts. The assembly worker receives the actual finished parts. If more information is needed about a part that cannot be obtained from the part itself, the detail drawing must be checked. Thus, the assembly drawing purports to show *relationships* of parts, *not shapes*. The view or views selected should be the minimum views or partial views that will show how the parts

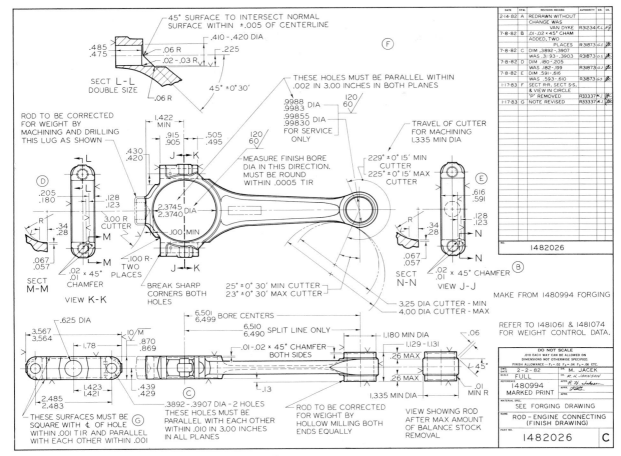

**Fig. 16.37** Detail Drawing of Connecting Rod. *Courtesy of Cadillac Motor Car Division*

fit together. In Fig. 16.38, only one view is needed, while in Fig. 16.39 only two views are necessary.

## 2. Sections

Since assemblies often have parts fitting into or overlapping other parts, hidden-line delineation is usually out of the question. Hence, in assemblies, sectioning can be used to great advantage. For example, in Fig. 16.39, try to imagine the right-side view drawn in elevation with interior parts represented by hidden lines. The result would be completely unintelligible.

Any kind of section may be used as needed. A broken-out section is shown in Fig. 16.39, a half section in Fig. 16.40, and several removed sections are shown in Fig. 16.36. For general information on

assembly sectioning, see §16.22. For methods of drawing threads in sections, see §15.15.

## 3. Hidden Lines

As a result of the extensive use of sectioning in assemblies, hidden lines are often not needed. However, they should be used wherever necessary for clearness.

## 4. Dimensions

As a rule, dimensions are not given on assembly drawings, since they are given completely on the detail drawings. If dimensions are given, they are limited to some function of the object as a whole, such as the maximum height of a jack, or the max-

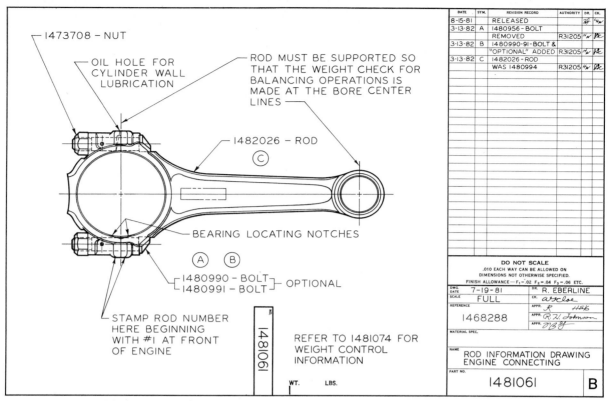

**Fig. 16.38** Assembly Drawing of Connecting Rod. *Courtesy of Cadillac Motor Car Division*

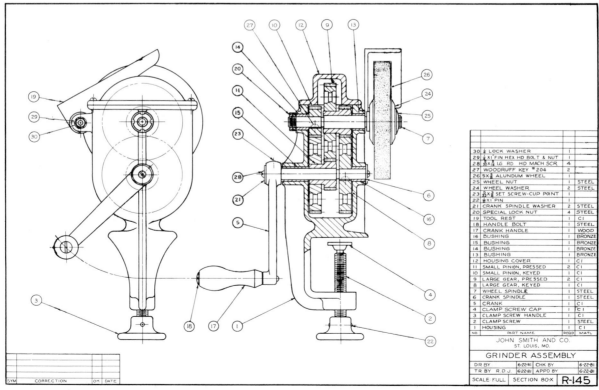

**Fig. 16.39** Assembly Drawing of Grinder.

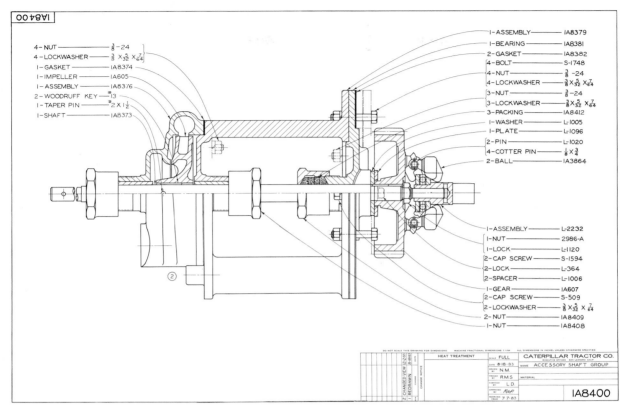

*Fig. 16.40* Subassembly of Accessory Shaft Group.

imum opening between the jaws of a vise. Or when machining is required in the assembly operation, the necessary dimensions and notes may be given on the assembly drawing.

## 5. Identification

The methods of identification of parts in an assembly are similar to those used in detail drawings where several details are shown on one sheet, as in Fig. 16.26. Circles containing the part numbers are placed adjacent to the parts, with leaders terminated by arrowheads touching the parts as in Fig. 16.39. The circles shown in Fig. 16.27 for detail drawings are, with the addition of radial leaders, satisfactory for assembly drawings. Note, in Fig. 16.39, that these circles are placed in orderly horizontal or vertical rows and not scattered over the sheet. Leaders are never allowed to cross, and adjacent leaders are parallel or nearly so.

The parts list includes the part numbers or symbols, a descriptive title of each part, the number required per machine or unit, the material specified, and frequently other information, such as pattern numbers, stock sizes, weights, and so on. Frequently the parts list is lettered or typed on a separate sheet.

Another method of identification is to letter the part names, numbers required, and part numbers, at the end of leaders as shown in Fig. 16.40. More commonly, however, only the part numbers are given, together with ANSI-approved straight-line leaders.

## 6. Drawing Revisions

Methods of recording changes are the same as those for detail drawings, Fig. 16.36, for example. See §16.17.

## 16.22 Assembly Sectioning

In assembly sections it is necessary not only to show the cut surfaces but to distinguish between adjacent parts. This is done by drawing the section lines in

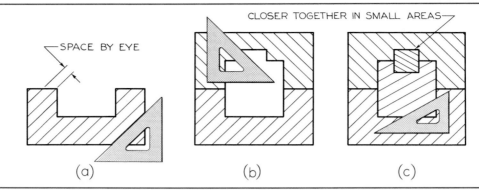

Fig. 16.41    Section Lining (Full Size).

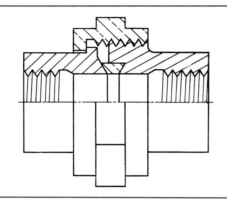

Fig. 16.42    Symbolic Section Lining.

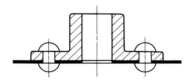

Fig. 16.43    Sectioning Thin Parts.

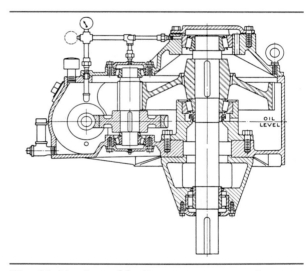

Fig. 16.44    Assembly Section. *Courtesy of Hewitt-Robins, Inc.*

opposing directions, as shown in Fig. 16.41. The first large area, (a), is section-lined at 45°. The next large area, (b), is section-lined at 45° in the opposite direction. Additional areas are then section-lined at other angles, as 30° or 60° with horizontal, as shown at (c). If necessary, "odd" angles may be used. Note at (c) that in small areas it is necessary to space the section lines closer together. The section lines in adjacent areas should not meet at the visible lines separating the areas.

For general use, the cast-iron general-purpose section lining is recommended for assemblies.

Wherever it is desired to give a general indication of the materials used, symbolic section lining may be used, as in Fig. 16.42.

In sectioning relatively thin parts in assembly, such as gaskets and sheet-metal parts, section lining is ineffective, and such parts should be shown in solid black, Fig. 16.43.

Often solid objects, or parts which themselves do not require sectioning, lie in the path of the cutting plane. It is customary and standard practice to show such parts unsectioned, or "in the round." These include bolts, nuts, shafts, keys, screws, pins,

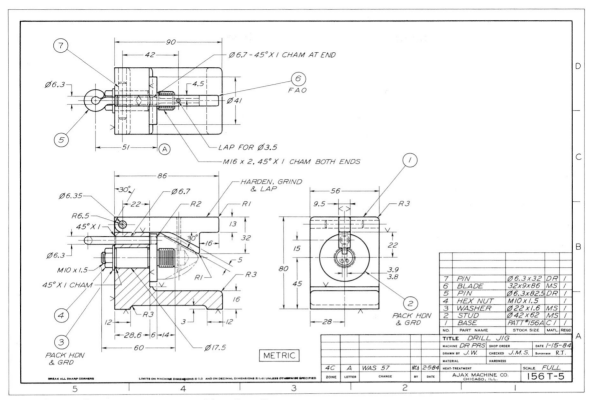

**Fig. 16.45** Working Drawing Assembly of Drill Jig.

ball or roller bearings, gear teeth, spokes, and ribs among others. Many are shown in Fig. 16.44, and similar examples are shown in Figs. 16.39 and 16.40.

### 16.23 Working Drawing Assembly
A working drawing assembly, Fig. 16.45, is a combined detail and assembly drawing. Such drawings are often used in place of separate detail and assembly drawings when the assembly is simple enough for all of its parts to be shown clearly in the single drawing. In some cases, all but one or two parts can be drawn and dimensioned clearly in the assembly drawing, in which event these parts are detailed separately on the same sheet. This type of drawing is common in valve drawings, locomotive subassemblies, aircraft subassemblies, and in drawings of jigs and fixtures.

### 16.24 Installation Assemblies
An assembly made specifically to show how to install or erect a machine or structure is an *installation assembly*. This type of drawing is also often called

an *outline assembly*, because it shows only the outlines and the relationships of exterior surfaces. A typical installation assembly is shown in Fig. 16.46. In aircraft drafting, an installation drawing (assembly) gives complete information for placing details or subassemblies in their final positions in the airplane.

### 16.25 Check Assemblies
After all detail drawings of a unit have been made, it may be necessary to make a *check assembly*, especially if a number of changes were made in the details. Such an assembly is drawn accurately to scale in order to check graphically the correctness of the details and their relationship in assembly. After the check assembly has served its purpose, it may be converted into a general assembly drawing.

### 16.26 Patent Drawings
The patent application for a machine or device must include drawings to illustrate and explain the invention. It is essential that all patent drawings be me-

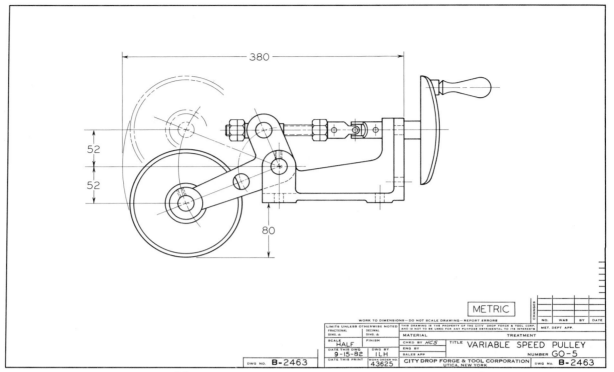

*Fig. 16.46*   Installation Assembly.

chanically correct and constitute complete illustrations of every feature of the invention claimed. The strict requirements of the U.S. Patent Office in this respect serve to facilitate the examination of applications and the interpretation of patents issued thereon. A typical patent drawing is shown in Fig. 16.47.

The drawings for patent applications are pictorial and explanatory in nature; hence, they are not detailed as are working drawings for production purposes. Center lines, dimensions, notes, and so forth are omitted. Views, features, and parts, for example, are identified by numbers that refer to the descriptions and explanations given in the specification section of the patent application.

Patent drawings are made with India ink on heavy, smooth, white paper, exactly 10.0″ × 15.0″ with 1.0″ borders on all sides. A space of not less than 1.25″ from the shorter border, which is the top of the drawing, is left blank for the heading of title, name, number, and other data to be added by the Patent Office.

All lines must be solid black and suitable for reproduction at a smaller size. Line shading is used whenever it improves readability.

The drawings must contain as many figures as necessary to show the invention clearly. There is no restriction on the number of sheets. The figures may be plan, elevation, section, pictorial, and detail views of portions or elements, and they may be drawn to an enlarged scale if necessary. The required signatures must be placed in the lower right-hand corner of the drawing, either inside or outside the border line.

Because of the strict requirements of the Patent Office, applicants are advised to employ competent drafters to make their drawings. To aid drafters in the preparation of drawing for submission in patent applications, the *Guide for Patent Draftsmen* was prepared and can be obtained from the Superintendent of Documents, U.S. Government Printing Office, Washington, D.C. 20402.

## 16.27 Reproduction and Control of Drawings

An essential part of the designer's or drafter's education is a thorough knowledge of reproduction techniques and processes. Specifically, they should be familiar with the various processes available for

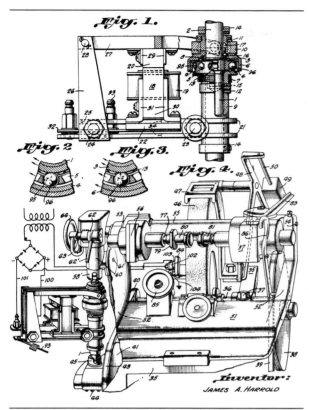

Fig. 16.47 A Well-Executed Patent Drawing.

the reproduction of drawings: *blueprint, diazo, microfilm, microfiche,* etc. Equally important is a knowledge of industrial printing and duplicating methods: *letterpress, lithography, xerography,* and so on.

Each of these processes has very definite advantages and disadvantages. In some, such as blueprinting, the original must be transparent, while in others, such as xerography, opaque or translucent originals may be used. In addition to copies, some reproducing equipment can make enlargements or reductions of the original, which often are extremely useful. A general familiarity with all of these repro-

duction processes is therefore absolutely necessary, since the reproduction method selected for a specific project will be dependent on the type of original used, number of copies required, size and appearance of copy desired, and the cost.

For a detailed explanation of these processes, we suggest that you consult your school or public library.

The average engineering drawing requires a considerable economic investment; therefore, adequate control and protection of the original drawing are mandatory. Such items as drawing numbers, methods of filing, microfilming, security files, print making and distribution, drawing changes, and retrieval of drawings are all inherent in proper drawing control.

A proper drawing-control system will enable those in charge of drawings (1) to know the location and status of the drawing at all times; (2) to minimize the damage to original drawings from the handling required for revisions, printing, and so on; and (3) to provide distribution of prints to proper persons.

Those organizations with large computer-aided design (CAD) systems use computer storage of finished drawings. In addition, digitized drawing information about frequently required components and elements, such as standard bolts, nuts, screws, pins, and piping valves, is stored in the computer for recall and placement on drawings as needed. CAD systems of this type are very expensive to purchase and maintain, but they have obvious advantages for production, storage, control, and recall of drawings. Most small and medium-size companies and a substantial number of large firms may not be able to justify the acquisition of high-capacity CAD systems and thus will continue to use the conventional methods of reproduction, storage, retrieval, and control of drawings described in this chapter. But CAD systems are becoming more compact and less expensive, and as advancing technology makes such systems more affordable, it is expected that many companies will make the change to computer graphics systems.

# _Design and Working Drawing Problems_

### Design Problems

The following suggestions for project assignments are of a general and very broad nature and it is expected that they will help generate many ideas for specific design projects. Much design work is undertaken to improve an existing product or system by utilization of new materials, new techniques, or new systems or procedures. In addition to the design of the product itself, another large amount of design work is essential for the tooling, production, and handling of the product. You are encouraged to discuss with your instructor any ideas you may have for a project.

1. Design new or improved playground, recreational, or sporting equipment. For example, a new child's toy could be both recreational and educational.
2. Design new or improved health equipment. For example, the physically handicapped have need for special equipment.
3. Design security or safety devices. Fire, theft, or poisonous gases are a threat to life and property.
4. Design devices and/or systems for waste handling. Home and factory waste disposal needs serious consideration.
5. Design new or improved educational equipment. Both teacher and student would welcome more efficient educational aids.
6. Design improvements in our land, sea, and air transportation systems. Vehicles, controls, highways, and airports need further refinement.
7. Design new or improved devices for material handling. A dispensing device for a powdered product is an example.
8. Improve the design of an existing device or system.
9. Design or redesign devices for improved portability.

Each solution to a design problem, whether prepared by an individual student or formulated by a group, should be in the form of a _report_, which should be typed or carefully lettered, assembled, and bound. Suitable folders or binders are usually available at the local school supply store. It is suggested that the report contain the following (or variations of them, as specified by your instructor):

1. A title sheet. The title of the design project should be placed in approximately the center of the sheet and in the lower right-hand corner place your name or the names of those in the group. The symbol PL should follow the name of the project leader.
2. Table of contents with page numbers.
3. Statement of the purpose of project with appropriate comments.
4. Preliminary design sketches, with comments on advantages and disadvantages of each, leading to the final selection of the _best_ solution. All work should be signed and dated.
5. An accurately made pictorial and/or assembly drawing(s), if more than one part is involved in the design.
6. Detail working drawings, freehand or mechanical as assigned. The 8.5″ × 11.0″ sheet size is preferred for convenient insertion in the report. Larger sizes may be bound in the report with appropriate folding.
7. A bibliography or credit for important sources of information, if applicable.

## Working Drawing Problems

The problems in Figs. 16.48 to 16.77 are presented to provide practice in making regular working drawings of the type used in industry. Many problems, especially those of the assemblies, offer an excellent opportunity for you to exercise your ability to redesign or improve upon the existing design. Owing to the variations in sizes and in scales that may be used, you are required to select the sheet sizes and scales, when these are not specified, subject to the approval of the instructor. Standard sheet layouts are shown inside the back cover of this book. See also §2.63.

The statements for each problem are intentionally brief, so that the instructor may amplify or vary the requirements when making assignments. Many problems lend themselves to the preferred metric system or the acceptable complete decimal-inch system, while others may be more suitable for a combination of fractional and decimal dimensions. Either the preferred unidirectional or acceptable aligned dimensioning may be assigned.

It should be clearly understood that in problems presented in pictorial form, the placement of dimensions and finish marks cannot always be followed in the drawing. *The dimensions given are in most cases those needed to make the parts, but owing to the limitations of pictorial drawings they are not in all cases the dimensions that should be shown on the working drawing.* In the pictorial problems the rough and finished surfaces are shown, but finish marks are usually omitted. You should add all necessary finish marks and place all dimensions in the preferred places in the final drawings.

Each problem should be preceded by a thumbnail sketch or a complete technical sketch, fully dimensioned. Any of the title blocks shown inside the back cover of this book may be used, with modification if desired, or you may design the title block if so assigned by the instructor.

Since many of the problems in this chapter are of a general nature, they can also be solved on most computer graphics systems. If a system is available, the instructor may choose to assign specific problems to be completed by this method.

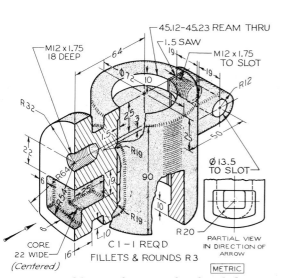

**Fig. 16.48** Table Bracket. Make detail drawing. Use Size B or A3 sheet.

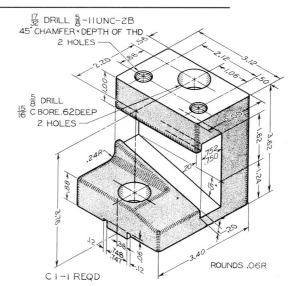

**Fig. 16.49** RH Tool Post. Make detail drawing. Use Size B or A3 sheet. If assigned, convert dimensions to metric system.

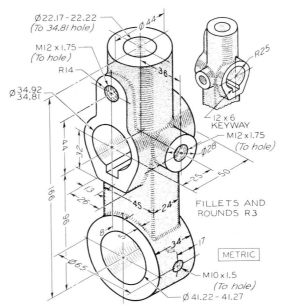

**Fig. 16.50** Idler Arm. Make detail drawing. Use Size B or A3 sheet. If assigned, convert dimensions to metric system.

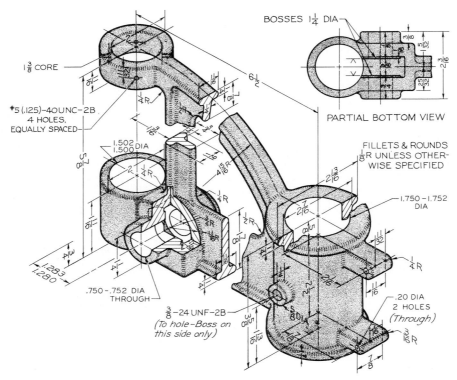

**Fig. 16.51** Drill Press Bracket. Make detail drawing. Use Size C or A2 sheet. If assigned, convert dimensions to decimal inches or redesign the part with metric dimensions.

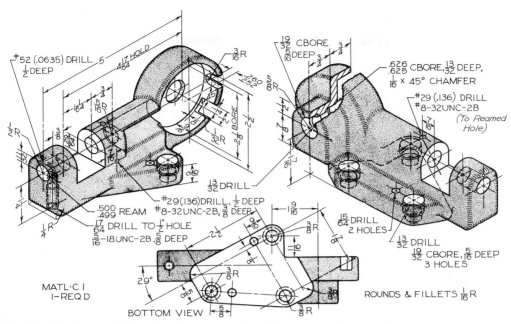

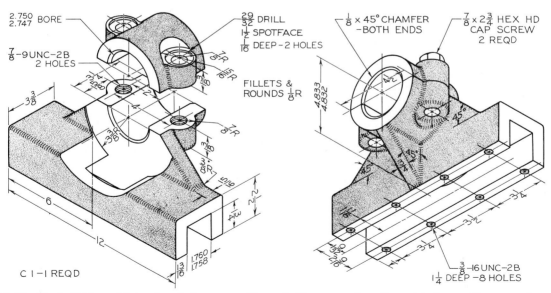

**Fig. 16.52** Dial Holder. Make detail drawing. Use Size C or A2 sheet. If assigned, convert dimensions to decimal inches or redesign the part with metric dimensions.

**Fig. 16.53** Rack Slide. Make detail drawings. Draw half size on Size B or A3 sheet. If assigned, convert dimensions to decimal inches or redesign the part with metric dimensions.

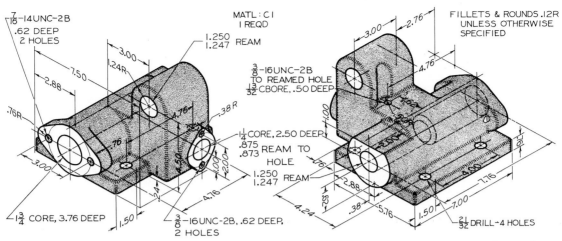

**Fig. 16.54** Automatic Stop Box. Make detail drawing. Draw half size on Size B or A3 sheet. If assigned, redesign the part with metric dimensions.

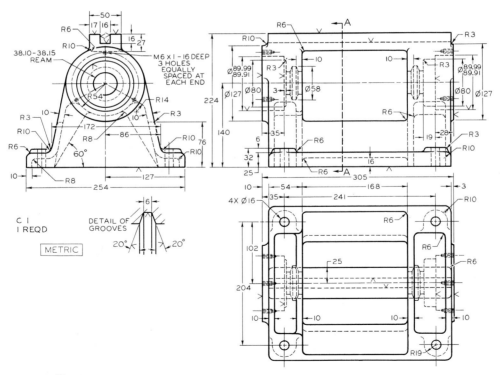

**Fig. 16.55** Spindle Housing.
Given: Front, left-side, and bottom views, and partial removed section.
Required: Front view in full section, top view, and right-side view in half section on **A–A**. Draw half size on Size C or A2 sheet. If assigned, dimension fully.

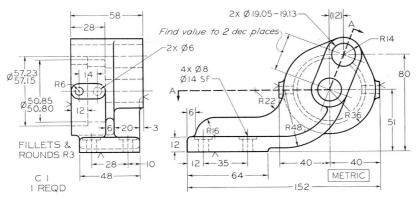

**Fig. 16.56** Pump Bracket for a Thread Milling Machine.
Given: Front and left-side views.
Required: Front and right-side views, and top view in section on A–A. Draw full size on Size B or A3 sheet. If assigned, dimension fully.

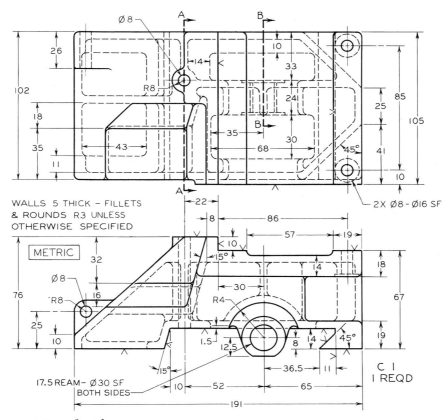

**Fig. 16.57** Support Base for Planer.
Given: Front and Top views.
Required: Front and top views, left-side view in full section A–A, and removed section B–B. Draw full size on Size C or A2 sheet. If assigned, dimension fully.

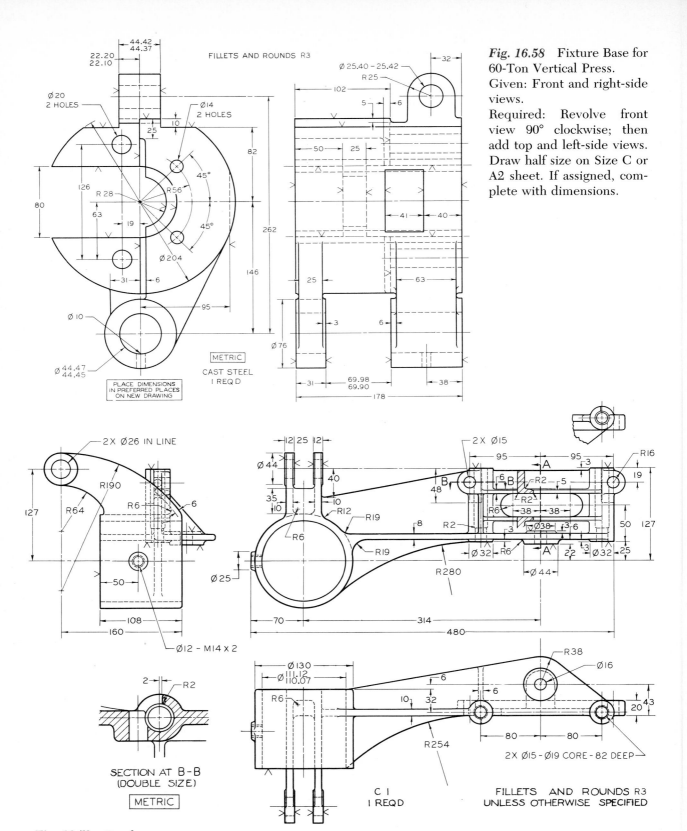

**Fig. 16.58** Fixture Base for 60-Ton Vertical Press.
Given: Front and right-side views.
Required: Revolve front view 90° clockwise; then add top and left-side views. Draw half size on Size C or A2 sheet. If assigned, complete with dimensions.

**Fig. 16.59** Bracket.
Given: Front, left-side, and bottom views, and partial removed section.
Required: Make detail drawing. Draw front, top, and right-side views, and removed sections A–A and B–B. Draw half size on size C or A2 sheet. Draw section B–B full size. If assigned, complete with dimensions.

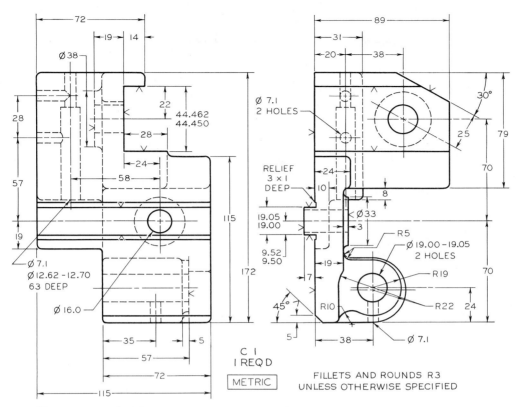

**Fig. 16.60** Roller Rest Bracket for Automatic Screw Machine.
Given: Front and left-side views.
Required: Revolve front view 90° clockwise; then add top and left-side views. Draw half size on Size C or A2 sheet. If assigned, complete with dimensions.

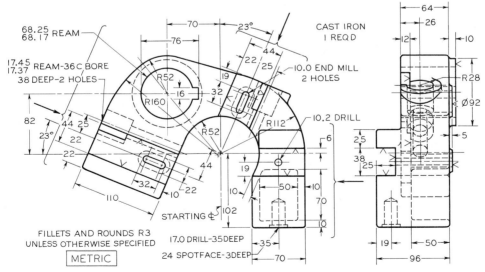

**Fig. 16.61** Guide Bracket for Gear Shaper.
Given: Front and right-side views.
Required: Front view, a partial right-side view, and two partial auxiliary views taken in direction of arrows. Draw half size on Size C or A2 sheet. If assigned, complete with unidirectional dimensions.

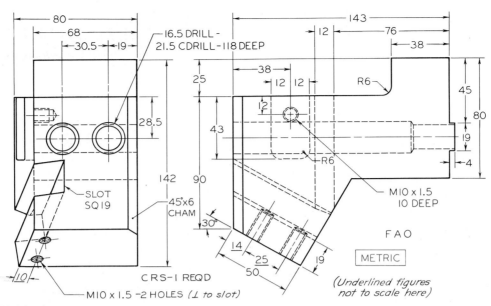

**Fig. 16.62**　Rear Tool Post.

Given: Front and left-side views.

Required: Take left-side view as new top view; add front and left-side views, approx. 215 mm apart, a primary auxiliary view, then a secondary view taken so as to show true end view of 19 mm slot. Complete all views, except show only necessary hidden lines in auxiliary views. Draw full size on Size C or A2 sheet. If assigned, complete with dimensions.

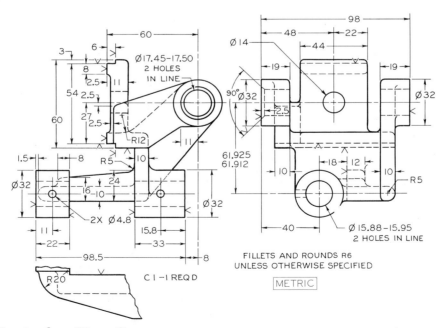

**Fig. 16.63**　Bearing for a Worm Gear.

Given: Front and right-side views.

Required: Front, top, and left-side views. Draw full size on Size C or A2 sheet. If assigned, complete with dimensions.

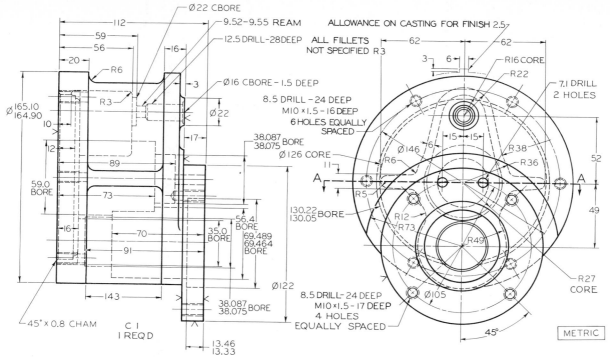

**Fig. 16.64** Generator Drive Housing.
Given: Front and left-side views.
Required: Front view, right-side in full section, and top view in full section on A–A. Draw full size on Size C or A2 sheet. If assigned, complete with dimensions.

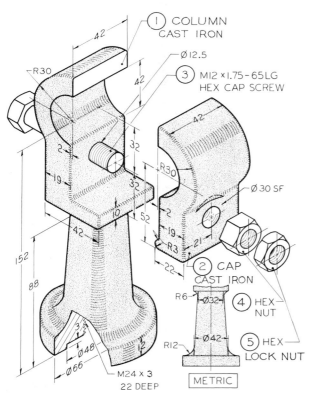

**Fig. 16.65** Hand Rail Column. (1) Draw details. If assigned, complete with dimensions. (2) Draw assembly.

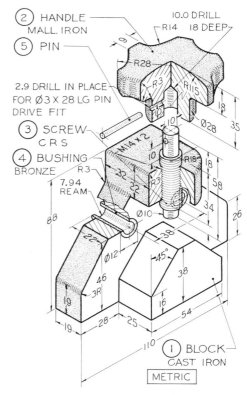

**Fig. 16.66** Drill Jig (1) Draw details. If assigned, complete with dimensions. (2) Draw assembly.

439

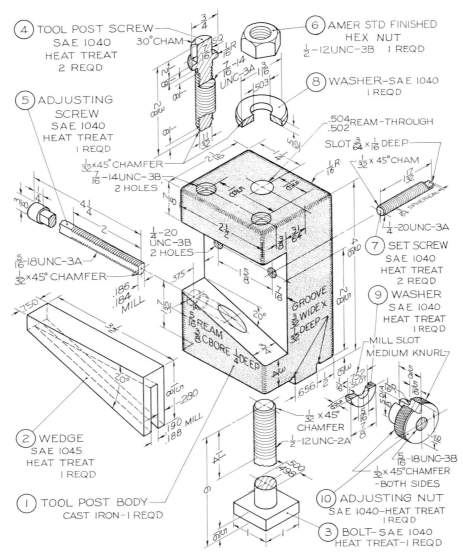

**Fig. 16.67** Tool Post. (1) Draw details. (2) Draw assembly. If assigned, use unidirectional two-place decimals for all fractional dimensions or redesign for all metric dimensions.

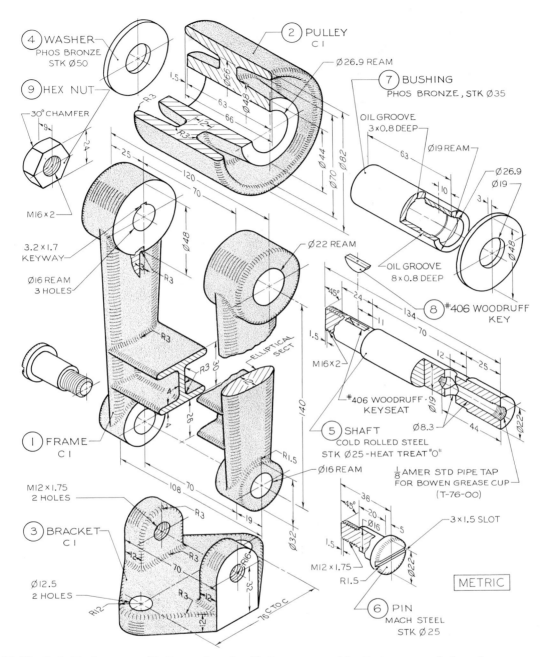

***Fig. 16.68*** Belt Tightener. (1) Draw details. (2) Draw assembly. It is assumed that the parts are to be made in quantity and they are to be dimensioned for interchangeability on the detail drawings. Use tables in Appendixes 11 to 14 for limit values. Design as follows.

a. Bushing fit in pulley: Locational interference fit.
b. Shaft fit in bushing: Free running fit.
c. Shaft fits in frame: Sliding fit.
d. Pin fit in frame: Free running fit.
e. Pulley hub length plus washers fit in frame: Allowance 0.13 and tolerances 0.10.
f. Make bushing 0.25mm shorter than pulley hub.
g. Bracket fit in frame: Same as e above.

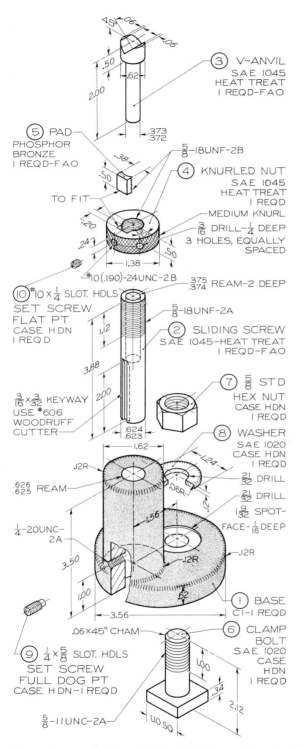

③ V-ANVIL
SAE 1045
HEAT TREAT
1 REQD-FAO

⑤ PAD
PHOSPHOR
BRONZE
1 REQD-FAO

TO FIT

⅝-18UNF-2B

④ KNURLED NUT
SAE 1045
HEAT TREAT
1 REQD

MEDIUM KNURL

3/16 DRILL-¼ DEEP
3 HOLES, EQUALLY
SPACED

#10(.190)-24UNC-2B

⑩ #10 × ¼ SLOT. HDLS
SET SCREW
FLAT PT
CASE HDN
1 REQD

.375 .374 REAM-2 DEEP

⅝-18UNF-2A

② SLIDING SCREW
SAE 1045-HEAT TREAT
1 REQD-FAO

3/16 × 3/32 KEYWAY
USE #606
WOODRUFF
CUTTER

⑦ ⅝ STD
HEX NUT
CASE HDN
1 REQD

⑧ WASHER
SAE 1020
CASE HDN
1 REQD

21/32 DRILL

21/32 DRILL
1 9/32 SPOT-
FACE-1/16 DEEP

.626 .625 REAM

¼-20UNC-2A

① BASE
CI-1 REQD

⑨ ¼ × ⅝ SLOT. HDLS
SET SCREW
FULL DOG PT
CASE HDN-1 REQD

.06×45° CHAM

⑥ CLAMP
BOLT
SAE 1020
CASE
HDN
1 REQD

⅝-11UNC-2A

**Fig. 16.69** Milling Jack. (1) Draw details. (2) Draw assembly. If assigned, convert dimensions to metric or decimal-inch system.

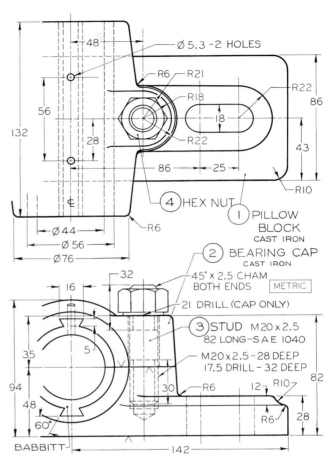

Ø 5.3 -2 HOLES

R6  R21
R18
R22
18
R22

R10

④ HEX NUT

① PILLOW
BLOCK
CAST IRON

② BEARING CAP
CAST IRON

45° × 2.5 CHAM
BOTH ENDS   METRIC

21 DRILL (CAP ONLY)

③ STUD M20 × 2.5
82 LONG-SAE 1040

M20×2.5-28 DEEP
17.5 DRILL - 32 DEEP

R6   R10
R6

BABBITT

**Fig. 16.70** Pillow Block Bearing. (1) Draw details. (2) Draw assembly. If assigned, complete with dimensions.

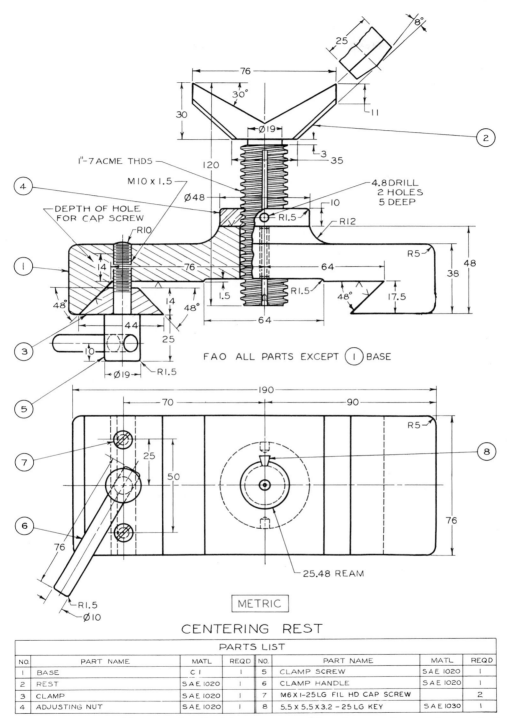

FAO ALL PARTS EXCEPT (1) BASE

METRIC

CENTERING REST

| NO. | PART NAME | MATL | REQD | NO. | PART NAME | MATL | REQD |
|---|---|---|---|---|---|---|---|
| | | | | | PARTS LIST | | |
| 1 | BASE | C I | 1 | 5 | CLAMP SCREW | SAE 1020 | 1 |
| 2 | REST | SAE 1020 | 1 | 6 | CLAMP HANDLE | SAE 1020 | 1 |
| 3 | CLAMP | SAE 1020 | 1 | 7 | M6X1-25LG FIL HD CAP SCREW | | 2 |
| 4 | ADJUSTING NUT | SAE 1020 | 1 | 8 | 5.5 X 5.5 X 3.2 - 25 LG KEY | SAE 1030 | 1 |

*Fig. 16.71* Centering Rest. (1) Draw details. (2) Draw assembly. If assigned, complete with dimensions.

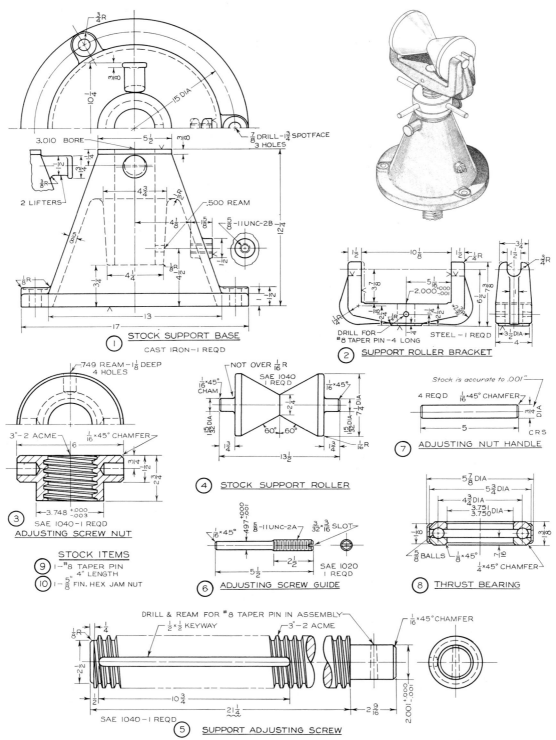

**Fig. 16.72** Stock Bracket for Cold Saw Machine. (1) Draw details. (2) Draw assembly. If assigned, use unidirectional decimal-inch dimensions or redesign for metric dimensions.

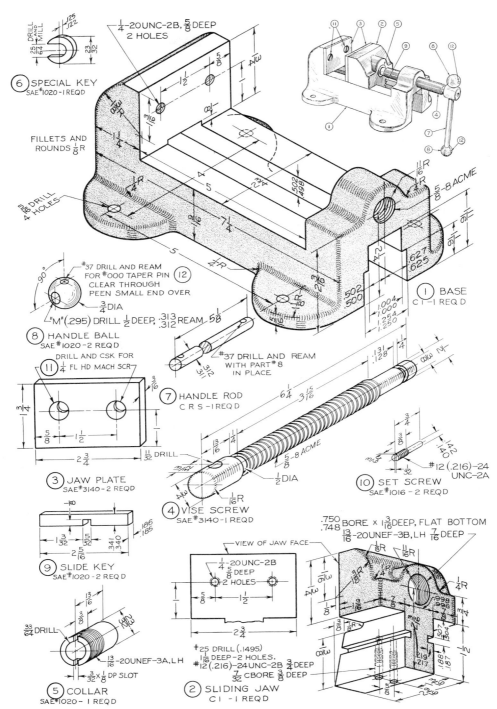

**Fig. 16.73  Machine Vise.**  (1) Draw details. (2) Draw assembly. If assigned, convert dimensions to the decimal-inch system or redesign with metric dimensions.

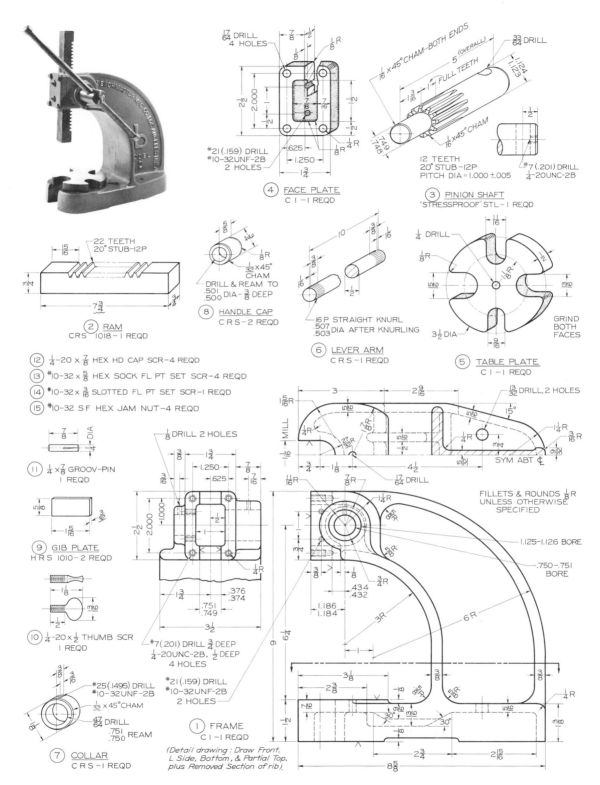

**Fig. 16.74** Arbor Press.   (1) Draw details. (2) Draw assembly. If assigned, convert dimensions to decimal inches or redesign for metric dimensions.

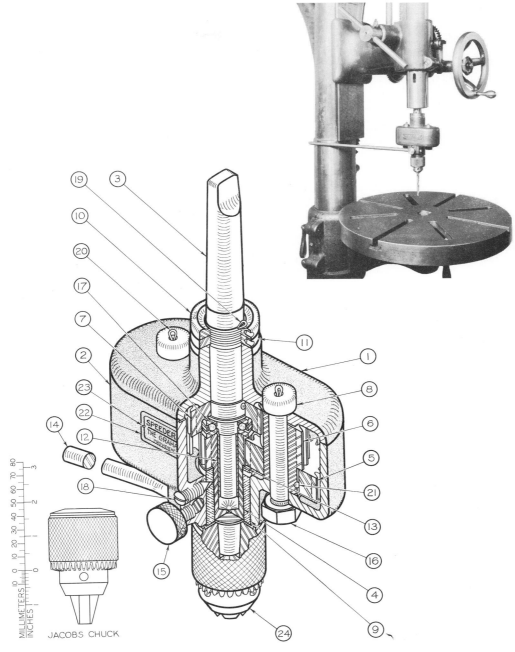

**MILLIMETERS**
**INCHES**

JACOBS CHUCK

*Fig. 16.75* Drill Speeder. See Figs. 16.76 and 16.77.

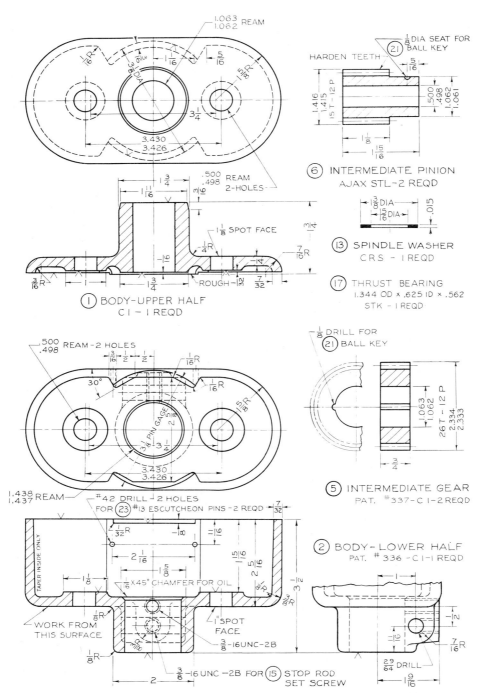

**Fig. 16.76** Drill Speeder (Continued). (1) Draw details. (2) Draw assembly. See Fig. 16.75. If assigned, convert dimensions to decimal inches or redesign with metric dimensions.

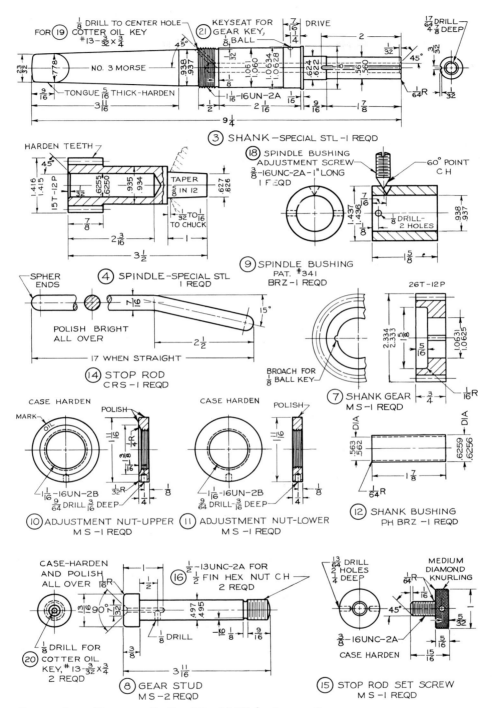

**Fig. 16.77** Drill Speeder (Continued). See Fig. 16.76 for instructions.

# CHAPTER
# 17

# Axonometric
# Projection

As described in Chapter 7, multiview drawing makes it possible to represent accurately the most complex forms of a design by showing a series of exterior views and sections. This type of representation has, however, two limitations: its execution requires a thorough understanding of the principles of multiview projection and its reading requires a definite exercise of the constructive imagination.

Frequently, it is necessary to prepare drawings for the presentation of a design idea that are accurate and scientifically correct and can be easily understood by persons without technical training. Such drawings show several faces of an object at once, approximately as they appear to the observer. This type of drawing is called a *pictorial drawing* [ANSI Y14.4–1957 (R1987)]. Since pictorial drawing shows only the appearances of parts or devices, it is not satisfactory for completely describing complex or detailed forms.

Pictorial drawing enables the person without technical training to visualize the design represented. It also enables the designer to visualize the successive stages of the design and to develop it in a satisfactory manner.

Various types of pictorial drawing are used extensively in catalogs, in general sales literature, and also in technical work,* to supplement and amplify multiview drawings. For example, pictorial drawing is used in Patent Office drawings, in piping diagrams, in machine, structural, and architectural designs, and in furniture design.

*Practically all of the pictorial drawings in this book were drawn by the methods described in Chapters 17 and 18. See especially Figs. 7.54 to 7.93 for examples.

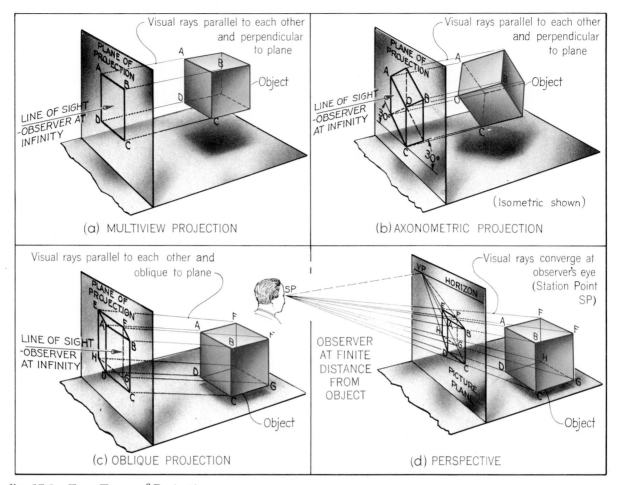

**Fig. 17.1**  Four Types of Projection.

## 17.1 Methods of Projection

The four principal types of projection are illustrated in Fig. 17.1, and all except the regular multiview projection, (a), are pictorial types since they show several sides of the object in a single view. In all cases the views, or projections, are formed by the piercing points in the plane of projection of an infinite number of visual rays or projectors.

In both multiview projection, (a), and *axonometric projection,* (b), the observer is considered to be at infinity, and the visual rays are parallel to each other and perpendicular to the plane of projection. Therefore, both are classified as *orthographic projections,* §1.10.

In *oblique projection,* (c) the observer is considered to be at infinity, and the visual rays are parallel to each other but oblique to the plane of projection. See Chapter 18.

In *perspective,* (d), the observer is considered to be at a finite distance from the object, and the visual rays extend from the observer's eye, or the station point (SP), to all points of the object to form a "cone of rays." See Fig. 1.10.

## 17.2 Types of Axonometric Projection

The distinguishing feature of axonometric projection, as compared to multiview projection, is the inclined position of the object with respect to the plane of projection. Since the principal edges and surfaces of the object are inclined to the plane of projection, the lengths of the lines, the sizes of the angles, and the general proportions of the object vary with the infinite number of possible positions in which the object may be placed with respect to the plane of projection. Three of these are shown in Fig. 17.2.

In these cases the edges of the cube are inclined to the plane of projection and are therefore foreshortened. See Fig. 7.20 (c). The degree of fore-

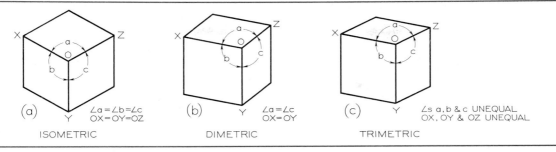

*Fig. 17.2* Axonometric Projections.

shortening of any line depends on its angle with the plane of projection; the greater the angle, the greater the foreshortening. If the degree of foreshortening is determined for each of the three edges of the cube that meet at one corner, scales can be easily constructed for measuring along these edges or any other edges parallel to them. See Figs. 17.41 (a) and 17.46.

It is customary to consider three edges of the cube that meet at the corner nearest the observer as the *axonometric axes.* In Fig. 17.1 (b), the axonometric axes, or simply the *axes,* are OA, OB, and OC. As shown in Fig. 17.2, axonometric projections are classified as (a) *isometric projection,* (b) *dimetric projection,* and (c) *trimetric projection,* depending upon the number of scales of reduction required.

# Isometric Projection

## 17.3 The Isometric Method of Projection

To produce an isometric projection (isometric means "equal measure"), it is necessary to place the object so that its principal edges, or axes, make equal angles with the plane of projection and are therefore foreshortened equally. See Fig. 6.24. In this position the edges of a cube would be projected equally and

would make equal angles with each other (120°), as shown in Fig. 17.2 (a).

In Fig. 17.3 (a) is shown a multiview drawing of a cube. At (b) the cube is shown revolved through 45° about an imaginary vertical axis. Now an auxiliary view in the direction of the arrow will show the cube diagonal ZW as a point, and the cube appears

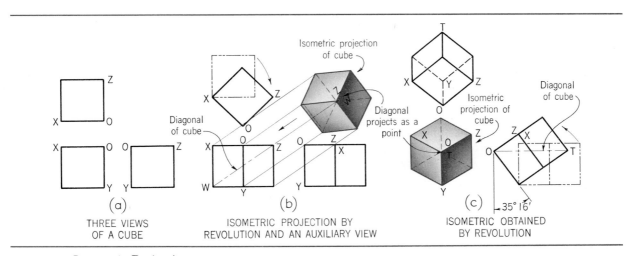

*Fig. 17.3* Isometric Projection.

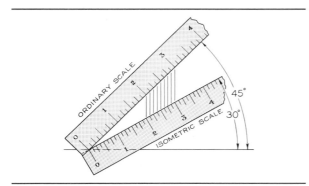

*Fig. 17.4*   Isometric Scale.

as a true isometric projection. However, instead of the auxiliary view at (b) being drawn, the cube may be further revolved as shown at (c), this time the cube being tilted forward about an imaginary horizontal axis until the three edges OX, OY, and OZ make equal angles with the frontal plane of projection and are therefore foreshortened equally. Here again, a diagonal of the cube, in this case OT, appears as a point in the isometric view. The front view thus obtained is a true isometric projection. In this projection the twelve edges of the cube make angles of about 35° 16′ with the frontal plane of projection. The lengths of their projections are equal to the lengths of the edges multiplied by $\sqrt{\frac{2}{3}}$, or by 0.816, approximately. Thus the projected lengths are about 80 percent of the true lengths, or still more roughly, about three-fourths of the true lengths. The projections of the axes OX, OY, and OZ make angles of 120° with each other and are called the *isometric axes.* Any line parallel to one of these is called an *isometric line;* a line that is not parallel

is called a *nonisometric line.* It should be noted that the angles in the isometric projection of the cube are either 120° or 60° and that all are projections of 90° angles. In an isometric projection of a cube, the faces of the cube, and any planes parallel to them, are called *isometric planes.*

## 17.4 The Isometric Scale

A correct isometric projection may be drawn with the use of a special isometric scale, prepared on a strip of paper or cardboard, Fig. 17.4. All distances in the isometric scale are $\sqrt{\frac{2}{3}}$ times true size, or approximately 80 percent of true size. The use of the isometric scale is illustrated in Fig. 17.5 (a). A scale of 9″ = 1′-0, or $\frac{3}{4}$-size scale (or metric equivalent), could be used to approximate the isometric scale.

## 17.5 Isometric Drawing

When a drawing is prepared with an isometric scale, or otherwise as the object is actually *projected* on a plane of projection, it is an *isometric projection,* as illustrated in Fig. 17.5 (a). When it is prepared with an ordinary scale, it is an *isometric drawing,* illustrated at (b). The isometric drawing, (b), is about 25 percent larger than the isometric projection (a), but the pictorial value is obviously the same in both.

Since the isometric projection is foreshortened and an isometric drawing is full-scale size, it is usually advantageous to make an isometric drawing rather than an isometric projection. The drawing is much easier to execute and, for all practical purposes, is just as satisfactory as the isometric projection.

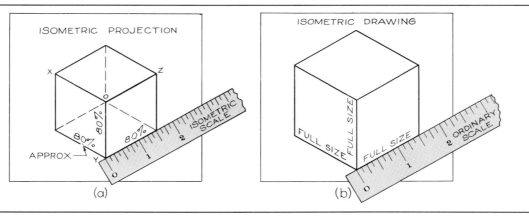

*Fig. 17.5*   Isometric and Ordinary Scales.

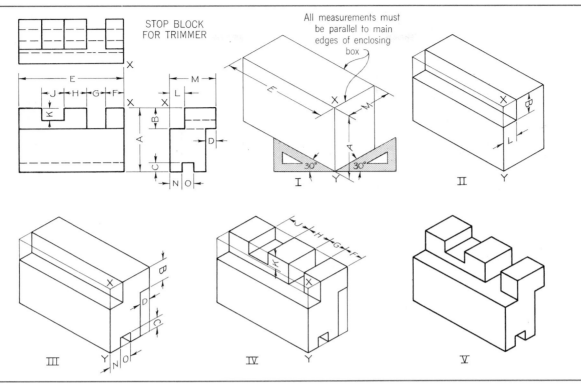

**Fig. 17.6** Isometric Drawing of Normal Surfaces.

## 17.6 Steps in Making an Isometric Drawing

The steps in constructing an isometric drawing of an object composed only of normal surfaces, §7.19, are illustrated in Fig. 17.6. Notice that all measurements are made parallel to the main edges of the enclosing box, that is, parallel to the isometric axes. No measurement along a diagonal (nonisometric line) on any surface or through the object can be set off directly with the scale. The object may be drawn in the same position by beginning at the corner Y, or any other corner, instead of at the corner X.

The method of constructing an isometric drawing of an object composed partly of inclined surfaces (and oblique edges) is shown in Fig. 17.7. Notice that inclined surfaces are located by *offset* or *coordinate measurements* along the isometric lines. For example, dimensions E and F are set off to locate

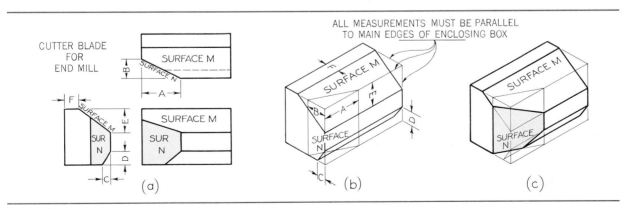

**Fig. 17.7** Inclined Surfaces in Isometric.

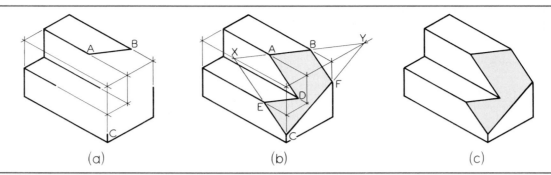

Fig. 17.8   Oblique Surfaces in Isometric.

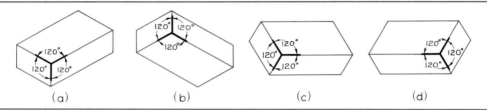

Fig. 17.9   Positions of Isometric Axes.

the inclined surface M, and dimensions A and B are used to locate surface N.

For sketching in isometric, see §§6.12 to 6.14.

## 17.7 Oblique Surfaces in Isometric

Oblique surfaces in isometric may be drawn by establishing the intersections of the oblique surface with the isometric planes. For example, in Fig. 17.8 (a), the oblique plane is known to contain points A, B, and C. To establish the plane, (b), line AB is extended to X and Y, which are in the same isometric planes as C. Lines XC and YC locate points E and F. Finally AD and ED are drawn, using the rule of parallelism of lines. The completed drawing is shown at (c).

## 17.8 Other Positions of the Isometric Axes

The isometric axes may be placed in any desired position according to the requirements of the problem, as shown in Fig. 17.9, but the angle between the axes must remain 120°. The choice of the directions of the axes is determined by the position from which the object is usually viewed, Fig. 17.10, or by the position that best describes the shape of the

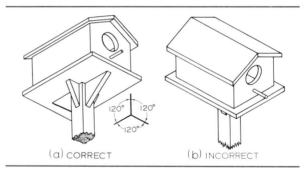

Fig. 17.10   An Object Naturally Viewed from Below.

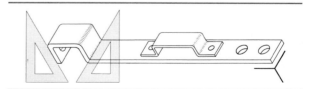

Fig. 17.11   Long Axis Horizontal.

object. If possible, both requirements should be met.

If the object is characterized by considerable length, the long axis may be placed horizontally for best effect, as shown in Fig. 17.11.

## 17.9 Offset Location Measurements

The method of locating one point with respect to another is illustrated in Figs. 17.12 and 17.13. In each case, after the main block has been drawn, the

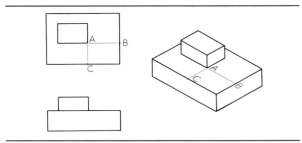

Fig. 17.12   Offset Location Measurements.

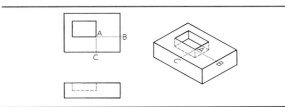

Fig. 17.13   Offset Location Measurements.

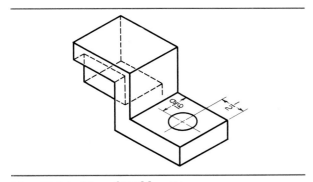

Fig. 17.14   Use of Hidden Lines.

offset lines CA and BA in the multiview drawing are drawn full size in the isometric drawing, thus, locating corner A of the small block or rectangular recess. These measurements are called *offset measurements*, and since they are parallel to certain edges of the main block in the multiview drawings, they will be parallel, respectively, to the same edges in the isometric drawings, §7.25.

## 17.10 Hidden Lines

The use of hidden lines in isometric drawing is governed by the same rules as in all other types of projection: *Hidden lines are omitted unless they are needed to make the drawing clear.* A case in which hidden lines are needed is illustrated in Fig. 17.14, in which a projecting part cannot be clearly shown without the use of hidden lines.

## 17.11 Center Lines

The use of center lines in isometric drawing is governed by the same rules as in multiview drawing: *Center lines are drawn if they are needed to indicate symmetry or if they are needed for dimensioning,* Fig. 17.14. In general, center lines should be used sparingly and omitted in cases of doubt. The use of too many center lines may produce a confusion of lines, which diminishes the clearness of the drawing. Examples in which center lines are not needed are shown in Figs. 17.10 and 17.11. Examples in which they are needed are seen in Figs. 17.14 and 17.39 (a).

## 17.12 Box Construction

Objects of rectangular shape may be more easily drawn by means of *box construction*, which consists simply in imagining the object to be enclosed in a rectangular box whose sides coincide with the main faces of the object. For example, in Fig. 17.15

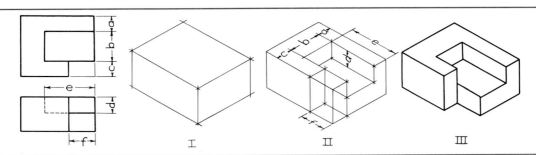

Fig. 17.15   Box Construction.

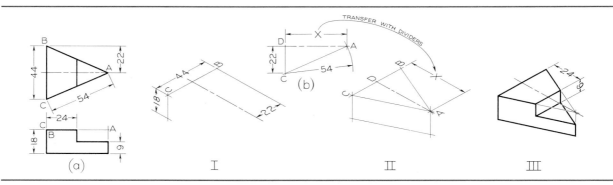

*Fig. 17.16*   Nonisometric Lines   (metric dimensions).

the object shown in two views is imagined to be enclosed in a construction box. This box is then drawn lightly with construction lines, as shown at I, the irregular features are then constructed as shown at II, and finally, as shown at III, the required lines are made heavy.

## 17.13 Nonisometric Lines

Since the only lines of an object that are drawn true length in an isometric drawing are the isometric axes or lines parallel to them, *nonisometric* lines cannot be set off directly with the scale. For example, in Fig. 17.16 (a), the inclined lines BA and CA are shown in their true lengths 54 mm in the top view, but since they are not parallel to the isometric axes, they will not be true length in the isometric. Such lines are drawn in isometric by means of box construction and offset measurements. First, as shown at I, the measurements 44 mm, 18 mm, and 22 mm can be set off directly since they are made along isometric lines. The nonisometric 54 mm dimension cannot be set off directly, but if one-half of the given top view is constructed full size to scale as shown at

(b), the dimension X can be determined. This dimension is parallel to an isometric axis and can be transferred with dividers to the isometric at II. The dimensions 24 mm and 9 mm are parallel to isometric lines and can be set off directly, as shown at III.

To realize the fact that nonisometric lines will not be true length in the isometric drawing, set your dividers on BA of II and then compare with BA on the given top view at (a). Do the same for line CA. It will be seen that BA is shorter and CA is longer in the isometric than the corresponding lines in the given views.

## 17.14 Angles in Isometric

As shown in §7.26, angles project true size only when the plane of the angle is parallel to the plane of projection. An angle may project larger or smaller than true size, depending upon its position. Since in isometric the various surfaces of the object are usually inclined to the plane of projection, it follows that angles generally will not be projected true size. For example, in the multiview drawing in Fig. 17.17

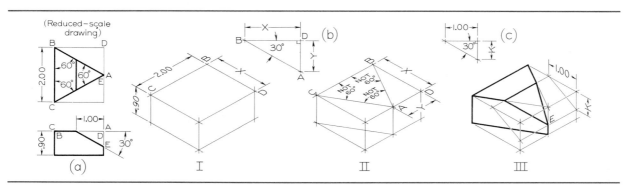

*Fig. 17.17*   Angles in Isometric.

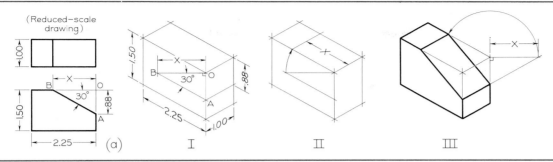

*Fig. 17.18*   Angle in Isometric.

(a), none of the three 60° angles will be 60° in the isometric drawing. To realize this fact, measure each angle in the isometric of II with the protractor and note the number of degrees compared to the true 60°. No two angles are the same; two are smaller and one larger than 60°.

As shown in I, the enclosing box can be drawn from the given dimensions, except for dimension **X**, which is not given. To find dimension **X**, draw triangle **BDA** from the top view full size, as shown at (b). Transfer dimension **X** to the isometric in I, to complete the enclosing box.

In order to locate point **A** in II, dimension **Y** must be used, but this is not given in the top view, (a). Dimension **Y** is found by the same construction, (b), and then transferred to the isometric, as shown. The completed isometric is shown at III where point **E** is located by using dimension **K**, as shown.

Thus, in order to set off angles in isometric, the regular protractor cannot be used.* *Angular measurements must be converted to linear measurements along isometric lines.*

In Fig. 17.18 (a) are two views of an object to be

*Isometric protractors for setting off angles on isometric surfaces are available from drafting supplies dealers.

drawn in isometric. Point **A** can easily be located in the isometric, step I, by measuring .88″ down from point **O**. However, in the given drawing at (a) the location of point **B** depends upon the 30° angle, and to locate **B** in the isometric linear dimension **X** must be known. This distance can be found graphically by drawing the right triangle **BOA** attached to the isometric, as shown. The distance **X** is then transferred to the isometric with the compass or dividers, as shown at II. Actually, the triangle could be attached in several different positions. One of these is shown at III.

When angles are given in degrees, it is necessary to convert the angular measurements into linear measurements. This is best done by drawing a right triangle separately, as in Fig. 17.17 (b), or attached to the isometric, as in Fig. 17.18.

## 17.15 Irregular Objects

If the general shape of an object does not conform somewhat to a rectangular pattern, as shown in Fig. 17.19, it may be drawn as shown at (a) by using the box construction discussed previously. Various points of the triangular base are located by means of offsets **a** and **b** along the edges of the bottom of

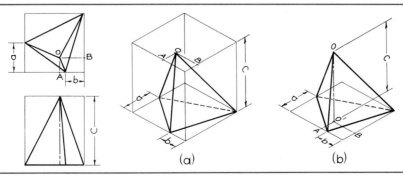

*Fig. 17.19*   Irregular Object in Isometric.

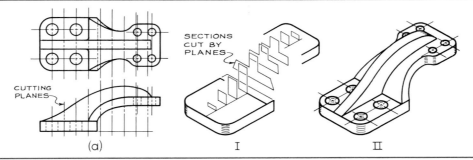

*Fig. 17.20* Use of Sections in Isometric.

the construction box. The vertex is located by means of offsets **OA** and **OB** on the top of the construction box.

However, it is not necessary to draw the complete construction box. If only the bottom of the box is drawn, as shown at (b), the triangular base can be constructed as before. The orthographic projection of the vertex O′ on the base can then be located by offsets O′A and O′B, as shown, and from this point, the vertical center line O′O can be erected, using measurement C.

An irregular object may be drawn by means of a series of sections, as illustrated in Fig. 17.20. The edge views of a series of imaginary cutting planes are shown in the top and front views of the multiview drawing at (a). At I the various sections are constructed in isometric, and at II the object is completed by drawing lines through the corners of the sections. In the isometric at I, all height dimensions are taken from the front view at (a), and all depth dimensions from the top view.

## 17.16 Curves in Isometric

Curves may be drawn in isometric by means of a series of offset measurements similar to those discussed in §17.9. In Fig. 17.21 any desired number of points, such as A, B, and C, are selected at random along the curve in the given top view at (a). Enough points should be chosen to fix accurately the path of the curve; the more points used, the greater the accuracy. Offset grid lines are then drawn from each point parallel to the isometric axes.

As shown at I, offset measurements a and b are laid off in the isometric to locate point A on the curve. Points B, C, and D are located in a similar manner, as shown at II. A light freehand curve is sketched smoothly through the points as shown at III. Points A′, B′, C′, and D′ are located directly under points A, B, C, and D, as shown at IV, by drawing vertical lines downward, making all equal to dimension C, the height of the block. A light freehand curve is then drawn through the points. The final curve is heavied in with the aid of the

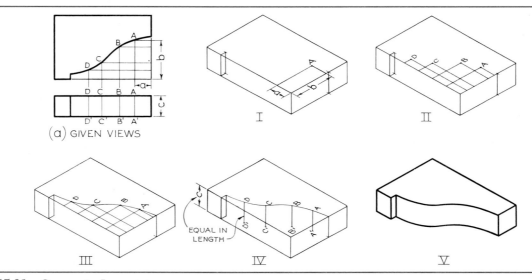

*Fig. 17.21* Curves in Isometric.

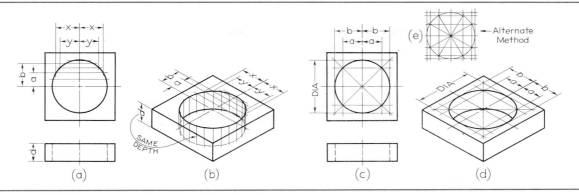

**Fig. 17.22** True Isometric Ellipse Construction.

irregular curve, §2.54, and all straight lines are darkened to complete the isometric at V.

## 17.17 True Ellipses in Isometric

As shown in §§5.51, 6.13, and 7.30, if a circle lies in a plane that is not parallel to the plane of projection, the circle will be projected as a true ellipse. The ellipse can be constructed by the method of offsets, §17.16. As shown in Fig. 17.22 (a), draw parallel lines, spaced at random, across the circle; then transfer these lines to the isometric as shown at (b), with the aid of the dividers. To locate points in the lower ellipse, transfer points of the upper ellipse down a distance equal to the height d of the block and draw the ellipse, part of which will be hidden, through these points. Draw the final ellipses with the aid of the irregular curve §2.54.

A variation of the method of offsets, which provides eight points on the ellipse, is illustrated at (c) and (d). If more points are desired, parallel lines, as at (a), can be added. As shown at (c), circumscribe a square around the given circle, and draw diagonals. Through the points of intersection of the diagonals and the circle, draw another square, as shown. Draw this construction in the isometric, as shown at (d), transferring distances a and b with the dividers.

A similar method that provides twelve points on the ellipse is shown at (e). The given circle is divided into twelve equal parts, using the 30° × 60° triangle, Fig. 2.23. Lines parallel to the sides of the square are drawn through these points. The entire construction is then drawn in isometric, and the ellipse is drawn through the points of intersection.

When the center lines shown in the top view at

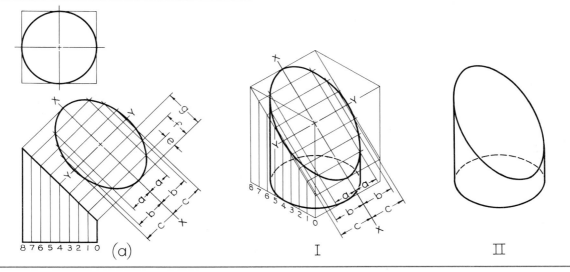

**Fig. 17.23** Ellipse in Inclined Plane.

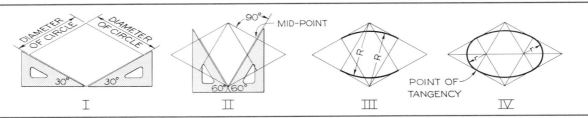

*Fig. 17.24*   Steps in Drawing Four-Center Ellipse.

(a) are drawn in isometric, (b), they become the *conjugate diameters* of the ellipse. The ellipse can then be constructed on the conjugate diameters by the methods of Figs. 5.51 and 5.52 (b).

When the 45° diagonals at (c) are drawn in isometric at (d), they coincide with the major and minor axes of the ellipse, respectively. Note that the minor axis is equal in length to the sides of the inscribed square at (c). The ellipse can be constructed upon the major and minor axes by any of the methods in §§5.48 and 5.51.

Remember the rule: *The major axis of the ellipse is always at right angles to the center line of the cylinder, and the minor axis is at right angles to the major axis and coincides with the center line.*

Accurate ellipses may be drawn with the aid of ellipse guides, §§5.56 and 17.21, or with a special *ellipsograph.*

If the curve lies in a nonisometric plane, all offset measurements cannot be applied directly. For example, in Fig. 17.23 (a) the elliptical face shown in the auxiliary view lies in an inclined nonisometric plane. The cylinder is enclosed in a construction box, and the box is then drawn in isometric, as shown at I. The base is drawn by the method of offsets, as shown in Fig. 17.22. The inclined ellipse is constructed by locating a number of points on the ellipse in the isometric and drawing the final curve by means of the irregular curve, §2.54.

Measurements a, b, c, and so on, are parallel to an isometric axis and can be set off in the isometric at I on each side of the center line X–X, as shown. Measurements e, f, g, and so on, are not parallel to any isometric axis and cannot be set off directly in isometric. However, when these measurements are projected to the front view and down to the base, as shown at (a), they can then be set off along the lower edge of the construction box, as shown at I. The completed isometric is shown at II.

The ellipse may also be drawn with the aid of an appropriate ellipse template selected to fit the major and minor axes established along X–X and Y–Y, respectively. See Fig. 5.55.

## 17.18 Approximate Four-Center Ellipse

An approximate ellipse is sufficiently accurate for nearly all isometric drawings. The method commonly used, called the *four-center ellipse*, is illustrated in Figs. 17.24, 17.25, and 17.26. It can be used only for ellipses in isometric planes.

To apply this method, Fig. 17.24, draw, or conceive to be drawn, a square around the given circle in the multiview drawing; then

I.   Draw the isometric of the square, which is an equilateral parallelogram whose sides are equal to the diameter of the circle.

II.   Erect perpendicular bisectors to each side, using the 30° × 60° triangle as shown. These perpendiculars will intersect at four points, which will be centers for the four circular arcs.

III.   Draw the two large arcs, with radius R, from the intersections of the perpendiculars in the two closest corners of the parallelogram, as shown.

IV.   Draw the two small arcs, with radius r, from the intersections of the perpendiculars within the parallelogram, to complete the ellipse. As a check on the accurate location of these centers, a long diagonal of the parallelogram may be drawn, as shown. The midpoints of the sides of the parallelogram are points of tangency for the four arcs.

A typical drawing with cylindrical shapes is illustrated in Fig. 17.25. Note that the centers of the larger ellipse cannot be used for the smaller ellipse, though the ellipses represent concentric circles. Each ellipse has its own parallelogram and its own centers. Observe also that the centers of the lower ellipse are obtained by projecting the centers of the upper large ellipse down a distance equal to the height of the cylinder.

The construction of the four-center ellipse upon the three visible faces of a cube is shown in Fig. 17.26, a study of which shows that all diagonals are horizontal or 60° with horizontal; hence, the entire construction is made with the T-square and 30° × 60° triangle.

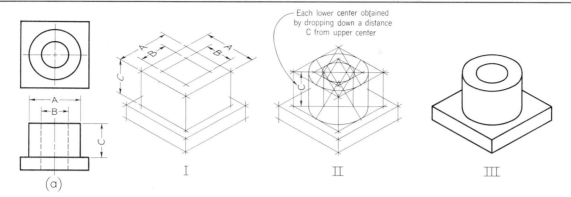

Fig. 17.25 Isometric Drawing of a Bearing.

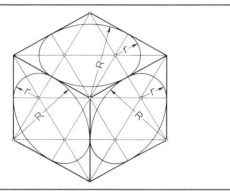

Fig. 17.26 Four-Center Ellipses.

Actually the four-center ellipse deviates considerably from the true ellipse. As shown in Fig. 17.27 (a), the four-center ellipse is somewhat shorter and "fatter" than the true ellipse. In constructions where tangencies or intersections with the four-center ellipse occur in the zones of error, the four-center ellipse is unsatisfactory, as shown at (b) and (c).

For a much closer approximation to the true ellipse, the Orth four-center ellipse, Fig. 17.28, which requires only one more step than the regular four-center ellipse, will be found sufficiently accurate for almost any problem.

When it is more convenient to start with the

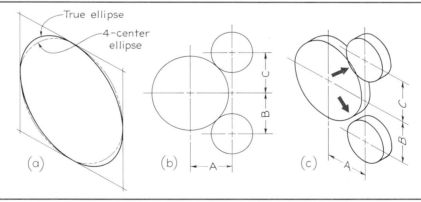

Fig. 17.27 Faults of Four-Center Ellipse.

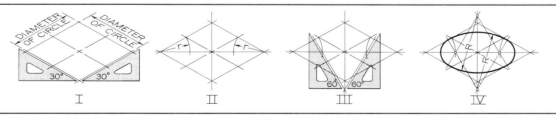

Fig. 17.28 Orth Four-Center Ellipse. *Courtesy of Professor H. D. Orth*

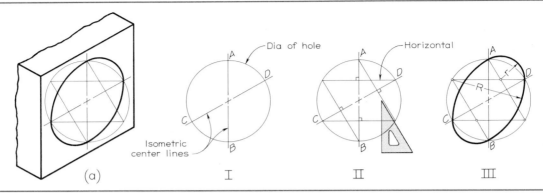

*Fig. 17.29* Alternate Four-Center Ellipse.

isometric center lines of a hole or cylinder in drawing the ellipse, rather than the enclosing parallelogram, the *alternate four-center ellipse* is recommended, Fig. 17.29. A completely constructed ellipse is shown at (a), and the steps followed are shown at the right in the figure.

I.    Draw the isometric center lines. From the center, draw a construction circle equal to the actual diameter of the hole or cylinder. The circle will intersect the center lines at four points A, B, C, and D.

II.    From the two intersection points on one center line, erect perpendiculars to the other center line; then from the two intersection points on the other center line, erect perpendiculars to the first center line.

III.    With the intersections of the perpendiculars as centers, draw two small arcs and two large arcs, as shown.

NOTE    The above steps are exactly the same as for the regular four-center ellipse of Fig. 17.24 except for the use of the isometric center lines instead of the enclosing parallelogram.

## 17.19 Screw Threads in Isometric

Parallel partial ellipses spaced equal to the symbolic thread pitch, Fig. 15.9 (a), are used to represent the crests only of a screw thread in isometric, Fig. 17.30. The ellipses may be drawn by the four-center method of §17.18, or with the ellipse template, which is much more convenient, §§5.56 and 17.21.

## 17.20 Arcs in Isometric

The four-center ellipse construction is used in drawing circular arcs in isometric, as shown in Fig. 17.31. At (a) the complete construction is shown. However, it is not necessary to draw the complete constructions for arcs, as shown at (b) and (c). In each case the radius R is set off from the construction corner; then at each point, perpendiculars to the lines are erected, their intersection being the center of the arc. Note that the R distances are equal in both cases, (b) and (c), but that the actual radii used are quite different.

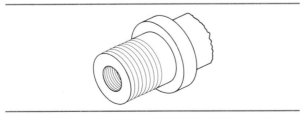

*Fig. 17.30*    Screw Threads in Isometric.

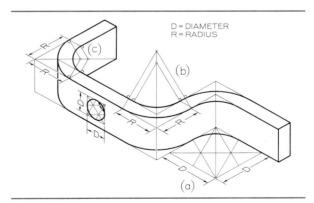

*Fig. 17.31*    Arcs in Isometric.

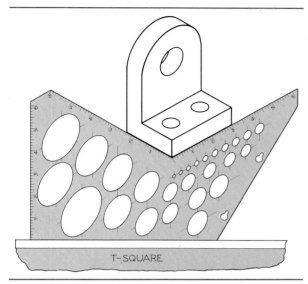

*Fig. 17.32*  Instrumaster Isometric Template.

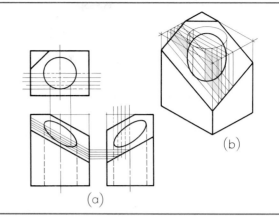

*Fig. 17.33*  Oblique Plane and Cylinder.

If a truer elliptic arc is required, the Orth construction, Fig. 17.28, can be used. Or a true elliptic arc may be drawn by the method of offsets, §17.17, or with the aid of an ellipse guide, §17.21.

### 17.21 Ellipse Guides

One of the principal time-consuming elements in pictorial drawing is the construction of ellipses. A wide variety of ellipse guides, or templates, is available for ellipses of various sizes and proportions. See §5.56. They are not available in every possible size, of course, and it may be necessary to "use the fudge factor," such as leaning the pencil or pen when inscribing the ellipse, or shifting the template slightly for drawing each quadrant of the ellipse.

The design of the ellipse template, Fig. 17.32, combines the angles, scales, and ellipses on the same instrument. The ellipses are provided with markings to coincide with the isometric center lines of the holes—a convenient feature in isometric drawing.

### 17.22 Intersections

To draw the elliptical intersection of a cylindrical hole in an oblique plane in isometric, Fig. 17.33, draw the ellipse in the isometric plane on top of the construction box, (b); then project points down to the oblique plane as shown. It will be seen that the construction for each point forms a trapezoid, which

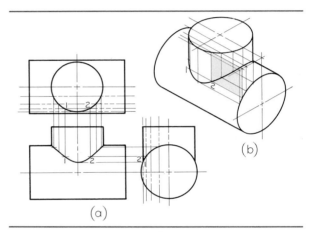

*Fig. 17.34*  Intersection of Cylinders.

is produced by a slicing plane parallel to a lateral surface of the block.

To draw the curve of intersection between two cylinders, Fig. 17.34, pass a series of imaginary cutting planes through the cylinders parallel to their axes, as shown. Each plane will cut elements on both cylinders that intersect at points on the curve of intersection, as shown at (b). As many points should be plotted as necessary to assure a smooth curve. For most accurate work, the ends of the cylinders should be drawn by the Orth construction, or with ellipse guides, or by one of the true-ellipse constructions.

### 17.23 The Sphere in Isometric

The isometric drawing of any curved surface is evidently the envelope of all lines that can be drawn on that surface. For the sphere, the great circles

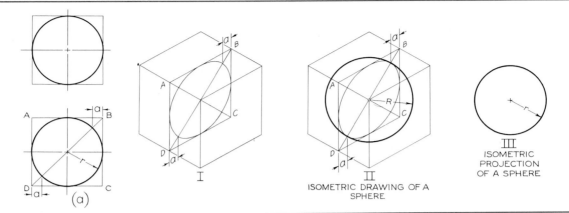

**Fig. 17.35**   Isometric of a Sphere.

(circles cut by any plane through the center) may be selected as the lines on the surface. Since all great circles, except those that are perpendicular or parallel to the plane of projection, are shown as ellipses having equal major axes, it follows that their envelope is a circle whose diameter is the major axis of the ellipses.

In Fig. 17.35 (a) two views of a sphere enclosed in a construction cube are shown. The cube is drawn at I, together with the isometric of a great circle that lies in a plane parallel to one face of the cube. Actually, the ellipse need not be drawn, for only the points on the diagonal located by measurements **a** are needed. These points establish the ends of the major axis from which the radius **R** of the sphere is determined. The resulting drawing shown at II is an *isometric drawing,* and its diameter is, therefore, $\sqrt{\frac{3}{2}}$ times the actual diameter of the sphere. The *isometric projection* of the sphere is simply a circle whose diameter is equal to the true diameter of the sphere, as shown at III.

### 17.24 Isometric Sectioning
In drawing objects characterized by open or irregular interior shapes, isometric sectioning is as appropriate as in multiview drawing. An *isometric full section* is shown in Fig. 17.36. In such cases it is usually best to draw the cut surface first and then to draw the portion of the object that lies behind the cutting plane. Other examples of isometric full sections are shown in Figs. 7.90, 9.11 (b), and 9.12 (d).

An *isometric half section* is shown in Fig. 17.37. The simplest procedure in this case is to make an isometric drawing of the entire object and then the cut surfaces. Since only a quarter of the object is

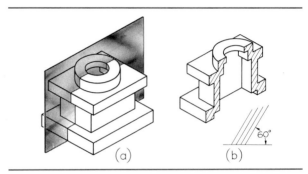

**Fig. 17.36**   Isometric Full Section.

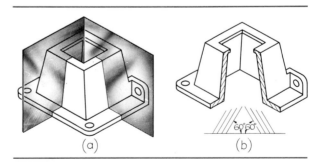

**Fig. 17.37**   Isometric Half Section.

removed in a half section, the resulting pictorial drawing is more useful than full sections in describing both exterior and interior shapes together. Other typical isometric half sections are shown in Figs. 9.13, 9.40, and 9.41.

*Isometric broken-out sections* are also sometimes used. Examples are shown in Figs. 7.97, 9.49, and 16.48.

Section lining in isometric drawing is similar to that in multiview drawing. Section lining at an angle

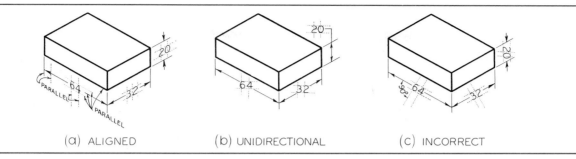

**Fig. 17.38**   Numerals and Arrowheads in Isometric (metric dimensions).

## 17.25 Isometric Dimensioning

Isometric dimensions are similar to ordinary dimensions used on multiview drawings, but are expressed in pictorial form. Two methods of dimensioning are approved by ANSI, namely, the pictorial plane (aligned) system and the unidirectional system, Fig. 17.38. Note that *vertical lettering* is used for either system of dimensioning. Inclined lettering is not recommended for pictorial dimensioning. The method of drawing numerals and arrowheads for the two systems is shown at (a) and (b). For the 64 mm dimension in the aligned system at (a), the extension lines, dimension lines, and lettering are all drawn in the isometric plane of one face of the object. The "horizontal" guide lines for the lettering are drawn parallel to the dimension line, and the "vertical" guide lines are drawn parallel to the extension lines.

of 60° with horizontal, Figs. 17.36 and 17.37, is recommended, but the direction should be changed if at this angle the lines would be parallel to a prominent visible line bounding the cut surface, or to other adjacent lines of the drawing.

The barbs of the arrowheads should line up parallel to the extension lines.

For the 64 mm dimension in the unidirectional system at (b), the extension lines and dimension lines are all drawn in the isometric plane of one face of the object and the barbs of the arrowheads should line up parallel to the extension lines, all exactly the same as at (a). However, the lettering for the dimensions is vertical and reads from the bottom of the drawing. This simpler system of dimensioning is often used on pictorials for production purposes.

As shown at (c), the vertical guide lines for the letters should not be perpendicular to the dimension lines. The example at (c) is incorrect because the 64 mm and 32 mm dimensions are lettered neither in the plane of the corresponding dimension and extension lines nor in a vertical position to read from the bottom of the drawing. The 20 mm dimension is awkward to read because of its position.

Correct and incorrect practice in isometric dimensioning using the aligned system of dimensioning is shown in Fig. 17.39. At (b) the $3\frac{1}{8}''$ dimension runs to a wrong extension line at the right, and con-

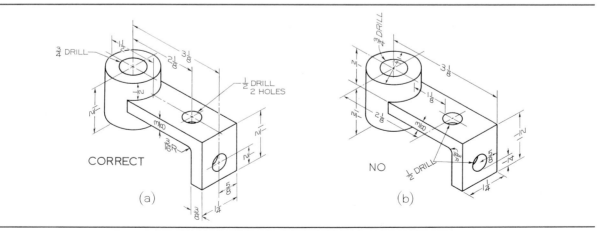

**Fig. 17.39**   Correct and Incorrect Isometric Dimensioning (Aligned System).

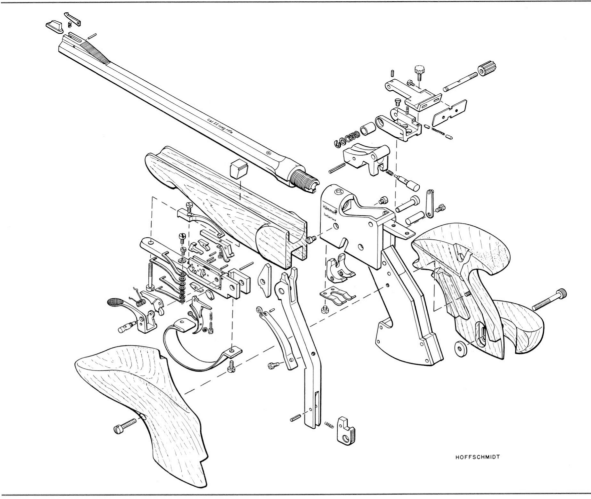

**Fig. 17.40** Isometric Exploded Assembly of Hammerli Match Pistol. *Courtesy of True Magazine, Fawcett Publications*

sequently the dimension does not lie in an isometric plane. Near the left side, a number of lines cross one another unnecessarily and terminate on the wrong lines. The upper ½″ drill hole is located from the edge of the cylinder when it should be dimensioned from its center line. Study these two drawings carefully to discover additional mistakes at (b).

The dimensioning methods described apply equally to fractional, decimal, and metric dimensions.

Many examples of isometric dimensioning are given in the problems at the end of Chapters 7, 9, 10, and 16, and you should study these to find samples of almost any special case you may encounter.

## 17.26 Exploded Assemblies

Exploded assemblies are often used in design presentations, catalogs, sales literature, and in the shop, to show all of the parts of an assembly and how they fit together. They may be drawn by any of the pictorial methods, including isometric, Fig. 17.40. Other isometric exploded assemblies are shown in Chapter 16.

## 17.27 Piping Diagrams

Isometric and oblique drawings are well suited for representation of piping layouts, as well as for all other structural work to be represented pictorially.

# Dimetric Projection

## 17.28 The Dimetric Method of Projection

A *dimetric projection* is an axonometric projection of an object so placed that two of its axes make equal angles with the plane of projection and the third axis makes either a smaller or a greater angle. Hence, the two axes making equal angles with the plane of projection are foreshortened equally, while the third axis is foreshortened in a different ratio.

Generally, the object is so placed that one axis will be projected in a vertical position. However, if the relative positions of the axes have been determined, the projection may be drawn in any revolved position, as in isometric drawing. See §17.8.

The angles between the *projection of the axes* must not be confused with the angles the *axes themselves* make with the plane projection.

The positions of the axes may be assumed such that any two angles between the axes are equal and over 90°, and the scales determined graphically, as shown in Fig. 17.41 (a), in which OP, OL, and OS are the projections of the axes or converging edges of a cube. In this case, angle POS = angle LOS. Lines PL, LS, and SP are the lines of intersection of the plane of projection with the three visible faces of the cube. From descriptive geometry we know that since line LO is perpendicular to the plane POS, in space, its projection LO is perpendicular to PS, the intersection of the plane POS and the plane of

projection. Similarly, OP is perpendicular to SL, and OS is perpendicular to PL.

If the triangle POS is revolved about the line PS as an axis into the plane of projection, it will be shown in its true size and shape as PO'S. If regular full-size scales are marked along the lines O'P and O'S, and the triangle is counterrevolved to its original position, the dimetric scales may be laid off on the axes OP and OS, as shown.

In order to avoid the preparation of special scales, use can be made of available scales on the architects' scale by assuming the scales and calculating the positions of the axes, as follows:

$$\cos a = -\frac{\sqrt{2h^2v^2 - v^4}}{2hv}$$

where *a* is one of the two equal angles between the projections of the axes, *h* is one of the two equal scales, and *v* is the third scale.

Examples are shown in the upper row of Fig. 17.42, in which the assumed scales, shown encircled, are taken from the architects' scale. One of these three positions of the axes will be found suitable for almost any practical drawing.

The Instrumaster Dimetric Template, Fig. 17.41 (b), has angles of approximately 11° and 39° with

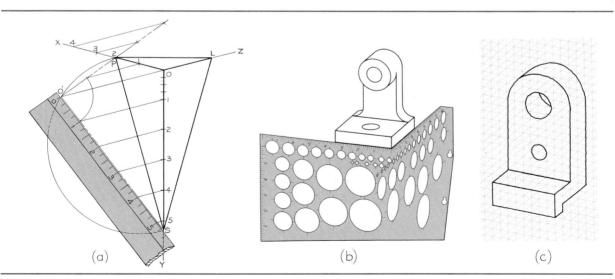

|  (a)  |  (b)  |  (c)  |

*Fig. 17.41*   Dimetric Projection.

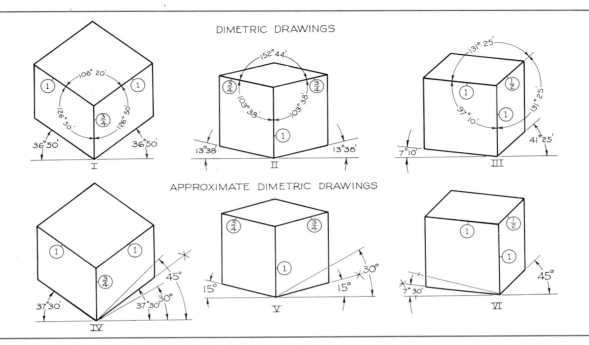

**Fig. 17.42**  Angles of Axes Determined by Assumed Scales.

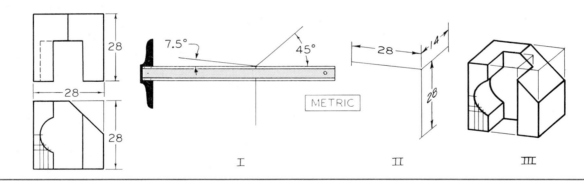

**Fig. 17.43**  Steps in Dimetric Drawing.

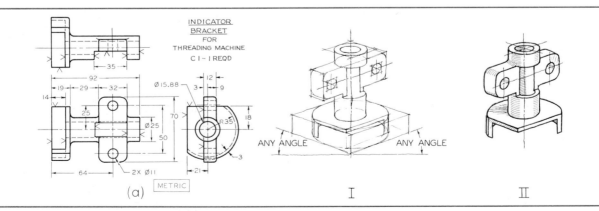

**Fig. 17.44**  Steps in Dimetric Sketching.

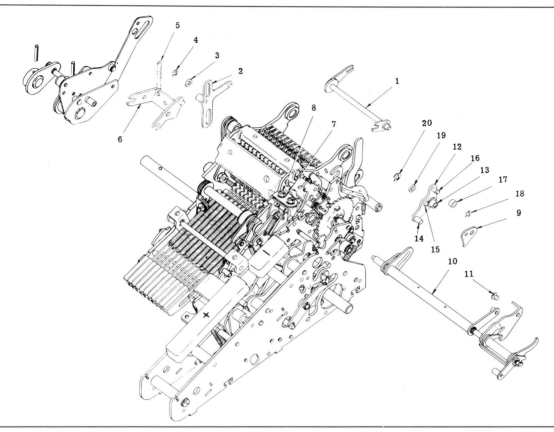

*Fig. 17.45* Exploded Dimetric of an Adding Machine. *Courtesy of Victor Adding Machine Co.*

horizontal, which provides a picture similar to that in Fig. 17.42 (III). In addition, the template has ellipses corresponding to the axes and accurate scales along the edges.

For other information on drawing of ellipses, see §17.33.

The Instrumaster Dimetric Graph paper, Fig. 17.41 (c), can be used to sketch in dimetric as easily as to sketch isometrics on isometric paper. The grid lines slope in conformity to the angles on the Dimetric Template at (b) and, when printed on vellum, the grid lines do not reproduce on prints.

## 17.29 Approximate Dimetric Drawing

Approximate dimetric drawings, which closely resemble true dimetrics, can be constructed by substituting for the true angles shown in the upper half

of Fig. 17.42, angles that can be obtained with the ordinary triangles and compass, as shown in the lower half of the figure. The resulting drawings will be sufficiently accurate for all practical purposes.

The procedure in preparing an approximate dimetric drawing, using the position of VI in Fig. 17.42, is shown in Fig. 17.43. The offset method of drawing a curve is shown in the figure. Other methods for drawing ellipses are the same as in trimetric drawing, §17.32.

The steps in making a dimetric sketch, using a position similar to that in Fig. 17.42 (V), are shown in Fig. 17.44. The two angles are equal and about 20° with horizontal for the most pleasing effect.

An exploded approximate dimetric drawing of an adding machine is shown in Fig. 17.45. The dimetric axes used are those in Fig. 17.42 (IV). Pictorials such as this are often used in service manuals.

# Trimetric Projection

### 17.30 The Trimetric Method of Projection

A *trimetric projection* is an axonometric projection of an object so placed that no two axes make equal angles with the plane of projection. In other words, each of the three axes and the lines parallel to them, respectively, have different ratios of foreshortening when projected to the plane of projection. If the three axes are assumed in any position on paper such that none of the angles is less than 90°, and if neither an isometric nor a dimetric position is deliberately arranged, the result will be a trimetric projection.

### 17.31 Trimetric Scales

Since the three axes are foreshortened differently, three different trimetric scales must be prepared and used. The scales are determined as shown in Fig. 17.46 (a), the method being the same as explained for the dimetric scales in §17.28. As shown at (a), any two of the three triangular faces can be revolved into the plane of projection to show the true lengths of the three axes. In the revolved position, the regular scale is used to set off inches or fractions thereof. When the axes have been counterrevolved to their original positions, the scales will

be correctly foreshortened, as shown. These dimensions should be transferred to the edges of three thin cards and marked OX, OZ, and OY for easy reference.

A special trimetric angle may be prepared from Bristol Board or plastic, as shown at (b). Perhaps six or seven such guides, using angles for a variety of positions of the axes, would be sufficient for all practical requirements.*

### 17.32 Trimetric Ellipses

The trimetric center lines of a hole, or on the end of a cylinder, become the conjugate diameters of the ellipse when drawn in trimetric. The ellipse may be drawn upon the conjugate diameters by the methods of Fig. 5.51 or 5.52 (b). Or the major and minor axes may be determined from the conjugate diameters, Fig. 5.53 (c), and the ellipse constructed upon them by any of the methods of Figs. 5.48 to 5.50, and 5.52 (a), or with the aid of an ellipse guide, Fig. 5.55.

One of the advantages of trimetric is the infinite number of positions of the object available. The an-

*Plastic templates of this type are available from drafting supplies dealers.

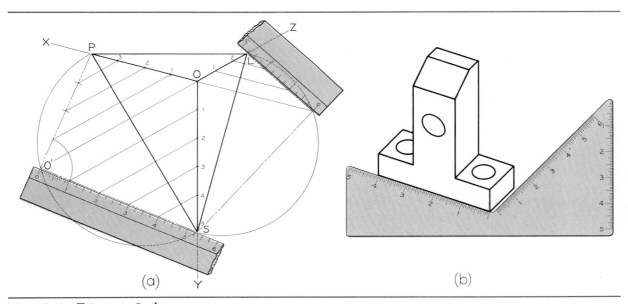

(a)                          (b)

*Fig. 17.46*    Trimetric Scales.

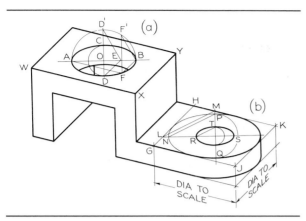

*Fig. 17.47*    Ellipses in Trimetric. *Method (b) courtesy of Professor H. E. Grant*

gles and scales can be handled without too much difficulty, as shown in §17.31. However, the infinite variety of ellipses has been a discouraging factor.

In drawing any axonometric ellipse, keep the following in mind:

1.  On the drawing, the major axis is always perpendicular to the center line, or axis, of the cylinder.

2.  The minor axis is always perpendicular to the major axis; that is, on the paper it coincides with the axis of the cylinder.

3.  The length of the major axis is equal to the actual diameter of the cylinder.

Thus we know at once the directions of both the major and minor axes, and the length of the major axis. *We do not know the length of the minor axis.* If we can find it, we can easily construct the ellipse with the aid of an ellipse guide or any of a number of ellipse constructions mentioned earlier.

In Fig. 17.47 (a), center O is located as desired, and horizontal and vertical construction lines that will contain the major and minor axes are drawn through O. Note that the major axis will be on the horizontal line perpendicular to the axis of the hole, and the minor axis will be perpendicular to it, or vertical.

Set the compass for the actual radius of the hole and draw the semicircle, as shown, to establish the ends A and B of the major axis. Draw AF and BF parallel to the axonometric edges WX and YX, respectively, to locate F, which lies on the ellipse.

Draw a vertical line through F to intersect the semicircle at F′ and join F′ to B as shown. From D′ where the minor axis, extended, intersects the semicircle, draw D′E and ED parallel to F′B and BF, respectively. Point D is one end of the minor axis. From center O, strike arc DC to locate C, the other end of the minor axis. Upon these axes, a true ellipse can be constructed, or drawn with the aid of an ellipse guide. A simple method for finding the "angle" of ellipse guide to use is shown in Fig. 5.55 (c). If an ellipse guide is not available, an approximate four-center ellipse, Fig. 5.56, will be found satisfactory in most cases.

In constructions where the enclosing parallelogram for an ellipse is available or easily constructed, the major and minor axes can be readily determined as shown in Fig. 17.47 (b). The directions of both axes, and the length of the major axis, are known. Extend the axes to intersect the sides of the parallelogram at L and M, and join the points with a straight line. From one end N of the major axis, draw a line NP parallel to LM. The point P is one end of the minor axis. To find one end T of the minor axis of the smaller ellipse, it is only necessary to draw RT parallel to LM or NP.

The method of constructing an ellipse on an oblique plane in trimetric is similar to that shown for isometric in Fig. 17.33.

## 17.33 Axonometric Projection by the Method of Intersections

Instead of constructing axonometric projections with the aid of specially prepared scales, as explained in the preceding paragraphs, an axonometric projection can be obtained directly by projection from two orthographic views of the object. This method is called the *method of intersections;* it was developed by Profs. L. Eckhart and T. Schmid of the Vienna College of Engineering and was published in 1937.

To understand this method, let us assume Fig. 17.48, that the axonometric projection of a rectangular object is given, and it is required to find its three orthographic projections: the top view, front view, and side view.

Assume that the object is placed so that its principal edges coincide with the coordinate axes, and assume that the plane of projection (the plane upon which the axonometric projection is drawn) intersects the three coordinate planes in the triangle ABC. From descriptive geometry, we know that lines BC, CA, and AB will be perpendicular, respec-

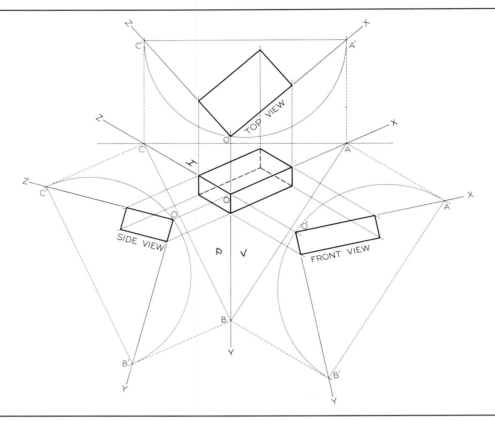

*Fig. 17.48*    Views from an Axonometric Projection.

tively, to axes **OX, OY,** and **OZ.** Any one of the three points **A, B,** or **C** may be assumed anywhere on one of the axes, and the triangle **ABC** drawn.

To find the true size and shape of the top view, revolve the triangular portion of the horizontal plane **AOC,** which is in front of the plane of projection, about its base **CA,** into the plane of projection. In this case, the triangle is revolved *inward* to the plane of projection through the smallest angle made with it. The triangle will then be shown in its true size and shape, and the top view of the object can be drawn in the triangle by projection from the axonometric projection, as shown, since all width dimensions remain the same. In the figure, the base **CA** of the triangle has been moved upward to **C′A′** so that the revolved position of the triangle will not overlap its projection.

In the same manner, the true sizes and shapes of the front view and side view can be found, as shown.

It is evident that if the three orthographic projections, or in most cases any two of them, are given in their relative positions, as shown in Fig. 17.48,

the directions of the projections could be reversed so that the intersections of the projecting lines would determine the required axonometric projection.

In order to draw an axonometric projection by the method of intersections, it is well to make a sketch, Fig. 17.49, of the desired general appearance of the projection. Even if the object is a complicated one, this sketch need not be complete, but may be only a sketch of an enclosing box. Draw the projections of the coordinate axes **OX, OY,** and **OZ,** parallel to the principal edges of the object as shown in the sketch, and the triangle **ABC** to represent the intersection of the three coordinate planes with the plane of projection.

Revolve the triangle **ABO** about its base **AB** as the axis into the plane of projection. Line **OA** will revolve to **O′A,** and this line, or one parallel to it, must be used as the base line of the front view of the object. The projecting lines from the front view to the axonometric must be drawn parallel to the projection of the unrevolved **Z**-axis, as indicated in the figure.

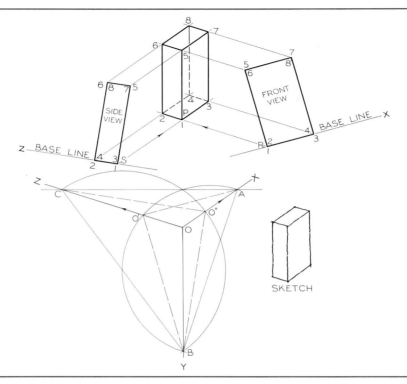

*Fig. 17.49*   Axonometric Projection.

Similarly, revolve the triangle **COB** about its base **CB** as the axis into the plane of projection. Line **CO** will revolve to **CO″**, and this line, or one parallel to it, must be used as the base line of the side view. The direction of the projecting lines must be parallel to the projection of the unrevolved X-axis, as shown.

Draw the front-view base line at a convenient location, but parallel to **O′X**, and with it as the base, draw the front view of the object. Draw the side-view base line also at a convenient location, but parallel to **O″C**, and with it as the base, draw the side view of the object, as shown. From the corners of the front view, draw projecting lines parallel to **OZ**, and from the corners of the side view, draw projecting lines parallel to **OX**. The intersections of these two sets of projecting lines determine the desired axonometric projection. It will be an isometric, dimetric, or a trimetric projection, depending upon the form of the sketch used as the basis for the projections, §17.2. If the sketch is drawn so that the three angles formed by the three coordinate axes are equal, the resulting projection will be an isometric projection; if two of the three angles are equal, the resulting projection will be a dimetric

projection; and if no two of the three angles are equal, the resulting projection will be a trimetric projection.

In order to place the desired projection on a specific location on the drawing, Fig. 17.49, select the desired projection **P** of the point **1**, for example, and draw two projecting lines **PR** and **PS** to intersect the two base lines and thereby to determine the locations of the two views on their base lines.

Another example of this method of axonometric projection is shown in Fig. 17.50. In this case, it was deemed necessary only to draw a sketch of the plan or base of the object in the desired position, as shown. The axes are then drawn with **OX** and **OZ** parallel, respectively, to the sides of the sketch plan, and the remaining axis **OY** is assumed in a vertical position. The triangles **COB** and **AOB** are revolved, and the two base lines drawn parallel to **O″C** and **O′A** as shown. Point **P**, the lower front corner of the axonometric drawing, was then chosen at a convenient place, and projecting lines drawn toward the base lines parallel to axes **OX** and **OZ** to locate the positions of the views on the base lines. The views are drawn upon the base lines or cut apart from

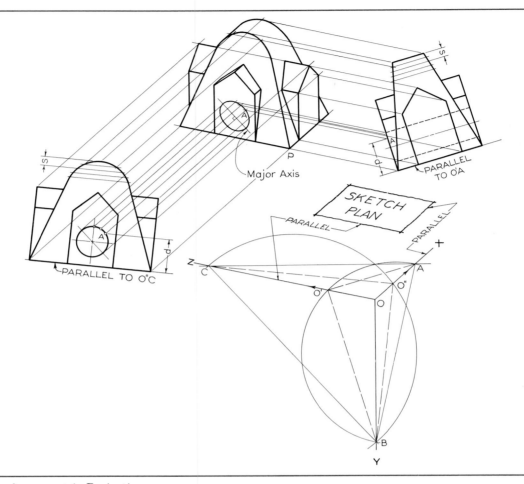

*Fig. 17.50*   Axonometric Projection.

another drawing and fastened in place with drafting tape.

To draw the elliptical projection of the circle, assume any points, such as A, on the circle in both front and side views. Note that point A is the same altitude d above the base line in both views. The axonometric projection of point A is found simply by drawing the projecting lines from the two views.

The major and minor axes may be easily found by projecting in this manner or by methods shown in Fig. 17.47, and the true ellipse drawn by any of the methods of Figs. 5.48 to 5.50 and 5.52 (a), or with the aid of an ellipse guide, §§5.56 and 17.21. Or an approximate ellipse, which is satisfactory for most drawings, may be used, Fig. 5.56.

# *Axonometric Problems*

A large number of problems to be drawn axonometrically are given in Figs. 17.51 to 17.54. The earlier isometric sketches may be drawn on isometric paper, §6.14; later sketches should be made on plain drawing paper. On drawings to be executed with instruments, show all construction lines required in the solutions.

For additional problems, see Figs. 7.51 to 7.53, and 18.23 to 18.25.

Since many of the problems in this chapter are of a general nature, they can also be solved on most computer graphics systems. If a system is available, the instructor may choose to assign specific problems to be completed by this method.

Axonometric problems in convenient form for solution may be found in *Principles of Engineering Graphics Problems* by Spencer, Hill, Loving, and Dygdon, a workbook designed to accompany this text that is also published by Macmillan Publishing Company.

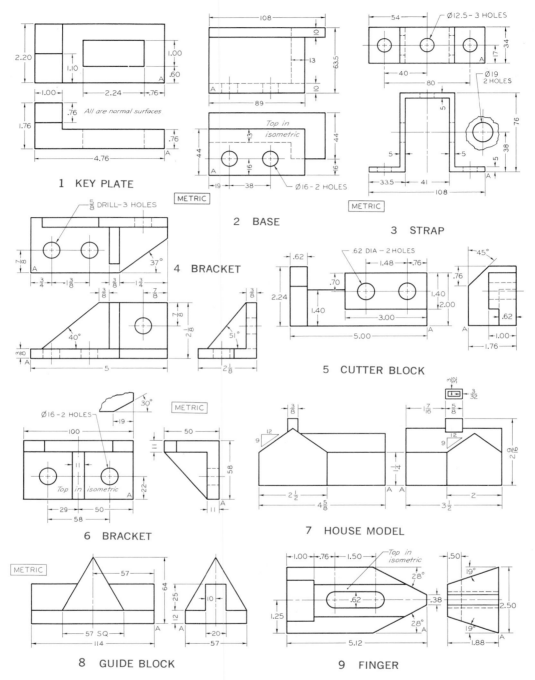

1 KEY PLATE

2 BASE

3 STRAP

4 BRACKET

5 CUTTER BLOCK

6 BRACKET

7 HOUSE MODEL

8 GUIDE BLOCK

9 FINGER

**Fig. 17.51** (1) Make freehand isometric sketches. (2) Make isometric drawings with instruments on Layout A–2 or A4–2 (adjusted). (3) Make dimetric drawings with instruments, using Layout A–2 or A4–2 (adjusted), and position assigned from Fig. 17.42. (4) Make trimetric drawings, using instruments, with axes chosen to show the objects to best advantage. If dimensions are required, study §17.25.

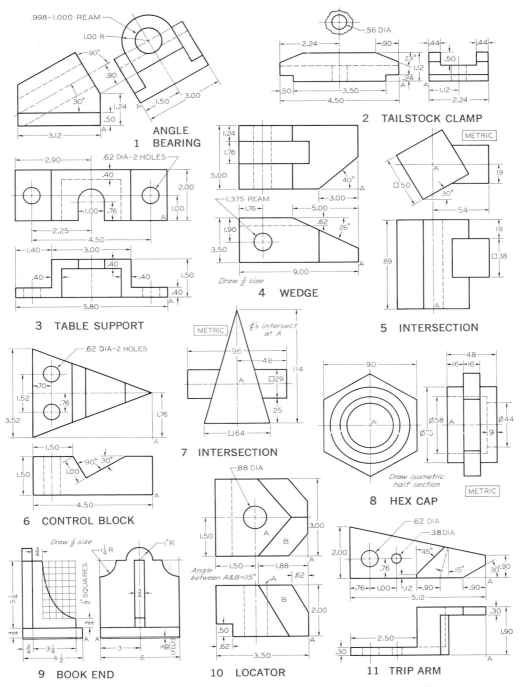

.998–1.000 REAM
1.00 R
90°
.90
30°
1.24
.50
3.12
A

**1  ANGLE BEARING**

.56 DIA
2.24   .90
23°
1.12
.24
A
.50   3.50
4.50
.44   .44
.50
1.12
2.24

**2  TAILSTOCK CLAMP**

2.90   .62 DIA–2 HOLES
.40
2.00
1.00   1.00   .76
2.25   4.50

1.40   3.00
.40
.40   .40   .40
5.80   1.50
.40

**3  TABLE SUPPORT**

1.24
1.76
5.00   40°
3.00

1.375 REAM
1.76   5.00
1.90   .62   26°
3.50
9.00
*Draw ½ size*

**4  WEDGE**

METRIC
A
□50   30°
54   19

89   □38   19

**5  INTERSECTION**

.62 DIA–2 HOLES
.70
1.52   .76
3.52   1.76
A

1.50
90°  30°
1.00
1.50
4.50   A

**6  CONTROL BLOCK**

METRIC   ℄'s intersect at A
96
48
A   □29   114
25
□64

**7  INTERSECTION**

90
48
16  16
Ø58  A   Ø44
Ø70   9
*Draw isometric half section*
METRIC

**8  HEX CAP**

*Draw ½ size*   1¼ R   1" R
¾
½" SQUARES
¾
5¼
¾  3¼  ¾   3¾
¾  3¼   3   6   (FELT)
4½

**9  BOOK END**

.88 DIA
1.50
A   3.00
B
1.50   1.88
Angle between A&B=115°   A   .62
.50   B   2.00
.62
3.50

**10  LOCATOR**

.62 DIA
.38 DIA
2.00   45°
.76   15°
.76  1.00  .12  .90   .90   30°  .90
5.12
.30
2.50   1.90
.30   A

**11  TRIP ARM**

***Fig. 17.52*** (1) Make freehand isometric sketches. (2) Make isometric drawings with instruments on Layout A–2 or A4–2 (adjusted). (3) Make dimetric drawings with instruments, using Layout A–2 or A4–2 (adjusted), and position assigned from Fig. 17.42. (4) Make trimetric drawings, using instruments, with axes chosen to show the objects to best advantage. If dimensions are required, study §17.25.

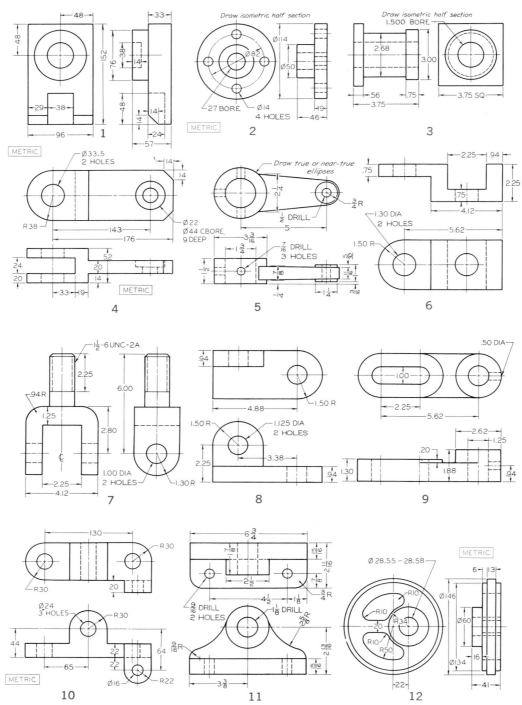

**Fig. 17.53** (1) Make isometric freehand sketches. (2) Make isometric drawings with instruments, using Size A or A4 sheet or Size B or A3 sheet, as assigned. (3) Make dimetric drawings with instruments, using Size A or A4 sheet or Size B or A3 sheet, as assigned, and position assigned from Fig. 17.42. (4) Make trimetric drawings, using instruments, with axes chosen to show the objects to best advantage. If dimensions are required, study §17.25.

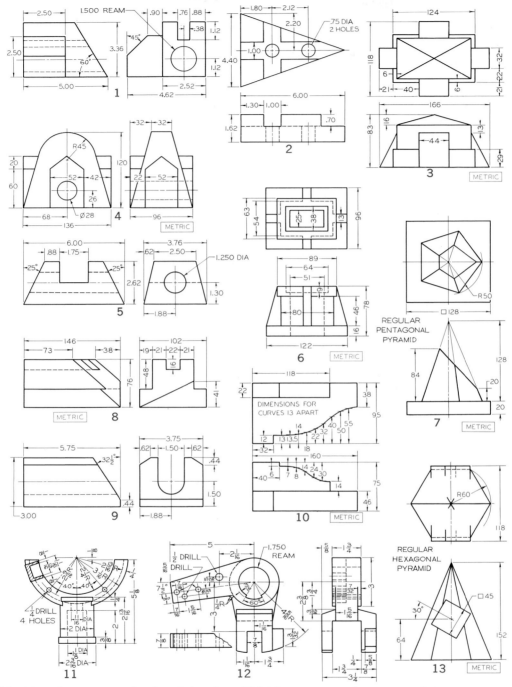

**Fig. 17.54**  (1) Make isometric freehand sketches. (2) Make isometric drawings with instruments, using Size A or A4 sheet or Size B or A3 sheet, as assigned. (3) Make dimetric drawings with instruments, using Size A or A4 sheet or Size B or A3 sheet, as assigned, and position assigned from Fig. 17.42. (4) Make trimetric drawings, using instruments, with axes chosen to show the objects to best advantage. If dimensions are required, study §17.25. For additional problems, assignments may be made from any of the problems in Fig. 18.23 to 18.25.

# CHAPTER
# 18

# Oblique Projection

If the observer is considered to be stationed at an infinite distance from the object, Fig. 17.1 (c), and looking toward the object so that the projectors are parallel to each other and oblique to the plane of projection, the resulting drawing is an *oblique projection*. As a rule, the object is placed with one of its principal faces parallel to the plane of projection. This is equivalent to holding the object in the hand and viewing it approximately as shown in Fig. 6.27.

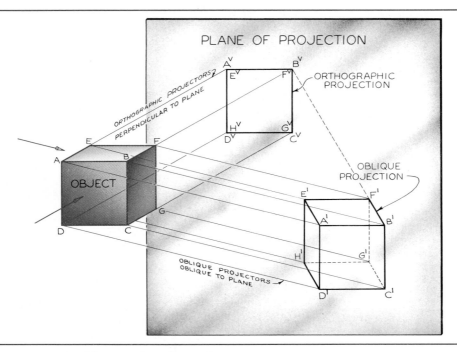

*Fig. 18.1*   Comparison of Oblique and Orthographic Projections.

## 18.1 Oblique and Other Projections Compared

A comparison of oblique projection and orthographic projection is shown in Fig. 18.1. The front face A′B′C′D′ in the oblique projection is identical with the front view, or orthographic projection, A$^V$B$^V$C$^V$D$^V$. Thus, if an object is placed with one of its faces parallel to the plane of projection, that face will be projected true size and shape in oblique projection as well as in orthographic or multiview projection. This is the reason why oblique projection is preferable to axonometric projection in representing certain objects pictorially. Note that surfaces of the object that are not parallel to the plane of projection will not project in true size and shape. For example, surface ABFE on the object (a square) projects as a parallelogram A′B′F′E′ in the oblique projection.

In axonometric projection, circles on the object nearly always lie in surfaces inclined to the plane of projection and project as ellipses. In oblique projection, the object may be positioned so that those surfaces are parallel to the plane of projection, in which case the circles will project as true circles, and can be easily drawn with the compass.

A comparison of the oblique and orthographic projections of a cylindrical object is shown in Fig. 18.2. In both cases, the circular shapes project as true circles. Note that although the observer, looking in the direction of the oblique arrow, does see these shapes as ellipses, the drawing, or projection, represents not what is seen but what is projected upon the plane of projection. This curious situation is peculiar to oblique projection.

Observe that the axis AB of the cylinder projects as a point A$^V$B$^V$ in the orthographic projection, since the line of sight is parallel to AB. But in the oblique projection, the axis projects as a line A′B′. The more nearly the direction of sight approaches the perpendicular with respect to the plane of projection — that is, the larger the angle between the projectors and the plane — the closer the oblique projection moves toward the orthographic projection, and the shorter A′B′ becomes.

## 18.2 Directions of Projectors

In Fig. 18.3, the projectors make an angle of 45° with the plane of projection; hence, the line CD′, which is perpendicular to the plane, projects true length at C′D′. If the projectors make a greater angle with the plane of projection, the oblique projection is shorter, and if the projectors make a smaller angle with the plane of projection, the oblique projection is longer. Theoretically, CD′ could project in any

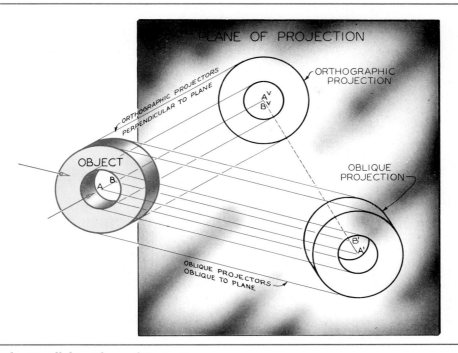

*Fig. 18.2* Circles Parallel to Plane of Projection.

length from zero to infinity. However, the line **AB** is parallel to the plane and will project in true length regardless of the angle the projectors make with the plane of projection.

In Fig. 18.1 the lines **AE, BF, CG,** and **DH** are perpendicular to the plane of projection, and project as parallel inclined lines **A'E', B'F', C'G',** and **D'H'** in the oblique projection. These lines on the drawing are called the *receding lines.* As we have seen,

they may be any length, from zero to infinity, depending upon the direction of the line of sight. Our next concern is: What angle do these lines make on paper with respect to horizontal?

In Fig. 18.4, the line **AO** is perpendicular to the plane of projection, and all the projectors make angles of 45° with it; therefore, all of the oblique projections **BO, CO, DO,** and so on, are equal in length to the line **AO.** It can be seen from the figure that

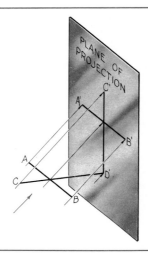

*Fig. 18.3* Lengths of Projections.

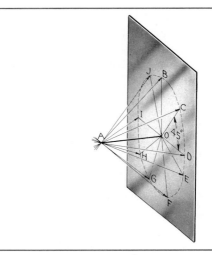

*Fig. 18.4* Directions of Projections.

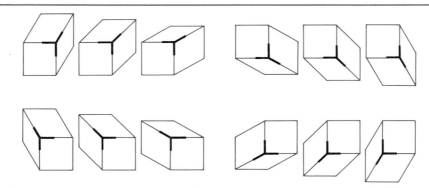

*Fig. 18.5*   Variation in Direction of Receding Axis.

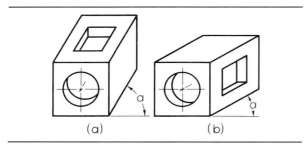

*Fig. 18.6*   Angle of Receding Axis.

## 18.3 Angles of Receding Lines

The receding lines may be drawn at any convenient angle. Some typical drawings with the receding lines in various directions are shown in Fig. 18.5. The angle that should be used in an oblique drawing depends upon the shape of the object and the location of its significant features. For example, in Fig. 18.6 (a) a large angle was used in order to obtain a better view of the rectangular recess on the top, while at (b) a small angle was chosen to show a similar feature on the side.

## 18.4 Length of Receding Lines

Since the eye is accustomed to seeing objects with all receding parallel lines appearing to converge, an oblique projection presents an unnatural appearance, with more or less serious distortion depending upon the object shown. For example, the object shown in Fig. 18.7 (a) is a cube, the receding lines being full length, but the receding lines appear to be too long and to diverge toward the rear of the block. A striking example of the unnatural appearance of an oblique drawing when compared with the natural appearance of a perspective is shown in Fig.

the projectors may be selected in any one of an infinite number of directions and yet maintain any desired angle with the plane of projection. It is also evident that the directions of the projections BO, CO, DO, and so on, are independent of the angles the projectors make with the plane of projection. Ordinarily, this inclination of the projection is 45° (CO in the figure), 30°, or 60° with horizontal, since these angles may be easily drawn with the triangles.

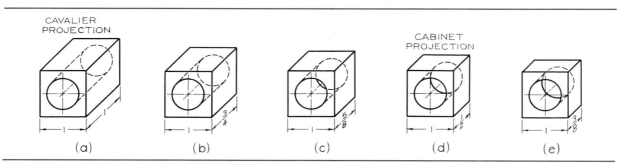

*Fig. 18.7*   Foreshortening of Receding Lines.

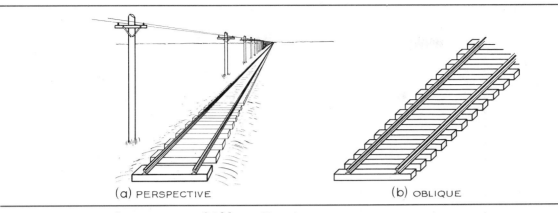

(a) PERSPECTIVE (b) OBLIQUE

*Fig. 18.8* Unnatural Appearance of Oblique Drawing.

18.8. This example points up one of the chief limitations of oblique projection: objects characterized by great length should not be drawn in oblique with the long dimension perpendicular to the plane of projection.

The appearance of distortion may be materially lessened by decreasing the length of the receding lines (remember, we established in §18.2 that they could be any length). In Fig. 18.7 a cube is shown in five oblique drawings with varying degrees of foreshortening of the receding lines. The range of scales chosen is sufficient for almost all problems, and most of the scales are available on the architects', engineers', or metric scales.

When the receding lines are true length—that

is, when the projectors make an angle of 45° with the plane of projection—the oblique drawing is called a *cavalier projection*, Fig. 18.7 (a). Cavalier projections originated in the drawing of medieval fortifications and were made upon horizontal planes of projection. On these fortifications the central portion was higher than the rest and it was called *cavalier* because of its dominating and commanding position.

When the receding lines are drawn to half size, as at (d), the drawing is commonly known as a *cabinet projection*. The term is attributed to the early use of this type of oblique drawing in the furniture industries. A comparison of cavalier projection and cabinet projection is shown in Fig. 18.9.

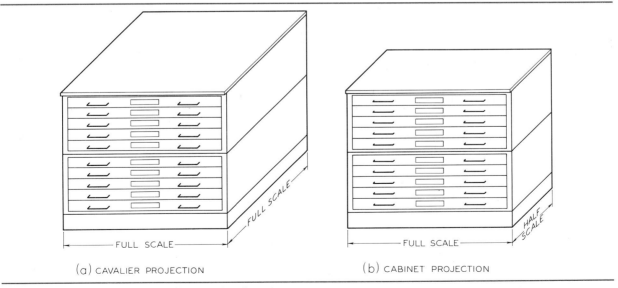

(a) CAVALIER PROJECTION (b) CABINET PROJECTION

*Fig. 18.9* Comparison of Cavalier and Cabinet Projections.

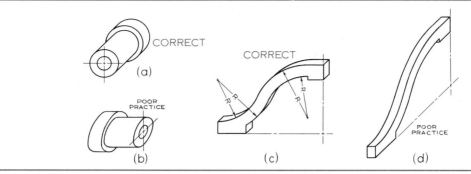

*Fig. 18.10*    Essential Contours Parallel to Plane of Projection.

## 18.5 Choice of Position

The face of an object showing the essential contours should generally be placed parallel to the plane of projection, Fig. 18.10. If this is done, distortion will be kept at a minimum and labor reduced. For example, at (a) and (c) the circles and circular arcs are shown in their true shapes and may be quickly

drawn with the compass, while at (b) and (d) these curves are not shown in their true shapes and must be plotted as free curves or in the form of ellipses.

The longest dimension of an object should generally be placed parallel to the plane of projection, as shown in Fig. 18.11 (b).

## 18.6 Steps in Oblique Drawing

The steps in drawing a cavalier drawing of a rectangular object are shown in Fig. 18.12. As shown in step I, draw the axes **OX** and **OY** perpendicular to each other and the receding axis **OZ** at any desired angle with horizontal. Upon these axes, construct an enclosing box, using the overall dimensions of the object.

As shown at II, block in the various shapes in detail, and as indicated at III, heavy in all final lines.

Many objects most adaptable to oblique representation are composed of cylindrical shapes built upon axes or center lines. In such cases, the oblique drawing is best constructed upon the projected center lines, as shown in Fig. 18.13. The object is positioned so that the circles shown in the given top view are parallel to the plane of projection and, hence, can be readily drawn with the compass in their true shapes. The general procedure is to draw

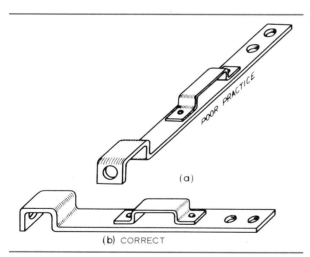

*Fig. 18.11*    Long Axis Parallel to Plane of Projection.

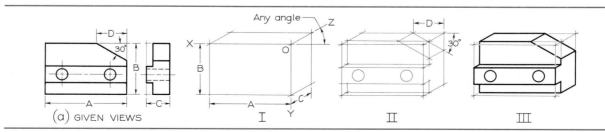

*Fig. 18.12*    Steps in Oblique Drawing—Box Construction.

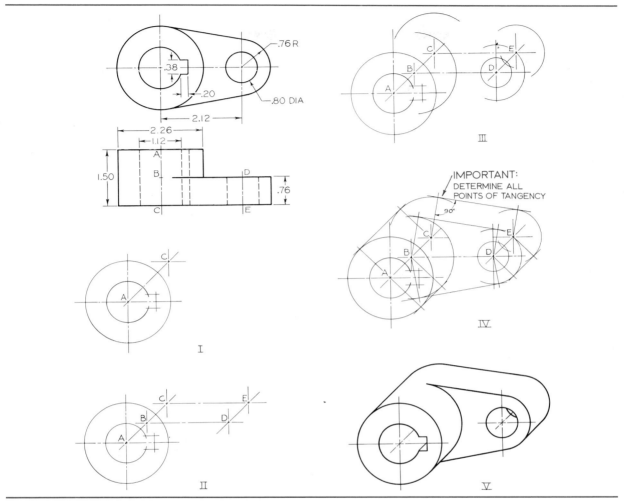

*Fig. 18.13*   Steps in Oblique Drawing—Skeleton Construction.

the center-line skeleton, as shown in steps I and II, and then to build the drawing upon these center lines.

It is very important to construct all points of tangency, as shown in step IV, especially if the drawing is to be inked. For a review of tangencies, see §§5.33 to 5.41. The final cavalier drawing is shown in step V.

## 18.7 Four-Center Ellipse

It is not always possible to place an object so that all of its significant contours are parallel to the plane of projection. For example, the object shown in Fig. 18.14 (a) has two sets of circular contours in different planes, and both cannot be placed parallel to the plane of projection.

In the oblique drawing at (b), the regular four-center method of Fig. 17.24 was used to construct ellipses representing circular curves not parallel to the plane of projection. This method can be used only in cavalier drawing in which case the enclosing parallelogram is equilateral—that is, the receding axis is drawn to full scale. The method is the same as in isometric: erect perpendicular bisectors to the four sides of the parallelogram; their intersections will be centers for the four circular arcs. If the angle of the receding lines is other than 30° with horizontal, as in this case, the centers of the two large arcs will not fall in the corners of the parallelogram.

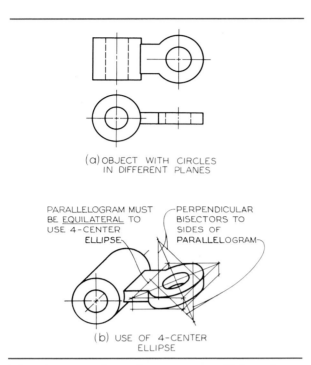

(a) OBJECT WITH CIRCLES
IN DIFFERENT PLANES

PARALLELOGRAM MUST
BE <u>EQUILATERAL</u> TO
USE 4-CENTER
ELLIPSE

PERPENDICULAR
BISECTORS TO
SIDES OF
PARALLELOGRAM

(b) USE OF 4-CENTER
ELLIPSE

***Fig. 18.14*** Circles and Arcs Not Parallel to Plane
of Projection.

The regular four-center method is not convenient in oblique drawing unless the receding lines make 30° with horizontal so that the perpendicular bisectors may be drawn easily with the 30° × 60° triangle and the T-square, parallel rule, or drafting machine without the necessity of first finding the midpoints of the sides. A more convenient method is the alternate four-center ellipse drawn upon the

two center lines, as shown in Fig. 18.15. This is the same method as used in isometric, Fig. 17.29, but in oblique drawing it varies slightly in appearance according to the different angles of the receding lines.

First, draw the two center lines. Then, from the center, draw a construction circle equal in diameter to the actual hole or cylinder. The circle will intersect each center line at two points. From the two points on one center line, erect perpendiculars to the other center line, then, from the two points on the other center line, erect perpendiculars to the first center line. From the intersections of the perpendiculars, draw four circular arcs, as shown.

It must be remembered that the four-center ellipse can be inscribed only in an *equilateral* parallelogram; hence, it cannot be used in any oblique drawing in which the receding axis is foreshortened. Its use is limited, therefore, to cavalier drawing.

## 18.8 Offset Measurements

Circles, circular arcs, and other curved or irregular lines may be drawn by means of offset measurements, as shown in Fig. 18.16. The offsets are first drawn on the multiview drawing of the curve, as shown at (a), and these are transferred to the oblique drawing, as shown at (b). In this case, the receding axis is full scale, and therefore all offsets can be drawn full scale. The four-center ellipse could be used, but the method here is more accurate. The final curve is drawn with the aid of the irregular curve, §2.54.

If the oblique drawing is a cabinet drawing, as shown at (c), or any oblique drawing in which the

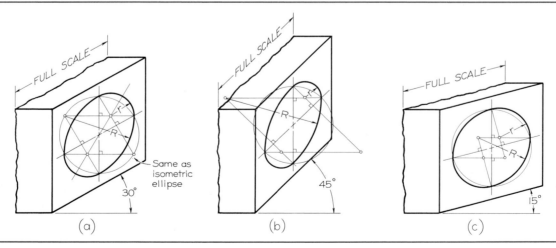

(a)    (b)    (c)

***Fig. 18.15*** Alternate Four-Center Ellipse.

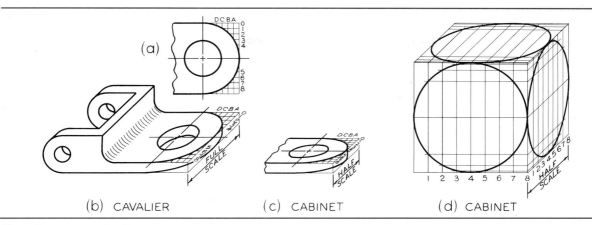

Fig. 18.16   Use of Offset Measurements.

receding axis is drawn to a reduced scale, the offset measurements parallel to the receding axis must be drawn to the same reduced scale. In this case, there is no choice of methods, since the four-center ellipse could not be used. A method of drawing ellipses in a cabinet drawing of a cube is shown at (d).

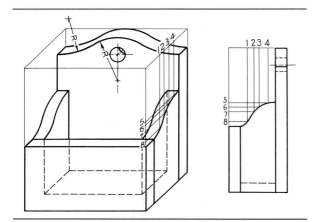

Fig. 18.17   Use of Offset Measurements.

As shown in Fig. 18.17, a free curve may be drawn in oblique by means of offset measurements. This figure also illustrates a case in which hidden lines are used to make the drawing clearer.

The use of offset measurements in drawing an ellipse in a plane inclined to the plane of projection is shown in Fig. 18.18. At (a) a number of parallel lines are drawn to represent imaginary cutting planes. Each plane will cut a rectangular surface between the front end of the cylinder and the inclined surface. These rectangles are drawn in oblique, as shown at (b), and the curve is drawn through corner points, as indicated. The final cavalier drawing is shown at (c).

## 18.9 Angles in Oblique Projection

If an angle that is specified in degrees lies in a receding plane, it is necessary to convert the angle into linear measurements in order to draw the angle in oblique. For example, in Fig. 18.19 (a) an angle of 30° is given. In order to draw the angle in oblique, we need to know dimensions AB and BC. The dis-

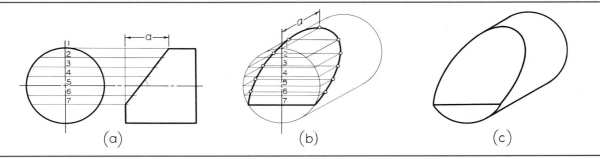

Fig. 18.18   Use of Offset Measurements.

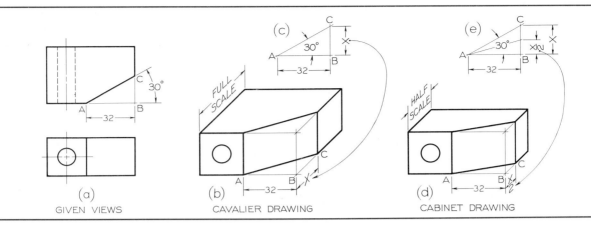

***Fig. 18.19***    Angles in Oblique Projection.

***Fig. 18.20***    Oblique Half Section.

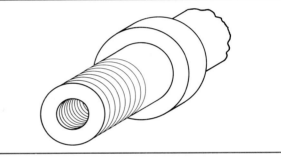

***Fig. 18.21***    Screw Threads in Oblique.

tance **AB** is given as $1\frac{1}{4}''$ and can be set off directly in the cavalier drawing, as shown at (b). Distance **BC** is not known, but can easily be found by constructing the right triangle **ABC** at (c) from the given dimensions in the top view at (a). The length **BC** is then transferred with the dividers to the cavalier drawing, as shown.

In cabinet drawing, it must be remembered that *all receding dimensions* must be reduced to half size. Thus, in the cabinet drawing at (d), the distance **BC** must be half the side **BC** of the right triangle at (e), as shown.

### 18.10 Oblique Sections

Sections are often useful in oblique drawing, especially in the representation of interior shapes. An *oblique half section* is shown in Fig. 18.20. Other examples are shown in Figs. 9.40, 9.42 and 9.44. *Oblique full sections*, in which the plane passes completely through the object, are seldom used because

they do not show enough of the exterior shapes. In general, all the types of sections discussed in §17.24 for isometric drawing may be applied equally to oblique drawing.

### 18.11 Screw Threads in Oblique

Parallel partial circles spaced equal to the symbolic thread pitch, Fig. 15.17 (b), are used to represent the crests only of a screw thread in a cavalier oblique, Fig. 18.21. For cabinet oblique the space would be one-half of the symbolic pitch. If the thread is so positioned to require ellipses, they may be drawn by the four-center method of §18.7.

### 18.12 Oblique Dimensioning

An oblique drawing may be dimensioned in a similar manner to that described in §17.25 for isometric drawing, as shown in Fig. 18.22. The general principles of dimensioning, as outlined in Chapter 13,

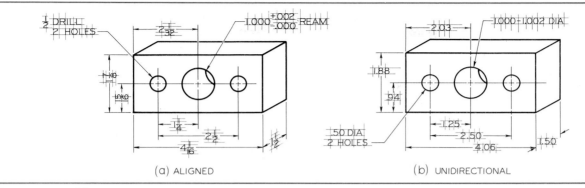

*Fig. 18.22*   Oblique Dimensioning.

must be followed. As shown in the figure, all dimension lines, extension lines, and arrowheads must lie in the planes of the object to which they apply. The dimension figures also will lie in the plane when the aligned dimensioning system is used as shown at (a). For the unidirectional system of dimensioning, (b), all dimension figures are set horizontal and read from the bottom of the drawing. This simpler system is often used on pictorials for production purposes. *Vertical lettering* should be used for all pictorial dimensioning.

Dimensions should be placed outside the outlines of the drawing except when greater clearness or directness of application results from placing the dimensions directly on the view. The dimensioning methods described apply equally to fractional, dec-

imal, and metric dimensions. For many other examples of oblique dimensioning, see Figs. 7.60, 7.64, 7.65, and others on following pages.

## 18.13 Oblique Sketching

Methods of sketching in oblique on plain paper are illustrated in Fig. 6.27. Ordinary graph paper is very useful in oblique sketching, Fig. 6.28. The height and width proportions can be easily controlled by simply counting the squares. A very pleasing depth proportion can be obtained by sketching the receding lines at 45° diagonally through the squares and through half as many squares as the actual depth would indicate.

# _Oblique Projection Problems_

A large number of problems to be drawn in oblique—either cavalier or cabinet—are given in Figs. 18.23 to 18.25. They may be drawn freehand, §6.15, using graph paper or plain drawing paper as assigned by the instructor, or they may be drawn with instruments. In the latter case, all construction lines should be shown on the completed drawing.

Many additional problems suitable for oblique projection will be found in Figs. 7.51 to 7.53, 10.27 to 10.29, and 17.51 to 17.54.

Since many of the problems in this chapter are of a general nature, they can also be solved on most computer graphics systems. If a system is available, the instructor may choose to assign specific problems to be completed by this method.

Oblique drawing problems in convenient form for solution may be found in _Principles of Engineering Graphics Problems_ by Spencer, Hill, Loving, and Dygdon, a workbook designed to accompany this text that is also published by Macmillan Publishing Company.

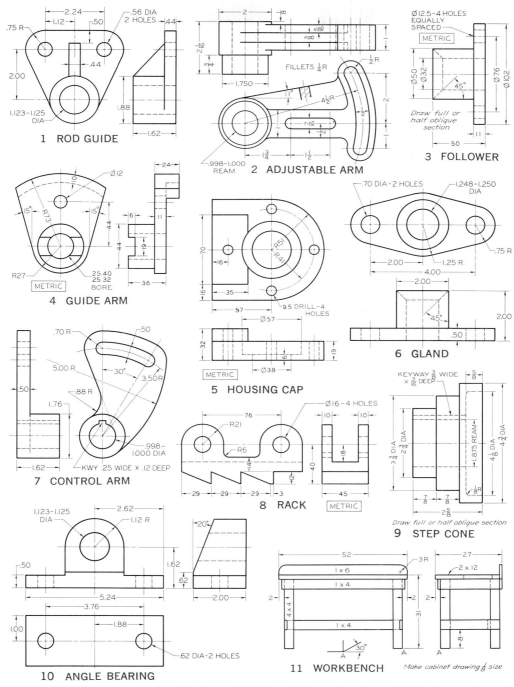

**Fig. 18.23** (1) Make freehand oblique sketches. (2) Make oblique drawings with instruments, using Size A or A4 sheet or Size B or A3 sheet, as assigned. If dimensions are required, study §18.12.

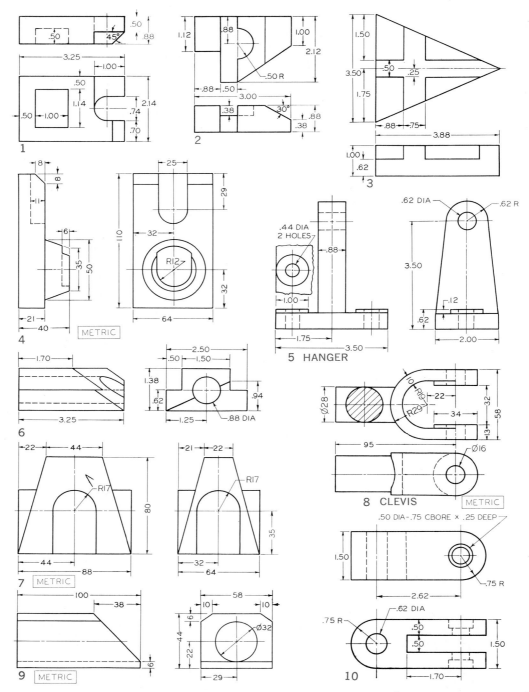

**Fig. 18.24** Make oblique drawings with instruments, using Size A or A4 sheet or Size B or A3 sheet, as assigned. If dimensions are required, study §18.12.

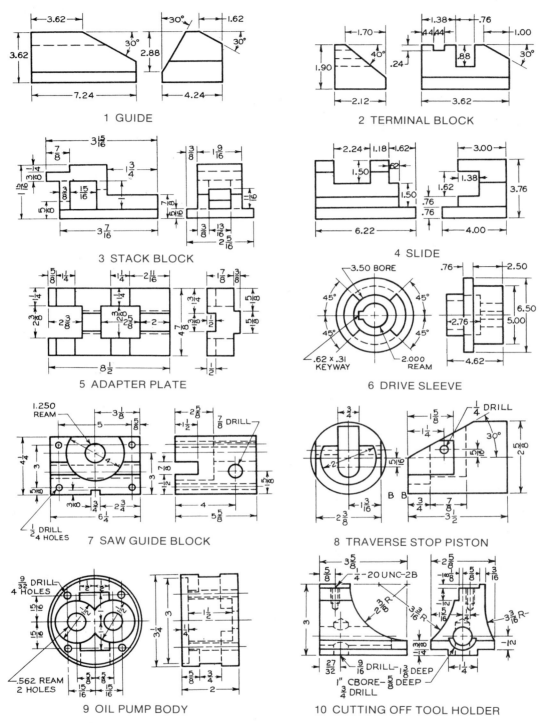

**Fig. 18.25** Make oblique drawings with instruments, using Size A or A4 sheet or Size B or A3 sheet, as assigned. If dimensions are required, study §18.12. For additional problems, see Figs 7.50 to 7.53, 10.27 to 10.29, and 17.51 to 17.54.

# CHAPTER
# 19

# Points, Lines,
# and Planes

The science of graphical representation and the solution of spatial relationships of points, lines, and planes by means of projections are the concerns of *descriptive geometry*. The methods of representation of solid objects, wherein the planes of projection of the "glass box" and views of the object were projected upon the planes, have been discussed in §§7.1, 7.2, 7.19 to 7.25. The elements of the objects—points, lines, and planes—now will be discussed and explained.

During the latter part of the eighteenth century the French mathematician Gaspard Monge developed the principles of descriptive geometry to solve spatial problems related to military structures. In France and Germany, Monge's descriptive geometry soon became a part of national education. In 1816 Claude Crozet introduced descriptive geometry into the curriculum of the United States Military Academy at West Point. In 1821 Crozet published his *Treatise on Descriptive Geometry*, the first important English work on descriptive geometry published in this country. Since then descriptive geometry has been taught in many engineering colleges, and today no study of engineering graphics is considered complete without a detailed study of descriptive geometry.

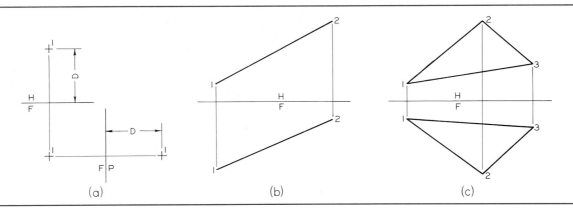

*Fig. 19.1*   Points, Lines, and Planes Individually Represented.

## 19.1 Basic Geometric Elements

We start with the representation of a single point. In Fig. 19.1(a) the front, top, and right-side views of point 1 are shown. The projections of the point are indicated by a small cross and the number 1. The folding line H/F is shown between the front and top views, and the folding line F/P is shown between the front and side views. Thus the names of the views are indicated by the letters H, F, and P (for horizontal, frontal, and profile planes of projection). See §7.3. As indicated by the dimensions D, the distance of the top view to the H/F folding line is equal to the distance of the side view to the F/P folding line.

In Fig. 19.1 (b) the front and top views, or projections, of two connected points (line 1–2) are shown. Note the thin projection lines between the views. (Projection lines are usually drawn in pencil as very light construction lines.) At (c) are shown the front and top views of three connected points, or

plane 1–2–3. Again, note the projection lines between the views.

In order to describe an object, lay out a mechanism, or begin the graphical solution of an engineering problem, the relative positions of two or more points must be specified. For example, in Fig. 19.2(a), the relative positions of points 1 and 2 could be described as follows: point 2 is 32 mm to the right of point 1, 12 mm below (or lower than) point 1, and 16 mm behind (or to the rear of) point 1.

When points 1 and 2 are connected, as at (b), observe that the preceding specifications have placed point 2 at a definite distance from point 1 along line 1–2 (more properly line *segment* 1–2, since line 1–2 could be extended).

When point 3 is introduced, as at (c), and is connected to point 2, line 2–3 is established. Since lines 1–2 and 2–3 have point 2 in common, they are *intersecting* lines.

If line 2–3 of Fig. 19.2 (c) is altered to position

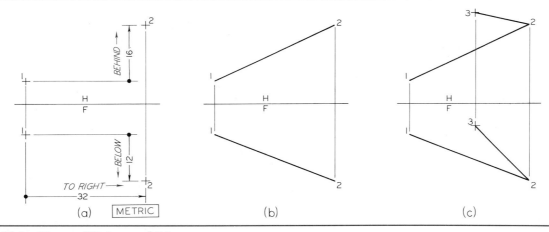

*Fig. 19.2*   Views of Points and Lines.

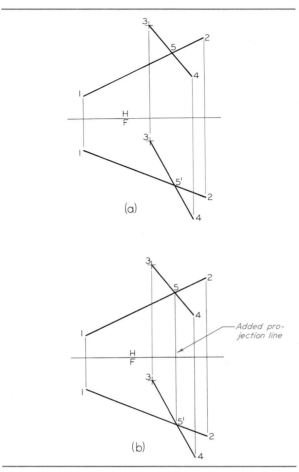

(a)

(b)

Added projection line

**Fig. 19.3** Intersecting Lines.

3–4, Fig. 19.3 (a), do we still have intersecting lines? The possible point in common could only be at **5** in the top view and **5′** in the front view. The question then becomes "Are **5** and **5′** in fact views of the same point?" Since adjacent views of a point must be aligned, §7.2, a vertical projection line is added at (b), and, since this line connects **5** and **5′**, it is evident that lines **1–2** and **3–4** are actually intersecting lines.

In Fig. 19.4 (a) another pair of lines, **1–2** and **3–4** is shown. In this case apparent points of intersection **5** and **5′** are not aligned with the projection lines between views and hence do not represent views of the same point. Therefore, these two lines do not intersect. Such nonintersecting, nonparallel lines are called *skew* lines. The relationship of these skew lines will now be considered in more detail.

Since the lines do not intersect, one must be above the other in the region of point **5**. At (b) this region has been assigned two numbers, **5** and **6**, in the top view, and it is arbitrarily decided that **5** is a point on line **1–2** and **6** is on line **3–4**. These points are then projected to the front view as shown. The direction of sight for the top view is downward toward the front view of the pair of points **5** and **6**. It is observed that point **5** on line **1–2** is *higher* in space or *nearer* to the observer than is point **6** on line **3–4**. Line **1–2** thus passes above line **3–4** in the vicinity of point **5**.

In like manner, at (c), numbers **7** and **8** are assigned to the apparent crossing point in the front view and projected to the top view, with **8** assigned to line **1–2** and **7** to line **3–4**. It is now noted that

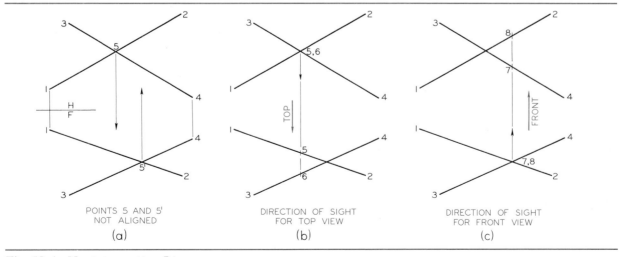

POINTS 5 AND 5′ NOT ALIGNED

(a)

DIRECTION OF SIGHT FOR TOP VIEW

(b)

DIRECTION OF SIGHT FOR FRONT VIEW

(c)

**Fig. 19.4** Nonintersecting Lines.

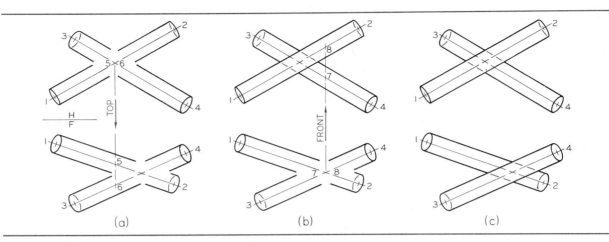

*Fig. 19.5*   Visibility of Nonintersecting Rods.

the direction of sight for a front view is upward (on the paper) toward the top view. With this in mind, it is observed that point 7, and therefore line 3–4, are *nearer* to the observer (in the front view) than are point 8 and line 1–2. Line 3–4 thus passes in front of line 1–2 in the vicinity of point 7.

The foregoing is useful in determining the visibility of nonintersecting members of a structure or of pipes and tubes, Fig. 19.5. At (a) the views are incomplete because it has not been determined which of the two rods is visible at the apparent crossover in each view. Only the relative positions of the center lines need be investigated. As before, concentration is limited temporarily to the apparent crossing point in the top view, with numbers 5 and 6 assigned to the region. Point 5 is projected to line 1–2 and point 6 to line 3–4 in the front view, where it is discovered that point 5 is above point 6. Line 1–2 therefore passes above line 3–4 and rod 1–2 is visible at the crossover in the top view, as shown at (b). Rod 3–4 is, of course, hidden where it passes below rod 1–2 and is completed accordingly, as shown.

Attention is now directed to the apparent crossing in the front view, Fig. 19.5 (b), and numbers 7 and 8 are assigned. Projected to the top view, these reveal that point 7 on line 3–4 is in front of point 8 on line 1–2. Rod 3–4 is therefore visible in the front view and rod 1–2 is hidden. The completed views are shown at (c).

This discussion has been in terms of front and top views, but it should be realized that the principle applies to *any pair of adjacent views* of the same structure. For example, study Figs. 19.4 and 19.5 with this book held upside down. Observe that the top views become the front views and vice versa but that the visibility is not altered.

Observe finally that *any* two adjacent views, Fig. 19.6, have this same fundamental relationship. The direction of sight for either view is always directed *toward* the adjacent view. Hence at (a), in view B, it is observed that point 5 on edge 1–3 is the nearer of the two assigned points 5 and 6. Thus edge 1–3 is visible in view B, and it follows that edge 2–4 is hidden. Note that this procedure reveals nothing about the visibility of the interior lines of view A of the tetrahedron. At (b), the positions of points 7 and 8 relative to the direction of sight for view A reveal that edge 1–3 is visible in view A.

## 19.2 Inclined Line and Angle with Plane of Projection

By definition, an inclined line appears true length on the plane to which it is parallel, §7.22. For convenience or precision, inclined lines are frequently classified as *frontal, horizontal,* or *profile,* Fig. 19.7.

Observe that the true-length view of an inclined line is always in an inclined position, while the foreshortened views are in either vertical or horizontal positions.

Note that additional information is available in the true-length views. The true angle between a line and a plane may be measured when the line is true length and the plane is in edge view in a single view. For example, in Fig. 19.7 (a), the horizontal and profile surfaces of the cube appear as lines (in edge

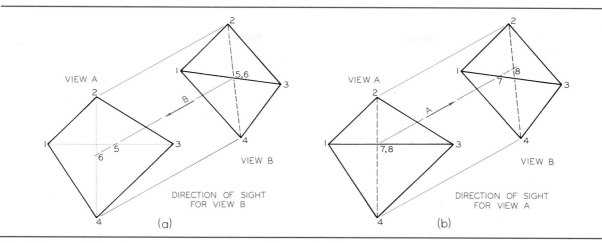

***Fig. 19.6***   Visibility of Nonintersecting Lines of a Tetrahedron.

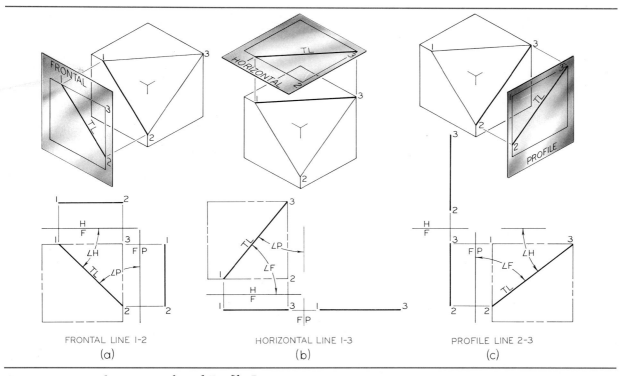

***Fig. 19.7***   Frontal, Horizontal, and Profile Lines.

view) in the front view, where edge 1–2 is true length. Thus the angles between edge 1–2 and the horizontal plane (∠H) and between edge 1–2 and the profile plane (∠P) may be measured in the front view. Similarly, ∠F and ∠P for edge 1–3 are measured in the top view, Fig. 19.7 (b), while ∠F and ∠H for edge 2–3 appear in the side view at (c).

## 19.3 True Length of Oblique Line and Angle with Plane of Projection

By definition, an oblique line does not appear true length in any principal view—front, top, or side, §7.24. It follows that the angles formed with the planes of projection also cannot be measured in the principal views.

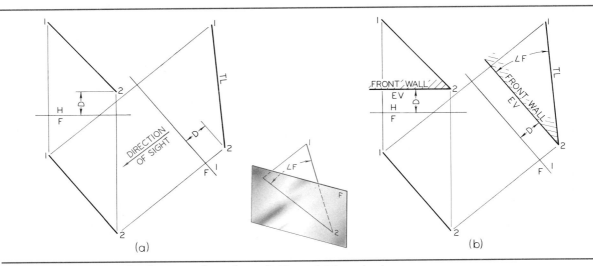

**Fig. 19.8**  True Length of Line and Angle with Frontal Plane (∠F).

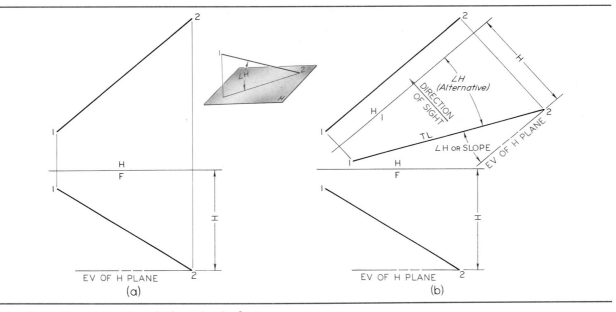

**Fig. 19.9**  True Length and Slope (∠H) of Line.

In §10.18, the true length (TL) of hip rafter 1–2 was obtained by assuming a direction of sight perpendicular to the front view of rafter 1–2 and constructing a depth auxiliary view. In Fig. 19.8 (a) this construction is repeated with the remainder of the roof omitted. At (b) a portion of the front wall of the building has been added, passing through point 2. Note that, because every point of the wall is at distance D from the folding line H/F in the top view, all points of the wall are at this same distance D from folding line F/1 in the auxiliary view. The entire wall

thus appears as a line (in edge view) in these views. Since line 1–2 appears true length in auxiliary view 1, the angle between line 1–2 and the edge view of the front wall, ∠F, can be measured.

In civil engineering, mining, and geology the most important principal plane is the horizontal plane because a map (of a relatively small area) is a horizontal projection and thus corresponds to a top view. The angle between a line, such as the center line of a highway, and a horizontal plane is a very important factor in the engineering description of

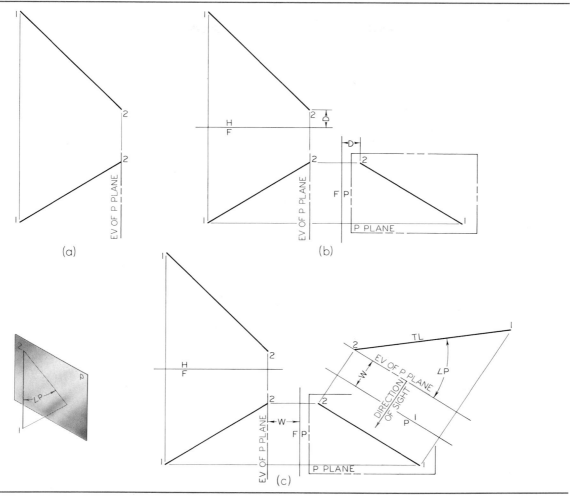

**Fig. 19.10**   True Length and Angle with Profile Plane (∠P).

the highway. If the angle (∠H) is measured in degrees, it is sometimes called the *slope* of the highway. More commonly it is measured by the ratio between the horizontal and vertical displacements and is called the *grade*. See §19.5.

To measure the slope (∠H) of line 1–2 in Fig. 19.9 (a), a view must be obtained in which line 1–2 appears true length and a horizontal plane appears in edge view. Any horizontal plane appears in edge view in the front view (parallel to the H/F folding line). Thus every point of a particular horizontal plane is at distance H (height) from H/F and any height auxiliary view, §10.8, will show the horizontal plane in edge view. At (b) a direction of sight perpendicular to the top view is chosen to obtain a true-length view of line 1–2. The resulting auxiliary view then shows the slope of line 1–2.

The observant student has probably noted that the angle could just as well be measured with respect to folding line H/1 in the auxiliary view. There is actually no need to introduce a special horizontal plane, providing the working space is suitable, since the H/F and H/1 folding lines represent edge views of the horizontal plane in the front view and in any height auxiliary view.

To obtain ∠P of line 1–2 in Fig. 19.10 (a), a view must be obtained in which line 1–2 appears true length and a profile plane appears in edge view.

At (b) side view P is constructed, which shows line 1–2 as it appears projected on a profile plane. At (c) the direction of sight is established perpendicular to the side view of line 1–2. The resulting auxiliary view shows a true-length view of line 1–2 and any profile plane in edge view and parallel to

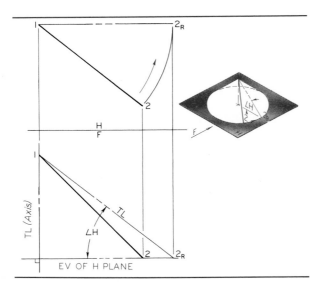

**Fig. 19.11**   True Length and Angle with Horizontal Plane by Revolution.

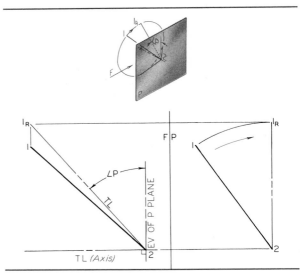

**Fig. 19.12**   True Length and Angle with Profile Plane by Revolution.

folding line P/1. The ∠P may then be measured with respect to the edge view of the profile plane, as indicated.

In summary, note that each of the angles ∠F, ∠H (slope), and ∠P is obtained by a separate auxiliary view, Figs. 19.8, 19.9, and 19.10, respectively; that is, an auxiliary view can show no more than one of these angles of an oblique line.

## 19.4 True Length and Angle with Plane of Projection by Revolution

The true length of a line may also be obtained by revolution, §11.10. In Fig. 19.11 a vertical axis of revolution is employed to find the true length of line 1–2. The path of revolution lies in a horizontal plane seen edgewise in the front view, as indicated. As the line revolves, its angle with horizontal (∠H) remains unchanged in space. Thus, in the true-length position this angle may be measured as indicated.

Note that as line 1–2 revolves about the chosen axis, its angles with the other two planes, frontal and profile, continually change. Hence this particular revolution, Fig. 19.11, cannot be used to find the angle the line forms with these planes.

The axis of revolution in Fig. 19.12 is perpendicular to a profile plane. Hence the angle revealed at the true-length position is ∠P.

To determine ∠F for line 1–2, it is necessary to establish the axis of revolution perpendicular to a frontal plane, Fig. 19.13. In practice, it is not necessary to show the axis of revolution, since the re-

maining construction makes the position of the axis obvious. Note that a separate revolution is needed for each angle of an oblique line with a projection plane.

## 19.5 Bearing and Grade

The position of a line in space, as is often found in geology, mining, and navigation, is described by its *bearing* and *grade* or by its bearing and slope. The

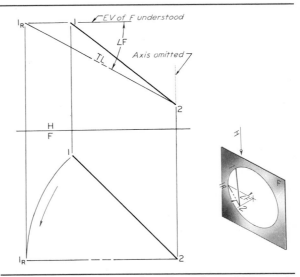

**Fig. 19.13**   True Length and Angle with Frontal Plane by Revolution.

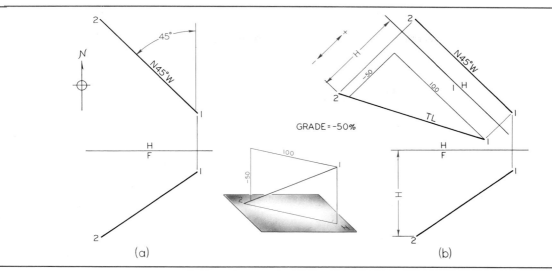

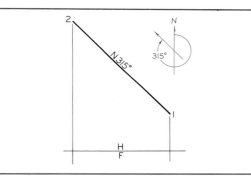

*Fig. 19.14*    Bearing and Percent Grade.

bearing of a line is the direction of a line on a map or horizontal projection. Since for practical purposes a limited area of the earth's surface may be considered a horizontal plane, a map is a top view of the area. Thus the bearing of a line is measured in degrees with respect to north or south in the *top view* of the line, Fig. 19.14 (a).

It is customary to consider north as being toward the top of the map unless information to the contrary is given. Hence the small symbol showing the directions of north, east, south, and west is not usually needed. These are the directions assumed in the absence of the symbol. Note that the bearing indicated, N 45° W, is that of line 1–2 (from 1 toward 2). The conventional practice is to give either the abbreviation for north or south first, chosen so that the angle is less than 90°, and then the angle, followed by the abbreviation for east or west, as appropriate. For example: N 45° W, as shown in the figure.*

At (b) a true-length auxiliary view is added, projected from the top view. As discussed in §19.3 this is the auxiliary view appropriate for measuring the slope of line 1–2 or ∠H. However, in this case another method, known as *grade*, is used to measure the inclination. The grade is the ratio of the vertical displacement (rise) to the horizonal displacement (run) expressed as a percentage.

A construction line horizontal in space (parallel to H/1) is drawn through a point of the line—point 1 in this example. Along this horizontal line 100 units of any appropriate scale are set off. In this

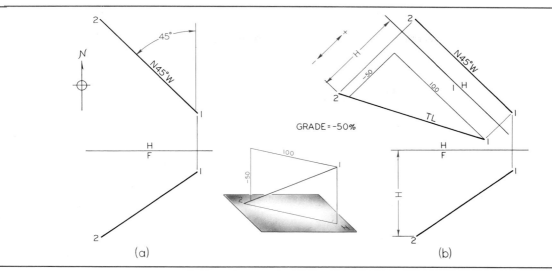

*Fig. 19.15*    Azimuth Bearing.

instance the 1/20 scale was used. At the 100th division a line is drawn perpendicular to the folding line H/1 and extended to intersect line 1–2 as shown. The length of this line, as measured by the previously used scale, becomes a numerical description of the inclination of line 1–2 expressed as −50%,* because

$$\frac{-50 \text{ units vertically}}{100 \text{ units horizontally}} \times 100 = -50\%†$$

This is the *percent grade*, or simply the *grade*, of line 1–2.

Another means of describing the bearing of a line is by its *azimuth* bearing, Fig. 19.15. Here the total

---

*Special cases are "Due north," "Due south," "Due east," and "Due west."

*It is common practice to designate a vertical distance as positive or negative according to whether it is measured upward or downward, respectively.

†The student familiar with trigonometry will recognize the ratio 50:100 as the *tangent* of ∠H.

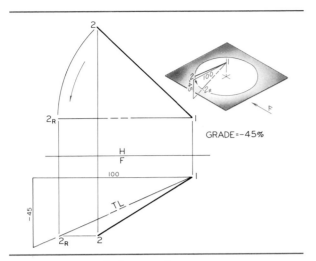

*Fig. 19.16*    Grade by Revolution.

*clockwise* angle from the base direction, usually north, is given. Line 1–2 here has the same direction as line 1–2 of Fig. 19.14 (point 1 toward 2) so that the clockwise angle is 360° −45° or 315°. If it is understood by all concerned that north is the base direction, the N may be omitted. Thus it is common for an aircraft pilot to describe his or her flight direction as "a course of 315°." On a drawing or map, however, it is best to retain the N to avoid possible confusion.

Grade may also be obtained by revolution. In Fig. 19.16, since the top view is revolved, the axis of revolution (not shown) projects as a point coincident with the top view of point 1. Hence the axis is vertical, and point 2 moves horizontally to the front view to 2$_R$. In the true-length position, the grade of −45% can be measured as shown.

## 19.6 Point View of Line

If a direction of sight for a view is parallel to a true-length view of a line, that line will appear as a point in the resulting view. See §7.15. In Fig. 19.17 (a) the vertical line 1–2 appears as a point in the top view, since a vertical line is true length in any height view.

At (b) line 3–4 appears true length in the front view. A direction of sight is chosen, as indicated by arrow 1, parallel to the true-length view. The resulting auxiliary view 1 is a point view, since all points of line 3–4 are the same distance D from the folding lines.

In Fig. 19.17 (c) auxiliary view 1 is necessary to show line 5–6 in true length. Direction of sight 2 is then introduced parallel to the true-length view. The resulting view 2, which is a secondary auxiliary view, shows the point view of line 5–6, §10.19.

Figure 19.17 (c) illustrates an important use of point views: finding the shortest distance from a point to a line. Since the shortest distance is measured along a perpendicular from the point to the line, the perpendicular will appear true length when the given line appears in point view. Observe point 7 in the illustration. An even more important use of the point view of a line is in obtaining an edge view of plane, §§10.11 and 19.9.

## 19.7 Representation of Planes

We have discussed planes as surfaces of objects, §7.15, bounded by straight lines or curves. Planes can be established or represented even more simply, Fig. 19.18, by intersecting lines, (a), parallel lines, (b), three points not in a straight line, (c), or a line and a point not on the line, (d). Careful study

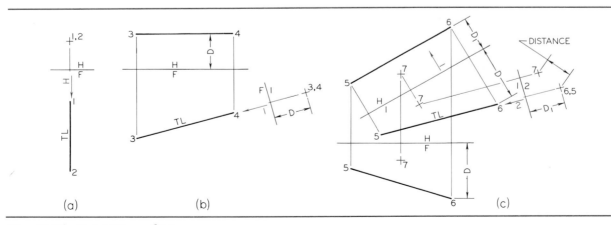

*Fig. 19.17*    Point View of Line.

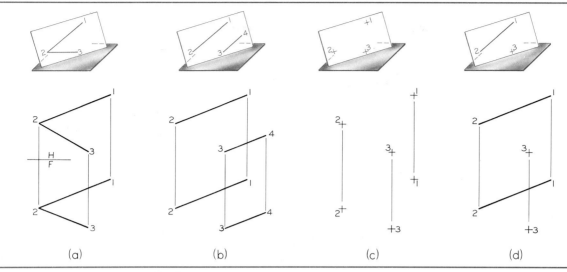

Fig. 19.18    Representation of a Plane.

of Fig. 19.18 will reveal that the same plane 1–2–3 is represented in all four examples. One method can be converted to another by adding or deleting appropriate lines without changing the position of the plane. Most problem solutions involving planes require adding lines at one stage or another, so that in practice the plane, regardless of its original representation, is in the end represented by intersecting lines.

## 19.8 Points and Lines in Planes
One formal definition of a plane is that it is a surface such that a straight line joining any two points of the surface lies in the surface. It follows that two

straight lines in the same plane must intersect, unless the lines are parallel. These concepts are used constantly in working with points and lines in planes.

Figure 19.19 (a) shows a typical elementary problem of this nature. The top view of a line 4–5 is given. The problem is to find the front view of line 4–5 that lies in plane 1–2–3. Since lines 4–5 and 1–2 are obviously not parallel, they must intersect at point 6 as shown at (b). Point 6 is then located by projecting vertically to the front view of line 1–2. Line 4–5 extended (top view) intersects line 2–3 at 7, which is projected to line 2–3 in the front view. Line segment 4–5 in the front view then lies along a construction line through points 6 and 7, and

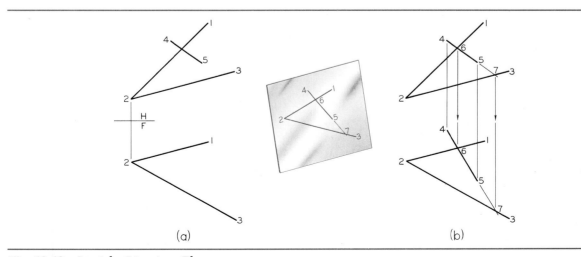

Fig. 19.19    Straight Line in a Plane.

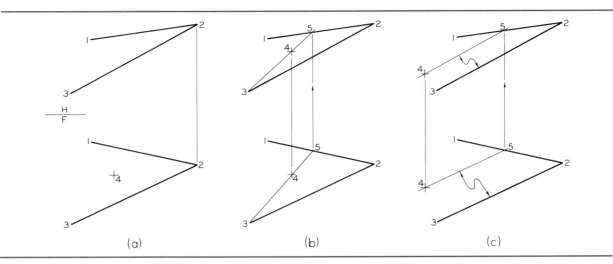

*Fig. 19.20*  Point in a Plane.

points 4 and 5 are established by projection from the top view as shown.

A point may be placed in a plane by locating it on a line known to be in the plane. In Fig. 19.20 (a) we are given the front view of a point 4 in plane 1–2–3 and desire to find the top view. At (b) a line is introduced through points 3 and 4 and, when extended, intersects line 1–2 at point 5. Point 5 is projected to the top view, establishing line 3–5 in that view. Point 4 is then projected from the front view to the top view of line 3–5. Theoretically any line could be drawn through point 4 to solve this problem. However, lines approaching parallel to the projection lines between views should be avoided as they may lead to significantly inaccurate results.

A different solution of a similar problem—using the principle of parallelism*—is shown in Fig. 19.20 (c). Here lines drawn through point 1 or point 3 and given point 4 lead to inconvenient intersections. If a line is drawn through point 4 in the front view parallel to line 2–3 and intersecting line 1–2 at point 5 as shown, it will not intersect line 2–3. Therefore, according to the principles stated at the beginning of this section, the line must be parallel to line 2–3. Thus, after intersection point 5 is projected to the top view, the new line through point 5 is drawn parallel to the top view of line 2–3, and point 4 is projected to it to complete the top view. Use of this parallelism principle requires minimum construction.

*Parallelism is discussed in more detail in Chapter 20.

Another example of locating a point on a plane is shown in Fig. 19.21. Here it is desired to locate in plane 1–2–3 a point P that is 10 mm above point 2 and 12 mm behind point 3.

At (a) a horizontal line 10 mm above (higher than) point 2 is added to the front view of plane 1–2–3. Its intersection points 4 and 5 with lines 1–2 and 2–3 are projected to the top view as shown. Any point along line 4–5 lies in plane 1–2–3 and is 10 mm above point 2. Line 4–5 is said to be the *locus* of such points.

At (b) a frontal line 12 mm behind (to the rear of) point 3 is added to the top view of the plane. Its front view is obtained by projection of intersection points 6 and 7. (Note the addition of line 1–3 to secure point 7.) Line 6–7 is the locus of points in plane 1–2–3 that are 12 mm behind point 3.

The intersection point of lines 4–5 and 6–7 is the required point P. The views of P at (b) are checked with a vertical projection line, as shown, to make sure that they are views of the same point.

## 19.9 Edge Views of Planes

In order to get the edge view of a plane, we must get the point view of a line in the plane, §10.11. For Fig. 19.22 (a) the edge view of plane 1–2–3 could be obtained by getting the true-length view and then the point view of any one of the three given lines of the plane. Since these lines are all oblique lines, obtaining their point views would each entail two successive auxiliary views, §19.6. It is easier to add a line that appears true length in one of the principal

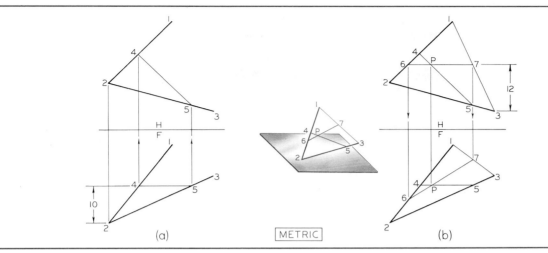

**Fig. 19.21** Locus Problem.

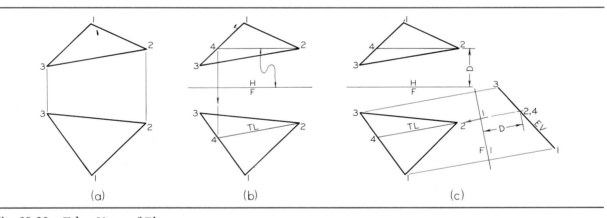

**Fig. 19.22** Edge View of Plane.

views, thus eliminating the need of a second auxiliary view.

At (b) line 2–4 is drawn parallel to the H/F folding line (horizontal on the paper) in the top view. Thus it is a frontal line, §19.2, and its front view, obtained by projecting point 4 to the front view of line 1–3, is true length, as indicated.

Thus a true-length line in plane 1–2–3 has been established without drawing an auxiliary view, and we may now proceed as at (c) by assuming a direction of sight 1 parallel to the true-length view of 2–4. The resulting auxiliary view is the desired edge view. Note that all points of the plane, not just a minimum two points, are actually projected to the auxiliary view. This provides a convenient check on

accuracy, since obviously the points must lie on a straight line in the auxiliary view.

Edge views are useful as the first step in obtaining the true-size view of an oblique plane, §§10.21 and 19.10. They are also employed in showing dihedral angles, §10.11. In Fig. 10.11 (c) it was possible to show the true angle by drawing a primary auxiliary view, since the line of intersection of the planes appeared in true length in the top view. In Fig. 19.23 (a) the line of intersection 1–2 between surfaces **A** and **B** is not shown true length in either view. Accordingly, at (b) auxiliary view 1 is constructed, with the direction of sight 1 perpendicular to line 1–2. At (c) secondary auxiliary view 2 is then added, with direction of sight 2 parallel to the true-length

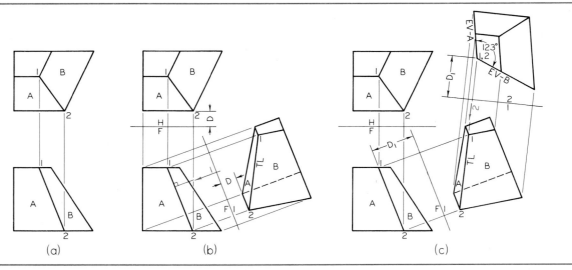

**Fig. 19.23**   Dihedral Angle with Oblique Line of Intersection.

view of line **1–2**. As a check, all points of the planes are located in the auxiliary views. When they fall on the respective straight-line (edge) views of the surfaces **A** and **B**, confidence in accuracy is established.

### 19.10 True-Size Views of Oblique Planes

The procedure for obtaining the true size of an oblique surface of an object is treated in §10.21. Many problems of a more abstract nature are also solved through obtaining true-size views of oblique planes.

For example, let it be required to find the center of the circle passing through points **1**, **2**, and **3** of Fig. 19.24. The plane geometry construction is that of §5.31. However, the construction illustrated in that section took place in the plane of the paper, which we now recognize as a form of a true-size view. Since this is not the situation at (a), it is necessary to proceed as at (b) with the edge view and

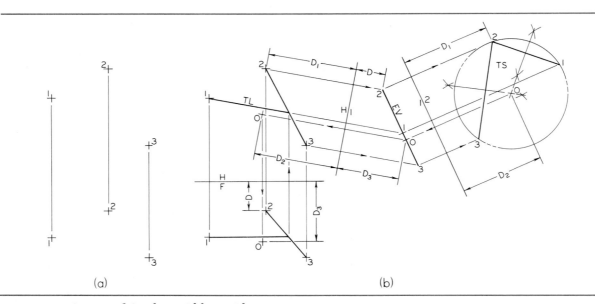

**Fig. 19.24**   Center of Circle in Oblique Plane.

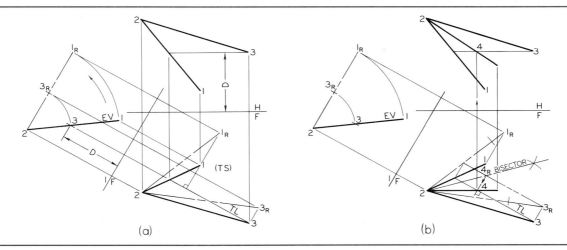

**Fig. 19.25** True Size of Oblique Plane—Revolution Method.

true-size views of plane **1–2–3**. In auxiliary view **2** the construction shown in §5.31 is performed, locating center **0** for a circle through points **1**, **2**, and **3**. If desired, point **0** can be located in the divider distances $D_2$ and $D_3$ as shown. The circle, if drawn, would appear elliptical in the front and top views, §10.24.

The revolution method explained in §11.11 may also be applied to finding the true-size view of an oblique surface for geometric construction. In Fig. 19.25, the problem is to bisect the plane angle **1–2–3**. The true bisector of a plane angle lies in the same plane as the angle, and only under special circumstances when the plane of the angle is not in true size will a view of the bisector actually bisect the corresponding view of the angle. To solve this problem, the edge view of plane **1–2–3** is first constructed at (a) and then is revolved until it is parallel to folding line **F/1**. The revolved front view is then true size and angle **$1_R$–2–$3_R$** is bisected as shown at (b). The revolution thus takes the place of a secondary auxiliary view. This has the major advantage of compactness of construction, but the overlapping front views may in some cases be confusing.

The front and top views of the bisector are obtained by selecting an additional point on the bisector in the true-size view and reversing the whole process—*counterrevolving*. Point **$4_R$** on the true-length line through point **$3_R$** is particularly convenient in this case because it will counterrevolve to the true-length line through point **3** and can then be projected to the top view, bypassing the auxiliary view. If desired, however, the selected point can be

projected to the revolved edge view, counter-revolved to the original edge view, and then returned to the front and top views by the usual methods.

## 19.11 Piercing Points

If a straight line is not parallel to a plane, it must intersect that plane in a single point called a *piercing point*. It may be necessary to extend the line, or plane, or both; but this is permissible, since the abstract terms *line* and *plane* do not imply any limits on their extent. There are two recommended methods for finding piercing points.

### Edge-View Method

All points of a plane are shown along its edge view. These, of course, include the piercing point of any lines that happen to be present. In Fig. 19.26 (a) the frontal line through point **3** is introduced to get a true-length line and thus the edge view of plane **1–2–3**. In this case it is necessary to extend line **4–5** to find the piercing point (encircled). At (b) the piercing point is projected first to the front view and then to the top view. Note the use of divider distance $D_1$ to check the accuracy of location of the top view. This procedure, under some circumstances, is more accurate than direct projection.

Note that a horizontal line could have been introduced into plane **1–2–3**, thereby establishing a different true-length line. This would have produced a different edge view, but would not give a different piercing point, as there is only one piercing

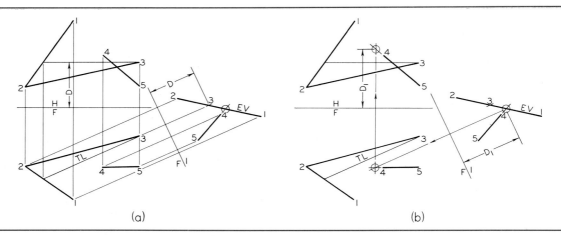

**Fig. 19.26**   Piercing Point—Edge-View Method.

point for a particular straight line and plane. This procedure would not be considered a different method but merely an alternative approach.

### Cutting-Plane Method or Given-View Method

If a cutting plane A–A containing line 4–5 is introduced, Fig. 19.27 (a), it will cut line 6–7 from given plane 1–2–3. Lines 6–7 and 4–5, being in the same plane A–A, must intersect at the piercing point (encircled). To make the method practical, an edge-view cutting plane is used, as shown at (b). To contain line 4–5, the cutting plane must coincide with a view of the line. At (b) it was chosen to have plane

A–A coincide with the top view of 4–5. The line cut from plane 1–2–3 is line 6–7. Projected to the front view, line 6–7 locates the front view of the piercing point, which is then projected to the top view as shown.

Actually, there is no need for some of the lettering shown at (b). At (c) the symbol **EV** adequately identifies the cutting plane, and the numbers 6 and 7 may be omitted as being of little value other than for purposes of discussion.

Note that this illustration is similar to Fig. 19.26, except that (1) the piercing point is within the line segment 4–5 and (2) the plane is *limited* or completely bounded. It is then feasible to consider the

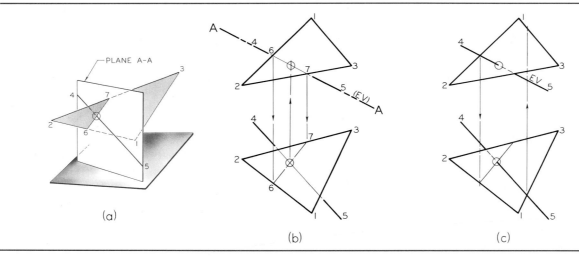

**Fig. 19.27**   Piercing Point—Cutting-Plane Method.

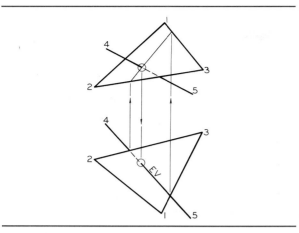

*Fig. 19.28* Piercing Point—Cutting-Plane Method (alternative solution).

bounded area to be opaque. The line then becomes hidden after it pierces the plane. The visibility displayed at (c) was determined by the methods of §19.1, investigating in each view any convenient point where line **4–5** crosses one of the boundary lines of the plane.

The problem in Fig. 19.27 is shown again in Fig. 19.28, this time with the edge view of the cutting plane introduced coincident with the front view of line **4–5**. Of course, the same answer is obtained, and it is a matter of personal choice and convenience as to which view is chosen for introduction of the edge-view cutting plane. For the convenience of the reader, always include the letters **EV** as shown when the problem solutions involve such cutting planes.

## 19.12 Intersections of Planes

The intersection of two planes is a straight line containing all points common to the two planes. Since planes are themselves represented by straight lines, §19.7, points common to two intersecting planes may be located by finding piercing points of lines of one plane with the other plane, by the use of edge-view method or the cutting-plane method of §19.11.

### Edge-View Method

In Fig. 19.29, two planes are given: **1–2–3** and **4–5–6**. If the edge view of either plane is constructed, the piercing points of the lines of the other plane will lie along the edge view. At (a) a horizontal line is introduced through point **3** of plane **1–2–3** in order to secure a true-length line in the top view. (A horizontal line in plane **4–5–6** or frontal lines in either plane would serve just as well in this problem.) Auxiliary view 1 is then constructed with its direction of sight parallel to the true-length view of the line. The completed auxiliary view 1 shows the edge view of plane **1–2–3** and the piercing points of lines **4–5** and **5–6** as indicated by the encircled points.

At (b) the piercing points are projected to the top view and then to the front view. (It is good practice to check accuracy by divider distances, as indicated by dimension $D_1$). Since the given planes are not completely bounded, there is no reason to restrict the drawn length of the segment of the line of intersection (LI). However, the views of the line of intersection should be compatible from view to view.

Since the **LI** is common to both planes, it must

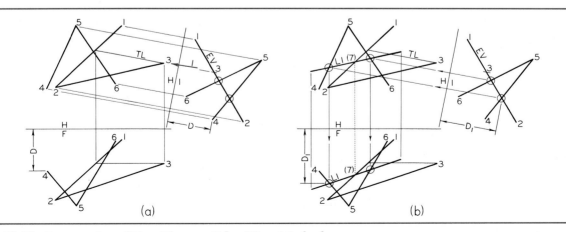

*Fig. 19.29* Intersection of Two Planes—Edge-View Method.

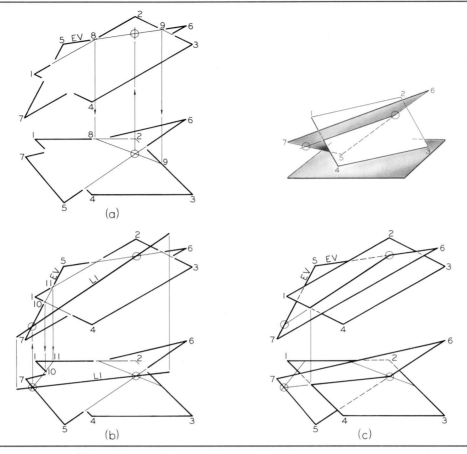

**Fig. 19.30**   Intersection of Two Planes—Cutting-Plane Method.

intersect or be parallel to each line of both planes. As a check on accuracy, observe in this case that the LI intersects line 1–2 at 7 and is parallel to line 2–3.

### Cutting-Plane Method

Because it requires no additional views, the cutting-plane method is frequently used to find the intersection of two planes, Fig. 19.30. At (a) it is arbitrarily decided to introduce an edge-view cutting plane coinciding with the top view of line 5–6, with the intention of finding the piercing point of line 5–6 in plane 1–2–3–4. The student should realize that one could introduce cutting planes in either view coinciding with any of the lines of the planes. With so many possibilities it is imperative that the choice be indicated with proper use of the symbol EV, both to avoid confusion on the student's part and as a courtesy to the person who must read the drawing.

In this case the introduced plane cuts line 8–9 from plane 1–2–3–4. Point 8 is on line 1–2 and point 9 is on line 2–3. Observe this carefully so as to avoid mistakes in projecting to the front view. The front view of line 8–9 intersects line 5–6 at the encircled piercing point which, after projection to the top view, represents one point common to the given planes.

At (b) another piercing point is located by introducing an edge-view cutting plane along line 5–7. The line of intersection, LI, passes through the two piercing points as shown.

In this illustration the given planes are bounded and can therefore be considered limited as at (c). The piercing point of line 5–7 falls outside plane 1–2–3–4 and is therefore not on the "real" portion of the line of intersection, which is drawn as a visible line only in the area common to the views of both planes. The termination of this segment is at the

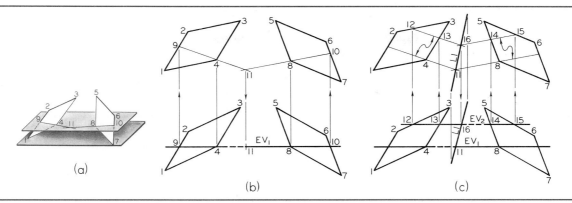

*Fig. 19.31*  Intersection of Two Planes—Special Cutting-Plane Method.

point that is actually the piercing point of line 1–4 in plane 5–6–7. However, this result was not obvious at the start of the construction. Visibility was determined by the method of §19.1. Usually it is necessary to examine only one apparent crossing point in each view. After visibility is established in one such region, the spatial relations of the remaining lines of that view are evident, since each boundary line in turn can change visibility only where it meets a piercing point or a boundary line of the other given plane.

### Special Cutting-Plane Method

The line of intersection of two planes also may be found through the use of cutting planes that do not coincide with the views of given lines. Any plane cutting the two given planes, Fig. 19.31 (a), cuts one line from each. Since these lines lie in the cutting plane as well as in the given planes, they will intersect at a point common to the given planes. A second cutting plane will establish a second common point, giving two points on the line of intersection. For convenience, the edge-view cutting planes employed are usually drawn parallel to a regular coordinate plane, but this is not necessary. It is suggested that the two planes be introduced in the same view for more control of the distance between the points secured.

At (b) horizontal plane EV₁ cuts lines 9–4 and 8–10 from the given planes. When these lines are projected to the top view, they intersect at point 11, which is then one point on the required line of intersection. The front view of point 11 is on line EV₁ as shown. At (c) a second horizontal plane EV₂ is introduced, cutting lines 12–13 and 14–15, which intersect in the top view at point 16. After point 16

is projected to line EV₂ the line of intersection 11–16 (LI) is drawn to any desired length. Note carefully the parallelism of the lines in the top view at (c). This affords a convenient and a very desirable check on accuracy.

This method involves more construction than did the previous methods and can be confusing when the given views occupy overlapping areas, as in Figs. 19.29 and 19.30. This particular method is therefore recommended primarily for problems in which the given views of the planes are separated.

### 19.13 Angle Between Line and Oblique Plane

The true angle between a line and a plane of projection (frontal, horizontal, or profile) is seen in the view in which the given line is true length and the plane in question is in edge view, §19.3. This is a general principle that applies to any plane: normal, inclined, or oblique.

In Fig. 19.32 (a) two views of a plane 1–2–3 and a line 5–6 are given. One cannot expect a primary auxiliary view to show plane 1–2–3 in edge view and also line 5–6 in true length, for generally the directions of sight for these two purposes will not be parallel. Note that in Fig. 19.8 the direction of sight for the auxiliary view is toward the front view, which shows the true-size view of all frontal planes. In Fig. 19.9 the direction of sight is toward the top view, which shows the true-size view of all horizontal planes. In summary, *any view projected from a true-size view of a plane shows an edge view of that plane.*

In Fig. 19.32 (a) frontal line 2–4 is added to given plane 1–2–3. We thus now have a true-length line

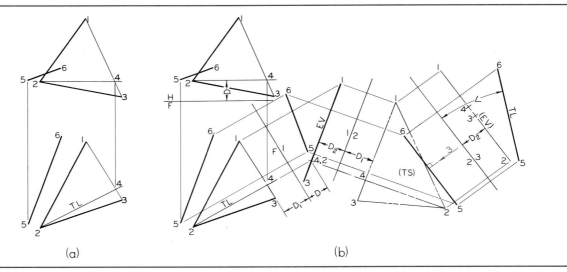

(a)                                        (b)

*Fig. 19.32*    Angle Between Line and Plane—True-Size Method.

in the front view and edge-view auxiliary view 1 is projected from it. The true-size secondary auxiliary view 2 is then constructed in the customary manner.

Any view projected from view 2 will show plane 1–2–3 in edge view. Therefore to show the true angle between line 5–6 and the plane, direction of sight 3 is established at right angles to view 2 of line 5–6. View 3 then shows the required angle (∠). Be-

cause in this chain of views the edge view of the plane in view 3 is always parallel to folding line 2/3 (note divider distance $D_2$), the construction can be simplified if desired by omitting plane 1–2–3 from auxiliary views 2 and 3. The required angle is then measured between line 4–5 and the folding line 2/3.

# *Point, Line, and Plane Problems*

The problems in Figs. 19.33 to 19.44 cover points and lines, intersecting and non-intersecting lines, visibility, true length and angles with principal planes, auxiliary-view method and revolution method, point views, points and lines in planes, dihedral angles, edge view and true size of planes, piercing points, intersection of planes, and angle between line and oblique plane.

Use Layout A–1 or A4–1 (adjusted) and divide the working area into four equal areas for problems to be assigned by the instructor. Some problems will require a single problem area and others will require two problem areas (half the sheet). See §2.63. Data for the layout for each problem are given by a coordinate system. For example, in Fig. 19.33, Prob. 1, point 1 is located by the full-scale coordinates, 22, 38, and 75 mm. The first coordinate locates the front view of the point from the left edge of the problem area. The second coordinate locates the front view of the point from the bottom edge of the problem area. The third coordinate locates either the top view of the point from the bottom edge of the problem area or the side view of the point from the left edge of the problem area. Inspection of the given problem layout will determine which application to use.

Since many of the problems in this chapter are of a general nature, they can also be solved on most computer graphics systems. If a system is available, the instructor may choose to assign specific problems to be completed by this method.

Additional problems, in convenient form for solution, are available in *Principles of Engineering Graphics Problems* by Spencer, Hill, Loving, and Dygdon, a workbook designed to accompany this text that is also published by Macmillan Publishing Company.

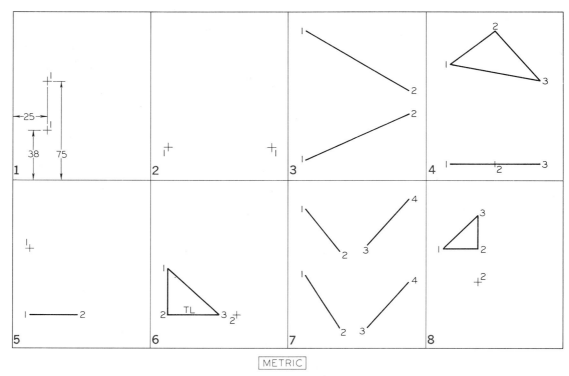

METRIC

***Fig. 19.33***   Lay out and solve four problems per sheet as assigned. Use Layout A–1 or A4–1 (adjusted) divided into four equal areas.

1.  Given point 1(25, 38, 75), locate the front and top views of point 2, which is 50 mm to the right of point 1, 25 mm below point 1, and 30 mm behind point 1.

2.  Given point 1(12, 25, 90), locate the front and side views of line 1–2 such that point 2 is 38 mm to the right of point 1, 45 mm above point 1, and 25 mm in front of point 1. Add the top view of line 1–2.

3.  Find the views of points 3, 4, and 5 on line 1(12, 15, 115)–2(90, 50, 70) that fit the following descriptions: point 3, 20 mm above point 1; point 4, 65 mm to the left of point 2; and point 5, 25 mm in front of point 1.

4.  Triangle 1(18, 12, 90)–2(50, 12, 115)–3(85, 12, 75) is the base of a pyramid. The vertex V is 8 mm behind point 1, 8 mm to the left of point 2, and 45 mm above point 3. Complete the front and top views of the pyramid.

5.  Line 1(12, 25, 75)–2(48, 25, ?) is 43 mm long (2 behind 1). Line 1–3 is a 50 mm frontal line, and line 2–3 is a profile line. Find the true length of line 2–3.

6.  Line 1(12, 60, ?)–2(12, 25, 64) is 45 mm long. The front view of line 2–3(50, 25, ?) is true length as indicated. Complete the front and side views and add a top view of triangle 1–2–3.

7.  Point 5 is on line 1(12, 56, 106)–2(38, 15, 74), 18 mm below point 1. Point 6 is on line 3(58, 15, 80)–4(90, 50, 115). Line 5–6 is frontal. Find the true length of line 5–6.

8.  Line 2(38, 50, 75)–3(38, ?, 100) is 38 mm long. Line 3–1(12, ?, 75) is horizontal. How long is line 1–2?

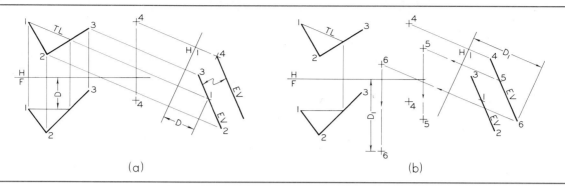

**Fig. 20.3** Parallel Planes by Parallel Edge Views.

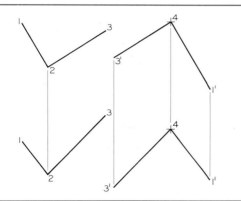

**Fig. 20.4** Parallel Planes by Parallel Lines.

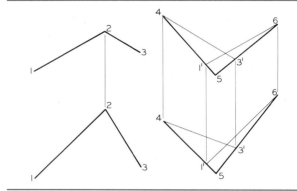

**Fig. 20.5** Checking Parallelism of Planes with Intersecting Lines.

jection lines. The front views are then located by projection from the top view and transfer of distances such as $D_1$.

A somewhat simpler and more commonly used procedure for drawing parallel planes is shown in Fig. 20.4. The method depends on this principle: *If a pair of intersecting lines in one plane is parallel to a pair of intersecting lines in a second plane, the planes are parallel.* Thus the preceding problem, Fig. 20.3 (a), is readily solved as in Fig. 20.4. Line 4–3′ is drawn parallel to line 2–3 in both views, and line 4–1′ is drawn parallel to line 1–2. Plane 3–4–1′ is parallel to plane 1–2–3. The lines 4–3′ and 4–1′ may be of any desired length.

The intersecting-line principle can also be used to check for parallelism. In Fig. 20.5, line 6–1′ was added in the top view of plane 4–5–6 and parallel to line 1–2 of plane 1–2–3. When the front view of point 1′ is located, line 6–1′ is found to be parallel to line 1–2. Continuing the investigation: In the top view of plane 4–5–6, line 4–3′ is added parallel to line 2–3. When the front view of line 4–3′ is located,

however, it is seen that it is *not* parallel to the front view of line 2–3. Hence planes 1–2–3 and 4–5–6 are *not* parallel in space.

## 20.3 Line Parallel to Plane

*A line is parallel to a plane if it is parallel to a line in the plane.* To establish line 4–5′ parallel to the plane 1–2–3, Fig. 20.6, line 3–5 is arbitrarily se-

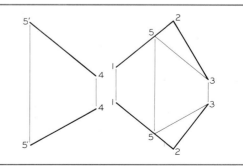

**Fig. 20.6** Line Parallel to Plane.

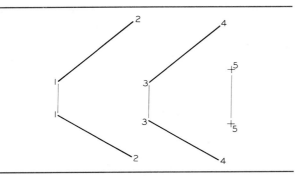

*Fig. 20.7*   Plane Parallel to Line.

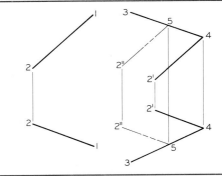

*Fig. 20.8*   Plane Through One or Two Skew Lines and Parallel to the Other.

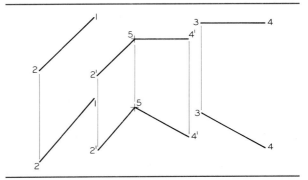

*Fig. 20.9*   Plane Through Point Parallel to Skew Lines.

lected and added to the plane 1–2–3. Line 4–5′ is then drawn parallel to line 3–5 in the plane. It is possible that line 4–5′ is *in* plane 1–2–3 (if extended). Even if true, this is not considered an exception to the general principle.

## 20.4 Plane Parallel to Line

*A plane is parallel to a line if it contains a line that is parallel to the given line.* Thus, if plane 1–2–3 is parallel to line 4–5′, Fig. 20.6, line 3–5 added to the plane 1–2–3 and parallel to line 4–5 will be parallel to line 4–5′ in all views. This is the converse of the principle of §20.3. Logically, if line 4–5′ in Fig. 20.6 is parallel to plane 1–2–3, it follows that plane 1–2–3 is parallel to line 4–5′.

Suppose we are given two parallel lines 1–2 and 3–4, Fig. 20.7. How many planes can be "passed through" line 3–4 parallel to line 1–2? If any random point 5 is added (not on 3–4), a plane 3–4–5 is established, §19.7. The plane is parallel to line 1–2 because it contains line 3–4. Thus it is seen that an infinite number of planes can be passed through one of two parallel lines and parallel to the other.

On the other hand, Fig. 20.8, let two *skew* lines 1–2 and 3–4 be given, and let it be required to pass a plane through line 3–4 and parallel to line 1–2. If a line is added, such as 4–2′, parallel to line 1–2, plane 3–4–2′ is parallel to line 1–2. Note that the added line could be made to intersect line 3–4 at any point, such as 5, resulting in plane 3–5–2″. However, this is merely a revised representation of plane 3–4–2′. It is evident, then, that through one of two skew lines only one plane can be passed parallel to the other line.

In Fig. 20.9, let skew lines 1–2 and 3–4 and point

5 be given, and let a plane be required through the point and parallel to the skew lines. If line 5–2′ is drawn parallel to line 1–2, and line 5–4′ is drawn parallel to line 3–4, plane 2′–5–4′ is parallel to both lines 1–2 and 3–4, even though nonintersecting lines 1–2 and 3–4 do not represent a plane.

## 20.5 Perpendicular Lines

In §7.26 it was observed that a 90° angle is projected in true size, even though it is in an inclined plane, provided one leg of the angle is a *normal* line, Fig. 7.29 (d). This principle can be restated in broader terms: *A 90° angle appears in true size in any view showing one leg true length,* provided the other leg does not appear as a point in the same view. Thus, in Fig. 20.10, lines 2–3, 2–4, 2–5, and 2–6 are all perpendicular to line 1–2, and they appear at 90° to the true-length front view of line 1–2. Note that the 90° angle is not observed in the top view where none of the lines is true length.

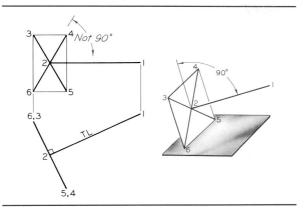

**Fig. 20.10** Lines Perpendicular to True-Length Line.

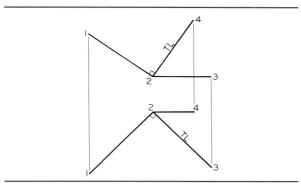

**Fig. 20.11** True-Length Lines Perpendicular to Oblique Line.

In Fig. 20.11 each of the lines 2–3 and 2–4 is perpendicular to oblique line 1–2 because their true-length views are perpendicular to the corresponding views of line 1–2. (Note that line 2–3 is a frontal line and line 2–4 is a horizontal line, §19.2.)

## 20.6 Plane Perpendicular to Line
### *Given-View Method*

To establish a plane perpendicular to a line, Fig. 20.12 (a), true-length lines are drawn perpendicular to given line 1–2 in the same manner as in Fig. 20.11. Now consider plane 3–2–4 in Fig. 20.12 (a). Since it is represented by intersecting lines, each of which is perpendicular to line 1–2, the plane is perpendicular to line 1–2.

Then consider the case at (b) where a plane through given point 5 and perpendicular to line 1–2 is desired. If plane 3′–5–4′ is constructed parallel to plane 3–2–4 of part (a), plane 3′–5–4′ will also be perpendicular to line 1–2. However, plane 3′–5–4′ could be drawn directly—without the use of plane 3–2–4—merely by drawing the true-length views of lines 3′–5 and 5–4′, respectively, perpendicular to the corresponding views of line 1–2, as indicated at (b).

Note that lines 3′–5 and 5–4′ do *not* intersect line 1–2. (You may prove this for yourself by extending the lines and checking vertical alignment of crossing points.) Thus, for present purposes, it is useful to regard the perpendicular true-length view *position principle* as indicating perpendicular lines, without regard to whether the lines intersect.

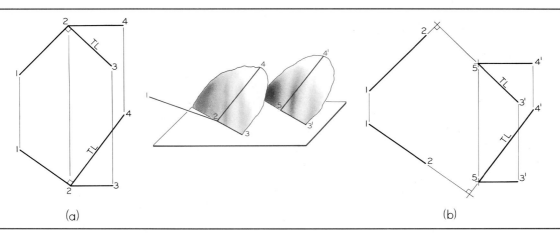

(a)                              (b)

**Fig. 20.12** Plane Perpendicular to Line—Given-View Method.

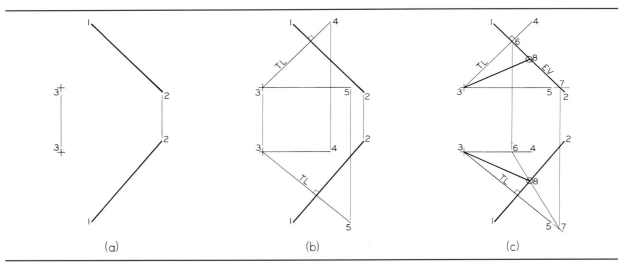

**Fig. 20.13** Plane Perpendicular to Line—Application.

### APPLICATION OF GIVEN-VIEW METHOD

*Given:* Views of a line 1–2 and a point 3, Fig. 20.13 (a).

*Req'd:* Find a line from point 3 perpendicular to and intersecting line 1–2. Use only the given views.

*Analysis and Procedure:* If a horizontal line 3–4 is drawn through point 3 with its true-length view perpendicular to the top view of 1–2 as at (b), it does not intersect line 1–2. Similarly, if a frontal line 3–5 is drawn perpendicular to line 1–2, it also does not intersect line 1–2. Neither of these lines is the required line. We conclude that the required line will not appear perpendicular in the given views. However, 4–3–5 represents a *plane* perpendicular to line 1–2, and all lines in the plane are perpendicular to line 1–2. Plane 4–3–5 is the locus of all lines through point 3 perpendicular to line 1–2. The required line belongs to this locus or family of lines.

By the cutting-plane method of §19.11, the piercing point 8 of line 1–2 in plane 4–3–5 is found, Fig. 20.13 (c). This is the only point on line 1–2 that is in the plane, and thus line 3–8 is the only possible solution to the problem.

### Auxiliary-View Method

A plane also may be constructed perpendicular to a line by drawing its edge view perpendicular to the true-length view of the line, Fig. 20.14. At (a) given line 1–2 and point 3 are projected to the true-length auxiliary view, where the edge view of the required plane is drawn through point 3 and perpendicular to the true-length view as indicated. If it is desired to represent the perpendicular plane in the top and front views, as at (b), any convenient pair of points, such as 4 and 5, is selected on the edge view. In the top view points 4 and 5 may be placed at any desired locations along the projection line from view 1. Points 4 and 5 are then located in the front view by the divider distances as indicated.

### APPLICATION OF AUXILIARY-VIEW METHOD

*Given:* A right square prism, Fig. 5.7 (6), has its axis along line 1–2 and one corner at point 3, Fig. 20.15 (a). The other base is centered at point 4.

*Req'd:* Find the views of the prism.

*Analysis and Procedure:* Auxiliary view 1 is added showing axis 1–2 in true length. Because a *right* prism is required, the bases must be perpendicular to axis 1–2 and appear in view 1 in edge view and perpendicular to axis 1–2. At (b) the point view of axis 1–2 is added. This view shows the true shape of the square bases and the size of the square is established by the position of corner 3. By the method of Fig. 5.23 (c), the square is constructed.

The projection process is now reversed. The corners are projected from view 2 to view 1, which is then completed. Next, the corners are projected to the top view and located with divider distances such as $D_2$. Observe that a square prism is composed of three sets of parallel lines. Check the view for parallelism and correct any errors before proceeding.

Finally, the front view is projected in similar fashion. Again, the construction work should be

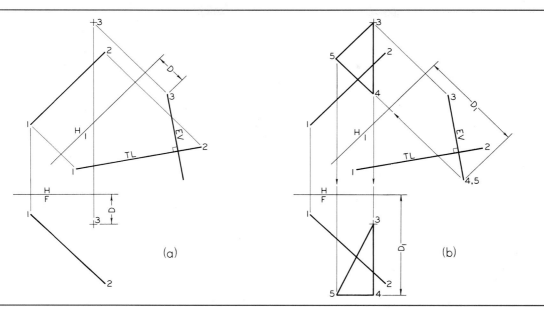

*Fig. 20.14*  Plane Perpendicular to Line—Auxiliary-View Method.

checked for accurate parallelism before the views are completed.

## 20.7 Line Perpendicular to Plane

*A line perpendicular to a plane is perpendicular to all lines in the plane.* In practice, it is sufficient to state that a line is perpendicular to a plane if it is perpendicular to at least two nonparallel lines in the plane. This line will also appear perpendicular (and in true length) to the edge view of the plane. Either principle may be used in the construction of a line perpendicular to a plane.

It is desired to construct by the given-view method, Fig. 20.16, a line from point 4 perpendicular to plane 1–2–3. Since the plane is oblique, the perpendicular line will be oblique and will not appear in true length. Although the required line will

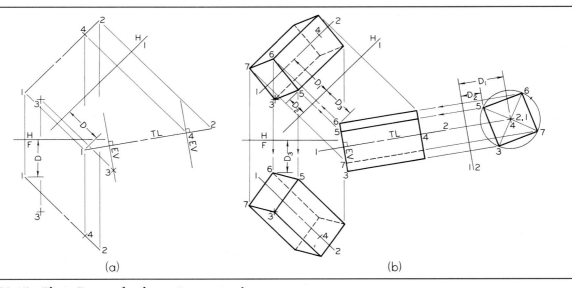

*Fig. 20.15*  Plane Perpendicular to Line—Application.

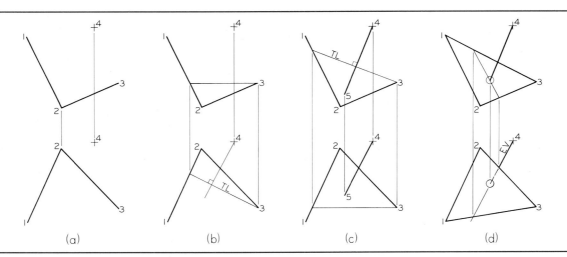

**Fig. 20.16**  Line Perpendicular to Plane—Given-View Method.

be perpendicular in space to lines 1–2 and 2–3, the 90° angles between the given lines 1–2 and 2–3 and the required line will not appear in the given views because lines 1–2 and 2–3 are not shown in true length.

If a frontal line is added to plane 1–2–3, as at (b), its front view is true length, and the front view of the required line will be perpendicular to the true-length view of the frontal line as shown. It must be realized that at this stage nothing has been determined about the direction of the perpendicular in the top view. Nor is the point at which the perpendicular strikes the plane—its piercing point—known. (Frequently the location of the piercing point is of no interest.)

At (c) a horizontal line is added. The true-length view of the horizontal line determines the direction of the top view of the required perpendicular. Again, the piercing point has not been determined, but two views have been established and thus a space description of the required perpendicular from point 4 with plane 1–2–3 has been constructed. Point 5 is arbitrarily selected as an end point of the line, not necessarily in the plane.

If it is desired to terminate the perpendicular in plane 1–2–3, it will be necessary to find the piercing point by one of the methods of §19.11. At (d) the cutting-plane method was used to minimize construction.

## 20.8 Perpendicular Planes

*If a line is perpendicular to a given plane, any plane containing the line is perpendicular to the given plane.* Since an infinite number of planes can be passed through such a perpendicular, it was chosen in Fig. 20.17 to illustrate perpendicular planes by a more restricted example: How to pass a plane through given line 4–5 and perpendicular to given plane 1–2–3.

At (a) a horizontal line is added to the plane. Since a horizontal line appears true length in the top view, it determines the direction of the required perpendicular in that view (but, remember, not in the front view). Point 5 is selected as being convenient for the origin of the perpendicular.

At (b) a frontal line is added to plane 1–2–3 to establish a true-length line in the front view, which is needed to determine the direction of the front view of the required perpendicular. It is assumed that the piercing point of the perpendicular is of no interest here. Accordingly, any arbitrary point, such as 6, is used to terminate the perpendicular and complete the representation of the required perpendicular plane 4–5–6.

## 20.9 Common Perpendicular Between Skew Lines

*The shortest distance, or clearance, between any two lines is measured along a line perpendicular to both lines.* Study Fig. 20.18. Suppose line 5–6 to be perpendicular to line 1–2, intersecting line 3–4 at 6. Now let line 5–6 move to position 5′–6′, still perpendicular to line 1–2 and intersecting line 3–4 at point 6′. Continue the process of moving line 5–6 upward along line 1–2. Note that it will gradually shorten until eventually it reaches a minimum length and will begin to lengthen if moved further. At its minimum length it will be perpendicular to

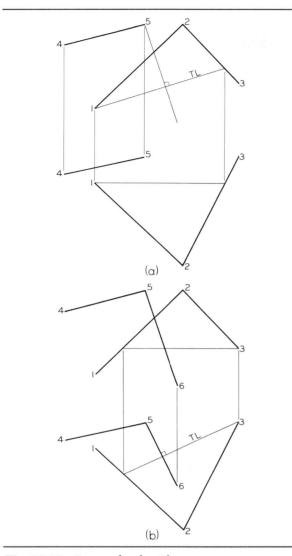

(a)

(b)

*Fig. 20.17*  Perpendicular Planes.

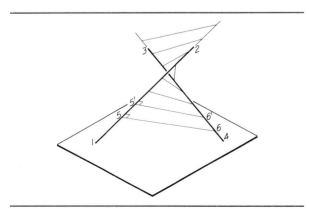

*Fig. 20.18*  Common Perpendicular.

line 3–4 also. This is the common perpendicular representing the shortest distance between the skew lines 1–2 and 3–4. Several procedures are available for locating the views of the common perpendicular in multiview projection, two of which will be discussed here.

### Point-View Method

If a point view of any given line is constructed, a line that is perpendicular to the given line will show in true length in the view showing the given line as a point. As noted in §19.6, a point view of a line must be preceded by a true-length view of the line. Accordingly, in Fig. 20.19 (a) line 3–4 is arbitrarily chosen to be shown in true length and view 1 is projected. As shown at (b), view 2 is then constructed showing line 3–4 as a point. In this view, any line perpendicular to line 3–4 (including the shortest connector) must appear in true length. Since the shortest connector is also perpendicular to line 1–2, it must appear at 90° to line 1–2 as shown, even though line 1–2 is not true length. If only the shortest distance is required, it is measured in view 2 and the construction is complete.

If, in addition, the views of the common perpendicular are required, we proceed as shown at (b). Point 5 is projected to line 1–2 in view 1. In this view the common perpendicular is not true length. Line 3–4 *is* true length, however, so line 5–6 is drawn perpendicular to line 3–4 in view 1 as shown. It is then routine to project line 5–6 to the top view and then to the front view. Divider distances such as $D_2$ and $D_3$ are used to check the accuracy of the construction.

### Plane Method

If a plane containing one of two skew lines is parallel to the second line, the perpendicular distance from the second line to the plane is the shortest distance between the two lines.

In Fig. 20.20 (a) a line is drawn through point 2 parallel to line 3–4, thus establishing a plane containing line 1–2 and parallel to line 3–4, §20.4. The edge view of the plane is then established in auxiliary view 1. In this view the shortest distance is measured as shown. The shortest distance being obtained in only one additional view is an advantage of the plane method over the point-view method.

If, in addition, the views of the common perpendicular are required, a second auxiliary view is necessary and the total amount of construction is slightly more than in the point-view method. At (b)

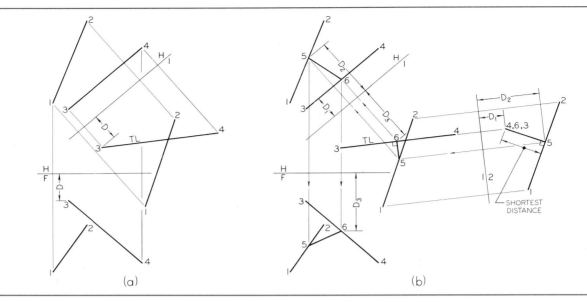

**Fig. 20.19**    Common Perpendicular—Point-View Method.

the second auxiliary view is constructed with its direction of sight parallel to the shortest distance (or perpendicular to the edge-view plane). In this view the shortest distance or common perpendicular must appear as a point. This can be only at the apparent crossing point 5, 6 of lines 1–2 and 3–4. This locates the common perpendicular, which is then projected back to the other views as shown.

## 20.10 Shortest Horizontal Line Connecting Skew Lines

Related to the preceding plane method is the problem of finding the shortest line at zero slope or grade, the shortest horizontal line connecting two skew lines.

In Fig. 20.21 (a) a plane is constructed containing line 1–2 and parallel to line 3–4. The edge view of

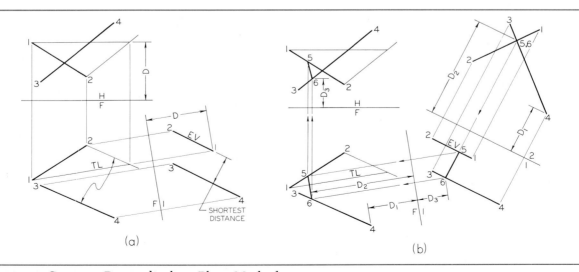

**Fig. 20.20**    Common Perpendicular—Plane Method.

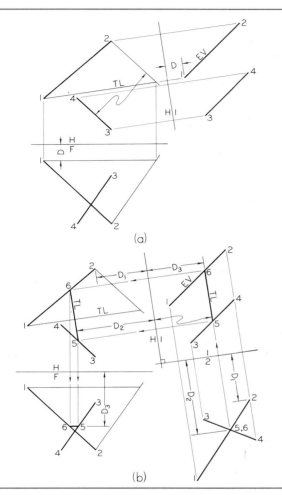

(a)

(b)

*Fig. 20.21*   Shortest Horizontal Line Connecting Two Skew Lines.

the plane is constructed through the use of a *horizontal* line added to the plane because in the auxiliary view **1** any horizontal connecting line will be parallel to folding line H/1.

Auxiliary view **2** is then constructed at (b) with its direction of sight parallel to folding line H/1. The shortest connector **5–6** appears in point view in view **2** at the apparent crossing point of lines **1–2** and **3–4**. It is then projected back to the other views.

## 20.11 Shortest Line at Specified Slope Connecting Skew Lines

For a specified slope other than zero (horizontal), the method of §20.10 requires only slight modification. In this example, it is assumed that 38° is specified for the slope of the connecting line. The first portion of the construction is the same, Fig. 20.22 (a): A plane is passed through line **1–2** and parallel to line **3–4**, and an edge view of the plane is constructed, as shown, in an auxiliary view adjacent to the top view.

At (b) projection lines for view **2** are drawn at the prescribed slope angle with folding line H/1. In view **2**, the apparent crossing point **5, 6** of lines **1–2** and **3–4** is the point view of the shortest connector **5–6** at the specified slope of **38°**. The other views of line **5–6** are then completed by projection as before.

It should be noted that the general procedure illustrated could be readily modified for other specifications, such as the shortest connecting line at a prescribed grade or the shortest frontal line connecting two skew lines.

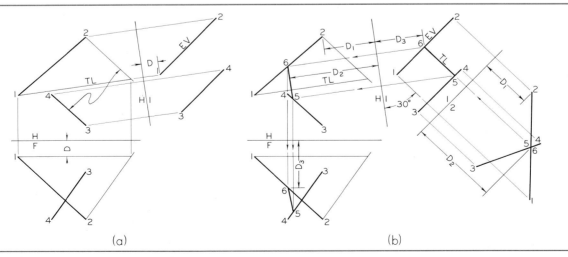

(a)                                     (b)

*Fig. 20.22*   Shortest Line at Specified Slope Connecting Two Skew Lines.

# _Problems in Parallelism and Perpendicularity_

In Figs. 20.23 to 20.26 are problems covering parallel lines, lines parallel to a plane, plane parallel to a line, plane parallel to a plane or skew lines, perpendicular lines, lines perpendicular to planes, common perpendicular, and shortest line at specified angle.

Use Layout A–1 or A4–1 (adjusted) and divide the working area into four equal areas for problems to be assigned by the instructor. Some problems will require a single problem area, and others will require two problem areas or one-half sheet. Data for the layout for each problem are given by a coordinate system using metric dimensions. For example, in Fig. 20.23, Prob. 1, point **6** is located by the full-scale coordinates (**35**, **20**, **50**). The first coordinate locates the front view of the point from the left edge of the problem area. The second coordinate locates the front view of the point from the bottom edge of the problem area. The third coordinate locates either the top view of the point from the bottom edge of the problem area or the side view of the point from the left edge of the problem area. Inspection of the given problem layout will determine which application to use.

Since many of the problems in this chapter are of a general nature, they can also be solved on most computer graphics systems. If a system is available, the instructor may choose to assign specific problems to be completed by this method.

Additional problems, in convenient form for solution, are available in _Principles of Engineering Graphics Problems_ by Spencer, Hill, Loving, and Dygdon, a workbook designed to accompany this text that is also published by Macmillan Publishing Company.

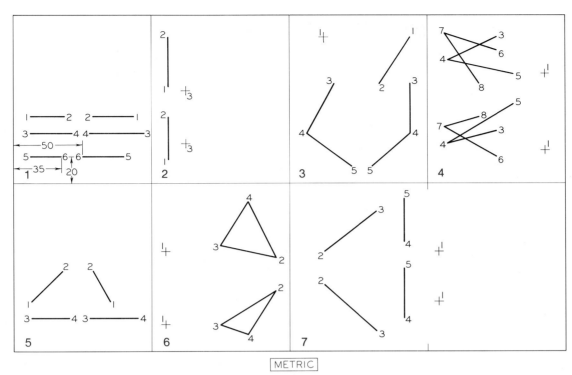

METRIC

***Fig. 20.23*** Lay out and solve problems as assigned. Use Layout A–1 or A4–1 (adjusted) divided into four equal areas.

1.  Determine which, if any, of the lines 1(12, 50, 88)–2(38, 50, 58), 3(12, 38, 96)–4(43, 38, 55), or 5(12, 20, 81)–6(35, 20, 50) are parallel.
2.  Draw a line 3(25, 30, 71)–4, 25 mm in length and parallel to line 1(12, 18, 73)–2(12, 50, 112).
3.  Complete the side view of line 1(25, 112, 88)–2(?, 75, 65) that is parallel to plane 3(33, 75, 88)–4(12, 38, 88)–5(45, 12, 60).
4.  Through point 1(86, 25, 83) draw a line 1–2 parallel to planes 3(50, 40, 112)–4(15, 30, 94)–5(63, 60, 84) and 6(50, 20, 100)–7(12, 43, 114)–8(38, 50, 75).
5.  Pass a plane through line 1(12, 38, 70)–2(35, 60, 58) parallel to line 3(12, 25, 56)–4(40, 25, 91). Add the top view.
6.  By means of a horizontal line and a frontal line represent a plane containing point 1(10, 20, 75) and parallel to plane 2(91, 45, 70)–3(50, 20, 81)–4(70, 12, 114).
7.  Pass a plane through point 1(109, 38, 75) parallel to lines 2(25, 50, 75)–3(63, 15, 108) and 4(85, 25, 84)–5(85, 63, 107).

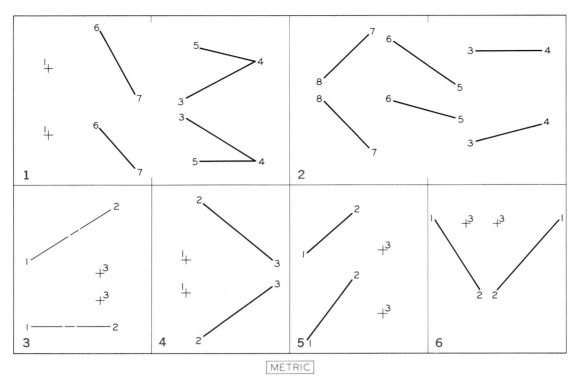

METRIC

*Fig. 20.24*   Lay out and solve problems as assigned. Use Layout A–1 or A4–1 (adjusted) divided into four equal areas.

1. Find a line 1(25, 38, 88)–2 that is parallel to plane 3(127, 50, 66)–4(178, 17, 94)–5(137, 17, 104) and intersects line 6(63, 43, 117)–7(88, 12, 60).

2. Establish a line 1–2 that is parallel to line 3(137, 33, 100)–4(185, 45, 100) and intersects lines 5(122, 50, 75)–6(75, 63, 109) and 7(58, 28, 114)–8(25, 63, 81).

3. Line 1(12, 20, 70)–2(70, 20, 109) is the center line of a pipe. Connect this pipe to point 3(63, 40, 60) with a 90° elbow (pipe fitting) at the juncture on 1–2. Find the true length of the center line of the connecting pipe. Scale: 1/10.

4. Draw a 50 mm frontal line from point 1(25, 45, 70) perpendicular to line 2(38, 12, 115)–3(88, 50, 70). Also draw a 50 mm horizontal line from point 1 perpendicular to line 2–3. Use only the given views. (Note that these lines do not intersect line 2–3.)

5. Using only the given views, find a line 3(68, 30, 78)–4 that is perpendicular to line 1(12, 10, 75)–2(45, 55, 106) and also intersects line 1–2.

6. Find the center of the smallest sphere that has its center on line 1(5,100, 96)–2(38, 48, 50) and has point 3(28, 88, 50) on its surface. Use only the given views. If assigned, find the diameter of the sphere.

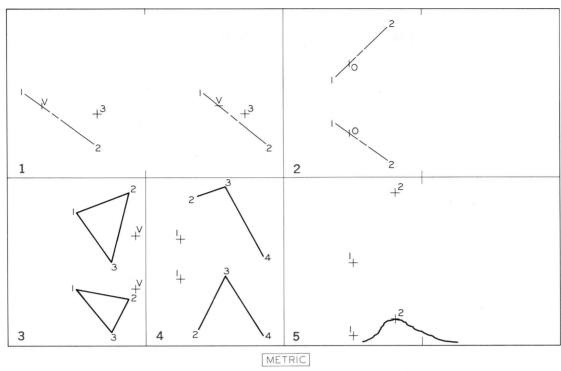

METRIC

*Fig. 20.25*   Lay out and solve problems as assigned. Use Layout A–1 or A4–1 (adjusted) divided into four equal areas.

1. The axis of a right square pyramid lies along center line 1(12, 63, 144)–2(63, 25, 190). One corner of the base is at point 3(66, 48, 175). The vertex is at point V(25, –, –). Find the front and side views of the pyramid.

2. The axis of a right prism lies along center line 1(38, 40, 75)–2(75, 12, 114). One base is centered at O(48, –, –). The bases are equilateral triangles inscribed in 36 mm diameter circles. The lowest side of each base is a horizontal line. The altitude of the prism is 35 mm. Complete the views.

3. If an oblique cone is drawn with its vertex at V(94, 43, 84) and its base in plane 1(50, 43, 100)–2(88, 35, 116)–3(75, 10, 63), what is its altitude in millimeters? Use an auxiliary view to solve this problem. Show the front and top views of the altitude.

4. Find the shadow of point 1(25, 50, 80) on plane 2(38, 12, 114)–3(57, 53, 122)–4(84, 7, 68) if light rays are perpendicular to the plane.

5. An aircraft on a landing approach course of N 45° passes 300 m above point 1(50, 7, 63). It is losing altitude at the rate of 200 m in 1000 m. Point 2(80, 20, 116) represents the peak of a hill. How close to point 2 does the aircraft pass? Scale: 1/10 000.

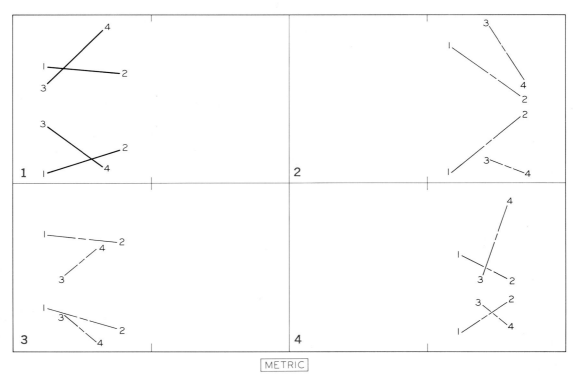

METRIC

**Fig. 20.26** Lay out and solve problems as assigned. Use Layout A–1 or A4–1 (adjusted) divided into four equal areas.

1. Find the clearance between high-voltage lines 1(25, 7, 88)–2(78, 25, 84) and 3(25, 43, 75)–4(66, 12, 117). Show the views of a line representing this clearance. Scale: 1/40.
2. Determine the bearing, grade, and length of the shortest shaft connecting tunnels 1(120, 7, 104)–2(170, 50, 65) and 3(147, 18, 120)–4(172, 7, 78). Scale: 1/80.
3. Ski slopes represented by lines 1(25, 33, 88)–2(75, 18, 84) and 3(38, 28, 58)–4(61, 7, 78) are to be connected by the shortest possible horizontal path. Find the views and measure the length of this path. Scale: 1/4000.
4. Tunnel 1(127, 15, 73)–2(160, 38, 56) is to be connected to tunnel 3(142, 35, 58)–4(162, 20, 112) with the shortest tunnel at a downgrade of 10%. Find the length and bearing of this connector. Scale: 1/2000.

# CHAPTER
# 21

# Intersections
# and Developments

A machine part or structure often consists of a number of geometric shapes so arranged as to produce the required form. Easily represented geometric forms frequently meet in lines of intersection that require knowledge and skill to produce in multiview projection. Accurate representation of the intersecting surfaces therefore becomes very important since precision fit is necessary for function and appearance. The development of these surfaces, such as those found in sheet-metal fabrication, is a flat pattern that represents the unfolded or unrolled surface of the form. The resulting plane figure gives the true size of each area of the form so connected that when fabricated the desired part or structure is produced.

## 21.1 Surfaces

A *surface* is a geometric magnitude having two dimensions. A surface may be generated by a line, called the *generatrix* of the surface. Any position of the generatrix is an *element* of the surface, Fig. 7.31 (a).

A *ruled surface* is one that may be generated by a straight line and may be a *plane,* a *single-curved surface,* or a *warped surface.*

A *plane* is a ruled surface that may be generated by a straight line one point of which moves along another straight line, while the generatrix remains parallel to its original position. Many of the geometric solids are bounded by plane surfaces, Fig. 5.7.

A *single-curved surface* is a developable ruled surface; that is, it can be unrolled to coincide with a plane. Any two adjacent positions of the generatrix lie in the same plane. Examples are the cylinder and the cone, Fig. 5.7.

A *warped surface* is a ruled surface that is not developable, Fig. 21.1. No two adjacent positions of the generatrix lie in the same plane. Many exterior surfaces on an airplane or automobile are warped surfaces.

A *double-curved surface* may be generated only by a curved line and has no straight-line elements. Such a surface, generated by revolving a curved line about a straight line in the plane of the curve, is called a *double-curved surface of revolution.* Common examples are the *sphere, torus, ellipsoid,* Fig. 5.7, and the *hyperboloid,* Fig. 21.1 (d).

A *developable surface* is one that may be unfolded or unrolled so as to coincide with a plane,

§21.4. Surfaces composed of single-curved surfaces, or of planes, or of combinations of these types, are developable. Warped surfaces and double-curved surfaces are not developable. They may be developed approximately by dividing them into sections and substituting for each section a developable surface, that is, a plane or a single-curved surface. If the material used is sufficiently pliable, the flat sheets may be stretched, pressed, stamped, spun, or otherwise forced to assume the desired shape. Nondevelopable surfaces are often produced by a combination of developable surfaces, which are then formed slightly to produce the required shape.

## 21.2 Solids

*Solids* bounded by plane surfaces are *polyhedra,* the most common of which are the pyramid and prism, Fig. 5.7. Convex solids whose faces are all equal regular polygons are *regular polyhedra.* The simple regular polyhedra are the *tetrahedron, cube, octahedron, dodecahedron,* and *icosahedron,* known as the five *Platonic solids.*

Plane surfaces that bound polyhedra are *faces* of the solids. Lines of intersection of faces are *edges* of the solids.

A solid generated by revolving a plane figure about an axis in the plane of the figure is a *solid of revolution.*

Solids bounded by warped surfaces have no group name. The most common example of such solids is the screw thread.

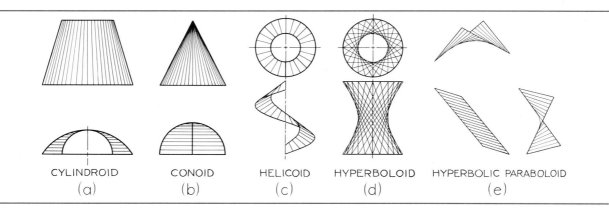

CYLINDROID (a)    CONOID (b)    HELICOID (c)    HYPERBOLOID (d)    HYPERBOLIC PARABOLOID (e)

*Fig. 21.1*   Warped Surfaces.

# Intersections and Developments of Planes and Solids

## 21.3 Principles of Intersections

The principles involved in intersections of planes and solids have their practical application in the cutting of openings in roof surfaces for flues and stacks and in wall surfaces for pipes and chutes, and so on, and in the building of sheet-metal structures (tanks, boilers, etc.).

In such cases, the problem is generally one of determining the true size and shape of the intersection of a plane and one of the more common geometric solids. The intersection of a plane and a solid is the locus of the points of intersection of the elements of the solid with the plane. For solids bounded by plane surfaces, it is necessary only to find the points of intersection of the edges of the solid with the plane and to join these points, in consecutive order, with straight lines. For solids bounded by curved surfaces, it is necessary to find the points of intersection of several elements of the solid with the plane and to trace a smooth curve through these points. The curve of intersection of a plane and a circular cone is a *conic section*. The various conic sections are defined and illustrated in §5.47 and Fig. 5.46.

## 21.4 Developments

The *development* of a surface is that surface laid out on a plane, Fig. 21.2. Practical applications of developments occur in sheet-metal work, stone cutting, pattern making, packaging, and package design.

Single-curved surfaces and the surfaces of polyhedra can be developed. Warped surfaces and double-curved surfaces can be developed only approximately. See §21.1.

In sheet-metal layout, extra material must be provided for laps or seams. If the material is heavy, the thickness may be a factor, and the crowding of metal in bends must be considered. See §13.39. The drafter must also take stock sizes into account and should make layouts so as to economize in the use of material and of labor. In preparing developments, it is best to put the seam at the shortest edge and to attach the bases at edges where they match, to economize in soldering, welding, or riveting.

It is common practice to draw development layouts with the *inside surfaces up*. In this way, all fold lines and other markings are related directly to inside measurements, which are the important di-

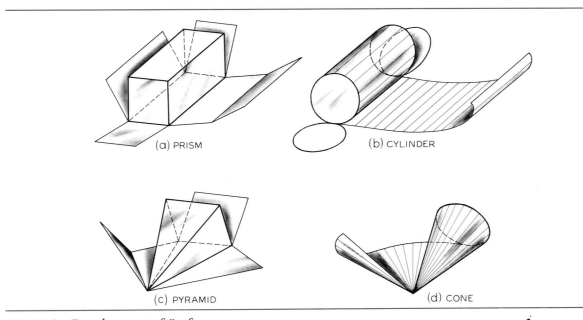

(a) PRISM          (b) CYLINDER

(c) PYRAMID          (d) CONE

*Fig. 21.2*   Development of Surfaces.

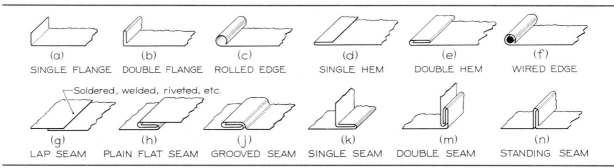

*Fig. 21.3*   Sheet-Metal Hems and Joints.

mensions in all ducts, pipes, tanks, and other vessels, and in this position they are also convenient for use in the fabricating shop.

## 21.5 Hems and Joints for Sheet Metal and Other Materials

A wide variety of hems and joints are used in the fabrication of sheet-metal developments and other materials, Fig. 21.3. Hems are used to eliminate the raw edge and also to stiffen the material. Joints and seams may be made for sheet metal by bending, welding, riveting, and soldering and by glueing and stapling for package materials.

Sufficient material as required for hems and joints must be added to the layout or development. The amount of allowance depends on the thickness of the material and the production equipment; therefore, no specific dimensions for allowances are given in this chapter. See §13.39.

## 21.6 To Find the Intersection of a Plane and a Prism and the Development of the Prism Fig. 21.4

*Intersection* Fig. 21.4 (a)

The true size and shape of the intersection is shown in the auxiliary view. See Chapter 10. The length **AB** is the same as **AB** in the front view, and the width **AD** is the same as **AD** in the top view.

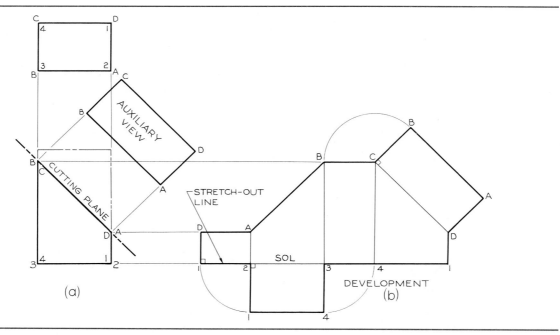

*Fig. 21.4*   Plane and Prism.

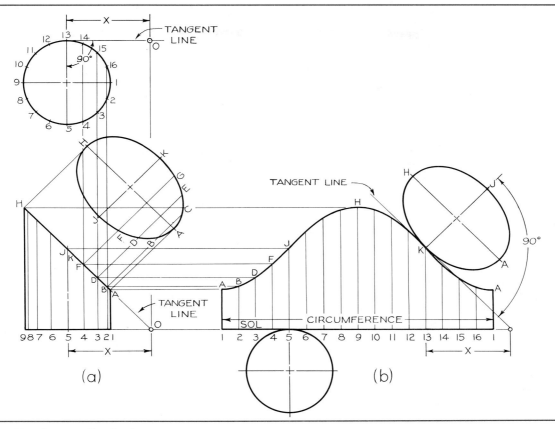

***Fig. 21.5***   Plane and Cylinder.

### *Development* Fig. 21.4 (b)

On the straight line 1–1, called the *stretchout line* (SOL), set off the widths of the faces 1–2, 2–3, . . ., taken from the top view. At the division points, erect perpendiculars to 1–1, and set off on each the length of the respective edge, taken from the front view. The lengths can be projected across from the front view, as shown. Join the points thus found by straight lines to complete the development of the lateral surface. Attach to this development the lower base and the upper base, or auxiliary view, to obtain the development of the entire surface of the frustum of the prism.

### 21.7 To Find the Intersection of a Plane and a Cylinder and the Development of the Cylinder Fig. 21.5

#### *Intersection* Fig. 21.5 (a)

The intersection is an ellipse whose points are the piercing points in the secant plane of the elements of the cylinder. In spacing the elements, it is best,

though not necessary, to divide the circumference of the base into *equal* parts and to draw an element at each division point. In the auxiliary view, the widths BC, DE, . . ., are taken from the top view at 2–16, 3–15, . . ., respectively, and the curve is traced through the points thus determined, with the aid of the irregular curve, §2.54.

The major axis AH and the minor axis JK are shown true length in the front view and the top view, respectively; therefore, the ellipse may also be constructed as explained in §§5.48 to 5.51 or with the aid of an ellipse template, §5.56.

#### *Development* Fig. 21.5 (b)

The base of the cylinder develops into a straight line 1–1, the stretchout line (SOL), equal to the circumference of the base, whose length may be determined by calculation ($\pi d$), by setting off with the bow dividers, or by rectifying the arcs of the base 1–2, 2–3, . . ., §5.45. Divide the stretchout line into the same number of equal parts as the circumference of the base, and draw an element through each

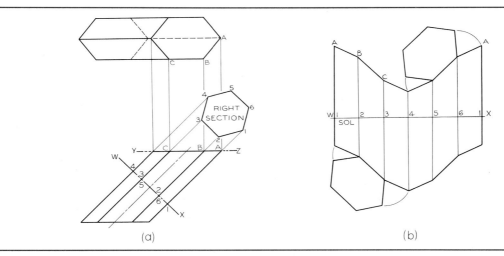

**Fig. 21.6** Plane and Oblique Prism.

division perpendicular to the line. Set off on each element its length, projected from the front view, as shown; then trace a smooth curve through the points A,B,D,, . . ., §2.54, and attach the bases.

## 21.8 To Find the Intersection of a Plane and an Oblique Prism and the Development of the Prism Fig. 21.6

**Intersection** Fig. 21.6 (a)

The right section cut by the plane WX is a regular hexagon, as shown in the auxiliary view; the oblique section, cut by the horizontal plane YZ, is shown in the top view.

**Development** Fig. 21.6 (b)

The right section develops into the straight line WX, the stretchout line (SOL). Set off, on the stretchout line, the widths of the faces 1–2, 2–3, . . ., taken from the auxiliary view, and draw a line through each division perpendicular to the line. Set off, from the stretchout line, the lengths of the respective edges measured from WX in the front view. Join the points A,B,C, . . ., with straight lines, and attach the bases, which are shown in their true sizes in the top view.

## 21.9 To Find the Intersection of a Plane and an Oblique Cylinder Fig. 21.7

**Intersection** Fig. 21.7 (a)

The right section cut by the plane WX is a circle, shown in the auxiliary view. The intersection of the horizontal plane YZ with the cylinder is an ellipse

shown in the top view, whose points are found as explained for the auxiliary view in Fig. 21.5 (a), §21.7. The major axis AH is shown true length in the top view, and the minor axis JK is equal to the diameter of the cylinder; therefore, the ellipse may be constructed as explained in §§5.48 to 5.51, or with the aid of an ellipse template, §5.56.

**Development** Fig. 21.7 (b)

The cylinder may be considered as a prism having an infinite number of edges; therefore, the development is found in a manner similar to that of the oblique prism shown in Fig. 21.6.

The circle of the right section cut by plane WX develops into a straight line 1–1, the stretchout line (SOL), equal in length to the circumference of the circle ($\pi d$). Divide the stretchout line into the same number of equal parts as the circumference of the circle as shown in the auxiliary view, and draw elements through these points perpendicular to the line. Set off on each element its length, taken from the front view with dividers, as shown; then trace a smooth curve through the points A, B, D, . . ., §2.54, and attach the bases.

## 21.10 To Find the Intersection of a Plane and a Pyramid and to Develop the Resulting Truncated Pyramid Fig. 21.8

**Intersection** Fig. 21.8 (a)

The intersection is a trapezoid whose vertices are the points in which the edges of the pyramid pierce the secant plane. In the auxiliary view, the altitude

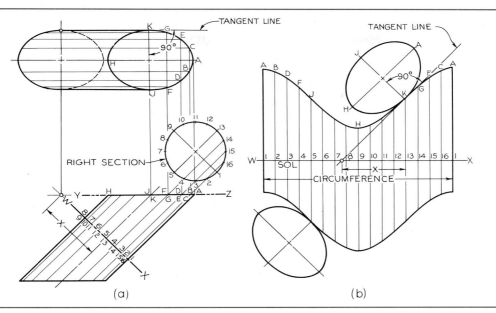

*Fig. 21.7*  Plane and Oblique Circular Cylinder.

of the trapezoid is projected from the front view, and the widths AD and BC are transferred from the top view with dividers.

### Development Fig. 21.8 (b)

With O in the development as center and O–1′ in the front view (the true length of one of the edges) as radius, draw the arc 1′–2′–3′. . . . Inscribe the cords 1′–2′, 2′–3′, . . ., equal, respectively, to the sides of the base, as shown in the top view. Draw the lines 1′–O, 2′–O, . . ., and set off the true lengths of the lines OD′, OA′, OB′, . . ., respectively, taken from the true lengths in the front view, §11.10.

To complete the development, join the points D′, A′, B′, . . ., by straight lines, and attach the bases to their corresponding edges. To transfer an irregular figure, such as the trapezoid shown here, refer to §§5.28 and 5.29.

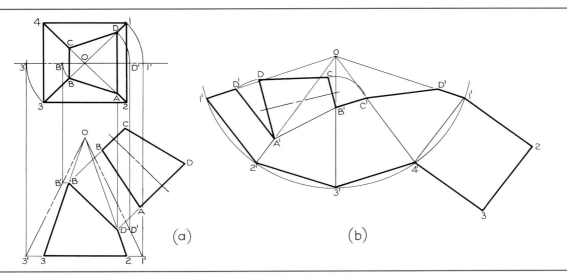

*Fig. 21.8*  Plane and Pyramid.

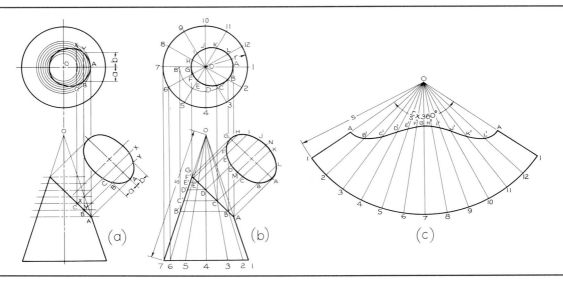

***Fig. 21.9***   Plane and Cone.

## 21.11 To Find the Intersection of a Plane and a Cone and to Develop the Lateral Surface of the Cone Fig. 21.9

***Intersection*** Fig. 21.9 (a)

The intersection is an ellipse. If a series of horizontal cutting planes is passed perpendicular to the axis, as shown, each plane will cut a circle from the cone that will show in true size and shape in the top view. Points in which these circles intersect the original secant plane are points on the ellipse. Since the secant plane is shown edgewise in the front view, all of these piercing points may be found in that view and projected to the others, as shown.

***Fig. 21.9 (b)***   This method is most suitable when a development also is required, since it utilizes elements that are also needed in the development. The piercing points of these elements in the secant plane are points on the intersection. Divide the base into any number of equal parts, and draw an element at each division point. These elements pierce the secant plane in points A, B, C, . . .. The top views of these points are found by projecting upward from the front view, as shown. In the auxiliary view, the widths BL, CK, . . ., are taken from the top view. The ellipse is then drawn with the aid of the irregular curve, §2.54.

The major axis of the ellipse, shown in the auxiliary view, is equal to AG in the front view. The minor axis MN bisects the major axis and is equal to

the minor axis of the ellipse in the top view. With these axes, the ellipse may also be constructed as explained in §§5.48 to 5.51 or with the aid of an ellipse template, §5.56.

***Development*** Fig. 21.9 (c)

The cone may be considered as a pyramid having an infinite number of edges; hence, the development is found in a manner similar to that explained for the pyramid in §21.10. The base of the cone develops into a circular arc, with the slant height of the cone as its radius and the circumference of the base as its length, §5.45. The lengths of the elements in the development are taken from the element O–7 or O–1 in the front view, (b). Instead of our finding the true circumference of the base, the vertical angle 1–O–1 in the development can be set off equal to $\frac{r}{s}\,360°$ (where r is the radius of the base and s is the slant height of the cone).

## 21.12 To Find the Development of a Hood and Flue Fig. 21.10

Since the hood is a conical surface, it may be developed as described in §21.11. The two end sections of the elbow are cylindrical surfaces and may be developed as described in §21.7. The two middle sections of the elbow are cylindrical surfaces, but since their bases are not perpendicular to the axes,

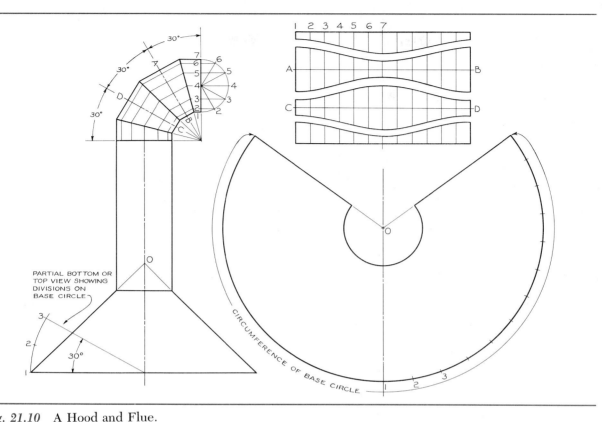

*Fig. 21.10*  A Hood and Flue.

they will not develop into straight lines. They will be developed in a manner similar to that for an oblique cylinder, §21.9, Fig. 21.7 (b). If the auxiliary planes **AB** and **DC** are passed perpendicular to the axes, they will cut right sections from the cylinders, which will develop into the straight lines **AB** and **CD** in the developments.

If the developments are arranged as shown in Fig. 21.10, the elbow can be constructed from a rectangular sheet of metal without wasting material. The patterns are shown separated after cutting. Before cutting, the adjacent curves coincided.

## 21.13 To Find the Development of a Truncated Oblique Rectangular Pyramid Fig. 21.11

None of the four lateral surfaces is shown in the multiview drawing in true size and shape. Using the method of §11.10, revolve each edge until it appears in true length in the front view, as shown. Thus, O–2 revolves to O–2′, O–3 revolves to O–3′, and so

on. These true lengths are transferred from the front view to the development with the compass, as shown. Notice that true lengths OD′, OA′, OB′, . . ., are found and transferred. The true lengths of the edges of the bases are given in the top view and are transferred directly to the development.

## 21.14 Triangulation

*Triangulation* is simply a method of dividing a surface into a number of triangles and transferring them to the development. A triangle is said to be "indestructible," because if its sides are of given lengths, it can be only one shape. A triangle can be easily transferred by transferring the sides with the aid of the compass, §5.28.

## 21.15 To Find the Development of an Oblique Cone by Triangulation Fig. 21.12

Divide the base, in the top view, into any number of equal parts, and draw an element at each division point. Find the true length of each element, §11.10.

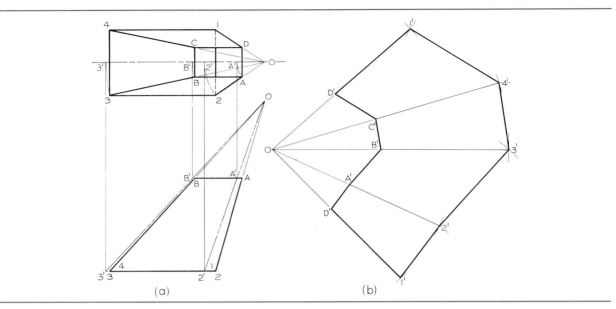

**Fig. 21.11**   Development of a Transition Piece.

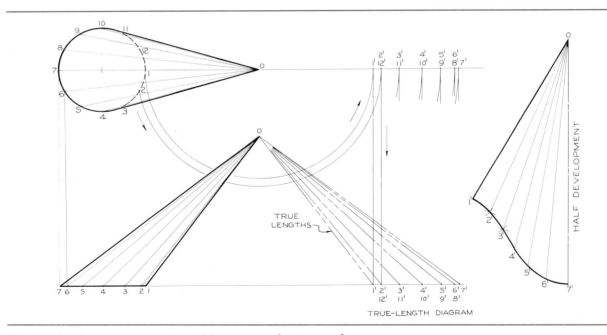

**Fig. 21.12**   Development of an Oblique Cone by Triangulation.

If the divisions of the base are comparatively small, the lengths of the chords may be set off in the development as representing the lengths of the respective subtending arcs. In the development, set off O–1′ equal to O–1 in the front view where it is shown true length. With 1′ in the development as center, and the chord 1–2 taken from the top view as radius, strike an arc at 2′. With O as center, and O–2′, the true-length of the element O–2 from the "true-length" diagram, as radius, draw the arc at 2′. The intersection of these arcs is a point on the development of the base of the cone. The points 3′, 4′,

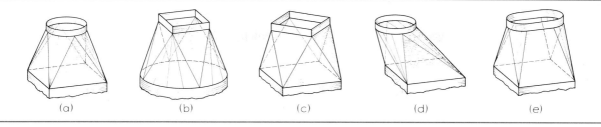

*Fig. 21.13*    Transition Pieces.

. . ., in the curve are found in a similar manner, and the curve is traced through these points with the aid of the irregular curve, §2.54.

Since the development is symmetrical about element O–7′, it is necessary to lay out only half the development, as shown.

## 21.16 Transition Pieces

A *transition piece* is one that connects two differently shaped, differently sized, or skewed-position openings, Fig. 21.13. In most cases, transition pieces are composed of plane surfaces and conical surfaces, the latter being developed by triangulation. Triangulation can also be used to develop, approximately, certain warped surfaces. Transition pieces are extensively used in air conditioning, heating, ventilating, and similar construction.

## 21.17 To Find the Development of a Transition Piece Connecting Rectangular Pipes on the Same Axis

The transition piece is a frustum of a pyramid, Fig. 21.14 (a). Find the vertex O of the pyramid by extending its edges to their intersection. Find the true lengths of the edges by any one of the methods explained in §11.10. The development can then be found as explained in §21.10.

If the transition piece is not a frustum of a pyramid, as in Fig. 21.14 (b), it can best be developed by triangulation, §21.14, as shown for the faces 1–5–8–4 and 2–6–7–3, or by extending the sides to form triangles, as shown for faces 1–2–6–5 and 3–4–8–7, and then finding the true lengths of the sides of the triangles, §11.10, and setting them off as shown.

As a check on the development, lines parallel on

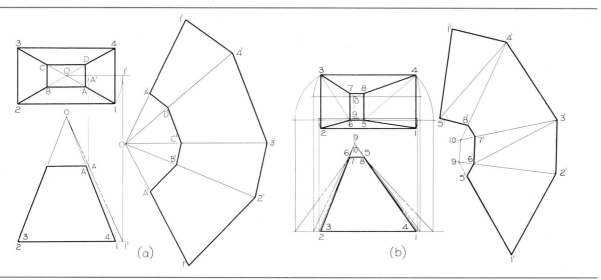

*Fig. 21.14*    Development of a Transition Piece.

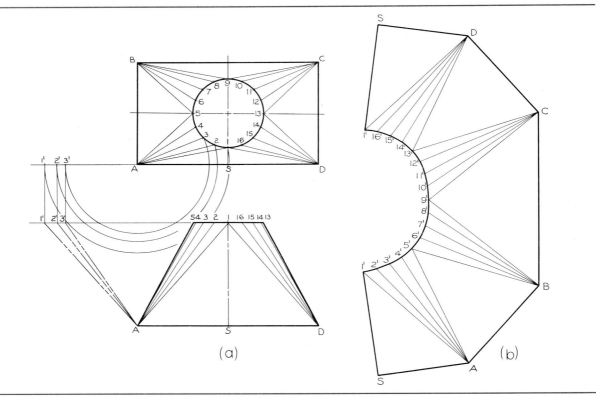

*Fig. 21.15*　Development of a Transition Piece.

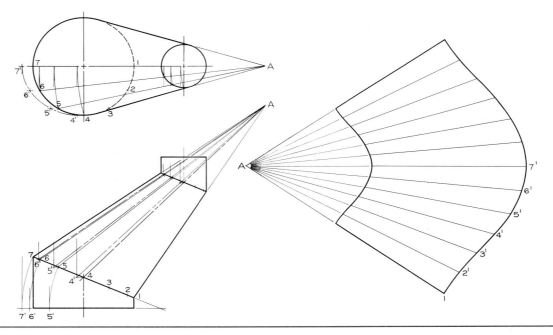

*Fig. 21.16*　Development of a Transition Piece.

the surface must also be parallel on the development; for example, 8′–5′ must be parallel to 4′–1′ on the development.

## 21.18 To Find the Development of a Transition Piece Connecting a Circular Pipe and a Rectangular Pipe on the Same Axis Fig. 21.15

The transition piece is composed of four isosceles triangles and four conical surfaces. The seam is along line S–1. Begin the development on the line 1′–S, and draw the right triangle 1′–S–A, whose base SA is equal to half the side AD and whose hypotenuse A–1′ is equal to the true length of side A–1.

The conical surfaces are developed by triangulation as explained in §§21.14 and 21.15.

## 21.19 To Find the Development of a Transition Piece Connecting Two Cylindrical Pipes on Different Axes Fig. 21.16

The transition piece is a frustum of a cone, the vertex of which may be found by extending the contour elements to their intersection A.

The development can be found by triangulation, as explained in §§21.14 and 21.15. The sides of each triangle are the true lengths of two adjacent elements of the cone, and the base is the true length of the curve of the base of the cone between the two elements. This curve is not shown in its true length in either view, and the plane of the base of the frustum must therefore be revolved until it is horizontal in order to find the distance from the foot of one element to the foot of the next. When the plane of the base is thus revolved, the foot of any element, such as 7, revolves to 7′, and the curve 6′–7′ (top view) is the true length of the curve of the base between the elements 6 and 7. In practice, the chord distances between these points are generally used to approximate the curved distances.

After the conical surface has been developed, the true lengths of the elements on the truncated section of the cone are set off from the vertex A of the development to secure points on the upper curve of the development.

If the transition piece is not a frustum of a cone, its development is found by another variation of triangulation, as shown in Fig. 21.17. The circular intersection with the large vertical pipe is shown

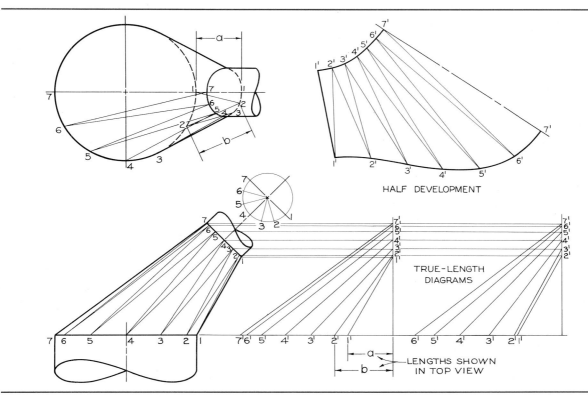

HALF DEVELOPMENT

TRUE-LENGTH DIAGRAMS

LENGTHS SHOWN IN TOP VIEW

*Fig. 21.17*   Development of a Transition Piece.

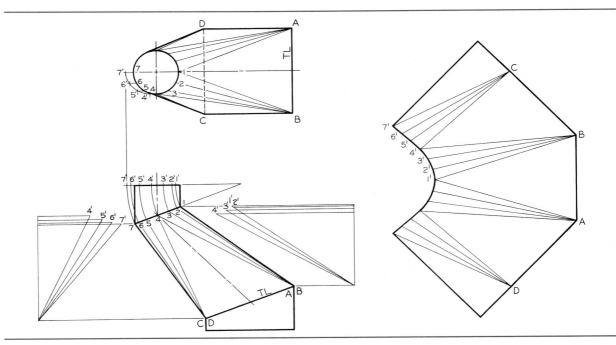

*Fig. 21.18*    Development of a Transition Piece.

true size in the top view, and the circular intersection with the small inclined pipe is shown true size in the auxiliary view. Since both intersections are true circles, and the planes containing them are not parallel, the lateral surface of the transition piece is a warped surface and not conical (single-curved). It is theoretically nondevelopable, but may be approximately developed by considering it to be made up of plane triangles, alternate ones of which are inverted, as shown in the development. The true lengths of the sides of the triangles are found by the method of Fig. 11.10 (d), but in a systematic manner so as to form true-length diagrams, as shown in Fig. 21.17.

### 21.20 To Find the Development of a Transition Piece Connecting a Square Pipe and a Cylindrical Pipe on Different Axes Fig. 21.18

The development of the transition piece is made up of five plane triangular surfaces and four triangular conical surfaces similar to those in Fig. 21.15. The development is made in a similar manner to those described in §§21.15 and 21.18.

### 21.21 To Find the Intersection of a Plane and a Sphere and to Find the Approximate Development of the Sphere Fig. 21.19

*Intersection* Fig. 21.19 (a)

The intersection of a plane and a sphere is a circle, as shown in the top views in Fig. 21.19, the diameter of the circle depending upon where the plane is passed. Any circle cut by a plane through the center of the sphere is called a *great circle*. If a plane passes through the center and perpendicular to the axis, the resulting great circle is called the *equator*. If a plane contains the axis, it will cut a great circle called a *meridian*.

*Development* Fig. 21.19 (a)

The surface of a sphere is a double-curved surface and is not developable, §21.1. The surface may be developed approximately by dividing it into a series of zones and substituting for each zone a frustum of a right-circular cone. The development of the conical surfaces is an approximate development of the spherical surface. If the conical surfaces are inscribed within the sphere, the development will be smaller than the spherical surface, while if the con-

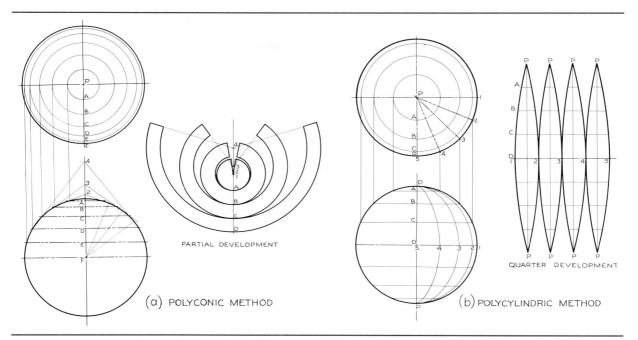

*Fig. 21.19*  Approximate Development of a Sphere.

ical surfaces are circumscribed about the sphere, the development will be larger. If the conical surfaces are partly within and partly without the sphere, as indicated in the figure, the resulting development very closely approximates the spherical surface.

This method of developing a spherical surface is the *polyconic* method. It is used on all government maps of the United States.

*Fig. 21.19 (b)*  Another method of making an approximate development of the double-curved sur-

face of a sphere is to divide the surface into equal sections with meridian planes and substitute cylindrical surfaces for the spherical sections. The cylindrical surfaces may be inscribed within the sphere, or circumscribed about it, or located partly within and partly without. The development of the series of cylindrical surfaces is an approximate development of the spherical surface. This method is the *polycylindric* method, sometimes designated as the *gore* method.

## Intersections and Developments of Solids

### 21.22 Principles of Intersections

Intersections of solids are generally regarded as in the province of descriptive geometry, and for information on the more complicated intersections the student is referred to any standard text on that subject. However, most of the intersections encountered in drafting practice do not require a knowledge of descriptive geometry, and some of the more common solutions may be found in the paragraphs that follow.

An intersection of two solids is referred to as a *figure of intersection*. Two plane surfaces intersect in a straight line; hence, if two solids that are composed of plane surfaces intersect, the figure of intersection will be composed of straight lines, as shown in Figs. 21.20 to 21.23. The method generally consists of finding the piercing points of the edges of one solid in the surfaces of the other solid and joining these points with straight lines.

If curved surfaces intersect, or if curved surfaces

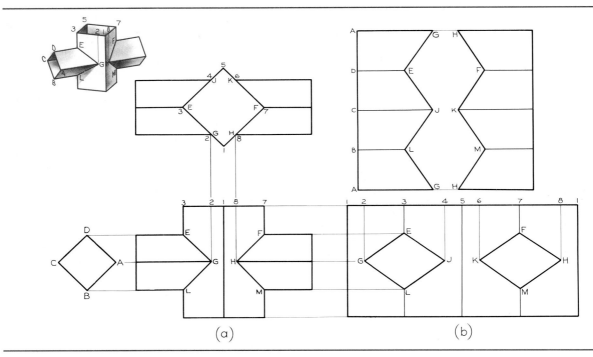

***Fig. 21.20***   Two Prisms at Right Angles to Each Other.

and plane surfaces intersect, the figure of intersection will be composed of curves, as shown in Figs. 21.5, 21.9, and 21.24 to 21.29. The method generally consists of finding the piercing points of *elements* of one solid in the surfaces of the other. A smooth curve is then traced through these points, with the aid of the irregular curve, §2.54.

### 21.23 To Find the Intersection and Developments of Two Prisms Fig. 21.20

***Intersection*** Fig. 21.20 (a)

The points in which the edges A, B, C, and D of the horizontal prism pierce the vertical prism are vertices of the intersection. The edges D and B of the horizontal prism intersect the edges 3 and 7 of the vertical prism at the points E, F, L, and M. The edges A and C of the horizontal prism intersect the faces of the vertical prism at the points G, H, J, and K. The intersection is completed by joining these points in order by straight lines.

***Developments*** Fig. 21.20 (b)

To develop the lateral surface of the horizontal prism, set off on the vertical stretchout line A–A the widths of the faces AB, BC, . . ., taken from the end

view, and draw the edges through these points, as shown. Set off, from the stretchout line, the lengths of the edges AG, BL, . . ., taken from the front view or from the top view, and join the points G, L, J, . . ., by straight lines.

To develop the lateral surface of the vertical prism, set off on the stretchout line 1–1 the widths of the faces 1–2, 3–5, . . ., taken from the top view, and draw the edges through these points, as shown. Set off on the stretchout line the distances 1–2, 5–4, 5–6, and 1–8, taken from the top view, and draw the intermediate elements parallel to the principal edges. Take the lengths of the principal edges and of the intermediate elements from the front view, and join the points E, G, L, . . ., in order with straight lines, to complete the development.

### 21.24 To Find the Intersection and Developments of Two Prisms Fig. 21.21

***Intersection*** Fig. 21.21 (a)

The points in which the edges ACEH of the horizontal prism pierce the surfaces of the vertical prism are found in the top view and are projected downward to the corresponding edges ACEH in the front view. The points in which the edges 5 and 11 of the

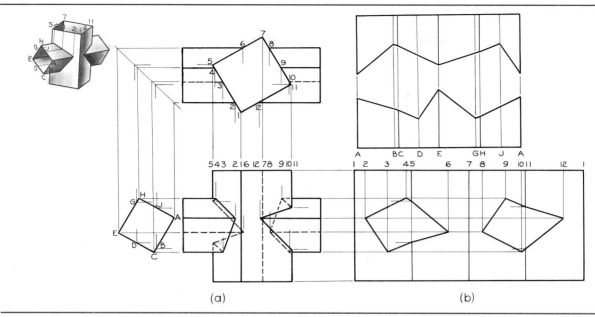

*Fig. 21.21*  Two Prisms at Right Angles to Each Other.

vertical prism pierce the surfaces of the horizontal prism are found in the left-side view at **G**, **D**, **J**, and **B** and are projected horizontally to the front view, intersecting the corresponding edges as shown. The intersection is completed by joining these points in order by straight lines.

### *Developments*  Fig. 21.21 (b)
The lateral surfaces of the two prisms are developed as explained in §21.23. True lengths of all lateral edges and lines parallel to them are shown in the front view of Fig. 21.21 at (a).

### 21.25 To Find the Intersection and Developments of Two Prisms  Fig. 21.22
*Intersection*  Fig. 21.22 (a)
The points in which edges 1–2–3–4 of the inclined prism pierce the surfaces of the vertical prism are vertices of the intersection. These points, found in the top view, are projected downward to the corresponding edges 1–2–3–4 in the front view, as shown. The intersection is completed by joining these points in order by straight lines.

### *Developments*  Fig. 21.22 (b)
The lateral surfaces of the two prisms are developed as explained in §21.23. True lengths of all edges of

both prisms are shown in the front view of Fig. 21.22 at (a).

### 21.26 To Find the Intersection and the Developments of Two Prisms  Fig. 21.23
In this case the edges of the oblique prism are oblique to the planes of projection, and in the front and top views none of the edges is shown true length, §7.24, and none of the faces is shown true size, §7.23. Furthermore, none of the angles, including the angle of inclination, is shown true size, §7.26. Therefore, it is necessary to draw a secondary auxiliary view, §10.19, to obtain the true size and shape of the right section of the oblique prism.

The direction of sight, indicated by arrow **A**, is assumed perpendicular to the end face 1–2–3, that is, parallel to the principal edges of the prism. The primary auxiliary view, taken in the direction of arrow **B**, shows the true lengths of the edges, the true inclination of the prism with respect to the horizontal and, incidentally, the true length and inclination of arrow **A**. In the secondary auxiliary view, arrow **A** is shown as a point, and the end face 1–2–3 is shown in its true size.

### *Intersection*  Fig. 21.23 (a)
The points in which the edges 1–2–3 of the oblique prism pierce the surfaces of the vertical prism are

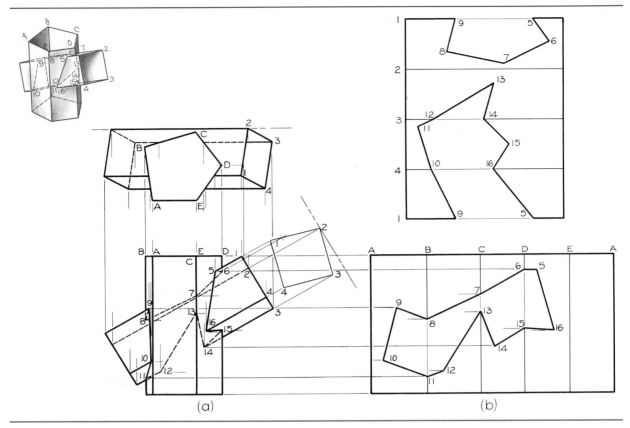

*Fig. 21.22*  Two Prisms Oblique to Each Other.

vertices of the intersection, found first in the top view and then projected downward to the front view.

### Developments Fig. 21.23 (b)
The lateral surfaces of the two prisms are developed as explained in §21.23. True lengths of the edges of the vertical prism are shown in the front view. True lengths of the edges of the oblique prism can be shown in the primary auxiliary view; true lengths to the vertices of the intersection may be found in this view, as shown for line X–5.

## 21.27 To Find the Intersection and Developments of Two Cylinders Fig. 21.24
### Intersection Fig. 21.24 (a)
Assume a series of elements (preferably equally spaced) on the horizontal cylinder, numbered 1, 2, 3, . . ., in the side view, and draw their top and front views. Their points of intersection with the surface

of the vertical cylinder are shown in the top view at A, B, C, . . ., and may be found in the front view by projecting downward to their intersections with the corresponding elements 1, 2, 3, . . ., in the front view. When a sufficient number of points have been found to determine the intersection, the curve is traced through the points with the aid of the irregular curve, §2.54. See also Fig. 7.38 (c).

### Developments Fig. 21.24 (b)
The lateral surfaces of the two cylinders are developed as explained in §21.7. True lengths of all elements of both cylinders are shown in the front view. Since both cylinders have bases at right angles to the center lines, the circles will develop as straight lines, and the developments will be rectangular, as shown. The length XY of the stretchout line for the development of the vertical cylinder, is equal to the circumference of the cylinder, or $\pi d$, and the length 1–1 of the stretchout line for the development of the horizontal cylinder is determined in the same way. Those elements of the large cylinder that pierce the small cylinder can be identified in the top view as

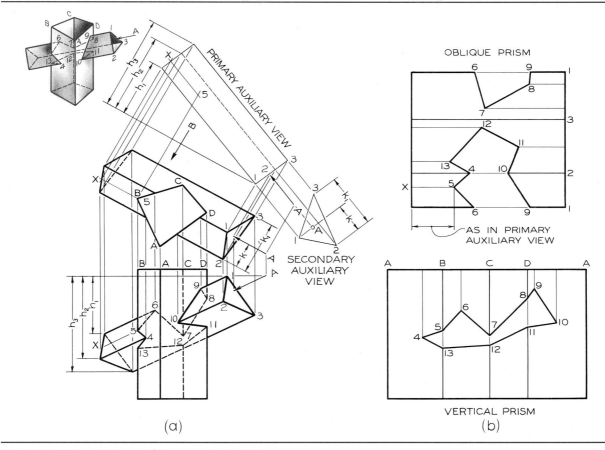

**Fig. 21.23** Two Prisms Oblique to Each Other.

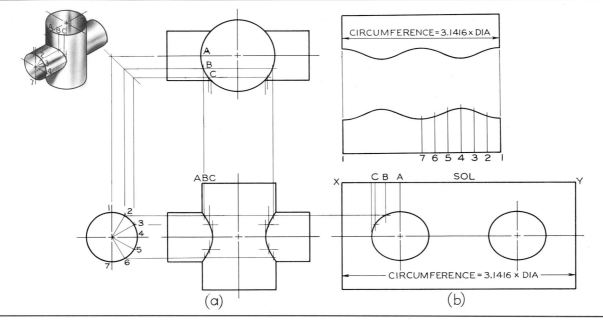

**Fig. 21.24** Two Cylinders at Right Angles to Each Other.

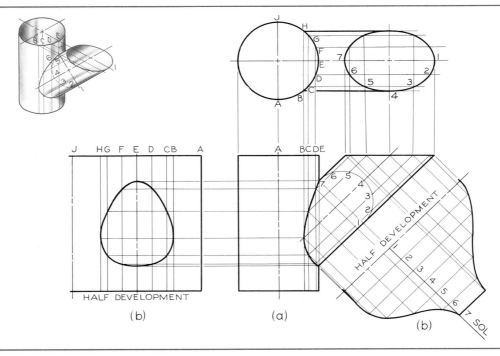

*Fig. 21.25*   Two Cylinders Oblique to Each Other.

elements A, B, C, . . . .. When these are drawn in the development, the points of intersections are found at their intersections with the corresponding elements of the horizontal cylinder taken from the front view, thus determining one of the figures of intersection, as shown in Fig. 21.24 (b).

### 21.28 To Find the Intersection and Developments of Two Cylinders Fig. 21.25

*Intersection* Fig. 21.25 (a)

A revolved right section of the inclined cylinder is divided into a number of equal parts, and an element is drawn at each of the division points, 1, 2, 3, . . .. The points of intersection of these elements with the surface of the vertical cylinder are shown in the top view at B, C, D, . . ., and are found in the front view by projecting downward to intersect the corresponding elements 1, 2, 3, . . .. The curve is traced through these points with the aid of the irregular curve, §2.54

### Developments Fig. 21.25 (b)

The lateral surfaces of the two cylinders are developed as explained in §§21.7 and 21.9. True lengths

of all elements of both cylinders are shown in the front view.

### 21.29 To Find the Intersection and Developments of a Prism and a Cone Fig. 21.26

*Intersection* Fig. 21.26 (a)

Points in which the edges of the prism intersect the surface of the cone are shown in the side view at A, C, and F. Intermediate points such as B, D, E, and G are piercing points of any lines on the lateral surface of the prism parallel to the edges. Through all of the piercing points in the side view, elements of the cone are drawn and then drawn in the top and front views. The intersections of the elements of the cone with the edges of the prism (and lines along the prism drawn parallel thereto) are points of the intersections. The figures of intersections are traced through these points with the aid of the irregular curve, §2.54.

The elements 6, 5, 4, . . ., in the side view of the cone may be regarded as the edge views of cutting planes that cut these elements on the cone and edges or elements on the prism. The intersection of

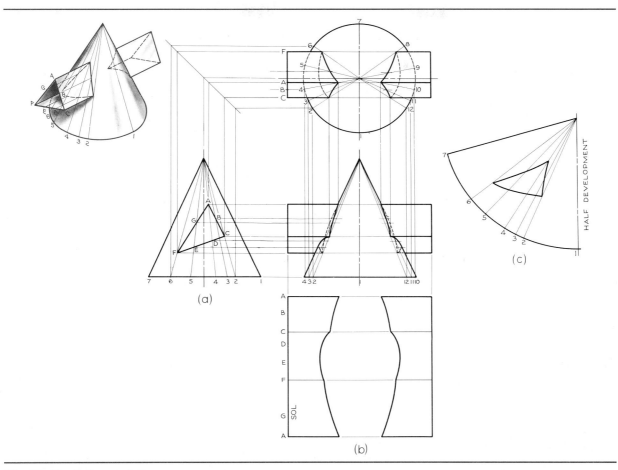

*Fig. 21.26*   Prism and Cone.

corresponding edges or elements on the two solids are points on the figure of intersection.

Another method of finding the figure of intersection is to pass a series of horizontal parallel planes through the solids in the manner of Fig. 21.9 (a). The plane will cut circles on the cone and straight lines on the prism, and their intersections will be points on the figure of intersection. See also Fig. 21.27 (b).

### *Developments* Fig. 21.26 (b)

The lateral surface of the prism is developed as explained in §21.23. True lengths of all edges and lines parallel thereto are shown in both the front and top views.

The lateral surface of the cone, Fig. 21.26 (c), is developed as explained in §21.11. True lengths of elements from the vertex to points on the intersections are found as shown in Fig. 11.10 (a).

## 21.30 To Find the Intersection of a Prism and a Cone with Edges of Prism Parallel to Axis of Cone Fig. 21.27

*Fig. 21.27 (a)*   Since the lateral surfaces of the prism are parallel to the axis of the cone, the figure of intersection will be composed of a series of hyperbolas, §§5.47 and 5.60. If a series of planes is assumed containing the axis of the cone, each plane will contain edges of the prism, or will cut lines parallel to them along the prism, and will cut elements on the cone that intersect these at points on the figure of intersection.

*Fig. 21.27 (b)*   The intersection is the same as at (a), but it is found in a different manner. Here a series of parallel planes perpendicular to the axis of the cone cut circles of varying diameters on the cone. These circles are shown true size in the top

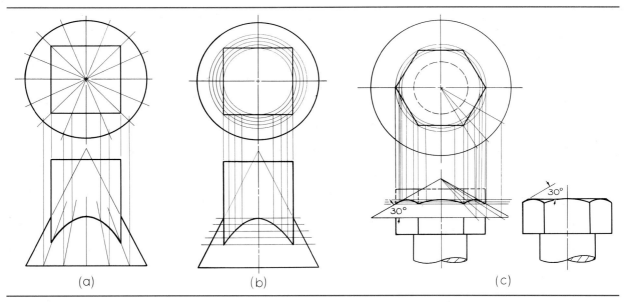

*Fig. 21.27*   Prisms and Cones.

view, where the piercing points of these circles in the vertical plane surfaces of the prism are also shown. The front views of these piercing points are found by projecting downward to the corresponding cutting-plane lines.

***Fig. 21.27 (c)***   The chamfer of an ordinary hexagon bolt head or hexagon nut is actually a conical surface that intersects the six vertical sides of a hexagonal prism to form hyperbolas. At (c) the methods of both (a) and (b) are shown to illustrate how points may be found by either method.

In machine drawings of bolts and nuts, these hyperbolic curves are approximated by means of circular arcs, as shown in Fig. 15.27.

## 21.31 To Find the Intersections and the Developments of a Cylinder and a Cone Fig. 21.28

***Intersections*** Fig. 21.28 (a)

Points in which elements of the cylinder (preferably equally spaced to facilitate the development) intersect the surface of the cone are shown in the side view at A, B, C, . . .. The elements of the cylinder are here shown as points. Elements of the cone are then drawn from the vertex through each of these points, and then drawn in their correct locations in the top and front views. The intersections of these elements with the elements A, B, C, . . ., of the cylinder are points on the figures of intersection. The

curves are then traced through these points with the aid of the irregular curve, §2.54.

As explained in §21.29, the elements 5, 6, 7, . . ., in the side view of Fig. 21.28 (a) could be regarded as edge views of cutting planes that cut elements from both the cone and the cylinder, the elements meeting at points on the figure of intersection. Or a series of horizontal parallel planes can be passed through the solids that will cut circles from the cone and elements from the cylinder that intersect at points on the figure of intersection.

### Developments Fig. 21.28 (b)

The lateral surface of the cylinder is developed as explained in §21.27. True lengths of all elements are shown in both the front and top views. The lateral surface of the cone is developed as explained in §21.11. True lengths of elements from the vertex to points on the intersections are found as shown in Fig. 11.10 (a).

## 21.32 To Find the Intersection of a Cylinder and a Sphere Fig. 21.29

Horizontal planes 1, 2, 3, . . ., which appear edgewise in the front and side views, cut elements A, B, C, . . ., from the cylinder and circular arcs 1′, 2′, 3′, . . ., from the sphere. The intersections of the elements with the arcs produced by the corresponding planes are points on the figure of intersection. Join the points with a smooth curve, §2.54.

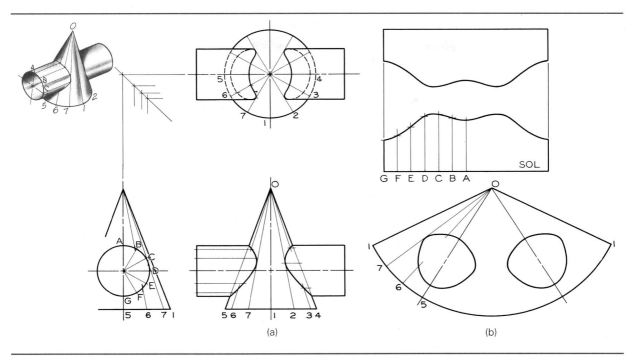

*Fig. 21.28*   Cone and Cylinder.

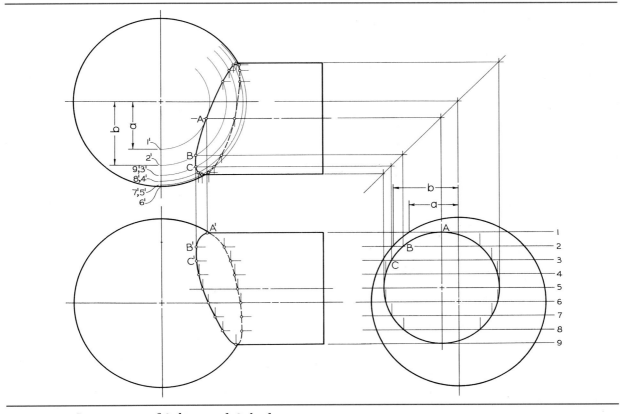

*Fig. 21.29*   Intersection of Sphere and Cylinder.

# _Intersection and Development Problems_

A wide selection of intersection and development problems is provided in Figs. 21.30 to 21.37. These problems are designed to fit size B (11.0″ × 17.0″) or A3 (297 mm × 420 mm) sheets. Dimensions should be included on the given views. The student is cautioned to take special pains to obtain accuracy in these drawings and to draw smooth curves as required.

Since many of the problems in this chapter are of a general nature, they can also be solved on most computer graphics systems. If a system is available, the instructor may choose to assign specific problems to be completed by this method.

Additional problems, in convenient form for solution, are available in _Principles of Engineering Graphics Problems_ by Spencer, Hill, Loving, and Dygdon, a workbook designed to accompany this text that is also published by Macmillan Publishing Company.

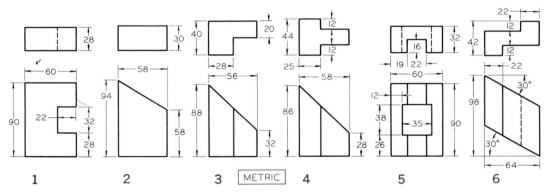

**_Fig. 21.30_**   Draw given views and develop lateral surface. Layout A3–3 or B–3.

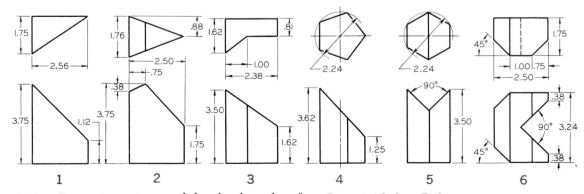

**_Fig. 21.31_**   Draw given views and develop lateral surface. Layout A3–3 or B–3.

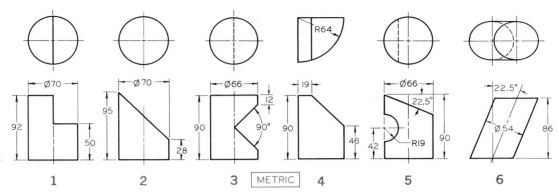

**Fig. 21.32** Draw given views and develop lateral surface. Layout A3–3 or B–3.

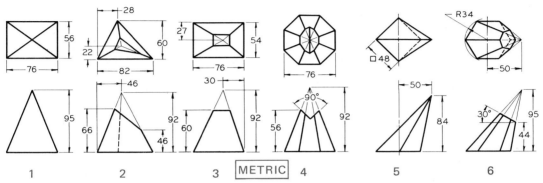

**Fig. 21.33** Draw given views and develop lateral surface. Layout A3–3 or B–3.

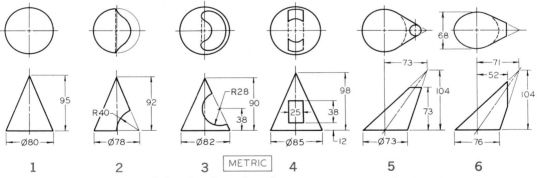

**Fig. 21.34** Draw given views and develop lateral surface. Layout A3–3 or B–3.

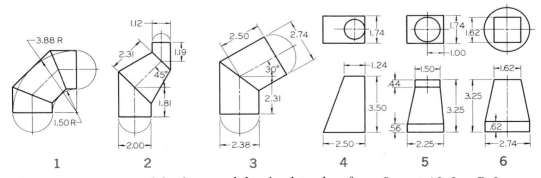

**Fig. 21.35** Draw given views of the forms and develop lateral surfaces. Layout A3–3 or B–3.

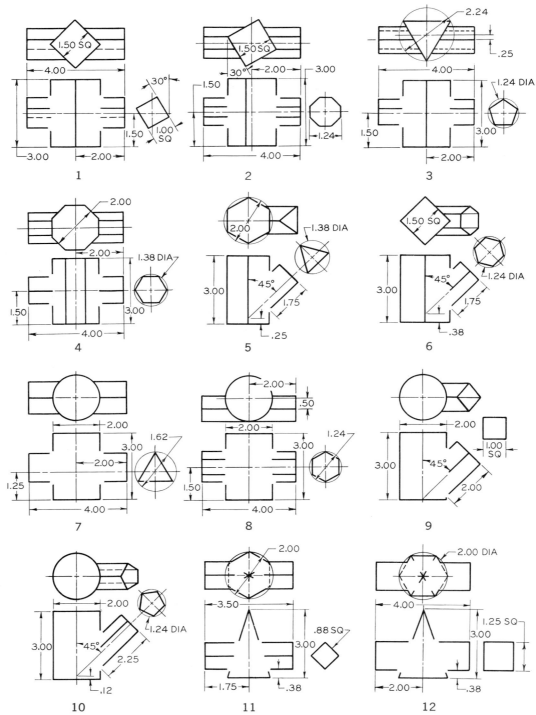

**Fig. 21.36**  Draw the given views of assigned form, and complete the intersection. Then develop lateral surfaces. Layout A3–3 or B–3.

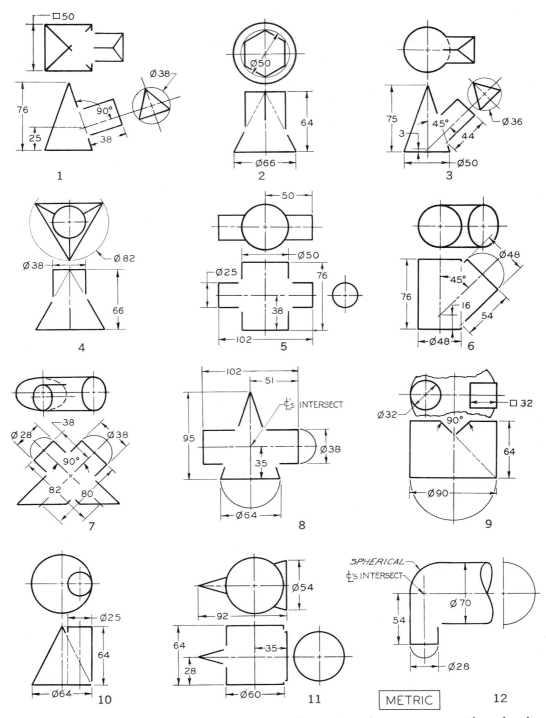

**Fig. 21.37** Draw the given views of assigned form, and complete the intersection. Then develop lateral surfaces. Layout A3–3 or B–3.

# Line and Plane Tangencies

A plane tangent to a ruled surface such as a cone or cylinder contains only one straight-line element of that surface. A plane tangent to a double-curved surface such as a sphere contains only one point in that surface. All lines tangent to a curved surface at a particular point or at points along the same straight-line element lie in a plane tangent at the point or element. A plane tangent to a ruled surface is conveniently represented by two straight lines, one an element and the other line tangent to the surface at a point on the element. For double-curved surfaces the plane is represented by two straight lines, both tangent at the same point on the double-curved surface.

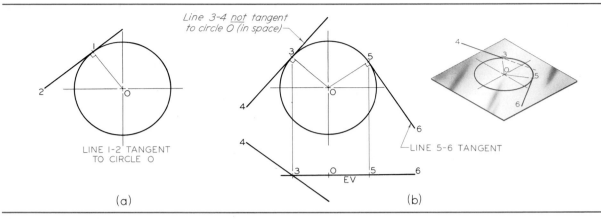

*Fig. 22.1*　Line Tangencies.

## 22.1 Line and Plane Tangencies

Line tangencies as encountered in plane geometry are discussed in Chapter 5, "Geometric Constructions." Methods such as those of §§5.33 to 5.35 are based on the principle that a line tangent to a circle is perpendicular to the radius drawn to the point of tangency. Thus in Fig. 22.1 (a) line 1–2 is perpendicular to radius O–1 at point 1 and is tangent to the circle O. If the assumption is made that line 1–2 and circle O are in the plane of the paper, a plane geometry construction suffices.

At (b) is given a multiview (or three-dimensional) drawing. Although the top view of line 3–4 appears to be tangent to the top view of the circle, line 3–4 is *not* tangent to the circle *in space* because line 3–4 is not in the plane of the circle, as is evident in the front view. By contrast, line 5–6 lies in the plane of the circle and *is* tangent in space to the given circle.

Because planes are easily represented by lines, §19.7, planes tangent to curved surfaces are often represented by suitable pairs of tangent lines, or by one tangent line and a line lying in the curved surface (a straight-line *element*, §7.28). It is sometimes convenient to represent a tangent plane by a tangent edge view of the plane, Fig. 22.10 (b).

## 22.2 Planes Tangent to Cones
### *Plane Tangent to a Right-Circular Cone Through Point on Surface* Fig. 22.2
Let the cone and the top view of point 1 on the surface of the cone be given as at (a).

At (b) the element through point 1 is drawn in the top view, establishing point 2 on the circular base. Point 2 is then projected to the front view, and point 1 is projected to the now-established front view of the element.

If the view of element O–2 is nearly parallel to the projectors, more dependable accuracy may be secured through the use of revolution, as at (c), which is an application of §11.6

At (d) line 2–3 is drawn tangent to the circular base by drawing its top view perpendicular to element O–2 at point 2 and by drawing its front view coinciding with the edge view of the base. Plane O–1–2–3 is the required tangent plane.

### *Plane Tangent to Oblique Cone Through Point Outside Its Surface* Fig. 22.3
Let the cone and point be given as at (a).

Because all elements of a cone pass through its vertex, a tangent plane, which must contain an element, will contain the vertex also. Hence line V–1 lies in the tangent plane. Any line tangent to the base circle lies in the plane of the base, §22.1, and thus can intersect line V–1 only at point 2, shown in the front view at (b).

At (c) point 2 is projected to the top view of line V–1, and from the top view of point 2 line 2–3 and 2–3′ may be drawn tangent to the base circle as shown. Either plane V–2–3 or V–2–3′ meets the requirements of the problem. In practice, it is usually evident which of two optional solutions is compatible with other features of the design.

### *Plane Tangent to Cone and Parallel to Given Line* Fig. 22.4
With the given cone and line 1–2 as shown at (a), a line is drawn through vertex V of the cone, parallel to the given line and intersecting the plane of the

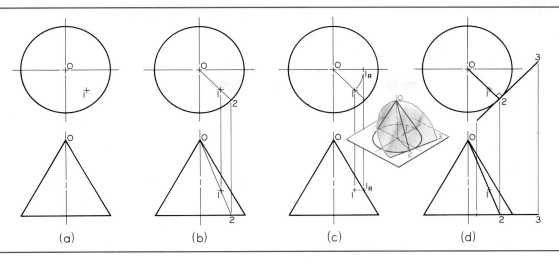

**Fig. 22.2** Plane Tangent to Right-Circular Cone Through Point on Surface.

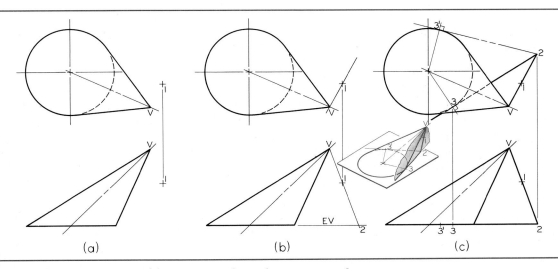

**Fig. 22.3** Plane Tangent to Oblique Cone Through Point Outside Cone.

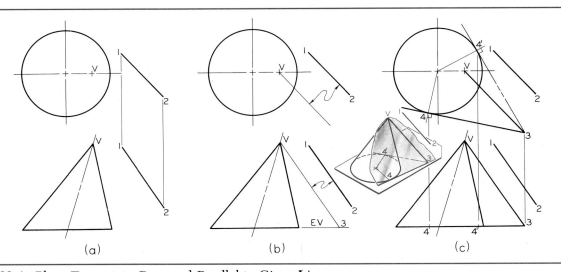

**Fig. 22.4** Plane Tangent to Cone and Parallel to Given Line.

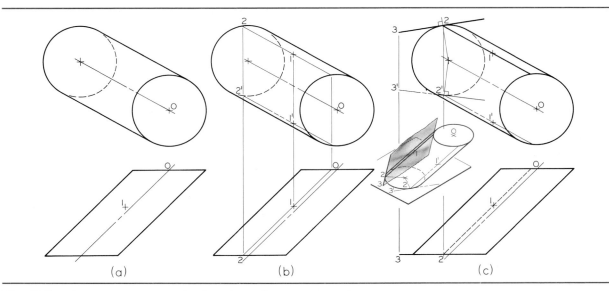

**Fig. 22.5**    Plane Tangent to Cylinder Through Point on Surface.

base at point **3**, as at (b), thus establishing a line in the required tangent plane. (Also see §20.4.)

At (c) point **3** is projected to the top view of the line through **V** and parallel to line **1–2**. From the top view of point **3** lines **3–4** and **3–4′** may be drawn tangent to the base circle as shown. Either plane **V–3–4** or **V–3–4′** satisfies the requirements of the problem.

## 22.3 Planes Tangent to Cylinders

By definition, §5.7, all elements of a cylinder are parallel to each other and to the axis of the cylinder. It follows from §20.4 that any plane tangent to a cylinder, and thus containing one element, is parallel to the remaining elements and to the axis.

### *Plane Tangent to Cylinder Through Point on Surface* Fig. 22.5.

With one view given of point **1** on the surface of an oblique cylinder as at (a), element **1–2** is introduced as at (b). When point **2** is projected to the top view, it is observed that point **2** may fall at either position **2** or position **2′**. There are thus alternative solutions, and point **1** may be at either of the locations **1** and **1′**, as shown. Addition of line **2–3** tangent to the base of the cylinder at point **2** (or line **2′–3′** at point **2′**) completes the representation of the required tangent plane.

### *Plane Tangent to Cylinder Through Point Outside Its Surface* Fig. 22.6

The cylinder and point **1** are given as at (a). As previously noted, any plane tangent to a cylinder must be parallel to the elements. Hence a line **1–2** drawn through point **1** and parallel to the elements, as at (b), must be common to all planes containing point **1** and parallel to the elements, §20.4. The representation of a tangent plane is completed by the addition of a line tangent to one base of the cylinder. As observed earlier, such a tangent line must lie in the same plane as the chosen base—in this example the lower base, (c). Therefore the tangent line could intersect line **1–2** only at point **2** located in the front view at the intersection of the (extended) edge view of the base plane with line **1–2**. Observe that lines could be drawn from point **2** tangent to the lower base at either point **3** or point **3′**, so that again there are alternative solutions, and in an application it would normally be apparent which solution is practical.

### *Plane Tangent to Cylinder and Parallel to Given Line Outside the Cylinder* Fig. 22.7

Let the cylinder and line **1–2** be given as at (a). A plane is constructed parallel to a given line by the method of §20.4. However, at this stage it is not known which element will be the line of tangency.

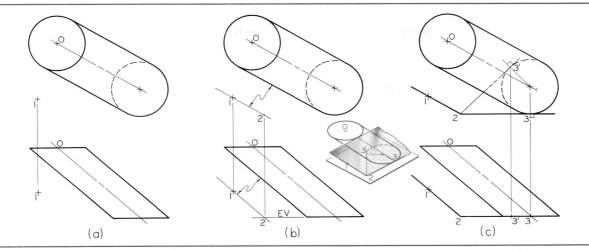

*Fig. 22.6* Plane Tangent to Cylinder Through Point Outside Surface.

The tangent plane must be parallel to all elements of the cylinder as well as to line 1–2. By the method of §20.4 a plane can be constructed at any convenient location and parallel both to the elements and to line 1–2. The required tangent plane will then be parallel to this plane.

A convenient representation for this preliminary plane includes the given line 1–2, as at (b). With line 1–3 drawn parallel to the cylinder as shown, plane 2–1–3 is established parallel to the cylinder.

Any line tangent to either given cylinder base must be a horizontal line, §22.1. Since all horizontal lines in the same oblique plane are parallel to each other, §19.8, it follows that horizontal lines in parallel oblique planes are likewise parallel to each other. Thus, if the direction of one such horizontal line is established, such as line 2–4 at (b), the direction of horizontal lines in planes parallel to plane 2–1–3, including the required plane, is also established.

At (c) line $2_1$–$4_1$ is drawn parallel to the top view of line 2–4 and tangent to either base—in this case the lower base—of the cylinder. The point of tangency is 5, and line 5–6 is the element of tangency.

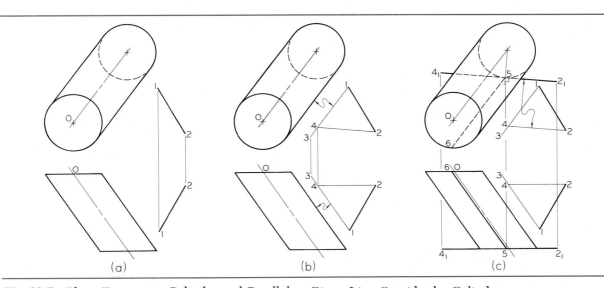

*Fig. 22.7* Plane Tangent to Cylinder and Parallel to Given Line Outside the Cylinder.

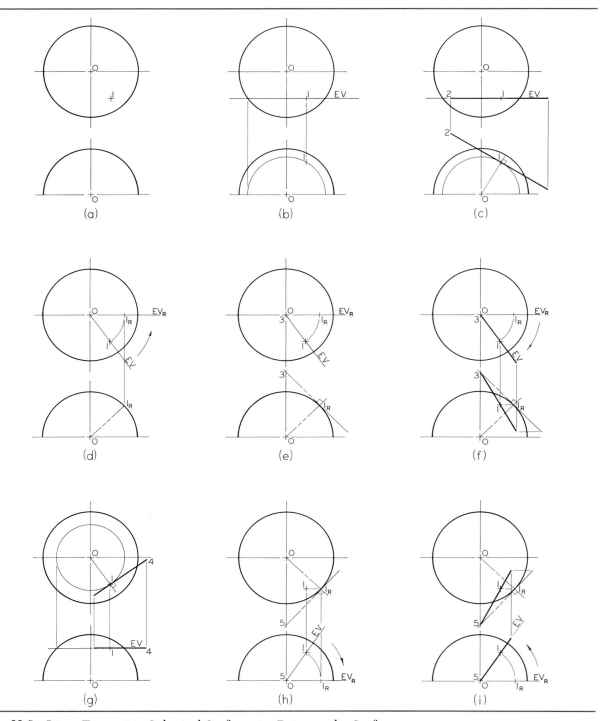

*Fig. 22.8*   Lines Tangent to Spherical Surface at a Point on the Surface.

Construction of the front view completes the representation of the required tangent plane parallel to given line 1–2.

Since tangent line $2_1$–$4_1$ could have been drawn on the opposite side of the base, there is an alternative tangent plane, which is not shown here.

## 22.4 Planes Tangent to Spheres

A plane tangent to a *double-curved surface*, §21.1, contains one and only one point of that surface, since it follows from the definition of such a surface that it contains no straight-line elements. Hence planes tangent to double-curved surfaces are represented by appropriate combinations of lines tangent at the desired point or points of tangency. Under suitable circumstances such a tangent line may represent an edge view of the required tangent plane. An example of this appears later in this section.

The sphere, §5.7, is by far the most practical, hence most common, form of the double-curved surface. This discussion will be limited to spherical surfaces.

### *Lines Tangent to a Sphere at a Given Point on Its Surface* Fig. 22.8

Let the front and top views of a hemisphere be given, as at (a), and let the top view of a point 1 on its surface be given also. The front view of point 1 can be located by passing a convenient cutting plane through the point and finding another view of the line (circle) of intersection. It follows that a "convenient" cutting plane is one in such position that the circle appears in one of the given views as a circle and not as an ellipse. As an example, at (b) a frontal, edge-view plane is introduced. The circle is located and constructed in the front view as shown, and point 1 is projected to it. A line tangent to this circle is tangent to the spherical surface. At (c) line 1–2 is constructed tangent to the circle at point 1 by drawing the front view of the tangent line perpendicular to radius O–1 and then drawing the top view coincident with the edge view of the cutting plane, §22.1.

At (d) point 1 is revolved, in the top view, to the frontal plane through center O. See §11.6. This amounts to revolving the edge view of a vertical cutting plane (EV), as indicated. The revolved view of the circle of intersection coincides with the circular front view of the sphere, and the revolved position $1_R$ of point 1 is projected to it. As shown at (e), line $1_R$–3 is now drawn tangent to this circle,

intersecting the vertical center line of the sphere at point 3. Since this vertical center line is also the axis of revolution, point 3 will not move as the cutting plane counterrevolves, as at (f). Line 1–3 is thus another line tangent to the spherical surface at given point 1.

At (g) a horizontal cutting plane is introduced by first drawing the top view of the circle of intersection passing through the top view of point 1. This in turn projects to the front view as shown, locating the edge of the cutting plane, to which point 1 is projected. Line 1–4, constructed tangent to the circle of intersection at point 1, is also tangent to the spherical surface.

Finally, at (h) and (i), line 1–5 is constructed tangent to the spherical surface in a variation of the method shown at (d) to (f). An edge-view cutting plane is introduced in the front view, through points 1 and O. After it revolves to horizontal, the circle of intersection coincides with the top view of the sphere, and line $1_R$–5 is drawn tangent to the cut circle of intersection. Counterrevolution establishes the views of the tangent line 1–5, as shown at (i).

### *Plane Tangent to a Sphere at a Given Point on Its Surface* Fig. 22.9

In Fig. 22.8 four lines, 1–2, 1–3, 1–4, and 1–5, were constructed tangent to the spherical surface at point 1. Any two of these constitute intersecting tangent lines and thus establish the plane tangent at point 1. As an example, tangent lines 1–2 and 1–4 are

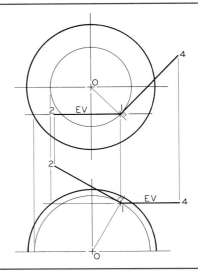

**Fig. 22.9** Plane Tangent to Sphere at Given Point on Surface.

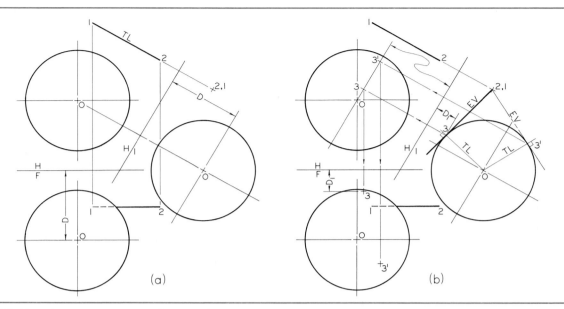

**Fig. 22.10**    Plane Tangent to Sphere and Containing Line Outside Sphere.

shown in Fig. 22.9. Plane 2–1–4 is one representation of the plane tangent to the spherical surface at point 1. Incidentally, study reveals that this is simply the construction of a plane perpendicular to radius O–1 at point 1 by the given-view method of §20.6. Analogous to the plane geometry description of a line tangent to a circle, §22.1, a plane tangent to a sphere may be defined as a plane perpendicular to the radius of the sphere drawn to the point of tangency.

### A Plane Tangent to a Sphere and Containing a Given Line That Does Not Intersect the Sphere Fig. 22.10

Let sphere O and line 1–2 be given in the front and top views, as at (a). If a point view of a line is constructed, any plane containing the line appears in edge view, §19.9. Also any orthographic projection of a sphere shows the true diameter of the sphere. Hence in the given problem, if a point view of line 1–2 is constructed, the required tangent plane will appear in edge view and tangent to the corresponding view of the sphere. Since line 1–2 appears in true length in the top view at (a), its point view may be constructed in primary auxiliary view 1 as shown.

At (b) alternative edge views of planes are drawn through point view 2, 1 and tangent to the sphere at point 3 or point 3′ as preferred. The front and top

views of point 3 or 3′ are then projected as shown, completing the representation of the tangent plane. In practice, other lines of the tangent plane could, and probably would, be drawn to establish a recognizable configuration. Theoretically, however, additional lines are not needed.

### 22.5 Applications of Right-Circular Cones

As pointed out in §19.3, all elements of a right-circular cone form the same angle with the base plane of the cone. This feature is the basis for the revolution constructions of §19.3 and for the constructions following.

### A Plane Containing an Oblique Line and Making a Specified Angle with Horizontal Fig. 22.11

Let line 1–2 be given, as at (a), and let it be required to construct a plane containing line 1–2 and forming an angle of 45° (or 135°) with horizontal.

A plane tangent to a right-circular cone contains one element and forms the same angle as does any element with the base plane of the cone. Because the vertex is common to all elements, it must lie in any tangent plane. Hence at (b) the vertex of a cone of suitable dimensions is placed at some chosen point 3 along given line 1–2. In this case the re-

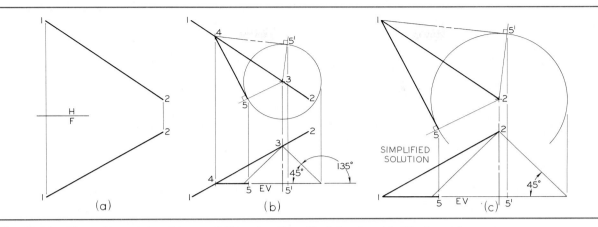

**Fig. 22.11**   Plane Containing Line and Forming Specified Angle with Horizontal.

quired angle is 45° with horizontal, so the cone is placed with its base horizontal (axis vertical) and with its elements at 45° with the base. This same construction is used for a specified angle of 135°, the supplement of 45°. Line 1–2 pierces the extended edge view of the base plane at point 4. See Fig. 22.3. Lines tangent to the base may be drawn alternatively from point 4 to point 5 or from point 4 to point 5′. Either of the two resulting tangent planes may be selected according to additional specifications, if any.

As shown at (c), the foregoing construction could be somewhat simplified in detail, not in principle, by placing the cone vertex at point 2 and the base plane at the same elevation as point 1.

### Line at Specified Angles
### with Given Planes Fig. 22.12
If two right-circular cones with the same vertex intersect, the common element or elements of the two cones form the same angles with the two base planes as do the respective sets of elements. To simplify determining which elements are common, the two cones should have elements of the same length so that their base lines intersect, as exemplified by points 1 and 2 in Fig. 22.12.

As an example of an application of the foregoing, Fig. 22.13, let it be required to construct a line 1–2, with point 1 given, such that line 1–2 forms an angle of 30° with a horizontal plane and an angle of 50° with a frontal plane. At (a) a right-circular cone is constructed with its vertex at point 1 and with its

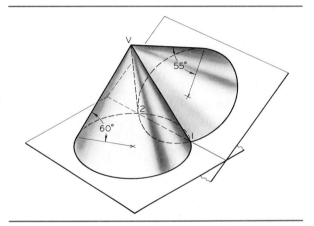

**Fig. 22.12**   Line at Specified Angles with Two Given Planes.

elements at 30° with horizontal. The length S of the elements can be any convenient or specified length.

At (b) a second cone is introduced with its vertex at point 1 but with its elements at 50° with a frontal plane. Note that the previously selected length S must also be used for the elements of the second cone. This selection results in the intersection of two base circles at points 2 and 2′. Thus the requirements of the problems are fulfilled by either line 1–2 or line 1–2′. There are additional alternative solutions. If we choose to reverse either cone, we find two more solutions. At (c) the cone with the 30° angle is drawn sloping upward from point 1. The two additional solutions are lines 1–3 and 1–3′.

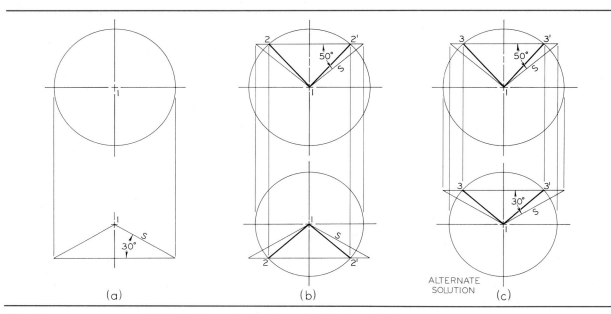

*Fig. 22.13*   Line at Specified Angles with Horizontal and Frontal Planes.

Additional reversals of the cones produce line-segment extensions of lines 1–2 and 1–2′ or 1–3 and 1–3′—not additional alternative solutions.

It is important to realize that there are limitations on the selection of the two angles. If their sum is greater than 90°, the cones do not intersect, and there is no solution.* If the sum equals 90°, the

*Assuming both specified angles are acute, or using supplements of obtuse angles.

cones are tangent and the element of tangency is the solution. (Two such single-element solutions are possible.) Only when the sum of the required angles is less than 90° do we have the four alternative possibilities shown in Fig. 22.13. For given planes that are not perpendicular, there are different but similar limitations dependent upon the dihedral angle, §10.11, between the two planes. In general, the sum of the required angles must be equal to or less than the dihedral angle between the given planes.

# *Line and Plane Tangency Problems*

In Figs. 22.14, 22.15, and 22.16 are problems involving planes tangent to cones, cylinders, and spheres, and applications of right-circular cones.

Use Layout A–1 or A4–1 (adjusted) and divide the working area into four equal areas for problems to be assigned by the instructor. Some problems require two problem areas or one-half sheet. Data for most problems are given by a coordinate system using metric dimensions. For example, in Fig. 22.14. Prob. 1, point O is located by the full-scale coordinates (60, 50, and 90 mm). The first coordinate locates the front view from the left edge of the problem area. The second coordinate locates the front view of the point from the bottom edge of the problem area. The third coordinate locates either the top view of the point from the bottom edge of the problem area or the side view of the point from the left edge of the problem area. Inspection of the given problem layout will determine which application to use.

Since many of the problems in this chapter are of a general nature, they can also be solved on most computer graphics systems. If a system is available, the instructor may choose to assign specific problems to be completed by this method.

Additional problems in convenient form for solution are available in *Principles of Engineering Graphics Problems* by Spencer, Hill, Loving, and Dygdon, a workbook designed to accompany this text that is also published by Macmillan Publishing Company.

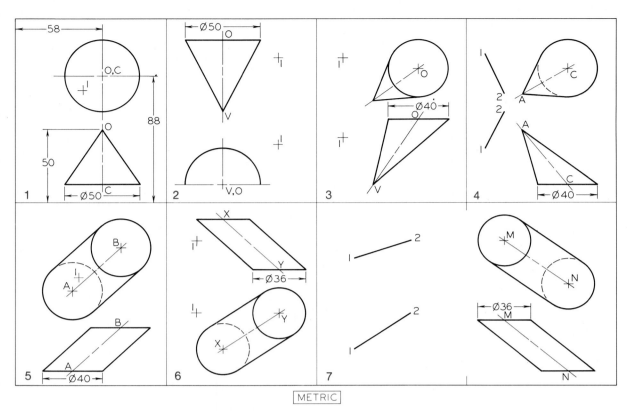

METRIC

**Fig. 22.14**  Lay out and solve problems as assigned. Use Layout A–1 or A4–1 (adjusted) divided into four equal areas.

1.  Point 1(46, –, 79) is on the surface of cone O(58, 30, 88)–C(58, 12, 88). Pass a plane tangent to the cone and containing point 1.
2.  Pass a plane through point 1(75, 40, 100) and tangent to cone V(38, 12, 63)–O(38, 12, 114).
3.  Pass a plane through point 1(18, 46, 100) and tangent to cone V(38, 12, 71)–O(69, 61, 94).
4.  Pass a plane tangent to cone A(38, 50, 75)–C(68, 12, 94) and parallel to line 1(12, 38, 104)–2(25, 63, 75).
5.  Pass a plane tangent to cylinder A(38, 10, 66)–B(71, 40, 96) and containing point 1(43, –, 75) on the surface of the cylinder.
6.  Pass a plane through point 1(20, 50, 100) and tangent to cylinder X(38, 25, 117)–Y(75, 50, 81).
7.  Pass a plane tangent to cylinder M(127, 46, 100)–N(170, 10, 70) and parallel to line 1(25, 25, 88)–2(63, 50, 100).

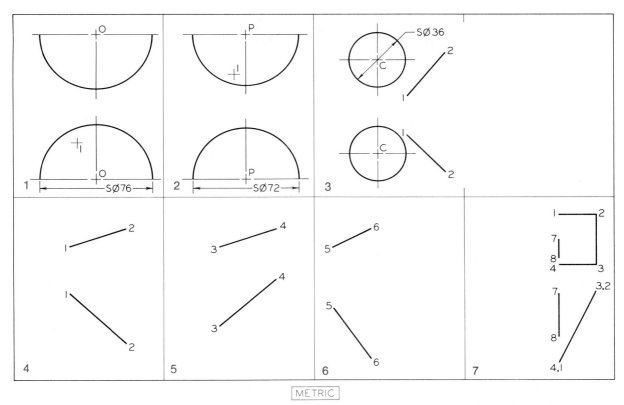

**Fig. 22.15**  Lay out and solve problems as assigned. Use Layout A–1 or A4–1 (adjusted) divided into four equal areas.

1. Pass a plane tangent to sphere O(56, 12, 114) and containing point 1(43, 38, –) on the surface of the sphere.
2. Draw three lines tangent to sphere P(56, 12, 114) and containing point 1(48, –, 86) on the surface of the sphere.
3. Pass a plane tangent to sphere C(43, 30, 96) and containing line 1(63, 43, 70)–2(88, 18, 100). Show the point of tangency in all views.
4. Pass a plane through line 1(38, 61, 94)–2(75, 25, 107) and making an angle of 30° with a frontal plane.
5. Pass a plane through line 3(38, 38, 94)–4(75, 71, 107) and making an angle of 60° with horizontal.
6. Pass a plane through line 5(12, 50, 94)–6(38, 15, 107) and making an angle of 135° with a profile plane.
7. Pass a plane through line 7(63, 61, 99)–8(63, 30, 86) and making an angle of 60° with plane 1(63, 12, 117)–2(88, 63, 117)–3(88, 63, 81)–4(63, 12, 81).

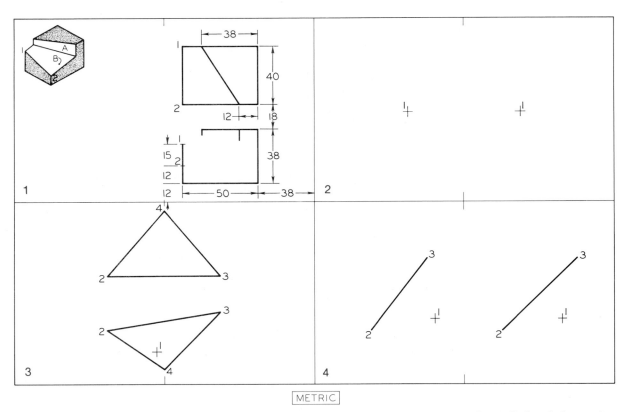

METRIC

*Fig. 22.16*  Lay out and solve problems as assigned. Use Layout A–1 or A4–1 (adjusted) divided into four equal areas.

1. Surfaces **A** and **B** form a dihedral angle of 130°. Complete the front view. Omit the pictorial in the layout.
2. Complete the views of a line 1(63, 63, 140)–2, which is 50 mm in length and forms angles of 45° with a profile plane and 35° with a frontal plane.
3. Point 1(96, 23, –) is in plane 2(63, 38, 75)–3(140, 50, 75)–4(100, 10, 122). Find in plane 2–3–4 a line 1–5 that forms an angle of 25° with a horizontal plane.
4. Find a line 1(81, 45, 167)–4 that is 40 mm in length, is perpendicular to line 2(38, 38, 127)–3(75, 88, 178), and makes an angle of 40° with a frontal plane.

# CHAPTER
# 23

# Graphs
# and Diagrams

by E. J. Mysiak*

In previous chapters we have seen how graphical representation is used instead of words to describe the size, shape, material, and fabrication methods for the manufacture of actual objects. Graphical representation is also used extensively to present and analyze data and to solve technical problems. A pictorial or graphical presentation is much more impressive and easier to understand than a numerical tabulation or a verbal description. These graphical descriptions are synonymously termed *graphs*, *charts*, or *diagrams*.

The term *chart* has two meanings. It is associated with maps and also describes any of the forms of graphical presentation described in this chapter. Therefore, the term *chart* includes all graphs and diagrams. A *graph* is a special form of *chart* with data plotted on some type of grid. A *diagram* is a *chart* without the use of a grid.

Tabulated data in Fig. 23.1 (a), showing the average weekly earnings of United States manufacturing workers for the years 1974–1984, are presented as a line graph at (b) and a bar graph at (c). The greater effectiveness of graphical representation is evident.

*Engineering Manager, Phoenix Company of Chicago, Wood Dale, IL.

| Year | Current Dollars | 1977 Dollars |
|------|-----------------|--------------|
| 1974 | 176.80 | 217.20 |
| 1975 | 190.79 | 214.85 |
| 1976 | 209.32 | 222.92 |
| 1977 | 228.90 | 228.90 |
| 1978 | 249.27 | 231.66 |
| 1979 | 268.94 | 224.64 |
| 1980 | 288.62 | 212.64 |
| 1981 | 318.00 | 212.00 |
| 1982 | 330.65 | 207.96 |
| 1983 | 354.08 | 216.03 |
| 1984 | 372.91 | 221.05 |

(a)

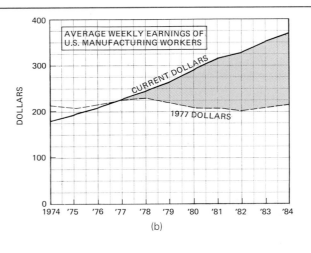

(b)

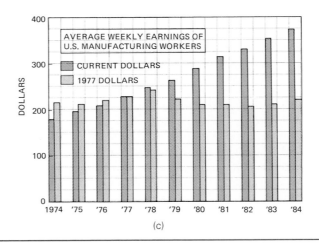

(c)

*Fig. 23.1*   Comparison of Tabulated and Graphical Presentations.

## 23.1 Uses of Graphical Representation

Graphs and diagrams can be classified into two broad categories depending upon whether their application or use is for technical purposes or for popular appeal.

Technically trained personnel communicate by means of graphs similar to Fig. 23.1 (b). When graphs are used to present data to lay personnel, a form similar to Fig. 23.1 (c) is preferred.

Scientific or technical personnel use graphs and diagrams to (a) present results of experimental investigations, (b) represent phenomena that follow natural laws, (c) represent equations for further computational purposes, and (d) derive equations to represent empirical test data. The forms of graphical representation most commonly used are

1. Rectangular coordinate line charts.

2. Semilogarithmic coordinate line charts.

3. Logarithmic coordinate line charts.

4. Trilinear coordinate line charts.

5. Polar coordinate line charts.

6. Nomographs or alignment charts.

7. Volume charts.

8. Rectangular coordinate distribution charts.

9. Flowcharts.

Often information and ideas must be presented in a way understandable to the layman in order to have the information or idea accepted. For popular appeal, graphs and diagrams frequently used are

1. Rectangular coordinate line charts.
2. Bar or column charts.
3. Rectangular coordinate surface or area charts.
4. Pie charts.
5. Volume (map) charts.
6. Flowcharts.
7. Map distribution charts.

## 23.2 Rectangular Coordinate Line Graphs

The rectangular coordinate line graph is the type in which values of two related variables are plotted on coordinate paper, and the points, joined together successively, form a continuous line or "curve."

The following are some of the purposes for which a line graph can be used to advantage.

1. Comparison of a large number of plotted values in a compact space.
2. Comparison of the relative movements (trends) of more than one set of data on the same graph. There should not be more than two or three curves on the same graph, and there should be some definite relationship among them.
3. Interpolation of intermediate values.

4. Representation of movement or overall trend (relative change) of a series of values rather than the difference between values (absolute amounts).

Line graphs are *not* well suited for (1) presenting relatively few plotted values in a series, (2) emphasizing changes or difference in absolute amounts, or (3) showing extreme or irregular movement of data.

Rectangular line graphs may be classified as (1) *mathematical graphs*, (2) *time-series charts*, or (3) *engineering graphs*. Any of these may have one or more curves on the same graph. If the values plotted along the axes are pure numbers (positive and negative), showing the relationship of an equation, the plot is commonly called a mathematical graph, Fig. 23.2 (a). When one of the variables is any unit of time, the chart is known as a time-series chart, (b). This is one of the most common forms, since time is frequently one of the variables. Line, bar, or surface chart forms may be used for time-series charts, line charts being the most widely used in engineering practice. The plotting of values of any two related physical variables on a rectangular coordinate grid is referred to as an engineering chart, graph, or diagram, (c).

Line curves are generally presented for any of three types of relationships.

1. Observed relationships, usually plotted with observed data points connected by straight, irregular lines, Fig. 23.3 (a).

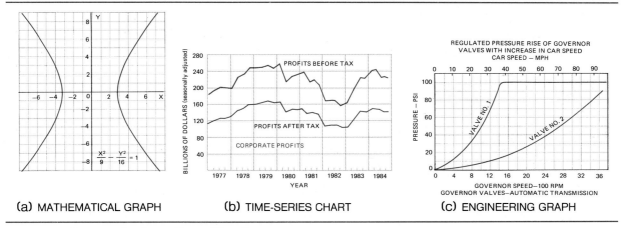

(a) MATHEMATICAL GRAPH  (b) TIME-SERIES CHART  (c) ENGINEERING GRAPH

*Fig. 23.2* Rectangular Line Chart Classification.

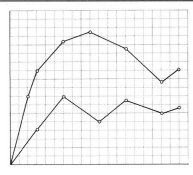 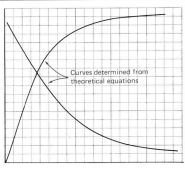

**(a) OBSERVED RELATIONSHIP**　　**(b) EMPIRICAL RELATIONSHIP**　　**(c) THEORETICAL RELATIONSHIP**

*Fig. 23.3*　Curve Fitting.

2. Empirical relationships, (b), normally reflecting the author's interpretation of his series of observations, represented as smooth curves or straight lines fitted to the data by eye or by formulas chosen empirically.

3. Theoretical relationships, (c), in which the curves are smooth and without point designations, though observed values may be plotted to compare them with a theoretical curve if desired. The curve thus drawn is based on theoretical considerations only, in which a theoretical function (equation) is used to compute values for the curve.

## 23.3 Design and Layout of Rectangular Coordinate Line Graphs

The steps in drawing a typical coordinate line graph are shown in Fig. 23.4.

I.　a. Compute and/or assemble data in a convenient arrangement.

b. Select the type of graph and coordinate paper most suitable, §23.4.

c. Determine the size of the paper and locate the axes.

d. Determine the variable for each axis and choose the appropriate scales, §23.5. Letter the unit values along the axes, §23.5.

II.　Plot the points representing the data and draw the curve or curves, §23.6.

III.　Identify the curves by lettering names or symbols, §23.6. Letter the title, §23.7. Ink in the graph, if desired.

The completed graph, which includes the curves, captions, and designations, should have a balanced arrangement relative to the axes. Much of the foregoing procedure is also applicable to the other forms of charts, graphs, and diagrams discussed in this chapter.

## 23.4 Grids and Composition

To simplify the plotting of values along the perpendicular axes and to eliminate the use of a special scale to locate them, *coordinate paper*, or "graph paper," ruled with grids, is generally used and can

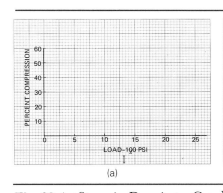 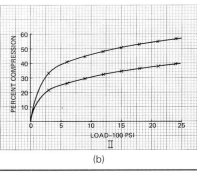 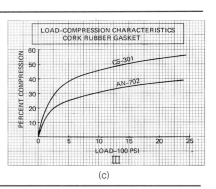

*Fig. 23.4*　Steps in Drawing a Graph.

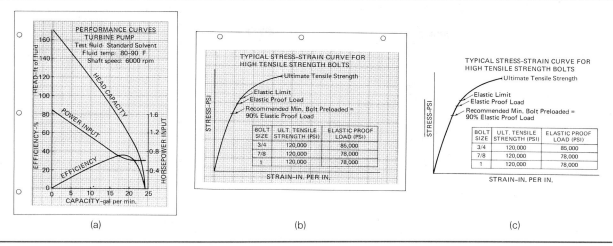

**Fig. 23.5** Printed and Prepared Coordinate Paper.

be purchased already printed. Alternatively, the grids can be drawn on blank paper.

Printed coordinate papers are available in various sizes and spacings of grids, $8\frac{1}{2}'' \times 11''$ being the most common paper size. The spacing of grid lines may be $\frac{1}{10}''$, $\frac{1}{20}''$, or multiples of $\frac{1}{16}''$. A spacing of $\frac{1}{8}''$ to $\frac{1}{4}''$ is preferred. Closely spaced coordinate ruling is generally avoided for publications, charts reduced in size, and charts used for slides. Much of engineering graphical analysis, however, requires (1) close study, (2) interpolation, and (3) only one copy, with possibly a few prints that can be readily prepared with little effort. Therefore, engineering graphs are usually plotted on the closely spaced, printed coordinate paper. Printed papers can be obtained in several colors of lines and in various weights and grades. A thin, translucent paper may also be used when prints are required. A special nonreproducible-grid coordinate paper is available for use when reproductions without visible grids are desired.

Scale values and captions should be placed outside the grid axes, if possible. Since printed papers do not have sufficient margins to accommodate the axes and nomenclature, the axes should be placed far enough inside the grid area to permit sufficient space for axes and lettering, as shown in Fig. 23.5 (a) and (b). As much of the remaining grid space as possible should be used for the curve—that is, the scale should be such as to spread the curve out over the available space. A title block (and tabular data, if any) should be placed in an open space on the chart, as shown. If only one copy of the chart is

required, tabular data may be placed on the back of the graph or on a separate sheet.

Charts prepared for printed publications, conferences, or projection (slides) generally do not require accurate or detailed interpolation and should emphasize the major facts presented. For such graphs, coordinate grids drawn or traced on blank paper, cloth, or film have definite advantages when compared to printed paper. The charts in Fig. 23.5 (b) and (c) show the same information plotted on printed coordinate paper and plain paper, respectively. The specially prepared sheet should have as few grid rulings as necessary—or none, as at (c)—to allow a clear interpretation of the curve. For ease of reading, lettering is not placed upon grid lines. The title and other data can be lettered in open areas, completely free of grid lines.

The layout of specially prepared grids is restricted by the overall paper size required, or space limitations for slides and other considerations. Space is first provided for margins and for axes nomenclature; the remaining space is then divided into the number of grid spaces needed for the range of values to be plotted. Another important consideration of composition that affects the spacing of grids is the slope or trend of the curve, as discussed in §23.5.

Since independent variable values are generally placed along the horizontal axis, especially in time-series charts, vertical rulings can be made for each value plotted, if uniformly spaced, Fig. 23.6 (a). If there are many values to plot, intermediate values can be designated by *ticks* on the curves or along the axis, (b) and (c), with the grid rulings omitted.

Fig. 23.6   Vertical Rulings—Specially Prepared Grids.

The horizontal lines are generally spaced according to the available space and the range of values.

The weight of the grid rulings should be thick enough to guide the eye in reading the values, but thin enough to provide contrast and emphasize the curve. The thickness of the lines generally should decrease as the number of rulings increases. As a general rule, as few rulings should be used as possible, but if a large number of rulings is necessary, major divisions should be drawn heavier than the subdivision rulings, for ease of reading.

## 23.5 Scales and Scale Designation

The choice of scale is the most important factor of composition and curve significance. Rectangular coordinate line graphs have values of the two related variables plotted with reference to two mutually perpendicular coordinate axes, meeting at a zero point or origin, Fig. 23.7 (a). The horizontal axis, normally designated as an *x*-axis, is called the *ab-*

*scissa*. The vertical axis is denoted a *y*-axis and is called the *ordinate*. It is common practice to place independent values along the abscissa and the dependent values along the ordinate. For example, if in an experiment at certain time intervals, related values are observed, recorded, or determined, the amount of these values is dependent upon the time intervals (independent or controlled) chosen. The values increase from the point of origin toward the right on the *x*-axis and upward on the *y*-axis.

Mathematical graphs, (b), quite often contain positive and negative values, which necessitates the division of the coordinate field into four quadrants, numbered counterclockwise as shown. Positive values increase toward the right on the *x*-axis and upward on the *y*-axis, from the origin. Negative values increase (negatively) to the left on the *x*-axis and downward on the *y*-axis.

Generally, a full range of values is desirable, beginning at zero and extending slightly beyond the largest value, to avoid crowding. The available co-

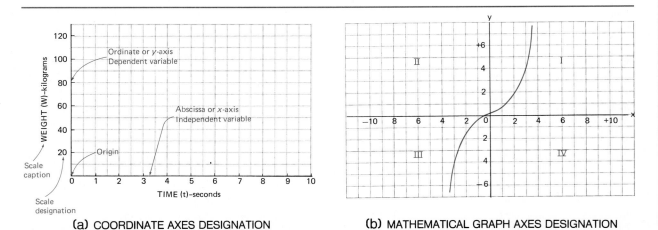

Fig. 23.7   Axes Designation.

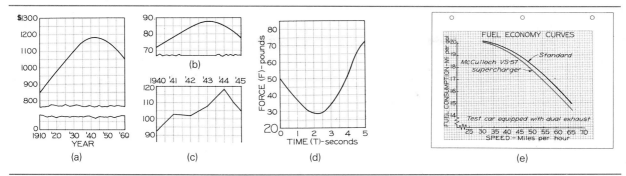

**Fig. 23.8** Axes Scale "Breaks."

ordinate area should be used as completely as possible. However, certain circumstances require special consideration to avoid wasted space. For example, if the values to be plotted along one of the axes do not range near zero, a "break" in the grid may be shown, as in Fig. 23.8 (a). However, when relative amount of change is required, as it is in Fig. 23.5 (a), the axes or grid should not be broken, and the zero line should not be omitted. If the absolute amount is the important consideration, the zero line may be omitted, as in Fig. 23.8 (b), (c), and (d). Time designations of years naturally are fixed and have no relation to zero.

If a few given values to be plotted are widely separated in amount from the others, the total range may be very great, and when this is compressed to fit on the sheet, the resulting curve will tend to be "flat," as shown in Fig. 23.9 (a). It is best to arrange for such values to fall off the sheet and to indicate them as "freak" values. The curve may then be drawn much more satisfactorily, as shown at (b).

A convenient manner in which to show related curves having the same units along the abscissa, but different ordinate units, is to place one or more sets of ordinate units along the left margin and another set of ordinate values along the right margin, as shown in Fig. 23.5 (a), using the same rulings. Multiple scales are also sometimes established along the abscissa, such as for time units of months covering multiple years, as in Fig. 23.6 (b). A more compact arrangement for the curve is shown in Fig. 23.9 (c), where the purpose is to compare the inventory/sales ratios for 1981 and 1982 on a monthly basis.

The choice of scales deserves careful consideration, since it has a controlling influence on the depicted rate of change of the dependent variable. The *slope* of the curve (trend) should be chosen to represent a true picture of the data or a correct impression of the trend.

The slope of a curve is affected by the spacing of the rulings and their designations. A slope or trend can be made to appear "steeper" by increasing the ordinate scale or decreasing the abscissa scale, Fig. 23.10 (a), and "flatter" by increasing the abscissa

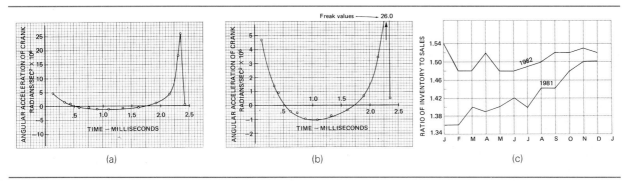

**Fig. 23.9** "Freak" Values and Combined Curves.

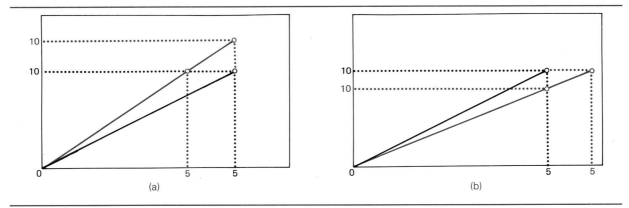

*Fig. 23.10*   Slopes.

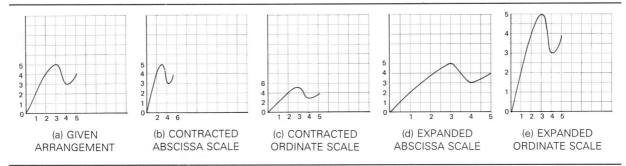

(a) GIVEN ARRANGEMENT     (b) CONTRACTED ABSCISSA SCALE     (c) CONTRACTED ORDINATE SCALE     (d) EXPANDED ABSCISSA SCALE     (e) EXPANDED ORDINATE SCALE

*Fig. 23.11*   Effects of Scale Designation.

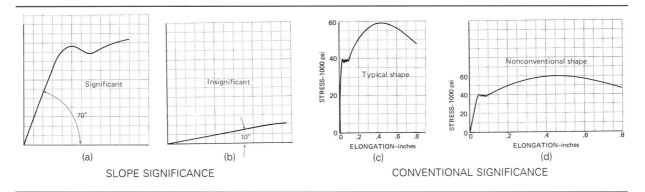

SLOPE SIGNIFICANCE          CONVENTIONAL SIGNIFICANCE

*Fig. 23.12*   Curve Shapes.

scale or decreasing the ordinate scale, (b). As shown in Fig. 23.11, a variety of slopes or shapes can be obtained by expanding or contracting the scales. A deciding factor is the impression desired to be conveyed graphically.

Normally, an angle greater than 40° with the horizontal gives an impression of a significant rise or increase of ordinate values, while an angle of 10° or less suggests an insignificant trend, Fig. 23.12 (a)

and (b). *The slope chosen should emphasize the significance of the data plotted.* Some relationships are customarily presented in a conventional shape, as shown at (c) and (d). In this case, an expanded abscissa scale, as shown at (d), should be avoided.

Scale designations should be placed outside the axes, where they can be shown clearly. Abscissa nomenclature is placed along the axis so that it can be read from the bottom of the graph. Ordinate values

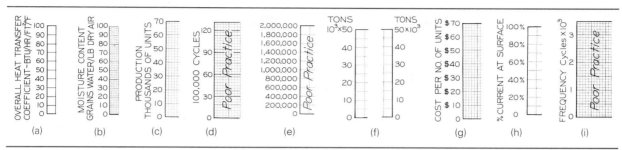

Fig. 23.13 Scale Designations.

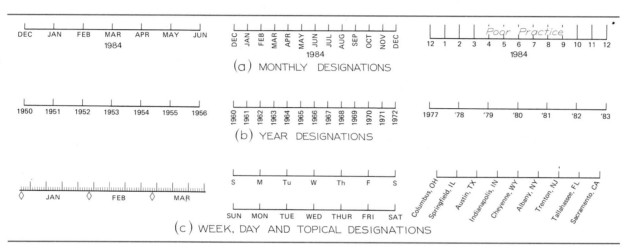

Fig. 23.14 Nonnumerical Designations.

are generally lettered so that they can also be read from the bottom; but ordinate captions, if lengthy, are lettered to be read from the right. The values can be shown on both the right and left sides of the graph, or along the top and bottom, if necessary for clearness, as when the graph is exceptionally wide or tall or when the rulings are closely spaced and hard to follow. When the major interest (e.g., maximum or minimum values) is situated at the right, the ordinate designations may be placed along the right, Fig. 23.33 (e). This arrangement also encourages reading the chart first and then the scale magnitudes.

When grid rulings are specially prepared on blank paper, every major division ruling should have its value designated, Fig. 23.13 (a). The labeled divisions should not be closer than .25″ and rarely more than 1″ apart. Intermediate values (rulings or ticks) should not be identified and should be spaced no closer than .05″. If the rulings are numerous and close together, as on printed graph paper, only the major values are noted, (b) and (c). The assigned values should be consistent with the minor divisions. For example, major divisions designated as 0, 5, 10, . . ., should not have 2 or 4 minor intervals, since resulting values of 1.25, 2.5, 3.75, . . ., are undesirable. Similarly, odd-numbered major divisions of 3, 5, 7, . . ., or multiples of odd numbers with an even number of minor divisions, should be avoided, as (d) indicates. The numbers, if three digits or smaller, can be fully given. If the numbers are larger than three digits, (e), dropping the ciphers is recommended, if the omission is indicated in the scale caption, as at (c). Values are shortened to even hundreds, thousands, or millions, in preference to tens of thousands, for example. Graphs for technical use can have the values shortened by indicating the shortened number times some power of 10, as at (f). In special cases, such as when giving values in dollars or percent, the symbols may be given adjacent to the numbers, as at (g) and (h).

Designations other than numbers usually require additional space; therefore, standard abbreviations should be used when available, Fig. 23.14.

These abscissa values may be lettered vertically, as in the center at (a) and (b), or inclined, as at the right in (c), to fit the designations along the axes.

Scale captions (or titles) should be placed along the scales so that they can be read from the bottom for the abscissa, and from the right for the ordinate. Captions include the name of the variable, symbol (if any), units of measurement, and any explanation of digit omission in the values. If space permits, the designations are lettered completely, but if necessary, standard abbreviations may be used for the units of measurement. Notations such as shown in Fig. 23.13 (i) should be avoided, since it is not clear whether the values shown are to be multiplied by the power of 10 or already have been. Short captions may be placed above the values, Fig. 23.13 (f), especially when graphs are prepared for projection slides, since reading from the right is difficult.

## 23.6 Points, Curves, and Curve Designations

In mathematical and popular-appeal graphs, curves without designated points are commonly used, since the purpose is to emphasize the general significance of the curves. On graphs prepared from observed data, as in laboratory experiments, points are usually designated by various symbols, Fig. 23.15. If more than one curve is plotted on the same grid, a combination of these symbols may be used, one type for each curve, although labels are preferable if clear. The use of open-point symbols is recommended, except in cases of "scatter" diagrams, Fig. 23.39, where the filled-in points are more visible. In general, filled-in symbols should be used only when more than three curves are plotted on the same graph and a different identification is required for each curve.

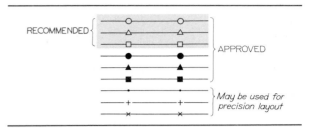

**Fig. 23.15**    Point Symbols.

The curve should not be drawn through the point symbols as they may be needed for reference later for additional information.

When several curves are to be plotted on the same grid, they can be distinguished by the use of various types of lines, Fig. 23.16 (a). However, solid lines are used for the curves wherever possible, while the dashed line is commonly used for *projections* (estimated values, such as future expectations), as shown at (b). The curve should be heavier in weight than the grid rulings, but a difference in weight can also be made between various curves to emphasize a preferred curve or a total value curve (sum of two or more curves), as shown at (c). A *key*, or *legend,* should be placed in an isolated portion of the grid, preferably enclosed by a border, to denote point symbols or line types that are used for the curves. If the grid lines are drawn on blank paper, a space should be left vacant for this information, Fig. 23.17. Keys may be placed off the grids below the title, if space permits. However, it is preferable to designate curves with labels, if possible, rather than letters, numbers, or keys, Fig. 23.2 (c). Colored lines are very effective for distinguishing the various curves on a grid, but they may not be suitable for multiple copies.

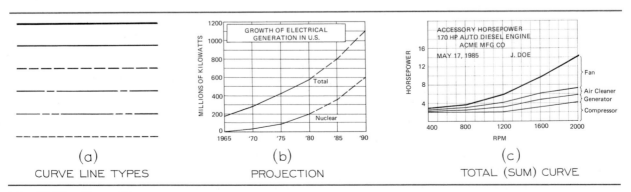

**Fig. 23.16**    Curve Lines.

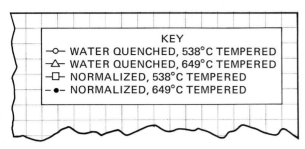

*Fig. 23.17*  Keys.

## 23.7 **Titles**

Titles for a graph may be placed on or off the grid surface. If placed on the grid, white space should be left for the title block, but if printed coordinate paper is used, a heavy border should enclose the title block. If further emphasis is desired, the title may be underlined. The contents of title blocks vary according to method of presentation. Typical title blocks include title, subtitle, institution or company, date of preparation, and name of the author, Fig. 23.16 (c). Some relationships may be given an appropriate name. For example, a number of curves showing the performance of an engine are commonly entitled "Performance Characteristics." If two variables plotted do not have a suitable title, "Dependent variable (name) vs. independent variable (name)" will suffice. For example, GOVERNOR PRESSURE vs SPEED.

Notes, when required, may be placed under the title for general information, Fig. 23.18 (a); labeled adjacent to the curve, (b), or along the curve, Fig. 23.20 (a); or referred to by means of reference symbols, Fig. 23.18 (c) and (d).

Any chart can be made more effective, whether it is drawn on blank paper or upon printed paper, if it is inked. For reproduction purposes, as for slides or for publications, inking is necessary.

## 23.8 **Semilogarithmic Coordinate Line Charts**

A semilog chart, also known as a *rate-of-change* or *ratio chart*, is a type in which two variables are plotted on semilogarithmic coordinate paper to form a continuous straight line or curve. Semilog paper contains uniformly spaced vertical rulings and logarithmically spaced horizontal rulings.

Semilog charts have the same advantages as rectangular coordinate line charts (arithmetic charts), §23.2. When rectangular coordinate line charts give a false impression of the trend of a curve, the semilog charts would be more effective in revealing whether the rate of change is increasing, decreasing, or constant. Semilog charts are also useful in the derivation of empirical equations.

Semilog charts, like rectangular coordinate line graphs, are not recommended for presenting only a few plotted values in a series, for emphasizing change in absolute amounts, or for showing extreme or irregular movement or trend of data.

In Fig. 23.19 (a) and (b), data are plotted on rectangular coordinate grids (arithmetic) and on semilogarithmic coordinate grids, respectively. The same data, which produce curves on the arithmetic graph, produce straight lines on the semilog grid. The straight lines permit an easier analysis of the trend or movements of the variables. If the logarithms of the ordinate values are plotted on a rectangular coordinate grid, instead of the actual values, straight lines will result on the arithmetic graph, as shown at (c). The straight lines produced on semilog grid provide a simple means of deriving empirical equations. Straight lines are not necessarily obtained on a semilog grid, but if they do occur, it means that the rate of change is constant, Fig. 23.20 (a). Irregular curves can be compared to constant-rate scales individually or between a series of curves, as shown at (b).

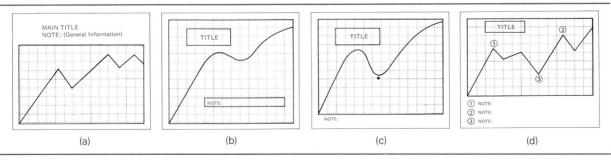

(a)  (b)  (c)  (d)

*Fig. 23.18*  Notes.

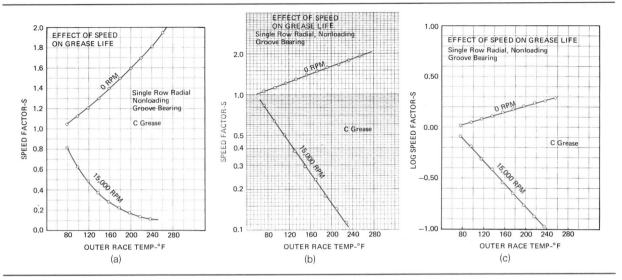

*Fig. 23.19*   Arithmetic and Semilogarithmic Plottings.

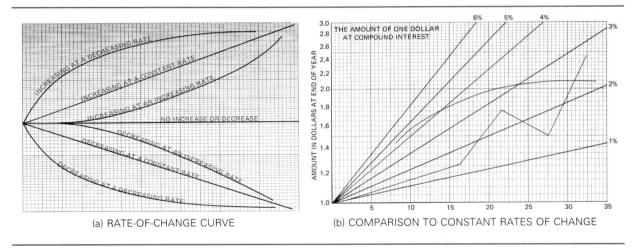

*Fig. 23.20*   Rates of Change.

## 23.9 Logarithmic Coordinate Line Charts

Logarithmic charts have two variables plotted on a logarithmic coordinate grid to form a continuous line or a "curve." Printed logarithmic paper contains logarithmically spaced horizontal and vertical rulings. As in the case of semilog charts, paper containing as many as five cycles on an axis can be purchased.

Log charts are applicable for the comparison of a large number of plotted values in a compact space and for the comparison of the relative trends of sev-

eral curves on the same chart. This form of graph is not the best form for presentation of relatively few plotted values in a series or for emphasizing change in absolute amounts. The designation of log cycles, however, permits the plotting of very extensive ranges of values.

Logarithmic charts are primarily used to determine empirical equations by fitting a single straight line to a series of plotted points. They are also used to obtain straight-line relationships when the data are suitable, as in Fig. 23.21.

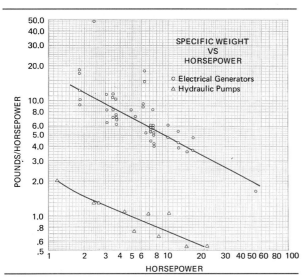

*Fig. 23.21* Logarithmic Chart.

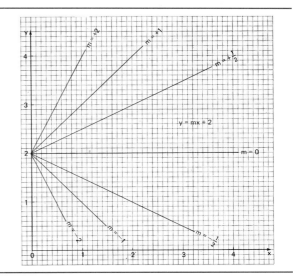

*Fig. 23.22* Equation of Straight Lines on Rectangular Grid.

## 23.10 Empirical Equations

Empirical equations by definition are equations derived from experimental data or experience, as distinguished from equations derived from logical reasoning or hypothesis (rational equations). At times, tabulated data or the analysis of a graph is inadequate, and an equation for the data is required. A graphical plot shows the trend of the data and the value of one variable relative to the corresponding second variable. The derivation of empirical equations is a more comprehensive method of analysis and can be used to calculate additional data not obtained in experimentation.

The derivations of equations are varied in methods. A basic procedure is to plot the data on rectangular, semilogarithmic, or logarithmic coordinate graph paper in an attempt to obtain a straight line. If the plot results in a reasonably straight line on one of these papers, an approximate (empirical) equation can be derived by geometric and algebraic methods. The reader is referred to texts in the school or local library for methods of derivation and forms of empirical equations.

## 23.11 Empirical Equations—Solution by Rectangular Coordinates

The equation for a straight line on a rectangular coordinate grid is $y = mx + b$. As shown in Fig. 23.22, $m$ represents the slope of the line (the tangent of the angle between the line and the $x$-axis) and $b$ is the intercept on the $y$-axis (when $x = 0$). A negative slope is inclined downward to the right. A positive slope has an upward trend to the right. The intercept may be positive or negative.

## 23.12 Empirical Equations—Semilog Coordinates: $y = b(10^{mx})$ or $y = be^{mx}$

Data plotted on rectangular coordinate paper, which do not result in a straight line, may rectify to an approximate straight-line graph on a semilogarithmic coordinate grid, if the rectangular coordinate curve resembles an exponential curve, as shown in Fig. 23.23 (a) or (b). The base of the exponent may be either $e$ ($e = 2.718$) or 10. An exponential curve intersects one of the axes at a steep angle and is asymptotic to the other axis. The same data used to plot some of the curves in Fig. 23.23 (b), when plotted on semilog coordinate paper, rectify to straight lines, as shown in Fig. 23.24.

Since semilog paper has logarithmic divisions along one of the axes (normally designated the $y$-axis), the equation for a straight line on this type of graph paper is

$$\log y = mx + \log b$$

or

$$\ln y = mx + \ln b$$

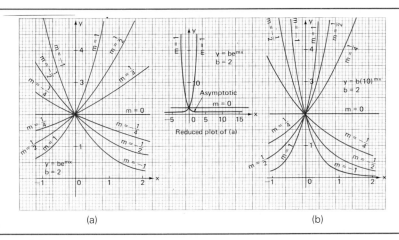

(a)                                 (b)

*Fig. 23.23*    Exponential Curves.

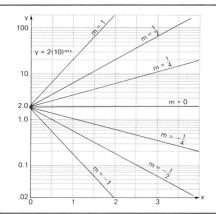

*Fig. 23.24*    Semilog Plot.

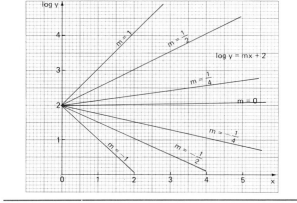

*Fig. 23.25*    Rectangular Grid Solution.

therefore,

$$y = b(10^{mx}) \qquad \text{or} \qquad y = be^{mx}$$

The derivation of the empirical equation requires the solution for the values $b$ ($y$-axis intercept, when $x = 0$) and $m$ (the slope of the straight line to the $x$-axis).

An alternate method is to plot log $y$ values on rectangular coordinate graph paper, if semilog coordinate paper is not readily available, Fig. 23.25.

When a rectangular coordinate plot results in an $x$-axis intercept and a curve appearing asymptotic to the $y$-axis, Fig. 23.26, the semilog plotting may rectify to a straight line if the logarithmic scale is placed on the $x$-axis. The equation becomes

$$x = a(10^{my}) \qquad \text{or} \qquad x = ae^{my}$$

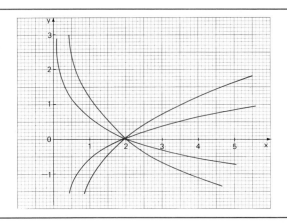

*Fig. 23.26*    Reverse Plot.

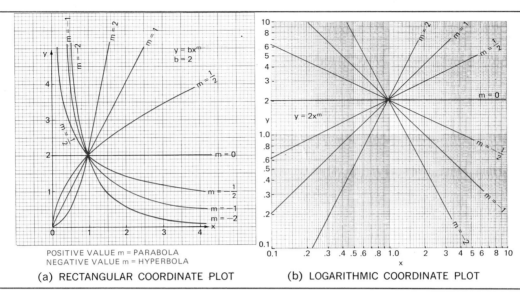

*Fig. 23.27* Power Curves.

## 23.13 Empirical Equations—Logarithmic Coordinates: $y = bx^m$

Curves that plot as a parabola through the origin or a hyperbola asymptotic to the $x$- and $y$-axes are known as power curves, Fig. 23.27 (a), and can be rectified to a straight line by plotting the same data on logarithmic coordinate paper, as shown at (b). The equation for a straight line on logarithmic paper is

$$\log y = m \log x + \log b$$

or

$$y = bx^m$$

The derivation of an empirical equation requires the determination of the values for the slope $m$ of the line and the intercept value $b$ ($y$ intercept when $x = 1$). The intercept is at the axis value of 1.0 since $\log 1.0 = 0$.

## 23.14 Trilinear Coordinate Line Charts
(Fig. 23.28)
Trilinear charts have three related variables plotted on a coordinate paper in the form of an equilateral triangle. The points joined together successively form a continuous straight line or "curve."

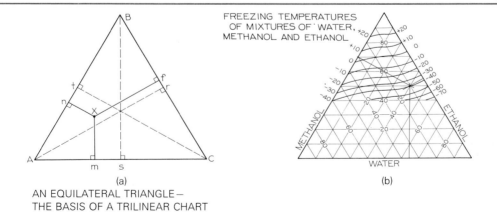

*Fig. 23.28* Trilinear Coordinate Line Charts.

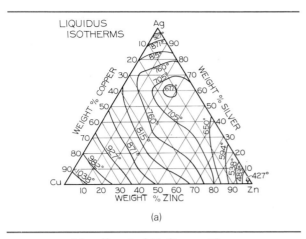

(a)

*Fig 23.29*　Metallurgical Trilinear Chart.

Trilinear charts are particularly suited for the following uses.

1.　Comparison of three related variables relative to their total composition (100 percent).

2.　Analysis of the composition structure by a combination of curves—for example, metallic microstructure of trinary alloys, Fig. 23.29.

3.　Emphasizing change in amount or differences between values.

The trilinear chart is *not* recommended for (1) emphasizing movement or trend of data or (2) comparing three related but dissimilar physical quantities—for example, force, acceleration, and time.

Trilinear charts are widely applied in the metallurgical and chemical fields because of the frequency of three variables in metallurgical and chemical composition. The basis of application is the geometric principle that the sum of the perpendiculars to the three sides from any point within an equilateral triangle is equal to the altitude of the triangle.

In an equilateral triangle ABC, Fig. 23.28 (a), the sum of the distances X–f, X–m, and X–n from the point X within the triangle is equal to the altitude A–r, B–s, or C–t of the triangle. For example, if the distances X–f, X–m, and X–n are, respectively, 50, 30, and 20 units, the altitude of the triangle is 100 units, and the point X will represent a quantity composed of 50, 30, and 20 parts of the three variables. At (b) is shown a chart for various freezing temperatures, with the mixture proportions by volume of water, methanol, and ethanol required. For example, a freezing temperature of $-40°F$ can be established by mixing 50 parts of water with 10 parts of ethanol and 40 parts of methanol.

## 23.15 Polar Coordinate Line Charts

Polar charts have two variables, one a *linear* magnitude and the other an *angular* quantity, plotted on a polar coordinate grid with respect to a pole (origin) to form a continuous line or "curve."

Polar charts are particularly applicable to

1.　Comparing two related variables, one being a linear magnitude (called a *radius vector*) and the second an angular value

2.　Indicating movement or trend or location with respect to a pole point

Polar charts are *not* suited for (1) emphasizing changes in amounts or differences between values or (2) interpolating intermediate values.

As shown in Fig. 23.30 (a), the zero degree line is the horizontal right axis. To locate a point P, it is necessary to know the radius vector r, and an angle

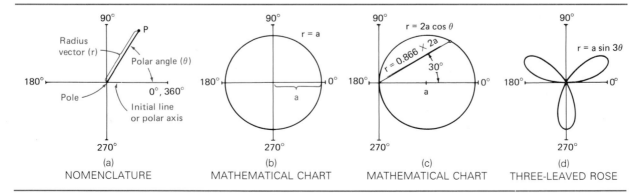

*Fig. 23.30*　Polar Coordinate Charts.

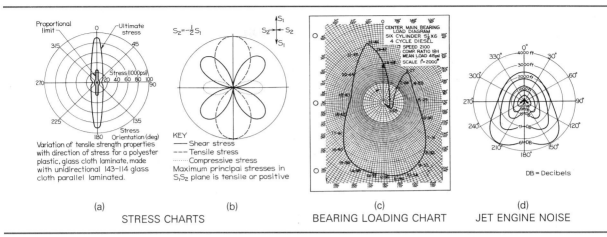

**Fig. 23.31** Polar Coordinate Charts. *(a) and (b) adapted from Charts by Robert L. Stedfield and F. W. Kinsman, respectively, with permission of* Machine Design. *(c) adapted from R. R. Slaymaker,* Bearing Lubrication Analysis, *copyright 1955 by John Wiley and Sons, with permission of the publisher and Clevite Corporation. (d) adapted from chart by G. S. Schairer, with permission of author and Society of Automotive Engineers.*

$\theta$ (e.g., 5, 70°). The point P could also be denoted as (5, 430°), (5, −290°), (−5, 250°), and (−5, −110°), for example. If we plot the equation r = a (no angular designation), we will obtain a circle with the center at the pole, as shown at (b). The value a is a constant value, which determines the relative size of the radius vector and the curve. The equation r = 2a cos $\theta$ produces a circle going through the pole point, with its center on the polar axis, (c). The plot of r = a sin 3$\theta$ produces a "three-leaved rose," as shown at (d). Polar charts also have many practical applications concerned with the magnitude of some values and their location with respect to a pole point. For example, Fig. 23.31 (a) and (b) illustrate stress charts from experimentation and for stress visualization, respectively. At (c) is shown a polar chart of a bearing load diagram, and the graph at (d) indicates the noise distribution from a jet engine.

## 23.16 Nomographs or Alignment Charts
Nomographs or alignment charts consist of straight or curved scales, arranged in various configurations so that a straight line drawn across the scales intersects them at values satisfying the equation represented.

Alignment charts can be used for analysis, but the predominant application is for computation.

Some of the more common forms of nomographs

are shown in Fig. 23.32. Basically, a nomograph is used to solve a three-variable equation. A straight line (*isopleth*) joining known or given values of two of the variables intersects the scale of the third variable at a value that satisfies the equation represented, as at (a), (c), (f), (g), and (h). For this reason they are also called *alignment charts*. All alignment charts are nomographs, but not all nomographs are alignment charts; a rectangular coordinate graph can be classified as a nomograph. Two or more such charts can sometimes be combined to solve an equation containing more than three variables as at (b), (d), and (e). The forms shown have fixed scales and a movable alignment line. However, movable-scale nomographs can be designed with a fixed direction alignment line, the slide rule being an example of this form.

Although alignment charts require time to construct, they have considerable popular appeal for the following reasons.

1. They save time when it is necessary to make repeated calculations of certain numerical relationships (equations).

2. They enable one unskilled or lacking a background in mathematics to handle analytical solutions.

3. When constructed properly, they are limited to the scale values for which the equation is valid.

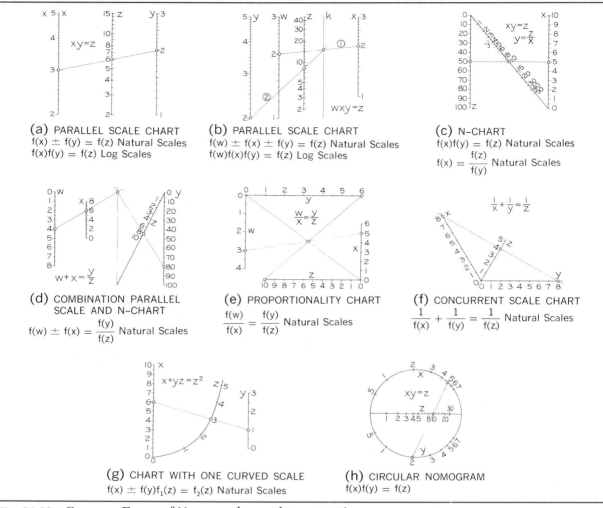

**(a)** PARALLEL SCALE CHART
$f(x) \pm f(y) = f(z)$ Natural Scales
$f(x)f(y) = f(z)$ Log Scales

**(b)** PARALLEL SCALE CHART
$f(w) \pm f(x) \pm f(y) = f(z)$ Natural Scales
$f(w)f(x)f(y) = f(z)$ Log Scales

**(c)** N-CHART
$f(x)f(y) = f(z)$ Natural Scales
$f(x) = \dfrac{f(z)}{f(y)}$ Natural Scales

**(d)** COMBINATION PARALLEL SCALE AND N-CHART
$f(w) \pm f(x) = \dfrac{f(y)}{f(z)}$ Natural Scales

**(e)** PROPORTIONALITY CHART
$\dfrac{f(w)}{f(x)} = \dfrac{f(y)}{f(z)}$ Natural Scales

**(f)** CONCURRENT SCALE CHART
$\dfrac{1}{f(x)} + \dfrac{1}{f(y)} = \dfrac{1}{f(z)}$ Natural Scales

**(g)** CHART WITH ONE CURVED SCALE
$f(x) \pm f(y)f_1(z) = f_2(z)$ Natural Scales

**(h)** CIRCULAR NOMOGRAM
$f(x)f(y) = f(z)$

*Fig. 21.32* Common Forms of Nomographs or Alignment Charts.

This prevents the use of the equation with values that are not applicable.

### 23.17 Bar or Column Charts

Bar charts are graphic representations of numerical values by lengths of bars, beginning at a base line, indicating the relationship between two or more related variables.

Bar charts are particularly suited for the following purposes.

1. Presentation for the nontechnical reader.

2. A simple comparison of two values along two axes.

3. Illustration of relatively few plotted values.

4. Representation of data for a total period of time in comparison to point data.

Bar or column charts are *not* recommended for (1) comparing several series of data or (2) plotting a comparatively large number of values.

The bar chart is effective for nontechnical use because it is most easily read and understood. Therefore, it is used extensively by newspapers, magazines, and similar publications. The bars may be placed horizontally or vertically; when they are placed vertically, the presentation is sometimes called a column chart.

Bar charts are plotted with reference to two mutually perpendicular coordinate axes, similar to those in rectangular coordinate line charts. The graph may be a simple bar chart with two related

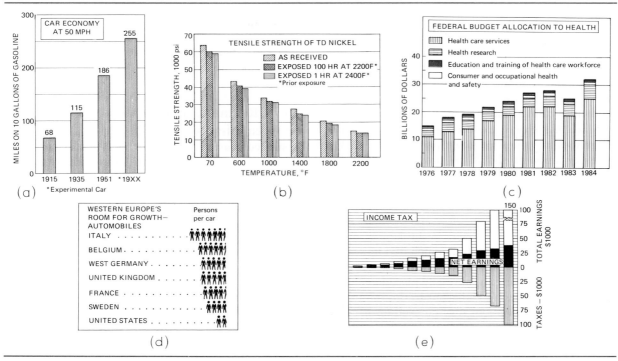

**Fig. 23.33** Bar or Column Charts.

variables, Fig. 23.33 (a); a grouped bar chart (three or more related variables), (b); or a combined bar chart (three or more related variables), (c). Another type of bar graph employs pictorial symbols, composed to form bars, as shown at (d). Bar charts can also effectively indicate a "deviation" or difference between values, (e).

## 23.18 Design and Layout of Bar Charts

If only a few values are to be represented by vertical bars, the chart should be higher than wide, Fig. 23.33 (a). When a relatively large number of values

are plotted, a chart wider than high is preferred, (b) and (c). Composition is dictated by the number of bars used, whether they are to be vertical or horizontal, and the available space.

A convenient method of spacing bars is to divide the available space into twice as many equal spaces as bars are required, Fig. 23.34 (a). Center the bars on every other division mark, beginning with the first division at each end, as shown at (b). When the series of data is incomplete, the missing bars should be indicated by the use of ticks, (c), indicating the lack of data. The bars should be spaced uniformly when the data used are distributed. When irregu-

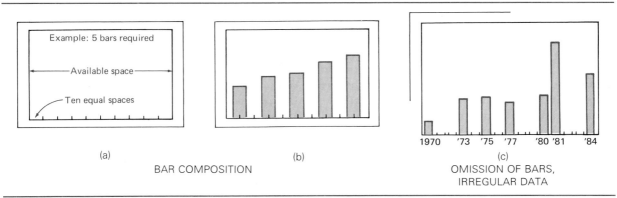

**Fig. 23.34** Bar Composition.

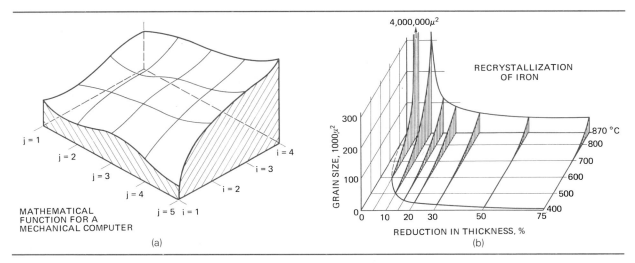

**Fig. 23.37**  Volume Charts. *(a) adapted from drawing by Eugene W. Pike and Thomas R. Silverberg, with permission of* Machine Design.

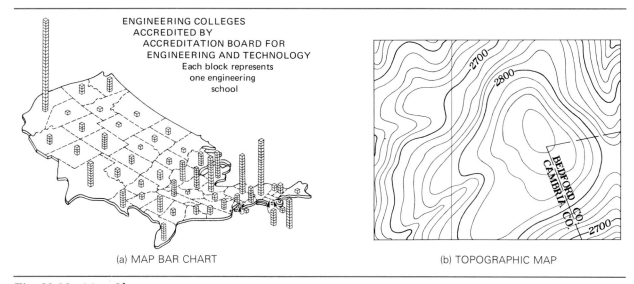

(a) MAP BAR CHART          (b) TOPOGRAPHIC MAP

**Fig. 23.38**  Map Charts.

direction), all drawn in one plane of the drawing paper, as shown at (b).

## 23.23 Rectangular Coordinate Distribution Charts

When the data observed or obtained vary greatly, they can be plotted on a rectangular coordinate grid for the purpose of observing the distribution or areas of major concentration, with no attempt to fit a curve, Fig. 23.39. Charts of this nature are commonly called *scatter diagrams*.

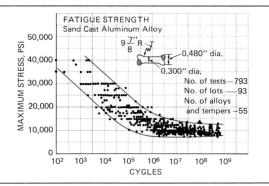

**Fig. 23.39**  "Scatter" Diagram. *Courtesy of Product Engineering*

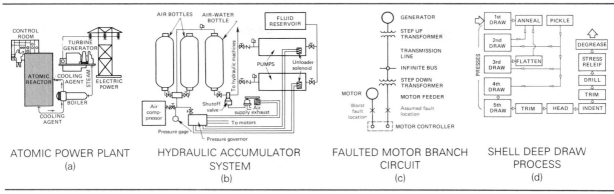

ATOMIC POWER PLANT
(a)

HYDRAULIC ACCUMULATOR
SYSTEM
(b)

FAULTED MOTOR BRANCH
CIRCUIT
(c)

SHELL DEEP DRAW
PROCESS
(d)

**Fig. 23.40** Flowcharts. *(b) adapted from chart by A. F. Welsh, with permission of* Machine Design.

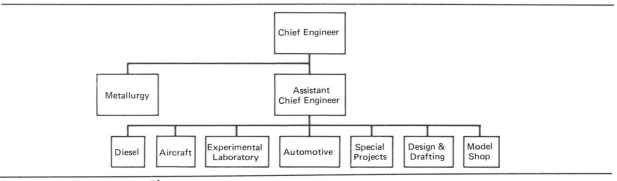

**Fig. 23.41** Organization Chart.

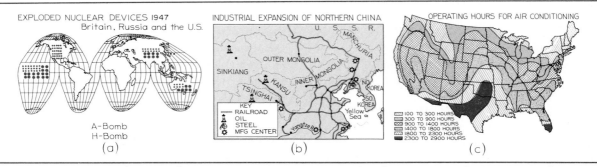

**Fig. 23.42** Map Distribution Charts. *(a), (b), and (c) courtesy of* Look *Magazine,* Life *Magazine, and* Heating, Piping and Air Conditioning, *respectively.*

## 23.24 Flowcharts

Flowcharts are predominantly schematic representations of the flow of a process—for example, manufacturing production processes and electric or hydraulic circuits, Fig. 23.40. Pictorial forms may be used as shown at (a). Schematic symbols are also used, if applicable, (b) and (c). Blocks with captions are used in the simplest form of flowchart, as illustrated at (d).

*Organization charts* are similar to flowcharts, except that they are usually representations of the arrangement of personnel and physical items of specific organizations, Fig. 23.41.

## 23.25 Map Distribution Charts

When it is desired to present data according to geographical distribution, maps are commonly used. Locations and emphasis of data can be shown by dots of various sizes, Fig. 23.42 (a); by the use of symbols, (b); by shading of areas, (c); and by the use of numbers or colors.

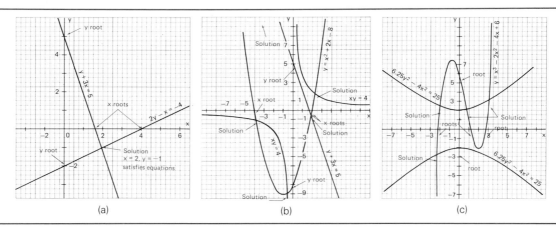

**Fig. 23.43**  Rectangular Coordinate Graphs—Algebraic Solutions.

## 23.26 Graphical Mathematics

In general, mathematical problems may be solved *algebraically* (using numerical and mathematical symbols) or *graphically* (using drawing techniques). The algebraic method is predominantly a verbal approach in comparison to the visual methods of graphics; therefore, errors and discrepancies are more evident and subject to detection in a graphical presentation. The advantages of graphics are quite evident in any mathematics text, since most writers in the field of mathematics supplement their algebraic notations with graphical illustrations to illuminate their writings and improve the visualization of the solutions.

The graphical methods cannot be used exclusively, but neither can the algebraic methods be used to the fullest degree of effectiveness without the use of graphics. The engineer should be familiar with both methods in order to convey a clearer understanding of his analyses and designs.

Since equations are mathematical expressions, mathematical operations can be (1) computations with equations, (2) derivation of equations, and (3) solutions of particular equations.

The reader is referred to texts in the school or local library for detailed methods of graphic solutions of particular equations for algebraic and calculus problems.

## 23.27 Rectangular Coordinate Algebraic Solution Graphs

Rectangular coordinate paper can be effectively used to visualize algebraic solutions. A common application is to solve two simultaneous polynomial equations with two unknowns. A graph of two linear equations, Fig. 23.43 (a), gives the solution at the intersection of the two lines. Linear equations are simple to solve algebraically; however, a graphical approach is advantageous in the solution of a quadratic equation for a parabola ($y = x^2 + 2x - 8$) with a linear equation for a straight line ($y + 3x = 5$) or an equation for an equilateral hyperbola ($xy = 4$), Fig. 23.43 (b). The graphical plot can also be used for further analysis. Note that the linear equation and the equation for the equilateral hyperbola do not have a real simultaneous equation solution. The ease of visualization and solution is evident as the equations become more complex. Figure 23.43 (c) is a graphical solution for the cubic equation $y = x^3 - 2x^2 - 4x + 6$ and the equation for a hyperbola $6.25y^2 - 4x^2 = 25$. If the data were empirical, without an equation, the graphical solution would be necessary.

Another graphic application to algebraic problems is the determination of the roots for the equations. The roots of the equation are determined by the intersection of the curve and the corresponding axis. The linear equation of Fig. 23.43 (a) has one root per variable, as denoted. The highest power of the quadratic equation for the x term of Fig. 23.43 (b) being 2, there are two roots on the x-axis. The y term has only one root. The equilateral hyperbola, in the same graph, is asymptotic to both axes and therefore has no real roots. The symmetrical hyperbola of Fig. 23.43 (c) has one plus and one minus root. A cubic equation can have one or three real roots, depending on the location of the curve relative to the axes. The cubic equation of Fig. 23.43 (c) has three roots.

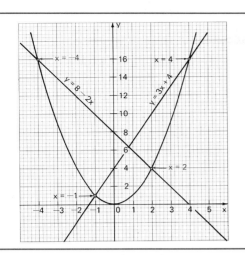

**Fig. 23.44**   Graphic Algebra Solutions.

A more convenient procedure to determine roots of an equation is to separate the equation and plot two separate curves. The quadratic equation of Fig. 23.43 (b), $y = x^2 + 2x - 8$, was separated to $y = x^2 = 8 - 2x$. The equation $y = x^2$ results in a symmetrical parabola, Fig. 23.44. The remaining portion of the quadratic equation, $y = 8 - 2x$, is a linear equation. The intersection of the parabola and the straight line provides the same root solution for the quadratic equation as in Fig. 23.43 (b). The parabolic curve can be used with other linear portions of similar quadratic equations to provide the respective root solutions on the same plot, (i.e., $y = 3x + 4$).

## 23.28 The Graphical Calculus

If two variables are so related that the value of one of them depends on the value assigned the other, then the first variable is said to be a function of the second. For example, the area of a circle is a function of the radius. *The calculus* is that branch of mathematics pertaining to the change of values of functions due to finite changes in the variables involved. It is a method of analysis called the *differential calculus* when concerned with the determination of the *rate of change* of one variable of a function with respect to a related variable of the same function. A second principal operation, called *the integral calculus*, is the inverse of the differential calculus and is defined as a *process of summation* (finding the total change).

## 23.29 The Differential Calculus

Fundamentally, the differential calculus is a means of determining for a given function the limit of the ratio of change in the dependent variable to the corresponding change in the independent variable, as this change approaches zero. This limit is the *derivative* of the function with respect to the independent variable. The derivative of a function $y = f(x)$ may be denoted by $dy/dx$, $f'(x)$, $y'$, or $D_x y$.

## 23.30 Geometric Interpretation of the Derivative

As illustrated in Fig. 23.45,

1.  The symbol $f(x)$ is used to denote a function of a single variable $x$ and is read "$f$ of $x$."

2.  The value of the function when $x$ has the value $x_0$ may be denoted by $f(x_0)$ or $F(x_0)$.

3.  The increment $\Delta y$ (read: "delta $y$") is the change in $y$ produced by increasing $x$ from $x_0$ to $x_0 + \Delta x$; therefore, $\Delta y = f(x_0 + \Delta x) - f(x_0)$.

4.  The differential, $dy$, of $y$ is the value that $\Delta y$ would have if the curve coincided with its tangent. The differential, $dx$, of $x$ is the same as $\Delta x$, when $x$ is the independent variable.

5.  The slope of the secant line is represented by the ratio $\Delta y/\Delta x$.

6.  Ratio $dy/dx$ represents the slope of the tangent line.

Draw a secant through $A$ (coordinates $x$, $y$) and a neighboring point $B$ on the curve (coordinates

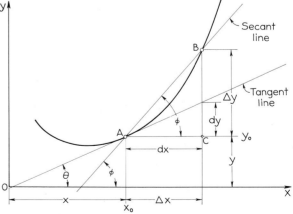

**Fig. 23.45**   Geometric Interpretation of the Derivative.

$x + \Delta x$, $y + \Delta y$). The slope of the secant line through $A$ and $B$ is

$$\frac{BC}{AC} = \frac{(y + \Delta y) - y}{(x + \Delta x) - x} = \frac{\Delta y}{\Delta x}$$
$$= \text{tangent of angle } BAC = \text{tangent } \phi$$

Let point $B$ move along the curve and approach point $A$ indefinitely. The secant line will revolve about point $A$, and its limiting position is the tangent line at $A$, when $\Delta x$, varying, approaches zero as a limit. The angle of inclination, $\theta$, of the tangent line at $A$ becomes the limit of the angle $\phi$ when $\Delta x$ approaches zero. Therefore,

$$\tan \phi = \tan \theta = \frac{dy}{dx}$$
$$= \text{slope of tangent line at } A$$

The result of this analysis is the following theorem:

> The value of the derivative at any point of a curve is equal to the slope of the tangent line to the curve at that point.

Accordingly, graphical differentiation is a process of drawing tangents or the equivalent (chords parallel to tangents) to a curve at any point and determining the value of a differential (value of the slope of the tangent line to the curve or to a parallel chord, at the point selected). The value of the slope is the corresponding ordinate value on the derived curve.

## 23.31 Graphical Differentiation— The Slope Law

The slope of the curve at any point is the tangent of the angle ($\theta$) between the $x$-axis of the graph and the tangent line to the curve at the selected point, Fig. 23.46. The slope is also the rise or fall of the tangent line in the $y$ direction per one unit of travel in the $x$ direction, the value of the slope being calculated in terms of the scale units of the $x$- and $y$-axes.

The *slope law* as applied to differentiation may be stated as follows.

> The slope at any point on a given curve is equal to the ordinate of the corresponding point on the next lower derived curve.

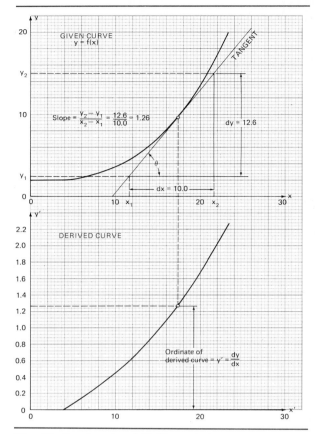

**Fig. 23.46** Graphical Differentiation.

The slope law is the graphical equivalent of differentiation of the calculus.

## 23.32 Area Law

Given a curve, Fig. 23.47 (upper portion), a tangent to the arc is to be constructed at $T$. A length of arc $AB$ is chosen so that $T$ is at the midpoint of the length of arc. A chord is constructed through $A$ and $B$, and the tangent is drawn through $T$ parallel to the chord. The coordinates of $A$ are $x_1$, $y_1$, and of $B$ are $x_2$, $y_2$. Since the chord and tangent are parallel, their slopes are equal. The slope of the tangent line at $T$ is equal to the mean ordinate $y'_m$ of $T'$ in the derived curve

$$y'_m = \frac{y_2 - y_1}{x_2 - x_1}$$

and

$$y_2 - y_1 = y_m'(x_2 - x_1)$$

but

$$y_m'(x_2 - x_1) = \text{area of rectangle } CDx_2x_1$$

Therefore

$$y_2 - y_1 = \text{area of rectangle } CDx_2x_1$$

The law derived from this analysis, the *area law*, may be stated as follows:

> The difference in the length of any two ordinates to a continuous curve equals the total net area between the corresponding ordinates of the next lower curve.

> The application of the area law as stated, which provides for dividing the given curve into short arcs, permits the determination of the derivative curve for the given curve. The law is also applicable to the process of integration, as discussed in §23.35.

## 23.33 Practical Applications— Differentiation

The application of the calculus, especially by graphical methods, is based on the practicality and knowledge of the subject matter. Table 23.1 summarizes only a few of the practical applications for the differential calculus. The derived curve also shows maximum and minimum values for the derivative and whether the rate is constant, variable, or zero.

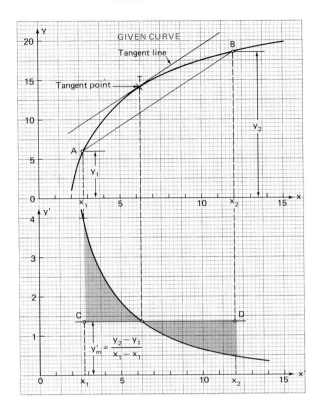

*Fig. 23.47*   Area Law.

**Table 23.1**   *Differentiation Applications*

| Plotted Data | | Derivative of Derived Curve |
| --- | --- | --- |
| **Independent** | **Dependent** | |
| Time | Displacement | Velocity |
| Time | Velocity | Acceleration |
| Time | Amount of | Rate of |
| | Population | Growth |
| | Inches | Growth |
| | Volume | Flow |
| | Temperature | Cooling, heating |
| Time | Energy or work | Power |
| Displacement | Energy or work | Force |
| Quantity | Cost | Rate of cost |
| Variable #2 | Related variable #1 | Rate of change of variable #1 as variable #2 changes |

## 23.34 The Integral Calculus

*Integration* is a process of summation, the integral calculus having been devised for the purpose of calculating the areas bounded by curves. If the given area is assumed to be divided into an infinite number of infinitesimal parts called *elements*, the sum of all these elements is the total area required. The integral sign ∫, the long S, was used by early writers and is still used in calculations to indicate "sum."

One of the most important applications of the integral calculus is the determination of a function from a given derivative, the inverse of differentiation. The process of determining such a function is *integration*, and the resulting function is called the *integral* of the given derivative or *integrand*. In many cases, however, graphical integration is used to determine the area bounded by curves, in which case the expression of the function is not necessary.

Graphical solutions may be classified into three general groups: graphical, semigraphical, and mechanical methods.

## 23.35 Graphical Solution—Area Law

Since integration is the inverse of differentiation, the area law as analyzed for differentiation, §23.32, is applicable to integration, but in reverse. The area law as applied to integration may be stated as follows.

> The area between any two ordinates of a given curve is equal to the difference in length between the corresponding ordinates of the next higher curve.

If an $\frac{acceleration}{time}$ curve were given, the integral of this derivative curve would be the $\frac{velocity}{time}$ curve. If the $\frac{velocity}{time}$ curve is integrated as shown in Fig. 23.48, the $\frac{displacement}{time}$ curve is obtained (the next higher curve).

## 23.36 Semigraphical Integration

When it becomes necessary to determine only an area and not an integral curve, many semigraphical methods are applicable, some of which are the rectangular rule, trapezoid rule, Simpson's rule, Durand's rule, Weddle's rule, method of Gauss, and the method of parabolas. All of these rules and methods may be found in texts at the school or local library. Simpson's rule is one of the more accurate methods. The solutions from these methods become

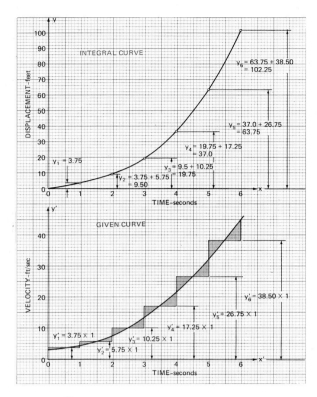

**Fig. 23.48**    Integration—Area Law.

partly graphical, since the data used are measured from the graphical plot.

## 23.37 Integration—Mechanical Methods

When the requirement for integration is the determination of the area, mechanical integrators called *planimeters* may be used. The operator manually traces the outline of the area, and the instrument automatically records on a dial the area circumnavigated.

A common type of planimeter is the polar planimeter, Fig. 23.49 (b). As illustrated at (a), by means of the polar arm $OM$, one end $M$ of the tracer arm is caused to move in a circle, and the other end $N$ is guided around a closed curve bounding the area measured. The area $M_1N_1NN_2M_2MM_1$ is "swept out" twice, but in opposite directions. The resulting displacement reading (difference in the two sweeps) on the integrating wheel indicates the amount of the area. The circumference of the wheel is graduated

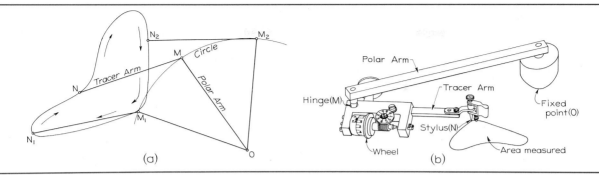

*Fig. 23.49*   Polar Planimeter.

so that one revolution corresponds to a definite number of square units of area.

The ordinary planimeter used to measure indicator diagrams has a length $L = 4''$, and a wheel circumference of 2.5″, so that one revolution of the wheel is $4 \times 2.5'' = 10$ sq in. The wheel is graduated into ten parts, each part being further subdivided into ten parts, and a vernier scale facilitates a further subdivision into ten parts, enabling a reading to the nearest hundredth of a square inch.

This same operation of mechanical integration can be performed on a typical CAD software program. Any geometric shape (closed configuration) entered into a computer will compute the area of that configuration, which will be presented upon the proper command.

### 23.38 Practical Applications of Integration

The definition of integration as being a summation process and the reverse operation of differentiation is not sufficient for a working knowledge of the subject matter. Table 23.2 summarizes a few of the practical applications for the integral calculus.

**Table 23.2**   *Integration Applications*

| Plotted Data | | Integral of Integration Curve |
| --- | --- | --- |
| **Independent** | **Dependent** | |
| Time | Acceleration | Velocity |
| Time | Velocity | Displacement |
| $y$ values | $x$ values | Area of $X$–$Y$ plot enclosed by $X$-axis and/or $Y$-axis and the curve |
| Area | Pressure | Force |
| Displacement | Force | Work or energy |
| Time | Power | Work or energy |
| Volume | Pressure | Work or energy |
| Time | Force | Momentum |
| Velocity | Momentum | Impulse |
| Time | Rate based on time | Total change or cumulative change |
| Quantity #2 | Rate of change of quantity #1 based on change of quantity #2 | Max. or min. quantity #1 for values of quantity #2 |

## 23.29 Computer Graphics

The graphic presentations discussed in this chapter have all been constructed using conventional drafting equipment. Graphs, charts, and diagrams can also be produced by using a modern computer graphics system. Most major computer manufacturers as well as independent software companies have developed graphics software programs to use with computers. These programs make it possible to store information and data that can be accessed when needed to create drawings and other graphic images in color or black and white. Text material can also be incorporated with graphics so that reports can be prepared, revised, and published in a reasonably short period of time. The images created may be viewed on a graphic display device known as a cathode ray tube (CRT), Fig. 23.50, and a record or copy of these images may be produced by an output device such as the plotter/printer shown in Fig. 23.51. For additional information, see Chapters 3 and 8.

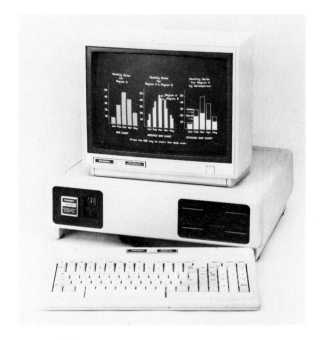

*Fig. 23.50*   Graphic Display Device. *Courtesy of Radio Shack.*

*Fig. 23.51*   Graphic Output Device. *Courtesy of Radio Shack.*

# *Graph Problems*

Construct an appropriate form of graph for each set of data listed. The determination of graph form (line, bar, surface, etc.) is left to the discretion of the instructor or the student and should be based on the nature of the data or the form of presentation desired. In some cases, more than one curve or more than one series of bars are required.

Since many of the problems in this chapter are of a general nature, they can also be solved on most computer graphics systems. If a system is available, the instructor may choose to assign specific problems to be completed by this method.

**Prob. 23.1**   Medicine for Profit: A Healthy Market (revenues generated by for-profit acute-care hospitals)

| Year | Gross Revenue (billions of dollars) |
|------|-------------------------------------|
| 1969 | 0.75 |
| 1970 | 1.00 |
| 1971 | 1.25 |
| 1972 | 1.60 |
| 1973 | 2.00 |
| 1974 | 2.50 |
| 1975 | 3.00 |
| 1976 | 3.50 |
| 1977 | 4.00 |
| 1978 | 4.50 |
| 1979 | 5.25 |
| 1980 | 6.25 |
| 1981 | 7.30 |
| 1982 | 9.00 |
| 1983 | 10.50[a] |
| 1984 | 13.00[a] |

[a]Estimated.

**Prob. 23.2**   Metal Hardness Comparison (Mohs' hardness scale)

| Metal | Comparative Degree of Hardness (diamond = 10) |
|-------|-----------------------------------------------|
| Lead | 1.5 |
| Tin | 1.8 |
| Cadmium | 2.0 |
| Zinc | 2.5 |
| Gold | 2.5 |
| Silver | 2.7 |
| Aluminum | 2.9 |
| Copper | 3.0 |
| Nickel | 3.5 |
| Platinum | 4.3 |
| Iron | 4.5 |
| Cobalt | 5.5 |
| Tungsten | 7.5 |
| Chromium | 9.0 |
| Diamond | 10.0 |

**Prob. 23.3**   Effect of Accuracy of Gear Manufacture on Available Strength in Terms of Horsepower (60 teeth gear and 30 teeth pinion of 6 diametral pitch, 1.5″ face width, $14\frac{1}{2}°$ pressure angle; 500 Brinell Case hardness)

| Pitch Line Velocity, ft/min | Horsepower | | | |
|-----------------------------|-------------|------------------------|------------------------|--------------------------|
| | Perfect Gear | Aircraft Quality Gear | Accurate Quality Gear | Commercial Quality Gear |
| 0 | 0 | 0 | 0 | 0 |
| 1000 | 140 | 100 | 75 | 50 |
| 2000 | 290 | 180 | 105 | 60 |
| 3000 | Straight-line | 250 | 120 | 70 |
| 4000 | curve through | 320 | 130 | 72 |
| 5000 | two points | 380 | 140 | 72 |

Noise limit: 2750 ft/min—accurate quality gear; 1400 ft/min—commercial quality gear.

*Prob. 23.4*    U.S. Population

| Year | Population |
|------|-----------|
| 1800 | 5,308,483 |
| 1810 | 7,239,881 |
| 1820 | 9,638,453 |
| 1830 | 12,866,020 |
| 1840 | 17,069,453 |
| 1850 | 23,191,876 |
| 1860 | 31,443,321 |
| 1870 | 38,558,371 |
| 1880 | 56,155,783 |
| 1890 | 62,947,714 |
| 1900 | 75,994,575 |
| 1910 | 96,977,266 |
| 1920 | 105,710,620 |
| 1930 | 122,775,046 |
| 1940 | 131,669,275 |
| 1950 | 151,325,798 |
| 1960 | 179,323,175 |
| 1970 | 203,302,031 |
| 1980 | 226,545,805 |
| 1990 | 249,000,000[a] |
| 2000 | 267,000,000[a] |

[a]Estimated.

*Prob. 23.5*    U.S. Pedestrian Deaths, 1975–1979

| Age | Number of Deaths Female | Number of Deaths Male | Age | Number of Deaths Female | Number of Deaths Male |
|-----|--------|------|-----|--------|------|
| 1 | 500 | 800 | 47 | 225 | 600 |
| 3 | 575 | 1000 | 50 | 225 | 600 |
| 5 | 600 | 1040 | 52 | 225 | 575 |
| 7 | 400 | 800 | 55 | 250 | 500 |
| 8 | 300 | 625 | 57 | 250 | 550 |
| 10 | 325 | 500 | 60 | 225 | 575 |
| 12 | 425 | 625 | 62 | 275 | 575 |
| 15 | 400 | 1100 | 65 | 250 | 500 |
| 17 | 375 | 1075 | 67 | 290 | 525 |
| 20 | 300 | 1000 | 70 | 300 | 525 |
| 22 | 275 | 900 | 72 | 350 | 475 |
| 25 | 200 | 750 | 75 | 340 | 450 |
| 27 | 225 | 700 | 77 | 300 | 475 |
| 30 | 200 | 600 | 80 | 300 | 400 |
| 32 | 175 | 500 | 82 | 225 | 375 |
| 35 | 150 | 450 | 85 | 150 | 300 |
| 37 | 160 | 475 | 87 | 100 | 200 |
| 40 | 175 | 475 | 90 | 50 | 150 |
| 42 | 200 | 500 | 92 | 25 | 50 |
| 45 | 175 | 500 | 95 | 0 | 0 |

*Prob. 23.6*    World Population

| Year | Population (billions) |
|------|----------------------|
| 1950 | 2.51 |
| 1960 | 3.03 |
| 1970 | 3.63 |
| 1980 | 4.42 |
| 1990 | 5.28[a] |
| 2000 | 6.20[a] |

[a]Projections.

*Prob. 23.7*    Worldwide Growth of Engineering Periodicals

| Years | Number of Periodicals |
|-------|----------------------|
| 1921–1930 | 166 |
| 1931–1940 | 215 |
| 1941–1950 | 301 |
| 1951–1960 | 485 |
| 1961–1970 | 705 |
| 1971–1980 | 927 |
| 1981–1985 | 991 |

*Prob. 23.8*    Lumber Prices—Softwoods

| Year | Index (1967 = 100) Mar. | June | Sept. | Dec. |
|------|------|------|-------|------|
| 1980 | 362.4 | 328.6 | 349.4 | 355.7 |
| 1981 | 346.0 | 357.0 | 335.3 | 321.8 |
| 1982 | 318.9 | 328.1 | 320.9 | 322.7 |
| 1983 | 369.4 | 397.9 | 361.4 | 360.3 |

*Prob. 23.9*    U.S. Personal Consumption Expenditures

| Year | Billions of Dollars Durables | Nondurables | Services |
|------|----------|-------------|----------|
| 1965 | 63.0 | 188.6 | 178.7 |
| 1970 | 85.2 | 265.7 | 270.8 |
| 1975 | 132.2 | 407.3 | 437.0 |
| 1980 | 214.7 | 668.8 | 784.5 |

**Prob. 23.10** Industrial Energy Sources

| Year | Percent of Total U.S. Industrial Use of Energy | | | |
|---|---|---|---|---|
| | Coal | Oil | Natural Gas | Electricity |
| 1969 | 23 | 23 | 45 | 9 |
| 1971 | 20 | 24 | 46 | 10 |
| 1973 | 17 | 36 | 39 | 8 |
| 1975 | 16 | 36 | 38 | 10 |
| 1977 | 15 | 39 | 35 | 11 |
| 1979 | 14 | 41 | 33 | 12 |
| 1981 | 15 | 37 | 35 | 13 |
| 1983 | 12[a] | 40[a] | 35[a] | 13[a] |
| 1985 | 13[a] | 38[a] | 35[a] | 14[a] |
| 1987 | 13[a] | 39[a] | 34[a] | 14[a] |

[a]Projections.

**Prob. 23.11** Psychological Analysis of Work Efficiency and Fatigue

| Hours of Work | Relative Production Index | |
|---|---|---|
| | Heavy Work | Light Work |
| 9–10 A.M. | 100 | 96 |
| 10–11 A.M. | 108 | 104 |
| 11–12 A.M. | 104 | 104 |
| 12–1 P.M. | 98 | 103 |
| Lunch | | |
| 2–3 P.M. | 103 | 100 |
| 3–4 P.M. | 99 | 102 |
| 4–5 P.M. | 98 | 101 |
| After 8-hour day | | |
| 5–6 P.M. | 91 | 94 |
| 6–7 P.M. | 86 | 93 |
| 7–8 P.M. | 68 | 83 |

**Prob. 23.12** The World Economy—Industrial Production

| Year | Index (1980 = 100) | | | | | |
|---|---|---|---|---|---|---|
| | United Kingdom | France | U.S. | West Germany | Italy | Japan |
| 1976 | 94 | 90 | 86 | 89 | 80 | 77 |
| 1977 | 100 | 96 | 90 | 94 | 92 | 84 |
| 1978 | 101 | 94 | 94 | 95 | 85 | 85 |
| 1979 | 104 | 99 | 103 | 98 | 95 | 92 |
| 1980 | 106 | 102 | 103 | 102 | 102 | 100 |
| 1981 | 96 | 98 | 102 | 98 | 99 | 99 |
| 1982 | 98 | 97 | 98 | 97 | 98 | 102 |
| 1983 | 100 | 95 | 93 | 93 | 94 | 101 |
| 1984 | 105 | 99 | 108 | 99 | 95 | 112 |

Plot on semilog shows rate of change of index.

**Prob. 23.13** Essential Qualities of a Successful Engineer (average estimate based on 1500 questionnaires from practicing engineers)

| Quality | Percent |
|---|---|
| Character | 41 |
| Judgment | $17\frac{1}{2}$ |
| Efficiency | $14\frac{1}{2}$ |
| Understanding human nature | 14 |
| Technical knowledge | 13 |
| Total | 100 |

**Prob. 23.14** Automobile Accident Analysis

| Type of Accident | Percent |
|---|---|
| Cross traffic (grade crossing, highway, railway) | 21 |
| Same direction | 30 |
| Head-on | 21 |
| Fixed object | 11 |
| Pedestrian | 10 |
| Miscellaneous | 7 |
| Total | 100 |

**Prob. 23.15**   Comparison of Horsepower at the Rear Wheels (as shown by dynamometer tests)

| Engine rpm | Horsepower at Rear Wheels | | | |
| --- | --- | --- | --- | --- |
| | McCulloch Supercharged with Dual Exhausts | McCulloch Supercharged | Unsupercharged —Dual Exhausts | Unsupercharged |
| 2000 | 77.0 | 73.0 | 64.5 | 59.5 |
| 2200 | 82.0 | 77.5 | 70.5 | 65.0 |
| 2400 | 88.0 | 83.5 | 75.0 | 69.0 |
| 2600 | 95.5 | 91.5 | 79.0 | 73.5 |
| 2800 | 105.0 | 99.0 | 82.5 | 76.5 |
| 3000 | 112.5 | 105.5 | 84.5 | 78.5 |
| 3200 | 117.0 | 109.5 | 85.5 | 79.5 |
| 3400 | 119.0 | 111.5 | 83.5 | 77.0 |
| 3600 | 118.5 | 111.5 | | |

**Prob. 23.16**   How the World Uses Its Work Force: Employment by Economic Sector, 1980

| Country | Percent of Workers | | | |
| --- | --- | --- | --- | --- |
| | Agriculture[a] | Mining and Construction | Manufacturing | Services[b] |
| United States | 3.6 | 7.2 | 22.1 | 67.1 |
| Canada | 5.4 | 7.7 | 19.7 | 67.2 |
| Australia | 6.5 | 9.1 | 19.9 | 64.5 |
| Japan | 10.1 | 10.1 | 25.0 | 54.8 |
| France | 8.7 | 9.3 | 25.8 | 56.2 |
| Great Britain | 2.7 | 8.3 | 28.4 | 60.6 |
| Italy | 14.2 | 11.2 | 26.9 | 47.7 |
| Netherlands | 6.0 | 9.6 | 21.3 | 63.1 |
| Sweden | 5.6 | 7.2 | 24.3 | 62.9 |

[a]Agriculture, forestry, hunting, and fishing.
[b]Transportation, communication, public utilities, trade, finance, public administration, private household services, and miscellaneous services.

Plot the data for Problems 23.17 and 23.18 on rectangular coordinate paper and on semilog paper.

**Prob. 23.17**   Rupture Strength of T. D. Nickel—High-Temperature Alloy

| Temperature $(T)$,°F | 100-hr Rupture Stress $(s_r)$, psi × 1000 (log scale) |
| --- | --- |
| 1200 | 24 |
| 1400 | 17 |
| 1600 | 12.5 |
| 1800 | 9 |
| 2000 | 6.5 |
| 2200 | 4.75 |
| 2400 | 3.5 |

**Prob. 23.18**  Nuisance Noise

| Frequency ($f$), cps (log scale) | Octave Band Level, db | | | |
| --- | --- | --- | --- | --- |
| | Hearing Loss Risk Region | | Power Lawn Mower at 3 ft | 5-hp Chainsaw at 3 ft |
| | Negligible | Serious | | |
| 53 | 104 | 122 | 84 | 93 |
| 106 | 93 | 113 | 93 | 103 |
| 220 | 87 | 107 | 94 | 103 |
| 425 | 85 | 105 | 90 | 111 |
| 850 | 85 | 105 | 84 | 112 |
| 1700 | 85 | 105 | 84 | 107 |
| 3400 | 85 | 105 | 82 | 104 |
| 6800 | 85 | 105 | 75 | 98 |

Plot the data for the following problems on rectangular coordinate paper and on logarithmic paper.

**Prob. 23.19**  Loss of Head for Water Flowing in Iron Pipes

| Velocity ($v$), ft/sec | Loss of Head, ft/1000 ft | | | |
| --- | --- | --- | --- | --- |
| | 1″ Pipe | 2″ Pipe | 4″ Pipe | 6″ Pipe |
| 0 | 0 | 0 | 0 | 0 |
| 1 | 6.0 | 2.9 | 1.6 | .7 |
| 2 | 23.5 | 9.5 | 4.2 | 2.4 |
| 3 | 50 | 20 | 8.2 | 4.7 |
| 4 | | 34 | 13.5 | 7.7 |
| 5 | | 51 | 20 | 11.4 |
| 6 | | | 28 | 15.5 |
| 7 | | | 37 | 20 |

**Prob. 23.20**  Material and Process Economics

| Weight of Steel Forging ($W$), lb | Unit Cost ($C$) for Forging, $ | | |
| --- | --- | --- | --- |
| | Simple | Average | Complex |
| 0.1 | 0.22 | | |
| 0.2 | 0.38 | | |
| 0.4 | 0.67 | | |
| 1.0 | 1.4 | | |
| 2.0 | 2.5 | 2.75 | 4.5 |
| 4.0 | 4.4 | 5.0 | 8.5 |
| 10.0 | 9.2 | 11.0 | 19.5 |
| 20.0 | 16.0 | 20.0 | 37.0 |
| 40.0 | 28.5 | 36.5 | 70.0 |
| 100 | 60.0 | 80.0 | 165.0 |

**Prob. 23.21**  Pressurized Square Tubing, Fig. 23.52.

| Orientation, degrees | Stress Ratio ($\sigma_1/P$) | |
| --- | --- | --- |
| | $D/a = 0.80$ | $D/a = 0.86$ |
| 0 | 2.0 | 2.6 |
| 15 | 3.2 | 4.3 |
| 30 | 4.2 | 6.3 |
| 45 | 4.3 | 5.0 |
| 60 | 4.2 | 6.3 |
| 75 | 3.2 | 4.3 |
| 90 | 2.0 | 2.6 |

Data are given for one quadrant; quadrants are identical.

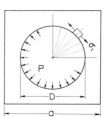

**Fig. 23.52**  (Prob. 23.21).

***Prob. 23.22*** Variation of Modulus of Elasticity (*E*) with Direction in Copper

| Orientation, degrees | $E \times 10^6$ psi | |
| --- | --- | --- |
| | As Rolled | Annealed |
| 0 | 20.0 | 9.5 |
| 15 | 18.5 | 11.0 |
| 30 | 16.5 | 14.5 |
| 45 | 15.0 | 17.5 |
| 60 | 16.5 | 14.5 |
| 75 | 18.5 | 11.0 |
| 90 | 20.0 | 9.5 |
| 105 | 18.5 | 11.0 |
| 120 | 16.5 | 14.5 |
| 135 | 15.0 | 17.5 |
| 150 | 16.5 | 14.5 |
| 165 | 18.5 | 11.0 |
| 180 | 20.0 | 9.5 |

Direction of rolling is from 0° to 180°.

***Prob. 23.23*** Light Distribution in a Vertical Plane for a Bulb Suspended from the Ceiling with the Filament at the Origin of the Polar Chart

| Orientation, degrees | Candle Power |
| --- | --- |
| 0 | 140 |
| 10 | 210 |
| 20 | 310 |
| 30 | 320 |
| 40 | 310 |
| 50 | 310 |
| 60 | 300 |
| 70 | 290 |
| 80 | 280 |
| 90 | 250 |
| 100 | 270 |
| 110 | 290 |
| 120 | 295 |
| 130 | 300 |
| 140 | 315 |
| 150 | 330 |
| 160 | 340 |
| 170 | 350 |
| 180 | 340 |

***Prob. 23.24*** Main Bearing Load Diagram (4000 rpm, no counterweight)

| Orientation, degrees | Load | | Orientation, degrees | Load | |
| --- | --- | --- | --- | --- | --- |
| | 1000 lb | (1000 kg) | | 1000 lb | (1000 kg) |
| 0 | 6.6 | (3.0) | 190 | 5.3 | (2.4) |
| 10 | 4.8 | (2.2) | 200 | 5.2 | (2.4) |
| 20 | 4.2 | (1.9) | 210 | 4.9 | (2.2) |
| 30 | 3.7 | (1.7) | 220 | 4.5 | (2.1) |
| 40 | 3.2 | (1.5) | 230 | 4.2 | (1.9) |
| 50 | 3.0 | (1.4) | 240 | 3.7 | (1.7) |
| 60 | 2.8 | (1.3) | 250 | 3.3 | (1.5) |
| 70 | 2.7 | (1.2) | 260 | 3.1 | (1.4) |
| 80 | 2.7 | (1.2) | 270 | 2.9 | (1.3) |
| 90 | 2.8 | (1.3) | 280 | 2.8 | (1.3) |
| 100 | 2.9 | (1.3) | 290 | 2.7 | (1.2) |
| 110 | 3.3 | (1.5) | 300 | 2.8 | (1.3) |
| 120 | 3.6 | (1.6) | 310 | 3.0 | (1.4) |
| 130 | 4.1 | (1.9) | 320 | 3.2 | (1.5) |
| 140 | 4.6 | (2.1) | 330 | 3.7 | (1.7) |
| 150 | 4.9 | (2.2) | 340 | 4.4 | (2.0) |
| 160 | 5.0 | (2.3) | 350 | 5.3 | (2.4) |
| 170 | 5.2 | (2.4) | 360 | 6.6 | (3.0) |
| 180 | 5.3 | (2.4) | | | |

Max. load = 6600 lb (3000 kg); mean load = 4530 lb (2059 kg).

# Appendix

## Contents of Appendix

# 1 Bibliography of American National Standards

American National Standards Institute, 1430 Broadway, New York, N.Y. 10018. For complete listing of standards, see ANSI catalog of American National Standards.

### Abbreviations

Abbreviations for Use on Drawings and in Text, ANSI Y1.1–1972 (R1984)

### Bolts, Screws, and Nuts

Bolts, Metric Heavy Hex, ANSI B18.2.3.6M–1979

Bolts, Metric Heavy Hex Structural, ANSI B18.2.3.7M–1979

Bolts, Metric Hex, ANSI B18.2.3.5M–1979

Bolts, Metric Round Head Short Square Neck, ANSI B18.5.2.1M–1981

Bolts, Metric Round Head Square Neck, ANSI/ASME B18.5.2.2M–1982

Hex Jam Nuts, Metric, ANSI B18.2.4.5M–1979

Hex Nuts, Heavy, Metric, ANSI B18.2.4.6M–1979

Hex Nuts, Slotted, Metric, ANSI B18.2.4.3M–1979

Hex Nuts, Style 1, Metric, ANSI B18.2.4.1M–1979

Hex Nuts, Style 2, Metric, ANSI B18.2.4.2M–1979

Hexagon Socket Flat Countersunk Head Cap Screws (Metric Series), ANSI/ASME B18.3.5M–1986

Mechanical Fasteners, Glossary of Terms, ANSI B18.12–1962 (R1981)

Miniature Screws, ANSI B18.11–1961 (R1983)

Nuts, Metric Hex Flange, ANSI B18.2.4.4M–1982

Plow Bolts, ANSI B18.9–1958 (R1977)

Round Head Bolts, ANSI B18.5–1978

Screws, Hexagon Socket Button Head Cap, Metric Series, ANSI/ASME B18.3.4M–1986

Screws, Hexagon Socket Head Shoulder, Metric Series, ANSI/ASME B18.3.3M–1986

Screws, Hexagon Socket Set, Metric Series, ANSI/ASME B18.3.6M–1986

Screws, Metric Formed Hex, ANSI B18.2.3.2M–1979

Screws, Metric Heavy Hex, ANSI B18.2.3.3M–1979

Screws, Metric Hex Cap, ANSI B18.2.3.1M–1979

Screws, Metric Hex Flange, ANSI/ASME B18.2.3.4M–1984

Screws, Metric Hex Lag, ANSI B18.2.3.8M–1981

Screws, Metric Machine, ANSI B18.6.7M–1985

Screws, Socket Head Cap, Metric Series, ANSI/ASME B18.3.1M–1986

Screws, Tapping and Metallic Drive, Inch Series, Thread Forming and Cutting, ANSI B18.6.4–1981

Slotted and Recessed Head Machine Screws and Machine Screw Nuts, ANSI B18.6.3–1972 (R1983)

Slotted Head Cap Screws, Square Head Set Screws, and

Slotted Headless Set Screws, ANSI B18.6.2–1972 (R1983)

Socket Cap, Shoulder, and Set Screws (Inch Series) ANSI/ASME B18.3–1986

Square and Hex Bolts and Screws, Inch Series, ANSI B18.2.1–1981

Square and Hex Nuts (Inch Series) ANSI/ASME B18.2.2–1987

Track Bolts and Nuts, ANSI/ASME B18.10–1982

Wing Nuts, Thumb Screws, and Wing Screws, ANSI B18.17–1968 (R1983)

Wood Screws, Inch Series, ANSI B18.6.1–1981

### Charts and Graphs

Illustrations for Publication and Projection, ANSI Y15.1M–1979 (R1986)

Process Charts, ANSI Y15.3M–1979 (R1986)

Time Series Charts, ANSI Y15.2M–1979 (R1986)

### Dimensioning and Surface Finish

General Tolerances for Metric Dimensioned Products, ANSI B4.3–1978 (R1984)

Preferred Limits and Fits for Cylindrical Parts, ANSI B4.1–1967 (R1987)

Preferred Metric Limits and Fits, ANSI B4.2–1978 (R1984)

Surface Texture, ANSI/ASME B46.1–1985

### Drafting Manual (Y14)

Sect. 1 Drawing Sheet Size and Format, ANSI Y14.1–1980 (R1987)

Sect. 2 Line Conventions and Lettering, ANSI Y14.2M–1979 (R1987)

Sect. 3 Multi and Sectional View Drawings, ANSI Y14.3–1975 (R1987)

Sect. 4 Pictorial Drawing, ANSI Y14.4–1957 (R1987)

Sect. 5 Dimensioning and Tolerancing, ANSI Y14.5M–1982 (R1988)

Sect. 6 Screw Thread Representation, ANSI Y14.6–1978 (R1987)

Sect. 6a Screw Thread Representation, Metric, ANSI Y14.6aM–1981 (R1987)

Sect. 7.1 Gear Drawing Standards—Part 1, for Spur, Helical, Double Helical, and Rack, ANSI Y14.7.1–1971 (R1988)

Sect. 7.2 Gear and Spline Drawing Standards—Part 2, Bevel and Hypoid Gears, ANSI Y14.7.2–1978 (R1984)

Sect. 13 Mechanical Spring Representation, ANSI Y14.13M–1981 (R1987)

Sect. 15 Electrical and Electronics Diagrams, ANSI Y14.15–1966 (R1988)

Sect. 15a Electrical and Electronics Diagrams—Supplement, ANSI Y14.15a–1971 (R1973)

Sect. 15b Electrical and Electronics Diagrams—Supplement, ANSI Y14.15b–1973

Sect. 17 Fluid Power Diagrams, ANSI Y14.17–1966 (R1987)

Sect. 34 Parts Lists, Data Lists, and Index Lists, ANSI Y14.34M–1982 (R1988)

Sect. 36 Surface Texture Symbols, ANSI Y14.36–1978 (R1987)

## Gears

Design for Fine-Pitch Worm Gearings, ANSI/AGMA 374.04–1977

Gear Nomenclature—Terms, Definitions, Symbols, and Abbreviations, ANSI/AGMA 112.05–1976

Nomenclature of Gear-Tooth Failure Modes, ANSI/AGMA 110.04–1980

System for Straight Bevel Gears, ANSI/AGMA 208.03–1979

Tooth Proportions for Coarse-Pitch Involute Spur Gears, ANSI/AGMA 201.02

Tooth Proportions for Fine-Pitch Involute Spur and Helical Gears, ANSI/AGMA 207.06–1977

## Graphic Symbols (Y32)

Fire-Protection Symbols for Architectural and Engineering Drawings, ANSI/NFPA 172–1986

Fire-Protection Symbols for Risk Analysis Diagrams, ANSI/NFPA 174–1986

Graphic Symbols for Electrical and Electronics Diagrams (Including 1986 Supplement) ANSI/IEEE 315–1975 (Y32.2–1975)

Graphic Symbols for Electrical Wiring and Layout Diagrams Used in Architecture and Building Construction, ANSI Y32.9–1972

Graphic Symbols for Fluid Power Diagrams, ANSI Y32.10–1967 (R1987)

Graphic Symbols for Grid and Mapping Used in Cable Television Systems, ANSE/IEEE 623–1976

Graphic Symbols for Heat-Power Apparatus, ANSI Y32.2.6M–1950 (R1984)

Graphic Symbols for Heating, Ventilating and Air Conditioning, ANSI Y32.2.4–1949 (R1984)

Graphic Symbols for Logic Functions, ANSI/IEEE 91–1984

Graphic Symbols for Pipe Fittings, Valves, and Piping, ANSI Y32.2.3–1949 (R1953)

Graphic Symbols for Plumbing Fixtures for Diagrams Used in Architecture and Building Construction, ANSI Y32.4–1977 (R1987)

Graphic Symbols for Process Flow Diagrams in the Petroleum and Chemical Industries, ANSI Y32.11–1961 (R1985)

Graphic Symbols for Railroad Maps and Profiles, ANSI Y32.7–1972 (R1987)

Instrumentation Symbols and Identification, ANSI/ISA S5.1–1984

Reference Designations for Electrical and Electronics Parts and Equipment, ANSI/IEEE 200–1975

Symbols for Fire Fighting Operations, ANSI/NFPA 178–1986

Symbols for Mechanical and Acoustical Elements as Used in Schematic Diagrams, ANSI Y32.18–1972 (R1985)

Symbols for Welding and Nondestructive Testing, Including Brazing, ANSI/AWS A2.4–86

## Keys and Pins

Clevis Pins and Cotter Pins, ANSI B18.8.1–1972 (R1983)

Hexagon Keys and Bits (Metric Series), ANSI B18.3.2M–1979 (R1986)

Keys and Keyseats, ANSI B17.1–1967 (R1973)

Pins-Taper Pins, Dowel Pins, Straight Pins, Grooved Pins and Spring Pins (Inch Series), ANSI B18.8.2–1978 (R1983)

Woodruff Keys and Keyseats, ANSI B17.2–1967 (R1978)

## Piping

Bronze Pipe Flanges and Flanged Fittings, Class 150 and 300, ANSI B16.24–1979

Cast Bronze Threaded Fittings, Class 125 and 250, ANSI/ASME B16.15–1985

Cast Iron Pipe Flanged and Flanged Fittings, Class 25, 125, 250 and 800, ANSI B16.1–1975

Cast Iron Threaded Fittings, Class 125 and 250, ANSI/ASME B16.4–1985

Ductile Iron Pipe, Centrifugally Cast, in Metal Molds or Sand-Lined Molds for Water or Other Liquids, ANSI/AWWA C151/A21.51–1986

Factory-Made Wrought Steel Buttwelding Fittings, ANSI/ASME B16.9–1986

Ferrous Pipe Plugs, Bushings, and Locknuts with Pipe Threads, ANSI B16.14–1983

Flanged Ductile-Iron and Gray Iron Pipe with Threaded Flanges, ANSI/AWWA C115/A21.15–83

Malleable-Iron Threaded Fittings, Class 150 and 300, ANSI/ASME B16.3–1985

Pipe Flanges and Flanged Fittings, Steel Nickel Alloy and Other Special Alloys, ANSI B16.5–1988

Stainless Steel Pipe, ANSI/ASME B36.19M–1985

Welded and Seamless Wrought Steel Pipe, ANSI/ASME B36.10M–1985

## Rivets

Large Rivets ($\frac{1}{2}$ Inch Nominal Diameter and Larger) ANSI B18.1.2–1972 (R1983)

Small Solid Rivets ($\frac{7}{16}$ Inch Nominal Diameter and Smaller) ANSI B18.1.1–1972 (R1983)

Small Solid Rivets, Metric, ANSI B18.1.3M–1983

### Small Tools and Machine Tool Elements

Jig Bushings, ANSI B94.33–1974 (R1986)
Machine Tapers, ANSI B5.10–1981 (R1987)
Milling Cutters and End Mills, ANSI/ASME B94.19–1985
Reamers, ANSI B94.2–1983 (R1988)
T-Slots—Their Bolts, Nuts and Tongues, ANSI/ASME B5.1M–1985
Twist Drills, Straight Shank and Taper Shank Combined Drills and Countersinks, ANSI B94.11M–1979 (R1987)

### Threads

Acme Screw Threads, ANSI/ASME B1.5–1988
Buttress Inch Screw Threads, ANSI B1.9–1973 (R1985)
Class 5 Interference-Fit Thread, ANSI/ASME B1.12–1987
Dryseal Pipe Threads (Inch), ANSI B1.20.3–1976 (R1982)
Dryseal Pipe Threads (Metric), ANSI B1.20.4–1976 (R1982)
Hose Coupling Screw Threads, ANSI/ASME B1.20.7–1966 (R1983)
Metric Screw Threads for Commercial Mechanical Fasteners—Boundary Profile Defined, ANSI B1.18M–1982 (R1987)
Metric Screw Threads—M Profile, ANSI/ASME B1.13M–1983
Metric Screw Threads—MJ Profile, ANSI B1.21M–1978
Nomenclature, Definitions and Letter Symbols for Screw Threads, ANSI/ASME B1.7M–1984
Pipe Threads, General Purpose (Inch), ANSI/ASME B1.20.1–1983

Stub Acme Threads, ANSI/ASME B1.8–1988
Unified Screw Threads (UN and UNR Thread Form), ANSI B1.1–1982
Unified Miniature Screw Threads, ANSI B1.10–1958 (R1988)

### Washers

Lock Washers, ANSI B18.21.1–1972 (R1983)
Plain Washers, ANSI B18.22.1–1965 (R1981)
Plain Washers, Metric, ANSI B18.22M–1981

### Miscellaneous

Knurling, ANSI/ASME B94.6–1984
Preferred Metric Sizes for Flat Metal Products, ANSI/ASME B32.3M–1984
Preferred Metric Equivalents of Inch Sizes for Tubular Metal Products Other Than Pipe, ANSI/ASME B32.6M–1984
Preferred Metric Sizes for Round, Square, Rectangle and Hexagon Metal Products, ANSI B32.4M–1980 (R1986)
Preferred Metric Sizes for Tubular Metal Products Other Than Pipe, ANSI B32.5–1977 (R1983)
Preferred Thickness for Uncoated Thin Flat Metals (Under 0.250 in.), ANSI B32.1–1952 (R1983)
Surface Texture (Surface Roughness, Waviness and Lay), ANSI/ASME B46.1–1985

## 2  Technical Terms

"The beginning of wisdom is to call things by their right names."
–CHINESE PROVERB

n *means a* noun; v *means a* verb

**acme** (*n*)  Screw thread form, §§15.3 and 15.13
**addendum** (*n*)  Radial distance from pitch circle to top of gear tooth.
**allen screw** (*n*)  Special set screw or cap screw with hexagon socket in head, §15.31.
**allowance** (*n*)  Minimum clearance between mating parts, §14.12.
**alloy** (*n*)  Two or more metals in combination, usually a fine metal with a baser metal.
**aluminum** (*n*)  A lightweight but relatively strong metal. Often alloyed with copper to increase hardness and strength.
**angle iron** (*n*)  A structural shape whose section is a right angle, §12.20.
**anneal** (*v*)  To heat and cool gradually, to reduce brittleness and increase ductility, §12.22.
**arc-weld** (*v*)  To weld by electric arc. The work is usually the positive terminal.

**babbitt** (*n*)  A soft alloy for bearings, mostly of tin with small amounts of copper and antimony.
**bearing** (*n*)  A supporting member for a rotating shaft.
**bevel** (*n*)  An inclined edge, not at right angle to joining surface.
**bolt circle** (*n*)  A circular center line on a drawing, containing the centers of holes about a common center, §13.25.
**bore** (*v*)  To enlarge a hole with a boring mill, Figs. 12.15 (b) and 12.18 (b).
**boss** (*n*)  A cylindrical projection on a casting or a forging.

BOSS

**brass** (*n*)  An alloy of copper and zinc.

**braze** (*v*)   To join with hard solder of brass or zinc.

**Brinell** (*n*)   A method of testing hardness of metal.

**broach** (*n*)   A long cutting tool with a series of teeth that gradually increase in size which is forced through a hole or over a surface to produce a desired shape, §12.15.

**bronze** (*n*)   An alloy of eight or nine parts of copper and one part of tin.

**buff** (*v*)   To finish or polish on a buffing wheel composed of fabric with abrasive powders.

**burnish** (*v*)   To finish or polish by pressure upon a smooth rolling or sliding tool.

**burr** (*n*)   A jagged edge on metal resulting from punching or cutting.

**bushing** (*n*)   A replaceable lining or sleeve for a bearing.

**calipers** (*n*)   Instrument (of several types) for measuring diameters, §12.17.

**cam** (*n*)   A rotating member for changing circular motion to reciprocating motion.

**carburize** (*v*)   To heat a low-carbon steel to approximately 2000°F in contact with material which adds carbon to the surface of the steel, and to cool slowly in preparation for heat treatment, §12.22.

**caseharden** (*v*)   To harden the outer surface of a carburized steel by heating and then quenching.

**castellate** (*v*)   To form like a castle, as a castellated shaft or nut.

**casting** (*n*)   A metal object produced by pouring molten metal into a mold, §12.3.

**cast iron** (*n*)   Iron melted and poured into molds, §12.3.

**center drill** (*n*)   A special drill to produce bearing holes in the ends of a workpiece to be mounted between centers. Also called a "combined drill and countersink," §13.35.

COMBINED DRILL
& C SINK

**chamfer** (*n*)   A narrow inclined surface along the intersection of two surfaces.

CHAMFER

**chase** (*v*)   To cut threads with an external cutting tool.

**cheek** (*n*)   The middle portion of a three-piece flask used in molding, §12.3.

**chill** (*v*)   To harden the outer surface of cast iron by quick cooling, as in a metal mold.

**chip** (*v*)   to cut away metal with a cold chisel.

**chuck** (*n*)   A mechanism for holding a rotating tool or workpiece.

**coin** (*v*)   To form a part in one stamping operation.

**cold-rolled steel** (CRS) (*n*)   Open hearth or Bessemer steel containing .12% to .20% carbon that has been rolled while cold to produce a smooth, quite accurate stock.

**collar** (*n*)   A round flange or ring fitted on a shaft to prevent sliding.

COLLAR

**colorharden** (*v*)   Same as *caseharden*, except that it is done to a shallower depth, usually for appearance only.

**cope** (*n*)   The upper portion of a flask used in molding, §12.3.

**core** (*v*)   To form a hollow portion in a casting by using a dry-sand core or a green-sand core in a mold, §12.3.

**coreprint** (*n*)   A projection on a pattern which forms an opening in the sand to hold the end of a core, §12.3.

**cotter pin** (*n*)   A split pin used as a fastener, usually to prevent a nut from unscrewing, Fig. 15.31 (g) and (h) and Appendix 30.

**counterbore** (*v*)   To enlarge an end of a hole cylindrically with a *counterbore*. §12.16.

COUNTERBORE

**countersink** (*v*)   To enlarge an end of a hole conically, usually with a *countersink*, §12.16.

COUNTERSINK

**crown** (*n*)   A raised contour, as on the surface of a pulley.

**cyanide** (*v*)   To surface-harden steel by heating in contact with a cyanide salt, followed by quenching.

**dedendum** (*n*)   Distance from pitch circle to bottom of tooth space.

**development** (*n*)   Drawing of the surface of an object unfolded or rolled out on a plane.

**diametral pitch** (*n*)   Number of gear teeth per inch of pitch diameter.

**die** (*n*)   (1) Hardened metal piece shaped to cut or form a required shape in a sheet of metal by pressing it against a mating die. (2) Also used for cutting small male threads. In a sense is opposite to a tap.

**die casting** (*n*)   Process of forcing molten metal under pressure into metal dies or molds, producing a very accurate and smooth casting.

**die stamping** (*n*)   Process of cutting or forming a piece of sheet metal with a die.

**dog** (*n*)   A small auxiliary clamp for preventing work from rotating in relation to the face plate of a lathe.

**dowel** (*n*)   A cylindrical pin, commonly used to prevent sliding between two contacting flat surfaces.

DOWEL

**draft** (*n*)   The tapered shape of the parts of a pattern to permit it to be easily withdrawn from the sand or, on a forging, to permit it to be easily withdrawn from the dies, §§12.3 and 12.21.

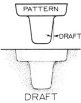

**drag** (*n*)   Lower portion of a flask used in molding, §12.3.

**draw** (*v*)   To stretch or otherwise to deform metal. Also to temper steel.

**drill** (*v*)   To cut a cylindrical hole with a drill. A *blind hole* does not go through the piece, §12.16.

**drill press** (*n*)   A machine for drilling and other hole-forming operations, §12.8.

**drop forge** (*v*)   To form a piece while hot between dies in a drop hammer or with great pressure, §12.21.

**face** (*v*)   To finish a surface at right angles, or nearly so, to the center line of rotation on a lathe.

**FAO**   Finish all over, §13.17.

**feather key** (*n*)   A flat key, which is partly sunk in a shaft and partly in a hub, permitting the hub to slide lengthwise of the shaft, §15.34.

**file** (*v*)   To finish or smooth with a file.

**fillet** (*n*)   An interior rounded intersection between two surfaces, §§7.34 and 12.5.

**fin** (*n*)   A thin extrusion of metal at the intersection of dies or sand molds.

**fit** (*n*)   Degree of tightness or looseness between two mating parts, as a *loose fit*, a *snug fit*, or a *tight fit*. §§13.9 and 14.1 to 14.3.

**fixture** (*n*)   A special device for holding the work in a machine tool, *but not for guiding the cutting tool*.

**flange** (*n*)   A relatively thin rim around a piece.

**flash** (*n*)   Same as *fin*.

**flask** (*n*)   A box made of two or more parts for holding the sand in sand molding, §12.3.

**flute** (*n*)   Groove, as on twist drills, reamers, and taps.

**forge** (*v*)   To force metal while it is hot to take on a desired shape by hammering or pressing, §12.21.

**galvanize** (*v*)   To cover a surface with a thin layer of molten alloy, composed mainly of zinc, to prevent rusting.

**gasket** (*n*)   A thin piece of rubber, metal, or some other material, placed between surfaces to make a tight joint.

**gate** (*n*)   The opening in a sand mold at the bottom of the *sprue* through which the molten metal passes to enter the cavity or mold, §12.3.

**graduate** (*v*)   To set off accurate divisions on a scale or dial.

**grind** (*v*)   To remove metal by means of an abrasive wheel, often made of carborundum. Use chiefly where accuracy is required, §12.14.

**harden** (*v*)   To heat steel above a critical temperature and then quench in water or oil, §12.22.

**heat-treat** (*v*)   To change the properties of metals by heating and then cooling, §12.22.

**interchangeable** (*adj.*)   Refers to a part made to limit dimensions so that it will fit any mating part similarly manufactured, §14.1.

**jig** (*n*)   A device *for guiding a tool* in cutting a piece. Usually it holds the work in position.

**journal** (*n*)   Portion of a rotating shaft supported by a bearing.

**kerf** (*n*)   Groove or cut made by a saw.

**key** (*n*)   A small piece of metal sunk partly into both shaft and hub to prevent rotation, §15.34.

**keyseat** (*n*)   A slot or recess in a shaft to hold a key, §15.34.

**keyway** (*n*)   A slot in a hub or portion surrounding a shaft to receive a key. §15.34.

**knurl** (*v*)   To impress a pattern of dents in a turned surface with a knurling tool to produce a better hand grip, §13.37.

**lap** (*v*)   To produce a very accurate finish by sliding contact with a *lap*, or piece of wood, leather, or soft metal impregnated with abrasive powder.

**lathe** (*n*)   A machine used to shape metal or other materials by rotating against a tool, §12.7.

**lug** (*n*)   An irregular projection of metal, but not round as in the case of a *boss*, usually with a hole in it for a bolt or screw.

**malleable casting** (*n*)   A casting that has been made less brittle and tougher by annealing.

**mill** (*v*)   To remove material by means of a rotating cutter on a milling machine, §12.9.

***mold*** (*n*)   The mass of sand or other material that forms the cavity into which molten metal is poured, §12.3.

***MS*** (*n*)   Machinery steel, sometimes called *mild steel* with a small percentage of carbon. Cannot be hardened.

***neck*** (*v*)   To cut a groove called a *neck* around a cylindrical piece.

NECK

***normalize*** (*v*)   To heat steel above its critical temperature and then to cool it in air, §12.22.

***pack-harden*** (*v*)   To *carburize*, then to *case-harden*. §12.22.

***pad*** (*n*)   A slight projection, usually to provide a bearing surface around one or more holes.

PAD

***pattern*** (*n*)   A model, usually of wood, used in forming a mold for a casting. In sheet metal work a pattern is called a *development*.

***peen*** (*v*)   To hammer into shape with a ballpeen hammer.

***pickle*** (*v*)   To clean forgings or castings in dilute sulphuric acid.

***pinion*** (*n*)   The smaller of two mating gears.

***pitch circle*** (*n*)   An imaginary circle corresponding to the circumference of the friction gear from which the spur gear was derived.

***plane*** (*v*)   To remove material by means of the *planer*, §12.11.

***planish*** (*v*)   To impart a planished surface to sheet metal by hammering with a smooth-surfaced hammer.

***plate*** (*v*)   To coat a metal piece with another metal, such as chrome or nickel, by electrochemical methods.

***polish*** (*v*)   To produce a highly finished or polished surface by friction, using a very fine abrasive.

***profile*** (*v*)   To cut any desired outline by moving a small rotating cutter, usually with a master template as a guide.

***punch*** (*v*)   To cut an opening of a desired shape with a rigid tool having the same shape, by pressing the tool through the work.

***quench*** (*v*)   To immerse a heated piece of metal in water or oil in order to harden it.

***rack*** (*n*)   A flat bar with gear teeth in a straight line to engage with teeth in a gear.

***ream*** (*v*)   To enlarge a finished hole slightly to give it greater accuracy, with a *reamer*, §12.16.

***relief*** (*n*)   An offset of surfaces to provide clearance for machining.

RELIEF

***rib*** (*n*)   A relatively thin flat member acting as a brace or support.

RIB

***rivet*** (*v*)   To connect with rivets or to clench over the end of a pin by hammering, §15.36.

***round*** (*n*)   An exterior rounded intersection of two surfaces, §§7.34 and 12.5.

***SAE***   Society of Automotive Engineers.

***sandblast*** (*v*)   To blow sand at high velocity with compressed air against castings or forgings to clean them.

***scleroscope*** (*n*)   An instrument for measuring hardness of metals.

***scrape*** (*v*)   To remove metal by scraping with a hand scraper, usually to fit a bearing.

***shape*** (*v*)   To remove metal from a piece with a *shaper*, §12.10.

***shear*** (*v*)   To cut metal by means of shearing with two blades in sliding contact.

***sherardize*** (*v*)   To galvanize a piece with a coating of zinc by heating it in a drum with zinc powder, to a temperature of 575 to 850 °F.

***shim*** (*n*)   A thin piece of metal or other material used as a spacer in adjusting two parts.

***solder*** (*v*)   To join with solder, usually composed of lead and tin.

***spin*** (*v*)   To form a rotating piece of sheet metal into a desired shape by pressing it with a smooth tool against a rotating form.

***spline*** (*n*)   A keyway, usually one of a series cut around a shaft or hole.

SPLINED HOLE

***spotface*** (*v*)   To produce a round *spot* or bearing surface around a hole, usually with a *spotfacer*. The spotface may be on top of a boss or it may be sunk into the surface, §§7.33 and 12.16.

SPOTFACE

***sprue*** (*n*)   A hole in the sand leading to the *gate* which leads to the mold, through which the metal enters, §12.3.

***steel casting*** (*n*)   Like cast-iron casting except that in the furnace scrap steel has been added to the casting.

*swage* (*v*)  To hammer metal into shape while it is held over a *swage*, or die, which fits in a hole in the *swage block*, or anvil.

*sweat* (*v*)  To fasten metal together by the use of solder between the pieces and by the application of heat and pressure.

*tap* (*v*)  To cut relatively small internal threads with a *tap*. §12.16.

*tape* (*n*)  Conical form given to a shaft or a hole. Also refers to the slope of a plane surface, §13.33.

*taper pin* (*n*)  A small tapered pin for fastening, usually to prevent a collar or hub from rotating on a shaft.

TAPER PIN

*taper reamer* (*n*)  A tapered reamer for producing accurate tapered holes, as for a taper pin. §§13.33 and 15.35.

*temper* (*v*)  To reheat hardened steel to bring it to a desired degree of hardness, §12.22.

*template* or *templet* (*n*)  A guide or pattern used to mark out the work, guide the tool in cutting it, or check the finished product.

*tin* (*n*)  A silvery metal used in alloys and for coating other metals, such as tin plate.

*tolerance* (*n*)  Total amount of variation permitted in limit dimension of a part, §14.2.

*trepan* (*v*)  To cut a circular groove in the flat surface at one end of a hole.

*tumble* (*v*)  To clean rough castings or forgings in a revolving drum filled with scrap metal.

*turn* (*v*)  To produce, on a lathe, a cylindrical surface parallel to the center line, §12.7.

*twist drill* (*n*)  A drill for use in a drill press, §12.16.

*undercut* (*n*)  A recessed cut or a cut with inwardly sloping sides.

UNDERCUT

*upset* (*v*)  To form a head or enlarged end on a bar or rod by pressure or by hammering between dies.

*web* (*n*)  A thin flat part joining larger parts. Also known as a *rib*.

*weld* (*v*)  Uniting metal pieces by pressure or fusion welding processes, §12.19.

*Woodruff key* (*n*)  A semicircular flat key, §15.34.

WOODRUFF KEYS

*wrought iron* (*n*)  Iron of low carbon content useful because of its toughness, ductility, and malleability.

# 3  CAD/CAM Glossary*

*access time* (or disk access time)  One measure of system response. The time interval between the instant that data is called for from storage and the instant that delivery is completed—i.e., read time. See also *response time*.

*alphanumeric* (or alphameric)  A term that encompasses letters, digits, and special characters that are machine-processable.

*alphanumeric display* (or alphameric display)  A work-station device consisting of a CRT on which text can be viewed. An alphanumeric display is capable of showing a fixed set of letters, digits, and special characters. It allows the designer to observe entered commands and to receive messages from the system.

*alphanumeric keyboard* (or alphameric keyboard)  A work-station device consisting of a typewriter-like keyboard that allows the designer to communicate with the system using an English-like command language.

*American Standard Code for Information Interchange* (ASCII)  An industry-standard character code widely used for information interchange among data processing systems, communications systems, and associated equipment.

*analog*  Applied to an electrical or computer system, this denotes the capability to represent data in continuously varying physical quantities.

*annotation*  Process of inserting text or a special note or identification (such as a flag) on a drawing, map, or diagram constructed on a CAD/CAM system. The text can be generated and positioned on the drawing using the system.

*application program* (or package)  A computer program or collection of programs to perform a task or tasks specific to a particular user's need or class of needs.

*archival storage*  Refers to memory (on magnetic tape, disks, printouts, or drums) used to store data on completed designs or elements outside of main memory.

*array* (*v*)  To create automatically on a CAD system an arrangement of identical elements or components. The designer defines the element once, then indicates the

*Extracted from *The CAD/CAM Glossary*, 1983 edition, published by the Computervision Corporation, Bedford, MA 01730; reproduced with permission of the publisher.

starting location and spacing for automatic generation of the array. (*n*) An arrangement created in the above manner. A series of elements or sets of elements arranged in a pattern—i.e., matrix.

**ASCII** See *American National Standard Code for Information Exchange.*

**assembler** A computer program that converts (i.e., translates) programmer-written symbolic instructions, usually in mnemonic form, into machine-executable (computer or binary-coded) instructions. This conversion is typically one-to-one (one symbolic instruction converts to one machine-executable instruction). A software programming aid.

**associative dimensioning** A CAD capabililty that links dimension entities to geometric entities being dimensioned. This allows the value of a dimension to be automatically updated as the geometry changes.

**attribute** A nongraphic characteristic of a part, component, or entity under design on a CAD system. Examples include: dimension entities associated with geometry, text with text nodes, and nodal lines with connect nodes. Changing one entity in an association can produce automatic changes by the system in the associated entity; e.g., moving one entity can cause moving or stretching of the other entity.

**automatic dimensioning** A CAD capability that computes the dimensions in a displayed design, or in a designated section, and automatically places dimensions, dimensional lines, and arrowheads where required. In the case of mapping, this capability labels the linear feature with length and azimuth.

**auxiliary storage** Storage that supplements main memory devices such as disk or drum storage. Contrast with *archival storage.*

**benchmark** The program(s) used to test, compare, and evaluate in real time the performance of various CAD/CAM systems prior to selection and purchase. A *synthetic* benchmark has preestablished parameters designed to exercise a set of system features and resources. A *live* benchmark is drawn from the prospective user's workload as a model of the entire workload.

**bit** The smallest unit of information that can be stored and processed by a digital computer. A bit may assume only one of two values: 0 or 1 (i.e., ON/OFF or YES/NO). Bits are organized into larger units called *words* for access by computer instructions.

    Computers are often categorized by word size in bits, i.e., the maximum word size that can be processed as a unit during an instruction cycle (e.g., 16-bit computers or 32-bit computers). The number of bits in a word is an indication of the processing power of the system, especially for calculations or for high-precision data.

**bit rate** The speed at which bits are transmitted, usually expressed in bits per second.

**bits per inch** (bpi) The number of bits that can be stored per inch of a magnetic tape. A measure of the data storage capacity of a magnetic tape.

**blank** A CAD command that causes a predefined entity to go temporarily blank on the CRT. The reversing command is *unblank.*

**blinking** A CAD design aid that makes a predefined graphic entity blink on the CRT to attract the attention of the designer.

**boot up** Start up (a system).

**B-spline** A sequence of parametric polynomial curves (typically quadratic or cubic polynomials) forming a smooth fit between a sequence of points in 3D space. The piece-wise defined curve maintains a level of mathematical continuity dependent upon the polynomial degree chosen. It is used extensively in mechanical design applications in the automotive and aerospace industries.

**bug** A flaw in the design or implementation of a software program or hardware design that causes erroneous results or malfunctions.

**bulk memory** A memory device for storing a large amount of data, e.g., disk, drum, or magnetic tape. It is not randomly accessible as main memory is.

**byte** A sequence of adjacent bits, usually eight, representing a character that is operated on as a unit. Usually shorter than a word. A measure of the memory capacity of a system, or of an individual storage unit (as a 300-million-byte disk).

**CAD** See *computer-aided design.*

**CAD/CAM** See *computer-aided design/computer-aided manufacturing.*

**CADDS®** Computervision Corporation's registered trademark for its prerecorded software programs.

**CAE** See *computer-aided engineering.*

**CAM** See *computer-aided manufacturing.*

**CAMACS™** (CAM Asynchronous Communications Software) Computervision's communications link, which enables users to exchange machine control data with other automation systems and devices, or to interact directly with local or remote manufacturing systems and machines. CAMACS tailors CAD/CAM data automatically for a wide range of machine tools, robots, coordinate measurement systems, and off-line storage-devices.

**cathode ray tube** (CRT) The principal component in a CAD display device. A CRT displays graphic representations of geometric entities and designs and can be of various types: storage tube, raster scan, or refresh. These tubes create images by means of a controllable beam of electrons striking a screen. The term *CRT* is often used to denote the entire display device.

**central processing unit** (CPU) The computer brain of a CAD/CAM system that controls the retrieval, decoding, and processing of information, as well as the interpretation and execution of operating instructions—the building blocks of application and other computer programs. A CPU comprises arithmetic, control, and logic elements.

**character**   An alphabetical, numerical, or special graphic symbol used as part of the organization, control, or representation of CAD/CAM data.

**characters per second** (cps)   A measure of the speed with which an alphanumeric terminal can process data.

**chip**   See *integrated circuit.*

**code**   A set of specific symbols and rules for representing data (usually instructions) so that the data can be understood and executed by a computer. A code can be in binary (machine) language, assembly language, or a high-level language. Frequently refers to an industry-standard code such as ANSI, ASCII, IPC, or Standard Code for Information Exchange. Many application codes for CAD/CAM are written in FORTRAN.

**color display**   A CAD/CAM display device. Color raster-scan displays offer a variety of user-selectable, contrasting colors to make it easier to discriminate among various groups of design elements on different layers of a large, complex design. Color speeds up the recognition of specific areas and subassemblies, helps the designer interpret complex surfaces, and highlights interference problems. Color displays can be of the penetration type, in which various phosphor layers give off different colors (refresh display) or the TV-type with red, blue, and green electron guns (raster-scan display).

**command**   A control signal or instruction to a CPU or graphics processor, commonly initiated by means of a menu/tablet and electronic pen or by an alphanumeric keyboard.

**command language**   A language for communicating with a CAD/CAM system in order to perform specific functions or tasks.

**communication link**   The physical means, such as a telephone line, for connecting one system module or peripheral to another in a different location in order to transmit and receive data. See also *data link.*

**compatibility**   The ability of a particular hardware module or software program, code, or language to be used in a CAD/CAM system without prior modification or special interfaces. *Upward compatible* denotes the ability of a system to interface with new hardware or software modules or enhancements (i.e., the system vendor provides with each new module a reasonable means of transferring data, programs, and operator skills from the user's present system to the new enhancements).

**compiler**   A computer program that converts or translates a high-level, user-written language (e.g., PASCAL, COBOL, VARPRO, or FORTRAN) or source, into a language that a computer can understand. The conversion is typically one to many (i.e., one user instruction to many machine-executable instructions). A software programming aid, the compiler allows the designer to write programs in an English-like language with relatively few statements, thus saving program development time.

**component**   A physical entity, or a symbol used in CAD to denote such an entity. Depending on the application, a component might refer to an IC or part of a wiring circuit (e.g., a resistor), or a valve, elbow, or vee in a plant layout, or a substation or cable in a utility map. Also applies to a subassembly or part that goes into higher level assemblies.

**computer-aided design** (CAD)   A process that uses a computer system to assist in the creation, modification, and display of a design.

**computer-aided design/computer-aided manufacturing** (CAD/CAM)   Refers to the integration of computers into the entire design-to-fabrication cycle of a product or plant.

**computer-aided engineering** (CAE)   Analysis of a design for basic error checking, or to optimize manufacturability, performance, and economy (for example, by comparing various possible materials or designs). Information drawn from the CAD/CAM design data base is used to analyze the functional characteristics of a part, product, or system under design and to simulate its performance under various conditions. In electronic design, CAE enables users of the Computervision Designer system to detect and correct potentially costly design flaws. CAE permits the execution of complex circuit loading analyses and simulation during the circuit definition stage. CAE can be used to determine section properties, moments of inertia, shear and bending moments, weight, volume, surface area, and center of gravity. CAE can precisely determine loads, vibration, noise, and service life early in the design cycle so that components can be optimized to meet those criteria. Perhaps the most powerful CAE technique is finite element modeling. See also *kinematics.*

**computer-aided manufacturing** (CAM)   The use of computer and digital technology to generate manufacturing-oriented data. Data drawn from a CAD/CAM data base can assist in or control a portion or all of a manufacturing process, including numerically controlled machines, computer-assisted parts programming, computer-assisted process planning, robotics, and programmable logic controllers, CAM can involve production programming, manufacturing engineering, industrial engineering, facilities engineering, and reliability engineering (quality control). CAM techniques can be used to produce process plans for fabricating a complete assembly, to program robots, and to coordinate plant operation.

**computer graphics**   A general term encompassing any discipline or activity that uses computers to generate, process and display graphic images. The essential technology of CAD/CAM systems. See also *computer-aided design.*

**computer network**   An interconnected complex (arrangement, or configuration) of two or more systems. See also *network.*

**computer program**   A specific set of software commands

in a form acceptable to a computer and used to achieve a desired result. Often called a *software program* or *package*.

**configuration** A particular combination of a computer, software and hardware modules, and peripherals at a single installation and interconnected in such a way as to support certain application(s).

**connector** A termination point for a signal entering or leaving a PC board or a cabling system.

**convention** Standardized methodology or accepted procedure for executing a computer program. In CAD, the term denotes a standard rule or mode of execution undertaken to provide consistency. For example, a drafting convention might require all dimensions to be in metric units.

**core** (core memory) A largely obsolete term for *main storage*.

**CPU** See *central processing unit*.

**CRT** See *cathode ray tube*.

**cursor** A visual tracking symbol, usually an underline or cross hairs, for indicating a location or entity selection on the CRT display. A text cursor indicates the alphanumeric input; a graphics cursor indicates the next geometric input. A cursor is guided by an electronic or light pen, joystick, keyboard, etc., and follows every movement of the input device.

**cycle** A preset sequence of events (hardware or software) initiated by a single command.

**data base** A comprehensive collection of interrelated information stored on some kind of mass data storage device, usually a disk. Generally consists of information organized into a number of fixed-format record types with logical links between associated records. Typically includes operating systems instructions, standard parts libraries, completed designs and documentation, source code, graphic and application programs, as well as current user tasks in progress.

**data communication** The transmission of data (usually digital) from one point (such as a CAD/CAM workstation or CPU) to another point via communication channels such as telephone lines.

**data link** The communication line(s), related controls, and interface(s) for the transmission of data between two or more computer systems. Can include modems, telephone lines, or dedicated transmission media such as cable or optical fiber.

**data tablet** A CAD/CAM input device that allows the designer to communicate with the system by placing an electronic pen or stylus on the tablet surface. There is a direct correspondence between positions on the tablet and addressable points on the display surface of the CRT. Typically used for indicating positions on the CRT, for digitizing input of drawings, or for menu selection. See also *graphic tablet*.

**debug** To detect, locate, and correct any bugs in a system's software or hardware.

**dedicated** Designed or intended for a single function or use. For example, a dedicated work station might be used exclusively for engineering calculations or plotting.

**default** The predetermined value of a parameter required in a CAD/CAM task or operation. It is automatically supplied by the system whenever that value (e.g., text, height, or grid size) is not specified.

**density** (1) A measure of the complexity of an electronic design. For example, IC density can be measured by the number of gates or transistors per unit area or by the number of square inches per component. (2) Magnetic tape storage capacity. High capacity might be 1600 bits/inch; low, 800 bits/inch.

**device** A system hardware module external to the CPU and designed to perform a specific function—i.e., a CRT, plotter, printer, hard-copy unit, etc. See also *peripheral*.

**diagnostics** Computer programs designed to test the status of a system or its key components and to detect and isolate malfunctions.

**dial up** To initiate station-to-station communication with a computer via a dial telephone, usually from a workstation to a computer.

**digital** Applied to an electrical or computer system, this denotes the capability to represent data in the form of digits.

**digitize** (1) General description: to convert a drawing into digital form (i.e., coordinate locations) so that it can be entered into the data base for later processing. A digitizer, available with many CAD systems, implements the conversion process. This is one of the primary ways of entering existing drawings, crude graphics, lines, and shapes into the system. (2) Computervision usage: to specify a coordinate location or entity using an electronic pen or other device; or a single coordinate value or entity pointer generated by a digitizing operation.

**digitizer** A CAD input device consisting of a data tablet on which is mounted the drawing or design to be digitized into the system. The designer moves a puck or electronic pen to selected points on the drawing and enters coordinate data for lines and shapes by simply pressing down the digitize button with the puck or pen.

**dimensioning, automatic** A CAD capability that will automatically compute and insert the dimensions of a design or drawing, or a designated section of it.

**direct access** (linkage) Retrieval or storage of data in the system by reference to its location on a tape, disk, or cartridge, without the need for processing on a CPU.

**direct-view storage tube** (DVST) One of the most widely used graphics display devices, DVST generates a long-lasting, flicker-tree image with high resolution and no refreshing. It handles an almost unlimited amount of data. However, display dynamics are limited since DVSTs do not permit selective erase. The image is not as bright as with refresh or raster. Also called *storage tube*.

*directory*  A named space on the disk or other mass storage device in which are stored the names of files and some summary information about them.

*discrete components*  Components with a single functional capability per package—for example, transistors and diodes.

*disk* (storage)  A device on which large amounts of information can be stored in the data base. Synonymous with *magnetic disk storage* or *magnetic disk memory.*

*display*  A CAD/CAM work station device for rapidly presenting a graphic image so that the designer can react to it, making changes interactively in real time. Usually refers to a CRT.

*dot-matrix plotter*  A CAD peripheral device for generating graphic plots. Consists of a combination of wire nibs (styli) spaced 100 to 200 styli per inch, which place dots where needed to generate a drawing. Because of its high speed, it is typically used in electronic design applications. Accuracy and resolution are not as great as with pen plotters. Also known as *electrostatic plotter.*

*drum plotter*  An electromechanical pen plotter that draws an image on paper or film mounted on a rotatable drum. In this CAD peripheral device a combination of plotting-head movement and drum rotation provides the motion.

*dynamic* (motion)  Simulation of movement using CAD software, so that the designer can see on the CRT screen 3D representations of the parts in a piece of machinery as they interact dynamically. Thus, any collision or interference problems are revealed at a glance.

*dynamic menuing*  This feature of Computervision's Instaview terminal allows a particular function or command to be initiated by touching an electronic pen to the appropriate key word displayed in the status text area on the screen.

*dynamics*  The capability of a CAD system to zoom, scroll, and rotate.

*edit*  To modify, refine, or update an emerging design or text on a CAD system. This can be done on-line interactively.

*electrostatic plotter*  See *dot-matrix plotter.*

*element*  The basic design entity in computer-aided design whose logical, positional, electrical, or mechanical function is identifiable.

*enhancements*  Software or hardware improvements, additions, or updates to a CAD/CAM system.

*entity*  A geometric primitive—the fundamental building block used in constructing a design or drawing, such as an arc, circle, line, text, point, spline, figure, or nodal line. Or a group of primitives processed as an identifiable unit. Thus, a square may be defined as a discrete entity consisting of four primitives (vectors), although each side of the square could be defined as an entity in its own right. See also *primitive.*

*feedback*  (1) The ability of a system to respond to an operator command in real time either visually or with a message on the alphanumeric display or CRT. This message registers the command, indicates any possible errors, and simultaneously displays the updated design on the CRT. (2) The signal or data fed back to a commanding unit from a controlled machine or process to denote its response to a command. (3) The signal representing the difference between actual response and desired response and used by the commanding unit to improve performance of the controlled machine or process. See also *prompt.*

*figure*  A symbol or a part that may contain primitive entities, other figures, nongraphic properties, and associations. A figure can be incorporated into other parts or figures.

*file*  A collection of related information in the system that may be accessed by a unique name. May be stored on a disk, tape, or other mass storage media.

*file protection*  A technique for preventing access to or accidental erasure of data within a file on the system.

*firmware*  Computer programs, instructions, or functions implemented in user-modifiable hardware, i.e., a microprocessor with read-only memory. Such programs or instructions, stored permanently in programmable read-only memories, constitute a fundamental part of system hardware. The advantage is that a frequently used program or routine can be invoked by a single command instead of multiple commands as in a software program.

*flatbed plotter*  A CAD/CAM peripheral device that draws an image on paper, glass, or film mounted on a flat table. The plotting head provides all the motion.

*flat-pattern generation*  A CAD/CAM capability for automatically unfolding a 3D design of a sheet metal part into its corresponding flat-pattern design. Calculations for material bending and stretching are performed automatically for any specified material. The reverse flat-pattern generation package automatically folds a flat-pattern design into its 3D version. Flat-pattern generation eliminates major bottlenecks for sheet metal fabricators.

*flicker*  An undesired visual effect on a CRT when the refresh rate is low.

*font, line*  Repetitive pattern used in CAD to give a displayed line appearance characteristics that make it more easily distinguishable, such as a solid, dashed, or dotted line. A line font can be applied to graphic images in order to provide meaning, either graphic (e.g., hidden lines) or functional (roads, tracks, wires, pipes, etc.). It can help a designer to identify and define specific graphic representations of entities that are view-dependent. For example, a line may be solid when drawn in the top view of an object but, when a line font is used, becomes dotted in the side view where it is not normally visible.

*font, text*  Sets of type faces of various styles and sizes.

In CAD, fonts are used to create text for drawings, special characters such as Greek letters, and mathematical symbols.

*FORTRAN*   *FOR*mula *TRAN*slation, a high-level programming language used primarily for scientific or engineering applications.

*fracturing*   The division of IC graphics by CAD into simple trapezoidal or rectangular areas for pattern-generation purposes.

*function key*   A specific square on a data tablet, or a key on a function key box, used by the designer to enter a particular command or other input. See also *data tablet*.

*function keyboard*   An input device located at a CAD/CAM workstation and containing a number of function keys.

*gap*   The gap between two entities on a computer-aided design is the length of the shortest line segment that can be drawn from the boundary of one entity to the other without intersecting the boundary of the other. CAD/CAM design-rules checking programs can automatically perform gap checks.

*graphic tablet*   A CAD/CAM input device that enables graphic and location instruments to be entered into the system using an electronic pen on the tablet. See also *data tablet*.

*gray scales*   In CAD systems with a monochromatic display, variations in brightness level (gray scale) are employed to enhance the contrast among various design elements. This feature is very useful in helping the designer discriminate among complex entities on different layers displayed concurrently on the CRT.

*grid*   A network of uniformly spaced points or crosshatch optionally displayed on the CRT and used for exactly locating and digitizing a position, inputting components to assist in the creation of a design layout, or constructing precise angles. For example, the coordinate data supplied by digitizers is automatically calculated by the CPU from the closest grid point. The grid determines the minimum accuracy with which design entities are described or connected. In the mapping environment, a grid is used to describe the distribution network of utility resources.

*hard copy*   A copy on paper of an image displayed on the CRT—for example, a drawing, printed report, plot, listing, or summary. Most CAD/CAM systems can automatically generate hard copy through an on-line printer or plotter.

*hardware*   The physical components, modules, and peripherals comprising a system—computer disk, magnetic tape, CRT terminal(s), plotter(s), etc.

*hard-wired link*   A technique of physically connecting two systems by fixed circuit interconnections using digital signals.

*high-level language*   A problem-oriented programming language using words, symbols, and command statements that closely resemble English-language statements. Each statement typically represents a series of computer instructions. Relatively easy to learn and use, a high-level language permits the execution of a number of subroutines through a simple command. Examples are BASIC, FORTRAN, PL/I, PASCAL, and COBOL.

A high-level language must be translated or compiled into machine language before it can be understood and processed by a computer. See also *assembler; low-level language*.

*host computer*   The primary or controlling computer in a multicomputer network. Large-scale host computers typically are equipped with mass memory and a variety of peripheral devices, including magnetic tape, line printers, card readers, and possibly hard-copy devices. Host computers may be used to support, with their own memory and processing capabilities, not only graphics programs running on a CAD/CAM system but also related engineering analysis.

*host-satellite system*   A CAD/CAM system configuration characterized by a graphic workstation with its own computer (typically holding the display file) that is connected to another, usually larger, computer for more extensive computation or data manipulation. The computer local to the display is a satellite to the larger host computer, and the two comprise a host-satellite system.

*IC*   See *integrated circuit*.

*IGES*   See *Initial Graphics Exchange Specification*.

*inches per second* (ips)   Measure of the speed of a device (i.e., the number of inches of magnetic tape that can be processed per second, or the speed of a pen plotter).

*Initial Graphics Exchange Specification* (IGES)   An interim CAD/CAM data base specification until the American National Standards Institute develops its own specification. IGES attempts to standardize communication of drawing and geometric product information between computer systems.

*initialize*   To set counters, switches, and addresses on a computer to zero or to other starting values at the beginning of, or at predetermined stages in, a program or routine.

*input* (data)   (1) The data supplied to a computer program for processing by the system. (2) The process of entering such data into the system.

*input devices*   A variety of devices (such as data tablets or keyboard devices) that allow the user to communicate with the CAD/CAM system, for example, to pick a function from many presented, to enter text and/or numerical data, to modify the picture shown on the CRT, or to construct the desired design.

*input/output* (I/O)   A term used to describe a CAD/CAM communications device as well as the process by which communications take place in a CAD/CAM system. An I/O device is one that makes possible communications between a device and a workstation operator or between devices on the system (such as workstations or controllers). By extension, input/output also denotes the process by which communications takes place. Input refers to the data transmitted to the processor for manipulation, and output refers to the data transmitted from the processor to the workstation operator or to another device (i.e., the results). Contrast with the other major parts of a CAD/CAM system: the CPU or central processing unit, which performs arithmetic and logical operations, and data storage devices (such as memories, disks, or tapes).

**insert** To create and place entities, figures, or information on a CRT or into an emerging design on the display.

**instruction set** *(1)* All the commands to which a CAD/CAM computer will respond. *(2)* The repertoire of functions the computer can perform.

**integrated circuit** (IC) A tiny complex of electronic components and interconnections comprising a circuit that may vary in functional complexity from a simple logic gate to a microprocessor. An IC is usually packaged in a single substrate such as a slice of silicon. The complexity of most IC designs and the many repetitive elements have made computer-aided design an economic necessity. Also called a *chip*.

**integrated system** A CAD/CAM system that integrates the entire product development cycle—analysis, design, and fabrication—so that all processes flow smoothly from concept to production.

**intelligent work station/terminal** A workstation in a system that can perform certain data processing functions in a stand-alone mode, independent of another computer. Contains a built-in computer, usually a microprocessor or minicomputer, and dedicated memory.

**interactive** Denotes two-way communications between a CAD/CAM system or workstation and its operators. An operator can modify or terminate a program and receive feedback from the system for guidance and verification. See also *feedback*.

**interactive graphics system** (IGS) or interactive computer graphics (ICG) A CAD/CAM system in which the workstations are used interactively for computer-aided design and/or drafting, as well as for CAM, all under full operator control, and possibly also for text-processing, generation of charts and graphs, or computer-aided engineering. The designer (operator) can intervene to enter data and direct the course of any program, receiving immediate visual feedback via the CRT. Bilateral communication is provided between the system and the designer(s). Often used synonymously with *CAD*.

**interface** *(n)* (1) A hardware and/or software link that enables two systems, or a system and its peripherals, to operate as a single, integrated system (2) The input devices and visual feedback capabilities that allow bilateral communication between the designer and the system. The interface to a large computer can be a communi-

cations link (hardware) or a combination of software and hard-wired connections. An interface might be a portion of storage accessed by two or more programs or a link between two subroutines in a program.

**I/O** See *input/output*.

**ips** See *inches per second*.

**jaggies** A CAD jargon term used to refer to straight or curved lines that appear to be jagged or saw-toothed on the CRT screen.

**joystick** A CAD data-entering device employing a hand-controlled lever to manually enter the coordinates of various points on a design being digitized into the system.

**key file** A disk file that provides user-defined definitions for a tablet menu. See *menu*.

**kinematics** A computer-aided engineering (CAE) process for plotting or animating the motion of parts in a machine or a structure under design on the system. CAE simulation programs allow the motion of mechanisms to be studied for interference, acceleration, and force determinations while still in the design stage.

**layering** A method of logically organizing data in a CAD/CAM data base. Functionally different classes of data (e.g., various graphic/geometric entities) are segregated on separate layers, each of which can be displayed individually or in any desired combination. Layering helps the designer distinguish among different kinds of data in creating a complex product such as a multilayered PC board or IC.

**layers** User-defined logical subdivisions of data in a CAD/CAM data base that may be viewed on the CRT individually or overlaid and viewed in groups.

**learning curve** A concept that projects the expected improvement in operator productivity over a period of time. Usually applied in the first 1 to $1\frac{1}{2}$ years of a new CAD/CAM facility as part of a cost-justification study, or when new operators are introduced. An accepted tool of management for predicting manpower requirements and evaluating training programs.

**library, graphics** (or parts library) A collection of standard, often-used symbols, components, shapes, or parts stored in the CAD data base as templates or building blocks to speed up future design work on the system. Generally an organization of files under a common library name.

**light pen** A hand-held photosensitive CAD input device used on a refreshed CRT screen for identifying display elements, or for designating a location on the screen where an action is to take place.

**line font** See *font, line*.

**line printer** A CAD/CAM peripheral device used for rapid printing of data.

**line smoothing** An automated mapping capability for the interpolation and insertion of additional points along a linear entity yielding a series of shorter linear segments to generate a smooth curved appearance to the original linear component. The additional points or segments are created only for display purposes and are interpolated from a relatively small set of stored representative points. Thus, data storage space is minimized.

**low-level language**  A programming language in which statements translate on a one-for-one basis. See also *machine language*.

**machine**  A computer, CPU, or other processor.

**machine instruction**  An instruction that a machine (computer) can recognize and execute.

**machine language**  The complete set of command instructions understandable to and used directly by a computer when it performs operations.

**macro**  (1) A sequence of computer instructions executable as a single command. A frequently used, multistep operation can be organized into a macro, given a new name, and remain in the system for easy use, thus shortening program development time. (2) In Computervision's IC design system, macro refers to macroexpansion of a cell. This system capability enables the designer to replicate the contents of a cell as primitives without the original cell grouping.

**magnetic disk**  A flat circular plate with a magnetic surface on which information can be stored by selective magnetization of portions of the flat surface. Commonly used for temporary working storage during computer-aided design. See also *disk*.

**magnetic tape**  A tape with a magnetic surface on which information can be stored by selective polarization of portions of the surface. Commonly used in CAD/CAM for off-line storage of completed design files and other archival material.

**mainframe** (computer)  A large central computer facility.

**main memory/storage**  The computer's general-purpose storage from which instructions may be executed and data loaded directly into operating registers.

**mass storage**  Auxiliary large-capacity memory for storing large amounts of data readily accessible by the computer. Commonly a disk or magnetic tape.

**matrix**  A 2D or 3D rectangular array (arrangement) of identical geometric or symbolic entities. A matrix can be generated automatically on a CAD system by specifying the building block entity and the desired locations. This process is used extensively in computer-aided electrical/electronic design.

**memory**  Any form of data storage where information can be read and written. Standard memories include RAM, ROM, and PROM. See also *programmable read-only memory; random access memory; read-only memory; storage*.

**menu**  A common CAD/CAM input device consisting of a checkerboard pattern of squares printed on a sheet of paper or plastic placed over a data tablet. These squares have been preprogrammed to represent a part of a command, a command, or a series of commands. Each square, when touched by an electronic pen, initiates the particular function or command indicated on that square. See also *data tablet, dynamic menuing*.

**merge**  To combine two or more sets of related data into one, usually in a specified sequence. This can be done automatically on a CAD/CAM system to generate lists and reports.

**microcomputer**  A smaller, lower-cost equivalent of a full-scale minicomputer. Includes a microprocessor (CPU), memory, and necessary interface circuits. Consists of one or more ICs (chips) comprising a chip set.

**microprocessor**  The central control element of a microcomputer, implemented in a single integrated circuit. It performs instruction sequencing and processing, as well as all required computations. It requires additional circuits to function as a microcomputer. See *microcomputer*.

**minicomputer**  A general-purpose, single processor computer of limited flexibility and memory performance.

**mirroring**  A CAD design aid that automatically creates a mirror image of a graphic entity on the CRT by flipping the entity or drawing on its $x$ or $y$ axis.

**mnemonic symbol**  An easily remembered symbol that assists the designer in communicating with the system (e.g., an abbreviation such as MPY for *multiply*).

**model, geometric**  A complete, geometrically accurate 3D or 2D representation of a shape, a part, a geographic area, a plant or any part of it, designed on a CAD system and stored in the data base. A mathematical or analytic model of a physical system used to determine the response of that system to a stimulus or load. See *modeling, geometric*.

**modeling, geometric**  Constructing a mathematical or analytic model of a physical object or system for the purpose of determining the response of that object or system to a stimulus or load. First, the designer describes the shape under design using a geometric model constructed on the system. The computer then converts this pictorial representation on the CRT into a mathematical model later used for other CAD functions such as design optimization.

**modeling, solid**  A type of 3D modeling in which the solid characteristics of an object under design are built into the data base so that complex internal structures and external shapes can be realistically represented. This makes computer-aided design and analysis of solid objects easier, clearer, and more accurate than with wireframe graphics.

**modem**  MOdulator-DEModulator, a device that converts digital signals to analog signals, and vice versa, for long-distance transmission over communications circuits such as telephone lines, dedicated wires, optical fiber, or microwave.

**module**  A separate and distinct unit of hardware or software that is part of a system.

**mouse**  A hand-held data-entering device used to position a cursor on a data tablet. See *cursor*.

**multiprocessor** A computer whose architecture consists of more than one processing unit. See *central processing unit; microcomputer.*

**network** An arrangement of two or more interconnected computer systems to facilitate the exchange of information in order to perform a specific function. For example, a CAD/CAM system might be connected to a mainframe computer to off-load heavy analytic tasks. Also refers to a piping network in computer-aided plant design.

**numerical control** (NC) A technique of operating machine tools or similar equipment in which motion is developed in response to numerically coded commands. These commands may be generated by a CAD/CAM system on punched tapes or other communications media. Also, the processes involved in generating the data or tapes necessary to guide a machine tool in the manufacture of a part.

**off-line** Refers to peripheral devices not currently connected to and under the direct control of the system's computer.

**on-line** Refers to peripheral devices connected to and under the direct control of the system's computer, so that operator-system interaction, feedback, and output are all in real time.

**operating system** A structured set of software programs that control the operation of the computer and associated peripheral devices in a CAD/CAM system, as well as the execution of computer programs and data flow to and from peripheral devices. May provide support for activities and programs such as scheduling, debugging, input/output control, accounting, editing, assembly, compilation, storage assignment, data management, and diagnostics. An operating system may assign task priority levels, support a file system, provide drives for I/O devices, support standard system commands or utilities for on-line programming, process commands, and support both networking and diagnostics.

**output** The end result of a particular CAD/CAM process or series of processes. The output of a CAD cycle can be artwork and hard-copy lists and reports. The output of a total design-to-manufacturing CAD/CAM system can also include numerical control tapes for manufacturing.

**overlay** A segment of code or data to be brought into the memory of a computer to replace existing code or data.

**paint** To fill in a bounded graphic figure on a raster display using a combination of repetitive patterns or line fonts to add meaning or clarity. See *font, line.*

**paper-tape punch/reader** A peripheral device that can read as well as punch a perforated paper tape generated by a CAD/CAM system. These tapes are the principal means of supplying/data to an NC machine.

**parallel processing** Executing more than one element of a single process concurrently on multiple processors in a computer system.

**password protection** A security feature of certain CAD/CAM systems that prevents access to the system or to files within the system without first entering a password, i.e., a special sequence of characters.

**PC board** See *printed circuit board.*

**pen plotter** An electromechanical CAD output device that generates hard copy of displayed graphic data by means of a ballpoint pen or liquid ink. Used when a very accurate final drawing is required. Provides exceptional uniformity and density of lines, precise positional accuracy, as well as various user-selectable colors.

**peripheral (device)** Any device, distinct from the basic system modules, that provides input to and/or output from the CPU. May include printers, keyboards, plotters, graphics display terminals, paper-tape reader/punches, analog-to-digital converters, disks, and tape drives.

**permanent storage** A method or device for storing the results of a completed program outside the CPU—usually in the form of magnetic tape or punched cards.

**photo plotter** A CAD output device that generates high-precision artwork masters photographically for PC board design and IC masks.

**pixel** The smallest portion of a CRT screen that can be individually referenced. An individual dot on a display image. Typically, pixels are evenly spaced, horizontally and vertically, on the display.

**plotter** A CAD peripheral device used to output for external use the image stored in the data base. Generally makes large, accurate drawings substantially better than what is displayed. Plotter types include pen, drum, electrostatic, and flatbed.

**postprocessor** A software program or procedure that formats graphic or other data processed on the system for some other purpose. For example, a postprocessor might format cutter centerline data into a form that a machine controller can interpret.

**precision** The degree of accuracy. Generally refers to the number of significant digits of information to the right of the decimal point for data represented within a computer system. Thus, the term denotes the degree of discrimination with which a design or design element can be described in the data base.

**preplaced line** (or bus) A run (or line) between a set of points on a PC board layout that has been predefined by the designer and must be avoided by a CAD automatic routing program.

**preprocessor** A computer program that takes a specific set of instructions from an external source and translates it into the format required by the system.

**primitive** A design element at the lowest stage of complexity. A fundamental graphic entity. It can be a vector, a point, or a text string. The smallest definable object in a display processor's instruction set.

**printed circuit (PC) board** A baseboard made of insulating materials and an etched copper-foil circuit pattern on which are mounted ICs and other components required to implement one or more electronic functions. PC boards plug into a rack or subassembly of electronic equipment to provide the brains or logic to control the operation of a computer, or a communications system, instrumentation, or other electronic systems. The name derives from the fact that the circuitry is connected not by wires but by copper-foil lines, paths, or traces actually etched onto the board surface. CAD/CAM is used extensively in PC board design, testing, and manufacture.

**process simulation** A program utilizing a mathematical model created on the system to try out numerous process design iterations with real-time visual and numerical feedback. Designers can see on the CRT what is taking place at every stage in the manufacturing process. They can therefore optimize a process and correct problems that could affect the actual manufacturing process downstream.

**processor** In CAD/CAM system hardware, any device that performs a specific function. Most often used to refer to the CPU. In software, it refers to a complex set of instructions to perform a general function. See also *central processing unit.*

**productivity ratio** A widely accepted means of measuring CAD/CAM productivity (throughput per hour) by comparing the productivity of a design/engineering group before and after installation of the system or relative to some standard norm or potential maximum. The most common way of recording productivity is Actual Manual Hours/Actual CAD Hours, expressed as 4:1, 6:1, etc.

**program** (n) A precise sequential set of instructions that direct a computer to perform a particular task or action or to solve a problem. A complete program includes plans for the transcription of data, coding for the computer, and plans for the absorption of the results into the system. (v) To develop a program. See also *computer program.*

**Programmable Read-Only Memory** (PROM) A memory that, once programmed with permanent data or instructions, becomes a ROM. See *read-only memory.*

**PROM** See *programmable read-only memory.*

**prompt** A message or symbol generated automatically by the system, and appearing on the CRT, to inform the user of (a) a procedural error or incorrect input to the program being executed or (b) the next expected action, option(s), or input. See also *tutorial.*

**puck** A hand-held, manually controlled input device that allows coordinate data to be digitized into the system from a drawing placed on the data tablet or digitizer surface. A puck has a transparent window containing cross hairs.

**RAM** See *random access memory.*

**random access memory** (RAM) A main memory read/write storage unit that provides the CAD/CAM operator direct access to the stored information. The time required to access any word stored in the memory is the same as for any other word.

**raster display** A CAD workstation display in which the entire CRT surface is scanned at a constant refresh rate. The bright, flicker-free image can be selectively written and erased. Also called a digital TV display.

**raster scan** (video) Currently, the dominant technology in CAD graphic displays. Similar to conventional television, it involves a line-by-line sweep across the entire CRT surface to generate the image. Raster-scan features include good brightness, accuracy, selective erase, dynamic motion capabilities, and the opportunity for unlimited color. The device can display a large amount of information without flicker, although resolution is not as good as with storage-tube displays.

**read-only memory** (ROM) A memory that cannot be modified or reprogrammed. Typically used for control and execute programs. See also *programmable read-only memory.*

**real time** Refers to tasks or functions executed so rapidly by a CAD/CAM system that the feedback at various stages in the process can be used to guide the designer in completing the task. Immediate visual feedback through the CRT makes possible real time, interactive operation of a CAD/CAM system.

**rectangular array** Insertion of the same entity at multiple locations on a CRT using the system's ability to copy design elements and place them at user-specified intervals to create a rectangular arrangement or matrix. A feature of PC and IC design systems.

**refresh** (or vector refresh) A CAD display technology that involves frequent redrawing of an image displayed on the CRT to keep it bright, crisp, and clear. Refresh permits a high degree of movement in the displayed image as well as high resolution. Selective erase or editing is possible at any time without erasing and repainting the entire image. Although substantial amounts of high-speed memory are required, large, complex images may flicker.

**refresh rate** The rate at which the graphic image on a CRT is redrawn in a refresh display, i.e., the time needed for one refresh of the displayed image.

**registration** The degree of accuracy in the positioning of one layer or overlay in a CAD display or artwork, relative to another layer, as reflected by the clarity and sharpness of the resulting image.

**repaint** A CAD feature that automatically redraws a design displayed on the CRT.

**resolution** The smallest spacing between two display elements that will allow the elements to be distinguished visually on the CRT. The ability to define very minute detail. For example, the resolution of Computervision's IC design system is one part in 33.5 million. As applied to an electrostatic plotter, resolution means the number of dots per square inch.

**response time** The elapsed time from initiation of an operation at a workstation to the receipt of the results at that workstation. Includes transmission of data to the CPU, processing, file access, and transmission of results back to the initiating workstation.

**restart**   To resume a computer program interrupted by operator intervention.

**restore**   To bring back to its original state a design currently being worked on in a CAD/CAM system after editing or modification that the designer now wants to cancel or rescind.

**resume**   A feature of some application programs that allows the designer to suspend the data-processing operation at some logical break point and restart it later from that point.

**reticle**   The photographic plate used to create an IC mask. See also *photo plotter*.

**rotate**   To turn a displayed 2D or 3D construction about an axis through a predefined angle relative to the original position.

**robotics**   The use of computer-controlled manipulators or arms to automate a variety of manufacturing processes such as welding, material handling, painting and assembly.

**ROM**   See *read-only memory*.

**routine**   A computer program, or a subroutine in the main program. The smallest separately compilable source code unit. See *computer program; source*.

**rubber banding**   A CAD capability that allows a component to be tracked (dragged) across the CRT screen, by means of an electronic pen, to a desired location, while simultaneously stretching all related interconnections to maintain signal continuity. During tracking the interconnections associated with the component stretch and bend, providing an excellent visual guide for optimizing the location of a component to best fit into the flow of the PC board, or other entity, minimizing total interconnect length and avoiding areas of congestion.

**satellite**   A remote system connected to another, usually larger, host system. A satellite differs from a remote intelligent work station in that it contains a full set of processors, memory, and mass storage resources to operate independently of the host. See *host-satellite system*.

**scale**   (*v*) To enlarge or diminish the size of a displayed entity without changing its shape, i.e., to bring it into a user-specified ratio to its original dimensions. Scaling can be done automatically by a CAD system. (*n*) Denotes the coordinate system for representing an object.

**scissoring**   The automatic erasing of all portions of a design on the CRT that lie outside user-specified boundaries.

**scroll**   To automatically roll up, as on a spool, a design or text message on a CRT to permit the sequential viewing of a message or drawing too large to be displayed all at once on the screen. New data appear on the CRT at one edge as other data disappear at the opposite edge. Graphics can be scrolled up, down, left, or right.

**selective erase**   A CAD feature for deleting portions of a display without affecting the remainder or having to repaint the entire CRT display.

**shape fill**   The automatic painting-in of an area, defined by user-specified boundaries, on an IC or PC board layout, for example, the area to be filled by copper when the PC board is manufactured. Can be done on-line by CAD.

**smoothing**   Fitting together curves and surfaces so that a smooth, continuous geometry results.

**software**   The collection of executable computer programs including application programs, operating systems, and languages.

**source**   A text file written in a high-level language and containing a computer program. It is easily read and understood by people but must be compiled or assembled to generate machine-recognizable instructions. Also known as *source code*. See also *high-level language*.

**source language**   A symbolic language comprised of statements and formulas used in computer processing. It is translated into object language (object code) by an assembler or compiler for execution by a computer.

**spline**   A subset of a B-spline where in a sequence of curves is restricted to a plane. An interpolation routine executed on a CAD/CAM system automatically adjusts a curve by design iteration until the curvature is continuous over the length of the curve. See also *B-spline*.

**storage**   The physical repository of all information relating to products designed on a CAD/CAM system. It is typically in the form of a magnetic tape or disk. Also called *memory*.

**storage tube**   A common type of CRT that retains an image continuously for a considerable period of time without redrawing (refreshing). The image will not flicker regardless of the amount of information displayed. However, the display tends to be slow relative to raster scan, the image is rather dim, and no single element by itself can be modified or deleted without redrawing. See also *direct view storage tube*.

**stretch**   A CAD design/editing aid that enables the designer to automatically expand a displayed entity beyond its original dimensions.

**string**   A linear sequence of entities, such as characters or physical elements, in a computer-aided design.

**stylus**   A hand-held pen used in conjunction with a data table to enter commands and coordinate input into the system. Also called an *electronic pen*.

**subfigure**   A part or a design element that may be extracted from a CAD library and inserted intact into another part displayed on the CRT.

**surface machining**   Automatic generation of NC tool paths to cut 3D shapes. Both the tool paths and the shapes may be constructed using the mechanical design capabilities of a CAD/CAM system.

**symbol**   Any recognizable sign, mark, shape or pattern

used as a building block for designing meaningful structures. A set of primitive graphic entities (line, point, arc, circle, text, etc.) that form a construction that can be expressed as one unit and assigned a meaning. Symbols may be combined or nested to form larger symbols and/or drawings. They can be as complex as an entire PC board or as simple as a single element, such as a pad. Symbols are commonly used to represent physical things. For example, a particular graphic shape may be used to represent a complete device or a certain kind of electrical component in a schematic. To simplify the preparation of drawings of piping systems and flow diagrams, standard symbols are used to represent various types of fittings and components in common use. Symbols are also basic units in a language. The recognizable sequence of characters END may inform a compiler that the routine it is compiling is completed. In computer-aided mapping, a symbol can be a diagram, design, letter, character, or abbreviation placed on maps and charts, that, by convention or reference to a legend, is understood to stand for or represent a specific characteristic or feature. In a CAD environment, symbol libraries contribute to the quick maintenance, placement, and interpretation of symbols.

**syntax** (1) A set of rules describing the structure of statements allowed in a computer language. To make grammatical sense, commands and routines must be written in conformity to these rules. (2) The structure of a computer command language, i.e., the English-sentence structure of a CAD/CAM command language, e.g., verb, noun, modifiers.

**system** An arrangement of CAD/CAM dataprocessing, memory, display, and plotting modules—coupled with appropriate software—to achieve specific objectives. The term CAD/CAM system implies both hardware and software. See also *operating system* (a purely software term).

**tablet** An input device on which a designer can digitize coordinate data or enter commands into a CAD/CAM system by means of an electronic pen. See also *data tablet*.

**task** (1) A specific project that can be executed by a CAD/CAM software program. (2) A specific portion of memory assigned to the user for executing that project.

**template** the pattern of a standard, commonly used component or part that serves as a design aid. Once created, it can be subsequently traced instead of redrawn whenever needed. The CAD equivalent of a designer's template might be a standard part in the data-base library that can be retrieved and inserted intact into an emerging drawing on the CRT.

**temporary storage** Memory locations for storing immediate and partial results obtained during the execution of a program on the system.

**terminal** See *workstation*.

**text editor** An operating system program used to create and modify text files on the system.

**text file** A file stored in the system in text format that can be printed and edited on-line as required.

**throughput** The number of units of work performed by a CAD/CAM system or a work station during a given period of time. A quantitative measure of system productivity.

**time-sharing** The use of a common CPU memory and processing capabilities by two or more CAD/CAM terminals to execute different tasks simultaneously.

**tool path** Centerline of the tip of an NC cutting tool as it moves over a part produced on a CAD/CAM system. Tool paths can be created and displayed interactively or automatically by a CAD/CAM system, and reformatted into NC tapes, by means of postprocessor, to guide or control machining equipment. See also *surface machining*.

**track ball** A CAD graphics input device consisting of a ball recessed into a surface. The designer can rotate it in any direction to control the position of the cursor used for entering coordinate data into the system.

**tracking** Moving a predefined (tracking) symbol across the surface of the CRT with a light pen or an electronic pen.

**transform** To change an image displayed on the CRT by, for example, scaling, rotating, translating, or mirroring.

**transformation** The process of transforming a CAD display image. Also the matrix representation of a geometric space.

**translate** (1) To convert CAD/CAM output from one language to another, for example, by means of a postprocessor such as Computervision's IPC-to-Numerics Translator program. (2) Also, by an editing command, to move a CAD display entity a specified distance in a specified direction.

**trap** The area that is searched around each digitize to find a hit on a graphics entity to be edited. See also *digitize*.

**turnaround time** The elapsed time between the moment a task or project is input into the CAD/CAM system and the moment the required output is obtained.

**turnkey** A CAD/CAM system for which the supplier/vendor assumes total responsibility for building, installing, and testing both hardware and software, and the training of user personnel. Also, loosely, a system that comes equipped with all the hardware and software required for a specific application or applications. Usually implies a commitment by the vendor to make the system work and to provide preventive and remedial maintenance of both hardware and software. Sometimes used interchangeably with stand-alone, although stand-alone applies more to system architecture than to terms of purchase.

**tutorial** A characteristic of CAD/CAM systems. If the

user is not sure how to execute a task, the system will show how. A message is displayed to provide information and guidance.

*utilities* Another term for system capabilities and/or features that enable the user to perform certain processes.

*vector* A quantity that has magnitude and direction and that, in CAD, is commonly represented by a directed line segment.

*verification* (1) A system-generated message to a work station acknowledging that a valid instruction or input has been received. (2) The process of checking the accuracy, viability, and/or manufacturability of an emerging design on the system.

*view port* A user-selected, rectangular view of a part, assembly, etc., that presents the contents of a window on the CRT. See also *window*.

*window* A temporary, usually rectangular, bounded area on the CRT that is user-specified to include particular entities for modification, editing, or deletion.

*wire-frame graphics* A computer-aided design technique for displaying a 3D object on the CRT screen as a series of lines outlining its surface.

*wiring diagram* (1) Graphic representation of all circuits and device elements of an electrical system and its associated apparatus or any clearly defined functional portion of that system. A wiring diagram may contain not only wiring system components and wires but also nongraphic information such as wire number, wire size, color, function, component label, and pin number. (2) Illustration of device elements and their interconnectivity as distinguished from their physical arrangement. (3) Drawing that shows how to hook things up.

Wiring diagrams can be constructed, annotated, and documented on a CAD system.

*word* A set of bits (typically 16 to 32) that occupies a single storage location and is treated by the computer as a unit. See also *bit*.

*working storage* That part of the system's internal storage reserved for intermediate results (i.e., while a computer program is still in progress). Also called *temporary storage*.

*workstation* The work area and equipment used for CAD/CAM operations. It is where the designer interacts (communicates) with the computer. Frequently consists of a CRT display and an input device as well as, possibly, a digitizer and a hard-copy device. In a distributed processing system, a work station would have local processing and mass storage capabilities. Also called a *terminal* or *design terminal*.

*write* To transfer information from CPU main memory to a peripheral device, such as a mass storage device.

*write-protect* A security feature in a CAD/CAM data storage device that prevents new data from being written over existing data.

*zero* The origin of all coordinate dimensions defined in an absolute system as the intersection of the baselines of the $x$, $y$, and $z$ axes.

*zero offset* On an NC unit, this features allows the zero point on an axis to be relocated anywhere within a specified range, thus temporarily redefining the coordinate frame of reference.

*zoom* A CAD capability that proportionately enlarges or reduces a figure displayed on a CRT screen.

# 4 Abbreviations for Use on Drawings and in Text— American National Standard

[Selected from ANSI Y1.1–1972 (R1984)]

**A**

| | | | | | |
|---|---|---|---|---|---|
| | | after | AFT. | American wire gage | AWG |
| | | aggregate | AGGR | amount | AMT |
| absolute | ABS | air condition | AIR COND | ampere | AMP |
| accelerate | ACCEL | airplane | APL | amplifier | AMPL |
| accessory | ACCESS. | allowance | ALLOW | anneal | ANL |
| account | ACCT | alloy | ALY | antenna | ANT. |
| accumulate | ACCUM | alteration | ALT | apartment | APT. |
| actual | ACT. | alternate | ALT | apparatus | APP |
| adapter | ADPT | alternating current | AC | appendix | APPX |
| addendum | ADD. | altitude | ALT | approved | APPD |
| addition | ADD. | aluminum | AL | approximate | APPROX |
| adjust | ADJ | American National | | arc weld | ARC/W |
| advance | ADV | Standard | AMER NATL STD | area | A |

| | | | | | | | |
|---|---|---|---|---|---|---|---|
| armature | ARM. | bolt circle | BC | circle | CIR |
| armor plate | ARM-PL | both faces | BF | circular | CIR |
| army navy | AN | both sides | BS | circular pitch | CP |
| arrange | ARR. | both ways | BW | circumference | CIRC |
| artificial | ART. | bottom | BOT | clear | CLR |
| asbestos | ASB | bottom chord | BC | clearance | CL |
| asphalt | ASPH | bottom face | BF | clockwise | CW |
| assemble | ASSEM | bracket | BRKT | coated | CTD |
| assembly | ASSY | brake | BK | cold drawn | CD |
| assistant | ASST | brake horsepower | BHP | cold-drawn steel | CDS |
| associate | ASSOC | brass | BRS | cold finish | CF |
| association | ASSN | brazing | BRZG | cold punched | CP |
| atomic | AT | break | BRK | cold rolled | CR |
| audible | AUD | Brinell hardness | BH | cold-rolled steel | CRS |
| audio frequency | AF | British Standard | BR STD | combination | COMB. |
| authorized | AUTH | British thermal unit | BTU | combustion | COMB |
| automatic | AUTO | broach | BRO | commercial | COML |
| auto-transformer | AUTO TR | bronze | BRZ | company | CO |
| auxiliary | AUX | Brown & Sharpe (wire gage, | | complete | COMPL |
| avenue | AVE | same as AWG) | B&S | compress | COMP |
| average | AVG | building | BLDG | concentric | CONC |
| aviation | AVI | bulkhead | BHD | concrete | CONC |
| azimuth | AZ | burnish | BNH | condition | COND |
| | | bushing | BUSH. | connect | CONN |
| | | button | BUT. | constant | CONST |
| | | | | construction | CONST |
| **B** | | | | contact | CONT |
| | | | | continue | CONT |
| Babbitt | BAB | | | copper | COP. |
| back feed | BF | | | corner | COR |
| back pressure | BP | cabinet | CAB. | corporation | CORP |
| back to back | B to B | calculate | CALC | correct | CORR |
| backface | BF | calibrate | CAL | corrugate | CORR |
| balance | BAL | cap screw | CAP SCR | cotter | COT |
| ball bearing | BB | capacity | CAP | counter | CTR |
| barometer | BAR | carburetor | CARB | counterbore | CBORE |
| base line | BL | carburize | CARB | counter clockwise | CCW |
| base plate | BP | carriage | CRG | counterdrill | CDRILL |
| bearing | BRG | case harden | CH | counterpunch | CPUNCH |
| bench mark | BM | cast iron | CI | countersink | CSK |
| bending moment | M | cast steel | CS | coupling | CPLG |
| bent | BT | casting | CSTG | cover | COV |
| bessemer | BESS | castle nut | CAS NUT | cross section | XSECT |
| between | BET. | catalogue | CAT. | cubic | CU |
| between centers | BC | cement | CEM | cubic foot | CU FT |
| between perpendiculars | BP | center | CTR | cubic inch | CU IN. |
| bevel | BEV | center line | CL | current | CUR |
| bill of material | B/M | center of gravity | CG | customer | CUST |
| Birmingham wire gage | BWG | center of pressure | CP | cyanide | CYN |
| blank | BLK | center to center | C to C | | |
| block | BLK | centering | CTR | | |
| blueprint | BP | chamfer | CHAM | | |
| board | BD | change | CHG | **D** | |
| boiler | BLR | channel | CHAN | | |
| boiler feed | BF | check | CHK | decimal | DEC |
| boiler horsepower | BHP | check valve | CV | dedendum | DED |
| boiling point | BP | chord | CHD | deflect | DEFL |

**C**

| | |
|---|---|
| degree | (°) DEG |
| density | D |
| department | DEPT |
| design | DSGN |
| detail | DET |
| develop | DEV |
| diagonal | DIAG |
| diagram | DIAG |
| diameter | DIA |
| diametral pitch | DP |
| dimension | DIM. |
| discharge | DISCH |
| distance | DIST |
| division | DIV |
| double | DBL |
| dovetail | DVTL |
| dowel | DWL |
| down | DN |
| dozen | DOZ |
| drafting | DFTG |
| drawing | DWG |
| drill or drill rod | DR |
| drive | DR |
| drive fit | DF |
| drop | D |
| drop forge | DF |
| duplicate | DUP |

**E**

| | |
|---|---|
| each | EA |
| east | E |
| eccentric | ECC |
| effective | EFF |
| elbow | ELL |
| electric | ELEC |
| elementary | ELEM |
| elevate | ELEV |
| elevation | EL |
| engine | ENG |
| engineer | ENGR |
| engineering | ENGRG |
| entrance | ENT |
| equal | EQ |
| equation | EQ |
| equipment | EQUIP |
| equivalent | EQUIV |
| estimate | EST |
| exchange | EXCH |
| exhaust | EXH |
| existing | EXIST. |
| exterior | EXT |
| extra heavy | X HVY |
| extra strong | X STR |
| extrude | EXTR |

**F**

| | |
|---|---|
| fabricate | FAB |
| face to face | F to F |
| Fahrenheit | F |
| far side | FS |
| federal | FED. |
| feed | FD |
| feet | (') FT |
| figure | FIG. |
| fillet | FIL |
| fillister | FIL |
| finish | FIN. |
| finish all over | FAO |
| flange | FLG |
| flat | F |
| flat head | FH |
| floor | FL |
| fluid | FL |
| focus | FOC |
| foot | (') FT |
| force | F |
| forged steel | FST |
| forging | FORG |
| forward | FWD |
| foundry | FDRY |
| frequency | FREQ |
| front | FR |
| furnish | FURN |

**G**

| | |
|---|---|
| gage or gauge | GA |
| gallon | GAL |
| galvanize | GALV |
| galvanized iron | GI |
| galvanized steel | GS |
| gasket | GSKT |
| general | GEN |
| glass | GL |
| government | GOVT |
| governor | GOV |
| grade | GR |
| graduation | GRAD |
| graphite | GPH |
| grind | GRD |
| groove | GRV |
| ground | GRD |

**H**

| | |
|---|---|
| half-round | $\frac{1}{2}$RD |
| handle | HDL |
| hanger | HGR |

| | |
|---|---|
| hard | H |
| harden | HDN |
| hardware | HDW |
| head | HD |
| headless | HDLS |
| heat | HT |
| heat-treat | HT TR |
| heavy | HVY |
| hexagon | HEX |
| high-pressure | HP |
| high-speed | HS |
| horizontal | HOR |
| horsepower | HP |
| hot rolled | HR |
| hot-rolled steel | HRS |
| hour | HR |
| housing | HSG |
| hydraulic | HYD |

**I**

| | |
|---|---|
| illustrate | ILLUS |
| inboard | INBD |
| inch | (") IN. |
| inches per second | IPS |
| inclosure | INCL |
| include | INCL |
| inside diameter | ID |
| instrument | INST |
| interior | INT |
| internal | INT |
| intersect | INT |
| iron | I |
| irregular | IREG |

**J**

| | |
|---|---|
| joint | JT |
| joint army-navy | JAN |
| journal | JNL |
| junction | JCT |

**K**

| | |
|---|---|
| key | K |
| keyseat | KST |
| Keyway | KWY |

**L**

| | |
|---|---|
| laboratory | LAB |

| | |
|---|---|
| laminate | LAM |
| lateral | LAT |
| left | L |
| left hand | LH |
| length | LG |
| length over all | LOA |
| letter | LTR |
| light | LT |
| line | L |
| locate | LOC |
| logarithm | LOG. |
| long | LG |
| lubricate | LUB |
| lumber | LBR |

**M**

| | |
|---|---|
| machine | MACH |
| machine steel | MS |
| maintenance | MAINT |
| malleable | MALL |
| malleable iron | MI |
| manual | MAN. |
| manufacture | MFR |
| manufactured | MFD |
| manufacturing | MFG |
| material | MATL |
| maximum | MAX |
| mechanical | MECH |
| mechanism | MECH |
| median | MED |
| metal | MET. |
| meter | M |
| miles | MI |
| miles per hour | MPH |
| millimeter | MM |
| minimum | MIN |
| minute | (') MIN |
| miscellaneous | MISC |
| month | MO |
| Morse taper | MOR T |
| motor | MOT |
| mounted | MTD |
| mounting | MTG |
| multiple | MULT |
| music wire gage | MWG |

**N**

| | |
|---|---|
| national | NATL |
| natural | NAT |
| near face | NF |
| near side | NS |
| negative | NEG |

| | |
|---|---|
| neutral | NEUT |
| nominal | NOM |
| normal | NOR |
| north | N |
| not to scale | NTS |
| number | NO. |

**O**

| | |
|---|---|
| obsolete | OBS |
| octagon | OCT |
| office | OFF. |
| on center | OC |
| opposite | OPP |
| optical | OPT |
| original | ORIG |
| outlet | OUT. |
| outside diameter | OD |
| outside face | OF |
| outside radius | OR |
| overall | OA |

**P**

| | |
|---|---|
| pack | PK |
| packing | PKG |
| page | P |
| paragraph | PAR. |
| part | PT |
| patent | PAT. |
| pattern | PATT |
| permanent | PERM |
| perpendicular | PERP |
| piece | PC |
| piece mark | PC MK |
| pint | PT |
| pitch | P |
| pitch circle | PC |
| pitch diameter | PD |
| plastic | PLSTC |
| plate | PL |
| plumbing | PLMB |
| point | PT |
| point of curve | PC |
| point of intersection | PI |
| point of tangent | PT |
| polish | POL |
| position | POS |
| potential | POT. |
| pound | LB |
| pounds per square inch | PSI |
| power | PWR |
| prefabricated | PREFAB |
| preferred | PFD |

| | |
|---|---|
| prepare | PREP |
| pressure | PRESS. |
| process | PROC |
| production | PROD |
| profile | PF |
| propeller | PROP |
| publication | PUB |
| push button | PB |

**Q**

| | |
|---|---|
| quadrant | QUAD |
| quality | QUAL |
| quarter | QTR |

**R**

| | |
|---|---|
| radial | RAD |
| radius | R |
| railroad | RR |
| ream | RM |
| received | RECD |
| record | REC |
| rectangle | RECT |
| reduce | RED. |
| reference line | REF L |
| reinforce | REINF |
| release | REL |
| relief | REL |
| remove | REM |
| require | REQ |
| required | REQD |
| return | RET. |
| reverse | REV |
| revolution | REV |
| revolutions per minute | RPM |
| right | R |
| right hand | RH |
| rivet | RIV |
| Rockwell hardness | RH |
| roller bearing | RB |
| room | RM |
| root diameter | RD |
| root mean square | RMS |
| rough | RGH |
| round | RD |

**S**

| | |
|---|---|
| schedule | SCH |
| schematic | SCHEM |
| scleroscope hardness | SH |

| | | | | | |
|---|---|---|---|---|---|
| screw | SCR | symbol | SYM | **V** | |
| second | SEC | system | SYS | | |
| section | SECT | | | vacuum | VAC |
| semi-steel | SS | | | valve | V |
| separate | SEP | **T** | | variable | VAR |
| set screw | SS | | | versus | VS |
| shaft | SFT | tangent | TAN. | vertical | VERT |
| sheet | SH | taper | TPR | volt | V |
| shoulder | SHLD | technical | TECH | volume | VOL |
| side | S | template | TEMP | | |
| single | S | tension | TENS. | | |
| sketch | SK | terminal | TERM. | **W** | |
| sleeve | SLV | thick | THK | | |
| slide | SL | thousand | M | wall | W |
| slotted | SLOT. | thread | THD | washer | WASH. |
| small | SM | threads per inch | TPI | watt | W |
| socket | SOC | through | THRU | week | WK |
| space | SP | time | T | weight | WT |
| special | SPL | tolerance | TOL | west | W |
| specific | SP | tongue & groove | T & G | width | W |
| spot faced | SF | tool steel | TS | wood | WD |
| spring | SPG | tooth | T | Woodruff | WDF |
| square | SQ | total | TOT | working point | WP |
| standard | STD | transfer | TRANS | working pressure | WP |
| station | STA | typical | TYP | wrought | WRT |
| stationary | STA | | | wrought iron | WI |
| steel | STL | | | | |
| stock | STK | | | | |
| straight | STR | **U** | | | |
| street | ST | | | **X, Y, Z** | |
| structural | STR | ultimate | ULT | | |
| substitute | SUB | unit | U | yard | YD |
| summary | SUM. | universal | UNIV | year | YR |
| support | SUP. | | | | |
| surface | SUR | | | | |

# 5   Running and Sliding Fits[a]—American National Standard

RC 1   *Close sliding fits* are intended for the accurate location of parts which must assemble without perceptible play.

RC 2   *Sliding fits* are intended for accurate location, but with greater maximum clearance than class RC 1. Parts made to this fit move and turn easily but are not intended to run freely, and in the larger sizes may seize with small temperature changes.

RC 3   *Precision running fits* are about the closest fits which can be expected to run freely, and are intended for precision work at slow speeds and light journal pressures, but are not suitable where appreciable temperature differences are likely to be encountered.

RC 4   *Close running fits* are intended chiefly for running fits on accurate machinery with moderate surface speeds and journal pressures, where accurate location and minimum play are desired.

Basic hole system. Limits are in thousandths of an inch. See §14.8.
Limits for hole and shaft are applied algebraically to the basic size to obtain the limits of size for the parts.
Data in **boldface** are in accordance with ABC agreements.
Symbols H5, g5, etc., are hole and shaft designations used in ABC System.

| Nominal Size Range, inches Over   To | Class RC 1 Limits of Clearance | Class RC 1 Standard Limits Hole H5 | Class RC 1 Standard Limits Shaft g4 | Class RC 2 Limits of Clearance | Class RC 2 Standard Limits Hole H6 | Class RC 2 Standard Limits Shaft g5 | Class RC 3 Limits of Clearance | Class RC 3 Standard Limits Hole H7 | Class RC 3 Standard Limits Shaft f6 | Class RC 4 Limits of Clearance | Class RC 4 Standard Limits Hole H8 | Class RC 4 Standard Limits Shaft f7 |
|---|---|---|---|---|---|---|---|---|---|---|---|---|
| 0   – 0.12 | 0.1 0.45 | +0.2 −0 | −0.1 −0.25 | 0.1 0.55 | +0.25 −0 | −0.1 −0.3 | 0.3 0.95 | +0.4 −0 | −0.3 −0.55 | 0.3 1.3 | +0.6 −0 | −0.3 −0.7 |
| 0.12– 0.24 | 0.15 0.5 | +0.2 −0 | −0.15 −0.3 | 0.15 0.65 | +0.3 −0 | −0.15 −0.35 | 0.4 1.12 | +0.5 −0 | −0.4 −0.7 | 0.4 1.6 | +0.7 −0 | −0.4 −0.9 |
| 0.24– 0.40 | 0.2 0.6 | +0.25 −0 | −0.2 −0.35 | 0.2 0.85 | +0.4 −0 | −0.2 −0.45 | 0.5 1.5 | +0.6 −0 | −0.5 −0.9 | 0.5 2.0 | +0.9 −0 | −0.5 −1.1 |
| 0.40– 0.71 | 0.25 0.75 | +0.3 −0 | −0.25 −0.45 | 0.25 0.95 | +0.4 −0 | −0.25 −0.55 | 0.6 1.7 | +0.7 −0 | −0.6 −1.0 | 0.6 2.3 | +1.0 −0 | −0.6 −1.3 |
| 0.71– 1.19 | 0.3 0.95 | +0.4 −0 | −0.3 −0.55 | 0.3 1.2 | +0.5 −0 | −0.3 −0.7 | 0.8 2.1 | +0.8 −0 | −0.8 −1.3 | 0.8 2.8 | +1.2 −0 | −0.8 −1.6 |
| 1.19– 1.97 | 0.4 1.1 | +0.4 −0 | −0.4 −0.7 | 0.4 1.4 | +0.6 −0 | −0.4 −0.8 | 1.0 2.6 | +1.0 −0 | −1.0 −1.6 | 1.0 3.6 | +1.6 −0 | −1.0 −2.0 |
| 1.97– 3.15 | 0.4 1.2 | +0.5 −0 | −0.4 −0.7 | 0.4 1.6 | +0.7 −0 | −0.4 −0.9 | 1.2 3.1 | +1.2 −0 | −1.2 −1.9 | 1.2 4.2 | +1.8 −0 | −1.2 −2.4 |
| 3.15– 4.73 | 0.5 1.5 | +0.6 −0 | −0.5 −0.9 | 0.5 2.0 | +0.9 −0 | −0.5 −1.1 | 1.4 3.7 | +1.4 −0 | −1.4 −2.3 | 1.4 5.0 | +2.2 −0 | −1.4 −2.8 |
| 4.73– 7.09 | 0.6 1.8 | +0.7 −0 | −0.6 −1.1 | 0.6 2.3 | +1.0 −0 | −0.6 −1.3 | 1.6 4.2 | +1.6 −0 | −1.6 −2.6 | 1.6 5.7 | +2.5 −0 | −1.6 −3.2 |
| 7.09– 9.85 | 0.6 2.0 | +0.8 −0 | −0.6 −1.2 | 0.6 2.6 | +1.2 −0 | −0.6 −1.4 | 2.0 5.0 | +1.8 −0 | −2.0 −3.2 | 2.0 6.6 | +2.8 −0 | −2.0 −3.8 |
| 9.85–12.41 | 0.8 2.3 | +0.9 −0 | −0.8 −1.4 | 0.8 2.9 | +1.2 −0 | −0.8 −1.7 | 2.5 5.7 | +2.0 −0 | −2.5 −3.7 | 2.5 7.5 | +3.0 −0 | −2.5 −4.5 |
| 12.41–15.75 | 1.0 2.7 | +1.0 −0 | −1.0 −1.7 | 1.0 3.4 | +1.4 −0 | −1.0 −2.0 | 3.0 6.6 | +2.2 −0 | −3.0 −4.4 | 3.0 8.7 | +3.5 −0 | −3.0 −5.2 |

[a]From ANSI B4.1—1967 (R1987). For larger diameters, see the standard.

# 5 Running and Sliding Fits[a]—American National Standard (continued)

RC 5 }
RC 6 }    *Medium running fits* are intended for higher running speeds, or heavy journal pressures, or both.

RC 7    *Free running fits* are intended for use where accuracy is not essential, or where large temperature variations are likely to be encountered, or under both these conditions.

RC 8 }
RC 9 }    *Loose running fits* are intended for use where wide commercial tolerances may be necessary, together with an allowance, on the external member.

| Nominal Size Range, inches Over — To | Class RC 5 Limits of Clearance | Class RC 5 Standard Limits Hole H8 | Class RC 5 Standard Limits Shaft e7 | Class RC 6 Limits of Clearance | Class RC 6 Standard Limits Hole H9 | Class RC 6 Standard Limits Shaft e8 | Class RC 7 Limits of Clearance | Class RC 7 Standard Limits Hole H9 | Class RC 7 Standard Limits Shaft d8 | Class RC 8 Limits of Clearance | Class RC 8 Standard Limits Hole H10 | Class RC 8 Standard Limits Shaft c9 | Class RC 9 Limits of Clearance | Class RC 9 Standard Limits Hole H11 | Class RC 9 Standard Limits Shaft |
|---|---|---|---|---|---|---|---|---|---|---|---|---|---|---|---|
| 0 – 0.12 | 0.6 1.6 | +0.6 −0 | −0.6 −1.0 | 0.6 2.2 | +1.0 −0 | −0.6 −1.2 | 1.0 2.6 | +1.0 −0 | − 1.0 − 1.6 | 2.5 5.1 | +1.6 −0 | − 2.5 − 3.5 | 4.0 8.1 | + 2.5 − 0 | − 4.0 − 5.6 |
| 0.12– 0.24 | 0.8 2.0 | +0.7 −0 | −0.8 −1.3 | 0.8 2.7 | +1.2 −0 | −0.8 −1.5 | 1.2 3.1 | +1.2 −0 | − 1.2 − 1.9 | 2.8 5.8 | +1.8 −0 | − 2.8 − 4.0 | 4.5 9.0 | + 3.0 − 0 | − 4.5 − 6.0 |
| 0.24– 0.40 | 1.0 2.5 | +0.9 −0 | −1.0 −1.6 | 1.0 3.3 | +1.4 −0 | −1.0 −1.9 | 1.6 3.9 | +1.4 −0 | − 1.6 − 2.5 | 3.0 6.6 | +2.2 −0 | − 3.0 − 4.4 | 5.0 10.7 | + 3.5 − 0 | − 5.0 − 7.2 |
| 0.40– 0.71 | 1.2 2.9 | +1.0 −0 | −1.2 −1.9 | 1.2 3.8 | +1.6 −0 | −1.2 −2.2 | 2.0 4.6 | +1.6 −0 | − 2.0 − 3.0 | 3.5 7.9 | +2.8 −0 | − 3.5 − 5.1 | 6.0 12.8 | + 4.0 − 0 | − 6.0 − 8.8 |
| 0.71– 1.19 | 1.6 3.6 | +1.2 −0 | −1.6 −2.4 | 1.6 4.8 | +2.0 −0 | −1.6 −2.8 | 2.5 5.7 | +2.0 −0 | − 2.5 − 3.7 | 4.5 10.0 | +3.5 −0 | − 4.5 − 6.5 | 7.0 15.5 | + 5.0 − 0 | − 7.0 −10.5 |
| 1.19– 1.97 | 2.0 4.6 | +1.6 −0 | −2.0 −3.0 | 2.0 6.1 | +2.5 −0 | −2.0 −3.6 | 3.0 7.1 | +2.5 −0 | − 3.0 − 4.6 | 5.0 11.5 | +4.0 −0 | − 5.0 − 7.5 | 8.0 18.0 | + 6.0 − 0 | − 8.0 −12.0 |
| 1.97– 3.15 | 2.5 5.5 | +1.8 −0 | −2.5 −3.7 | 2.5 7.3 | +3.0 −0 | −2.5 −4.3 | 4.0 8.8 | +3.0 −0 | − 4.0 − 5.8 | 6.0 13.5 | +4.5 −0 | − 6.0 − 9.0 | 9.0 20.5 | + 7.0 − 0 | − 9.0 −13.5 |
| 3.15– 4.73 | 3.0 6.6 | +2.2 −0 | −3.0 −4.4 | 3.0 8.7 | +3.5 −0 | −3.0 −5.2 | 5.0 10.7 | +3.5 −0 | − 5.0 − 7.2 | 7.0 15.5 | +5.0 −0 | − 7.0 −10.5 | 10.0 24.0 | + 9.0 − 0 | −10.0 −15.0 |
| 4.73– 7.09 | 3.5 7.6 | +2.5 −0 | −3.5 −5.1 | 3.5 10.0 | +4.0 −0 | −3.5 −6.0 | 6.0 12.5 | +4.0 −0 | − 6.0 − 8.5 | 8.0 18.0 | +6.0 −0 | − 8.0 −12.0 | 12.0 28.0 | +10.0 − 0 | −12.0 −18.0 |
| 7.09– 9.85 | 4.0 8.6 | +2.8 −0 | −4.0 −5.8 | 4.0 11.3 | +4.5 −0 | −4.0 −6.8 | 7.0 14.3 | +4.5 −0 | − 7.0 − 9.8 | 10.0 21.5 | +7.0 −0 | −10.0 −14.5 | 15.0 34.0 | +12.0 − 0 | −15.0 −22.0 |
| 9.85–12.41 | 5.0 10.0 | +3.0 −0 | −5.0 −7.0 | 5.0 13.0 | +5.0 −0 | −5.0 −8.0 | 8.0 16.0 | +5.0 −0 | − 8.0 −11.0 | 12.0 25.0 | +8.0 −0 | −12.0 −17.0 | 18.0 38.0 | +12.0 − 0 | −18.0 −26.0 |
| 12.41–15.75 | 6.0 11.7 | +3.5 −0 | −6.0 −8.2 | 6.0 15.5 | +6.0 −0 | −6.0 −9.5 | 10.0 19.5 | +6.0 −0 | −10.0 13.5 | 14.0 29.0 | +9.0 −0 | −14.0 −20.0 | 22.0 45.0 | +14.0 − 0 | −22.0 −31.0 |

[a]From ANSI B4.1—1967 (R1987). For larger diameters, see the standard.

# 6  Clearance Locational Fits[a]—American National Standard

LC  Locational clearance fits are intended for parts which are normally stationary, but which can be freely assembled or disassembled. They run from snug fits for parts requiring accuracy of location, through the medium clearance fits for parts such as spigots, to the looser fastener fits where freedom of assembly is of prime importance.

Basic hole system. Limits are in thousandths of an inch. See §14.8.
Limits for hole and shaft are applied algebraically to the basic size to obtain the limits of size for the parts.
Data in **boldface** are in accordance with ABC agreements.
Symbols H6, h5, etc., are hole and shaft designations used in ABC System.

| Nominal Size Range, inches Over | To | Class LC 1 Limits of Clearance | Class LC 1 Hole H6 | Class LC 1 Shaft h5 | Class LC 2 Limits of Clearance | Class LC 2 Hole H7 | Class LC 2 Shaft h6 | Class LC 3 Limits of Clearance | Class LC 3 Hole H8 | Class LC 3 Shaft h7 | Class LC 4 Limits of Clearance | Class LC 4 Hole H10 | Class LC 4 Shaft h9 | Class LC 5 Limits of Clearance | Class LC 5 Hole H7 | Class LC 5 Shaft g6 |
|---|---|---|---|---|---|---|---|---|---|---|---|---|---|---|---|---|
| 0 | 0.12 | 0, 0.45 | +0.25, −0 | +0, −0.2 | 0, 0.65 | +0.4, −0 | +0, −0.25 | 0, 1 | +0.6, −0 | +0, −0.4 | 0, 2.6 | +1.6, −0 | +0, −1.0 | 0.1, 0.75 | +0.4, −0 | −0.1, −0.35 |
| 0.12 | 0.24 | 0, 0.5 | +0.3, −0 | +0, −0.2 | 0, 0.8 | +0.5, −0 | +0, −0.3 | 0, 1.2 | +0.7, −0 | +0, −0.5 | 0, 3.0 | +1.8, −0 | +0, −1.2 | 0.15, 0.95 | +0.5, −0 | −0.15, −0.45 |
| 0.24 | 0.40 | 0, 0.65 | +0.4, −0 | +0, −0.25 | 0, 1.0 | +0.6, −0 | +0, −0.4 | 0, 1.5 | +0.9, −0 | +0, −0.6 | 0, 3.6 | +2.2, −0 | +0, −1.4 | 0.2, 1.2 | +0.6, −0 | −0.2, −0.6 |
| 0.40 | 0.71 | 0, 0.7 | +0.4, −0 | +0, −0.3 | 0, 1.1 | +0.7, −0 | +0, −0.4 | 0, 1.7 | +1.0, −0 | +0, −0.7 | 0, 4.4 | +2.8, −0 | +0, −1.6 | 0.25, 1.35 | +0.7, −0 | −0.25, −0.65 |
| 0.71 | 1.19 | 0, 0.9 | +0.5, −0 | +0, −0.4 | 0, 1.3 | +0.8, −0 | +0, −0.5 | 0, 2 | +1.2, −0 | +0, −0.8 | 0, 5.5 | +3.5, −0 | +0, −2.0 | 0.3, 1.6 | +0.8, −0 | −0.3, −0.8 |
| 1.19 | 1.97 | 0, 1.0 | +0.6, −0 | +0, −0.4 | 0, 1.6 | +1.0, −0 | +0, −0.6 | 0, 2.6 | +1.6, −0 | +0, −1 | 0, 6.5 | +4.0, −0 | +0, −2.5 | 0.4, 2.0 | +1.0, −0 | −0.4, −1.0 |
| 1.97 | 3.15 | 0, 1.2 | +0.7, −0 | +0, −0.5 | 0, 1.9 | +1.2, −0 | +0, −0.7 | 0, 3 | +1.8, −0 | +0, −1.2 | 0, 7.5 | +4.5, −0 | +0, −3 | 0.4, 2.3 | +1.2, −0 | −0.4, −1.1 |
| 3.15 | 4.73 | 0, 1.5 | +0.9, −0 | +0, −0.6 | 0, 2.3 | +1.4, −0 | +0, −0.9 | 0, 3.6 | +2.2, −0 | +0, −1.4 | 0, 8.5 | +5.0, −0 | +0, −3.5 | 0.5, 2.8 | +1.4, −0 | −0.5, −1.4 |
| 4.73 | 7.09 | 0, 1.7 | +1.0, −0 | +0, −0.7 | 0, 2.6 | +1.6, −0 | +0, −1.0 | 0, 4.1 | +2.5, −0 | +0, −1.6 | 0, 10 | +6.0, −0 | +0, −4 | 0.6, 3.2 | +1.6, −0 | −0.6, −1.6 |
| 7.09 | 9.85 | 0, 2.0 | +1.2, −0 | +0, −0.8 | 0, 3.0 | +1.8, −0 | +0, −1.2 | 0, 4.6 | +2.8, −0 | +0, −1.8 | 0, 11.5 | +7.0, −0 | +0, −4.5 | 0.6, 3.6 | +1.8, −0 | −0.6, −1.8 |
| 9.85 | 12.41 | 0, 2.1 | +1.2, −0 | +0, −0.9 | 0, 3.2 | +2.0, −0 | +0, −1.2 | 0, 5 | +3.0, −0 | +0, −2.0 | 0, 13 | +8.0, −0 | +0, −5 | 0.7, 3.9 | +2.0, −0 | −0.7, −1.9 |
| 12.41 | 15.75 | 0, 2.4 | +1.4, −0 | +0, −1.0 | 0, 3.6 | +2.2, −0 | +0, −1.4 | 0, 5.7 | +3.5, −0 | +0, −2.2 | 0, 15 | +9.0, −0 | +0, −6 | 0.7, 4.3 | +2.2, −0 | −0.7, −2.1 |

[a]From ANSI B4.1—1967 (R1987). For larger diameters, see the standard.

# 6  Clearance Locational Fits[a]—American National Standard (continued)

| Nominal Size Range, inches (Over – To) | LC 6 Limits of Clearance | LC 6 Hole H9 | LC 6 Shaft f8 | LC 7 Limits of Clearance | LC 7 Hole H10 | LC 7 Shaft e9 | LC 8 Limits of Clearance | LC 8 Hole H10 | LC 8 Shaft d9 | LC 9 Limits of Clearance | LC 9 Hole H11 | LC 9 Shaft c10 | LC 10 Limits of Clearance | LC 10 Hole H12 | LC 10 Shaft | LC 11 Limits of Clearance | LC 11 Hole H13 | LC 11 Shaft |
|---|---|---|---|---|---|---|---|---|---|---|---|---|---|---|---|---|---|---|
| 0 – 0.12 | 0.3 / 1.9 | +1.0 / −0 | −0.3 / −0.9 | 0.6 / 3.2 | +1.6 / −0 | −0.6 / −1.6 | 1.0 / 3.6 | +1.6 / −0 | −1.0 / −2.0 | 2.5 / 6.6 | +2.5 / −0 | −2.5 / −4.1 | 4 / 12 | +4 / −0 | −4 / −8 | 5 / 17 | +6 / −0 | −5 / −11 |
| 0.12– 0.24 | 0.4 / 2.3 | +1.2 / −0 | −0.4 / −1.1 | 0.8 / 3.8 | +1.8 / −0 | −0.8 / −2.0 | 1.2 / 4.2 | +1.8 / −0 | −1.2 / −2.4 | 2.8 / 7.6 | +3.0 / −0 | −2.8 / −4.6 | 4.5 / 14.5 | +5 / −0 | −4.5 / −9.5 | 6 / 20 | +7 / −0 | −6 / −13 |
| 0.24– 0.40 | 0.5 / 2.8 | +1.4 / −0 | −0.5 / −1.4 | 1.0 / 4.6 | +2.2 / −0 | −1.0 / −2.4 | 1.6 / 5.2 | +2.2 / −0 | −1.6 / −3.0 | 3.0 / 8.7 | +3.5 / −0 | −3.0 / −5.2 | 5 / 17 | +6 / −0 | −5 / −11 | 7 / 25 | +9 / −0 | −7 / −16 |
| 0.40– 0.71 | 0.6 / 3.2 | +1.6 / −0 | −0.6 / −1.6 | 1.2 / 5.6 | +2.8 / −0 | −1.2 / −2.8 | 2.0 / 6.4 | +2.8 / −0 | −2.0 / −3.6 | 3.5 / 10.3 | +4.0 / −0 | −3.5 / −6.3 | 6 / 20 | +7 / −0 | −6 / −13 | 8 / 28 | +10 / −0 | −8 / −18 |
| 0.71– 1.19 | 0.8 / 4.0 | +2.0 / −0 | −0.8 / −2.0 | 1.6 / 7.1 | +3.5 / −0 | −1.6 / −3.6 | 2.5 / 8.0 | +3.5 / −0 | −2.5 / −4.5 | 4.5 / 13.0 | +5.0 / −0 | −4.5 / −8.0 | 7 / 23 | +8 / −0 | −7 / −15 | 10 / 34 | +12 / −0 | −10 / −22 |
| 1.19– 1.97 | 1.0 / 5.1 | +2.5 / −0 | −1.0 / −2.6 | 2.0 / 8.5 | +4.0 / −0 | −2.0 / −4.5 | 3.0 / 9.5 | +4.0 / −0 | −3.0 / −5.5 | 5 / 15 | +6 / −0 | −5 / −9 | 8 / 28 | +10 / −0 | −8 / −18 | 12 / 44 | +16 / −0 | −12 / −28 |
| 1.97– 3.15 | 1.2 / 6.0 | +3.0 / −0 | −1.2 / −3.0 | 2.5 / 10.0 | +4.5 / −0 | −2.5 / −5.5 | 4.0 / 11.5 | +4.5 / −0 | −4.0 / −7.0 | 6 / 17.5 | +7 / −0 | −6 / −10.5 | 10 / 34 | +12 / −0 | −10 / −22 | 14 / 50 | +18 / −0 | −14 / −32 |
| 3.15– 4.73 | 1.4 / 7.1 | +3.5 / −0 | −1.4 / −3.6 | 3.0 / 11.5 | +5.0 / −0 | −3.0 / −6.5 | 5.0 / 13.5 | +5.0 / −0 | −5.0 / −8.5 | 7 / 21 | +9 / −0 | −7 / −12 | 11 / 39 | +14 / −0 | −11 / −25 | 16 / 60 | +22 / −0 | −16 / −38 |
| 4.73– 7.09 | 1.6 / 8.1 | +4.0 / −0 | −1.6 / −4.1 | 3.5 / 13.5 | +6.0 / −0 | −3.5 / −7.5 | 6 / 16 | +6 / −0 | −6 / −10 | 8 / 24 | +10 / −0 | −8 / −14 | 12 / 44 | +16 / −0 | −12 / −28 | 18 / 68 | +25 / −0 | −18 / −43 |
| 7.09– 9.85 | 2.0 / 9.3 | +4.5 / −0 | −2.0 / −4.8 | 4.0 / 15.5 | +7.0 / −0 | −4.0 / −8.5 | 7 / 18.5 | +7 / −0 | −7 / −11.5 | 10 / 29 | +12 / −0 | −10 / −17 | 16 / 52 | +18 / −0 | −16 / −34 | 22 / 78 | +28 / −0 | −22 / −50 |
| 9.85–12.41 | 2.2 / 10.2 | +5.0 / −0 | −2.2 / −5.2 | 4.5 / 17.5 | +8.0 / −0 | −4.5 / −9.5 | 7 / 20 | +8 / −0 | −7 / −12 | 12 / 32 | +12 / −0 | −12 / −20 | 20 / 60 | +20 / −0 | −20 / −40 | 28 / 88 | +30 / −0 | −28 / −58 |
| 12.41–15.75 | 2.5 / 12.0 | +6.0 / −0 | −2.5 / −6.0 | 5.0 / 20.0 | +9.0 / −0 | −5 / −11 | 8 / 23 | +9 / −0 | −8 / −14 | 14 / 37 | +14 / −0 | −14 / −23 | 22 / 66 | +22 / −0 | −22 / −44 | 30 / 100 | +35 / −0 | −30 / −65 |

[a]From ANSI B4.1−1967 (R1987). For larger diameters, see the standard.

# 7 Transition Locational Fits[a]—American National Standard

LT  *Transition fits* are a compromise between clearance and interference fits, for application where accuracy of location is important, but either a small amount of clearance or interference is permissible.

Basic hole system. Limits are in thousandths of an inch. See §14.8.

Limits for hole and shaft are applied algebraically to the basic size to obtain the limits of size for the mating parts.

Data in **boldface** are in accordance with ABC agreements.

"Fit" represents the maximum interference (minus values) and the maximum clearance (plus values).

Symbols H7, js6, etc., are hole and shaft designations used in ABC System.

| Nominal Size Range, inches Over–To | Class LT 1 Fit | Class LT 1 Hole H7 | Class LT 1 Shaft js6 | Class LT 2 Fit | Class LT 2 Hole H8 | Class LT 2 Shaft js7 | Class LT 3 Fit | Class LT 3 Hole H7 | Class LT 3 Shaft k6 | Class LT 4 Fit | Class LT 4 Hole H8 | Class LT 4 Shaft k7 | Class LT 5 Fit | Class LT 5 Hole H7 | Class LT 5 Shaft n6 | Class LT 6 Fit | Class LT 6 Hole H7 | Class LT 6 Shaft n7 |
|---|---|---|---|---|---|---|---|---|---|---|---|---|---|---|---|---|---|---|
| 0 – 0.12 | −0.10 / +0.50 | +0.4 / −0 | +0.10 / −0.10 | −0.2 / +0.8 | +0.6 / −0 | +0.2 / −0.2 | | | | | | | −0.5 / +0.15 | +0.4 / −0 | +0.5 / +0.25 | −0.65 / +0.15 | +0.4 / −0 | +0.65 / +0.25 |
| 0.12– 0.24 | −0.15 / +0.65 | +0.5 / −0 | +0.15 / −0.15 | −0.25 / +0.95 | +0.7 / −0 | +0.25 / −0.25 | | | | | | | −0.6 / +0.2 | +0.5 / −0 | +0.6 / +0.3 | −0.8 / +0.2 | +0.5 / −0 | +0.8 / +0.3 |
| 0.24– 0.40 | −0.2 / +0.8 | +0.6 / −0 | +0.2 / −0.2 | −0.3 / +1.2 | +0.9 / −0 | +0.3 / −0.3 | −0.5 / +0.5 | +0.6 / −0 | +0.5 / +0.1 | −0.7 / +0.8 | +0.9 / −0 | +0.7 / +0.1 | −0.8 / +0.2 | +0.6 / −0 | +0.8 / +0.4 | −1.0 / +0.2 | +0.6 / −0 | +1.0 / +0.4 |
| 0.40– 0.71 | −0.2 / +0.9 | +0.7 / −0 | +0.2 / −0.2 | −0.35 / +1.35 | +1.0 / −0 | +0.35 / −0.35 | −0.5 / +0.6 | +0.7 / −0 | +0.5 / +0.1 | −0.8 / +0.9 | +1.0 / −0 | +0.8 / +0.1 | −0.9 / +0.2 | +0.7 / −0 | +0.9 / +0.5 | −1.2 / +0.2 | +0.7 / −0 | +1.2 / +0.5 |
| 0.71– 1.19 | −0.25 / +1.05 | +0.8 / −0 | +0.25 / −0.25 | −0.4 / +1.6 | +1.2 / −0 | +0.4 / −0.4 | −0.6 / +0.7 | +0.8 / −0 | +0.6 / +0.1 | −0.9 / +1.1 | +1.2 / −0 | +0.9 / +0.1 | −1.1 / +0.2 | +0.8 / −0 | +1.1 / +0.6 | −1.4 / +0.2 | +0.8 / −0 | +1.4 / +0.6 |
| 1.19– 1.97 | −0.3 / +1.3 | +1.0 / −0 | +0.3 / −0.3 | −0.5 / +2.1 | +1.6 / −0 | +0.5 / −0.5 | −0.7 / +0.9 | +1.0 / −0 | +0.7 / +0.1 | −1.1 / +1.5 | +1.6 / −0 | +1.1 / +0.1 | −1.3 / +0.3 | +1.0 / −0 | +1.3 / +0.7 | −1.7 / +0.3 | +1.0 / −0 | +1.7 / +0.7 |
| 1.97– 3.15 | −0.3 / +1.5 | +1.2 / −0 | +0.3 / −0.3 | −0.6 / +2.4 | +1.8 / −0 | +0.6 / −0.6 | −0.8 / +1.1 | +1.2 / −0 | +0.8 / +0.1 | −1.3 / +1.7 | +1.8 / −0 | +1.3 / +0.1 | −1.5 / +0.4 | +1.2 / −0 | +1.5 / +0.8 | −2.0 / +0.4 | +1.2 / −0 | +2.0 / +0.8 |
| 3.15– 4.73 | −0.4 / +1.8 | +1.4 / −0 | +0.4 / −0.4 | −0.7 / +2.9 | +2.2 / −0 | +0.7 / −0.7 | −1.0 / +1.3 | +1.4 / −0 | +1.0 / +0.1 | −1.5 / +2.1 | +2.2 / −0 | +1.5 / +0.1 | −1.9 / +0.4 | +1.4 / −0 | +1.9 / +1.0 | −2.4 / +0.4 | +1.4 / −0 | +2.4 / +1.0 |
| 4.73– 7.09 | −0.5 / +2.1 | +1.6 / −0 | +0.5 / −0.5 | −0.8 / +3.3 | +2.5 / −0 | +0.8 / −0.8 | −1.1 / +1.5 | +1.6 / −0 | +1.1 / +0.1 | −1.7 / +2.4 | +2.5 / −0 | +1.7 / +0.1 | −2.2 / +0.4 | +1.6 / −0 | +2.2 / +1.2 | −2.8 / +0.4 | +1.6 / −0 | +2.8 / +1.2 |
| 7.09– 9.85 | −0.6 / +2.4 | +1.8 / −0 | +0.6 / −0.6 | −0.9 / +3.7 | +2.8 / −0 | +0.9 / −0.9 | −1.4 / +1.6 | +1.8 / −0 | +1.4 / +0.2 | −2.0 / +2.6 | +2.8 / −0 | +2.0 / +0.2 | −2.6 / +0.4 | +1.8 / −0 | +2.6 / +1.4 | −3.2 / +0.4 | +1.8 / −0 | +3.2 / +1.4 |
| 9.85–12.41 | −0.6 / +2.6 | +2.0 / −0 | +0.6 / −0.6 | −1.0 / +4.0 | +3.0 / −0 | +1.0 / −1.0 | −1.4 / +1.8 | +2.0 / −0 | +1.4 / +0.2 | −2.2 / +2.8 | +3.0 / −0 | +2.2 / +0.2 | −2.6 / +0.6 | +2.0 / −0 | +2.6 / +1.4 | −3.4 / +0.6 | +2.0 / −0 | +3.4 / +1.4 |
| 12.41–15.75 | −0.7 / +2.9 | +2.2 / −0 | +0.7 / −0.7 | −1.0 / +4.5 | +3.5 / −0 | +1.0 / −1.0 | −1.6 / +2.0 | +2.2 / −0 | +1.6 / +0.2 | −2.4 / +3.3 | +3.5 / −0 | +2.4 / +0.2 | −3.0 / +0.6 | +2.2 / −0 | +3.0 / +1.6 | −3.8 / +0.6 | +2.2 / −0 | +3.8 / +1.6 |

[a]From ANSI B4.1–1967 (R1987). For larger diameters, see the standard.

# 8   Interference Locational Fits[a]—American National Standard

LN   *Locational interference fits* are used where accuracy of location is of prime importance, and for parts requiring rigidity and alignment with no special requirements for bore pressure. Such fits are not intended for parts designed to transmit frictional loads from one part to another by virtue of the tightness of fit, as these conditions are covered by force fits.

Basic hole system. Limits are in thousandths of an inch. See §14.8.
Limits for hole and shaft are applied algebraically to the basic size to obtain the limits of size for the parts.
Data in **boldface** are in accordance with ABC agreements.
Symbols H7, p6, etc., are hole and shaft designations used in ABC System.

| Nominal Size Range, inches Over    To | Class LN 1 | | | Class LN 2 | | | Class LN 3 | | |
| --- | --- | --- | --- | --- | --- | --- | --- | --- | --- |
| | Limits of Interference | Standard Limits | | Limits of Interference | Standard Limits | | Limits of Interference | Standard Limits | |
| | | Hole H6 | Shaft n5 | | Hole H7 | Shaft p6 | | Hole H7 | Shaft r6 |
| 0   – 0.12 | **0** **0.45** | **+0.25** **−0** | **+0.45** **+0.25** | **0** **0.65** | **+0.4** **−0** | **+0.65** **+0.4** | **0.1** **0.75** | **+0.4** **−0** | **+0.75** **+0.5** |
| 0.12– 0.24 | **0** **0.5** | **+0.3** **−0** | **+0.5** **+0.3** | **0** **0.8** | **+0.5** **−0** | **+0.8** **+0.5** | **0.1** **0.9** | **+0.5** **0** | **+0.9** **+0.6** |
| 0.24– 0.40 | **0** **0.65** | **+0.4** **−0** | **+0.65** **+0.4** | **0** **1.0** | **+0.6** **−0** | **+1.0** **+0.6** | **0.2** **1.2** | **+0.6** **−0** | **+1.2** **+0.8** |
| 0.40– 0.71 | **0** **0.8** | **+0.4** **−0** | **+0.8** **+0.4** | **0** **1.1** | **+0.7** **−0** | **+1.1** **+0.7** | **0.3** **1.4** | **+0.7** **−0** | **+1.4** **+1.0** |
| 0.71– 1.19 | **0** **1.0** | **+0.5** **−0** | **+1.0** **+0.5** | **0** **1.3** | **+0.8** **−0** | **+1.3** **+0.8** | **0.4** **1.7** | **+0.8** **−0** | **+1.7** **+1.2** |
| 1.19– 1.97 | **0** **1.1** | **+0.6** **−0** | **+1.1** **+0.6** | **0** **1.6** | **+1.0** **−0** | **+1.6** **+1.0** | **0.4** **2.0** | **+1.0** **−0** | **+2.0** **+1.4** |
| 1.97– 3.15 | **0.1** **1.3** | **+0.7** **−0** | **+1.3** **+0.7** | **0.2** **2.1** | **+1.2** **−0** | **+2.1** **+1.4** | **0.4** **2.3** | **+1.2** **−0** | **+2.3** **+1.6** |
| 3.15– 4.73 | **0.1** **1.6** | **+0.9** **−0** | **+1.6** **+1.0** | **0.2** **2.5** | **+1.4** **−0** | **+2.5** **+1.6** | **0.6** **2.9** | **+1.4** **−0** | **+2.9** **+2.0** |
| 4.73– 7.09 | **0.2** **1.9** | **+1.0** **−0** | **+1.9** **+1.2** | **0.2** **2.8** | **+1.6** **−0** | **+2.8** **+1.8** | **0.9** **3.5** | **+1.6** **−0** | **+3.5** **+2.5** |
| 7.09– 9.85 | **0.2** **2.2** | **+1.2** **−0** | **+2.2** **+1.4** | **0.2** **3.2** | **+1.8** **−0** | **+3.2** **+2.0** | **1.2** **4.2** | **+1.8** **−0** | **+4.2** **+3.0** |
| 9.85–12.41 | **0.2** **2.3** | **+1.2** **−0** | **+2.3** **+1.4** | **0.2** **3.4** | **+2.0** **−0** | **+3.4** **+2.2** | **1.5** **4.7** | **+2.0** **−0** | **+4.7** **+3.5** |

[a]From ANSI B4.1—1967 (R1987). For larger diameters, see the standard.

# 9    Force and Shrink Fits[a]—American National Standard

FN 1    Light *drive* fits are those requiring light assembly pressures, and produce more or less permanent assemblies. They are suitable for thin sections or long fits, or in cast-iron external members.

FN 2    Medium *drive* fits are suitable for ordinary steel parts, or for shrink fits on light sections. They are about the tightest fits that can be used with high-grade cast-iron external members.

FN 3    Heavy *drive* fits are suitable for heavier steel parts or for shrink fits in medium sections.

FN 4)
FN 5)    Force *fits* are suitable for parts which can be highly stressed, or for shrink fits where the heavy pressing forces required are impractical.

Basic hole system. Limits are in thousandths of an inch. See §14.8.

Limits for hole and shaft are applied algebraically to the basic size to obtain the limits of size for the parts.

Data in **boldface** are in accordance with ABC agreements.

Symbols H7, s6, etc., are hole and shaft designations used in ABC System.

| Nominal Size Range, inches Over — To | Class FN 1 | | | Class FN 2 | | | Class FN 3 | | | Class FN 4 | | | Class FN 5 | | |
|---|---|---|---|---|---|---|---|---|---|---|---|---|---|---|---|
| | Limits of Interference | Standard Limits Hole H6 | Standard Limits Shaft | Limits of Interference | Standard Limits Hole H7 | Standard Limits Shaft s6 | Limits of Interference | Standard Limits Hole H7 | Standard Limits Shaft t6 | Limits of Interference | Standard Limits Hole H7 | Standard Limits Shaft u6 | Limits of Interference | Standard Limits Hole H8 | Standard Limits Shaft x7 |
| 0 – 0.12 | 0.05 / 0.5 | +0.25 / –0 | +0.5 / +0.3 | 0.2 / 0.85 | +0.4 / –0 | +0.85 / +0.6 | | | | 0.3 / 0.95 | +0.4 / –0 | +0.95 / +0.7 | 0.3 / 1.3 | +0.6 / –0 | +1.3 / +0.9 |
| 0.12– 0.24 | 0.1 / 0.6 | +0.3 / –0 | +0.6 / +0.4 | 0.2 / 1.0 | +0.5 / –0 | +1.0 / +0.7 | | | | 0.4 / 1.2 | +0.5 / –0 | +1.2 / +0.9 | 0.5 / 1.7 | +0.7 / –0 | +1.7 / +1.2 |
| 0.24– 0.40 | 0.1 / 0.75 | +0.4 / –0 | +0.75 / +0.5 | 0.4 / 1.4 | +0.6 / –0 | +1.4 / +1.0 | | | | 0.6 / 1.6 | +0.6 / –0 | +1.6 / +1.2 | 0.5 / 2.0 | +0.9 / –0 | +2.0 / +1.4 |
| 0.40– 0.56 | 0.1 / 0.8 | +0.4 / –0 | +0.8 / +0.5 | 0.5 / 1.6 | +0.7 / –0 | +1.6 / +1.2 | | | | 0.7 / 1.8 | +0.7 / –0 | +1.8 / +1.4 | 0.6 / 2.3 | +1.0 / –0 | +2.3 / +1.6 |
| 0.56– 0.71 | 0.2 / 0.9 | +0.4 / –0 | +0.9 / +0.6 | 0.5 / 1.6 | +0.7 / –0 | +1.6 / +1.2 | | | | 0.7 / 1.8 | +0.7 / –0 | +1.8 / +1.4 | 0.8 / 2.5 | +1.0 / –0 | +2.5 / +1.8 |
| 0.71– 0.95 | 0.2 / 1.1 | +0.5 / –0 | +1.1 / +0.7 | 0.6 / 1.9 | +0.8 / –0 | +1.9 / +1.4 | 0.8 / 2.1 | +0.8 / –0 | +2.1 / +1.6 | 0.8 / 2.1 | +0.8 / –0 | +2.1 / +1.6 | 1.0 / 3.0 | +1.2 / –0 | +3.0 / +2.2 |
| 0.95– 1.19 | 0.3 / 1.2 | +0.5 / –0 | +1.2 / +0.8 | 0.6 / 1.9 | +0.8 / –0 | +1.9 / +1.4 | | | | 1.0 / 2.3 | +0.8 / –0 | +2.3 / +1.8 | 1.3 / 3.3 | +1.2 / –0 | +3.3 / +2.5 |
| 1.19– 1.58 | 0.3 / 1.3 | +0.6 / –0 | +1.3 / +0.9 | 0.8 / 2.4 | +1.0 / –0 | +2.4 / +1.8 | 1.0 / 2.6 | +1.0 / –0 | +2.6 / +2.0 | 1.5 / 3.1 | +1.0 / –0 | +3.1 / +2.5 | 1.4 / 4.0 | +1.6 / –0 | +4.0 / +3.0 |

## 9 Force and Shrink Fits[a]—American National Standard (continued)

| Nominal Size Range, inches Over / To | Class FN 1 Limits of Interference | Class FN 1 Hole H6 | Class FN 1 Shaft | Class FN 2 Limits of Interference | Class FN 2 Hole H7 | Class FN 2 Shaft s6 | Class FN 3 Limits of Interference | Class FN 3 Hole H7 | Class FN 3 Shaft t6 | Class FN 4 Limits of Interference | Class FN 4 Hole H7 | Class FN 4 Shaft u6 | Class FN 5 Limits of Interference | Class FN 5 Hole H8 | Class FN 5 Shaft x7 |
|---|---|---|---|---|---|---|---|---|---|---|---|---|---|---|---|
| 1.58–1.97 | 0.4 / 1.4 | +0.6 / −0 | +1.4 / +1.0 | 0.8 / 2.4 | +1.0 / −0 | +2.4 / +1.8 | 1.2 / 2.8 | +1.0 / −0 | +2.8 / +2.2 | 1.8 / 3.4 | +1.0 / −0 | +3.4 / +2.8 | 2.4 / 5.0 | +1.6 / −0 | +5.0 / +4.0 |
| 1.97–2.56 | 0.6 / 1.8 | +0.7 / −0 | +1.8 / +1.3 | 0.8 / 2.7 | +1.2 / −0 | +2.7 / +2.0 | 1.3 / 3.2 | +1.2 / −0 | +3.2 / +2.5 | 2.3 / 4.2 | +1.2 / −0 | +4.2 / +3.5 | 3.2 / 6.2 | +1.8 / −0 | +6.2 / +5.0 |
| 2.56–3.15 | 0.7 / 1.9 | +0.7 / −0 | +1.9 / +1.4 | 1.0 / 2.9 | +1.2 / −0 | +2.9 / +2.2 | 1.8 / 3.7 | +1.2 / −0 | +3.7 / +3.0 | 2.8 / 4.7 | +1.2 / −0 | +4.7 / +4.0 | 4.2 / 7.2 | +1.8 / −0 | +7.2 / +6.0 |
| 3.15–3.94 | 0.9 / 2.4 | +0.9 / −0 | +2.4 / +1.8 | 1.4 / 3.7 | +1.4 / −0 | +3.7 / +2.8 | 2.1 / 4.4 | +1.4 / −0 | +4.4 / +3.5 | 3.6 / 5.9 | +1.4 / −0 | +5.9 / +5.0 | 4.8 / 8.4 | +2.2 / −0 | +8.4 / +7.0 |
| 3.94–4.73 | 1.1 / 2.6 | +0.9 / −0 | +2.6 / +2.0 | 1.6 / 3.9 | +1.4 / −0 | +3.9 / +3.0 | 2.6 / 4.9 | +1.4 / −0 | +4.9 / +4.0 | 4.6 / 6.9 | +1.4 / −0 | +6.9 / +6.0 | 5.8 / 9.4 | +2.2 / −0 | +9.4 / +8.0 |
| 4.73–5.52 | 1.2 / 2.9 | +1.0 / −0 | +2.9 / +2.2 | 1.9 / 4.5 | +1.6 / −0 | +4.5 / +3.5 | 3.4 / 6.0 | +1.6 / −0 | +6.0 / +5.0 | 5.4 / 8.0 | +1.6 / −0 | +8.0 / +7.0 | 7.5 / 11.6 | +2.5 / −0 | +11.6 / +10.0 |
| 5.52–6.30 | 1.5 / 3.2 | +1.0 / −0 | +3.2 / +2.5 | 2.4 / 5.0 | +1.6 / −0 | +5.0 / +4.0 | 3.4 / 6.0 | +1.6 / −0 | +6.0 / +5.0 | 5.4 / 8.0 | +1.6 / −0 | +8.0 / +7.0 | 9.5 / 13.6 | +2.5 / −0 | +13.6 / +12.0 |
| 6.30–7.09 | 1.8 / 3.5 | +1.0 / −0 | +3.5 / +2.8 | 2.9 / 5.5 | +1.6 / −0 | +5.5 / +4.5 | 4.4 / 7.0 | +1.6 / −0 | +7.0 / +6.0 | 6.4 / 9.0 | +1.6 / −0 | +9.0 / +8.0 | 9.5 / 13.6 | +2.5 / −0 | +13.6 / +12.0 |
| 7.09–7.88 | 1.8 / 3.8 | +1.2 / −0 | +3.8 / +3.0 | 3.2 / 6.2 | +1.8 / −0 | +6.2 / +5.0 | 5.2 / 8.2 | +1.8 / −0 | +8.2 / +7.0 | 7.2 / 10.2 | +1.8 / −0 | +10.2 / +9.0 | 11.2 / 15.8 | +2.8 / −0 | +15.8 / +14.0 |
| 7.88–8.86 | 2.3 / 4.3 | +1.2 / −0 | +4.3 / +3.5 | 3.2 / 6.2 | +1.8 / −0 | +6.2 / +5.0 | 5.2 / 8.2 | +1.8 / −0 | +8.2 / +7.0 | 8.2 / 11.2 | +1.8 / −0 | +11.2 / +10.0 | 13.2 / 17.8 | +2.8 / −0 | +17.8 / +16.0 |
| 8.86–9.85 | 2.3 / 4.3 | +1.2 / −0 | +4.3 / +3.5 | 4.2 / 7.2 | +1.8 / −0 | +7.2 / +6.0 | 6.2 / 9.2 | +1.8 / −0 | +9.2 / +8.0 | 10.2 / 13.2 | +1.8 / −0 | +13.2 / +12.0 | 13.2 / 17.8 | +2.8 / −0 | +17.8 / +16.0 |
| 9.85–11.03 | 2.8 / 4.9 | +1.2 / −0 | +4.9 / +4.0 | 4.0 / 7.2 | +2.0 / −0 | +7.2 / +6.0 | 7.0 / 10.2 | +2.0 / −0 | +10.2 / +9.0 | 10.0 / 13.2 | +2.0 / −0 | +13.2 / +12.0 | 15.0 / 20.0 | +3.0 / −0 | +20.0 / +18.0 |
| 11.03–12.41 | 2.8 / 4.9 | +1.2 / −0 | +4.9 / +4.0 | 5.0 / 8.2 | +2.0 / −0 | +8.2 / +7.0 | 7.0 / 10.2 | +2.0 / −0 | +10.2 / +9.0 | 12.0 / 15.2 | +2.0 / −0 | +15.2 / +14.0 | 17.0 / 22.0 | +3.0 / −0 | +22.0 / +20.0 |
| 12.41–13.98 | 3.1 / 5.5 | +1.4 / −0 | +5.5 / +4.5 | 5.8 / 9.4 | +2.2 / −0 | +9.4 / +8.0 | 7.8 / 11.4 | +2.2 / −0 | +11.4 / +10.0 | 13.8 / 17.4 | +2.2 / −0 | +17.4 / +16.0 | 18.5 / 24.2 | +3.5 / +0 | +24.2 / +22.0 |

[a]From ANSI B4.1–1967 (R1987). For larger diameters, see the standard.

# 10    International Tolerance Grades[a]

Dimensions are in millimeters.

| Basic sizes | | Tolerance grades[b] | | | | | | | | | | | | | | | | | | |
|---|---|---|---|---|---|---|---|---|---|---|---|---|---|---|---|---|---|---|---|---|
| Over | Up to and Including | IT01 | IT0 | IT1 | IT2 | IT3 | IT4 | IT5 | IT6 | IT7 | IT8 | IT9 | IT10 | IT11 | IT12 | IT13 | IT14 | IT15 | IT16 |
| 0 | 3 | 0.0003 | 0.0005 | 0.0008 | 0.0012 | 0.002 | 0.003 | 0.004 | 0.006 | 0.010 | 0.014 | 0.025 | 0.040 | 0.060 | 0.100 | 0.140 | 0.250 | 0.400 | 0.600 |
| 3 | 6 | 0.0004 | 0.0006 | 0.001 | 0.0015 | 0.0025 | 0.004 | 0.005 | 0.008 | 0.012 | 0.018 | 0.030 | 0.048 | 0.075 | 0.120 | 0.180 | 0.300 | 0.480 | 0.750 |
| 6 | 10 | 0.0004 | 0.0006 | 0.001 | 0.0015 | 0.0025 | 0.004 | 0.006 | 0.009 | 0.015 | 0.022 | 0.036 | 0.058 | 0.090 | 0.150 | 0.220 | 0.360 | 0.580 | 0.900 |
| 10 | 18 | 0.0005 | 0.0008 | 0.0012 | 0.002 | 0.003 | 0.005 | 0.008 | 0.011 | 0.018 | 0.027 | 0.043 | 0.070 | 0.110 | 0.180 | 0.270 | 0.430 | 0.700 | 1.100 |
| 18 | 30 | 0.0006 | 0.001 | 0.0015 | 0.0025 | 0.004 | 0.006 | 0.009 | 0.013 | 0.021 | 0.033 | 0.052 | 0.084 | 0.130 | 0.210 | 0.330 | 0.520 | 0.840 | 1.300 |
| 30 | 50 | 0.0006 | 0.001 | 0.0015 | 0.0025 | 0.004 | 0.007 | 0.011 | 0.016 | 0.025 | 0.039 | 0.062 | 0.100 | 0.160 | 0.250 | 0.390 | 0.620 | 1.000 | 1.600 |
| 50 | 80 | 0.0008 | 0.0012 | 0.002 | 0.003 | 0.005 | 0.008 | 0.013 | 0.019 | 0.030 | 0.046 | 0.074 | 0.120 | 0.190 | 0.300 | 0.460 | 0.740 | 1.200 | 1.900 |
| 80 | 120 | 0.001 | 0.0015 | 0.0025 | 0.004 | 0.006 | 0.010 | 0.015 | 0.022 | 0.035 | 0.054 | 0.087 | 0.140 | 0.220 | 0.350 | 0.540 | 0.870 | 1.400 | 2.200 |
| 120 | 180 | 0.0012 | 0.002 | 0.0035 | 0.005 | 0.008 | 0.012 | 0.018 | 0.025 | 0.040 | 0.063 | 0.100 | 0.160 | 0.250 | 0.400 | 0.630 | 1.000 | 1.600 | 2.500 |
| 180 | 250 | 0.002 | 0.003 | 0.0045 | 0.007 | 0.010 | 0.014 | 0.020 | 0.029 | 0.046 | 0.072 | 0.115 | 0.185 | 0.290 | 0.460 | 0.720 | 1.150 | 1.850 | 2.900 |
| 250 | 315 | 0.0025 | 0.004 | 0.006 | 0.008 | 0.012 | 0.016 | 0.023 | 0.032 | 0.052 | 0.081 | 0.130 | 0.210 | 0.320 | 0.520 | 0.810 | 1.300 | 2.100 | 3.200 |
| 315 | 400 | 0.003 | 0.005 | 0.007 | 0.009 | 0.013 | 0.018 | 0.025 | 0.036 | 0.057 | 0.089 | 0.140 | 0.230 | 0.360 | 0.570 | 0.890 | 1.400 | 2.300 | 3.600 |
| 400 | 500 | 0.004 | 0.006 | 0.008 | 0.010 | 0.015 | 0.020 | 0.027 | 0.040 | 0.063 | 0.097 | 0.155 | 0.250 | 0.400 | 0.630 | 0.970 | 1.550 | 2.500 | 4.000 |
| 500 | 630 | 0.0045 | 0.006 | 0.009 | 0.011 | 0.016 | 0.022 | 0.030 | 0.044 | 0.070 | 0.110 | 0.175 | 0.280 | 0.440 | 0.700 | 1.100 | 1.750 | 2.800 | 4.400 |
| 630 | 800 | 0.005 | 0.007 | 0.010 | 0.013 | 0.018 | 0.025 | 0.035 | 0.050 | 0.080 | 0.125 | 0.200 | 0.320 | 0.500 | 0.800 | 1.250 | 2.000 | 3.200 | 5.000 |
| 800 | 1000 | 0.0055 | 0.008 | 0.011 | 0.015 | 0.021 | 0.029 | 0.040 | 0.056 | 0.090 | 0.140 | 0.230 | 0.360 | 0.560 | 0.900 | 1.400 | 2.300 | 3.600 | 5.600 |
| 1000 | 1250 | 0.0065 | 0.009 | 0.013 | 0.018 | 0.024 | 0.034 | 0.046 | 0.066 | 0.105 | 0.165 | 0.260 | 0.420 | 0.660 | 1.050 | 1.650 | 2.600 | 4.200 | 6.600 |
| 1250 | 1600 | 0.008 | 0.011 | 0.015 | 0.021 | 0.029 | 0.040 | 0.054 | 0.078 | 0.125 | 0.195 | 0.310 | 0.500 | 0.780 | 1.250 | 1.950 | 3.100 | 5.000 | 7.800 |
| 1600 | 2000 | 0.009 | 0.013 | 0.018 | 0.025 | 0.035 | 0.048 | 0.065 | 0.092 | 0.150 | 0.230 | 0.370 | 0.600 | 0.920 | 1.500 | 2.300 | 3.700 | 6.000 | 9.200 |
| 2000 | 2500 | 0.011 | 0.015 | 0.022 | 0.030 | 0.041 | 0.057 | 0.077 | 0.110 | 0.175 | 0.280 | 0.440 | 0.700 | 1.100 | 1.750 | 2.800 | 4.400 | 7.000 | 11.000 |
| 2500 | 3150 | 0.013 | 0.018 | 0.026 | 0.036 | 0.050 | 0.069 | 0.093 | 0.135 | 0.210 | 0.330 | 0.540 | 0.860 | 1.350 | 2.100 | 3.300 | 5.400 | 8.600 | 13.500 |

[a]From ANSI B4.2−1978 (R1984).

[b]IT Values for tolerance grades larger than IT16 can be calculated by using the formulas: IT17 = IT × 10, IT18 = IT13 × 10, etc.

## 11   Preferred Metric Hole Basis Clearance Fits[a]— American National Standard

Dimensions are in millimeters.

| Basic Size | | Loose Running | | | Free Running | | | Close Running | | | Sliding | | | Locational Clearance | | |
|---|---|---|---|---|---|---|---|---|---|---|---|---|---|---|---|---|
| | | Hole H11 | Shaft c11 | Fit | Hole H9 | Shaft d9 | Fit | Hole H8 | f7 | Fit | Hole H7 | Shaft g6 | Fit | Hole H7 | Shaft h6 | Fit |
| 1 | Max | 1.060 | 0.940 | 0.180 | 1.025 | 0.980 | 0.070 | 1.014 | 0.994 | 0.030 | 1.010 | 0.998 | 0.018 | 1.010 | 1.000 | 0.016 |
| | Min | 1.060 | 0.880 | 0.060 | 1.000 | 0.955 | 0.020 | 1.000 | 0.984 | 0.006 | 1.000 | 0.992 | 0.002 | 1.000 | 0.994 | 0.000 |
| 1.2 | Max | 1.260 | 1.140 | 0.180 | 1.225 | 1.180 | 0.070 | 1.214 | 1.194 | 0.030 | 1.210 | 1.198 | 0.018 | 1.210 | 1.200 | 0.016 |
| | Min | 1.200 | 1.080 | 0.060 | 1.200 | 1.155 | 0.020 | 1.200 | 1.184 | 0.036 | 1.200 | 1.192 | 0.002 | 1.200 | 1.194 | 0.000 |
| 1.6 | Max | 1.660 | 1.540 | 0.180 | 1.625 | 1.580 | 0.070 | 1.614 | 1.594 | 0.030 | 1.610 | 1.598 | 0.018 | 1.610 | 1.600 | 0.016 |
| | Min | 1.600 | 1.480 | 0.060 | 1.600 | 1.555 | 0.020 | 1.600 | 1.584 | 0.006 | 1.600 | 1.592 | 0.002 | 1.600 | 1.594 | 0.000 |
| 2 | Max | 2.060 | 1.940 | 0.180 | 2.025 | 1.980 | 0.070 | 2.014 | 1.994 | 0.030 | 2.010 | 1.998 | 0.018 | 2.010 | 2.000 | 0.016 |
| | Min | 2.000 | 1.880 | 0.060 | 2.000 | 1.955 | 0.020 | 2.000 | 1.984 | 0.006 | 2.000 | 1.992 | 0.002 | 2.000 | 1.994 | 0.000 |
| 2.5 | Max | 2.560 | 2.440 | 0.180 | 2.525 | 2.480 | 0.070 | 2.514 | 2.494 | 0.030 | 2.510 | 2.498 | 0.018 | 2.510 | 2.500 | 0.016 |
| | Min | 2.500 | 2.380 | 0.060 | 2.500 | 2.455 | 0.020 | 2.500 | 2.484 | 0.006 | 2.500 | 2.492 | 0.002 | 2.500 | 2.494 | 0.000 |
| 3 | Max | 3.060 | 2.940 | 0.180 | 3.025 | 2.980 | 0.070 | 3.014 | 2.994 | 0.030 | 3.010 | 2.998 | 0.018 | 3.010 | 3.000 | 0.016 |
| | Min | 3.000 | 2.880 | 0.060 | 3.000 | 2.955 | 0.020 | 3.000 | 2.984 | 0.006 | 3.000 | 2.992 | 0.002 | 3.000 | 2.994 | 0.000 |
| 4 | Max | 4.075 | 3.930 | 0.220 | 4.030 | 3.970 | 0.090 | 4.018 | 3.990 | 0.040 | 4.012 | 3.996 | 0.024 | 4.012 | 4.000 | 0.020 |
| | Min | 4.000 | 3.855 | 0.070 | 4.000 | 3.940 | 0.030 | 4.000 | 3.978 | 0.010 | 4.000 | 3.988 | 0.004 | 4.000 | 3.992 | 0.000 |
| 5 | Max | 5.075 | 4.930 | 0.220 | 5.030 | 4.970 | 0.090 | 5.018 | 4.990 | 0.040 | 5.012 | 4.996 | 0.024 | 5.012 | 5.000 | 0.020 |
| | Min | 5.000 | 4.855 | 0.070 | 5.000 | 4.940 | 0.030 | 5.000 | 4.978 | 0.010 | 5.000 | 4.988 | 0.004 | 5.000 | 4.992 | 0.000 |
| 6 | Max | 6.075 | 5.930 | 0.220 | 6.030 | 5.970 | 0.090 | 6.018 | 5.990 | 0.040 | 6.012 | 5.996 | 0.024 | 6.012 | 6.000 | 0.020 |
| | Min | 6.000 | 5.855 | 0.070 | 6.000 | 5.940 | 0.030 | 6.000 | 5.978 | 0.010 | 6.000 | 5.988 | 0.004 | 6.000 | 5.992 | 0.000 |
| 8 | Max | 8.090 | 7.920 | 0.260 | 8.036 | 7.960 | 0.112 | 8.022 | 7.987 | 0.050 | 8.015 | 7.995 | 0.029 | 8.015 | 8.000 | 0.024 |
| | Min | 8.000 | 7.830 | 0.080 | 8.000 | 7.924 | 0.040 | 8.000 | 7.972 | 0.013 | 8.000 | 7.986 | 0.005 | 8.000 | 7.991 | 0.000 |
| 10 | Max | 10.090 | 9.920 | 0.260 | 10.036 | 9.960 | 0.112 | 10.022 | 9.987 | 0.050 | 10.015 | 9.995 | 0.029 | 10.015 | 10.000 | 0.024 |
| | Min | 10.000 | 9.830 | 0.080 | 10.000 | 9.924 | 0.040 | 10.000 | 9.972 | 0.013 | 10.000 | 9.986 | 0.005 | 10.000 | 9.991 | 0.000 |
| 12 | Max | 12.110 | 11.905 | 0.315 | 12.043 | 11.950 | 0.136 | 12.027 | 11.984 | 0.061 | 12.018 | 11.994 | 0.035 | 12.018 | 12.000 | 0.029 |
| | Min | 12.000 | 11.795 | 0.095 | 12.000 | 11.907 | 0.050 | 12.000 | 11.966 | 0.016 | 12.000 | 11.983 | 0.006 | 12.000 | 11.989 | 0.000 |
| 16 | Max | 16.110 | 15.905 | 0.315 | 16.043 | 15.950 | 0.136 | 16.027 | 15.984 | 0.061 | 16.018 | 15.994 | 0.035 | 16.018 | 16.000 | 0.029 |
| | Min | 16.000 | 15.795 | 0.095 | 16.000 | 15.907 | 0.050 | 16.000 | 15.966 | 0.016 | 16.000 | 15.983 | 0.006 | 16.000 | 15.989 | 0.000 |
| 20 | Max | 20.130 | 19.890 | 0.370 | 20.052 | 19.935 | 0.169 | 20.033 | 19.980 | 0.074 | 20.021 | 19.993 | 0.041 | 20.021 | 20.000 | 0.034 |
| | Min | 20.000 | 19.760 | 0.110 | 20.000 | 19.883 | 0.065 | 20.000 | 19.959 | 0.020 | 20.000 | 19.980 | 0.007 | 20.000 | 19.987 | 0.000 |
| 25 | Max | 25.130 | 24.890 | 0.370 | 25.052 | 24.935 | 0.169 | 25.033 | 24.980 | 0.074 | 25.021 | 24.993 | 0.041 | 25.021 | 25.000 | 0.034 |
| | Min | 25.000 | 24.760 | 0.110 | 25.000 | 24.883 | 0.065 | 25.000 | 24.959 | 0.020 | 25.000 | 24.980 | 0.007 | 25.000 | 24.987 | 0.000 |
| 30 | Max | 30.130 | 29.890 | 0.370 | 30.052 | 29.935 | 0.169 | 30.033 | 29.980 | 0.074 | 30.021 | 29.993 | 0.041 | 30.021 | 30.000 | 0.034 |
| | Min | 30.000 | 29.760 | 0.110 | 30.000 | 29.883 | 0.065 | 30.000 | 29.959 | 0.020 | 30.000 | 29.980 | 0.007 | 30.000 | 29.987 | 0.000 |

[a]From ANSI B4.2—1978 (R1984). For description of preferred fits, see Table 14.2.

## 11 Preferred Metric Hole Basis Clearance Fits[a]— American National Standard (continued)

Dimensions are in millimeters.

| Basic Size | | Loose Running | | | Free Running | | | Close Running | | | Sliding | | | Locational Clearance | | |
|---|---|---|---|---|---|---|---|---|---|---|---|---|---|---|---|---|
| | | Hole H11 | Shaft c11 | Fit | Hole H9 | Shaft d9 | Fit | Hole H8 | Shaft f7 | Fit | Hole H7 | Shaft g6 | Fit | Hole H7 | Shaft h6 | Fit |
| 40 | Max | 40.160 | 39.880 | 0.440 | 40.062 | 39.920 | 0.204 | 40.039 | 39.975 | 0.089 | 40.025 | 39.991 | 0.050 | 40.025 | 40.000 | 0.041 |
| | Min | 40.000 | 39.720 | 0.120 | 40.000 | 39.858 | 0.080 | 40.000 | 39.950 | 0.025 | 40.000 | 39.975 | 0.009 | 40.000 | 39.984 | 0.000 |
| 50 | Max | 50.160 | 49.870 | 0.450 | 50.062 | 49.920 | 0.204 | 50.039 | 49.975 | 0.089 | 50.025 | 49.991 | 0.050 | 50.025 | 50.000 | 0.041 |
| | Min | 50.000 | 49.710 | 0.130 | 50.000 | 49.858 | 0.080 | 50.000 | 49.950 | 0.025 | 50.000 | 49.975 | 0.009 | 50.000 | 49.984 | 0.000 |
| 60 | Max | 60.190 | 59.860 | 0.520 | 60.074 | 59.900 | 0.248 | 60.046 | 59.970 | 0.106 | 60.030 | 59.990 | 0.059 | 60.030 | 60.000 | 0.049 |
| | Min | 60.000 | 59.670 | 0.140 | 60.000 | 59.826 | 0.100 | 60.000 | 59.940 | 0.030 | 60.000 | 59.971 | 0.010 | 60.000 | 59.981 | 0.000 |
| 80 | Max | 80.190 | 79.950 | 0.530 | 80.074 | 79.900 | 0.248 | 80.046 | 79.970 | 0.106 | 80.030 | 79.990 | 0.059 | 80.030 | 80.000 | 0.049 |
| | Min | 80.000 | 79.660 | 0.150 | 80.000 | 79.826 | 0.100 | 80.000 | 79.940 | 0.030 | 80.000 | 79.971 | 0.010 | 80.000 | 79.981 | 0.000 |
| 100 | Max | 100.220 | 99.830 | 0.610 | 100.087 | 99.880 | 0.294 | 100.054 | 99.964 | 0.125 | 100.035 | 99.988 | 0.069 | 100.035 | 100.000 | 0.057 |
| | Min | 100.000 | 99.610 | 0.170 | 100.000 | 99.793 | 0.120 | 100.000 | 99.929 | 0.036 | 100.000 | 99.966 | 0.012 | 100.000 | 99.978 | 0.000 |
| 120 | Max | 120.220 | 119.820 | 0.620 | 120.087 | 119.880 | 0.294 | 120.054 | 119.964 | 0.125 | 120.035 | 119.988 | 0.069 | 120.035 | 120.000 | 0.057 |
| | Min | 120.000 | 119.600 | 0.180 | 120.000 | 119.793 | 0.120 | 120.000 | 119.929 | 0.036 | 120.000 | 119.966 | 0.012 | 120.000 | 119.978 | 0.000 |
| 160 | Max | 160.250 | 159.790 | 0.710 | 160.100 | 159.855 | 0.345 | 160.063 | 159.957 | 0.146 | 160.040 | 159.986 | 0.079 | 160.040 | 160.000 | 0.065 |
| | Min | 160.000 | 159.540 | 0.210 | 160.000 | 159.755 | 0.145 | 160.000 | 159.917 | 0.043 | 160.000 | 159.961 | 0.014 | 160.000 | 159.975 | 0.000 |
| 200 | Max | 200.290 | 199.760 | 0.820 | 200.115 | 199.830 | 0.400 | 200.072 | 199.950 | 0.168 | 200.046 | 199.985 | 0.090 | 200.046 | 200.000 | 0.075 |
| | Min | 200.000 | 199.470 | 0.240 | 200.000 | 199.715 | 0.170 | 200.000 | 199.904 | 0.050 | 200.000 | 199.956 | 0.015 | 200.000 | 199.971 | 0.000 |
| 250 | Max | 250.290 | 249.720 | 0.860 | 250.115 | 249.830 | 0.400 | 250.072 | 249.950 | 0.168 | 250.046 | 249.985 | 0.090 | 250.046 | 250.000 | 0.075 |
| | Min | 250.000 | 249.430 | 0.280 | 250.000 | 249.715 | 0.170 | 250.000 | 249.904 | 0.050 | 250.000 | 249.956 | 0.015 | 250.000 | 249.971 | 0.000 |
| 300 | Max | 300.320 | 299.670 | 0.970 | 300.130 | 299.810 | 0.450 | 300.081 | 299.944 | 0.189 | 300.052 | 299.983 | 0.101 | 300.052 | 300.000 | 0.084 |
| | Min | 300.000 | 299.350 | 0.330 | 300.000 | 299.680 | 0.190 | 300.000 | 299.892 | 0.056 | 300.000 | 299.951 | 0.017 | 300.000 | 299.968 | 0.000 |
| 400 | Max | 400.360 | 399.600 | 1.120 | 400.140 | 399.790 | 0.490 | 400.089 | 399.938 | 0.208 | 400.057 | 399.982 | 0.111 | 400.057 | 400.000 | 0.093 |
| | Min | 400.000 | 399.240 | 0.400 | 400.000 | 399.650 | 0.210 | 400.000 | 399.881 | 0.062 | 400.000 | 399.946 | 0.018 | 400.000 | 399.964 | 0.000 |
| 500 | Max | 500.400 | 499.520 | 1.280 | 500.155 | 499.770 | 0.540 | 500.097 | 499.932 | 0.228 | 500.063 | 499.980 | 0.123 | 500.063 | 500.000 | 0.103 |
| | Min | 500.000 | 499.120 | 0.480 | 500.000 | 499.615 | 0.230 | 500.000 | 499.869 | 0.068 | 500.000 | 499.940 | 0.020 | 500.000 | 499.960 | 0.000 |

[a]From ANSI B4.2–1978 (R1984). For description of preferred fits, see Table 14.2.

## 12   Preferred Metric Hole Basis Transition and Interference Fits[a] — American National Standard

Dimensions are in millimeters.

| Basic Size | | Locational Transn. Hole H7 | Shaft k6 | Fit | Locational Transn. Hole H7 | Shaft n6 | Fit | Locational Interf. Hole H7 | Shaft p6 | Fit | Medium Drive Hole H7 | Shaft s6 | Fit | Force Hole H7 | Shaft u6 | Fit |
|---|---|---|---|---|---|---|---|---|---|---|---|---|---|---|---|---|
| 1 | Max | 1.010 | 1.006 | 0.010 | 1.010 | 1.010 | 0.006 | 1.010 | 1.012 | 0.004 | 1.010 | 1.020 | −0.004 | 1.010 | 1.024 | −0.008 |
|   | Min | 1.000 | 1.000 | −0.006 | 1.000 | 1.004 | −0.010 | 1.000 | 1.006 | −0.012 | 1.000 | 1.014 | −0.020 | 1.000 | 1.018 | −0.024 |
| 1.2 | Max | 1.210 | 1.206 | 0.010 | 1.210 | 1.210 | 0.006 | 1.210 | 1.212 | 0.004 | 1.210 | 1.220 | −0.004 | 1.210 | 1.224 | −0.008 |
|   | Min | 1.200 | 1.200 | −0.006 | 1.200 | 1.204 | −0.010 | 1.200 | 1.206 | −0.012 | 1.200 | 1.214 | −0.020 | 1.200 | 1.218 | −0.024 |
| 1.6 | Max | 1.610 | 1.606 | 0.010 | 1.610 | 1.610 | 0.006 | 1.610 | 1.612 | 0.004 | 1.610 | 1.620 | −0.004 | 1.610 | 1.624 | −0.008 |
|   | Min | 1.600 | 1.600 | −0.006 | 1.600 | 1.604 | −0.010 | 1.600 | 1.606 | −0.012 | 1.600 | 1.614 | −0.020 | 1.600 | 1.618 | −0.024 |
| 2 | Max | 2.010 | 2.006 | 0.010 | 2.010 | 2.010 | 0.006 | 2.010 | 2.012 | 0.004 | 2.010 | 2.020 | −0.004 | 2.010 | 2.024 | −0.008 |
|   | Min | 2.000 | 2.000 | −0.006 | 2.000 | 2.004 | −0.010 | 2.000 | 2.006 | −0.012 | 2.000 | 2.014 | −0.020 | 2.000 | 2.018 | −0.024 |
| 2.5 | Max | 2.510 | 2.506 | 0.010 | 2.510 | 2.510 | 0.006 | 2.510 | 2.512 | 0.004 | 2.510 | 2.520 | −0.004 | 2.510 | 2.524 | −0.008 |
|   | Min | 2.500 | 2.500 | −0.006 | 2.500 | 2.504 | −0.010 | 2.500 | 2.506 | −0.012 | 2.500 | 2.514 | −0.020 | 2.500 | 2.518 | −0.024 |
| 3 | Max | 3.010 | 3.006 | 0.010 | 3.010 | 3.010 | 0.006 | 3.010 | 3.012 | 0.004 | 3.010 | 3.020 | −0.004 | 3.010 | 3.024 | −0.008 |
|   | Min | 3.000 | 3.000 | −0.006 | 3.000 | 3.004 | −0.010 | 3.000 | 3.006 | −0.012 | 3.000 | 3.014 | −0.020 | 3.000 | 3.018 | −0.024 |
| 4 | Max | 4.012 | 4.009 | 0.011 | 4.012 | 4.016 | 0.004 | 4.012 | 4.020 | 0.000 | 4.012 | 4.027 | −0.007 | 4.012 | 4.031 | −0.011 |
|   | Min | 4.000 | 4.001 | −0.009 | 4.000 | 4.008 | −0.016 | 4.000 | 4.012 | −0.020 | 4.000 | 4.019 | −0.027 | 4.000 | 4.023 | −0.031 |
| 5 | Max | 5.012 | 5.009 | 0.011 | 5.012 | 5.016 | 0.004 | 5.012 | 5.020 | 0.000 | 5.012 | 5.027 | −0.007 | 5.012 | 5.031 | −0.011 |
|   | Min | 5.000 | 5.001 | −0.009 | 5.000 | 5.008 | −0.016 | 5.000 | 5.012 | −0.020 | 5.000 | 5.019 | −0.027 | 5.000 | 5.023 | −0.031 |
| 6 | Max | 6.012 | 6.009 | 0.011 | 6.012 | 6.016 | 0.004 | 6.012 | 6.020 | 0.000 | 6.012 | 6.027 | −0.007 | 6.012 | 6.031 | −0.011 |
|   | Min | 6.000 | 6.001 | −0.009 | 6.000 | 6.008 | −0.016 | 6.000 | 6.012 | −0.020 | 6.000 | 6.019 | −0.027 | 6.000 | 6.023 | −0.031 |
| 8 | Max | 8.015 | 8.010 | 0.014 | 8.015 | 8.019 | 0.005 | 8.015 | 8.024 | 0.000 | 8.015 | 8.032 | −0.008 | 8.015 | 8.037 | −0.013 |
|   | Min | 8.000 | 8.001 | −0.010 | 8.000 | 8.010 | −0.019 | 8.000 | 8.015 | −0.024 | 8.000 | 8.023 | −0.032 | 8.000 | 8.028 | −0.037 |
| 10 | Max | 10.015 | 10.010 | 0.014 | 10.015 | 10.019 | 0.005 | 10.015 | 10.024 | 0.000 | 10.015 | 10.032 | −0.008 | 10.015 | 10.037 | −0.013 |
|   | Min | 10.000 | 10.001 | −0.010 | 10.000 | 10.010 | −0.019 | 10.000 | 10.015 | −0.024 | 10.000 | 10.023 | −0.032 | 10.000 | 10.028 | −0.037 |
| 12 | Max | 12.018 | 12.012 | 0.017 | 12.018 | 12.023 | 0.006 | 12.018 | 12.029 | 0.000 | 12.018 | 12.039 | −0.010 | 12.018 | 12.044 | −0.015 |
|   | Min | 12.000 | 12.001 | −0.012 | 12.000 | 12.012 | −0.023 | 12.000 | 12.018 | −0.029 | 12.000 | 12.028 | −0.039 | 12.000 | 12.033 | −0.044 |
| 16 | Max | 16.018 | 16.012 | 0.017 | 16.018 | 16.023 | 0.006 | 16.018 | 16.029 | 0.000 | 16.018 | 16.039 | −0.010 | 16.018 | 16.044 | −0.015 |
|   | Min | 16.000 | 16.001 | −0.012 | 16.000 | 16.012 | −0.023 | 16.000 | 16.018 | −0.029 | 16.000 | 16.028 | −0.039 | 16.000 | 16.033 | −0.044 |
| 20 | Max | 20.021 | 20.015 | 0.019 | 20.021 | 20.028 | 0.006 | 20.021 | 20.035 | −0.001 | 20.021 | 20.048 | −0.014 | 20.021 | 20.054 | −0.020 |
|   | Min | 20.000 | 20.002 | −0.015 | 20.000 | 20.015 | −0.028 | 20.000 | 20.022 | −0.035 | 20.000 | 20.035 | −0.048 | 20.000 | 20.041 | −0.054 |
| 25 | Max | 25.021 | 25.015 | 0.019 | 25.021 | 25.028 | 0.006 | 25.021 | 25.035 | −0.001 | 25.021 | 25.048 | −0.014 | 25.021 | 25.061 | −0.027 |
|   | Min | 25.000 | 25.002 | −0.015 | 25.000 | 25.015 | −0.028 | 25.000 | 25.022 | −0.035 | 25.000 | 25.035 | −0.048 | 25.000 | 25.048 | −0.061 |
| 30 | Max | 30.021 | 30.015 | 0.019 | 30.021 | 30.028 | 0.006 | 30.021 | 30.035 | −0.001 | 30.021 | 30.048 | −0.014 | 30.021 | 30.061 | −0.027 |
|   | Min | 30.000 | 30.002 | −0.015 | 30.000 | 30.015 | −0.028 | 30.000 | 30.022 | −0.035 | 30.000 | 30.035 | −0.048 | 30.000 | 30.048 | −0.061 |

[a]From ANSI B4.2—1978 (R1984). For description of preferred fits, see Table 14.2.

## 12  Preferred Metric Hole Basis Transition and Interference Fits[a]— American National Standard (continued)

Dimensions are in millimeters.

| Basic Size | | Locational Transn. Hole H7 | Shaft k6 | Fit | Locational Transn. Hole H7 | Shaft n6 | Fit | Locational Interf. Hole H7 | Shaft p6 | Fit | Medium Drive Hole H7 | Shaft s6 | Fit | Force Hole H7 | Shaft u6 | Fit |
|---|---|---|---|---|---|---|---|---|---|---|---|---|---|---|---|---|
| 40 | Max | 40.025 | 40.018 | 0.023 | 40.025 | 40.033 | 0.008 | 40.025 | 40.042 | −0.001 | 40.025 | 40.059 | −0.018 | 40.025 | 40.076 | −0.035 |
|    | Min | 40.000 | 40.002 | −0.018 | 40.000 | 40.017 | −0.033 | 40.000 | 40.026 | −0.042 | 40.000 | 40.043 | −0.059 | 40.000 | 40.060 | −0.076 |
| 50 | Max | 50.025 | 50.018 | 0.023 | 50.025 | 50.033 | 0.008 | 50.025 | 50.042 | −0.001 | 50.025 | 50.059 | −0.018 | 50.025 | 50.086 | −0.045 |
|    | Min | 50.000 | 50.002 | −0.018 | 50.000 | 50.017 | −0.033 | 50.000 | 50.026 | −0.042 | 50.000 | 50.043 | −0.059 | 50.000 | 50.070 | −0.086 |
| 60 | Max | 60.030 | 60.021 | 0.028 | 60.030 | 60.039 | 0.010 | 60.030 | 60.051 | −0.002 | 60.030 | 60.072 | −0.023 | 60.030 | 60.106 | −0.057 |
|    | Min | 60.000 | 60.002 | −0.021 | 60.000 | 60.020 | −0.039 | 60.000 | 60.032 | −0.051 | 60.000 | 60.053 | −0.072 | 60.000 | 60.087 | −0.106 |
| 80 | Max | 80.030 | 80.021 | 0.028 | 80.030 | 80.039 | 0.010 | 80.030 | 80.051 | −0.002 | 80.030 | 80.078 | −0.029 | 80.030 | 80.121 | −0.072 |
|    | Min | 80.000 | 80.002 | −0.021 | 80.000 | 80.020 | −0.039 | 80.000 | 80.032 | −0.051 | 80.000 | 80.059 | −0.078 | 80.000 | 80.102 | −0.121 |
| 100 | Max | 100.035 | 100.025 | 0.032 | 100.035 | 100.045 | 0.012 | 100.035 | 100.059 | −0.002 | 100.035 | 100.093 | −0.036 | 100.035 | 100.146 | −0.089 |
|     | Min | 100.000 | 100.003 | −0.025 | 100.000 | 100.023 | −0.045 | 100.000 | 100.037 | −0.059 | 100.000 | 100.071 | −0.093 | 100.000 | 100.124 | −0.146 |
| 120 | Max | 120.035 | 120.025 | 0.032 | 120.035 | 120.045 | 0.012 | 120.035 | 120.059 | −0.002 | 120.035 | 120.101 | −0.044 | 120.035 | 120.166 | −0.109 |
|     | Min | 120.000 | 120.003 | −0.025 | 120.000 | 120.023 | −0.045 | 120.000 | 120.037 | −0.059 | 120.000 | 120.079 | −0.101 | 120.000 | 120.144 | −0.166 |
| 160 | Max | 160.040 | 160.028 | 0.037 | 160.040 | 160.052 | 0.013 | 160.040 | 160.068 | −0.003 | 160.040 | 160.125 | −0.060 | 160.040 | 160.215 | −0.150 |
|     | Min | 160.000 | 160.003 | −0.028 | 160.000 | 160.027 | −0.052 | 160.000 | 160.043 | −0.068 | 160.000 | 160.100 | −0.125 | 160.000 | 160.190 | −0.215 |
| 200 | Max | 200.046 | 200.033 | 0.042 | 200.046 | 200.060 | 0.015 | 200.046 | 200.079 | −0.004 | 200.046 | 200.151 | −0.076 | 200.046 | 200.265 | −0.190 |
|     | Min | 200.000 | 200.004 | −0.033 | 200.000 | 200.031 | −0.060 | 200.000 | 200.050 | −0.079 | 200.000 | 200.122 | −0.151 | 200.000 | 200.236 | −0.265 |
| 250 | Max | 250.046 | 250.033 | 0.042 | 250.046 | 250.060 | 0.015 | 250.046 | 250.079 | −0.004 | 250.046 | 250.169 | −0.094 | 250.046 | 250.313 | −0.238 |
|     | Min | 250.000 | 250.004 | −0.033 | 250.000 | 250.031 | −0.060 | 250.000 | 250.050 | −0.079 | 250.000 | 250.140 | −0.169 | 250.000 | 250.284 | −0.313 |
| 300 | Max | 300.052 | 300.036 | 0.048 | 300.052 | 300.066 | 0.018 | 300.052 | 300.088 | −0.004 | 300.052 | 300.202 | −0.118 | 300.052 | 300.382 | −0.298 |
|     | Min | 300.000 | 300.004 | −0.036 | 300.000 | 300.034 | −0.066 | 300.000 | 300.056 | −0.088 | 300.000 | 300.170 | −0.202 | 300.000 | 300.350 | −0.382 |
| 400 | Max | 400.057 | 400.040 | 0.053 | 400.057 | 400.073 | 0.020 | 400.057 | 400.098 | −0.005 | 400.057 | 400.244 | −0.151 | 400.057 | 400.471 | −0.378 |
|     | Min | 400.000 | 400.004 | −0.040 | 400.000 | 400.037 | −0.073 | 400.000 | 400.062 | −0.098 | 400.000 | 400.208 | −0.244 | 400.000 | 400.435 | −0.471 |
| 500 | Max | 500.063 | 500.045 | 0.058 | 500.063 | 500.080 | 0.023 | 500.063 | 500.108 | −0.005 | 500.063 | 500.292 | −0.189 | 500.063 | 500.580 | −0.477 |
|     | Min | 500.000 | 500.005 | −0.045 | 500.000 | 500.040 | −0.080 | 500.000 | 500.068 | −0.108 | 500.000 | 500.252 | −0.292 | 500.000 | 500.540 | −0.580 |

[a]From ANSI B4.2−1978 (R1984). For description of preferred fits, see Table 14.2.

# 13 Preferred Metric Shaft Basis Clearance Fits[a]— American National Standard

Dimensions are in millimeters.

| Basic Size | | Loose Running | | | Free Running | | | Close Running | | | Sliding | | | Locational Clearance | | |
|---|---|---|---|---|---|---|---|---|---|---|---|---|---|---|---|---|
| | | Hole C11 | Shaft h11 | Fit | Hole D9 | Shaft h9 | Fit | Hole F8 | Shaft h7 | Fit | Hole G7 | Shaft h6 | Fit | Hole H7 | Shaft h6 | Fit |
| 1 | Max | 1.120 | 1.000 | 0.180 | 1.045 | 1.000 | 0.070 | 1.020 | 1.000 | 0.030 | 1.012 | 1.000 | 0.018 | 1.010 | 1.000 | 0.016 |
| | Min | 1.060 | 0.940 | 0.060 | 1.020 | 0.975 | 0.020 | 1.006 | 0.990 | 0.006 | 1.002 | 0.994 | 0.002 | 1.000 | 0.994 | 0.000 |
| 1.2 | Max | 1.320 | 1.200 | 0.180 | 1.245 | 1.200 | 0.070 | 1.220 | 1.200 | 0.030 | 1.212 | 1.200 | 0.018 | 1.210 | 1.200 | 0.016 |
| | Min | 1.260 | 0.140 | 0.060 | 1.220 | 0.175 | 0.020 | 1.206 | 1.190 | 0.006 | 1.202 | 1.194 | 0.002 | 1.200 | 1.194 | 0.000 |
| 1.6 | Max | 1.720 | 1.600 | 0.180 | 1.645 | 1.600 | 0.070 | 1.620 | 1.600 | 0.030 | 1.612 | 1.600 | 0.018 | 1.610 | 1.600 | 0.016 |
| | Min | 1.660 | 0.540 | 0.060 | 1.620 | 0.575 | 0.020 | 1.606 | 1.590 | 0.006 | 1.602 | 1.594 | 0.002 | 1.600 | 1.594 | 0.000 |
| 2 | Max | 2.120 | 2.000 | 0.180 | 2.045 | 2.000 | 0.070 | 2.020 | 2.000 | 0.030 | 2.012 | 2.000 | 0.018 | 2.010 | 2.000 | 0.016 |
| | Min | 2.060 | 1.940 | 0.060 | 2.020 | 1.975 | 0.020 | 2.006 | 1.990 | 0.006 | 2.002 | 1.994 | 0.002 | 2.000 | 1.994 | 0.000 |
| 2.5 | Max | 2.620 | 2.500 | 0.180 | 2.545 | 2.500 | 0.070 | 2.520 | 2.500 | 0.030 | 2.512 | 2.500 | 0.018 | 2.510 | 2.500 | 0.016 |
| | Min | 2.560 | 2.440 | 0.060 | 2.520 | 2.475 | 0.020 | 2.506 | 2.490 | 0.006 | 2.502 | 2.494 | 0.002 | 2.500 | 2.494 | 0.000 |
| 3 | Max | 3.120 | 3.000 | 0.180 | 3.045 | 3.000 | 0.070 | 3.020 | 3.000 | 0.030 | 3.012 | 3.000 | 0.018 | 3.010 | 3.000 | 0.016 |
| | Min | 3.060 | 2.940 | 0.060 | 3.020 | 2.975 | 0.020 | 3.006 | 2.990 | 0.006 | 3.002 | 2.994 | 0.002 | 3.000 | 2.994 | 0.000 |
| 4 | Max | 4.145 | 4.000 | 0.220 | 4.060 | 4.000 | 0.090 | 4.028 | 4.000 | 0.040 | 4.016 | 4.000 | 0.024 | 4.012 | 4.000 | 0.020 |
| | Min | 4.070 | 3.925 | 0.070 | 4.030 | 3.970 | 0.030 | 4.010 | 3.988 | 0.010 | 4.004 | 3.992 | 0.004 | 4.000 | 3.992 | 0.000 |
| 5 | Max | 5.145 | 5.000 | 0.220 | 5.060 | 5.000 | 0.090 | 5.028 | 5.000 | 0.040 | 5.016 | 5.000 | 0.024 | 5.012 | 5.000 | 0.020 |
| | Min | 5.070 | 4.925 | 0.070 | 5.030 | 4.970 | 0.030 | 5.010 | 4.988 | 0.010 | 5.004 | 4.992 | 0.004 | 5.000 | 4.992 | 0.000 |
| 6 | Max | 6.145 | 6.000 | 0.220 | 6.060 | 6.000 | 0.090 | 6.028 | 6.000 | 0.040 | 6.016 | 6.000 | 0.024 | 6.012 | 6.000 | 0.020 |
| | Min | 6.070 | 5.925 | 0.070 | 6.030 | 5.970 | 0.030 | 6.010 | 5.988 | 0.010 | 6.004 | 5.992 | 0.004 | 6.000 | 5.992 | 0.000 |
| 8 | Max | 8.170 | 8.000 | 0.260 | 8.076 | 8.000 | 0.112 | 8.035 | 8.000 | 0.050 | 8.020 | 8.000 | 0.029 | 8.015 | 8.000 | 0.024 |
| | Min | 8.080 | 7.910 | 0.080 | 8.040 | 7.964 | 0.040 | 8.013 | 7.985 | 0.013 | 8.005 | 7.991 | 0.005 | 8.000 | 7.991 | 0.000 |
| 10 | Max | 10.170 | 10.000 | 0.260 | 10.076 | 10.000 | 0.112 | 10.035 | 10.000 | 0.050 | 10.020 | 10.000 | 0.029 | 10.015 | 10.000 | 0.024 |
| | Min | 10.080 | 9.910 | 0.080 | 10.040 | 9.964 | 0.040 | 10.013 | 9.985 | 0.013 | 10.005 | 9.991 | 0.005 | 10.000 | 9.991 | 0.000 |
| 12 | Max | 12.205 | 12.000 | 0.315 | 12.093 | 12.000 | 0.136 | 12.043 | 12.000 | 0.061 | 12.024 | 12.000 | 0.035 | 12.018 | 12.000 | 0.029 |
| | Min | 12.095 | 11.890 | 0.095 | 12.050 | 11.957 | 0.050 | 12.016 | 11.982 | 0.016 | 12.006 | 11.989 | 0.006 | 12.000 | 11.989 | 0.000 |
| 16 | Max | 16.205 | 16.000 | 0.315 | 16.093 | 16.000 | 0.136 | 16.043 | 16.000 | 0.061 | 16.024 | 16.000 | 0.035 | 16.018 | 16.000 | 0.029 |
| | Min | 16.095 | 15.890 | 0.095 | 16.050 | 15.957 | 0.050 | 16.016 | 15.982 | 0.016 | 16.006 | 15.989 | 0.006 | 16.000 | 15.989 | 0.000 |
| 20 | Max | 20.240 | 20.000 | 0.370 | 20.117 | 20.000 | 0.169 | 20.053 | 20.000 | 0.074 | 20.028 | 20.000 | 0.041 | 20.021 | 20.000 | 0.034 |
| | Min | 20.110 | 19.870 | 0.110 | 20.065 | 19.948 | 0.065 | 20.020 | 19.979 | 0.020 | 20.007 | 19.987 | 0.007 | 20.000 | 19.987 | 0.000 |
| 25 | Max | 25.240 | 25.000 | 0.370 | 25.117 | 25.000 | 0.169 | 25.053 | 25.000 | 0.074 | 25.028 | 25.000 | 0.041 | 25.021 | 25.000 | 0.034 |
| | Min | 25.110 | 24.870 | 0.110 | 25.065 | 24.948 | 0.065 | 25.020 | 24.979 | 0.020 | 25.007 | 24.987 | 0.007 | 25.000 | 24.987 | 0.000 |
| 30 | Max | 30.240 | 30.000 | 0.370 | 30.117 | 30.000 | 0.169 | 30.053 | 30.000 | 0.074 | 30.028 | 30.000 | 0.041 | 30.021 | 30.000 | 0.034 |
| | Min | 30.110 | 29.870 | 0.110 | 30.065 | 29.948 | 0.065 | 30.020 | 29.979 | 0.020 | 30.007 | 29.987 | 0.007 | 30.000 | 29.987 | 0.000 |

[a]From ANSI B4.2—1978 (R1984). For description of preferred fits, see Table 14.2.

# 13 Preferred Metric Shaft Basis Clearance Fits[a]— American National Standard (continued)

Dimensions are in millimeters.

| Basic Size | | Loose Running | | | Free Running | | | Close Running | | | Sliding | | | Locational Clearance | | |
|---|---|---|---|---|---|---|---|---|---|---|---|---|---|---|---|---|
| | | Hole C11 | Shaft h11 | Fit | Hole D9 | Shaft h9 | Fit | Hole F8 | Shaft h7 | Fit | Hole G7 | Shaft h6 | Fit | Hole H7 | Shaft h6 | Fit |
| 40 | Max | 40.280 | 40.000 | 0.440 | 40.142 | 40.000 | 0.204 | 40.064 | 40.000 | 0.089 | 40.034 | 40.000 | 0.050 | 40.025 | 40.000 | 0.041 |
| | Min | 40.120 | 39.840 | 0.120 | 40.080 | 39.938 | 0.080 | 40.025 | 39.975 | 0.025 | 40.009 | 39.984 | 0.009 | 40.000 | 39.984 | 0.000 |
| 50 | Max | 50.290 | 50.000 | 0.450 | 50.142 | 50.000 | 0.204 | 50.064 | 50.000 | 0.089 | 50.034 | 50.000 | 0.050 | 50.025 | 50.000 | 0.041 |
| | Min | 50.130 | 49.840 | 0.130 | 50.080 | 49.938 | 0.080 | 50.025 | 49.975 | 0.025 | 50.009 | 49.984 | 0.009 | 50.000 | 49.984 | 0.000 |
| 60 | Max | 60.330 | 60.000 | 0.520 | 60.174 | 60.000 | 0.248 | 60.076 | 60.000 | 0.106 | 60.040 | 60.000 | 0.059 | 60.030 | 60.000 | 0.049 |
| | Min | 60.140 | 59.810 | 0.140 | 60.100 | 59.926 | 0.100 | 60.030 | 59.970 | 0.030 | 60.010 | 59.981 | 0.010 | 60.000 | 59.981 | 0.000 |
| 80 | Max | 80.340 | 80.000 | 0.530 | 80.174 | 80.000 | 0.248 | 80.076 | 80.000 | 0.106 | 80.040 | 80.000 | 0.059 | 80.030 | 80.000 | 0.049 |
| | Min | 80.150 | 79.810 | 0.150 | 80.100 | 79.926 | 0.100 | 80.030 | 79.970 | 0.030 | 80.010 | 79.981 | 0.010 | 80.000 | 79.981 | 0.000 |
| 100 | Max | 100.390 | 100.000 | 0.610 | 100.207 | 100.000 | 0.294 | 100.090 | 100.000 | 0.125 | 100.047 | 100.000 | 0.069 | 100.035 | 100.000 | 0.057 |
| | Min | 100.170 | 99.780 | 0.170 | 100.120 | 99.913 | 0.120 | 100.036 | 99.965 | 0.036 | 100.012 | 99.978 | 0.012 | 100.000 | 99.978 | 0.000 |
| 120 | Max | 120.400 | 120.000 | 0.620 | 120.207 | 120.000 | 0.294 | 120.090 | 120.000 | 0.125 | 120.047 | 120.000 | 0.069 | 120.035 | 120.000 | 0.057 |
| | Min | 120.180 | 119.780 | 0.180 | 120.120 | 119.913 | 0.120 | 120.036 | 119.965 | 0.036 | 120.012 | 119.978 | 0.012 | 120.000 | 119.978 | 0.000 |
| 160 | Max | 160.460 | 160.000 | 0.710 | 160.245 | 160.000 | 0.345 | 160.106 | 160.000 | 0.146 | 160.054 | 160.000 | 0.079 | 160.040 | 160.000 | 0.065 |
| | Min | 160.210 | 159.750 | 0.210 | 160.145 | 159.900 | 0.145 | 160.043 | 159.960 | 0.043 | 160.014 | 159.975 | 0.014 | 160.000 | 159.975 | 0.000 |
| 200 | Max | 200.530 | 200.000 | 0.820 | 200.285 | 200.000 | 0.400 | 200.122 | 200.000 | 0.168 | 200.061 | 200.000 | 0.090 | 200.046 | 200.000 | 0.075 |
| | Min | 200.240 | 199.710 | 0.240 | 200.170 | 199.885 | 0.170 | 200.050 | 199.954 | 0.050 | 200.015 | 199.971 | 0.015 | 200.000 | 199.971 | 0.000 |
| 250 | Max | 250.570 | 250.000 | 0.860 | 250.285 | 250.000 | 0.400 | 250.122 | 250.000 | 0.168 | 250.061 | 250.000 | 0.090 | 250.046 | 250.000 | 0.075 |
| | Min | 250.280 | 249.710 | 0.280 | 250.170 | 249.885 | 0.170 | 250.050 | 249.954 | 0.050 | 250.015 | 249.971 | 0.015 | 250.000 | 249.971 | 0.000 |
| 300 | Max | 300.650 | 300.000 | 0.970 | 300.320 | 300.000 | 0.450 | 300.137 | 300.000 | 0.189 | 300.069 | 300.000 | 0.101 | 300.052 | 300.000 | 0.084 |
| | Min | 300.330 | 299.680 | 0.330 | 300.190 | 299.870 | 0.190 | 300.056 | 299.948 | 0.056 | 300.017 | 299.968 | 0.017 | 300.000 | 299.968 | 0.000 |
| 400 | Max | 400.760 | 400.000 | 1.120 | 400.350 | 400.000 | 0.490 | 400.151 | 400.000 | 0.208 | 400.075 | 400.000 | 0.111 | 400.057 | 400.000 | 0.093 |
| | Min | 400.400 | 399.640 | 0.400 | 400.210 | 399.860 | 0.210 | 400.062 | 399.943 | 0.062 | 400.018 | 399.964 | 0.018 | 400.000 | 399.964 | 0.000 |
| 500 | Max | 500.880 | 500.000 | 1.280 | 500.385 | 500.000 | 0.540 | 500.165 | 500.000 | 0.228 | 500.083 | 500.000 | 0.123 | 500.063 | 500.000 | 0.103 |
| | Min | 500.480 | 499.600 | 0.480 | 500.230 | 499.845 | 0.230 | 500.068 | 499.937 | 0.068 | 500.020 | 499.960 | 5.020 | 500.000 | 499.960 | 0.000 |

[a]From ANSI B4.2–1978 (R1984). For description of preferred fits, see Table 14.2.

## 14   Preferred Metric Shaft Basis Transition and Interference Fits[a] — American National Standard

Dimensions are in millimeters.

| Basic Size | | Locational Transn. Hole K7 | Shaft h6 | Fit | Locational Transn. Hole N7 | Shaft h6 | Fit | Locational Interf. Hole P7 | Shaft h6 | Fit | Medium Drive Hole S7 | Shaft h6 | Fit | Force Hole U7 | Shaft h6 | Fit |
|---|---|---|---|---|---|---|---|---|---|---|---|---|---|---|---|---|
| 1 | Max | 1.000 | 1.000 | 0.006 | 0.996 | 1.000 | 0.002 | 0.994 | 1.000 | 0.000 | 0.986 | 1.000 | -0.008 | 0.982 | 1.000 | -0.012 |
|   | Min | 0.990 | 0.994 | -0.010 | 0.986 | 0.994 | -0.014 | 0.984 | 0.994 | -0.016 | 0.976 | 0.994 | -0.024 | 0.972 | 0.994 | -0.028 |
| 1.2 | Max | 1.200 | 1.200 | 0.006 | 1.196 | 1.200 | 0.002 | 1.194 | 1.200 | 0.000 | 1.186 | 1.200 | -0.008 | 1.182 | 1.200 | -0.012 |
|   | Min | 1.190 | 1.194 | -0.010 | 1.186 | 1.194 | -0.014 | 1.184 | 1.194 | -0.016 | 1.176 | 1.194 | -0.024 | 1.172 | 1.194 | -0.028 |
| 1.6 | Max | 1.600 | 1.600 | 0.006 | 1.596 | 1.600 | 0.002 | 1.594 | 1.600 | 0.000 | 1.586 | 1.600 | -0.008 | 1.582 | 1.600 | -0.012 |
|   | Min | 1.590 | 1.594 | -0.010 | 1.586 | 1.594 | -0.014 | 1.584 | 1.594 | -0.016 | 1.576 | 1.594 | -0.024 | 1.572 | 1.594 | -0.028 |
| 2 | Max | 2.000 | 2.000 | 0.006 | 1.996 | 2.000 | 0.002 | 1.994 | 2.000 | 0.000 | 1.986 | 2.000 | -0.008 | 1.982 | 2.000 | -0.012 |
|   | Min | 1.990 | 1.994 | -0.010 | 1.986 | 1.994 | -0.014 | 1.984 | 1.994 | -0.016 | 1.976 | 1.994 | -0.024 | 1.972 | 1.994 | -0.028 |
| 2.5 | Max | 2.500 | 2.500 | 0.006 | 2.496 | 2.500 | 0.002 | 2.494 | 2.500 | 0.000 | 2.486 | 2.500 | -0.008 | 2.482 | 2.500 | -0.012 |
|   | Min | 2.490 | 2.494 | -0.010 | 2.486 | 2.494 | -0.014 | 2.484 | 2.494 | -0.016 | 2.476 | 2.494 | -0.024 | 2.472 | 2.494 | -0.028 |
| 3 | Max | 3.000 | 3.000 | 0.006 | 2.996 | 3.000 | 0.002 | 2.994 | 3.000 | 0.000 | 2.986 | 3.000 | -0.008 | 2.982 | 3.000 | -0.012 |
|   | Min | 2.990 | 2.994 | -0.010 | 2.986 | 2.994 | -0.014 | 2.984 | 2.994 | -0.016 | 2.976 | 2.994 | -0.024 | 2.972 | 2.994 | -0.028 |
| 4 | Max | 4.003 | 4.000 | 0.011 | 3.996 | 4.000 | 0.004 | 3.992 | 4.000 | 0.000 | 3.985 | 4.000 | -0.007 | 3.981 | 4.000 | -0.011 |
|   | Min | 3.991 | 3.992 | -0.009 | 3.984 | 3.992 | -0.016 | 3.980 | 3.992 | -0.020 | 3.973 | 3.992 | -0.027 | 3.969 | 3.992 | -0.031 |
| 5 | Max | 5.003 | 5.000 | 0.011 | 4.996 | 5.000 | 0.004 | 4.992 | 5.000 | 0.000 | 4.985 | 5.000 | -0.007 | 4.981 | 5.000 | -0.011 |
|   | Min | 4.991 | 4.992 | -0.009 | 4.984 | 4.992 | -0.016 | 4.980 | 4.992 | -0.020 | 4.973 | 4.992 | -0.027 | 4.969 | 4.992 | -0.031 |
| 6 | Max | 6.003 | 6.000 | 0.011 | 5.996 | 6.000 | 0.004 | 5.992 | 6.000 | 0.000 | 5.985 | 6.000 | -0.007 | 5.981 | 6.000 | -0.011 |
|   | Min | 5.991 | 5.992 | -0.009 | 5.984 | 5.992 | -0.016 | 5.980 | 5.992 | -0.020 | 5.973 | 5.992 | -0.027 | 5.969 | 5.992 | -0.031 |
| 8 | Max | 8.005 | 8.000 | 0.014 | 7.996 | 8.000 | 0.005 | 7.991 | 8.000 | 0.000 | 7.983 | 8.000 | -0.008 | 7.978 | 8.000 | -0.013 |
|   | Min | 7.990 | 7.991 | -0.010 | 7.981 | 7.991 | -0.019 | 7.976 | 7.991 | -0.024 | 7.968 | 7.991 | -0.032 | 7.963 | 7.991 | -0.037 |
| 10 | Max | 10.005 | 10.000 | 0.014 | 9.996 | 10.000 | 0.005 | 9.991 | 10.000 | 0.000 | 9.983 | 10.000 | -0.008 | 9.978 | 10.000 | -0.013 |
|   | Min | 9.990 | 9.991 | -0.010 | 9.981 | 9.991 | -0.019 | 9.976 | 9.991 | -0.024 | 9.968 | 9.991 | -0.032 | 9.963 | 9.991 | -0.037 |
| 12 | Max | 12.006 | 12.000 | 0.017 | 11.995 | 12.000 | 0.006 | 11.989 | 12.000 | 0.000 | 11.979 | 12.000 | -0.010 | 11.974 | 12.000 | -0.015 |
|   | Min | 11.988 | 11.989 | -0.012 | 11.977 | 11.989 | -0.023 | 11.971 | 11.989 | -0.029 | 11.961 | 11.989 | -0.039 | 11.956 | 11.989 | -0.044 |
| 16 | Max | 16.006 | 16.000 | 0.017 | 15.995 | 16.000 | 0.006 | 15.989 | 16.000 | 0.000 | 15.979 | 16.000 | -0.010 | 15.974 | 16.000 | -0.015 |
|   | Min | 15.988 | 15.989 | -0.012 | 15.977 | 15.989 | -0.023 | 15.971 | 15.989 | -0.029 | 15.961 | 15.989 | -0.039 | 15.956 | 15.989 | -0.044 |
| 20 | Max | 20.006 | 20.000 | 0.019 | 19.993 | 20.000 | 0.006 | 19.986 | 20.000 | -0.001 | 19.973 | 20.000 | -0.014 | 19.967 | 20.000 | -0.020 |
|   | Min | 19.985 | 19.987 | -0.015 | 19.972 | 19.987 | -0.028 | 19.965 | 19.987 | -0.035 | 19.952 | 19.987 | -0.048 | 19.946 | 19.987 | -0.054 |
| 25 | Max | 25.006 | 25.000 | 0.019 | 24.993 | 25.000 | 0.006 | 24.986 | 25.000 | -0.001 | 24.973 | 25.000 | -0.014 | 24.960 | 25.000 | -0.027 |
|   | Min | 24.985 | 24.987 | -0.015 | 24.972 | 24.987 | -0.028 | 24.965 | 24.987 | -0.035 | 24.952 | 24.987 | -0.048 | 24.939 | 24.987 | -0.061 |
| 30 | Max | 30.006 | 30.000 | 0.019 | 29.993 | 30.000 | 0.006 | 29.986 | 30.000 | -0.001 | 29.973 | 30.000 | -0.014 | 29.960 | 30.000 | -0.027 |
|   | Min | 29.985 | 29.987 | -0.015 | 29.972 | 29.987 | -0.028 | 29.965 | 29.987 | -0.035 | 29.952 | 29.987 | -0.048 | 29.939 | 29.987 | -0.061 |

[a]From ANSI B4.2–1978 (R1984). For description of preferred fits, see Table 14.2.

# 14 Preferred Metric Basis Transition and Interference Fits[a]—American National Standard (continued)

Dimensions are in millimeters.

| Basic Size | | Locational Transn. Hole K7 | Shaft h6 | Fit | Locational Transn. Hole N7 | Shaft h6 | Fit | Locational Interf. Hole P7 | Shaft h6 | Fit | Medium Drive Hole S7 | Shaft h6 | Fit | Force Hole U7 | Shaft h6 | Fit |
|---|---|---|---|---|---|---|---|---|---|---|---|---|---|---|---|---|
| 40 | Max | 40.007 | 40.000 | 0.023 | 39.992 | 40.000 | 0.008 | 39.983 | 40.000 | -0.001 | 39.966 | 40.000 | -0.018 | 39.949 | 40.000 | -0.035 |
|    | Min | 39.982 | 39.984 | -0.018 | 39.967 | 39.984 | -0.033 | 39.958 | 39.984 | -0.042 | 39.941 | 39.984 | -0.059 | 39.924 | 39.984 | -0.076 |
| 50 | Max | 50.007 | 50.000 | 0.023 | 49.992 | 50.000 | 0.008 | 49.983 | 50.000 | -0.001 | 49.966 | 50.000 | -0.018 | 49.939 | 50.000 | -0.045 |
|    | Min | 49.982 | 49.984 | -0.018 | 49.967 | 49.984 | -0.033 | 49.958 | 49.984 | -0.042 | 49.941 | 49.984 | -0.059 | 49.914 | 49.984 | -0.086 |
| 60 | Max | 60.009 | 60.000 | 0.028 | 59.991 | 60.000 | 0.010 | 59.979 | 60.000 | -0.002 | 59.958 | 60.000 | -0.023 | 59.924 | 60.000 | -0.057 |
|    | Min | 59.979 | 59.981 | -0.021 | 59.961 | 59.981 | -0.039 | 59.949 | 59.981 | -0.051 | 59.928 | 59.981 | -0.072 | 59.894 | 59.981 | -0.106 |
| 80 | Max | 80.009 | 80.000 | 0.028 | 79.991 | 80.000 | 0.010 | 79.979 | 80.000 | -0.002 | 79.952 | 80.000 | -0.029 | 79.909 | 80.000 | -0.072 |
|    | Min | 79.979 | 79.981 | -0.021 | 79.961 | 79.981 | -0.039 | 79.949 | 79.981 | -0.051 | 79.922 | 79.981 | -0.078 | 79.879 | 79.981 | -0.121 |
| 100 | Max | 100.010 | 100.000 | 0.032 | 99.990 | 100.000 | 0.012 | 99.976 | 100.000 | -0.002 | 99.942 | 100.000 | -0.036 | 99.889 | 100.000 | -0.089 |
|     | Min | 99.975 | 99.978 | -0.025 | 99.955 | 99.978 | -0.045 | 99.941 | 99.978 | -0.059 | 99.907 | 99.978 | -0.093 | 99.854 | 99.978 | -0.146 |
| 120 | Max | 120.010 | 120.000 | 0.032 | 119.990 | 120.000 | 0.012 | 119.976 | 120.000 | -0.002 | 119.934 | 120.000 | -0.044 | 119.869 | 120.000 | -0.109 |
|     | Min | 119.975 | 119.978 | -0.025 | 119.955 | 199.978 | -0.025 | 119.941 | 119.978 | -0.059 | 119.899 | 119.978 | -0.101 | 119.834 | 119.978 | -0.166 |
| 160 | Max | 160.012 | 160.000 | 0.037 | 159.988 | 160.000 | 0.013 | 159.972 | 160.000 | -0.003 | 159.915 | 160.000 | -0.060 | 159.825 | 160.000 | -0.150 |
|     | Min | 159.972 | 159.975 | -0.028 | 159.948 | 159.975 | -0.052 | 159.932 | 159.975 | -0.068 | 159.875 | 159.975 | -0.125 | 159.785 | 159.975 | -0.215 |
| 200 | Max | 200.013 | 200.000 | 0.042 | 199.986 | 200.000 | 0.015 | 199.967 | 200.000 | -0.004 | 199.895 | 200.000 | -0.076 | 199.781 | 200.000 | -0.190 |
|     | Min | 199.967 | 199.971 | -0.033 | 199.940 | 199.971 | -0.060 | 199.921 | 199.971 | -0.079 | 199.849 | 199.971 | -0.151 | 199.735 | 199.971 | -0.265 |
| 250 | Max | 250.013 | 250.000 | 0.042 | 249.986 | 250.000 | 0.015 | 249.967 | 250.000 | -0.004 | 249.877 | 250.000 | -0.094 | 249.733 | 250.000 | -0.238 |
|     | Min | 249.967 | 249.971 | -0.033 | 249.940 | 249.971 | -0.060 | 249.921 | 249.971 | -0.079 | 249.831 | 249.971 | -0.169 | 249.687 | 249.971 | -0.313 |
| 300 | Max | 300.016 | 300.000 | 0.048 | 299.986 | 300.000 | 0.018 | 299.964 | 300.000 | -0.004 | 299.850 | 300.000 | -0.118 | 299.670 | 300.000 | -0.298 |
|     | Min | 299.964 | 299.968 | -0.036 | 299.934 | 299.968 | -0.066 | 299.912 | 299.968 | -0.088 | 299.798 | 299.968 | -0.202 | 299.618 | 299.968 | -0.382 |
| 400 | Max | 400.017 | 400.000 | 0.053 | 399.984 | 400.000 | 0.020 | 399.959 | 400.000 | -0.005 | 399.813 | 400.000 | -0.151 | 399.586 | 400.000 | -0.378 |
|     | Min | 399.960 | 399.964 | -0.040 | 399.927 | 399.964 | -0.073 | 399.902 | 399.964 | -0.098 | 399.756 | 399.964 | -0.244 | 399.529 | 399.964 | -0.471 |
| 500 | Max | 500.018 | 500.000 | 0.058 | 499.983 | 500.000 | 0.023 | 499.955 | 500.000 | -0.005 | 499.771 | 500.000 | -0.189 | 499.483 | 500.000 | -0.477 |
|     | Min | 499.955 | 499.960 | -0.045 | 499.920 | 499.960 | -0.080 | 499.892 | 499.960 | -0.108 | 499.708 | 499.960 | -0.292 | 499.420 | 499.960 | -0.580 |

[a] From ANSI B4.2–1978 (R1984). For description of preferred fits, see Table 14.2.

# 15 Screw Threads, American National, Unified, and Metric

## AMERICAN NATIONAL STANDARD UNIFIED AND AMERICAN NATIONAL SCREW THREADS[a]

| Nominal Diameter | Coarse[b] NC UNC Thds. per Inch | Tap Drill[d] | Fine[b] NF UNF Thds. per Inch | Tap Drill[d] | Extra Fine[c] NEF UNEF Thds. per Inch | Tap Drill[d] | Nominal Diameter | Coarse[b] NC UNC Thds. per Inch | Tap Drill[d] | Fine[b] NF UNF Thds. per Inch | Tap Drill[d] | Extra Fine[c] NEF UNEF Thds. per Inch | Tap Drill[d] |
|---|---|---|---|---|---|---|---|---|---|---|---|---|---|
| 0 (.060) | | | 80 | 3/64 | | | 1 | 8 | 7/8 | 12 | 59/64 | 20 | 61/64 |
| 1 (.073) | 64 | No. 53 | 72 | No. 53 | .... | .... | 1 1/16 | .... | .... | .... | .... | 18 | 1 |
| 2 (.086) | 56 | No. 50 | 64 | No. 50 | .... | .... | 1 1/8 | 7 | 63/64 | 12 | 1 3/64 | 18 | 1 5/64 |
| 3 (.099) | 48 | No. 47 | 56 | No. 45 | .... | .... | 1 3/16 | .... | .... | .... | .... | 18 | 1 9/64 |
| 4 (.112) | 40 | No. 43 | 48 | No. 42 | .... | .... | 1 1/4 | 7 | 1 7/64 | 12 | 1 11/64 | 18 | 1 3/16 |
| 5 (.125) | 40 | No. 38 | 44 | No. 37 | .... | .... | 1 5/16 | .... | .... | .... | .... | 18 | 1 17/64 |
| 6 (.138) | 32 | No. 36 | 40 | No. 33 | .... | .... | 1 3/8 | 6 | 1 7/32 | 12 | 1 19/64 | 18 | 1 5/16 |
| 8 (.164) | 32 | No. 29 | 36 | No. 29 | .... | .... | 1 7/16 | .... | .... | .... | .... | 18 | 1 3/8 |
| 10 (.190) | 24 | No. 25 | 32 | No. 21 | .... | .... | 1 1/2 | 6 | 1 11/32 | 12 | 1 27/64 | 18 | 1 7/16 |
| 12 (.216) | 24 | No. 16 | 28 | No. 14 | 32 | No. 13 | 1 9/16 | .... | .... | .... | .... | 18 | 1 1/2 |
| 1/4 | 20 | No. 7 | 28 | No. 3 | 32 | 7/32 | 1 5/8 | .... | .... | .... | .... | 18 | 1 9/16 |
| 5/16 | 18 | F | 24 | I | 32 | 9/32 | 1 11/16 | .... | .... | .... | .... | 18 | 1 5/8 |
| 3/8 | 16 | 5/16 | 24 | Q | 32 | 11/32 | 1 3/4 | 5 | 1 9/16 | .... | .... | .... | .... |
| 7/16 | 14 | U | 20 | 25/64 | 28 | 13/32 | 2 | 4 1/2 | 1 25/32 | .... | .... | .... | .... |
| 1/2 | 13 | 27/64 | 20 | 29/64 | 28 | 15/32 | 2 1/4 | 4 1/2 | 2 1/32 | .... | .... | .... | .... |
| 9/16 | 12 | 31/64 | 18 | 33/64 | 24 | 33/64 | 2 1/2 | 4 | 2 1/4 | .... | .... | .... | .... |
| 5/8 | 11 | 17/32 | 18 | 37/64 | 24 | 37/64 | 2 3/4 | 4 | 2 1/2 | .... | .... | .... | .... |
| 11/16 | .... | .... | .... | .... | 24 | 41/64 | 3 | 4 | 2 3/4 | .... | .... | .... | .... |
| 3/4 | 10 | 21/32 | 16 | 11/16 | 20 | 45/64 | 3 1/4 | 4 | .... | .... | .... | .... | .... |
| 13/16 | .... | .... | .... | .... | 20 | 49/64 | 3 1/2 | 4 | .... | .... | .... | .... | .... |
| 7/8 | 9 | 49/64 | 14 | 13/16 | 20 | 53/64 | 3 3/4 | 4 | .... | .... | .... | .... | .... |
| 15/16 | .... | .... | .... | .... | 20 | 57/64 | 4 | 4 | .... | .... | .... | .... | .... |

[a]ANSI B1.1. For 8-, 12-, and 16-pitch thread series, see next page.
[b]Classes 1A, 2A, 3A, 1B, 2B, 3B, 2, and 3.
[c]Classes 2A, 2B, 2, and 3.
[d]For approximate 75% full depth of thread. For decimal sizes of numbered and lettered drills, see Appendix 16.

# 15  Screw Threads, American National, Unified, and Metric (continued)

AMERICAN NATIONAL STANDARD UNIFIED AND AMERICAN NATIONAL SCREW THREADS[a] (continued)

| Nominal Diameter | 8-Pitch[b] Series 8N and 8UN | | 12-Pitch[b] Series 12N and 12UN | | 16-Pitch[b] Series 16N and 16UN | | Nominal Diameter | 8-Pitch[b] Series 8N and 8UN | | 12-Pitch[b] Series 12N and 12UN | | 16-Pitch[b] Series 16N and 16UN | |
|---|---|---|---|---|---|---|---|---|---|---|---|---|---|
| | Thds. per Inch | Tap Drill[c] | Thds. per Inch | Tap Drill[c] | Thds. per Inch | Tap Drill[c] | | Thds. per Inch | Tap Drill[c] | Thds. per Inch | Tap Drill[c] | Thds. per Inch | Tap Drill[c] |
| 1/2 | .... | .... | 12 | 27/64 | .... | .... | 2 1/16 | .... | .... | .... | .... | **16** | 2 |
| 9/16 | .... | .... | 12[e] | 31/64 | .... | .... | 2 1/8 | .... | .... | 12 | 2 3/64 | 16 | 2 1/16 |
| 5/8 | .... | .... | 12 | 35/64 | .... | .... | 2 3/16 | .... | .... | .... | .... | **16** | 2 1/8 |
| 11/16 | .... | .... | 12 | 39/64 | .... | .... | 2 1/4 | 8 | 2 1/8 | 12 | 2 11/64 | 16 | 2 3/16. |
| 3/4 | .... | .... | 12 | 43/64 | 16[e] | 11/16 | 2 5/16 | .... | .... | .... | .... | **16** | 2 1/4 |
| 13/16 | .... | .... | 12 | 47/64 | 16 | 3/4 | 2 3/8 | .... | .... | 12 | 2 19/64 | 16 | 2 5/16 |
| 7/8 | .... | .... | 12 | 51/64 | 16 | 13/16 | 2 7/16 | .... | .... | .... | .... | **16** | 2 3/8 |
| 15/16 | .... | .... | 12 | 55/64 | 16 | 7/8 | 2 1/2 | 8 | 2 3/8 | 12 | 2 27/64 | 16 | 2 7/16 |
| 1 | 8[e] | 7/8 | 12 | 59/64 | 16 | 15/16 | 2 5/8 | .... | .... | 12 | 2 35/64 | 16 | 2 9/16 |
| 1 1/16 | .... | .... | 12 | 63/64 | 16 | 1 | 2 3/4 | 8 | 2 5/8 | 12 | 2 43/64 | 16 | 2 11/16 |
| 1 1/8 | 8 | 1 | 12[e] | 1 3/64 | 16 | 1 1/16 | 2 7/8 | .... | .... | 12 | .... | 16 | .... |
| 1 3/16 | .... | .... | 12 | 1 7/64 | 16 | 1 1/8 | 3 | 8 | 2 7/8 | 12 | .... | 16 | .... |
| 1 1/4 | 8 | 1 1/8 | 12 | 1 11/64 | 16 | 1 3/16 | 3 1/8 | .... | .... | 12 | .... | 16 | .... |
| 1 5/16 | .... | .... | 12 | 1 15/64 | 16 | 1 1/4 | 3 1/4 | 8 | .... | 12 | .... | 16 | .... |
| 1 3/8 | 8 | 1 1/4 | 12[e] | 1 19/64 | 16 | 1 5/16 | 3 3/8 | .... | .... | 12 | .... | 16 | .... |
| 1 7/16 | .... | .... | 12 | 1 23/64 | 16 | 1 3/8 | 3 1/2 | 8 | .... | 12 | .... | 16 | .... |
| 1 1/2 | 8 | 1 3/8 | 12[e] | 1 27/64 | 16 | 1 7/16 | 3 5/8 | .... | .... | 12 | .... | 16 | .... |
| 1 9/16 | .... | .... | .... | .... | 16 | 1 1/2 | 3 3/4 | 8 | .... | 12 | .... | 16 | .... |
| 1 5/8 | 8 | 1 1/2 | 12 | 1 35/64 | 16 | 1 9/16 | 3 7/8 | .... | .... | 12 | .... | 16 | .... |
| 1 11/16 | .... | .... | .... | .... | 16 | 1 5/8 | 4 | 8 | .... | 12 | .... | 16 | .... |
| 1 3/4 | 8 | 1 5/8 | 12 | 1 43/64 | 16[e] | 1 11/16 | 4 1/4 | 8 | .... | 12 | .... | 16 | .... |
| 1 13/16 | .... | .... | .... | .... | 16 | 1 3/4 | 4 1/2 | 8 | .... | 12 | .... | 16 | .... |
| 1 7/8 | 8 | 1 3/4 | 12 | 1 51/64 | 16 | 1 13/16 | 4 3/4 | 8 | .... | 12 | .... | 16 | .... |
| 1 15/16 | .... | .... | .... | .... | 16 | 1 7/8 | 5 | 8 | .... | 12 | .... | 16 | .... |
| 2 | 8 | 1 7/8 | 12 | 1 59/64 | 16[e] | 1 15/16 | 5 1/4 | 8 | .... | 12 | .... | 16 | .... |

[a] ANSI B1.1.
[b] Classes 2A, 3A, 2B, 3B, 2, and 3.
[c] For approximate 75% full depth of thread.
[d] Boldface type indicates American National threads only.
[e] This is a standard size of the Unified or American National threads of the coarse, fine, or extra fine series. See preceding page.

## 15   Screw Threads, American National, Unified, and Metric (continued)

METRIC SCREW THREADS[a]

Preferred sizes for commercial threads and fasteners are shown in **boldface** type.

| Coarse (general purpose) | | Fine | |
|---|---|---|---|
| Nominal Size & Thd Pitch | Tap Drill Diameter, mm | Nominal Size & Thd Pitch | Tap Drill Diameter, mm |
| **M1.6 × 0.35** | 1.25 | — | — |
| M1.8 × 0.35 | 1.45 | — | — |
| **M2 × 0.4** | 1.6 | — | — |
| M2.2 × 0.45 | 1.75 | — | — |
| **M2.5 × 0.45** | 2.05 | — | — |
| **M3 × 0.5** | 2.5 | — | — |
| M3.5 × 0.6 | 2.9 | — | — |
| **M4 × 0.7** | 3.3 | — | — |
| M4.5 × 0.75 | 3.75 | — | — |
| **M5 × 0.8** | 4.2 | — | — |
| **M6 × 1** | 5.0 | — | — |
| M7 × 1 | 6.0 | — | — |
| **M8 × 1.25** | 6.8 | **M8 × 1** | 7.0 |
| M9 × 1.25 | 7.75 | — | — |
| **M10 × 1.5** | 8.5 | **M10 × 1.25** | 8.75 |
| M11 × 1.5 | 9.50 | — | — |
| **M12 × 1.75** | 10.30 | **M12 × 1.25** | 10.5 |
| M14 × 2 | 12.00 | **M14 × 1.5** | 12.5 |
| **M16 × 2** | 14.00 | **M16 × 1.5** | 14.5 |
| M18 × 2.5 | 15.50 | **M18 × 1.5** | 16.5 |
| **M20 × 2.5** | 17.5 | **M20 × 1.5** | 18.5 |
| M22 × 2.5[b] | 19.5 | **M22 × 1.5** | 20.5 |
| **M24 × 3** | 21.0 | **M24 × 2** | 22.0 |
| M27 × 3[b] | 24.0 | **M27 × 2** | 25.0 |
| **M30 × 3.5** | 26.5 | **M30 × 2** | 28.0 |
| M33 × 3.5 | 29.5 | **M30 × 2** | 31.0 |
| **M36 × 4** | 32.0 | **M36 × 2** | 33.0 |
| M39 × 4 | 35.0 | M39 × 2 | 36.0 |
| **M42 × 4.5** | 37.5 | **M42 × 2** | 39.0 |
| M45 × 4.5 | 40.5 | M45 × 1.5 | 42.0 |
| **M48 × 5** | 43.0 | **M48 × 2** | 45.0 |
| M52 × 5 | 47.0 | M52 × 2 | 49.0 |
| **M56 × 5.5** | 50.5 | **M56 × 2** | 52.0 |
| M60 × 5.5 | 54.5 | M60 × 1.5 | 56.0 |
| **M64 × 6** | 58.0 | **M64 × 2** | 60.0 |
| M68 × 6 | 62.0 | M68 × 2 | 64.0 |
| **M72 × 6** | 66.0 | **M72 × 2** | 68.0 |
| **M80 × 6** | 74.0 | **M80 × 2** | 76.0 |
| **M90 × 6** | 84.0 | **M90 × 2** | 86.0 |
| **M100 × 6** | 94.0 | **M100 × 2** | 96.0 |

[a]Metric Fasteners Standard, IFI-500 (1983) and ANSI/ASME B1.13M—1983.
[b]Only for high strength structural steel fasteners.

# 16 Twist Drill Sizes—American National Standard and Metric

AMERICAN NATIONAL STANDARD DRILL SIZES[a]

All dimensions are in inches.
Drills designated in common fractions are available in diameters $\frac{1}{64}''$ to $1\frac{3}{4}''$ in $\frac{1}{64}''$ increments, $1\frac{3}{4}''$ to $2\frac{1}{4}''$ in $\frac{1}{32}''$ increments. $2\frac{1}{4}''$ to $3''$ in $\frac{1}{16}''$ increments and $3''$ to $3\frac{1}{2}''$ in $\frac{1}{8}''$ increments. Drills larger than $3\frac{1}{2}''$ are seldom used, and are regarded as special drills.

| Size | Drill Diameter | Size | Drill Diameter | Size | Drill Diameter | Size | Drill Diameter | Size | Drill Diameter | Size | Drill Diameter |
|------|------|------|------|------|------|------|------|------|------|------|------|
| 1 | .2280 | 17 | .1730 | 33 | .1130 | 49 | .0730 | 65 | .0350 | 81 | .0130 |
| 2 | .2210 | 18 | .1695 | 34 | .1110 | 50 | .0700 | 66 | .0330 | 82 | .0125 |
| 3 | .2130 | 19 | .1660 | 35 | .1100 | 51 | .0670 | 67 | .0320 | 83 | .0120 |
| 4 | .2090 | 20 | .1610 | 36 | .1065 | 52 | .0635 | 68 | .0310 | 84 | .0115 |
| 5 | .2055 | 21 | .1590 | 37 | .1040 | 53 | .0595 | 69 | .0292 | 85 | .0110 |
| 6 | .2040 | 22 | .1570 | 38 | .1015 | 54 | .0550 | 70 | .0280 | 86 | .0105 |
| 7 | .2010 | 23 | .1540 | 39 | .0995 | 55 | .0520 | 71 | .0260 | 87 | .0100 |
| 8 | .1990 | 24 | .1520 | 40 | .0980 | 56 | .0465 | 72 | .0250 | 88 | .0095 |
| 9 | .1960 | 25 | .1495 | 41 | .0960 | 57 | .0430 | 73 | .0240 | 89 | .0091 |
| 10 | .1935 | 26 | .1470 | 42 | .0935 | 58 | .0420 | 74 | .0225 | 90 | .0087 |
| 11 | .1910 | 27 | .1440 | 43 | .0890 | 59 | .0410 | 75 | .0210 | 91 | .0083 |
| 12 | .1890 | 28 | .1405 | 44 | .0860 | 60 | .0400 | 76 | .0200 | 92 | .0079 |
| 13 | .1850 | 29 | .1360 | 45 | .0820 | 61 | .0390 | 77 | .0180 | 93 | .0075 |
| 14 | .1820 | 30 | .1285 | 46 | .0810 | 62 | .0380 | 78 | .0160 | 94 | .0071 |
| 15 | .1800 | 31 | .1200 | 47 | .0785 | 63 | .0370 | 79 | .0145 | 95 | .0067 |
| 16 | .1770 | 32 | .1160 | 48 | .0760 | 64 | .0360 | 80 | .0135 | 96 | .0063 |
| | | | | | | | | | | 97 | .0059 |

LETTER SIZES

| | | | | | | | | | | | |
|------|------|------|------|------|------|------|------|------|------|------|------|
| A | .234 | G | .261 | L | .290 | Q | .332 | V | .377 | | |
| B | .238 | H | .266 | M | .295 | R | .339 | W | .386 | | |
| C | .242 | I | .272 | N | .302 | S | .348 | X | .397 | | |
| D | .246 | J | .277 | O | .316 | T | .358 | Y | .404 | | |
| E | .250 | K | .281 | P | .323 | U | .368 | Z | .413 | | |
| F | .257 | | | | | | | | | | |

[a] ANSI B94.11M—1979 (R1987).

# 16   Twist Drill Sizes—American National Standard and Metric (continued)

METRIC DRILL SIZES
Decimal-inch equivalents are for reference only.

| Drill Diameter | | Drill Diameter | | Drill Diameter | | Drill Diameter | | Drill Diameter | | Drill Diameter | |
|---|---|---|---|---|---|---|---|---|---|---|---|
| mm | in. | mm | in. | mm | in. | mm | in. | mm | in. | mm | in. |
| 0.40 | .0157 | 1.95 | .0768 | 4.70 | .1850 | 8.00 | .3150 | 13.20 | .5197 | 25.50 | 1.0039 |
| 0.42 | .0165 | 2.00 | .0787 | 4.80 | .1890 | 8.10 | .3189 | 13.50 | .5315 | 26.00 | 1.0236 |
| 0.45 | .0177 | 2.05 | .0807 | 4.90 | .1929 | 8.20 | .3228 | 13.80 | .5433 | 26.50 | 1.0433 |
| 0.48 | .0189 | 2.10 | .0827 | 5.00 | .1969 | 8.30 | .3268 | 14.00 | .5512 | 27.00 | 1.0630 |
| 0.50 | .0197 | 2.15 | .0846 | 5.10 | .2008 | 8.40 | .3307 | 14.25 | .5610 | 27.50 | 1.0827 |
| 0.55 | .0217 | 2.20 | .0866 | 5.20 | .2047 | 8.50 | .3346 | 14.50 | .5709 | 28.00 | 1.1024 |
| 0.60 | .0236 | 2.25 | .0886 | 5.30 | .2087 | 8.60 | .3386 | 14.75 | .5807 | 28.50 | 1.1220 |
| 0.65 | .0256 | 2.30 | .0906 | 5.40 | .2126 | 8.70 | .3425 | 15.00 | .5906 | 29.00 | 1.1417 |
| 0.70 | .0276 | 2.35 | .0925 | 5.50 | .2165 | 8.80 | .3465 | 15.25 | .6004 | 29.50 | 1.1614 |
| 0.75 | .0295 | 2.40 | .0945 | 5.60 | .2205 | 8.90 | .3504 | 15.50 | .6102 | 30.00 | 1.1811 |
| 0.80 | .0315 | 2.45 | .0965 | 5.70 | .2244 | 9.00 | .3543 | 15.75 | .6201 | 30.50 | 1.2008 |
| 0.85 | .0335 | 2.50 | .0984 | 5.80 | .2283 | 9.10 | .3583 | 16.00 | .6299 | 31.00 | 1.2205 |
| 0.90 | .0354 | 2.60 | .1024 | 5.90 | .2323 | 9.20 | .3622 | 16.25 | .6398 | 31.50 | 1.2402 |
| 0.95 | .0374 | 2.70 | .1063 | 6.00 | .2362 | 9.30 | .3661 | 16.50 | .6496 | 32.00 | 1.2598 |
| 1.00 | .0394 | 2.80 | .1102 | 6.10 | .2402 | 9.40 | .3701 | 16.75 | .6594 | 32.50 | 1.2795 |
| 1.05 | .0413 | 2.90 | .1142 | 6.20 | .2441 | 9.50 | .3740 | 17.00 | .6693 | 33.00 | 1.2992 |
| 1.10 | .0433 | 3.00 | .1181 | 6.30 | .2480 | 9.60 | .3780 | 17.25 | .6791 | 33.50 | 1.3189 |
| 1.15 | .0453 | 3.10 | .1220 | 6.40 | .2520 | 9.70 | .3819 | 17.50 | .6890 | 34.00 | 1.3386 |
| 1.20 | .0472 | 3.20 | .1260 | 6.50 | .2559 | 9.80 | .3858 | 18.00 | .7087 | 34.50 | 1.3583 |
| 1.25 | .0492 | 3.30 | .1299 | 6.60 | .2598 | 9.90 | .3898 | 18.50 | .7283 | 35.00 | 1.3780 |
| 1.30 | .0512 | 3.40 | .1339 | 6.70 | .2638 | 10.00 | .3937 | 19.00 | .7480 | 35.50 | 1.3976 |
| 1.35 | .0531 | 3.50 | .1378 | 6.80 | .2677 | 10.20 | .4016 | 19.50 | .7677 | 36.00 | 1.4173 |
| 1.40 | .0551 | 3.60 | .1417 | 6.90 | .2717 | 10.50 | .4134 | 20.00 | .7874 | 36.50 | 1.4370 |
| 1.45 | .0571 | 3.70 | .1457 | 7.00 | .2756 | 10.80 | .4252 | 20.50 | .8071 | 37.00 | 1.4567 |
| 1.50 | .0591 | 3.80 | .1496 | 7.10 | .2795 | 11.00 | .4331 | 21.00 | .8268 | 37.50 | 1.4764 |
| 1.55 | .0610 | 3.90 | .1535 | 7.20 | .2835 | 11.20 | .4409 | 21.50 | .8465 | 38.00 | 1.4961 |
| 1.60 | .0630 | 4.00 | .1575 | 7.30 | .2874 | 11.50 | .4528 | 22.00 | .8661 | 40.00 | 1.5748 |
| 1.65 | .0650 | 4.10 | .1614 | 7.40 | .2913 | 11.80 | .4646 | 22.50 | .8858 | 42.00 | 1.6535 |
| 1.70 | .0669 | 4.20 | .1654 | 7.50 | .2953 | 12.00 | .4724 | 23.00 | .9055 | 44.00 | 1.7323 |
| 1.75 | .0689 | 4.30 | .1693 | 7.60 | .2992 | 12.20 | .4803 | 23.50 | .9252 | 46.00 | 1.8110 |
| 1.80 | .0709 | 4.40 | .1732 | 7.70 | .3031 | 12.50 | .4921 | 24.00 | .9449 | 48.00 | 1.8898 |
| 1.85 | .0728 | 4.50 | .1772 | 7.80 | .3071 | 12.50 | .5039 | 24.50 | .9646 | 50.00 | 1.9685 |
| 1.90 | .0748 | 4.60 | .1811 | 7.90 | .3110 | 13.00 | .5118 | 25.00 | .9843 | | |

# 17   Acme Threads, General-Purpose[a]

| Size | Threads per Inch | Size | Threads per Inch | Size | Threads per Inch | Size | Threads per Inch |
|---|---|---|---|---|---|---|---|
| ¼ | 16 | ¾ | 6 | 1½ | 4 | 3 | 2 |
| ⁵⁄₁₆ | 14 | ⅞ | 6 | 1¾ | 4 | 3½ | 2 |
| ⅜ | 12 | 1 | 5 | 2 | 4 | 4 | 2 |
| ⁷⁄₁₆ | 12 | 1⅛ | 5 | 2¼ | 3 | 4½ | 2 |
| ½ | 10 | 1¼ | 5 | 2½ | 3 | 5 | 2 |
| ⅝ | 8 | 1⅜ | 4 | 2¾ | 3 | . . . | . . |

[a]ANSI/ASME B1.5—1988.

# 18 Bolts, Nuts, and Cap Screws—Square and Hexagon— American National Standard and Metric

## AMERICAN NATIONAL STANDARD SQUARE AND HEXAGON BOLTS[a] AND NUTS[b] AND HEXAGON CAP SCREWS[c]

**Boldface type** indicates product features unified dimensionally with British and Canadian standards.
All dimensions are in inches.
For thread series, minimum thread lengths, and bolt lengths, see §15.25.

| Nominal Size D Body Diameter of Bolt | | Regular Bolts | | | | | Heavy Bolts | | |
|---|---|---|---|---|---|---|---|---|---|
| | | Width Across Flats W | | Height H | | | Width Across Flats W | Height H | |
| | | Sq. | Hex. | Sq. (Unfin.) | Hex. (Unfin.) | Hex. Cap Scr.[c] (Fin.) | | Hex. (Unfin.) | Hex. Screw (Fin.) |
| ¼ | 0.2500 | ⅜ | ⁷⁄₁₆ | ¹¹⁄₆₄ | ¹¹⁄₆₄ | ⁵⁄₃₂ | .... | .... | .... |
| ⁵⁄₁₆ | 0.3125 | ½ | ½ | ¹³⁄₆₄ | ⁷⁄₃₂ | ¹³⁄₆₄ | .... | .... | .... |
| ⅜ | 0.3750 | ⁹⁄₁₆ | ⁹⁄₁₆ | ¼ | ¼ | ¹⁵⁄₆₄ | .... | .... | .... |
| ⁷⁄₁₆ | 0.4375 | ⅝ | ⅝ | ¹⁹⁄₆₄ | ¹⁹⁄₆₄ | ⁹⁄₃₂ | .... | .... | .... |
| ½ | 0.5000 | ¾ | ¾ | ²¹⁄₆₄ | ¹¹⁄₃₂ | ⁵⁄₁₆ | ⅞ | ¹¹⁄₃₂ | ⁵⁄₁₆ |
| ⁹⁄₁₆ | 0.5625 | .... | ¹³⁄₁₆ | .... | .... | ²³⁄₆₄ | .... | .... | .... |
| ⅝ | 0.6250 | ¹⁵⁄₁₆ | ¹⁵⁄₁₆ | ²⁷⁄₆₄ | ²⁷⁄₆₄ | ²⁵⁄₆₄ | 1¹⁄₁₆ | ²⁷⁄₆₄ | ²⁵⁄₆₄ |
| ¾ | 0.7500 | 1⅛ | 1⅛ | ½ | ½ | ¹⁵⁄₃₂ | 1¼ | ½ | ¹⁵⁄₃₂ |
| ⅞ | 0.8750 | 1⁵⁄₁₆ | 1⁵⁄₁₆ | ¹⁹⁄₃₂ | ³⁷⁄₆₄ | ³⁵⁄₆₄ | 1⁷⁄₁₆ | ³⁷⁄₆₄ | ³⁵⁄₆₄ |
| 1 | 1.000 | 1½ | 1½ | ²¹⁄₃₂ | ⁴³⁄₆₄ | ³⁹⁄₆₄ | 1⅝ | ⁴³⁄₆₄ | ³⁹⁄₆₄ |
| 1⅛ | 1.1250 | 1¹¹⁄₁₆ | 1¹¹⁄₁₆ | ¾ | ¾ | ¹¹⁄₁₆ | 1¹³⁄₁₆ | ¾ | ¹¹⁄₁₆ |
| 1¼ | 1.2500 | 1⅞ | 1⅞ | ²⁷⁄₃₂ | ²⁷⁄₃₂ | ²⁵⁄₃₂ | 2 | ²⁷⁄₃₂ | ²⁵⁄₃₂ |
| 1⅜ | 1.3750 | 2¹⁄₁₆ | 2¹⁄₁₆ | ²⁹⁄₃₂ | ²⁹⁄₃₂ | ²⁷⁄₃₂ | 2³⁄₁₆ | ²⁹⁄₃₂ | ²⁷⁄₃₂ |
| 1½ | 1.5000 | 2¼ | 2¼ | 1 | 1 | ¹⁵⁄₁₆ | 2⅜ | 1 | ¹⁵⁄₁₆ |
| 1¾ | 1.7500 | .... | 2⅝ | .... | 1⁵⁄₃₂ | 1³⁄₃₂ | 2¾ | 1⁵⁄₃₂ | 1³⁄₃₂ |
| 2 | 2.0000 | .... | 3 | .... | 1¹¹⁄₃₂ | 1⁷⁄₃₂ | 3⅛ | 1¹¹⁄₃₂ | 1⁷⁄₃₂ |
| 2¼ | 2.2500 | .... | 3⅜ | .... | 1½ | 1⅜ | 3½ | 1½ | 1⅜ |
| 2½ | 2.5000 | .... | 3¾ | .... | 1²¹⁄₃₂ | 1¹⁷⁄₃₂ | 3⅞ | 1²¹⁄₃₂ | 1¹⁷⁄₃₂ |
| 2¾ | 2.7500 | .... | 4⅛ | .... | 1¹³⁄₁₆ | 1¹¹⁄₁₆ | 4¼ | 1¹³⁄₁₆ | 1¹¹⁄₁₆ |
| 3 | 3.0000 | .... | 4½ | .... | 2 | 1⅞ | 4⅝ | 2 | 1⅞ |
| 3¼ | 3.2500 | .... | 4⅞ | .... | 2³⁄₁₆ | .... | .... | .... | .... |
| 3½ | 3.5000 | .... | 5¼ | .... | 2⁵⁄₁₆ | .... | .... | .... | .... |
| 3¾ | 3.7500 | .... | 5⅝ | .... | 2½ | .... | .... | .... | .... |
| 4 | 4.0000 | .... | 6 | .... | 2¹¹⁄₁₆ | .... | .... | .... | .... |

[a]ANSI B18.2.1—1981.
[b]ANSI B18.2.2—1987.
[c]Hexagon cap screws and finished hexagon bolts are combined as a single product.

## 18    Bolts, Nuts, and Cap Screws—Square and Hexagon— American National Standard and Metric (continued)

### AMERICAN NATIONAL STANDARD SQUARE AND HEXAGON BOLTS AND NUTS AND HEXAGON CAP SCREWS (continued)

See ANSI B18.2.2 for jam nuts, slotted nuts, thick nuts, thick slotted nuts, and castle nuts.
For methods of drawing bolts and nuts and hexagon-head cap screws, see Figs. 15.29, 15.30, and 15.32.

| Nominal Size D Body Diameter of Bolt | | Regular Nuts | | | | | Heavy Nuts | | | |
|---|---|---|---|---|---|---|---|---|---|---|
| | | Width Across Flats W | | Thickness T | | | Width Across Flats W | Thickness T | | |
| | | Sq. | Hex. | Sq. (Unfin.) | Hex. Flat (Unfin.) | Hex. (Fin.) | | Sq. (Unfin.) | Hex. Flat (Unfin.) | Hex. (Fin.) |
| ¼ | 0.2500 | $\frac{7}{16}$ | $\frac{7}{16}$ | $\frac{7}{32}$ | $\frac{7}{32}$ | $\frac{7}{32}$ | ½ | ¼ | $\frac{15}{64}$ | $\frac{15}{64}$ |
| $\frac{5}{16}$ | 0.3125 | $\frac{9}{16}$ | ½ | $\frac{17}{64}$ | $\frac{17}{64}$ | $\frac{17}{64}$ | $\frac{9}{16}$ | $\frac{5}{16}$ | $\frac{19}{64}$ | $\frac{19}{64}$ |
| ⅜ | 0.3750 | ⅝ | $\frac{9}{16}$ | $\frac{21}{64}$ | $\frac{21}{64}$ | $\frac{21}{64}$ | $\frac{11}{16}$ | ⅜ | $\frac{23}{64}$ | $\frac{23}{64}$ |
| $\frac{7}{16}$ | 0.4375 | ¾ | $\frac{11}{16}$ | ⅜ | ⅜ | ⅜ | ¾ | $\frac{7}{16}$ | $\frac{27}{64}$ | $\frac{27}{64}$ |
| ½ | 0.5000 | $\frac{13}{16}$ | ¾ | $\frac{7}{16}$ | $\frac{7}{16}$ | $\frac{7}{16}$ | ⅞$^a$ | ½ | $\frac{31}{64}$ | $\frac{31}{64}$ |
| $\frac{9}{16}$ | 0.5625 | . . . . | ⅞ | . . . . | $\frac{31}{64}$ | $\frac{31}{64}$ | $\frac{15}{16}$ | . . . . | $\frac{35}{64}$ | $\frac{35}{64}$ |
| ⅝ | 0.6250 | 1 | $\frac{15}{16}$ | $\frac{35}{64}$ | $\frac{35}{64}$ | $\frac{35}{64}$ | $1\frac{1}{16}$$^a$ | ⅝ | $\frac{39}{64}$ | $\frac{39}{64}$ |
| ¾ | 0.7500 | 1⅛ | 1⅛ | $\frac{21}{32}$ | $\frac{41}{64}$ | $\frac{41}{64}$ | $1\frac{1}{4}$$^a$ | ¾ | $\frac{47}{64}$ | $\frac{47}{64}$ |
| ⅞ | 0.8750 | $1\frac{5}{16}$ | $1\frac{5}{16}$ | $\frac{49}{64}$ | ¾ | ¾ | $1\frac{7}{16}$$^a$ | ⅞ | $\frac{55}{64}$ | $\frac{55}{64}$ |
| 1 | 1.0000 | 1½ | 1½ | ⅞ | $\frac{55}{64}$ | $\frac{55}{64}$ | $1\frac{5}{8}$$^a$ | 1 | $\frac{63}{64}$ | $\frac{63}{64}$ |
| 1⅛ | 1.1250 | $1\frac{11}{16}$ | $1\frac{11}{16}$ | 1 | 1 | $\frac{31}{32}$ | $1\frac{13}{16}$$^a$ | 1⅛ | 1⅛ | $1\frac{7}{64}$ |
| 1¼ | 1.2500 | 1⅞ | 1⅞ | $1\frac{3}{32}$ | $1\frac{3}{32}$ | $1\frac{1}{16}$ | 2$^a$ | 1¼ | 1¼ | $1\frac{7}{32}$ |
| 1⅜ | 1.3750 | $2\frac{1}{16}$ | $2\frac{1}{16}$ | $1\frac{13}{64}$ | $1\frac{13}{64}$ | $1\frac{11}{64}$ | $2\frac{3}{16}$$^a$ | 1⅜ | 1⅜ | $1\frac{11}{32}$ |
| 1½ | 1.5000 | 2¼ | 2¼ | $1\frac{5}{16}$ | $1\frac{5}{16}$ | $1\frac{9}{32}$ | 2⅜$^a$ | 1½ | 1½ | $1\frac{15}{32}$ |
| 1⅝ | 1.6250 | . . . . | . . . . | . . . . | . . . . | . . . . | $2\frac{9}{16}$ | . . . . | . . . . | $1\frac{19}{32}$ |
| 1¾ | 1.7500 | . . . . | . . . . | . . . . | . . . . | . . . . | 2¾ | . . . . | 1¾ | $1\frac{23}{32}$ |
| 1⅞ | 1.8750 | . . . . | . . . . | . . . . | . . . . | . . . . | $2\frac{15}{16}$ | . . . . | . . . . | $1\frac{27}{32}$ |
| 2 | 2.0000 | . . . . | . . . . | . . . . | . . . . | . . . . | 3⅛ | . . . . | 2 | $1\frac{31}{32}$ |
| 2¼ | 2.2500 | . . . . | . . . . | . . . . | . . . . | . . . . | 3½ | . . . . | 2¼ | $2\frac{13}{64}$ |
| 2½ | 2.5000 | . . . . | . . . . | . . . . | . . . . | . . . . | 3⅞ | . . . . | 2½ | $2\frac{29}{64}$ |
| 2¾ | 2.7500 | . . . . | . . . . | . . . . | . . . . | . . . . | 4¼ | . . . . | 2¾ | $2\frac{45}{64}$ |
| 3 | 3.0000 | . . . . | . . . . | . . . . | . . . . | . . . . | 4⅝ | . . . . | 3 | $2\frac{61}{64}$ |
| 3¼ | 3.2500 | . . . . | . . . . | . . . . | . . . . | . . . . | 5 | . . . . | 3¼ | $3\frac{3}{16}$ |
| 3½ | 3.5000 | . . . . | . . . . | . . . . | . . . . | . . . . | 5⅜ | . . . . | 3½ | $3\frac{7}{16}$ |
| 3¾ | 3.7500 | . . . . | . . . . | . . . . | . . . . | . . . . | 5¾ | . . . . | 3¾ | $3\frac{11}{16}$ |
| 4 | 4.0000 | . . . . | . . . . | . . . . | . . . . | . . . . | 6⅛ | . . . . | 4 | $3\frac{15}{16}$ |

$^a$ Product feature not unified for heavy square nut.

# 18   Bolts, Nuts, and Cap Screws—Square and Hexagon—American National Standard and Metric (continued)

METRIC HEXAGON BOLTS, HEXAGON CAP SCREWS,
HEXAGON STRUCTURAL BOLTS, AND HEXAGON NUTS

| Nominal Size D, mm | Width Across Flats W (max) | | Thickness T (max) | | | |
| --- | --- | --- | --- | --- | --- | --- |
| Body Dia and Thd Pitch | Bolts,[a] Cap Screws,[b] and Nuts[c] | Heavy Hex & Hex Structural Bolts[a] & Nuts[c] | Bolts (Unfin.) | Cap Screw (Fin.) | Nut (Fin. or Unfin.) | |
| | | | | | Style 1 | Style 2 |
| M5 × 0.8 | 8.0 | | 3.88 | 3.65 | 4.7 | 5.1 |
| M6 × 1 | 10.0 | | 4.38 | 4.47 | 5.2 | 5.7 |
| M8 × 1.25 | 13.0 | | 5.68 | 5.50 | 6.8 | 7.5 |
| M10 × 1.5 | 16.0 | | 6.85 | 6.63 | 8.4 | 9.3 |
| M12 × 1.75 | 18.0 | 21.0 | 7.95 | 7.76 | 10.8 | 12.0 |
| M14 × 2 | 21.0 | 24.0 | 9.25 | 9.09 | 12.8 | 14.1 |
| M16 × 2 | 24.0 | 27.0 | 10.75 | 10.32 | 14.8 | 16.4 |
| M20 × 2.5 | 30.0 | 34.0 | 13.40 | 12.88 | 18.0 | 20.3 |
| M24 × 3 | 36.0 | 41.0 | 15.90 | 15.44 | 21.5 | 23.9 |
| M30 × 3.5 | 46.0 | 50.0 | 19.75 | 19.48 | 25.6 | 28.6 |
| M36 × 4 | 55.0 | 60.0 | 23.55 | 23.38 | 31.0 | 34.7 |
| M42 × 4.5 | 65.0 | | 27.05 | 26.97 | .... | .... |
| M48 × 5 | 75.0 | | 31.07 | 31.07 | .... | .... |
| M56 × 5.5 | 85.0 | | 36.20 | 36.20 | .... | .... |
| M64 × 6 | 95.0 | | 41.32 | 41.32 | .... | .... |
| M72 × 6 | 105.0 | | 46.45 | 46.45 | .... | .... |
| M80 × 6 | 115.0 | | 51.58 | 51.58 | .... | .... |
| M90 × 6 | 130.0 | | 57.74 | 57.74 | .... | .... |
| M100 × 6 | 145.0 | | 63.90 | 63.90 | .... | .... |

HIGH STRENGTH STRUCTURAL HEXAGON BOLTS[a] (Fin.) AND HEXAGON NUTS[c]

| Nominal Size | W (max) | Heavy Hex | Bolts (Unfin.) | Cap Screw | Style 1 | Style 2 |
| --- | --- | --- | --- | --- | --- | --- |
| M16 × 2 | 27.0 | .... | 10.75 | .... | .... | 17.1 |
| M20 × 2.5 | 34.0 | .... | 13.40 | .... | .... | 20.7 |
| M22 × 2.5 | 36.0 | .... | 14.9 | .... | .... | 23.6 |
| M24 × 3 | 41.0 | .... | 15.9 | .... | .... | 24.2 |
| M27 × 3 | 46.0 | .... | 17.9 | .... | .... | 27.6 |
| M30 × 3.5 | 50.0 | .... | 19.75 | .... | .... | 31.7 |
| M36 × 4 | 60.0 | .... | 23.55 | .... | .... | 36.6 |

[a]ANSI B18.2.3.5M—1979, B18.2.3.6M—1979, B18.2.3.7M—1979.
[b]ANSI B18.2.3.1M—1979.
[c]ANSI B18.2.4.1M—1979, B18.2.4.2M—1979.

# 19　Cap Screws, Slotted[a] and Socket Head[b] — American National Standard and Metric

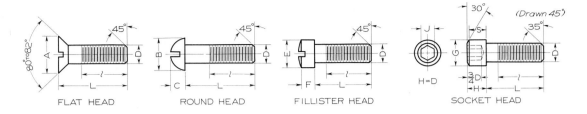

FLAT HEAD　　　ROUND HEAD　　　FILLISTER HEAD　　　SOCKET HEAD

For methods of drawing cap screws, screw lengths, and thread data, see Fig. 15.32.

| Nominal Size D | Flat Head[a] | Round Head[a] | | Fillister Head[a] | | Socket Head[b] | | |
|---|---|---|---|---|---|---|---|---|
| | A | B | C | E | F | G | J | S |
| 0 (.060) | .... | .... | .... | .... | .... | .096 | .05 | .054 |
| 1 (.073) | .... | .... | .... | .... | .... | .118 | 1⁄16 | .066 |
| 2 (.086) | .... | .... | .... | .... | .... | .140 | 5⁄64 | .077 |
| 3 (.099) | .... | .... | .... | .... | .... | .161 | 5⁄64 | .089 |
| 4 (.112) | .... | .... | .... | .... | .... | .183 | 3⁄32 | .101 |
| 5 (.125) | .... | .... | .... | .... | .... | .205 | 3⁄32 | .112 |
| 6 (.138) | .... | .... | .... | .... | .... | .226 | 7⁄64 | .124 |
| 8 (.164) | .... | .... | .... | .... | .... | .270 | 9⁄64 | .148 |
| 10 (.190) | .... | .... | .... | .... | .... | .312 | 5⁄32 | .171 |
| 1⁄4 | .500 | .437 | .191 | .375 | .172 | .375 | 3⁄16 | .225 |
| 5⁄16 | .625 | .562 | .245 | .437 | .203 | .469 | 1⁄4 | .281 |
| 3⁄8 | .750 | .675 | .273 | .562 | .250 | .562 | 5⁄16 | .337 |
| 7⁄16 | .812 | .750 | .328 | .625 | .297 | .656 | 3⁄8 | .394 |
| 1⁄2 | .875 | .812 | .354 | .750 | .328 | .750 | 3⁄8 | .450 |
| 9⁄16 | 1.000 | .937 | .409 | .812 | .375 | .... | .... | .... |
| 5⁄8 | 1.125 | 1.000 | .437 | .875 | .422 | .938 | 1⁄2 | .562 |
| 3⁄4 | 1.375 | 1.250 | .546 | 1.000 | .500 | 1.125 | 5⁄8 | .675 |
| 7⁄8 | 1.625 | .... | .... | 1.125 | .594 | 1.312 | 3⁄4 | .787 |
| 1 | 1.875 | .... | .... | 1.312 | .656 | 1.500 | 3⁄4 | .900 |
| 1 1⁄8 | 2.062 | .... | .... | .... | .... | 1.688 | 7⁄8 | 1.012 |
| 1 1⁄4 | 2.312 | .... | .... | .... | .... | 1.875 | 7⁄8 | 1.125 |
| 1 3⁄8 | 2.562 | .... | .... | .... | .... | 2.062 | 1 | 1.237 |
| 1 1⁄2 | 2.812 | .... | .... | .... | .... | 2.250 | 1 | 1.350 |

[a]ANSI B18.6.2–1972 (R1983).
[b]ANSI/ASME B18.3—1986. For hexagon-head screws, see §15.29 and Appendix 18.

# 19 Cap Screws, Slotted[a] and Socket Head[b]—American National Standard and Metric (continued)

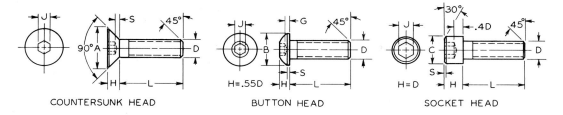

COUNTERSUNK HEAD     BUTTON HEAD     SOCKET HEAD

For methods of drawing cap screws, screw lengths, and thread data, see Fig. 15.32.

| | METRIC SOCKET HEAD CAP SCREWS | | | | | | | | |
|---|---|---|---|---|---|---|---|---|---|
| Nominal Size D | Countersunk Head[a] | | | Button Head[b] | | | Socket Head[c] | | Hex Socket Size |
| | A (max) | H | S | B | S | G | C | S | J |
| M1.6 × 0.35 | .... | .... | .... | .... | .... | .... | 3.0 | 0.16 | 1.5 |
| M2 × 0.4 | .... | .... | .... | .... | .... | .... | 3.8 | 0.2 | 1.5 |
| M2.5 × 0.45 | .... | .... | .... | .... | .... | .... | 4.5 | 0.25 | 2.0 |
| M3 × 0.5 | 6.72 | 1.86 | 0.25 | 5.70 | 0.38 | 0.2 | 5.5 | 0.3 | 2.5 |
| M4 × 0.7 | 8.96 | 2.48 | 0.45 | 7.6 | 0.38 | 0.3 | 7.0 | 0.4 | 3.0 |
| M5 × 0.8 | 11.2 | 3.1 | 0.66 | 9.5 | 0.5 | 0.38 | 8.5 | 0.5 | 4.0 |
| M6 × 1 | 13.44 | 3.72 | 0.7 | 10.5 | 0.8 | 0.74 | 10.0 | 0.6 | 5.0 |
| M8 × 1.25 | 17.92 | 4.96 | 1.16 | 14.0 | 0.8 | 1.05 | 13.0 | 0.8 | 6.0 |
| M10 × 1.5 | 22.4 | 6.2 | 1.62 | 17.5 | 0.8 | 1.45 | 16.0 | 1.0 | 8.0 |
| M12 × 1.75 | 26.88 | 7.44 | 1.8 | 21.0 | 0.8 | 1.63 | 18.0 | 1.2 | 10.0 |
| M14 × 2 | 30.24 | 8.12 | 2.0 | .... | .... | .... | 21.0 | 1.4 | 12.0 |
| M16 × 2 | 33.6 | 8.8 | 2.2 | 28.0 | 1.5 | 2.25 | 24.0 | 1.6 | 14.0 |
| M20 × 2.5 | 19.67 | 10.16 | 2.2 | .... | .... | .... | 30.0 | 2.0 | 17.0 |
| M24 × 3 | .... | .... | .... | .... | .... | .... | 36.0 | 2.4 | 19.0 |
| M30 × 3.5 | .... | .... | .... | .... | .... | .... | 45.0 | 3.0 | 22.0 |
| M36 × 4 | .... | .... | .... | .... | .... | .... | 54.0 | 3.6 | 27.0 |
| M42 × 4.5 | .... | .... | .... | .... | .... | .... | 63.0 | 4.2 | 32.0 |
| M48 · × 5 | .... | .... | .... | .... | .... | .... | 72.0 | 4.8 | 36.0 |

[a]IFI 535—1982.
[b]ANSI/ASME 18.3.4M—1986.
[c]ANSI/ASME 18.3.1M—1986.

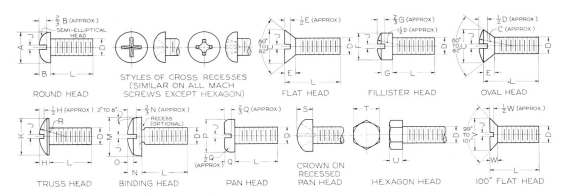

ROUND HEAD    STYLES OF CROSS RECESSES (SIMILAR ON ALL MACH SCREWS EXCEPT HEXAGON)    FLAT HEAD    FILLISTER HEAD    OVAL HEAD

TRUSS HEAD    BINDING HEAD    PAN HEAD    CROWN ON RECESSED PAN HEAD    HEXAGON HEAD    100° FLAT HEAD

### AMERICAN NATIONAL STANDARD MACHINE SCREWS[a]

*Length of Thread:* On screws 2″ long and shorter, the threads extend to within two threads of the head and closer if practicable; longer screws have minimum thread length of 1¾″.

*Points:* Machine screws are regularly made with plain sheared ends, not chamfered.

*Threads:* Either Coarse or Fine Thread Series, Class 2 fit.

*Recessed Heads:* Two styles of cross recesses are available on all screws except hexagon head.

| Nominal Size | Max. Diameter D | Round Head | | Flat Heads & Oval Head | | Fillister Head | | Truss Head | | | Slot Width |
|---|---|---|---|---|---|---|---|---|---|---|---|
| | | A | B | C | E | F | G | K | H | R | J |
| 0 | 0.060 | 0.113 | 0.053 | 0.119 | 0.035 | 0.096 | 0.045 | 0.131 | 0.037 | 0.087 | 0.023 |
| 1 | 0.073 | 0.138 | 0.061 | 0.146 | 0.043 | 0.118 | 0.053 | 0.164 | 0.045 | 0.107 | 0.026 |
| 2 | 0.086 | 0.162 | 0.069 | 0.172 | 0.051 | 0.140 | 0.062 | 0.194 | 0.053 | 0.129 | 0.031 |
| 3 | 0.099 | 0.187 | 0.078 | 0.199 | 0.059 | 0.161 | 0.070 | 0.226 | 0.061 | 0.151 | 0.035 |
| 4 | 0.112 | 0.211 | 0.086 | 0.225 | 0.067 | 0.183 | 0.079 | 0.257 | 0.069 | 0.169 | 0.039 |
| 5 | 0.125 | 0.236 | 0.095 | 0.252 | 0.075 | 0.205 | 0.088 | 0.289 | 0.078 | 0.191 | 0.043 |
| 6 | 0.138 | 0.260 | 0.103 | 0.279 | 0.083 | 0.226 | 0.096 | 0.321 | 0.086 | 0.211 | 0.048 |
| 8 | 0.164 | 0.309 | 0.120 | 0.332 | 0.100 | 0.270 | 0.113 | 0.384 | 0.102 | 0.254 | 0.054 |
| 10 | 0.190 | 0.359 | 0.137 | 0.385 | 0.116 | 0.313 | 0.130 | 0.448 | 0.118 | 0.283 | 0.060 |
| 12 | 0.216 | 0.408 | 0.153 | 0.438 | 0.132 | 0.357 | 0.148 | 0.511 | 0.134 | 0.336 | 0.067 |
| ¼ | 0.250 | 0.472 | 0.175 | 0.507 | 0.153 | 0.414 | 0.170 | 0.573 | 0.150 | 0.375 | 0.075 |
| ⁵⁄₁₆ | 0.3125 | 0.590 | 0.216 | 0.635 | 0.191 | 0.518 | 0.211 | 0.698 | 0.183 | 0.457 | 0.084 |
| ⅜ | 0.375 | 0.708 | 0.256 | 0.762 | 0.230 | 0.622 | 0.253 | 0.823 | 0.215 | 0.538 | 0.094 |
| ⁷⁄₁₆ | 0.4375 | 0.750 | 0.328 | 0.812 | 0.223 | 0.625 | 0.265 | 0.948 | 0.248 | 0.619 | 0.094 |
| ½ | 0.500 | 0.813 | 0.355 | 0.875 | 0.223 | 0.750 | 0.297 | 1.073 | 0.280 | 0.701 | 0.106 |
| ⁹⁄₁₆ | 0.5625 | 0.938 | 0.410 | 1.000 | 0.260 | 0.812 | 0.336 | 1.198 | 0.312 | 0.783 | 0.118 |
| ⅝ | 0.625 | 1.000 | 0.438 | 1.125 | 0.298 | 0.875 | 0.375 | 1.323 | 0.345 | 0.863 | 0.133 |
| ¾ | 0.750 | 1.250 | 0.547 | 1.375 | 0.372 | 1.000 | 0.441 | 1.573 | 0.410 | 1.024 | 0.149 |

| Nominal Size | Max. Diameter D | Binding Head | | | Pan Head | | | Hexagon Head | | 100° Flat Head | | Slot Width |
|---|---|---|---|---|---|---|---|---|---|---|---|---|
| | | M | N | O | P | Q | S | T | U | V | W | J |
| 2 | 0.086 | 0.181 | 0.050 | 0.018 | 0.167 | 0.053 | 0.062 | 0.125 | 0.050 | . . . . | . . . . | 0.031 |
| 3 | 0.099 | 0.208 | 0.059 | 0.022 | 0.193 | 0.060 | 0.071 | 0.187 | 0.055 | . . . . | . . . . | 0.035 |
| 4 | 0.112 | 0.235 | 0.068 | 0.025 | 0.219 | 0.068 | 0.080 | 0.187 | 0.060 | 0.225 | 0.049 | 0.039 |
| 5 | 0.125 | 0.263 | 0.078 | 0.029 | 0.245 | 0.075 | 0.089 | 0.187 | 0.070 | . . . . | . . . . | 0.043 |
| 6 | 0.138 | 0.290 | 0.087 | 0.032 | 0.270 | 0.082 | 0.097 | 0.250 | 0.080 | 0.279 | 0.060 | 0.048 |
| 8 | 0.164 | 0.344 | 0.105 | 0.039 | 0.322 | 0.096 | 0.115 | 0.250 | 0.110 | 0.332 | 0.072 | 0.054 |
| 10 | 0.190 | 0.399 | 0.123 | 0.045 | 0.373 | 0.110 | 0.133 | 0.312 | 0.120 | 0.385 | 0.083 | 0.060 |
| 12 | 0.216 | 0.454 | 0.141 | 0.052 | 0.425 | 0.125 | 0.151 | 0.312 | 0.155 | . . . . | . . . . | 0.067 |
| ¼ | 0.250 | 0.513 | 0.165 | 0.061 | 0.492 | 0.144 | 0.175 | 0.375 | 0.190 | 0.507 | 0.110 | 0.075 |
| ⁵⁄₁₆ | 0.3125 | 0.641 | 0.209 | 0.077 | 0.615 | 0.178 | 0.218 | 0.500 | 0.230 | 0.635 | 0.138 | 0.084 |
| ⅜ | 0.375 | 0.769 | 0.253 | 0.094 | 0.740 | 0.212 | 0.261 | 0.562 | 0.295 | 0.762 | 0.165 | 0.094 |
| ⁷⁄₁₆ | .4375 | . . . . | . . . . | . . . . | .865 | .247 | .305 | . . . . | . . . . | . . . . | . . . . | .094 |
| ½ | .500 | . . . . | . . . . | . . . . | .987 | .281 | .348 | . . . . | . . . . | . . . . | . . . . | .106 |
| ⁹⁄₁₆ | .5625 | . . . . | . . . . | . . . . | 1.041 | .315 | .391 | . . . . | . . . . | . . . . | . . . . | .118 |
| ⅝ | .625 | . . . . | . . . . | . . . . | 1.172 | .350 | .434 | . . . . | . . . . | . . . . | . . . . | .133 |
| ¾ | .750 | . . . . | . . . . | . . . . | 1.435 | .419 | .521 | . . . . | . . . . | . . . . | . . . . | .149 |

# 20 Machine Screws—American National Standard and Metric
## (continued)

### METRIC MACHINE SCREWS[b]

*Length of Thread:* On screws 36 mm long or shorter, the threads extend to within one thread of the head: on longer screws the thread extends to within two threads of the head.
*Points:* Machine screws are regularly made with sheared ends, not chamfered.
*Threads:* Coarse (general-purpose) threads series are given.
*Recessed Heads:* Two styles of cross-recesses are available on all screws except hexagon head.

| Nominal Size & Thd Pitch | Max. Dia. D, mm | Flat Heads & Oval Head | | Pan Heads | | | Hex Head | | Slot Width |
|---|---|---|---|---|---|---|---|---|---|
| | | C | E | P | Q | S | T | U | J |
| M2 × 0.4 | 2.0 | 3.5 | 1.2 | 4.0 | 1.3 | 1.6 | 3.2 | 1.6 | 0.7 |
| M2.5 × 0.45 | 2.5 | 4.4 | 1.5 | 5.0 | 1.5 | 2.1 | 4.0 | 2.1 | 0.8 |
| M3 × 0.5 | 3.0 | 5.2 | 1.7 | 5.6 | 1.8 | 2.4 | 5.0 | 2.3 | 1.0 |
| M3.5 × 0.6 | 3.5 | 6.9 | 2.3 | 7.0 | 2.1 | 2.6 | 5.5 | 2.6 | 1.2 |
| M4 × 0.7 | 4.0 | 8.0 | 2.7 | 8.0 | 2.4 | 3.1 | 7.0 | 3.0 | 1.5 |
| M5 × 0.8 | 5.0 | 8.9 | 2.7 | 9.5 | 3.0 | 3.7 | 8.0 | 3.8 | 1.5 |
| M6 × 1 | 6.0 | 10.9 | 3.3 | 12.0 | 3.6 | 4.6 | 10.0 | 4.7 | 1.9 |
| M8 × 1.25 | 8.0 | 15.14 | 4.6 | 16.0 | 4.8 | 6.0 | 13.0 | 6.0 | 2.3 |
| M10 × 1.5 | 10.0 | 17.8 | 5.0 | 20.0 | 6.0 | 7.5 | 15.0 | 7.5 | 2.8 |
| M12 × 1.75 | 12.0 | .... | .... | .... | .... | .... | 18.0 | 9.0 | .... |

| Nominal Size | Metric Machine Screw Lengths—L[c] | | | | | | | | | | | | | | | | | | | | | |
|---|---|---|---|---|---|---|---|---|---|---|---|---|---|---|---|---|---|---|---|---|---|---|
| | 2.5 | 3 | 4 | 5 | 6 | 8 | 10 | 13 | 16 | 20 | 25 | 30 | 35 | 40 | 45 | 50 | 55 | 60 | 65 | 70 | 80 | 90 |
| M2 × 0.4 | PH | A | A | A | A | A | A | A | A | A | | | | | | | | | | | | |
| M2.5 × 0.45 | | PH | A | A | A | A | A | A | A | A | A | | | | | | | | | | | |
| M3 × 0.5 | | | PH | A | A | A | A | A | A | A | A | A | | | | | | | | | | |
| M3.5 × 0.6 | | | | PH | A | A | A | A | A | A | A | A | A | | | | | | | | | |
| M4 × 0.7 | | | | PH | A | A | A | A | A | A | A | A | A | A | | | | | | | | |
| M5 × 0.8 | | | | | PH | A | A | A | A | A | A | A | A | A | A | A | | | | | | |
| M6 × 1 | | | | | | A | A | A | A | A | A | A | A | A | A | A | A | A | | | | |
| M8 × 1.25 | | | | | | | A | A | A | A | A | A | A | A | A | A | A | A | A | A | A | |
| M10 × 1.5 | | | | | | | A | A | A | A | A | A | A | A | A | A | A | A | A | A | A | A |
| M12 × 1.75 | | | | | | | | A | A | A | A | A | A | A | A | A | A | A | A | A | A | A |

Min. Thd Length—28 mm

Min. Thd Length—38 mm

[b]Metric Fasteners Standard. IFI-513 (1982).
[c]PH = recommended lengths for only pan and hex head metric screws;
    A = recommended lengths for all metric screw head-styles.

## 21  Keys—Square, Flat, Plain Taper,[a] and Gib Head

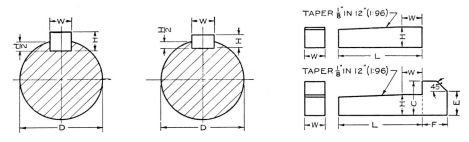

| Shaft Diameters | Square Stock Key | Flat Stock Key | Gib Head Taper Stock Key | | | | | |
|---|---|---|---|---|---|---|---|---|
| | | | Square | | | Flat | | |
| | | | Height | Length | Height to Chamfer | Height | Length | Height to Chamfer |
| D | W = H | W × H | C | F | E | C | F | E |
| ½ to ⁹⁄₁₆ | ⅛ | ⅛ × ³⁄₃₂ | ¼ | ⁷⁄₃₂ | ⁵⁄₃₂ | ³⁄₁₆ | ⅛ | ⅛ |
| ⅝ to ⅞ | ³⁄₁₆ | ³⁄₁₆ × ⅛ | ⁵⁄₁₆ | ⁹⁄₃₂ | ⁷⁄₃₂ | ¼ | ³⁄₁₆ | ⁵⁄₃₂ |
| ¹⁵⁄₁₆ to 1¼ | ¼ | ¼ × ³⁄₁₆ | ⁷⁄₁₆ | ¹¹⁄₃₂ | ¹¹⁄₃₂ | ⁵⁄₁₆ | ¼ | ³⁄₁₆ |
| 1⁵⁄₁₆ to 1⅜ | ⁵⁄₁₆ | ⁵⁄₁₆ × ¼ | ⁹⁄₁₆ | ¹³⁄₃₂ | ¹³⁄₃₂ | ⅜ | ⁵⁄₁₆ | ¼ |
| 1⁷⁄₁₆ to 1¾ | ⅜ | ⅜ × ¼ | ¹¹⁄₁₆ | ¹⁵⁄₃₂ | ¹⁵⁄₃₂ | ⁷⁄₁₆ | ⅜ | ⁵⁄₁₆ |
| 1¹³⁄₁₆ to 2¼ | ½ | ½ × ⅜ | ⅞ | ¹⁹⁄₃₂ | ⅝ | ⅝ | ½ | ⁷⁄₁₆ |
| 2⁵⁄₁₆ to 2¾ | ⅝ | ⅝ × ⁷⁄₁₆ | 1¹⁄₁₆ | ²³⁄₃₂ | ¾ | ¾ | ⅝ | ½ |
| 2⅞ to 3¼ | ¾ | ¾ × ½ | 1¼ | ⅞ | ⅞ | ⅞ | ¾ | ⅝ |
| 3⅜ to 3¾ | ⅞ | ⅞ × ⅝ | 1½ | 1 | 1 | 1¹⁄₁₆ | ⅞ | ¾ |
| 3⅞ to 4½ | 1 | 1 × ¾ | 1¾ | 1³⁄₁₆ | 1³⁄₁₆ | 1¼ | 1 | 1³⁄₁₆ |
| 4¾ to 5½ | 1¼ | 1¼ × ⅞ | 2 | 1⁷⁄₁₆ | 1⁷⁄₁₆ | 1½ | 1¼ | 1 |
| 5¾ to 6 | 1½ | 1½ × 1 | 2½ | 1¾ | 1¾ | 1¾ | 1½ | 1¼ |

[a] Plain taper square and flat keys have the same dimensions as the plain parallel stock keys, with the addition of the taper on top. Gib head taper square and flat keys have the same dimensions as the plain taper keys, with the addition of the gib head.

*Stock lengths for plain taper and gib head taper keys:* The minimum stock length equals 4W, and the maximum equals 16W. The increments of increase of length equal 2W.

## 22  Screw Threads,[a] Square and Acme

| Size | Threads per Inch | Size | Threads per Inch | Size | Threads per Inch | Size | Threads per Inch |
|---|---|---|---|---|---|---|---|
| ⅜ | 12 | ⅞ | 5 | 2 | 2 | 2½ | 3½ | 1⅓ |
| ⁷⁄₁₆ | 10 | 1 | 5 | 2¼ | 2 | 3¾ | 1⅓ |
| ½ | 10 | 1⅛ | 4 | 2½ | 2 | 4 | 1⅓ |
| ⁹⁄₁₆ | 8 | 1¼ | 4 | 2¾ | 2 | 4¼ | 1⅓ |
| ⅝ | 8 | 1½ | 3 | 3 | 1½ | 4½ | 1 |
| ¾ | 6 | 1¾ | 2½ | 3¼ | 1½ | over 4½ | 1 |

[a] See Appendix 17 for General-Purpose Acme Threads.

## 23   Woodruff Keys[a]—American National Standard

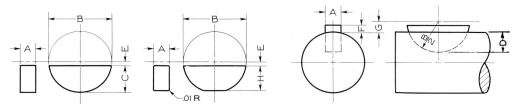

| Key No.[b] | Nominal Sizes | | | | Maximum Sizes | | | Key No.[b] | Nominal Sizes | | | | Maximum Sizes | | |
|---|---|---|---|---|---|---|---|---|---|---|---|---|---|---|---|
| | $A \times B$ | E | F | G | H | D | C | | $A \times B$ | E | F | G | H | D | C |
| 204 | $\frac{1}{16} \times \frac{1}{2}$ | $\frac{3}{64}$ | $\frac{1}{32}$ | $\frac{5}{64}$ | .194 | .1718 | .203 | 808 | $\frac{1}{4} \times 1$ | $\frac{1}{16}$ | $\frac{1}{8}$ | $\frac{3}{16}$ | .428 | .3130 | .438 |
| 304 | $\frac{3}{32} \times \frac{1}{2}$ | $\frac{3}{64}$ | $\frac{3}{64}$ | $\frac{3}{32}$ | .194 | .1561 | .203 | 809 | $\frac{1}{4} \times 1\frac{1}{8}$ | $\frac{5}{64}$ | $\frac{1}{8}$ | $\frac{13}{64}$ | .475 | .3590 | .484 |
| 305 | $\frac{3}{32} \times \frac{5}{8}$ | $\frac{1}{16}$ | $\frac{3}{64}$ | $\frac{7}{64}$ | .240 | .2031 | .250 | 810 | $\frac{1}{4} \times 1\frac{1}{4}$ | $\frac{5}{64}$ | $\frac{1}{8}$ | $\frac{13}{64}$ | .537 | .4220 | .547 |
| 404 | $\frac{1}{8} \times \frac{1}{2}$ | $\frac{3}{64}$ | $\frac{1}{16}$ | $\frac{7}{64}$ | .194 | .1405 | .203 | 811 | $\frac{1}{4} \times 1\frac{3}{8}$ | $\frac{3}{32}$ | $\frac{1}{8}$ | $\frac{7}{32}$ | .584 | .4690 | .594 |
| 405 | $\frac{1}{8} \times \frac{5}{8}$ | $\frac{1}{16}$ | $\frac{1}{16}$ | $\frac{1}{8}$ | .240 | .1875 | .250 | 812 | $\frac{1}{4} \times 1\frac{1}{2}$ | $\frac{7}{64}$ | $\frac{1}{8}$ | $\frac{15}{64}$ | .631 | .5160 | .641 |
| 406 | $\frac{1}{8} \times \frac{3}{4}$ | $\frac{1}{16}$ | $\frac{1}{16}$ | $\frac{1}{8}$ | .303 | .2505 | .313 | 1008 | $\frac{5}{16} \times 1$ | $\frac{1}{16}$ | $\frac{5}{32}$ | $\frac{7}{32}$ | .428 | .2818 | .438 |
| 505 | $\frac{5}{32} \times \frac{5}{8}$ | $\frac{1}{16}$ | $\frac{5}{64}$ | $\frac{9}{64}$ | .240 | .1719 | .250 | 1009 | $\frac{5}{16} \times 1\frac{1}{8}$ | $\frac{5}{64}$ | $\frac{5}{32}$ | $\frac{15}{64}$ | .475 | .3278 | .484 |
| 506 | $\frac{5}{32} \times \frac{3}{4}$ | $\frac{1}{16}$ | $\frac{5}{64}$ | $\frac{9}{64}$ | .303 | .2349 | .313 | 1010 | $\frac{5}{16} \times 1\frac{1}{4}$ | $\frac{5}{64}$ | $\frac{5}{32}$ | $\frac{15}{64}$ | .537 | .3908 | .547 |
| 507 | $\frac{5}{32} \times \frac{7}{8}$ | $\frac{1}{16}$ | $\frac{5}{64}$ | $\frac{9}{64}$ | .365 | .2969 | .375 | 1011 | $\frac{5}{16} \times 1\frac{3}{8}$ | $\frac{3}{32}$ | $\frac{5}{32}$ | $\frac{8}{32}$ | .584 | .4378 | .594 |
| 606 | $\frac{3}{16} \times \frac{3}{4}$ | $\frac{1}{16}$ | $\frac{3}{32}$ | $\frac{5}{32}$ | .303 | .2193 | .313 | 1012 | $\frac{5}{16} \times 1\frac{1}{2}$ | $\frac{7}{64}$ | $\frac{5}{32}$ | $\frac{17}{64}$ | .631 | .4848 | .641 |
| 607 | $\frac{3}{16} \times \frac{7}{8}$ | $\frac{1}{16}$ | $\frac{3}{32}$ | $\frac{5}{32}$ | .365 | .2813 | .375 | 1210 | $\frac{3}{8} \times 1\frac{1}{4}$ | $\frac{5}{64}$ | $\frac{3}{16}$ | $\frac{17}{64}$ | .537 | .3595 | .547 |
| 608 | $\frac{3}{16} \times 1$ | $\frac{1}{16}$ | $\frac{3}{32}$ | $\frac{5}{32}$ | .428 | .3443 | .438 | 1211 | $\frac{3}{8} \times 1\frac{3}{8}$ | $\frac{3}{32}$ | $\frac{3}{16}$ | $\frac{9}{32}$ | .584 | .4065 | .594 |
| 609 | $\frac{3}{16} \times 1\frac{1}{8}$ | $\frac{5}{64}$ | $\frac{3}{32}$ | $\frac{11}{64}$ | .475 | .3903 | .484 | 1212 | $\frac{3}{8} \times 1\frac{1}{2}$ | $\frac{7}{64}$ | $\frac{3}{16}$ | $\frac{19}{64}$ | .631 | .4535 | .641 |
| 807 | $\frac{1}{4} \times \frac{7}{8}$ | $\frac{1}{16}$ | $\frac{1}{8}$ | $\frac{3}{16}$ | .365 | .2500 | .375 | . . . . | . . . . . . . . . | . . | . . | . . | . . . . | . . . . | . . . . |

[a]ANSI B17.2–1967 (R1978).
[b]Key numbers indicate nominal key dimensions. The last two digits give the nominal diameter B in eighths of an inch, and the digits before the last two give the nominal width A in thirty-seconds of an inch.

## 24   Woodruff Key Sizes for Different Shaft Diameters[a]

| Shaft Diameter | $\frac{5}{16}$ to $\frac{3}{8}$ | $\frac{7}{16}$ to $\frac{1}{2}$ | $\frac{9}{16}$ to $\frac{3}{4}$ | $\frac{13}{16}$ to $\frac{15}{16}$ | 1 to $1\frac{3}{16}$ | $1\frac{1}{4}$ to $1\frac{7}{16}$ | $1\frac{1}{2}$ to $1\frac{3}{4}$ | $1\frac{13}{16}$ to $2\frac{1}{8}$ | $2\frac{3}{16}$ to $2\frac{1}{2}$ |
|---|---|---|---|---|---|---|---|---|---|
| Key Numbers | 204 | 304 305 | 404 405 406 | 505 506 507 | 606 607 608 609 | 807 808 809 | 810 811 812 | 1011 1012 | 1211 1212 |

[a]Suggested sizes; not standard.

# 25 Pratt and Whitney Round-End Keys

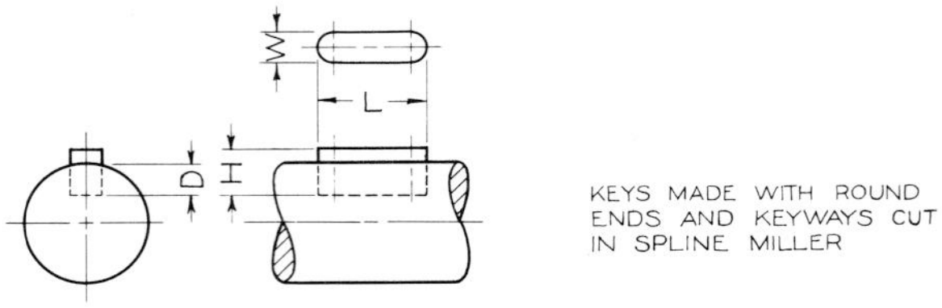
W
L
D
H
KEYS MADE WITH ROUND
ENDS AND KEYWAYS CUT
IN SPLINE MILLER

Maximum length of slot is 4″ + W. Note that key is sunk two-thirds into shaft in all cases.

| Key No. | L[a] | W or D | H | Key No. | L[a] | W or D | H |
|---|---|---|---|---|---|---|---|
| 1 | 1/2 | 1/16 | 3/32 | 22 | 1 3/8 | 1/4 | 3/8 |
| 2 | 1/2 | 3/32 | 9/64 | 23 | 1 1/38 | 5/16 | 15/32 |
| 3 | 1/2 | 1/8 | 3/16 | F | 1 3/8 | 3/8 | 9/16 |
| 4 | 5/8 | 3/32 | 9/64 | 24 | 1 1/2 | 1/4 | 3/8 |
| 5 | 5/8 | 1/8 | 3/16 | 25 | 1 1/2 | 5/16 | 15/32 |
| 6 | 5/8 | 5/32 | 15/64 | G | 1 1/2 | 3/8 | 9/16 |
| 7 | 3/4 | 1/8 | 3/16 | 51 | 1 3/4 | 1/4 | 3/8 |
| 8 | 3/4 | 5/32 | 15/64 | 52 | 1 3/4 | 5/16 | 15/32 |
| 9 | 3/4 | 3/16 | 9/32 | 53 | 1 3/4 | 3/8 | 9/16 |
| 10 | 7/8 | 5/32 | 15/64 | 26 | 2 | 3/16 | 9/32 |
| 11 | 7/8 | 3/16 | 9/32 | 27 | 2 | 1/4 | 3/8 |
| 12 | 7/8 | 7/32 | 21/64 | 28 | 2 | 5/16 | 15/32 |
| A | 7/8 | 1/4 | 3/8 | 29 | 2 | 3/8 | 9/16 |
| 13 | 1 | 3/16 | 9/32 | 54 | 2 1/4 | 1/4 | 3/8 |
| 14 | 1 | 7/32 | 21/64 | 55 | 2 1/4 | 5/16 | 15/32 |
| 15 | 1 | 1/4 | 3/8 | 56 | 2 1/4 | 3/8 | 9/16 |
| B | 1 | 5/16 | 15/32 | 57 | 2 1/4 | 7/16 | 21/32 |
| 16 | 1 1/8 | 3/16 | 9/32 | 58 | 2 1/2 | 5/16 | 15/32 |
| 17 | 1 1/8 | 7/32 | 21/64 | 59 | 2 1/2 | 3/8 | 9/16 |
| 18 | 1 1/8 | 1/4 | 3/8 | 60 | 2 1/2 | 7/16 | 21/32 |
| C | 1 1/8 | 5/16 | 15/32 | 61 | 2 1/2 | 1/2 | 3/4 |
| 19 | 1 1/4 | 3/16 | 9/32 | 30 | 3 | 3/8 | 9/16 |
| 20 | 1 1/4 | 7/32 | 21/64 | 31 | 3 | 7/16 | 21/32 |
| 21 | 1 1/4 | 1/4 | 3/8 | 32 | 3 | 1/2 | 3/4 |
| D | 1 1/4 | 5/16 | 15/32 | 33 | 3 | 9/16 | 27/32 |
| E | 1 1/4 | 3/8 | 9/16 | 34 | 3 | 5/8 | 15/16 |

[a]The length L may vary from the table, but equals at least 2W.

# 26 Washers,[a] Plain—American National Standard

For parts lists, etc., give inside diameter, outside
diameter, and the thickness; for example,
.344 × .688 × .065 TYPE A PLAIN WASHER.

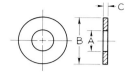

## PREFERRED SIZES OF TYPE A PLAIN WASHERS[b]

| Nominal Washer Size[c] | | | Inside Diameter<br>A | Outside Diameter<br>B | Nominal Thickness<br>C |
|---|---|---|---|---|---|
| . . . . | . . . . | | 0.078 | 0.188 | 0.020 |
| . . . . | . . . . | | 0.094 | 0.250 | 0.020 |
| . . . . | . . . . | | 0.125 | 0.312 | 0.032 |
| No. 6 | 0.138 | | 0.156 | 0.375 | 0.049 |
| No. 8 | 0.164 | | 0.188 | 0.438 | 0.049 |
| No. 10 | 0.190 | | 0.219 | 0.500 | 0.049 |
| ³⁄₁₆ | 0.188 | | 0.250 | 0.562 | 0.049 |
| No. 12 | 0.216 | | 0.250 | 0.562 | 0.065 |
| ¼ | 0.250 | N | 0.281 | 0.625 | 0.065 |
| ¼ | 0.250 | W | 0.312 | 0.734 | 0.065 |
| ⁵⁄₁₆ | 0.312 | N | 0.344 | 0.688 | 0.065 |
| ⁵⁄₁₆ | 0.312 | W | 0.375 | 0.875 | 0.083 |
| ⅜ | 0.375 | N | 0.406 | 0.812 | 0.065 |
| ⅜ | 0.375 | W | 0.438 | 1.000 | 0.083 |
| ⁷⁄₁₆ | 0.438 | N | 0.469 | 0.922 | 0.065 |
| ⁷⁄₁₆ | 0.438 | W | 0.500 | 1.250 | 0.083 |
| ½ | 0.500 | N | 0.531 | 1.062 | 0.095 |
| ½ | 0.500 | W | 0.562 | 1.375 | 0.109 |
| ⁹⁄₁₆ | 0.562 | N | 0.594 | 1.156 | 0.095 |
| ⁹⁄₁₆ | 0.562 | W | 0.625 | 1.469 | 0.109 |
| ⅝ | 0.625 | N | 0.656 | 1.312 | 0.095 |
| ⅝ | 0.625 | W | 0.688 | 1.750 | 0.134 |
| ¾ | 0.750 | N | 0.812 | 1.469 | 0.134 |
| ¾ | 0.750 | W | 0.812 | 2.000 | 0.148 |
| ⅞ | 0.875 | N | 0.938 | 1.750 | 0.134 |
| ⅞ | 0.875 | W | 0.938 | 2.250 | 0.165 |
| 1 | 1.000 | N | 1.062 | 2.000 | 0.134 |
| 1 | 1.000 | W | 1.062 | 2.500 | 0.165 |
| 1⅛ | 1.125 | N | 1.250 | 2.250 | 0.134 |
| 1⅛ | 1.125 | W | 1.250 | 2.750 | 0.165 |
| 1¼ | 1.250 | N | 1.375 | 2.500 | 0.165 |
| 1¼ | 1.250 | W | 1.375 | 3.000 | 0.165 |
| 1⅜ | 1.375 | N | 1.500 | 2.750 | 0.165 |
| 1⅜ | 1.375 | W | 1.500 | 3.250 | 0.180 |
| 1½ | 1.500 | N | 1.625 | 3.000 | 0.165 |
| 1½ | 1.500 | W | 1.625 | 3.500 | 0.180 |
| 1⅝ | 1.625 | | 1.750 | 3.750 | 0.180 |
| 1¾ | 1.750 | | 1.875 | 4.000 | 0.180 |
| 1⅞ | 1.875 | | 2.000 | 4.250 | 0.180 |
| 2 | 2.000 | | 2.125 | 4.500 | 0.180 |
| 2¼ | 2.250 | | 2.375 | 4.750 | 0.220 |
| 2½ | 2.500 | | 2.625 | 5.000 | 0.238 |
| 2¾ | 2.750 | | 2.875 | 5.250 | 0.259 |
| 3 | 3.000 | | 3.125 | 5.500 | 0.284 |

[a]From ANSI B18.22.1–1965 (R1981). For complete listings, see the standard.
[b]Preferred sizes are for the most part from series previously designated "Standard Plate" and "SAE." Where common sizes existed in the two series, the SAE size is designated "N" (narrow) and the Standard Plate "W" (wide).
[c]Nominal washer sizes are intended for use with comparable nominal screw or bolt sizes.

# 27 Washers,ᵃ Lock—American National Standard

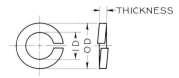

←THICKNESS

For parts lists, etc., give nominal size and series; for example, ¼ REGULAR LOCK WASHER

## PREFERRED SERIES

| Nominal Washer Size[b] | | Inside Diameter, Min. | Regular | | Extra Duty | | Hi-Collar | |
|---|---|---|---|---|---|---|---|---|
| | | | Outside Diameter, Max. | Thickness, Min. | Outside Diameter, Max. | Thickness, Min. | Outside Diameter, Max. | Thickness, Min. |
| No. 2 | 0.086 | 0.088 | 0.172 | 0.020 | 0.208 | 0.027 | . . . . | . . . . |
| No. 3 | 0.099 | 0.101 | 0.195 | 0.025 | 0.239 | 0.034 | . . . . | . . . . |
| No. 4 | 0.112 | 0.115 | 0.209 | 0.025 | 0.253 | 0.034 | 0.173 | 0.022 |
| No. 5 | 0.125 | 0.128 | 0.236 | 0.031 | 0.300 | 0.045 | 0.202 | 0.030 |
| No. 6 | 0.138 | 0.141 | 0.250 | 0.031 | 0.314 | 0.045 | 0.216 | 0.030 |
| No. 8 | 0.164 | 0.168 | 0.293 | 0.040 | 0.375 | 0.057 | 0.267 | 0.047 |
| No. 10 | 0.190 | 0.194 | 0.334 | 0.047 | 0.434 | 0.068 | 0.294 | 0.047 |
| No. 12 | 0.216 | 0.221 | 0.377 | 0.056 | 0.497 | 0.080 | . . . . | . . . . |
| ¼ | 0.250 | 0.255 | 0.489 | 0.062 | 0.535 | 0.084 | 0.365 | 0.078 |
| ⁵⁄₁₆ | 0.312 | 0.318 | 0.586 | 0.078 | 0.622 | 0.108 | 0.460 | 0.093 |
| ⅜ | 0.375 | 0.382 | 0.683 | 0.094 | 0.741 | 0.123 | 0.553 | 0.125 |
| ⁷⁄₁₆ | 0.438 | 0.446 | 0.779 | 0.109 | 0.839 | 0.143 | 0.647 | 0.140 |
| ½ | 0.500 | 0.509 | 0.873 | 0.125 | 0.939 | 0.162 | 0.737 | 0.172 |
| ⁹⁄₁₆ | 0.562 | 0.572 | 0.971 | 0.141 | 1.041 | 0.182 | . . . . | . . . . |
| ⅝ | 0.625 | 0.636 | 1.079 | 0.156 | 1.157 | 0.202 | 0.923 | 0.203 |
| ¹¹⁄₁₆ | 0.688 | 0.700 | 1.176 | 0.172 | 1.258 | 0.221 | . . . | . . . . |
| ¾ | 0.750 | 0.763 | 1.271 | 0.188 | 1.361 | 0.241 | 1.111 | 0.218 |
| ¹³⁄₁₆ | 0.812 | 0.826 | 1.367 | 0.203 | 1.463 | 0.261 | . . . . | . . . . |
| ⅞ | 0.875 | 0.890 | 1.464 | 0.219 | 1.576 | 0.285 | 1.296 | 0.234 |
| ¹⁵⁄₁₆ | 0.938 | 0.954 | 1.560 | 0.234 | 1.688 | 0.308 | . . . . | . . . . |
| 1 | 1.000 | 1.017 | 1.661 | 0.250 | 1.799 | 0.330 | 1.483 | 0.250 |
| 1¹⁄₁₆ | 1.062 | 1.080 | 1.756 | 0.266 | 1.910 | 0.352 | . . . . | . . . . |
| 1⅛ | 1.125 | 1.144 | 1.853 | 0.281 | 2.019 | 0.375 | 1.669 | 0.313 |
| 1³⁄₁₆ | 1.188 | 1.208 | 1.950 | 0.297 | 2.124 | 0.396 | . . . . | . . . . |
| 1¼ | 1.250 | 1.271 | 2.045 | 0.312 | 2.231 | 0.417 | 1.799 | 0.313 |
| 1⁵⁄₁₆ | 1.312 | 1.334 | 2.141 | 0.328 | 2.335 | 0.438 | . . . . | . . . . |
| 1⅜ | 1.375 | 1.398 | 2.239 | 0.344 | 2.439 | 0.458 | 2.041 | 0.375 |
| 1⁷⁄₁₆ | 1.438 | 1.462 | 2.334 | 0.359 | 2.540 | 0.478 | . . . . | . . . . |
| 1½ | 1.500 | 1.525 | 2.430 | 0.375 | 2.638 | 0.496 | 2.170 | 0.375 |

ᵃFrom ANSI B18.21.1–1972 (R1983). For complete listing, see the standard.
ᵇNominal washer sizes are intended for use with comparable nominal screw or bolt sizes.

# 28 Wire Gage Standards[a]

Dimensions of sizes in decimal parts of an inch.[b]

| No. of Wire | American or Brown & Sharpe for Non-ferrous Metals | Birming-ham, or Stubs' Iron Wire[c] | American S. & W. Co.'s (Washburn & Moen) Std. Steel Wire | American S. & W. Co.'s Music Wire | Imperial Wire | Stubs' Steel Wire[c] | Steel Manu-facturers' Sheet Gage[b] | No. of Wire |
|---|---|---|---|---|---|---|---|---|
| 7–0's | .651354 | . . . . | .4900 | . . . . | .500 | . . . . | . . . . . | 7–0's |
| 6–0's | .580049 | . . . . | .4615 | .004 | .464 | . . . . | . . . . . | 6–0's |
| 5–0's | .516549 | .500 | .4305 | .005 | .432 | . . . . | . . . . . | 5–0's |
| 4–0's | .460 | .454 | .3938 | .006 | .400 | . . . . | . . . . . | 4–0's |
| 000 | .40964 | .425 | .3625 | .007 | .372 | . . . . | . . . . . | 000 |
| 00 | .3648 | .380 | .3310 | .008 | .348 | . . . . | . . . . . | 00 |
| 0 | .32486 | .340 | .3065 | .009 | .324 | . . . . | . . . . . | 0 |
| 1 | .2893 | .300 | .2830 | .010 | .300 | .227 | . . . . . | 1 |
| 2 | .25763 | .284 | .2625 | .011 | .276 | .219 | . . . . . | 2 |
| 3 | .22942 | .259 | .2437 | .012 | .252 | .212 | .2391 | 3 |
| 4 | .20431 | .238 | .2253 | .013 | .232 | .207 | .2242 | 4 |
| 5 | .18194 | .220 | .2070 | .014 | .212 | .204 | .2092 | 5 |
| 6 | .16202 | .203 | .1920 | .016 | .192 | .201 | .1943 | 6 |
| 7 | .14428 | .180 | .1770 | .018 | .176 | .199 | .1793 | 7 |
| 8 | .12849 | .165 | .1620 | .020 | .160 | .197 | .1644 | 8 |
| 9 | .11443 | .148 | .1483 | .022 | .144 | .194 | .1495 | 9 |
| 10 | .10189 | .134 | .1350 | .024 | .128 | .191 | .1345 | 10 |
| 11 | .090742 | .120 | .1205 | .026 | .116 | .188 | .1196 | 11 |
| 12 | .080808 | .109 | .1055 | .029 | .104 | .185 | .1046 | 12 |
| 13 | .071961 | .095 | .0915 | .031 | .092 | .182 | .0897 | 13 |
| 14 | .064084 | .083 | .0800 | .033 | .080 | .180 | .0747 | 14 |
| 15 | .057068 | .072 | .0720 | .035 | .072 | .178 | .0763 | 15 |
| 16 | .05082 | .065 | .0625 | .037 | .064 | .175 | .0598 | 16 |
| 17 | .045257 | .058 | .0540 | .039 | .056 | .172 | .0538 | 17 |
| 18 | .040303 | .049 | .0475 | .041 | .048 | .168 | .0478 | 18 |
| 19 | .03589 | .042 | .0410 | .043 | .040 | .164 | .0418 | 19 |
| 20 | .031961 | .035 | .0348 | .045 | .036 | .161 | .0359 | 20 |
| 21 | .028462 | .032 | .0317 | .047 | .032 | .157 | .0329 | 21 |
| 22 | .025347 | .028 | .0286 | .049 | .028 | .155 | .0299 | 22 |
| 23 | .022571 | .025 | .0258 | .051 | .024 | .153 | .0269 | 23 |
| 24 | .0201 | .022 | .0230 | .055 | .022 | .151 | .0239 | 24 |
| 25 | .0179 | .020 | .0204 | .059 | .020 | .148 | .0209 | 25 |
| 26 | .01594 | .018 | .0181 | .063 | .018 | .146 | .0179 | 26 |
| 27 | .014195 | .016 | .0173 | .067 | .0164 | .143 | .0164 | 27 |
| 28 | .012641 | .014 | .0162 | .071 | .0149 | .139 | .0149 | 28 |
| 29 | .011257 | .013 | .0150 | .075 | .0136 | .134 | .0135 | 29 |
| 30 | .010025 | .012 | .0140 | .080 | .0124 | .127 | .0120 | 30 |
| 31 | .008928 | .010 | .0132 | .085 | .0116 | .120 | .0105 | 31 |
| 32 | .00795 | .009 | .0128 | .090 | .0108 | .115 | .0097 | 32 |
| 33 | .00708 | .008 | .0118 | .095 | .0100 | .112 | .0090 | 33 |
| 34 | .006304 | .007 | .0104 | . . . . | .0092 | .110 | .0082 | 34 |
| 35 | .005614 | .005 | .0095 | . . . . | .0084 | .108 | .0075 | 35 |
| 36 | .005 | .004 | .0090 | . . . . | .0076 | .106 | .0067 | 36 |
| 37 | .004453 | . . . . | .0085 | . . . . | .0068 | .103 | .0064 | 37 |
| 38 | .003965 | . . . . | .0080 | . . . . | .0060 | .101 | .0060 | 38 |
| 39 | .003531 | . . . . | .0075 | . . . . | .0052 | .099 | . . . . . | 39 |
| 40 | .003144 | . . . . | .0070 | . . . . | .0048 | .097 | . . . . . | 40 |

[a]Courtesy Brown & Sharpe Mfg. Co.

[b]Now used by steel manufacturers in place of old U.S. Standard Gage.

[c]The difference between the Stubs' Iron Wire Gage and the Stubs' Steel Wire Gage should be noted, the first being commonly known as the English Standard Wire, or Birmingham Gage, which designates the Stubs' soft wire sizes and the second being used in measuring drawn steel wire or drill rods of Stubs' make.

# 29   Taper Pins[a]—American National Standard

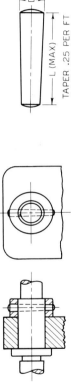

TAPER .25 PER FT

To find small diameter of pin, multiply the length by .02083 and subtract the result from the larger diameter.
All dimensions are given in inches.
Standard reamers are available for pins given above the heavy line.

| Number | 7/0 | 6/0 | 5/0 | 4/0 | 3/0 | 2/0 | 0 | 1 | 2 | 3 | 4 | 5 | 6 | 7 | 8 |
|---|---|---|---|---|---|---|---|---|---|---|---|---|---|---|---|
| Size (Large End) | .0625 | .0780 | .0940 | .1090 | .1250 | .1410 | .1560 | .1720 | .1930 | .2190 | .2500 | .2890 | .3410 | .4090 | .4920 |
| Shaft Diameter (Approx)[b] | | 7/32 | 1/4 | 5/16 | 3/8 | 7/16 | 1/2 | 9/16 | 5/8 | 3/4 | 13/16 | 7/8 | 1 | 1 1/4 | 1 1/2 |
| Drill Size (Before Reamer)[b] | .0312 | .0312 | .0625 | .0625 | .0781 | .0938 | .0938 | .1094 | .1250 | .1250 | .1562 | .1562 | .2188 | .2344 | .3125 |
| **Length L** | | | | | | | | | | | | | | | |
| .250 | X | X | X | X | X | | | | | | | | | | |
| .375 | X | X | X | X | X | X | | | | | | | | | |
| .500 | X | X | X | X | X | X | | | | | | | | | |
| .625 | X | X | X | X | X | X | X | | | | | | | | |
| .750 | X | X | X | X | X | X | X | X | | | | | | | |
| .875 | X | X | X | ... | X | X | X | X | X | | | | | | |
| 1.000 | X | X | X | X | X | X | X | X | X | X | X | | | | |
| 1.250 | ... | X | X | X | X | X | X | X | X | X | X | X | | | |
| 1.500 | ... | X | X | X | X | X | X | X | X | X | X | X | X | | |
| 1.750 | | ... | ... | X | X | X | X | X | X | X | X | X | X | X | X |
| 2.000 | | | ... | ... | X | X | X | X | X | X | X | X | X | X | X |
| 2.250 | | | ... | ... | X | X | X | X | X | X | X | X | X | X | X |
| 2.500 | | | | ... | ... | X | X | X | X | X | X | X | X | X | X |
| 2.750 | | | | | ... | ... | X | X | X | X | X | X | X | X | X |
| 3.000 | | | | | | ... | X | X | X | X | X | X | X | X | X |
| 3.250 | | | | | | | ... | X | X | X | X | X | X | X | X |
| 3.500 | | | | | | | ... | X | X | X | X | X | X | X | X |
| 3.750 | | | | | | | | ... | ... | X | X | X | X | X | X |
| 4.000 | | | | | | | | ... | ... | ... | ... | X | X | X | X |
| 4.250 | | | | | | | | ... | ... | ... | ... | X | X | X | X |
| 4.500 | | | | | | | | ... | ... | ... | ... | X | X | X | X |

[a]ANSI B18.8.2–1978 (R1983). For Nos. 9 and 10, see the standard. Pins Nos 11 (size .8600), 12 (size 1.032), 13 (size 1.241), and 14 (size 1.523) are special sizes; hence their lengths are special.
[b]Suggested sizes; not American National Standard.

# 30    Cotter Pins[a]—American National Standard

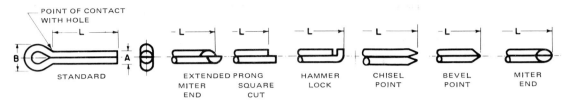

PREFERRED POINT TYPES

All dimensions are given in inches.

| Nominal Size or Pin Diameter | Diameter A | | Outside Eye Diameter B Min. | Extended Prong Length Min. | Hole Sizes Recommended |
|---|---|---|---|---|---|
| | Max. | Min. | | | |
| 1/32 | .031 | .032 | .028 | .06 | .01 | .047 |
| 3/64 | .047 | .048 | .044 | .09 | .02 | .062 |
| 1/16 | .062 | .060 | .056 | .12 | .03 | .078 |
| 5/64 | .078 | .076 | .072 | .16 | .04 | .094 |
| 3/32 | .094 | .090 | .086 | .19 | .04 | .109 |
| 7/64 | .109 | .104 | .100 | .22 | .05 | .125 |
| 1/8 | .125 | .120 | .116 | .25 | .06 | .141 |
| 9/64 | .141 | .134 | .130 | .28 | .06 | .156 |
| 5/32 | .156 | .150 | .146 | .31 | .07 | .172 |
| 3/16 | .188 | .176 | .172 | .38 | .09 | .203 |
| 7/32 | .219 | .207 | .202 | .44 | .10 | .234 |
| 1/4 | .250 | .225 | .220 | .50 | .11 | .266 |
| 5/16 | .312 | .280 | .275 | .62 | .14 | .312 |
| 3/8 | .375 | .335 | .329 | .75 | .16 | .375 |
| 7/16 | .438 | .406 | .400 | .88 | .20 | .438 |
| 1/2 | .500 | .473 | .467 | 1.00 | .23 | .500 |
| 5/8 | .625 | .598 | .590 | 1.25 | .30 | .625 |
| 3/4 | .750 | .723 | .715 | 1.50 | .36 | .750 |

[a]ANSI B18.8.1—1972 (R1983).

# 31 Metric Equivalents

## Length

| U.S. to Metric | Metric to U.S. |
| --- | --- |
| 1 inch = 2.540 centimeters | 1 millimeter = .039 inch |
| 1 foot = .305 meter | 1 centimeter = .394 inch |
| 1 yard = .914 meter | 1 meter = 3.281 feet or 1.094 yards |
| 1 mile = 1.609 kilometers | 1 kilometer = .621 mile |

## Area

| U.S. to Metric | Metric to U.S. |
| --- | --- |
| 1 inch$^2$ = 6.451 centimeter$^2$ | 1 millimeter$^2$ = .00155 inch$^2$ |
| 1 foot$^2$ = .093 meter$^2$ | 1 centimeter$^2$ = .155 inch$^2$ |
| 1 yard$^2$ = .836 meter$^2$ | 1 meter$^2$ = 10.764 foot$^2$ or 1.196 yard$^2$ |
| 1 acre$^2$ = 4,046.873 meter$^2$ | 1 kilometer$^2$ = .386 mile$^2$ or 247.04 acre$^2$ |

## Volume

| U.S. to Metric | Metric to U.S. |
| --- | --- |
| 1 inch$^3$ = 16.387 centimeter$^3$ | 1 centimeter$^3$ = .061 inch$^3$ |
| 1 foot$^3$ = .028 meter$^3$ | 1 meter$^3$ = 35.314 foot$^3$ or 1.308 yard$^3$ |
| 1 yard$^3$ = .764 meter$^3$ | 1 liter = .2642 gallons |
| 1 quart = .946 liter | 1 liter = 1.057 quarts |
| 1 gallon = .003785 meter$^3$ | 1 meter$^3$ = 264.02 gallons |

## Weight

| U.S. to Metric | Metric to U.S. |
| --- | --- |
| 1 ounce = 28.349 grams | 1 gram = .035 ounce |
| 1 pound = .454 kilogram | 1 kilogram = 2.205 pounds |
| 1 ton = .907 metric ton | 1 metric ton = 1.102 tons |

## Velocity

| U.S. to Metric | Metric to U.S. |
| --- | --- |
| 1 foot/second = .305 meter/second | 1 meter/second = 3.281 feet/second |
| 1 mile/hour = .447 meter/second | 1 kilometer/hour = .621 mile/second |

## Acceleration

| U.S. to Metric | Metric to U.S. |
| --- | --- |
| 1 inch/second$^2$ = .0254 meter/second$^2$ | 1 meter/second$^2$ = 3.278 feet/second$^2$ |
| 1 foot/second$^2$ = .305 meter/second$^2$ | |

## Force

N (newton) = basic unit of force, kg-m/s$^2$. A mass of one kilogram (1 kg) exerts a gravitational force of 9.8 N (theoretically 9.80665 N) at mean sea level.

# 32  Form and Proportion of Geometric Tolerancing Symbols[a]

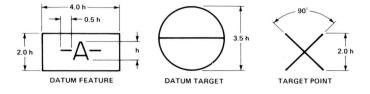

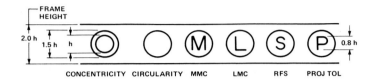

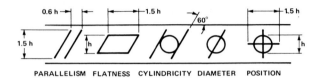

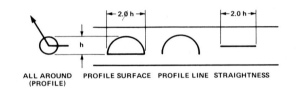

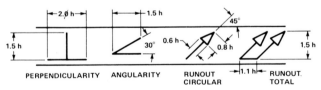

[a]ANSI Y14.5M—1982 (R1988).

# Index

691

# Sheet Layouts

A convenient code to identify American National Standard sheet sizes and forms suggested by the authors for title, parts or material list, and revision blocks, for use of instructors in making assignments, is shown here. All dimensions are in inches.

Three **sizes** of sheets are illustrated: **Size A**, Fig. I, **Size B**, Fig. V, and **Size C**, Fig. VI. Metric size sheets are not shown.

Eight **forms** of lettering arrangements are suggested, known as **Forms 1, 2, 3, 4, 5, 6, 7, and 8**, as shown below. The total length of **Forms 1, 2, 3,** and **4**, may be adjusted to fit **Sizes A4, A3,** and **A2**.

The term **layout** designates a sheet of certain size plus a certain arrangement of lettering. Thus **Layout A-1** is a combination of **Size A**, Fig. I, and **Form 1**, Fig. II. **Layout C-678** is a combination of **Size C**, Fig. VI, and **Forms 6, 7** and **8**, Figs. IX to XI inclusive. **Layout A4-2** (adjusted is a combination of **Size A4** and **Form 2**, Fig. III, adjusted to fit between the borders. Other combinations may be employed as assigned by the instructor.

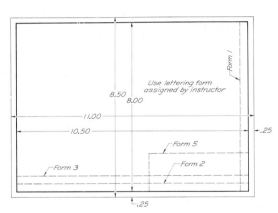

Fig. I  Size A Sheet (8.50″ × 11.00)

Fig. II  Form 1. Title Block

Fig. III  Form 2. Title Block

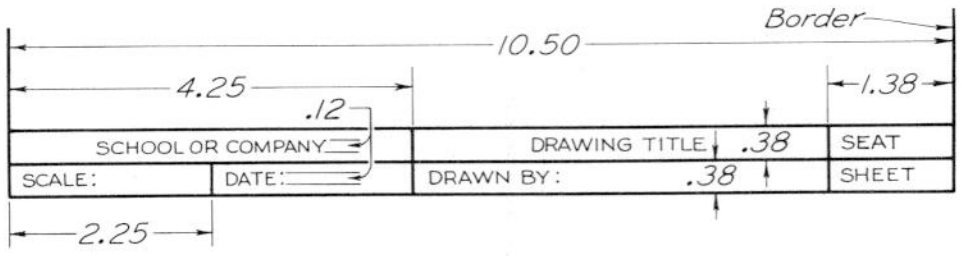

Fig. IV  Form 3. Title Block

## Sheet Sizes

### American National Standard

A -  8.50″ × 11.00″
B - 11.00″ × 17.00″
C - 17.00″ × 22.00″
D - 22.00″ × 34.00″
E - 34.00″ × 44.00″

### International Standard

A4 - 210 mm ×  297 mm
A3 - 297 mm ×  420 mm
A2 - 420 mm ×  594 mm
A1 - 594 mm ×  841 mm
A0 - 841 mm × 1189 mm
(25.4 mm = 1.00″)

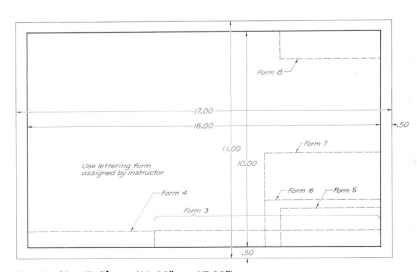

Fig. V  Size B Sheet (11.00″ × 17.00″)